Subcellular Biochemistry

Volume 93

The book series SUBCELLULAR BIOCHEMISTRY is a renowned and well recognized forum for disseminating advances of emerging topics in Cell Biology and related subjects. All volumes are edited by established scientists and the individual chapters are written by experts on the relevant topic. The individual chapters of each volume are fully citable and indexed in Medline/Pubmed to ensure maximum visibility of the work.

More information about this series at http://www.springer.com/series/6515

J. Robin Harris · Jon Marles-Wright
Editors

Macromolecular Protein Complexes II: Structure and Function

Springer

Editors
J. Robin Harris
Institute of Zoology
University of Mainz
Mainz, Germany

Jon Marles-Wright
School of Natural and Environmental Sciences
Newcastle University
Newcastle upon Tyne, UK

ISSN 0306-0225 ISSN 2542-8810 (electronic)
Subcellular Biochemistry
ISBN 978-3-030-28150-2 ISBN 978-3-030-28151-9 (eBook)
https://doi.org/10.1007/978-3-030-28151-9

This Springer imprint is published by the registered company Springer Nature Switzerland AG
The registered company address is: Gewerbestrasse 11, 6330 Cham, Switzerland

Preface

The compilation of this volume of the *Subcellular Biochemistry* series follows on from several volumes already devoted to the Macromolecular Protein Complex theme. The present, and indeed subsequent, volumes are justified because so many interesting relevant topics, deriving from X-ray crystallographic, cryo-EM and other structural studies, have appeared in the scientific literature over recent years. While already building up to be an almost encyclopaedic coverage, there is still much that could be added.

This present volume of the series contains a diverse collection of 19 fascinating chapters, as can be seen from the content list. While we will not expand here on any one chapter, it is clear that all the chapters stand alone, with some chapters complementing each other. The overall field of Macromolecular Protein Complexes continues to expand due to developments in cryogenic transmission electron microscopy and computational methods for reconstructing single particles and volumes. It therefore represents an important continuing component of structural biology, without in anyway detracting from the importance of X-ray crystallographic and NMR studies on an increasingly large number of smaller, usually monomeric, protein molecules.

The receipt and compilation of the chapters in this book have been a true excitement. Chapter authors have provided much exciting material for inclusion in the book, many containing impressive figures. As so often happens when editing a multi-author book, several planned chapters were either withdrawn or failed to appear in time for inclusion. Nevertheless, ample material is available to create a strong book, to supplement Volumes 82, 83, 84, 87 and 88 of the series (see https://www.springer.com/series/6515?detailsPage=titles). With two further related volumes at an earlier stage of compilation, continuation has been established, and it is hoped that many more interesting related topics will eventually be included in the *Subcellular Biochemistry* series.

The book is directed towards advanced undergraduates, postgraduates, research workers and academics who have a specialism or broad interest in molecular structural biology. It is hoped that the book, available in print and as an e-book, will be of interest and value to many.

Mainz, Germany J. Robin Harris
Newcastle upon Tyne, UK Jon Marles-Wright
June 2019

Contents

Chapter 1
Introduction: Protein Oligomerization and the Formation of Macromolecular Assemblies

J. Robin Harris and Jon Marles-Wright

Abstract The ability of biomolecules to link together to form higher order assemblies underlies much of cellular structure and function. Here we emphasise protein oligomerisation and discuss some of the principles of molecular interaction, from early considerations through to the present day. A few protein examples are presented, selected from our research interests, to highlight assembly features, ranging from the hemoglobins, the hemocyanins to the peroxiredoxins, collagen, the encapsulins and ferritins.

Keywords Protein · Oligomerization · Assembly · Hemoglobin · Hemocyanin · Peroxiredoxin · Collagen · Encapsulin · Ferritin

Fundamentals

A fundamental property of the organic molecules present in biological organisms is their ability to group together to form polymer chains, as found in DNA, proteins and carbohydrates, or bilayer sheets as for the lipids. Furthermore, these four principal biomolecules also interact with one another. The ability of biological molecules to link or associate together is implicit in the structure of DNA, with anti-parallel helices containing covalently linked nucleotide phosphates, dimerized as a *double helix* stabilized by hydrogen bonds between the complementary A-T, G-C base pairs. Indeed be it DNA or a protein, the dimer can be looked upon as the simplest oligomeric form; in the case of the helix it does not expand beyond the dimer (except under crystallization conditions, and in Holliday junctions and

J. R. Harris (✉)
Institute of Molecular Physiology, University of Mainz, Joachim-Becher-Weg 7, 55099 Mainz, Germany
e-mail: rharris@uni-mainz.de

J. Marles-Wright
School of Biology, Newcastle University, Newcastle upon Tyne NE2 4HH, UK
e-mail: Jon.marles-wright1@ncl.ac.uk

J. R. Harris and J. Marles-Wright (eds.), *Macromolecular Protein Complexes II: Structure and Function*, Subcellular Biochemistry 93,
https://doi.org/10.1007/978-3-030-28151-9_1

G-quadruplexes). For protein molecules the situation is somewhat different, with a multitude of structural possibilities for the creation of higher oligomeric forms.

Starting with the sequence of the 22 proteinogenic amino acids in a polypeptide chain, it is immediately apparent that the number of variations of primary structure is enormous, in turn reflected in the secondary structure with its combination of alpha helices, beta sheets, and connecting loops, that then determine the folding of the polypeptide chain and tertiary structure, stabilized by electrostatic/ionic, hydrogen bonds and in some cases by disulphide bonds. The arrangement of protein chains into multimeric assemblies gives the quaternary structure.

There are cases where secondary structure elements from individual chains in a multimeric-assembly combine to form tertiary structures across subunits. For instance, in some fibrous proteins, the tertiary structure includes the coming together of a number of closely associated polypeptide chains in the form of double or triple coiled-coil alpha helices, as in the collagen heterotrimer. The coming together of beta sheets from a number beta barrels has also emerged as a significant structural feature. This is especially apparent in the amyloid proteins, with their extensive cross-beta sheet and beta-helix structures formed by Tau filaments (Fitzpatrick et al. 2017) and amyloid fibres of beta-macroglobulin (Iadanza et al. 2018).

The definition of individual *domains* within a protein also has significance in relation to dimerization and multimerization. To some extent these features immediately indicate the possibility of creating sequence-defined protein-protein "subunit" interactions, that can then determine the widely occurring quaternary structure of multimeric macromolecules. Within protein multimers, the overall array of inter-subunit contacts and interactions is always determined by accessible ionic and hydrophobic surface groups rather than those of internally *buried* amino acids; nevertheless these latter determine the internal and overall surface geometry of the subunit. Whatever the protein-protein interaction involved in determining tertiary and quaternary structure, it generally does not involve covalent bonding (unlike in primary structure, and the relatively labile disulphide bonds in tertiary structure). It hardly needs to be emphasised that this leads to an increasing number of structural possibilities, exemplified by the expanding number of protein structures that have been determined by X-ray crystallography over the past 50 years (www.rcsb.org/pdb/statistics/contentGrowthChart.do?content=tota).

Among these are numerous large multimeric (homo- and hetero-oligomeric) protein complexes, many of which were *discovered* or defined by transmission electron microscopy, that have until relatively recently presented difficulties for crystallization and X-ray crystallography. The coming together of conventional transmission electron microscopy and cryo-electron microscopy with X-ray crystallography has led to much progress in the structural analysis of many large oligomeric proteins and macromolecular assemblies. Nevertheless, these structural tools are greatly supported by several essential biochemical and biophysical techniques, such as electrophoresis, chromatography, centrifugation, light scattering, spectroscopy and others, for the purification and analysis of proteins (see three recent texts "Protein Purification and Analysis", iConcept Press). The chapters in the present book and related volumes of the *Subcellular Biochemistry* series present detailed reviews on the structure and function of many interesting protein complexes, and thereby create an almost encyclopaedic coverage of this important aspect of protein biochemistry.

In a class of their own are the non-covalent antigen-antibody complexes, so important for the immune response. Furthermore, the polyclonality of in vivo antibody production contrasts with the monoclonality of in vitro cellular production, along with the varying affinity of antibodies for their highly specific protein epitope. Whilst being an exciting area of study, this is beyond the scope of the present chapter and this particular volume. Likewise, a consideration of covalent protein-protein and protein-nucleic acid complexes, whilst of great interest, cannot be dealt with in this chapter.

Here we present a limited selection of macromolecules in which we are interested and have studied in recent years, purely as examples of the many macromolecular assemblies that could be considered. Undoubtedly each protein complex possesses unique structural features responsible for the quaternary structure created by the subunit protein-protein association. Nevertheless, some principles of assembly can be defined and have been advanced by protein chemists, virologists and others. In the main, however, symmetrical and asymmetrical protein complexes remain as individual items for study, with an ever increasing number under investigation as present-day biological and biomedical research advances.

Protein Oligomeric Structure: Background

Ahnert et al. (2015) advanced a classification model for protein complexes based upon theoretically possible quaternary structure organization. Although a *periodic table* of protein subunit organization was established as a useful predictive model, it serves primarily to emphasise the incredible number of quaternary structures that spontaneously occur in nature. There can be no doubt that protein complex formation occurred during early evolution (Marsh and Teichmann 2014) and that the accrued functional benefits resulting from stability, allostery, and the conservation and compartmentalisation of reaction intermediates, have been retained, carried forward and *evolved* within all species through to the present day.

The late N. Michael Green, who worked at the UK MRC National Institute for Medical Research, Mill Hill (http://www.bridgesmathart.org/art-exhibits/bridges06/green.html), illustrated the diversity of protein multimerization through "plastic monkey" models, which made 'asymmetric units' in his sculptures. His wide interest in protein structure, including immunoglobulins, avidin, glutamine synthase, and transcarbamylase, informed his ideas and art (Valentine and Green 1967; Green et al. 1972; Green 1972, 1990).

"*Every living organism produces a variety of complex structures by self-assembly of identical building blocks of one or more types. These structural units are usually protein molecules which have evolved to assemble spontaneously, using multiple weak bonds. The principles of thermodynamics ensure that the most stable links are used. The number of such links is maximised in symmetrical structures, in which all the units are identically bonded. Natural selection ensures that assemblies with useful biological properties are perpetuated* (e.g. *virus coats, muscle fibres, mitotic spindles, intercellular junctions and many enzymes*)."

Whilst being an over simplification of the complex inter-molecular bonding present in any symmetrical oligomer, along with asymmetrical oligomers, Green's early concepts did indicate the basic principle of homo-oligomeric and hetero-oligomeric assembly, insight that has been advanced greatly in recent years. Further examples of early application of negative staining to the enzymes complexes and macromolecules was presented by Van Bruggen et al. (1960, 1981), Matthews and Bernhard (1973), and Oliver (1973).

The Structural Techniques

Biochemical and biophysical techniques have contributed toward the study of protein structure. Whilst analytical ultracentrifugation played a useful role in defining monomers and their higher mass assemblies during the first half of the 20th century, its contribution has diminished. However, first and foremost must be native- and SDS-polyacrylamide gel electrophoresis (PAGE), techniques valuable for the assessment of protein purity and subunit composition (Laemelli 1970). These electrophoretic techniques go hand in hand with the various chromatographic, preparative centrifugation and spectroscopic procedures used to achieve and assess protein purity from cells and biological tissues, or more usually in recent years from proteins expressed in bacteria, yeasts and animal cells. In this context, gene sequencing and cloning, along with protein amino acid sequencing, have played an important role. Indeed, the exponential increase in protein structure depositions in the protein databank has been a direct consequence of the technological advances from the major structural genomics initiatives over the last twenty years (Grabowski et al. 2016) and technological advances in X-ray sources, both in the laboratory and at synchrotron facilities (Owen et al. 2016). Laboratory automation and robotics have been of particular impact, with nano-scale protein crystallisation robotics reducing the amount of precious protein required for crystallisation, and automated sample-changers at synchrotrons speeding up data collection from protein crystals.

For the study of small proteins (<250 kDa), in addition to protein sequencing, X-ray crystallography (and to a lesser extent NMR spectroscopy) (Bernal and Crowfoot 1934) has been the predominant technique, expanding seemingly exponentially ever since the atomic structures of myoglobin and haemoglobin were determined more than 50 years ago (Kendrew et al. 1960; Perutz et al. 1960). With the polypeptide hormone insulin, the presence of two polypeptide chains, containing three covalent disulphide bridges was defined by Fred Sanger and colleagues (Ryle et al. 1955). Insulin Zinc-crystals contain hexamers as the unit cell (Harding et al. 1966), indicating the property of molecules to non-covalently associate under appropriate solution conditions. Indeed, the repeating unit cell within a protein crystal is often an oligomer, thereby indicating the strong tendency of like molecules to associate. It is abundantly clear that the progressive increase in computing power through the recent decades has greatly helped protein crystallography for the analysis of protein structure, as it has in many areas of science and technology. However, for an increasing number of large protein molecules the production of suitably diffracting crystals and subsequent generation of crystallographic data has proved a stumbling block.

It is here that transmission electron microscopy has made a significant contribution. Starting with negative staining and to a lesser extent metal shadowing that reveal the protein surface profile (reviewed by Harris 2015), followed by higher resolution cryo-electron microscopy of unstained vitrified specimens (Adrian et al. 1984) (that also has an increasing potential to reveal protein internal and surface structure to the low Å atomic level), the structure of many protein oligomers and assemblies has been defined.

The low resolution achieved by negative staining ($\sim$20 Å) has been steadily improved with the increasing availability of cryo-electron microscopic data, derived from the technical advances at the instrumental level, particularly in the development of direct electron detectors (McMullan et al. 2016), and massively increased image processing capability of recent decades, to achieve near-to atomic resolution ($\sim$2 Å) from single particle analysis and crystallographic analysis of 2D membrane crystals (Henderson 2015; Vinothkumar and Henderson 2016; Stahlberg et al. 2015). The development of software for protein digital image analysis and 3D reconstruction has played a significant role in this (Zivanov et al. 2018), along with the fitting/docking of higher resolution X-ray data of subunits into lower resolution electron microscopy data and homology modelling (Nicholls et al. 2018). It is appropriate to acknowledge the contribution of numerous physical/biophysical and computer-literate scientists and electron microscopists to the development of this field; it probably would not have happened in the hands of biologist, biochemists and protein chemists alone! Indeed, the continuing impact of computer refinement, image analysis and 3D modelling on the understanding of protein assemblies is steadily increasing, but will not be dealt with in the present chapter, and it does not represent a major component of the book as a whole.

Some Oligomeric Proteins and Complexes

Hemoglobin

Following the seminal X-ray crystallographic studies on myoglobin and haemoglobin by Kendrew and Perutz protein biochemists must have immediately contemplated as to why myoglobin is a monomer and the evolutionary-related haemoglobin A is an $\alpha 2ß2$ globin tetramer? And, why is it not and $\alpha 4$ or $ß4$ tetramer? The expanding sequence data on the embryonic, foetal and adult hemoglobins and the range of hemoglobinopathies has generated much information as to the range of tetramer formation, all indicating that this is the predominant and stable oligomeric form adopted by the haemoglobin monomers. The amino acid substitution of valine in the place of glutamine, at position 6 of the ß-globin chain due to the single codon missense mutation GAG to GTG, is responsible for sickle cell haemoglobin (HbS). This valine substitution creates a more hydrophobic environment and increases the tetramer instability when deoxygenated, resulting in decreased solubility of the HbS, with formation of helical polymers/fibres that bundle within the erythrocyte, essentially in a *quasi* crystalline manner. Due to the restricted space inside the erythrocyte, the HbS bundles bend, creating the characteristic sickle cells. Repeated cycles of oxygenation of HbS oxygenation and deoxygenation lead to irreversible sickling. This intracellular *quasi* crystallization of HbS is one of the few examples of natural 3D crystallization found in nature, other examples include the potato proteinase inhibitor (Rodis and Hoff 1984), the *Bacillus thuriengensis* toxin (Li et al. 1991) and the polyhedrin proteins from insect viruses (Ji et al. 2015). There are rather more examples of naturally occurring 2D membrane protein crystals (*Halobacterium halobum* purple membrane and the chloroplast thylakoid light harvesting complex).

The annelid, crustacean and insect hemoglobins all exhibit rather different oligomerization properties to those from vertebrates, exemplified most strongly by the earthworm (*Lumbricus tearrestris*) and ragworm (*Nereis virens*) hexagonally-ordered high molecular mass hemoglobins (erythrocruorins) (Fig. 1.1). Although earthworm haemoglobin was one of the first proteins from which 3D crystals were produced (and indeed it readily produces 2D crystals), only a low resolution 5.5 Å X-ray structure has been produced (Royer et al. 2000). and it has fallen to cryoEM of single molecules to generate singe particle reconstructions. Chen et al. (2015) produced an 8.1 Å structure and Afanasyev et al. (2017) achieved a higher resolution structure (at $\sim$3.8 Å). In this hexameric double-layer macromolecule a significant feature is the 1/12th unit containing 12 haem-containing globins and a "stem" containing the triple coiled-coil linker chain (Fig. 1.2) directed to the centre of the hexagonal structure (PDBID: 5M3L). Other invertebrate haemoglobins are apparently less well-ordered oligomers (Rousselot et al. 2006) and even occur as linear haemoglobin polymers (Borhani et al. 2012).

The Haemocyanins

For the megadalton copper-containing cephalopod and gastropod respiratory proteins, the haemocyanins, ring-like decameric $\sim$3.8 MDa oligomers are formed (Fig. 1.3), and in some species these form as didecamers and multidecamers (Markl 2013; Kato et al. 2017). Negative staining in the presence of PEG potentiates the formation of 2D hemocyanin crystals (see Fig. 1.3) (Harris and De Carlo 2014). An antiparallel subunit dimer (protamer) is the repeating unit within the decamer (Harris et al. 2004; Meissner et al. 2007a). The individual elongated haemocyanin

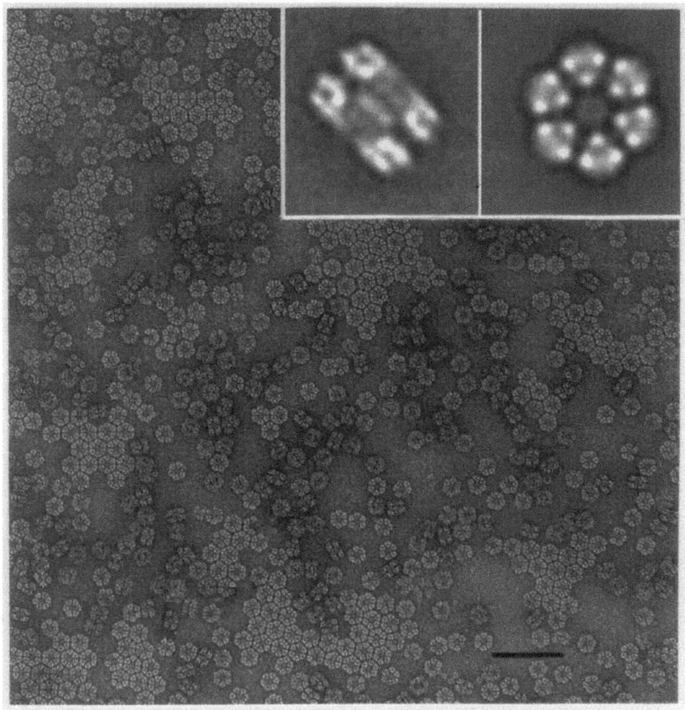

Fig. 1.1 Haemoglobin from the ragworm *Nereis virens*. The hexameric protein complex is shown by cryo-negative staining (courtesy of the late Marc Adrian), with the molecules oriented in the two predominant orientations. The insets show 2D molecular averages

subunits contain seven or eight covalently-linked genetically individual "functional units" (FUs), each with a copper-containing oxygen binding site. Gene duplication during evolution is believed to have generated these distinct FUs. Indeed, the polypeptide linker regions between the FUs can be considered to create an "oligomeric" subunit containing multiple covalent links. Most molluscan hemocyanins will dissociate to the single subunit in an alkaline buffer, but the hemocyanin subunit dimer from the chiton *Acanthochiton fasicularis* remains stable (Harris et al. 2004). An understanding of the formation of hemocyanin multidecamers of varying length (and indeed the proteolytically derived helical tubules), together with the deviant tridecameric mega-hemocyanin (Gatsogiannis et al. 2015) remain to be established.

Recent, 3 Å, crystallographic data from the *T. Pacificus* haemocyanin decamer (Gia et al. 2015) has revealed considerable structural detail. The subunit organization and copper-containing FUs were defined, in particular the hydrophobic FU interaction sites, responsible for creation of the protomer (dimer). Calcium/ionic bridging between dimers is also thought to be of importance for the creation of the higher order decamers.

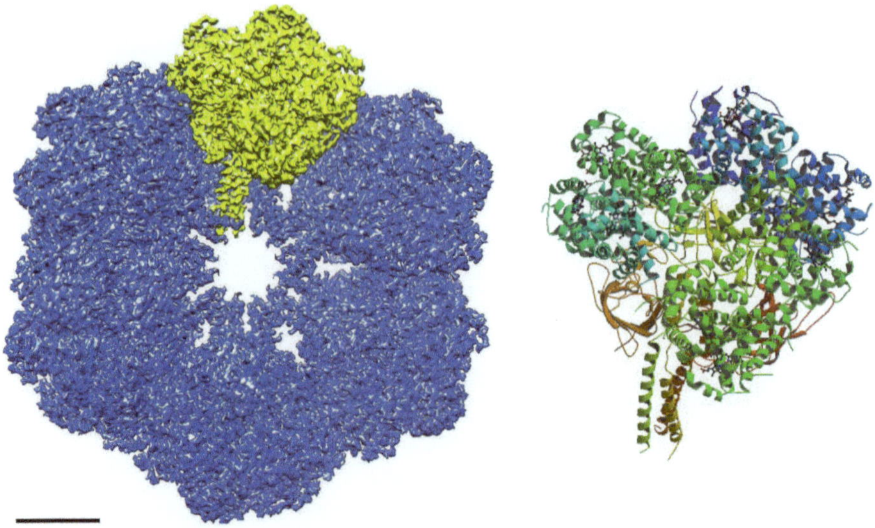

Fig. 1.2 The asymmetric unit ('protomer' or '1/12th unit') of the earthworm *Lumbricus terrestris* haemoglobin at 3.8 Å containing 12 haem-containing globins and a "stem" containing the triple coiled-coil linker chain (PDB entry 5m3l) from cryoEM (Afanasyev et al. 2017). Reproduced with permission of the International Union of Crystallography. (https://journals.iucr.org/)

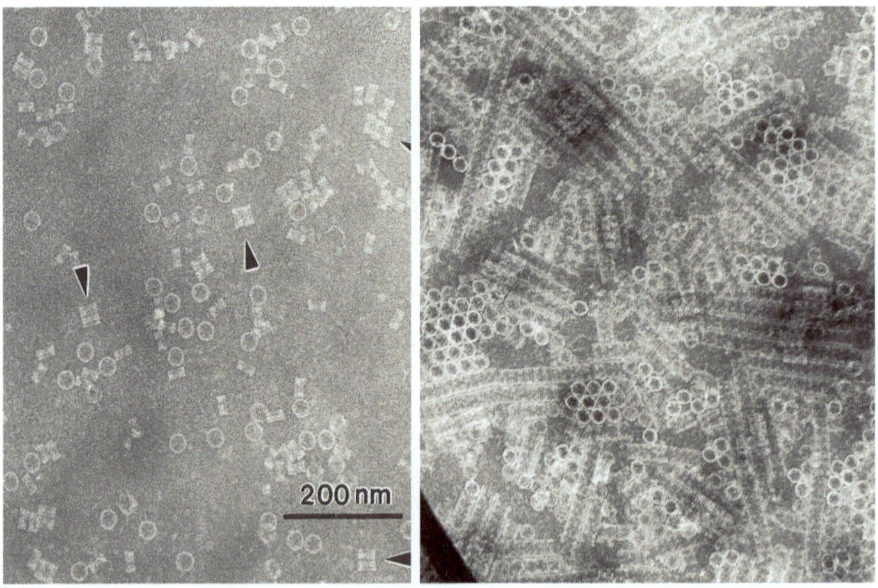

Fig. 1.3 TEM images of octopus haemocyanin decamers negatively stained with ammonium molybdate-trehalose (LHS) and ammonium molybdate-trehalose in the presence of PEG 1000 (RHS) across a hole. Note the ring-like face-on decamers and the side-on rectangular molecular images that have a tendency to form stacks (arrows). The presence of PEG during the negative staining procedure with a higher protein concentration promotes the formation of 2D crystals (De Carlo and Harris 2011)

The oligomeric haemocyanins from numerous other arthropods, products of a different gene family to the molluscan haemocyanins, are currently under active investigation and present an equally fascinating structural story, yet to be fully revealed (Martin et al. 2007).

The Peroxiredoxins

The large family of thiol-specific antioxidant and redox signalling enzymes, generally termed the peroxiredoxins, has emerged as a significant component of the overall enzymic antioxidant protection. The discovery of the erythrocyte peroxiredoxin-II (Prx-II_as a ring-like protein oligomer goes back to the late 1960s, when a ring-like decameric protein of then unknown function was defined by negative stain electron microscopy of protein extracts produced from erythrocyte ghosts (Harris 1968, 1969). Subsequently, from SDS-PAGE the single subunit molecular mass of ~20 kDa was defined (Harris and Naeem 1978) and from X-ray crystallography (Schröder et al. 2000) it became clear that the oligomer is indeed a pentamer of domain swapped dimers (D5 symmetry). Correlation of the then available TEM (Fig. 1.4) and X-ray data (Harris et al. 2001) produced a high degree of unity with a meaningful molecular superimposition (Fig. 1.5). The topic was

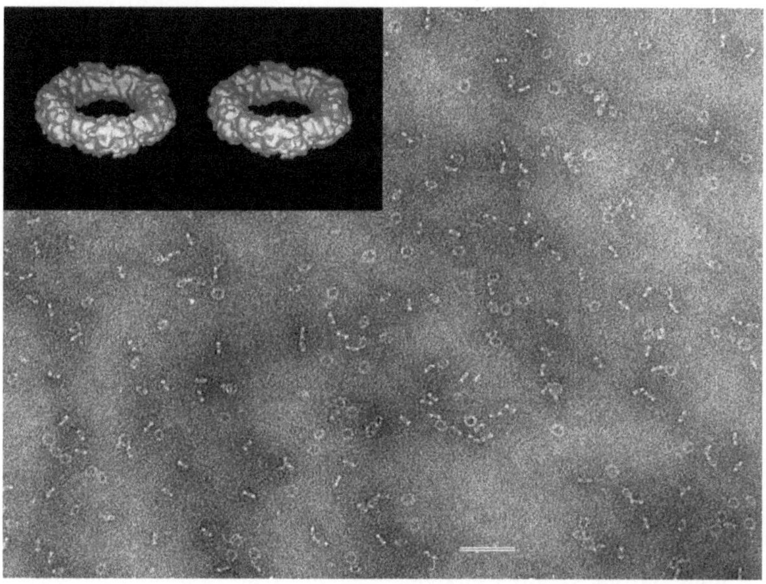

Fig. 1.4 Human erythrocyte peroxiredoxin II imaged by TEM negative staining with ammonium molybdate in the presence of trehalose (Harris et al. 2001), with a stereo pair of the 19 Å image reconstruction (inset)

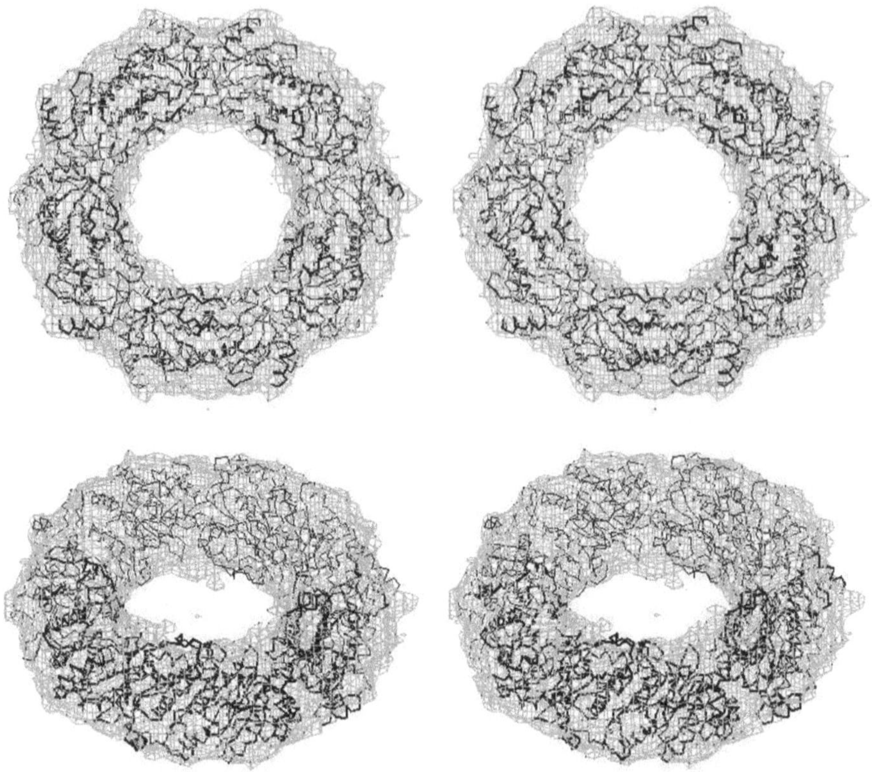

Fig. 1.5 Fitting of the 2.7 Å X-ray data of peroxiredoxin II (Schöeder et al. 2000) into the ~19 Å TEM envelope (Harris et al. 2001), as stereo pairs

reviewed in depth from a structural and functional perspective by Wood et al. (2003) and also by Cao and Lindsay (2017) in Volume 83 of this series, and Volume 44 was devoted to Peroxiredoxin Systems (Flohé and Harris 2007). Figure 1.6 depicts the historical data progression from negative stain TEM of the Prx-2 decamer to the X-ray crystal structure, within which the monomer-dimer and the dimer-pentamer interfaces have been defined (Schröder et al. 2000). Unexpectedly, when trying to produce 2-D crystals of the erythrocyte Prx-2 decamer in the presence of ammonium molybdate-PEG for TEM study, Meissner et al. (2007a) created a higher-order dodecahedron of decamers (Fig. 1.7). This assembly resembles virus-like particles and other protein cages. As a homo-oligomeric dodecahedron it emphasises the likely creation of specific non-covalent protein-protein interactions mediated by compounds used for protein crystallisation in this instance, although the creation and stabilization of this structure in conventional buffer solutions, or other crystallization conditions has yet to be demonstrated. Nevertheless, the ordered charge distribution on the peroxiredoxin decamer (Fig. 1.8) may contribute to the likelihood of dodecahedron creation. Interest in the

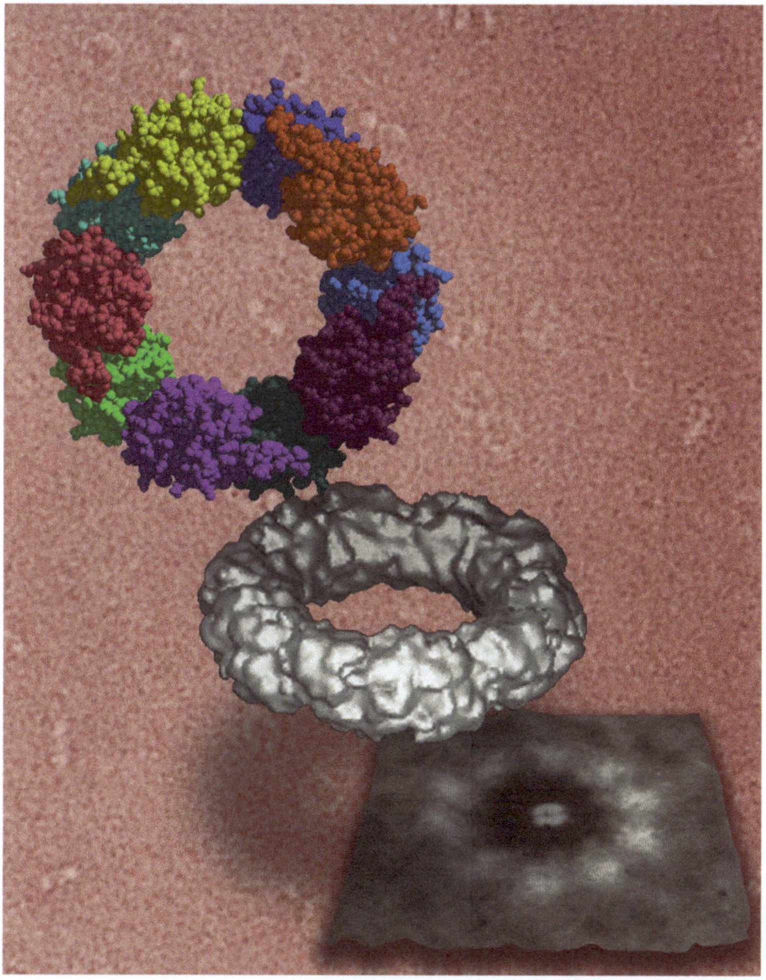

Fig. 1.6 A composite showing the first EM definition of 10 subunits with the Prx II toroid Å (Harris 1969), the Prx II molecular reconstruction from TEM| (Harris et al. 2001) and the 2.7 Å Prx II X-ray model (Schröder et al. 2000), with a background TEM image. (Courtesy of Zachary A Wood)

formation of cage-like protein assemblies and virus-like particles is expanding, revealing by TEM structures similar to the peroxiredoxin II decamer (Bale et al. 2016). Similar considerations undoubtedly apply during the formation of tubular stacks and 2D crystals of Prx II (Harris et al. 2001) and tubules of Prx III (Yewdall et al. 2018a).

Prx III is a ring of six dimers and has been shown to create double/catenated dodecamers (Cao et al. 2005, 2015). The ready formation of stacked ring tubules (nanotubules), a characteristic of Prx III, is a feature expanded upon by Phillips

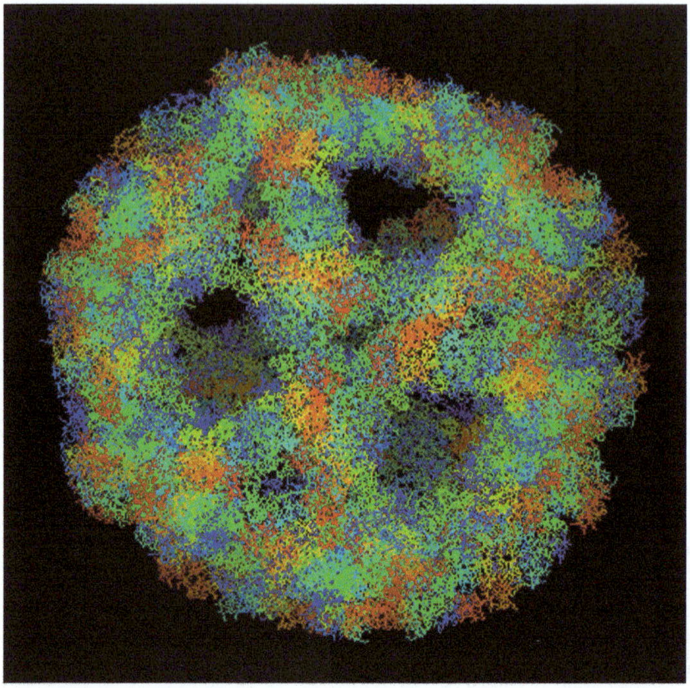

Fig. 1.7 The peroxiredoxin II dodecahedral higher-order assembly (i.e. containing 12 decamers) viewed down the three-fold axis (see Meissner et al. 2007a, b)

et al. (2014) and subsequently within the sphere of nanotechnology by Yewdall et al. (2018b) and Manuguri et al. (2018). Using the recently developed Volta phase plate has found application for the structural study of Prx-III (Khoshouei et al. 2016), with the claim that there is benefit for the study of smaller protein molecules.

Other members of the peroxiredoxin family have also received the attention of electron microscopists and more extensively X-ray crystallographers. The PDB contains many examples of peroxiredoxins, monomers, dimers and oligomers, in oxidised and reduced states and molecules from mutant genes. Thus, it is probable that these proteins and their oligomers will continue to be of structural interest.

Collagen Assemblies

Collagen Fibrils and SLS Crystallites

The fibrous proteins present special classes of protein association, in some cases bordering upon linear crystallization. Although not included in the present volume, this group of proteins were discussed in detail in Volume 82 of the series, edited by

David Parry and John Squire. Here we will consider briefly only fibrillar collagen, as a special example of protein self-association. (Other extracellular fibrils, such as the intermediate filaments, elastin, amyloid, and intracellular fibrils and tubules, including actin, myosin, titin, tropomyosin, microtubules and others, have also received the attention of electron microscopists and X-ray crystallographers. Perhaps most significant is the work on skeletal muscle myofibrils and the sliding filament mechanism of contraction (Huxley and Hanson 1953), induced by the action potential, calcium entry and control of ATP/energy utilization).

The collagen molecule is an elongated heterotrimer ($\sim$300 nm triple-helix) hydroxyproline-rich structural protein; most collagen family members possesses unique self-assembly properties in vivo and in vitro in physiological and experimental solutions, generating quasi-crystalline banded fibres aligned as bundles that possess inherent longitudinal strength and flexibility (Holmes and Kadler 2006; Harris and Reiber 2007; Harris et al. 2013; Harris 2017). The imaging and modelling of collagen fibril assembly has been considered extensively since the early days of transmission electron microscopy through to the present day, resulting in the widely accepted linear gap ($\sim$40 nm) and partial overlap ($\sim$27 nm) model with lateral alignment of heterotrimers that creates the $\sim$67 nm "D-banded" repeating structure (Fig. 1.9). Detailed molecular aspects of the fibrillar collagens were presented by Bella and Hulmes (2017).

What is less widely appreciated is the fact that under in vitro experimental conditions the collagen molecule can also create a deviant collagen assembly, as discrete rod-like structures termed segment long spacing (SLS) bundles or *crystallites* (Fig. 1.10), containing many parallel N-/C-terminus laterally aligned

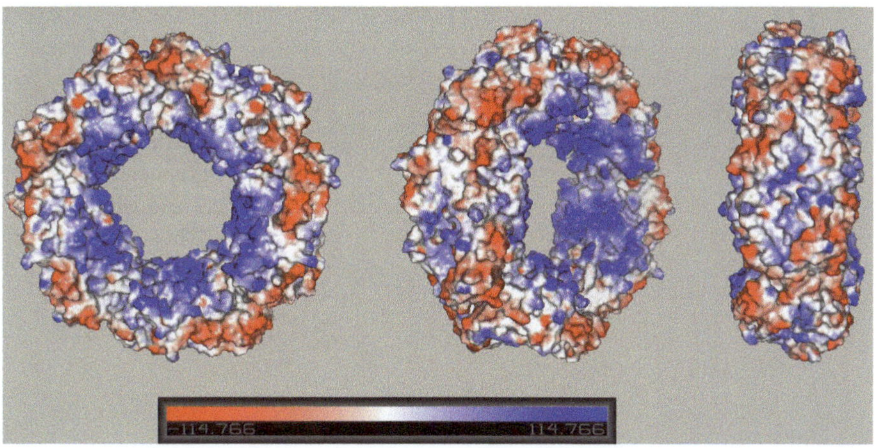

Fig. 1.8 The charge distribution on the surface of the peroxiredoxin II decamer (red negatively charged residues, blue positive residues). Note the central predominance of positive surface charge and the repeating distribution of positive and negative charge on the outer surface, available for possible inter-decamer linkage

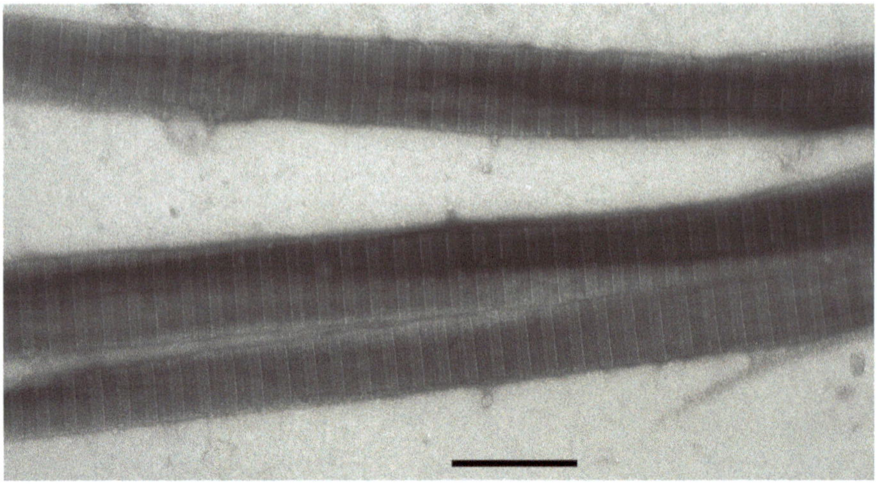

Fig. 1.9 Collagen type I fibres produced by in vitro assembly from the acid-soluble heterotrimer (Harris 2017). Each fibre contains a number of thinner fibres aligned with respect to their underlying molecular 67 nm repeating pattern of linear molecules (i.e. the established gap/overlap model). The collagen fibres are negatively stained with uranyl acetate. The scale bar indicates 400 nm

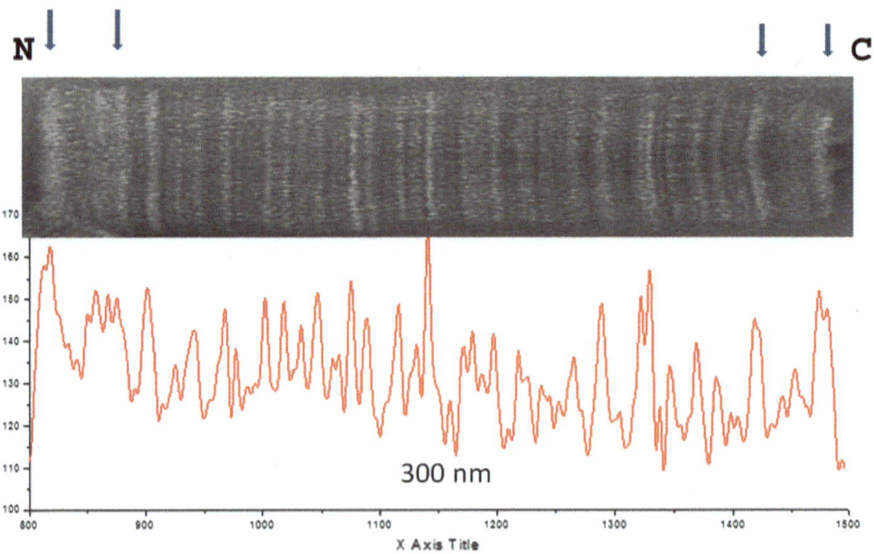

Fig. 1.10 A single collagen type I segment long spacing (SLS) crystallite (length ~300 nm), negatively stained with uranyl acetate (top). A densitometric scan (bottom) reveals the zones of increased protein thickness, indicative of the varying thickness along the individual aligned heterotromers oligomerized within the SLS bundle

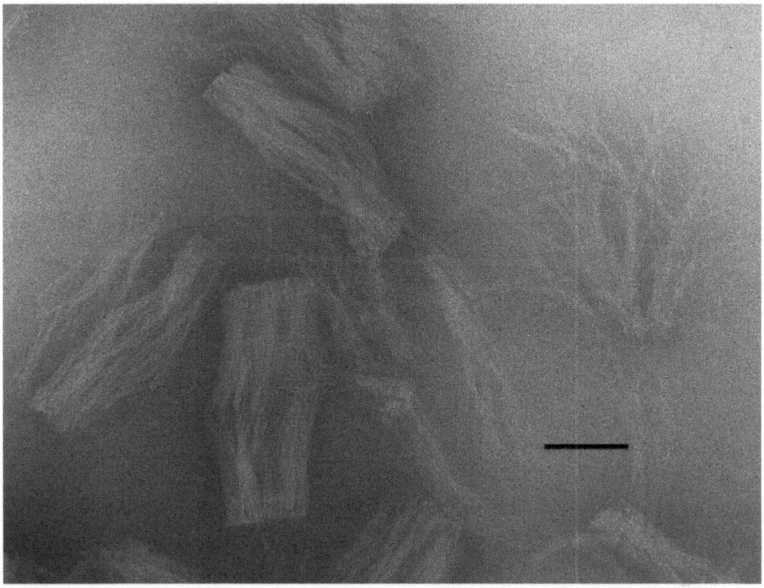

Fig. 1.11 SLS crystallites negatively stained in ammonium molybdate for TEM study undergoing spontaneous dissociation during specimen preparation. The individual hetetrotrimers can be seen to splay out from the unstable SLS bundles that have lost their characteristic banding pattern (Harris and Lewis 2016). From the original acidic collagen solution the heterotrimers have been imaged as flexible rather than straight elongated molecules (Harris et al. 2013) (The transfer to neutral pH saline solutions imparts structural linearity/rigidity during fibril formation, also during mildly acidic ATP conditions that generate SLS crystallites)

heterotrimers. (For the other deviant fibrous form of collagen, termed fibrous long spacing (FLS), there is only a ~27 nm molecular overlap, with no linear spacing; see Doyle et al. 1975). The band pattern exhibited by collagen type I SLS crystallite indicates the summated variation in protein thickness along its length (densitometric scan, Fig. 1.9), in turn indicating the periodic thickness variation along a single triple helix. It is this feature of the collagen molecule that underlies and creates the more complex banding pattern exhibited by the collagen fibril. The stability of the collagen SLS crystallite, formed under mildly acidic conditions in the presence of ATP is not great, indicated by the partial dissociation back to individual hetereotrimers when negatively stained with neutral pH ammonium molybdate (Fig. 1.11) (see Harris and Lewis 2016). However, negative staining with uranyl acetate stabilizes the SLS structures, although stability can be imparted by glutaraldehyde crosslinking when negative staining with ammonium molybdate. The subtleties of fibrous protein-protein interactions undoubtedly play numerous important roles in filament creation.

High-Resolution Studies of Model Proteins

With the now widespread installation of highly-automated 300 kV cryo-electron microscopes with direct electron detectors, much recent interest has focused on pushing the effective resolution of single-particle reconstructions of protein complexes and with this, a need for well-behaved and readily available protein complex samples has grown. In an analogous manner to the utility that lysozyme and thaumatin have found for testing synchrotron beamlines and X-ray diffraction processing software, ferritin and beta-galastosidase have become the mainstay test protein complexes for cryo-electron microscopy. The *Escherichia coli* beta-galactosidase complex fulfils this role, weighing in at nearly 500 kDa the homotetramer, with its mixed alpha- and beta- secondary structure and D2 symmetry is widely used to test new microscope installations and software. A recent reconstruction using Relion-3.0 achieved a resolution of 1.9 Å (Zivanov et al. 2018). The four-helix bundle of ferritin forms a highly ordered and stable 24-meric nanocage of around half a megadalton. Horse spleen ferritin is readily available from chemical suppliers and recombinant ferritins are simple to purify, as they are usually thermostable. Using Relion, a 1.65 Å resolution reconstruction was produced of the human apo-ferritin complex, which is higher than many ferritin structures determined by X-ray crystallography. These records for published high resolution cryo-EM single particle reconstructions have already been broken on test data collected and current anecdotally reported resolutions are close to what an X-ray crystsllographer would consider to be a true ‘atomic resolution’ structure. We will certainly see many more single particle reconstructions at sub-2 Å resolution as new microscopes and detectors are introduced in laboratories across the world. Much like the crystallographic test samples, lysozyme and thaumatin, our knowledge of the biochemistry and structure of ferritin and beta-galactosidase has not been significantly advanced by these new high-resolution reconstructions. They do however provide an important benchmark and show what can be done by modern microscopes, detectors, and software (Fig. 1.12).

Encapsulins

The encapsulins are a class of protein complexes that are of growing interest, from both a basic biological perspective and as tools in applied biotechnology. The encapsulins were originally identified as a 30 nm diameter virus-like particles in the hyperthermophilic archaeaon *Pyrococcus furiosus* (Namba et al. 2005), a later crystal structure of these particles highlighted their structural relationship to bacteriophage capsids (Tatur et al. 2007). An accidental discovery of a related virus-like particle in *Thermotoga maritima*, with an encapsulated ferritin-like protein led to the coining of the term ‘encapsulin’ for the proteins forming these bacterial nanocompartments (Sutter et al. 2008). Close analysis of the crystal

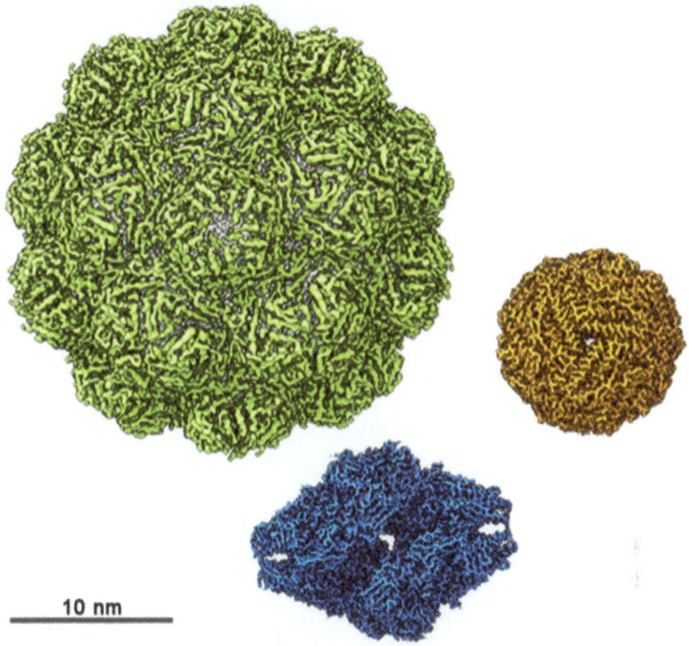

Fig. 1.12 Depictions of the electron potential maps of some recent high-resolution single-particle reconstructions of protein complexes. The T = 3 icosahedral *Myxococcus xanthus* encapsulin is shown in lime yellow (EMD: 5917) (McHugh et al. 2014); octahedral horse spleen ferritin in orange (EMD: 0263) (Zivanov et al. 2018); and the tetrameric D2-symmetry *E. coli* beta-galactosidase in blue (EMD: 0153) (Zivanov et al. 2018). The scale bar represents 10 nm

structure of the *T. maritima* encapsulin showed a short peptide sequence from the encapsulated enzyme bound to the interior wall of the nanocage. This peptide is responsible for directing enzyme cargoes to the encapsulin cage and it has been shown to be sufficient to direct heterologous proteins to recombinant encapsulins (Tamura et al. 2014).

Interest in encapsulins intensified after 2013, when primary research began appearing more frequently on the structure and function of these fascinating complexes. Since then, a single-particle cryo-EM reconstruction of the *Myxococcus xanthus* encapsulin has been published (Fig. 1.12) (McHugh et al. 2014), along with a detailed study of the biochemistry of the new family of ferritin-like proteins found within many of these proteins (He et al. 2016). What is striking about these encapsulated-ferritins is their annular decameric structure, which is in contrast to the 24-meric nanocages formed by the classical ferritins. The encapsulated ferritins function as ferroxidase enzymes to oxidise iron(II) to iron(III) within the encapsulin nanocage, where it is stored as inert ferrihydrite, and iron-phosphate minerals. What is striking about these complexes, is their ability to hold up to ten times more iron than the classical ferritin nanocages (He et al. 2016; McHugh et al. 2014). The role that such a massive iron store plays in bacteria is still open to speculation. A recent survey of bacterial and archaeal species highlighted the widespread distribution of encapsulins across diverse environmental niches and identified a number of new cargo proteins found within them, including new ferroxidase enzymes found only in bacteria of the firmicutes phylum, and enzymes involved in the response to nitrosative stress (Giessen and Silver 2017). It is clear from this survey that the primary role of encapsulins is to either protect the host cell from oxidative damage caused by the redox reactions carried out by their enzyme cargoes, or to protect the host from oxidative damage caused by the substrates of these cargo enzymes.

The ability to target non-native proteins to the interior of the encapsulin cage, coupled with its stability and amenability to surface modification, has led to the adoption of encapsulins for biotechnological applications that range from the targeted delivery of drugs (Moon et al. 2014), to imaging tools for studying mammalian cells (Sigmund et al. 2018). While these applications are still in the early stages of development and implementation in medicine and biotechnology, there is still much to be learnt about the basic biology and function of these complexes in their host organisms.

This introductory chapter cannot cover all interesting protein complexes and represents only a selection of specific interest to the editors of this volume. Indeed, in this series of books, we can only hope to cover a small selection of the recent advances in the study of macromolecular complexes. The importance of higher-order structure in biology and the coming together of multiple proteins, is central to the function of all organisms. As our technology for the high-resolution study of proteins advances, in particular our ability to image and generate 3D reconstructions of proteins by cryo-electron microscopy, more interesting macromolecular complex structures are being elucidated and their functions illuminated through the combination of structural and functional data. We hope that these selected chapters are of interest to our readers and that authors whose work we have not been able to cover in this introduction, and who have not contributed to this volume, are not offended by the omission of their work. Finally, we would like to thank all of our contributors, who have all provided interesting considerations of their work and its context within the wider biochemistry and biology of their host systems and organisms.

References

Adrian M, Dubochet J, Lepault J, McDowall AW (1984) Cryo-electron microscopy of viruses. Nature 308:32–36

Afanasyev P, Seer-Linnemayr C, Ravelli RBG, Matadeen R, De Carlo S, Alewijnse B, Rodrigo V. Portugal RV, Pannu NS, Schatz M, van Heel M (2017) Single-particle cryo-EM using alignment by classification (ABC): the structure of *Lumbricus terrestris* haemoglobin. IUCrJ 4: 678–694

Ahnert SE, Marsh JA, Hernández H, Robinson CV, Teichmann SA (2015) Principles of assembly reveal a periodic table of protein complexes. Science 350. https://doi.org/10.1126/science.aaa2245

Bale JB, Gonen S, Liu Y, Sheffler W, Ellis D, Thomas C, Cascio D, Yeates TO, Gonen T, King NP, Baker D (2016) Accurate design of megadalton-scale two-component icosahedral protein complexes. Science 353:389–394

Bella J, Hulmes DJS (2017) Fibrillar collagens. In: Parry DAD, Squire JM (eds) Fibrous proteins: structures and mechanisms. Subcellular biochemistry vol. 82. Springer Nature, pp 457–490m

Bernal JD, Crowfoot D (1934) X-ray photographs of crystalline pepsin. Nature 133:794–795

Borhani HA, Berghmans H, Trashin S, DeWael K, Fago A, Moens L, Habibi-Rezaei M, Dewilde S (2012) Kinetic properties and heme pocket structure of two domains of the polymeric hemoglobin of Artemia in comparison with the native molecule. Biochim Biophys Acta 1854:1307–1316

Cao Z, Roszak AW, Gourlay LJ, Lindsay JG, Isaacs NW (2005) Bovine mitochondrial peroxiredoxin III forms a two-ring catenane. Structure 13:1661–1664

Cao Z, McGow DP, Shepherd C, Lindsay JG (2015) Improved catenated structures of bovine Peroxiredoxin III F190L reveal details of ring-ring interactions and a novel conformational state. PLoS ONE 10(4):e0123303. https://doi.org/10.1371/journal.pone.0123303

Cao Z, Lindsay JG (2017) The peroxiredoxin family: an unfolding story. In: Harris JR, Marles-Wright J (eds) Macromolecular protein complexes. Springer Nature, pp 127–147

Chen W-T, Chen Y-C, Liou H-H, Chao C-Y (2015) Structural basis for cooperative oxygen binding and bracelet-assisted assembly of *Lumbricus terrestris* haemoglobin. Scientific Reports 5:9494. https://doi.org/10.1038/srep09494

De Carlo S, Harris JR (2011) Negative staining and cryo-negative staining of macromolecules and viruses for TEM. Micron 42:117–131

Doyle BB, Hukins DWL, Hulmes DJS, Miller A, Woodhead-Galloway J (1975) Collagen polymorphism: its origins in the amino acid sequence. J Mol Biol 91:79–99

Fitzpatrick AWP, Falcon, He S, Murzin AG, Murshudov G, Garringer HJ,- RA, Ghetti B, Goedert M, Scheres SHW (2017) Cryo-EM structures of Tau filaments from Alzheimer's disease brain. Nature. 547(7662):185–190. https://doi.org/10.1038/nature23002

Flohé L, Harris JR (2007) Peroxiredoxin systems, subcellular biochemistry, vol. 44. Springer, New York

Gatsogiannis C, Hofnagel O, Markl J, Raunser S (2015) Structure of mega-hemocyanin reveals protein origami in snails. Structure 23:93–103

Gia Z, Matsuno A, Kato K, Kato S, Khan MRI, Shimizu T,Yoshioka T,Kato Y, Kishimura H, Kanno G, Miyabe Y, Terada T, Tanaka Y, Yao M (2015) Crystal structure of the 3.8-MDa respiratory supermolecule hemocyanin at 3.0 Å. Resolution Structure 23:2204–2212

Giessen TW, Silver PA (2017) Widespread distribution of encapsulin nanocompartments reveals functional diversity. Nature Microbiol 2 Article number: 17029. https://doi.org/10.1038/nmicrobiol.2017.29

Grabowski M, Niedzialkowska E, Minor W (2016) The impact of structural genomics: the first quindecennial. J Struct Funct Genomics 17:1–16

Green NM (1972) Analysis of the Structure of Complex Proteins by Electron Microscopy. In: Jaenicke R, Helmreich E (eds) Protein-protein interactions., Springer| Verlag, pp 183–211

Green MN, Valentine R, Wrigley NG, Ahmad F, Jacobson B, Wood HG (1972) Transcarboxylase XI. Electron microscopy and subunit structure. J Biol Chem 247:6284–6298

Green NM (1990) Avidin and streptavidin. Methods Enzymol 84:51–67

Harding MM, Crowfoot Hodgkin D, Kennedy AF (1966) The crystal structure of insulin ll. An investigation of rhombohedral zinc insulin crystals and a report of other crystalline forms. J Mol Biol 16:212–226

Harris JR (1968) Release of a macromolecular protein component from human erythrocyte ghosts. Biochim Biophys Acta 150:534–537

Harris JR (1969) The isolation and purification of a macromolecular protein component from the human erythrocyte ghost. Biochim Biophys Acta 188:31–42

Harris JR (2015) Transmission electron microscopy in molecular structural biology: a historical survey. Arch Biochem Biophys 581:3–18

Harris JR (2017) Visualizing in vitro type I collagen fibrillogenesis by transmission electron microscopy. In: Rittié L (ed) Fibrosis: methods and protocols. Methods in Molecular Biology, vol. 1627. Springer Science + Business Media LLC, pp 367–383

Harris JR, De Carlo S (2014) Negative staining and Cryonegative staining. In: John Kuo (ed.), Electron microscopy: methods and protocols. Methods in Molecular Biology, vol. 1117. © Springer Science + Business Media, New York, pp 215–258. https://doi.org/10.1007/978-1-62703-776-1_11

Harris JR, Naeem I (1978) The subunit composition of two high molecular weight extrinsic proteins from human erythrocyte membranes. Biochim Biophys Acta 537:495–500

Harris JR, Schroder E, Isupov MN, Scheffler D, Kristensen P, Littlechild JA, Vagin AA, Meissner U (2001) Comparison of the decameric structure of peroxiredoxin II by transmission electron microscopy and X-ray crystallography. Biochim Biophys Acta 1547:221–234

Harris JR, Meissner U, Gebauer W, Markl J (2004) 3D reconstruction of the hemocyanin subunit dimer from the chiton *Acanthochiton fascicularis*. Micron 35:23–26

Harris JR, Lewis RJ (2016) The collagen type I segment long spacing (SLS) and fibrillar forms: formation by ATP and sulphonated diazo dyes. Micron 86:36–47

Harris JR, Reiber A (2007) Influence of saline and pH on collagen type I fibrillogenesis in vitro: fibrin polymorphism and colloidal gold binding. Micron 38:115–521

Harris JR, Soliakov A, Lewis RJ (2013) In vitro fibrillogenesis of collagen type I in varying ionic and pH conditions. Micron 49:60–68

He D, Hughes S, Vanden-Hehjir S, Georgiev A, Altenbach K,Tarrant E, MackaY cl, Waldron KJ, Clarke DJ (2016) Structural characterization of encapsulated ferritin provides insight into iron storage in bacterial nanocompartments. eLIFE. https://doi.org/10.7554/eLife.18972.001

Henderson R (2015) Overview and future of single particle electron cryomicroscopy. Arch Biochem Biophys 581:19–24

Holmes DF, Kadler KE (2006) Collagen Fibril assembly in vitro. In: Harris R, Graham J, Rickwood D (eds) Cell biology protocols. Wiley, pp 375–378

Huxley HE, Hanson EJ (1953) Structural basis of the cross-striations in muscle. Nature 172: 530–532

Iadanza MG, Silvers R, Boardman J, Smith HI, Karamanos TK, Deb elouchina GT, Su Y, Griffin RG, Ranson NA, Radford SE (2018) The structure of a β_2-microglobulin fibril suggests a molecular basis for its amyloid polymorphism. Nat Commun. https://doi.org/10.1038/s41467-018-06761-6

Ji X, Axford D, Owen R, Evans G, Ginn HM, Sutton G, Stuart DI (2015) Polyhedra structures and the evolution of the insect viruses. J Struct Biol 192:88–99

Kato S, Matsui T, Gatsogiannis C, Tanaka Y (2017) Molluscan hemocyanin: structure, evolution, and physiology Biophys Rev. https://doi.org/10.1007/s12551-017-0349-4

Kendrew JC, Dickerson RE, Hart RG, Davies DR, Phillips DC, Shore VG (1960) Structure of myoglobin: a three dimensional Fourier synthesis at 2 Å. Nature 185:422–427

Khoshouei M, Radjainia M, Phillips AJ, Gerrard JA, Mitra AK, Plitzko JM, Baumeister W, Danev R(2016) Volta phase plate cryo-EM of the small protein complex Prx3. Nature. https://doi.org/10.1038/ncomms10534|www.nature.com/naturecommunications

Laemli UK (1970) Cleavage of structural proteins during the assembly of the head of bacteriophage T4. Nature 227:680–685

Li J, Carroll J, Ellar DJ (1991) Crystal structure of insecticidal δ-endotoxin from *Bacillus thuringiensis* at 2.5 Å resolution Nature 353:815–821

Manuguri V, Webster K, Yewdall NA, An Y, Venugopal H, Bhugra V, Turner A, Domigan LJ, Gerrard JA, Williams DE, Malamoström (2018) Assembly of protein stacks with in situ synthesized nanoparticle cargo. Nano Lett. 18:5138–5145. https://doi.org/10.1021/acs.nanolett.8b02055

Markl J (2013) Evolution of molluscan hemocyanin structures. Biochim Biophys Acta 1834:1840–1852

Martin AG, Depoix F, Stohr M, Meissner U, Hagner-Holler S, Hammouti K, Burmester T, Heyd J, Wriggers W, Markl J (2007) Limulus polyphemus Hemocyanin: 10 Å Cryo-EM structure, sequence analysis, molecular modelling and rigid-body fitting reveal the interfaces between the eight hexamers. J Mol Biol 366:1332–1350

Marsh JA, Teichmann SA (2014) Structure, dynamics, assembly, and evolution of protein complexes. Ann Rev Biochem https://doi.org/10.1146/annurev-biochem-060614-034142

Matthews BW, Bernhard SA (1973) Structure and symmetry of oligomeric enzymes. Ann Rev Biophys Bioeng 2:257–317

McHugh CA, Fontana J, Nemecek D, Cheng N, Aksyuk AA, Heymann JB, Winkler DC, Lam AS, Wall JS, Steven AC, Hoiczyk E (2014) A virus capsid-like nanocompartment that stores iron and protects bacteria from oxidative stress. EMBO J 33:1891–1917

McMullan G, Faruqi AR, Henderson R (2016) Direct electron detectors. Meth Enzymol 579:1–16

Meissner U, Gatsogiannis C, Moeller A, Depoix F, Harris JR, Markl J (2007a) Comparative 11 Å structure of two molluscan hemocyanins from 3D cryo-electron microscopy. Micron 38:754–768

Meissner U, Schröder E, Scheffler D, Martin AG, Harris JR (2007b) Formation, TEM study and 3D reconstruction n of the human erythrocyte peroxiredoxin-2 dodecahedral higher-order assembly. Micron 38:29–39

Moon H, Lee J, Min J, Kang S (2014) Developing genetically engineered encapsulin protein cage nanoparticles as a targeted delivery nanoplatform. Biomacromol 15:3794–3801

Namba K, Hagiwara K, Tanaka H, Nakaishji Y (2005) Expression and molecular characterization of spherical particles derived from the genome of the hyperthermophilic euryarchaeote *Pyrococcus furiosus*. J Biochem 138:193–199

Nicholls RA, Tykac M, Kovalevskiy O, Murshudov GN (2018) Current approaches for the fitting and refinement of atomic models into cryo-EM maps using CCP-E. Acta Crystallog D74 https://doi.org/10.1107/s2059798318007313

Oliver RM (1973) Negative stain electron microscopy of protein macromolecules. Meth Enzymol 27:617–672

Owen RL, Juanhuix J, Fuchs M (2016) Current advances in synchrotron radiation instrumentation for MX experiments. Arch Biochem Biophys 602:21–31

Perutz MF, Rossmann MG, Cullis AF, Muirhead H, Will G, North ACT (1960) Structure of haemoglobin: a three-dimensional Fourier synthesis at 5.5-A. resolution, obtained by X-ray analysis. Nature 185:416–422

Phillips AJ, Littlejohn J, Yewdall NA, Zhu T, Valéry C, Pearce FG, Mitra AK, Radjainia M, Gerrard JA (2014) Peroxiredoxin is a versatile self-assembling tecton for protein nanotechnology. Biomacromol 15:1871–1881

Rodis P, Hoff JE (1984) Naturally occurring protein crystals in the potato. Plant Physiol 74:907–911

Rousselot T, Jaenicke E, Lamkemeyer T, Harris JR, Pirow R (2006) Native and subunit molecular mass and quarternary structure of the hemoglobin from the primitive branchiopod crustacean *Triops cancriformis.* Febs J. https://doi.org/10.1111/j.1742-4658.2006.05408.x

Royer WE, Kristen Strand K, van Heel Hendrickson WA (2000) Structural hierarchy in erythrocruorin, the giant respiratory assemblage of annelids. Proc Natl Acad Sci 97:7107–7111

Ryle AP, Sanger F, Smith LF, Kitai R (1955) The disulphide bonds of insulin. Biochem J 60:541–556

Schroder E, Littlechild JA, Lebedev AA, Errington N, Vagin AA, Isupov MN (2000) Crystal structure of decameric 2-Cys peroxiredoxin from human erythrocytes at 1.7 A resolution. Structure 8:605–615

Sigmund F, Massner C, Erdmann P, Stelzl A, Rolbieski H, Desai M, Bricault S, Wörner TP, Snijder J, Geerlof A, Fuchs H, Hrabě de Angelis M, Heck AJR, Jasanoff A, Ntziachristos V, Plitzko J, Westmeyer GG (2018) Bacterial encapsulins as orthogonal compartments for mammalian cell engineering. Nat Commun. https://doi.org/10.1038/s41467-018-04227-3

Stahlberg H, Biyani N, Engel A (2015) 3D reconstruction if two-dimensional crystals. Arch Biochem Biophys 581:68–77

Sutter M, Boehringer D, Gutmann S, Günther S, Prangishvili D, Loessner MJ, Stetter KO, Weber-Ban E, Ban N (2008) Structural basis of enzyme encapsulation into a bacterial nanocompartment. Nat Struct Mol Biol 15:939–947. https://doi.org/10.1038/nsmb.1473

Tamura A, Fukutani Y, Takami T, Fujii M, Nakaguchi Y, Murakami Y, Noguchi K, Yohda M, Odaka M (2014) Packaging guest proteins into the encapsulin nanocompartment from *Rhodococcus erythropolis* N771. Biotechnol Bioeng 112:13–20

Tatur J, Hagen WR, Matias PM (2007) Crystal structure of the ferritin from the hyperthermophilic archaeal anaerobe *Pyrococcus furiosus.* J Biol Inorg Chem 12:615–630

Van Bruggen EFJ, Wiebenga EH, Gruber M (1960) Negative staining electron microscopy of proteins at pH values below their isoelectric points; its application to haemocyanin. Biochim Biophys Acta 42:171–172

Van Bruggen EFJ, Schutter WG, van Breemen JFL, Bijlholt MMC, Wichertjes T (1981) Arthropodan and Molluscan Hemocyanins. In Harris jr (ed) Electron microscopy of proteins, vol. 1. Academic Press, London, pp 3–38

Valentine RC, Green NM (1967) Electron microscopy of an antibody-hapten complex. J Mol Biol 27:615–617

Vinothkumar KR, Henderson R (2016) Single particle electron cryomicroscopy: trends, issues and future perspective. Quart Rev Biophys 49:e13, page 1 of 25. https://doi.org/10.1017/s0033583516000068

Wood Z, Harris JR, Schröder E, Poole L (2003) Structure, mechanism and regulation of peroxiredoxins. Trends Biochem Sci 28:32–40

Yewdall NA, Peskin AV, Hampton MB, Goldstone DC, Pearce FG, Gerrard JA (2018a) Niochem Biophys Res Commun 497:558–563

Yewdall NA, Allison TM, Pearce FG, Robinson CV, Gerrard JA (2018b) Self-assembly of toroidal proteins explored using native mass spectrometry. Chem Sci. https://doi.org/10.1039/x0xx00000x

Zivanov J, Nakane T, Forsberg BO, Kimanius D, Lindahl E, Scheres SHW (2018) New tools for automated high-resolution cryo-EM structure determination in RELION-3 eLife 7:e42166. https://doi.org/10.7554/elife.42166

Chapter 2
Antibody Complexes

Reetesh Raj Akhouri, Lars-Göran Öfverstedt, Gunnar Wilken and Ulf Skoglund

Abstract Monoclonal based therapeutics have always been looked at as a futuristic natural way we could take care of pathogens and many diseases. However, in order to develop, establish and realize monoclonal based therapy we need to understand how the immune system contains or kill pathogens. Antibody complexes serve the means to decode this black box. We have discussed examples of antibody complexes both at biochemical and structural levels to understand and appreciate how discoveries in the field of antibody complexes have started to decoded mechanism of viral invasion and create potential vaccine targets against many pathogens. Antibody complexes have made advancement in our knowledge about the molecular interaction between antibody and antigen. It has also led to identification of potent protective monoclonal antibodies. Further use of selective combination of monoclonal antibodies have provided improved protection against deadly diseases. The administration of newly designed and improved immunogen has been used as potential vaccine. Therefore, antibody complexes are important tools to develop new vaccine targets and design an improved combination of monoclonal antibodies for passive immunization or protection with very little or no side effects.

Keywords Antibody · Therapeutics · Monoclonal antibody · Vaccine · Immunization · Immunogen

R. R. Akhouri · L.-G. Öfverstedt · G. Wilken · U. Skoglund (✉)
Okinawa Institute of Science and Technology Graduate University, Okinawa, Japan
e-mail: ulf.skoglund@oist.jp

R. R. Akhouri
e-mail: reetesh.akhouri@oist.jp

L.-G. Öfverstedt
e-mail: lg.ofverstedt@oist.jp

G. Wilken
e-mail: wilken@oist.jp

J. R. Harris and J. Marles-Wright (eds.), *Macromolecular Protein Complexes II: Structure and Function*, Subcellular Biochemistry 93,
https://doi.org/10.1007/978-3-030-28151-9_2

Introduction

Antibodies, also known as Immunoglobulins, are responsible for functionality of the immune system in an organism. The characteristic features of the immune system, such as protection from diverse range of pathogens (Wec et al. 2019; Arunkumar et al. 2019; Lu et al. 2019; Tan et al. 2018; Pauthner et al. 2018; Rudkin et al. 2018; Abreu-Mota et al. 2018; Micoli et al. 2018) and toxins (Babcock et al. 2006; Aboudola et al. 2003), are due to antibodies. The specificity and clearance of a pathogen is achieved by the recognition and formation of complexes with a variety of epitopes with diverse origins, such as nucleic acid (Pisetsky 1997, 1998), carbohydrates (Amon et al. 2014), proteins (Arunkumar et al. 2019), lipids (Wong-Baeza et al. 2016) or synthetic compounds (Marquardt et al. 1990), followed by neutralization (Zhang et al. 2017; Flyak et al. 2015) or opsonization (Chehadeh et al. 2009). The successful neutralizing antibodies are further amplified and stored as memories to prevent from future infections from those pathogens (Katzelnick et al. 2016; Robbiani et al. 2017). Besides opsonization, the antibodies also activate the complement system that forms a membrane attack complex causing lysis of pathogens and inflammation reactions (Janeway Jr et al. 2001; Noris and Remuzzi 2013).

The power of antibodies to recognize a diverse set of molecules as antigens is due to its ability to be present as soluble or membrane bound antibodies because of alternate splicing (Hippocrates 1959; Borghesi et al. 2006). The membrane bound form has a transmembrane region and is expressed on the surface of B lymphocytes. Soluble antibodies can travel to different tissues and mucosal surfaces for antigen recognition and neutralization. The membrane bound antibody is important for antigen-specific recognition and an activation process that would further lead to development of memory plasma cells.

Structure of Antibodies

Antibodies are heterodimers forming a 'Y' shaped structure. The two arms of the Y-shaped structure consist of two heavy and two light chains (Janeway et al. 2001; Woof and Burton 2004), each consisting of a variable (V) and a constant (C) immunoglobulin domain. The light chain contains one variable (V_L) and one constant region (C_L). The heavy chain consists of one variable (V_H) and three constant regions (C_H1, C_H2 and C_H3). The hinge or spacer region present in between C_H1 and C_H2 makes it sensitive to enzymes like papain that is used to obtain F_{ab} fragments of antibodies. The heavy and light chains are linked together by disulfide bonds through C_H1 and C_L domain that upon reduction gives rise to 55 kDa heavy chain and 25 kDa molecular weight light chain. The distal C_H2 and C_H3 forms the F_c region and are responsible for an effector function of the antibody. Thus, soluble antibodies can bind to non-lymphoid cells expressing F_c receptors and help them function as heterologous cell surface antigens. In order for the antibody to be membrane bound, alternative splicing removes the secretory sequence and replaces it with transmembrane/cytoplasmic tail domains (M1 and M2) (Schroeder and Cavacini 2010).

The feature of the antibody to recognize diverse epitopes originates at the sequence variability in the V region and enables the immune system to bind to many antigens or to differentiate between different antigens by as little as one atom. The diversity in paratopes is created at the genetic level in the heavy chain (encoded at chromosome 14) and the light chain (encoded at chromosome 22). V_H is encoded by three genetic elements V (variable), D (diversity) and J (joining elements). The V_H can be approx. 100 elements with more than 10 D elements and a small number of J elements, giving rise to variety of combinations. The VDJ elements are followed by constant regions consisting of μ, δ, γ3, γ1, γ2b, γ2a, ε and α, where any of V region can be expressed with each of C regions. In the germline the V, D and J are far from each other, and with multiple forms. However, later, during the lymphoid development process, the translocation of one of the D elements is synchronized with the J_H elements after splicing out the intervening DNA. In a following second translocation event one of V_H elements joins with DJ_H to form complete V region (V_HDJ_H). Besides the probability of combinations between different V_HDJ_H elements, addition and deletion of bases during translocation can further increase the diversity with restricted genetic repertoire (Janeway et al. 2001).

The L chains, in the light chain, consist of only V_L and J_L regions with an additional diversity brought about by κ and λ types of L chains. The variability (V) region consists of three hypervariable regions or complementary-determining regions (CDR) interspersed with four framework regions that are responsible for the antibody‘s combining site. Similar features are observed in the V region of the heavy chain and thus the pairing between different elements of IgH and IgL chains generates the diversity towards various antigens (Schroeder and Cavacini 2010). The enzymes responsible for these recombination events are RAG-1 and RAG-2 where they nick DNA for translocation. A Cryo-EM structure of RAG-1 reveals that it works as a dimer (Kim et al. 2018).

Based on the C region of each H chain, antibodies can be classified into different isotypes, providing distinct biological functions to each isotype.

IgM

IgM contains a μ-chain at the C region and is presented on the B-lymphocyte surface as monomers. On the B cell surface upon activation, it oligomerizes and transduces the signal for immune responses. IgM is also secreted into serum as pentamers or hexamers that are linked together by disulfide bonds in C_H4 domains (Randall et al. 1990; Petrušic et al. 2011) and by a J-chain (Vollmers and Brandlein 2006). This pentameric structure helps in increasing the valency of interaction with an antigen. Therefore, even after having low affinity with antigen, it provides increased avidity for generating a strong immune response (Boes 2000). The major functions of IgM is the activation of membrane attack complex that are highly effective mechanism for lysis and clearance of pathogens (Murphy et al. 2008).

IgA

IgA contains the α-chain in the C region of heavy chain, and is mostly secreted in body fluids and mucosal surfaces (Germain 1994). IgA can exist in both monomers and a secretory dimeric form adjoined by a J-chain. The monomer IgA is mostly predominant in serum. IgA can exist in two subclasses IgA1 and IgA2 where IgA1 is characterized by the presence of a longer hinge region than A2. Because of the presence of short hinge region in A2, IgA2 is not easily cleaved by proteases and thus it is mostly present in mucosa as it leads to protection against bacterial proteases. IgA1 is sensitive to proteases, but due to heavy glycosylation it is somewhat protected and can thus provide high multivalent interactions for better immune responses (Roche and Cresswell 1991).

IgE

IgE contains ε-chain in the heavy chain C region and has the shortest half-life (Lawrence et al. 2017). It is mostly responsible for allergy inflammatory responses (Pier et al. 2004).

IgD

IgD contains the δ-chain type in the heavy chain and is mostly present in a membrane bound form on B-cells. Thus, it works in a similar manner to IgM in passing on signals to cells for immune activation (Schlosstein et al. 1973). It has shorter half-life than IgM because of a longer hinge region that makes it sensitive to proteolysis (Schroeder and Cavacini 2010). However, secretory IgD are mostly present in mucosa and works against pathogen attacks.

IgG

IgG contains γ type heavy chains and has a maximum half-life of around 21 days. Due to this property, it is the most predominant type of antibody and possesses a major role in the immunological memory. Another important feature of IgG is the activation of the complement system, either directly, or through its Fc region on FcR bearing cells. Among its four subclasses (IgG1, IgG2, IgG3 and IgG4), IgG1 and 3 are effective activators, while IgG2 and 4 are non-or weak activators of the complement system (Kindred and Shreffler 1972; Rosenthal and Shevach 1973).

Diversity and Flexibility in Antibodies

A variety of strategies in antigen-antibody recognition are an outcome of exceptionally large possibilities of combinations of linkages, modifications and heterogeneous levels of glycosylation of antibodies as well as its epitopes in biological systems (Dudley et al. 2005; Maizels 2005). A large number of repertoire of antibodies are generated due to recombination of genes, that code for VH, VL and J gene segments for the heavy and the VJ of light chains, that further come from lambda or kappa type chromosomes. Therefore, the recombination and substitutions add a level of complexity in the antibody classes by post-translational modifications of antibodies that modulate effector functions (Mimura et al. 2001; Jefferis 2009).

3D Structure of IgG Antibodies and Its Conformational Diversity

The efficient function of antibodies has prompted a large number of two- and three-dimensional structure determinations of antibodies using a diverse set of physical modalities like X-ray Crystallography, Atomic Force Microscopy (AFM), Small-angle X-ray scattering (SAXS), Nuclear Magnetic Resonance (NMR), Transmission Electron Microscope (TEM) methods and Surface Plasmon Resonance methods. These experimental methods all convey different information of the molecules and thus basically answer different questions.

The modalities that traditionally have given the highest resolution, X-ray crystallography and NMR, have been used for hundreds of antibody-related structures resolved to atomic resolution and these structures are deposited in the Protein Data Bank repository. However, these are generally only fragments of an entire antibody. So far, we know of three structures of a complete and functional antibody, type IgG, resolved to atomic resolution and available for general downloading from the PDB repository (Fig. 2.1a–b). They have the PDB entry names 1IGT, 1IGY, 1HZH, and are human IgG1 (Harris et al. 1992), murine IgG1 (Harris et al. 1998) and murine IgG2(a) (Saphire et al. 2001) respectively. There is a fourth one, IgG4, which is complete and to atomic resolution, but it was stabilized by a mutation (PDB entry 5DK3) to enable crystallization (Scapin et al. 2015).

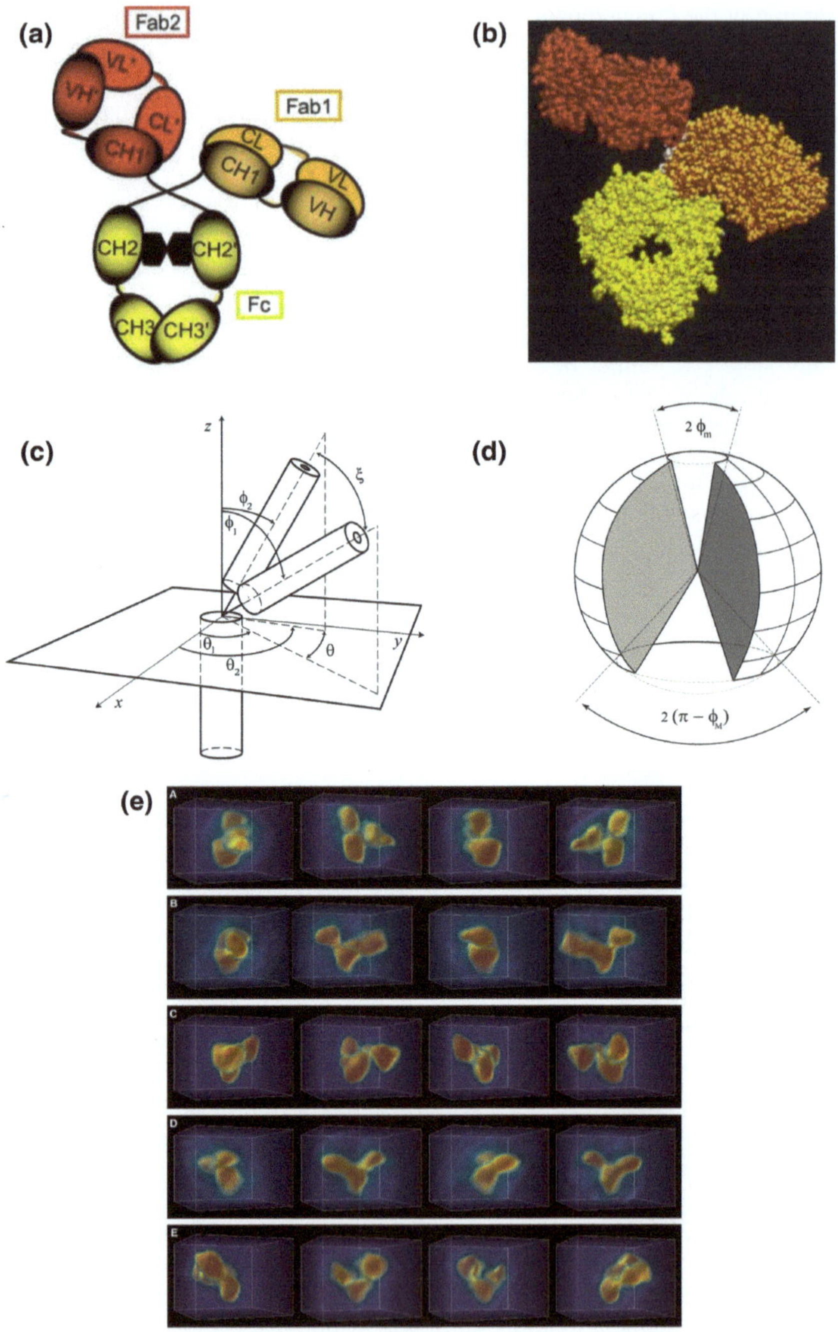
(a)
Fab2
VL'
VH'
CL'
CH1
Fab1
CL
CH1
VL
VH
CH2
CH2'
Fc
CH3
CH3'
(b)
(c)
z
ϕ_2
ϕ_1
ξ
θ_1
θ_2
θ
x
y
(d)
$2\,\phi_m$
$2\,(\pi - \phi_M)$
(e)
A
B
C
D
E

◂**Fig. 2.1** Overall structure of IgG2a. **a** A schematic illustration and **b** A space filling depiction of the crystal structure (Harris et al. 1992) of an intact IgG2a molecule. IgG consists of two light chains and two heavy chains, divided into twelve domains (named constant heavy [CH1-3], variable heavy [VH], constant light [CL], and variable light [VL]) of 110 amino acids residues each, with a similar immunoglobulin fold. **c** Coarse-grained model of an immunoglobulin molecule. The two Fabs **d** Representation of the IgG configurations space. The vertical axis of the sphere coincides with the F_C stem; the hinge region lies at the center. **e** Gallery of refined IgG tomograms (A-E) A gallery of five individual IgG molecules, visualize by volume rendering, with intensities colored from red to blue. Four views separated by 90° rotations around the vertical axis are shown. The box is 50X40X50 voxels

Generally, these structures all clarify and support the domain partitioning outline above. NMR is mainly used for the determination of structures in the mass range up to around 40 kDa (Paterson et al. 1990), so it has been more used to study structural details of a fragment of the antibody that is in complex with another molecule, *e.g.* an antigen. SAXS usage in antibody analysis has so far been very limited. The method allows for the approximate calculation of an average shape, an envelope for the protein, in solution. If the protein is reasonably stable and does not simultaneously exists in too many different structural states one finds a good correspondence between X-ray crystallographic results, SAXS results and that from Cryo-TEM Tomography (Gherardi et al. 2006).

SAXS results from IgD (Sun et al. 2005) display a wider and more diffuse spread of the molecular envelope than when individual antibodies are imaged, e.g. by AFM (Vilhena et al. 2016). A SAXS generated envelope of IgG is also quite wide (Rayner et al. 2013). The AFM method suggests quite a spread of possible conformations of antibodies, as does SAXS. Using X-ray crystallography to show more conformations has proven very difficult. Thus, it's not surprising that TEM methods have been used to understand more of the possible native conformations in solution attainable by an individual antibody. In fact, the three native IgG's determined with X-ray crystallographic techniques also indicate a remarkably large conformational space, showing Fab rotations and angular spread between the Fab's relative to the Fc 'stem' (Fig. 2.1d).

Since the antigen binding site is positioned near the peripheral parts of the Fab, relative to the hinge coupling between Fab and Fc, a large conformational space would allow the Fab to 'scan' effectively the surrounding space for antigens.

Using Cryo-EM with tomographic three-dimensional reconstructions of native IgG2(a) it could be shown that the two Fabs could basically occupy any position in space relative to each other and the Fc defined by the flexibility allowance from the hinge region (19 amino acids in this case) (Sandin et al. 2004). The occupancy of different large-scale conformations of the IgG2(a) indicated that a very simple model for the dynamics of the molecule could be used mainly describing the flexible position of the Fabs relative to a fixed Fc. This simplified model allowed for a solution of the Schrödinger equation in a buffer solution for this molecule, using as input data all the various tomograms of the individual IgG2(a)'s available, and the various rotational speeds of the Fabs as well as the rotation of the entire antibody could be calculated (Bongini et al. 2004). It was found that the antibody rotated around its Fc-axis at about 1.6 × 109 times/s and the Fabs were 'flapping' up and down between 2.9 × 109 and 1.0 × 1010 times/s (Fig. 2.1c–e).

Apart from Cryo-EM, antibodies have also been studied with the earlier ways of using TEM's, on stained molecules, either with negative stain or with positive stain. The advantage is the higher contrast of the molecules during the imaging. Basically, these stained images confirm the conformational space findings from the native IgG2(a) studied in Cryo-EM conditions using no stain. Trying to establish the large-scale dynamics from stained images can lead to errors, however, since the mass of the molecule is vital in calculating the flexing speeds, and the stained molecules have a rather unknown actual mass. Additionally, we don't know if the chemical agent in the stain alters the flexibility per se of the molecules.

Further, knowing the rotation and flexing speeds of the Fabs enables many optimizing parameters for the size of the antigen and the probability for antigen-Fab interaction to be calculated, helping the design of antibody-carriers, like nanoparticles, for efficient medical treatments (Bongini et al. 2005, 2007; Piazza et al. 2005; Galanti et al. 2016).

3D Structure of IgM

Soluble IgM in serum is primarily pentameric, with 70 immunoglobulin domains and a J chain (Fuentes-Panana et al. 2004; Fellah et al. 1992), but is also occasionally hexameric (Randall et al. 1992). The pentameric structure helps IgM to form strong interactions with several binding partners due to its high valency. Consequently, while monomeric IgM is membrane attached that does not bind to C1q due to the lack of binding pocket, soluble IgM binds complement component C1q with much higher affinity than large aggregated deposits of IgG, thus acts as a potent activator of the complement pathway (Quartier et al. 2005; Ogden et al. 2005; Chen et al. 2009). This leads to either opsonization or lysis through the membrane attack complex. Overall, there have been three serious efforts to resolve the structure of whole IgM. Aaron Feinstein and colleagues made the first attempt to visualize IgM using negative stained samples by electron microscopy (Feinstein and Munn 1969). Using this technique, they showed that IgM is star shaped. Perkins and colleagues followed this up with an in-solution study using SAXS to derive a 3D structure of IgM. Based on the fitting models, they found that IgM has a maximum dimension of 35-37 nm. Alone, Fc5 was found to be between 17.9 and 18.7 nm, which further confirms that the IgM core is highly flexible (Perkins et al. 1991). These authors suggested that electron microscopy data appeared to underestimate lengths, probably because of limitation of stain penetration. Fitting electron microscopy images in the SAXS model suggested that about one-quarter of images would fit a circle of diameter 37 nm, while the remaining would fit an ellipse of 39 nm × 35 nm. The Fc, discs in these images of IgM are such that 20% would fit a circle of 14 nm diameter, while large population would fit an ellipse of axes 15 nm × 13 nm.

However, in 2009 Czajkowsky and colleagues found that human IgM is mushroom-shaped (Czajkowsky and Shao 2009). This was mainly based on the fact that Fc5 or central core of the images looked protruded. Even more interesting were the energetics calculations and prediction that suggested three forms or conformations of IgM, that needed to be validated. However, better image processing and study at low electron dose was needed to achieve this. In order to avoid use of stain and overcome limits of stain penetration, and without use of averaging techniques, so that all the possible conformations could be visualized, Akhouri et al. (2016) used cryo-electron tomography to analyze individual IgM molecules. Reconstruction and characterization based on IgM diameter could be used to bin IgM into three major classes: extended, turtle shaped and bell shaped. The extended form of IgM had a planar core or Fc5 with an overall diameter of 29–30 nm. The turtle form has a protrusion in the core, with distinct convex and concave surfaces, and an overall diameter of 24–25 nm. In an extreme closed state IgM also has bell shaped structure with an overall diameter of 15–16 nm. These measurements coincides with the predicted conformations of IgM in Czajkowsky and Shao (2009). All these conformations of IgM could also be verified by negative stained images. The core of IgM is ~ 17 nm, and the Fab_2s pairs are between 7 and 11 nm in length (Fig. 2.2a and b). Turtle-shaped IgMs have a diameter in the range of 24–25 nm (Akhouri et al. 2016). The mass of the core is ~ 400 kDa, and the Fab_2s units are 100–120 kDa which are comparable with the estimated volume, or size, using small-angle X-ray scattering (SAXS); however, they are considerably different from measurements reported using AFM on a mica-surface. Given that all IgM structure still lacks resolution, these structures together explain why IgM detailed structure has been elusive for almost 40 years since it was first visualized. We also need to understand that averaging of IgM image contrast is not yet possible until a new software develops that could reconstruct higher resolution images from its 2D projection. Detailed information on the structure of IgM will remain elusive until this is achieved.

Secretory IgM is found in germ- or foreign antigen-free mice (Coutinho et al. 1995) and is polyreactive, but high affinity binding antibodies are reactive IgG. It is understood that internal or endogenous antigens do not drive IgM production (Haury et al. 1997). Most of the immunity that is imparted with the reactive antibodies are IgG and when we talk about antibody complexes we generally mean antigen bound to reactive IgG of one class or its fragment.

In our discussion below, we will consider a few well studied examples to show and appreciate the research that has revealed a great deal and has given us hope to provide good medicinal care. While there are various methods of research that have impacted this development, we will mainly focus on how antibody complexes have helped and advanced our knowledge regarding these disease and potential therapeutics. The main purpose of using antibody complexes in this discussion is to visualize and understand the natural process of how our immune system recognizes foreign material. Beyond this we develop an understanding of how naturally occurring antibodies either neutralize or mediate antibody mediated cellular cytotoxicity, to clear pathogens from the human body. This whole process serves the purpose of leading towards developing therapeutics that could be used for the treatment directly as monoclonal antibodies, or

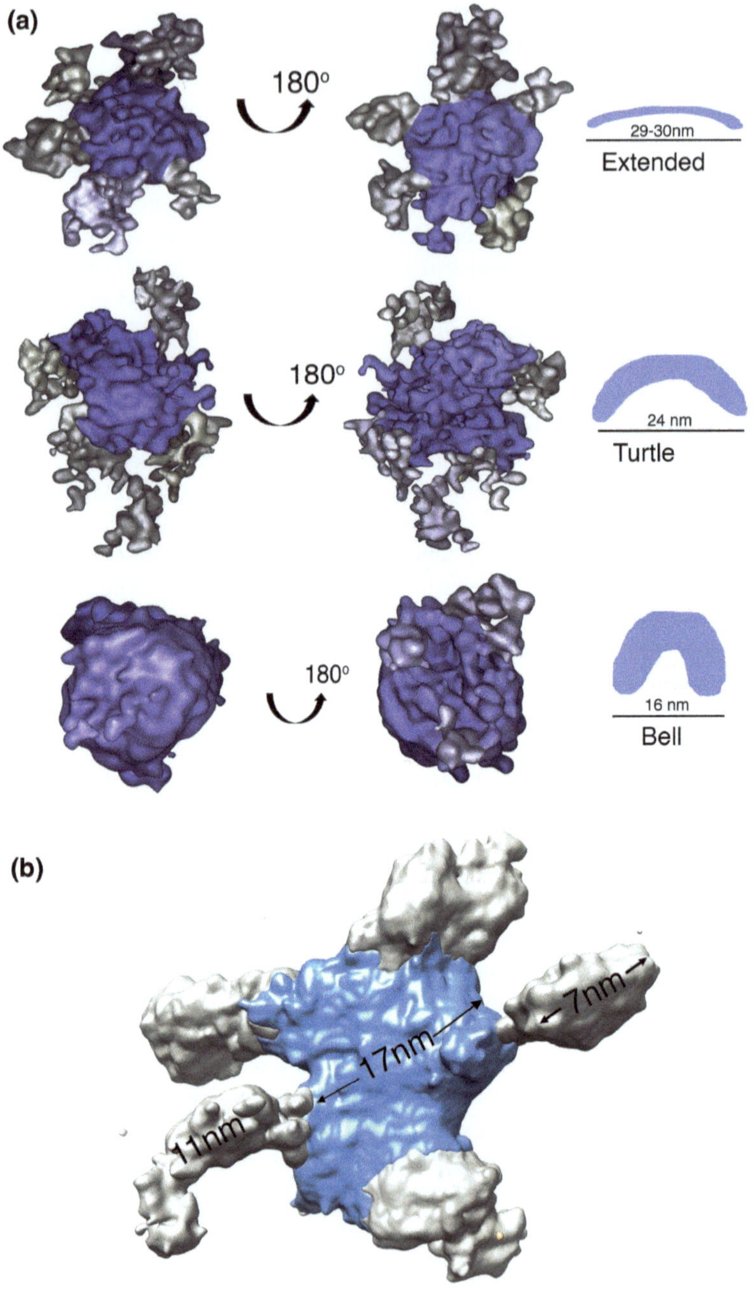

Fig. 2.2 Cryo-ET Architecture of non-immune human IgM. **a** IgM has three major conformations; Extended, Turtle and Bell shaped. Core of IgM is shown in blue and Fabs in grey. **b** Core of Extended form of IgM is 17 nm and Fabs are approximately 7–11 nm. Figure was adapted from Akhouri et al. (2016) and generated using Chimera visualization software (Pettersen et al. 2004)

to develop our understanding how we should target an antigen to generate a functional inhibitor to fight off viruses and pathogens.

Ebola Virus

Ebola is not a major killer in terms of number but it demands no explanation that the nature of disease and the threat is poses when an outbreak really occurs, as to why Ebola virus is considered to be one of most dangerous viruses. The recent outbreaks and number of deaths has created an urgent need for a better functional vaccine for protection of infected patients as well as care providers like doctors, nurses and volunteers, and to work as an efficient prophylactic. Monoclonal antibodies have generated much interest due to their efficiency and could be used as successful therapeutics. In the field of infectious diseases, especially those as dangerous as Ebola virus, there is an urgent need and to focus on generating neutralizing antibodies that could impart protection. Due to precision of target, almost negligible side effect and toxicity, there is an increasing demand for monoclonal antibody-based therapeutics. Antibody complexes have provided enough clues to develop efficient inhibitors that could mimic an antibody-antigen/pathogen complex that mediates efficient neutralization or complete protection due to antibody mediated cell cytotoxicity.

A well characterized mAb, KZ52 from a human survivor, showed promise due to its neutralization capability, and for being directly effective against Ebola virus glycoprotein. It imparted protection in mice and guinea pig. However, it was not effective in non-human primates (NHP) (Maruyama et al. 1999; Parren et al. 2002; Oswald et al. 2007). The crystal structure of EBOV GP trimer, in a pre-fusion conformation in complex with Fab_2 of KZ52 revealed at 3.4 Å resolution that the trimer is held together in a shape of a chalice or goblet that has a dimension of 9.5 nm × 9.5 nm × 7.0 nm. In this complex of trimer, GP1 is linked to GP2 with a disulfide, and GP2 are linked to each other without overlap from neighboring GP1. While GP1 forms a sauce-pan, three GP2 hold them together. KZ52 is an antibody against Zaire ebolavirus and directly neutralizes it. KZ52 binds to a more exposed part of GP which is not glycosylated and binds at the base of the trimer, mostly between amino acids 42 and 43 in GP1, 505–514 and again between 594 and 556 in GP2. It is believed that although KZ52 has a large contact with GP2, it is clear that GP1 can keep GP2 in a conformation that would favor recognition of the epitope in a pre-fusion conformation. Furthermore it is strongly believed that KZ52 successfully neutralizes the virus by blocking the GP2 $HR1_A/HR1_B$ rearrangement, that further blocks host membrane insertion of the internal fusion loop (IFL) that is formed by GP2 (Lee et al. 2008). KZ52 antibody interacts with 15 aa residues through van der Waal interaction, out of which 10 residues are only specific to Zaire ebolavirus (between the furin cleavage site and the HR helix bundle). It is highly likely that these interactions guide KZ52 specificity to only this virus (Fig. 2.3a, b) (Lee et al. 2008).

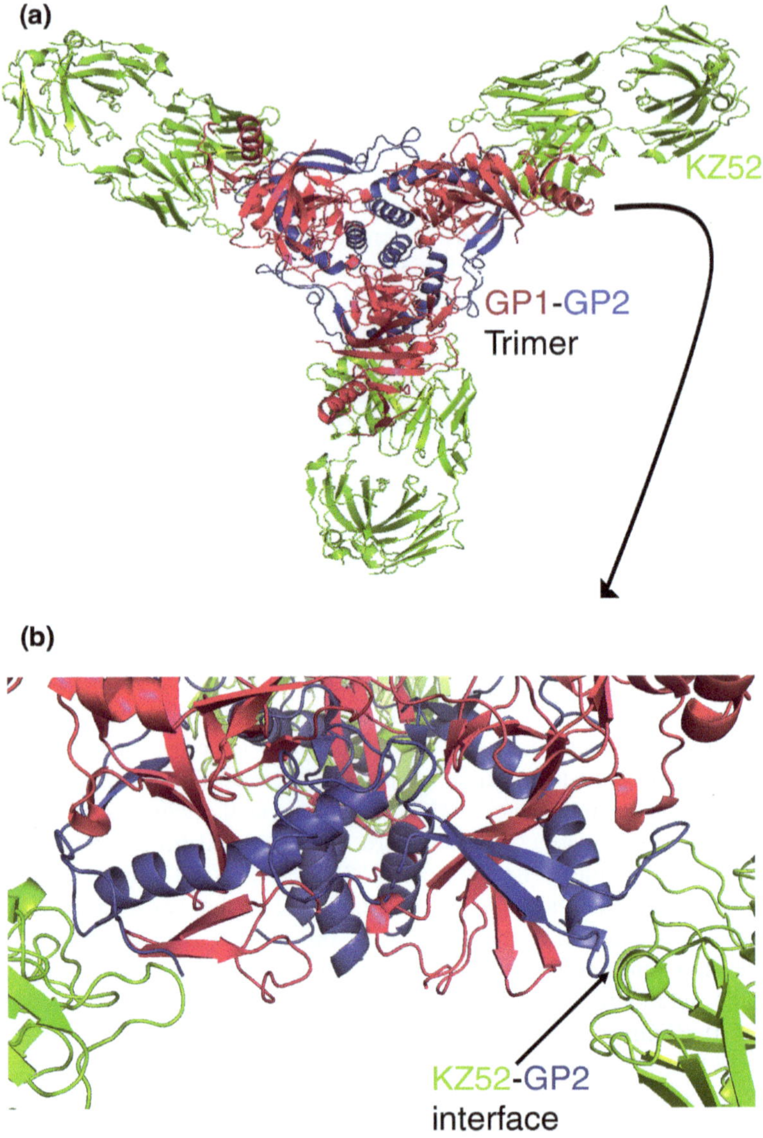

Fig. 2.3 Crystal structure model of Ebola Glycoprotein (GP) protein in complex with KZ52. **a** Overall architecture of Ebola GP trimer (GP1 in pink and GP2 in blue) in complex with broadly neutralizing monoclonal antibody KZ52 (bright green). **b** Magnified view of interacting GP2 and KZ52 coils from one of the trimer in packaged assembly. PyMOL was used for figures

In another example, biochemical characterization of mAbs 13F6 and 6D8 (both bind with a mucin-like domain) led to a greater understanding of how these antibodies recognize virus. Wilson and colleagues found that both antibodies recognize only the GP1 linear epitope and protect challenged mice (Wilson et al. 2000). They also found that 13C6 binds a quaternary epitope that involves both GP1 and sGP. When these three antibodies were prepared in a cocktail it was called MB-003. Structural analysis with a stained complex of GP trimer that contains the mucin-like domain but lacks trans-membrane part, revealed how these antibodies bind on the surface of soluble trimer (Murin et al. 2014). It was revealed that the c4G7 and c2G4 binding surface, and both of them bind to a GP2 ordered region that consists of both a helical and beta sheet region. However, these do not involve the IFL loop (Murin et al. 2014). Similarly, it has been established that c13C6 and c1H3 compete and hence may have an overlapping binding surface. The binding surface for c13C6 supersedes the binding surface of c1H3, or their binding affinity is the reason that c14C6 abrogates binding of c1H3, but not vice versa. And, as discussed earlier 13F6 and CD8 both bind to the mucin-like domain, but do not compete for the same surface. While 13C6 binds to a region of glycan cap, KZ52 binds to the base of GP that mainly involves GP2 and blocks rearrangement of the helix. It is also believed that deletion of the mucin-like domain facilitated the crystallization process. The inclusion of a mucin-like domain to visualize this domain and the antibody complex to this domain has not yielded any clear images from cryo-EM averaging. This could be because of flexibility of this region as well as the heavy glycosylation.

Based on a time course study on viral entry it has been established that epitopes in heavily glycosylated regions of glycoprotein are cleaved from the virion during entry, and hence, antibody against this region of virus should not be very effective. Interestingly, results confirmed the hypothesis that the glycoprotein is indeed cleaved. However, an even more interesting observation established that mAbs against this region of antigen impart full protection, probably due to an Fc-mediated effector function. Similarly, antibody complexes that impart protection against Ebola virus with mAb 114 or VRC608 show both neutralizing antibody and cell targeting antibodies (Corti et al. 2016). Thus, antibody-based complex analysis, along with biochemical understanding of the viral entry, have revealed that antibody-complexes could play an instrumental role in the understanding of how we could develop functionally relevant antibodies for therapeutics. Of the available therapeutic antibodies against Ebola, REGN-EB3 that is a combination of REGN3470, REGN3471 and REGN3479, has been able to impart protection due to the reason that it provides complimentary activity. REGN3479 binds conserved GP2 IFL and helps neutralization just like KZ52. REGN3470 makes a complex with GP1 head and mediate both neutralization as well as cell targeting activities, including FcγRIIIa and other FcγR-related functions. The third component REGN3471 mediates a cell targeting function by binding to the outer glycan cap that has similarity to 13C6. A set of functional models have been proposed as to how the combination McAb would interact with the epitopes. However, further studies are required to validate it. Therefore, taken together, antibody-complexes that target both neutralization as well as cell-targeting function are required for the complete protection against deadly viruses like Ebola.

It has also been found that a cluster of antibodies that were fucosylated have greater functionality towards activating FcR mediated cell activation and protection (Zeitlin et al. 2011; Olinger et al. 2012).

Human Immunmodefficiency Virus: HIV

HIV remains a major health concern due to its capability to compromise CD4+T cells and deplete them completely, thereby facilitating the establishment of secondary infections. Our understanding of HIV and its biology has come a long way in last decade. There are many lines and generations of drugs available for treatment and extend the life of an infected individual to a great extent. Yet, the social stigma attached to HIV infection has led to more behavioral and psychological issues. Due to the large population that are infected worldwide and the huge risk of HIV spreading due to patients being unaware and undetected for a long period of time, we desperately need an effective vaccine that could neutralize HIV. We also know that the error prone replication of HIV makes this a near impossible task. In spite of this, immune complex studies have led to great understanding of how neutralizing antibodies function and whether we could use them as a future therapeutics. In 1997 three research groups derived the first post-fusion conformation of gp41 crystal structure (Chan et al. 1997; Weissenhorn et al. 1997; Tan et al. 1997). In the following year a crystal structure of gp120 was derived in complex with the CD4 receptor and neutralizing antibody (Kwong et al. 1998). This complex revealed how CD4 receptor interacts with gp120. However, the details of the structure biologists needed in order to start proposing a hypothesis was delayed due to the flexible nature of the molecule and also due to heavy glycosylation on the surface of gp120 and gp41. It needed an out of the box approach to break the resolution barrier. Relentless effort from HIV researchers and innovative thinking, along with a huge amount of interactive experiments to understand conformational change during HIV interaction with CD4 and various antibodies, revealed the flexibility of the envelope protein. This inspired modification of both gp120 and gp41 in such a way that it would be more native like, allowing for a conformation change into pre-fusion and post-fusion states and also allowing crystallization with and without stabilizing antibodies or ligands. As a result, a disulfide bond between subunits was designed that was called SOS (A501C-T605C). Further, to add post-fusion flexibility and conformation change, another mutation was engineered to insert a proline at I559P. This allowed for trimer flexibility similar to the original trimer; this was called IP. Together, the engineering of this epitope was called SOSIP (Sanders et al. 2000; Sanders and Moore 2017). The results were very encouraging as it furnished better crystal contacts and improved crystal features, but did not yet produce a high-resolution structure (Binley et al. 2002; Sanders et al. 2000, 2002). It took deletion of hydrophobic residues close to the membrane (665–681) to allow for better solubility of the complex (Khayat et al. 2013; Klasse et al. 2013). Trimer complex with antibody PGT123 Fab revealed the structure of the soluble gp120-gp41 complex at 7.7A resolution (for clade A KNH1144 strain). These engineering steps were subsequently employed as a strategy on many other strains (Julien et al. 2013a, b). What it allowed for, along with the structural details, are improved melting temperatures of the whole complex to be screened for various neutralizing antibodies, and to be visualized by electron microscopy to better characterize all broad neutralizing antibodies. It also underscored the need for cleavage between gp120 and gp41 to allow for conformational flexibility to attain a native fold. While structural flexibility was under study for attaining the native fold, structures derived using cryo-ET revealed that the base of the complex, that would attach it to the virus, existed in more than one conformation, as trimeric or tripod (Zhu et al. 2003, 2006; Zanetti et al. 2006; Liu et al. 2008). It was further established, using cryo-EM, that when CD4 and a neutralizing antibody bind to the glycoprotein it drives a conformational change and to prove that, the trimer was shown to exist in various conformations (Tran et al. 2012; Bartesaghi et al. 2013). These two findings appear contradictory, but it is very important to understand that while flexibility was necessary to achieve pre- and post-fusion conformation, the use of mAb used in the structure could bind to the epitope and hence allow better structural resolution. While BG505 SOSIP with PGT122 yielded low resolution structure of the complex, PGV04 antibody with BG505 SOSIP trimer, enabled the structure to reach 5.8Å resolution. In a crystal structure, gp140 trimer with PGT122 yielded a structure at very high resolution. It established the interaction of the V1 and V3 loop of gp120 and several glycans with antibody, which acted as stabilizing factor (Fig. 2.4a, b). This structure exposed the importance of glycan at residue N322 (Lyumkis et al. 2013; Julien et al. 2013a, b). Taken together, it revealed that gp120 alone cannot act as a good immunogen. So, the purpose of studying antibody complexes to understand how we should generate an immunogen to generate antibodies that would be functionally relevant was becoming a reality. A detailed high-resolution structure of BG505 SOSIP in complex with PGT122 and 35O22 was derived by Pancera et al. (2014). Authors revealed the trimer complex in the pre-fusion state and how the viral protein could exist before they make a final committed attachment and then entry into CD4+cells. This is how we should design the vaccine to target free viruses in our system. They reveal that gp41 HR1 and HR2 undergo structural changes in the post-fusion state. This structure is in complex with neutralizing antibodies, and thus it further reveals how they bind to this epitope. In order to show how affinity maturation, generates antibodies that have better access to epitopes on glycosylated trimer Garces and colleagues derived the structure for BG505 SOSIP in complex with PGT121 and 35O22 (Garces et al. 2015). It shows that a mannose patch on the env glycoprotein is essential for binding of neutralizing antibody PGT121. They showed through elegant crystallization and binding experiments that several rounds of affinity maturation are integral to induce B cells to produce higher affinity antibodies that would take care of the minimizing blocking effect of neighboring glycans (Garces et al. 2015). It revealed heavy glycosylation of the trimer surface and showed that how the virus surface is almost impenetrable to starting lineages of antibodies, mainly due to selection of B cells and resultant antibodies that would not overcome steric clash

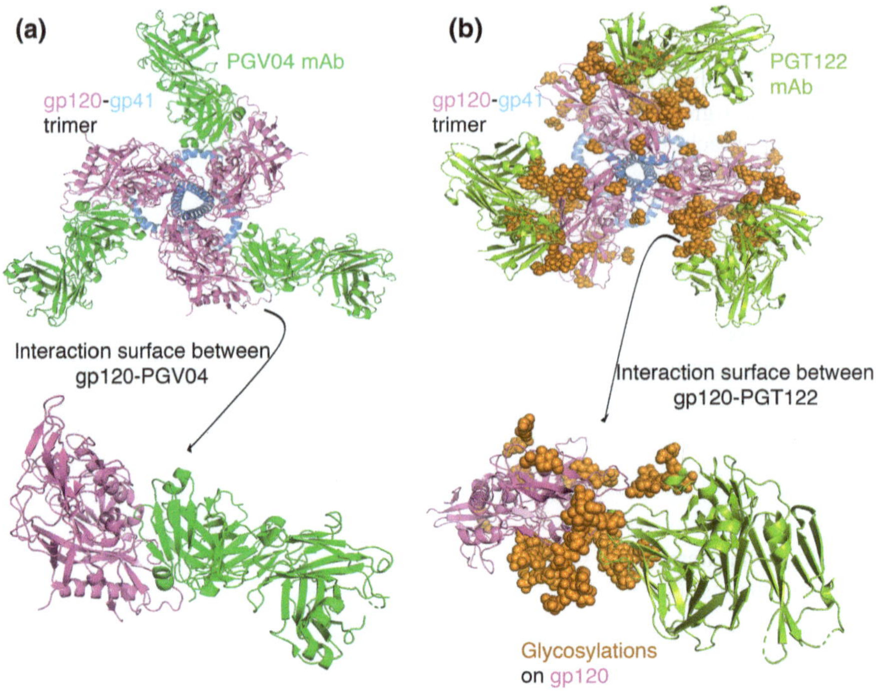

Fig. 2.4 Structure envelope trimer in complex with neutralizing antibodies. **a** Cryo-EM Structure of HIV envelope consisting of gp120 (pink)-gp41 (blue) trimer with PGV04 mAb (green). An enlarged view shown below highlights that neutralizing Antibody PGV04 interacts with gp120. **b** crystal structure of HIV envelope consisting of gp120 (pink)-gp41 (blue) trimer in complex with monoclonal antibodies PGT122 (Lime green). An enlarged view reveals interaction between glycosylated (orange) surface of gp120 with PGT122. PyMOL was used for figures

from various glycans. This explains why few neutralizing antibodies, like 2F5 and 4E10, have long CDRH3 loops (Haynes et al. 2005) which generally bind to the base gp41. Two more mutations in the BG505·SOSIP at G459C in gp120 and A60C in VRC01 helped crystallize glycosylated trimer from new clades. Lee and colleagues used PGT151 to extract and stabilize a near native envelope protein (del cytoplasmic domain) trimer and analyzed it using cryo-EM (Lee et al. 2016). Authors proposed that PGT151 stabilized the prefusion envelop trimer for a prolonged time period by contact with N611 and N637 glycans, which in turn prevent re-arrangement of HR1-HR2 upon fusion. Previously it was postulated by Pancera et al. (2014) that the pre-fusion trimer state of BG505·SOSIP.664 could be derived, because it was crystallized in complex with PGT122 and 35O22.

From all these exceptional studies, we learn that protein modification through disulfide linkage, mutation at I559 and removal of the hydrophobic region were essential to achieve a high resolution structure of envelope protein in complex with neutralizing antibodies. But Lee et al. (2016) structure revealed that to hold the complex in the prefusion state, which is vital for designing vaccine, can be achieved without such changes. One important lesson is that to generate a better immunogen we need to avoid looking at gp120 or gp41 in isolation. Since the ultimate goal of these advancements in our knowledge is aimed to serve the purpose of generating a greater vaccine or immunogen target, we should now also discuss two new approaches to that simplified generation of immunogen. Two research group showed that by inserting gly-ser linker immunogen production could be simplified so that this will avoid the need for cleavage by furin to gain a native fold (Sharma et al. 2015; Georgiev et al. 2015). Overall this sums up the importance of antibody complexes in the field of HIV research, and establishes an understanding of how viruses enter into CD4+cells, how neutralizing antibodies function against viruses and most importantly the designing of potent vaccine immunogens for future therapeutic use.

Hepatitis C Virus: HCV

Much focus has been on HIV and HBV due to the constant flow of new literature on these two viruses. However, we must not forget that HCV is a silent killer as only one in every twenty infected patients are aware of the infection. A large population being unaware of their infection can act as source for transmission and can infect a healthy population, and also cause reinfection in those who have sometimes cleared infection through available treatment, such as direct-acting antiviral therapy (Franco et al. 2014). Due to this unawareness, and without treatment, HCV is estimated to have infected more than 71 million people worldwide (WHO Global Hepatitis Report 2017). Thus, HCV is also a silent killer, as it is largely asymptomatic and can only be detected when causes an end stage liver disease, carcinoma and cirrhosis (Lagging et al. 2002; Hoofnagle 2002). Since it is largely asymptomatic, it would be highly desirable that a prophylactic vaccine be developed (Holmberg et al. 2013; Gravitz 2011). However, major challenges towards vaccine development against HCV are the 7 genotypes of HCV that show up to 30% genetic variability at the amino acid level around regions that code for the proteins E1 and E2 (Smith et al. 2014). Subsequently, due to a further 15% variation among all these seven genotypes of HCV, this problem is further compounded. Like almost all viruses, HCV envelope glycoproteins are major targets of neutralizing antibodies. It has been found that the E2 domain of the virus generates broadly neutralizing antibodies. Through a large number of studies it has been established that broadly neutralizing antibodies, for example AR3A, HEPC3 HEPC74 that bind to Antigenic region 3 (AR3), almost all bind to the same surface of the E2 domain that interacts with CD81 (Law et al. 2008; Bailey et al. 2017; Pierce et al. 2016). The groups of Bailey, Law and Gopal have together defined that yellow color residues together are a site where most of the neutralizing antibodies bind. It is also promising to see that antibodies against conserved region of E2, that interacts with CD81, are also neutralizing regions around domains B, D and AR3, and are especially important as frontal layer residues that interact with CD81. To name a few, antibodies like AP33, 3/11 and H77.39 along with HCV-1 and HCV-33.1, show promise targeting the residues between 412 and 423 in domain E2 (Gopal et al. 2017; Broering et al. 2009). It is further interesting to see that though there is huge genetic diversity, the hypervariable region-specific antibodies (HVR-1 monoclonal antibodies) , isolated from patients, do not show strain specificity and hence do not limit binding to specific strains of HCV. The junction of E1 and E2 also show reactivity in the region AR4 and AR5 that also mediate moderate to broad neutralization. In a clear demarcation of regions that promises great future, it has been found that antibodies like AR1A and AR2A that bind E2 domain are weak or non-neutralizing, and antibodies that bind the opposite surface of where the CD81 interaction occurs are also found to be non-productive in the process of neutralization (Fig. 2.5a, b) (Law et al. 2008; Gopal et al. 2017). Antibodies that have been found neutralizing have produced enough evidence that their functional capability is limited by the conformation and the angle at which they interact. For example, E2 interactions with AP33, 3/11 and H77.39 along with HCV-1 and HCV-33.1 are at different angles. HC-84-1 and HC84-27 are effective because they interact with the E2 in its frontal alpha helix which participates in interactions with CD81 (Kong et al. 2012a, b; Potter et al. 2012; Meola et al. 2015; Li et al. 2015).

Since, these antibodies were isolated from patients and therefore offer huge promise for the future, small molecules or inhibitors that would block residues of E2, that are required for its interaction with CD81, will also be useful. In all probability this treatment will need to be combined with another therapeutic approach, as again genetic diversity will limit its efficacy against all the strains/genotypes. For example, few antibodies against E1 domain that have been isolated seem to neutralize HCV from genotypes 1 (a and b), 4, 5 and 6, but fail to neutralize genotype 2 and 3 efficiently (Meunier et al. 2008). In spite of all these limitations we cannot stress enough how informative these antibody complexes are in terms of developing our understanding of disease and pathogens to fight against deadly diseases.

Malaria

Plasmodium falciparum causes the most severe form of malaria. The severe form of malaria is due to the ability of parasites to sequester in the microvasculature (Roberts et al. 2000). This causes blockage of blood flow causing severe symptoms like anemia, multiple organ failure, coma and death (Miller et al. 2002). The sequestration ability of the parasite is due to antigenically variant protein families, PfEMP1 (Scherf et al. 2008; Baruch et al. 1995; Su et al. 1995) and RIFINs present as 60 and 150 variant genes respectively (Gardner et al. 2002). The protein of these family bind to receptors such as CD36, ICAM-1, VCAM-1, EPCR, heparin sulfate proteoglycans, blood group A and sialic acid leading to sequestration, followed by severe outcomes in humans (Wahlgren et al. 2017; Goel et al. 2015; Biswas et al. 2007). However, these severe outcomes are prominent in children, but adults are protected against severe malaria (Doolan et al. 2009; O'Meara et al. 2008;

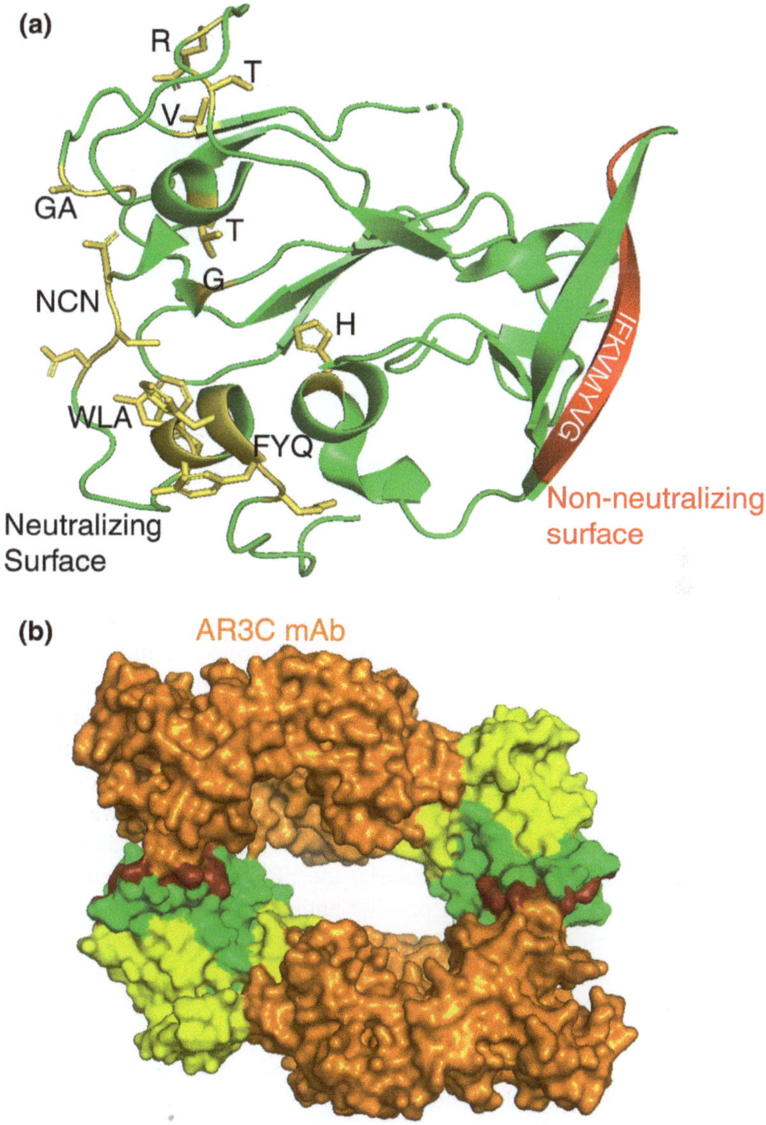

Fig. 2.5 Overall crystal structure of HCV envelope E2 protein. **a** Residues mapped on E2 crystal structure recognized by various antibodies shows polarity of neutralizing (yellow) and non-neutralizing surface (red). **b** Crystal structure of E2 in complex with neutralizing monoclonal antibodies AR3C (orange) shows neutralizing and non-neutralizing surfaces. PyMOL was used for figures

Crowley et al. 2010). This is clearly shown in areas endemic with *P. falciparum*, where only adults that acquire immunity against malaria have the capability to recognize multiple variants of PfEMP1 and RIFIN proteins. Due to the ability of parasites to develop resistance against drugs, monoclonal antibody have the potential to be used as therapeutic agents for treatment of severe malaria.

Quite interestingly, RIFIN binds to a leucocyte immunoglobulin-like receptor B1 (LILRB1). The binding of LILRB1 to RIFINs on infected RBC protects the parasite from host immune responses. It will not be inappropriate to diverge from the current topic of antibody complexes here, because, in contrast to the RIFIN-LILRB1 interaction that protects parasite from host immune response, the host immune system has also developed multiple antibodies that could recognize multiple field isolates (Saito et al. 2017). This holds great value in terms of therapeutics, as drug resistance has posed a major problem in *P. falciparum* malaria treatment. In another interesting study, Monoclonal antibodies (MGC34, MGD21 and MGD39) that recognized multiple field strains were isolated from plasma of infected individuals. Interestingly, sequencing of monoclonal antibodies identified insertion of 100 amino acids of the LAIR1 extracellular domain, a collagen binding inhibitory factor between the V and DJ region. This region was highly prone to somatic mutations and deeper analysis showed that mutations in T67L, N69S and A77T increased the binding to the infected erythrocyte surface, with reduced collagen binding. Besides being broadly cross-reactive, it showed high opsonizing capacity and effectively promoted phagocytosis (Tan et al. 2016). This transposition of a domain to bind to RIFINs that causes severe malaria is a novel mechanism used by the host to generate cross-reactive antibodies and protect against severe malaria. The transposition events are suggested to be independent of RAG mediated insertion, thus it will be interesting to see how research in this area may have implications in developing antibody-mediated protective immunity, which will be an important area of research in terms of therapeutics.

The above antibodies are responsible for reducing severe malaria or protecting from host immune responses once the infection has settled inside the human host. However, in order to protect from malaria, the nRTSS vaccine has been shown partially successful. It consists of a parasite antigen, the Cirumsporozoite protein (rsCSP), that also inhibits invasion into liver. In order to understand their mechanism of interaction and increase their efficacy in parasite inhibition, monoclonal antibodies that provide complete protection (mAb311) have been structurally analyzed in CSP in complex with the protein mAB311, bound to the repeat region NANP with in ratio of Fab311-$(NANP)_3$. It was observed in the cryo-EM map that 11 Fabs are arranged in pseudo helical arrangement that is bound to repeat regions of CSP forming an extended spiral structure. However, the complex of CSP with another protective antibody Fab317 with similar germline, showed up to 5Fabs bound to rsCSP, suggesting that the number of Fabs bound to rsCSP did not determine the protection capabilities of the antigen. Although inter-Fab contacts were observed when Fab311 was bound to PfCSP, but a change in amino acids involved in binding did not perturb the interaction, suggesting that inter-Fab contacts are encoded in the germ line (Oyen et al. 2017, 2018).

In a new approach, live attenuated sporozoites were inoculated and human monoclonals were screened for reactivity for PfCSP. After screening, monoclonal antibody CIS43 was found to provide sterile protection in two different mouse

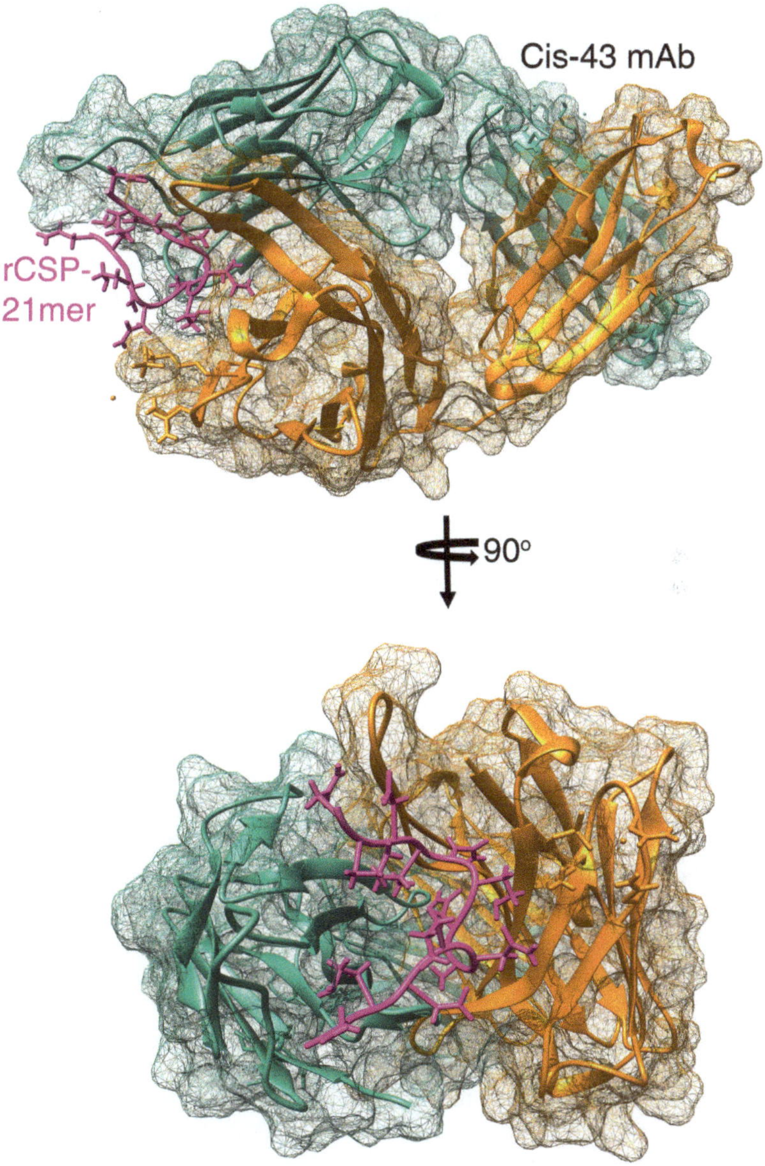

Fig. 2.6 Crystal structure of peptide (21mer, pink) of Plasmodium falciparum sporozoite stage protein Circumsporozoite Protein (CSP) in complex with Cis43 monoclonal antibody that blocks invasion of parasites to liver. Cis 43 heavy chain is represented as blue and light chain in orange. Chimera visualization software was used to generate this figure

models (Fig. 2.6). CIS43 recognized the junction of the N-terminus and the central repeat domain of PfCSP, suggesting that the N-terminus that was absent in RTSS was important for providing complete protection. Further, CIS43 bound to PfCSP in a sequential manner with two binding events, where the first binding is low affinity with a single binding site per antibody and second one with high affinity with five binding sites per antibody. The crystal structure of CIS43 in complex with corresponding peptide showed the binding of CIS43 at a specific angle to a rare confirmation at the N-terminus-repeat junction, where NAPN is the structural epitope in the NANP repeat sequences. This antibody complex leads to sterile infection as it inhibits processing of CSP on the sporozoite surface that is required for successful invasion of sporozoites in liver (Kisalu et al. 2018).

In contrast to the fact that antibody complexes are increasing our understanding about many diseases, the malaria parasite produces a very interesting example where the parasite uses IgM to mediate severe malaria. We have shown that IT4Var60 (a PfEMP1) that is expressed on the surface of the parasitized RBC binds to IgM. It was shown that IgM acts like a flower vase to cluster many PfEMP1s and help form a very high affinity binding pocket for the receptor on host cells and hence mediates an 8-fold tighter interaction and a highly robust host parasite interaction (Akhouri et al. 2016). This interaction could further explain how infants (usually with high IgM) in an endemic region who also lack any protective antibodies against various PfEMP1, could be exposed to a severe form of malaria and hence could be more prone to death.

Conclusion and Perspective

The contest between our immune system and pathogens is under constant pressure and helps both evolve for better survival. The biochemical and structural information through research on antibody complexes reveals how only a small fraction of the antibodies among the many antibodies that we generate against pathogens successfully neutralize them and protect our system from being highjacked completely. It gives us a clue about how our immune system does many trial experiments, but only few of them generated the desired results. Imagine a situation where we needed to probe the whole molecular surface of antigens for the development of inhibitors and the amount of chemistry we would be forced to undertake in the absence of clues from antibody complexes. Therefore, these antibodies complexes are of immense importance as they narrow down our area of interest for drug development. For example, the outcome of the mapping of mAb for the Ebola virus GP trimer revealed the combination of antibodies that could be used for protection. On the other hand, it highlighted a potential new combination of antibodies c4G7, c2G4, 16F6 and KZ52 that could be tried, as they bind to the region of Ebola virus that are not cleaved off and hence may impart better protection. It is not certain that these combinations would yield better results than the current or existing combination as not only neutralization that matters, it also depends on how they activate cell mediated opsonization and Fc dependent cell cytotoxicity. Therefore, it further needs validation through extensive challenge studies would be highly valuable. Antibody complexes have indeed established that Ebola is vulnerable at the base of

GP1 and GP2. Further design of new inhibitors that would harness this advantage and may need lower dose for protection and could also be used as a potential drug and prophylactic. As we discussed antibody complexes against HIV epitopes we learned that this field of research has started to deliver the goal that research community hoped to achieve. It simplifies the design and production of immunogen production that may deliver the desired result without depending on the mAb derived from patients and hence avoid passive immunity. In the field of malaria research monoclonal antibody screens have led to discovery of antibody that could neutralize a large number of micro-organisms isolated from an endemic area, which is definitely the first step towards the desired result that we could expect from antibody complex use. Therefore, all these experiments that nature does it for us leaves a foot print of how to design a successful drug against deadly viruses and many other pathogens, that can be visualized through research related to antibody complexes. Therefore, studies about antibody complexes and the collective information that we accumulate after extensive research, helps us prepare a library of information/database that would serve towards development of potent combinatorial drugs that might prepare us better against invading micro-organisms. Among all antibodies that target both neutralization as well as cell-targeting function give us indications which way our research should bend to give us next generation monoclonal antibody-based therapeutics and clues to design inhibitors.

References

Aboudola S, Kotloff KL, Kyne L, Sougioultzis S et al (2003) *Clostridium difficile* vaccine and serum immunoglobulin G antibody response to toxin A. Infect Immun. 71:1608–1610

Abreu-Mota T, Hagen KR, Cooper K et al (2018) Non-neutralizing antibodies elicited by recombinant Lassa-Rabies vaccine are critical for protection against Lassa fever. Nat Commun 9(1):4223. https://doi.org/10.1038/s41467-018-06741-w

Akhouri RR, Goel S, Furusho H et al (2016) Architecture of human IgM in complex with *P. falciparum* erythrocyte membrane protein 1. Cell Rep. 14:723–736

Amon R, Reuven EM, Leviatan Ben-Arye S et al (2014) Carbohydr Res 389:115–22. https://doi.org/10.1016/j.carres.2014.02.004

Arunkumar GA, Ioannou A, Wohlbold TJ et al (2019) Broadly cross-reactive, non-neutralizing antibodies against the influenza B virus hemagglutinin demonstrate effector function dependent protection against lethal viral challenge in mice. J Virol pii: JVI.01696-18. https://doi.org/10.1128/jvi.01696-18

Babcock GJ, Broering Teresa J et al (2006) Human monoclonal antibodies directed against Toxins A and B Prevent *Clostridium difficile*-induced mortality in hamsters. Infect Immun 74:6339–6347

Bailey JR, Flyak AI, Cohen VJ et al (2017) Broadly neutralizing antibodies with few somatic mutations and hepatitis C virus clearance. JCI Insight. 2:92872

Bartesaghi A, Merk A, Borgnia MJ et al (2013) Prefusion structure of trimeric HIV-1 envelope glycoprotein determined by cryo-electron microscopy. Nat Struct Mol Biol 20:1352–1357

Baruch DI, Pasloske B, Singh HB et al (1995) Cloning the *P. falciparum* gene encoding PfEMP1, a malarial variant antigen and adherence receptor on the surface of parasitized human erythrocytes. Cell 82:77–87

Binley JM, Sanders RW, Master A et al (2002) Enhancing the proteolytic maturation of human immunodeficiency virus type 1 envelope glycoproteins. J Virol 76:2606–2616
Biswas AK, Hafiz A, Banerjee B et al (2007) Plasmodium falciparum uses gC1qR/HABP1/p32 as a receptor to bind to vascular endothelium and for platelet-mediated clumping. PLoS Pathog 3:1271–1280
Boes M (2000) Role of natural and immune IgM antibodies in immune responses. Mol Immunol 37:1141–1149
Bongini L, Fanelli D, Piazza F et al (2007) A dynamical study of antibody-antigen encounter reactions. Phys Biol 4:172–180
Bongini L, Fanelli D, Piazza F et al (2004) Freezing immunoglobulins to see them move. Proc Natl Acad Sci 101:6466–6471
Bongini L, Fanelli D, Piazza F et al (2005) Dynamics of antibodies from cryo-electron tomography. Biophys Chem 115:235–240
Borghesi L, Milcarek C (2006) From B cell to plasma cell: regulation of V(D)J recombination and antibody secretion. Immunol Res 36:27–32
Broering TJ, Garrity KA, Boatright NK et al (2009) Identification and characterization of broadly neutralizing human monoclonal antibodiesdirected against the E2 envelope glycoprotein of hepatitis C virus. J Virol 83:12473–12482
Chan DC, Fass D, Berger JM et al (1997) Core structure of gp41 from the HIV envelope glycoprotein. Cell 89:263–273
Chehadeh W, Halim MA, Al-Nakib W (2009) Antibody mediated opsonization of red blood cells in parvovirus B19 infection. Virology 390:56–63. https://doi.org/10.1016/j.virol.2009.04.016
Chen Y, Park YB, Patel E et al (2009) IgM antibodies to apoptosis-associated determinants recruit C1q and enhance dendritic cell phagocytosis of apoptotic cells. J Immunol 182:6031–6043. https://doi.org/10.4049/jimmunol.0804191
Corti D, Misasi J, Mulangu S et al (2016) Protective monotherapy against lethal Ebola virus infection by a potently neutralizing antibody. Science 351:1339–1342
Coutinho A, Kazatchkine MD, Avrameas S (1995) Natural autoantibodies. Curr Opin Immunol 7:812–818
Crowley J, Chu C, Love GM et al (2010) Malaria in children. Lancet 375:1468–1481
Czajkowsky DM, Shao Z (2009) The human IgM pentamer is a mushroom-shaped molecule with a flexural bias. Proc Natl Acad Sci U S A. 106:14960–14965
Doolan DL, Dobaño C, Baird JK (2009) Acquired immunity to malaria. Clin Microbiol Rev 22:13–36
Dudley DD, Chaudhuri J, Bassing CH et al (2005) Mechanism and control of V(D)J recombination versus class switch recombination: similarities and differences. Adv Immunol 86:43–112
Feinstein A, Munn EA (1969) Conformation of the free and antigen-bound IgM antibody molecules. Nature 224:1307–1309
Fellah JS, Wiles MV, Charlemagne J et al (1992) Evolution of vertebrate IgM: complete amino acid sequence of the constant region of *Ambystoma mexicanum* mu chain deduced from cDNA sequence. Eur J Immunol 22:2595–2601
Flyak AI, Ilinykh PA, Murin CD et al (2015) Mechanism of human antibody mediated neutralization of Marburg virus. Cell 160:893–903. https://doi.org/10.1016/j.cell.2015.01.031
Franco S, Tural C, Nevot M et al (2014) Detection of a sexually transmitted hepatitis C virus protease inhibitor-resistance variant in a human immunodeficiency virus-infected homosexual man. Gastroenterology 147:599–601
Fuentes-Panana EM, Bannish G, Monroe JG (2004) Basal B-cell receptor signaling in B lymphocytes: mechanisms of regulation and role in positive selection, differentiation, and peripheral survival. Immunol Rev 197:26–40
Galanti M, Fanelli D, Piazza F (2016) Conformation-controlled binding kinetics of antibodies. Sci Rep 6:18976. https://doi.org/10.1038/srep18976
Garces F, Lee JH, de Val N et al (2015) Affinity maturation of a potent family of HIV antibodies is primarily focused on accommodating or avoiding Glycans. Immunity 43:1053–1063

Gardner MJ, Hall N, Fung E et al (2002) Genome sequence of the human malaria parasite *Plasmodium falciparum*. Nature 419:498–511

Georgiev IS, Joyce MG, Yang Y et al (2015) Single-chain soluble BG505·SOSIP gp140 trimers as structural and antigenic mimics of mature closed HIV-1 Env. J Virol 89:5318–5329

Germain RN (1994) MHC-dependent antigen processing and peptide presentation: providing ligands for T lymphocyte activation. Cell 76:287–299

Gherardi E, Sandin S, Petoukhov MV et al (2006) Structural basis of hepatocyte growth factor/ scatter factor and MET signalling. Proc Natl Acad Sci 103:4046–4051

Goel S, Pamkvist M, Moll K et al (2015). RIFINs are adhesins implicated in severe *Plasmodium falciparum* malaria. Nat Med 21:314–317

Gopal R, Jackson K, Tzarum N et al (2017) Probing the antigenicity of hepatitis C virus envelope glycoprotein complex by high-throughputmutagenesis. PLoS Pathog 13:e1006735

Gravitz L (2011) Introduction: a smouldering public-health crisis. Nature 474:2–4

Harris LJ, Larson SB, Hasel KW et al (1992) The three-dimensional structure of an intact monoclonal antibody for canine lymphoma. Nature 360:369–372

Harris LJ, Larson SB, Skaletsky E et al (1998) Comparison of the conformations of two intact monoclonal antibodies with hinges. Immunol Rev 163:35–43

Haury M, Sundblad A, Grandien A et al (1997) The repertoire of serum IgM in normal mice is largely independent of external antigenic contact. Eur J Immunol 27:1557–1563

Haynes BF, Moody MA, Verkoczy L et al (2005) Antibody polyspecificity and neutralization of HIV-1: a hypothesis. Hum Antibodies. 14:59–67

Hippocrates (1959) On the nature of man. Harvard University Press, Cambridge, MA

Holmberg SD, Spradling PR, Moorman AC et al (2013) Hepatitis C in the United States. N Engl J Med 368:1859–1861

Hoofnagle JH (2002) Course and outcome of hepatitis C. Hepatology 36:21–29

Janeway CA Jr, Travers P, Walport M et al (2001) Immunobiology: the immune system in health and disease, 5th edn. Garland Science, New York

Jefferis R (2009) Glycosylation as a strategy to improve antibody-based therapeutics. Nat Rev Drug Discov 8:226–234. https://doi.org/10.1038/nrd2804

Julien JP, Cupo A, Sok D et al (2013a) Crystal structure of a soluble cleaved HIV-1 envelope trimer. Science 342:1477–1483

Julien JP, Lee JH, Cupo A et al (2013b) Asymmetric recognition of the HIV-1 trimer by broadly neutralizing antibody PG9. Proc Natl Acad Sci U S A. 110:4351–4356

Katzelnick LC, Montoya M, Gresh L et al (2016) Neutralizing antibody titers against dengue virus correlate with protection from symptomatic infection in a longitudinal cohort. Proc Natl Acad Sci USA 113:728–733. https://doi.org/10.1073/pnas.1522136113

Khayat R, Lee JH, Julien JP et al (2013) Structural characterization of cleaved, soluble HIV-1 envelope glycoprotein trimers. J Virol 87:9865–9872

Kim MS, Chuenchor W, Chen X et al (2018) Cracking the DNA code for V(D)J recombination. Mol Cell 70:358–370. https://doi.org/10.1016/j.molcel.2018.03.008

Kindred B, Shreffler DC (1972) H-2 dependence of co-operation between T and B cells in vivo. J Immunol 109:940–943

Kisalu NK, Idris AH, Weidle C et al (2018) A human monoclonal antibody prevents malaria infection by targeting a new site of vulnerability on the parasite. Nat Med 24:408–416

Klasse PJ, Depetris RS, Pejchal R et al (2013) Influences on trimerization and aggregation of soluble, cleaved HIV-1 SOSIP envelope glycoprotein. J Virol 87:9873–9885

Kong L, Giang E, Nieusma T et al (2012a) Structure of hepatitis C virus envelope glycoprotein E2 antigenic site 412 to 423 in complex with antibody AP33. J Virol 86:13085–13088

Kong L, Giang E, Robbins JB et al (2012b) Structural basis of hepatitis C virus neutralization by broadly neutralizing antibody HCV1. Proc Natl Acad Sci U S A. 109:9499–9504

Kwong PD, Wyatt R, Robinson J et al (1998) Structure of an HIV gp120 envelope glycoprotein in complex with the CD4 receptor and a neutralizing human antibody. Nature 393:648–659

Lagging LM, Westin J, Svensson E et al (2002) Progression of fibrosis in untreated patients with hepatitis C virus infection. Liver 22:136–144

Law M, Maruyama T, Lewis J et al (2008) Broadly neutralizing antibodies protect against hepatitis C virus quasispecies challenge. Nat Med 14:25–27

Lawrence MG, Woodfolk JA, Schuyler AJ et al (2017) Half-life of IgE in serum and skin: consequences for anti IgE therapy in patients with allergic disease. J Allergy Clin Immunol. 139:422–428

Lee JE, Fusco ML, Hessell AJ et al (2008) Structure of the Ebola virus glycoprotein bound to an antibody from a human survivor. Nature 454:177–182

Lee JH, Ozorowski G, Ward AB (2016) Cryo-EM structure of a native, fully glycosylated, cleaved HIV-1 envelope trimer. Science 351:1043–1048

Li Y, Pierce BG, Wang Q et al (2015) Structural basis for penetration of the glycan shield of hepatitis C virus E2 glycoprotein by a broadly neutralizing human antibody. J Biol Chem 290:10117–10125

Liu J, Bartesaghi A, Borgnia MJ et al (2008) Molecular architecture of native HIV-1 gp120 trimers. Nature 455:109–113

Lu X, Xiao H, Li S, Pang X, Song J et al (2019) Double lock of a human neutralizing and protective monoclonal antibody targeting the yellow fever virus envelope. Cell Rep. 26:438–446. https://doi.org/10.1016/j.celrep.2018.12.065

Lyumkis D, Julien JP, de Val N et al (2013) Cryo-EM structure of a fully glycosylated soluble cleaved HIV-1 envelope trimer. Science 342:1484–1490

Maizels N (2005) Immunoglobulin gene diversification. Annu Rev Genet 39:23–46

Marquardt D, McCrone S, Center MS (1990) Mechanisms of multidrug resistance in HL60 cells: detection of resistance-associated proteins with antibodies against synthetic peptides that correspond to the deduced sequence of P-glycoprotein. Cancer Res 50:1426–1430

Maruyama T, Rodriguez LL, Jahrling PB et al (1999) Ebola virus can be effectively neutralized by antibody produced in natural human infection. J Virol 73:6024–6030

Meola A, Tarr AW, England P et al (2015) Structural flexibility of a conserved antigenic region in hepatitisvirus glycoprotein E2 recognized by broadly neutralizing antibodies. J Virol 89:2170–2181

Meunier JC, Russell RS, Goossens V et al (2008) Isolation and characterization of broadly neutralizing human monoclonal antibodies to the e1glycoprotein of hepatitis C virus. J Virol 82:966–973

Micoli F, Rondini S, Alfini R et al (2018) Comparative immunogenicity and efficacy of equivalent outer membrane vesicle and glycoconjugate vaccines against nontyphoidal *Salmonella*. Proc Natl Acad Sci U S A. 115:10428–10433. https://doi.org/10.1073/pnas.1807655115

Miller LH, Baruch DI, Marsh K et al (2002) The pathogenic basis of malaria. Nature 415:673–679

Mimura Y, Lund J, Church S et al (2001) Butyrate increases production of human chimeric IgG in CHO-K1 cells whilst maintaining function and glycoform profile. J Immunol Methods 247:205–216

Murin CD, Fusco ML, Bornholdt ZA et al (2014) Structures of protective antibodies reveal sites of vulnerability on Ebola virus. Proc Natl Acad Sci U S A. 111:17182–17187

Murphy KP, Travers P, Walport M (2008) Janeway's immunobiology,7th edn. Garland Science, New York

Noris M, Remuzzi G (2013) Overview of complement activation and regulation. Semin Nephrol 33(6):479–492. https://doi.org/10.1016/j.semnephrol.2013.08.001

O'Meara WP, Mwangi TW, Williams et al (2008) Relationship between exposure, clinical malaria, and age in an area of changing transmission intensity. Am J Trop Med Hyg 79:185–191

Ogden CA, Kowalewski R, Peng Y et al (2005) IGM is required for efficient complement mediated phagocytosis of apoptotic cells in vivo. Autoimmunity 38:259–264

Olinger GG Jr, Pettitt J, Kim D et al (2012) Delayed treatment of Ebola virus infection with plant-derived monoclonal antibodies provides protection in rhesus macaques. Proc Natl Acad Sci U S A. 109:18030–18035

Oswald WB, Geisbert TW, Davis KJ et al (2007) Neutralizing antibody fails to impact the course of Ebola virus infection in monkeys. PLoS Pathog 3:e9

Oyen D, Torres JL, Cottrell CA et al (2018) Cryo-EM structure of *P. falciparum* circumsporozoite protein with a vaccine-elicited antibody is stabilized by somatically mutated inter-Fab contacts. Sci Adv 4:eaau8529

Oyen D, Torres JL, Wille-Reece U et al (2017) Structural basis for antibody recognition of the NANP repeats in *Plasmodium falciparum* circumsporozoite protein. Proc Natl Acad Sci U S A. 114:E10438–E10445

Pancera M, Zhou T, Druz A et al (2014) Structure and immune recognition of trimeric pre-fusion HIV-1 Env. Nature 514:455–461

Parren PW, Geisbert TW, Maruyama T et al (2002) Pre- and postexposure prophylaxis of Ebola virus infection in an animal model by passive transfer of a neutralizing human antibody. J Virol 76:6408–6412

Paterson Y, Englander SW, Roder H (1990) An antibody binding site on cytochrome c defined by hydrogen exchange and two-dimensional NMR. Science 249:755–759

Pauthner MG, Nkolola JP, Havenar-Daughton C et al (2018) Vaccine-induced protection from homologous Tier 2 SHIV challenge in nonhuman primates depends on serum-neutralizing antibody titers. Immunity 50:241–252. https://doi.org/10.1016/j.immuni.2018.11.011

Perkins SJ, Nealis AS, Sutton BJ et al (1991) Solution structure of human and mouse immunoglobulin M by synchrotron X-ray scattering and molecular graphics modelling. A possible mechanism for complement activation. J Mol Biol 221:1345–1366

Petrušić V, Živković I, Stojanović M et al (2011) Hexameric immunoglobulin M in humans: desired or unwanted? Med Hypotheses 77:959–961

Pettersen EF, Goddard TD, Huang CC, Couch GS, Greenblatt DM, Meng EC, Ferrin TE (2004) UCSF Chimera? A visualization system for exploratory research and analysis. J Comput Chem 25(13):1605–1612

Piazza F, De Los Rios P, Fanelli D et al (2005) Anti-cooperativity in diffusion-controlled reactions with pairs of anisotropic domains: a model for the antigen-antibody encounter. Eur Biophys J 34:899–911

Pier GB, Lyczak JB, Wetzler LM (2004) Immunology, infection, and immunity. ASM Press. ISBN 1-55581-246-5

Pierce BG, Keck ZY, Lau P et al (2016) Global mapping of antibody recognition of the hepatitis C virus E2 glycoprotein: implications for vaccine design. Proc Natl Acad Sci U S A. 113:E6946–E6954

Pisetsky DS (1997) Specificity and immunochemical properties of antibodies to bacterial DNA. Methods 11:55–61

Pisetsky DS (1998) Antibody responses to DNA in normal immunity and aberrant immunity. Clin Diagn Lab Immunol 5:1–6

Potter JA, Owsianka AM, Jeffery N et al (2012) Toward a hepatitis C virus vaccine: the structural basis of hepatitis C virus neutralization by AP33, a broadly neutralizing antibody. J Virol 86:12923–12932

PyMOL software. http://www.pymol.org

Quartier P, Potter PK, Ehrenstein MR et al (2005) Predominant role of IgM-dependent activation of the classical pathway in the clearance of dying cells by murine bone marrow-derived macrophages in vitro. Eur J Immunol 35:252–260

Randall TD, Brewer JW, Corley RB (1992) Direct evidence that J chain regulates the polymeric structure of IgM in antibody-secreting B cells. J Biol Chem 267:18002–18007

Randall TD, King LB, Corley RB (1990) The biological effects of IgM hexamer formation. Eur J Immunol 20:1971–1979

Rayner LE, Kadkhodayi-Kholghi N, Heenan et al (2013) The solution structure of rabbit IgG accounts for its interactions with the Fc receptor and complement C1q and its conformational stability. J Mol Biol 425:506–523

Robbiani DF, Bozzacco L, Keeffe JR et al (2017) Recurrent potent human neutralizing antibodies to Zika Virus in Brazil and Mexico. Cell 169:597–609.e11. https://doi.org/10.1016/j.cell.2017.04.024

Roberts DJ, Pain A, Kai O et al (2000) Autoagglutination of malaria-infected red blood cells and malaria severity. Lancet 355:1427–1428

Roche PA, Cresswell P (1991) Proteolysis of the class II-associated invariant chain generates a peptide binding site in intracellular HLA-DR molecules. Proc Natl Acad Sci U S A. 88:3150–3154

Rosenthal AS, Shevach EM (1973) Function of macrophages in antigen recognition by guinea pig T lymphocytes. I. Requirement for histocompatible macrophages and lymphocytes. J Exp Med 138:1194–1212

Rudkin FM, Raziunaite I, Workman H et al (2018) Single human B cell-derived monoclonal anti-Candida antibodies enhance phagocytosis and protect against disseminated candidiasis. Nat Commun. 9:5288. https://doi.org/10.1038/s41467-018-07738-1

Saito F, Hirayasu K, Satoh T et al (2017) Immune evasion of Plasmodium falciparum by RIFIN via inhibitory receptors. Nature 552:101–105

Sanders RW, de Jong EC, Baldwin CE et al (2002) Differential transmission of human immunodeficiency virus type 1 by distinct subsets of effector dendritic cells. J Virol 76:7812–7821

Sanders RW, Moore JP (2017) Native-like Env trimers as a platform for HIV-1 vaccine design. Immunol Rev 275:161–182. https://doi.org/10.1111/imr.12481

Sanders RW, Schiffner L, Master et al (2000) Variable-loop-deleted variants of the human immunodeficiency virus type 1 envelope glycoprotein can be stabilized by an intermolecular disulfide bond between the gp120 and gp41 subunits. J Virol 74:5091–100

Sandin S, Öfverstedt LG, Wikström et al (2004). Structure and flexibility of individual immunoglobulin G molecules in solution. Structure 12: 409–415

Saphire EO, Parren PW, Pantophlet R et al (2001) Crystal structure of a neutralizing human IgG against HIV-1: template for vaccine design. Science 293:1155–1159

Scapin G, Yang X, Prosise WW et al (2015) Structure of full-length human anti-PD1 therapeutic IgG4 antibody pembrolizumab. Nat Struct Mol Biol 22:953–958

Scherf A, Lopez-Rubio JJ, Riviere L (2008) Antigenic variation in *Plasmodium falciparum*. Annu Rev Microbiol 62:445–470

Schlosstein L, Terasaki PI, Bluestone R et al (1973) High association of an HL-A antigen, W27, with ankylosing spondylitis. N Engl J Med 288:704–706

Schroeder HW Jr, Cavacini L (2010) Structure and function of immunoglobulins. J Allergy Clin Immunol. 125:41–52. https://doi.org/10.1016/j.jaci.2009.09.046

Sharma SK, de Val N, Bale S et al (2015) Cleavage-independent HIV-1 Env trimers engineered as soluble native spike mimetics for vaccine design. Cell Rep. 11:539–550

Smith DB, Bukh J, Kuiken C et al (2014) Expanded classification of hepatitis C virus into 7 genotypes and 67 subtypes: updated criteria and genotype assignment web resource. Hepatology 59:318–327

Su XZ, Heatwole VM, Wertheimer SP et al (1995) The large diverse gene family *var* encodes proteins involved in cytoadherence and antigenic variation of *Plasmodium falciparum*-infected erythrocytes. Cell 82:89–100

Sun Z, Almogren A, Furtado PB et al (2005) Semi-extended solution structure of human myeloma immunoglobulin D determined by constrained X-ray scattering. J Mol Biol 353:155–173

Tan J, Piccoli L, Lanzavecchia A (2018) The Antibody Response to *Plasmodium falciparum*: cues for vaccine design and the discovery of receptor-based antibodies. Annu Rev Immunol. https://doi.org/10.1146/annurev-immunol-042617-053301

Tan J, Pieper K, Piccoli L et al (2016) A LAIR1 insertion generates broadly reactive antibodies against malaria variant antigens. Nature 529:105–109

Tan K, Liu J, Wang J et al (1997) Atomic structure of a thermostable subdomain of HIV-1 gp41. Proc Natl Acad Sci U S A. 94:12303–12308

Tran EE, Borgnia MJ, Kuybeda O et al (2012) Structural mechanism of trimeric HIV-1 envelope glycoprotein activation. PLoS Pathog 8:e1002797

Vilhena JG, Dumitru AC, Herruzo ET et al (2016) Adsorption orientations and immunological recognition of antibodies on graphene. Nanoscale 8:13463–13475

Vollmers HP, Brandlein S (2006) Natural IgM antibodies: the orphaned molecules in immune surveillance. Adv Drug Deliv Rev 58:755–765

Wahlgren M, Goel S, Akhouri RR (2017) Variant surface antigens of *Plasmodium falciparum* and their roles in severe malaria. Nat Rev Microbiolgy. 15-479-491

Wec AZ, Bornholdt ZA, He S, Herbert AS et al (2019) Development of a human antibody cocktail that deploys multiple functions to confer pan-ebolavirus protection. Cell Host Microbe 25(39–48):e5. https://doi.org/10.1016/j.chom.2018.12.004

Weissenhorn W, Dessen A, Harrison SC et al (1997) Atomic structure of the ectodomain from HIV-1 gp41. Nature 387:426–430

WHO Global Hepatitis Report (2017) https://apps.who.int/iris/rest/bitstreams/1082592/retrieve

Wilson JA, Hevey M, Bakken R et al (2000) Epitopes involved in antibody-mediated protection from Ebola virus. Science 287:1664–1666

Wong-Baeza C, Reséndiz-Mora A, Donis-Maturano L et al (2016) Anti-lipid IgG antibodies are produced *via* Germinal Centers in a murine model resembling human lupus. Front Immunol 7:396

Woof J, Burton D (2004) Human antibody-Fc receptor interactions illuminated by crystal structures. Nat Rev Immunol 4:89–99

Zanetti G, Briggs JA, Grünewald K et al (2006) Cryo-electron tomographic structure of an immunodeficiency virus envelope complex in situ. PLoS Pathog 2:e83

Zeitlin L, Pettitt J, Scully C et al (2011) Enhanced potency of a fucose-free monoclonal antibody being developed as an Ebola virus immunoprotectant. Proc Natl Acad Sci U S A. 108:20690–20694

Zhang J, Liu D, Li G et al (2017) Antibody mediated neutralization of soluble MIC significantly enhances CTLA4 blockade therapy. Sci Adv 3(5):e1602133. https://doi.org/10.1126/sciadv.1602133

Zhu P, Chertova E, Bess J Jr et al (2003) Electron tomography analysis of envelope glycoprotein trimers on HIV and simian immunodeficiency virus virions. Proc Natl Acad Sci U S A. 100:15812–15817

Zhu P, Liu J, Bess J Jr et al (2006) Distribution and three-dimensional structure of AIDS virus envelope spikes. Nature 441:847–852

Chapter 3
Unravelling Ribosome Function Through Structural Studies

Abid Javed and Elena V. Orlova

Abstract Ribosomes are biological nanomachine that synthesise all proteins within a cell. It took decades to reveal the architecture of this essential cellular component. To understand the structure-function relationship of this nanomachine needed the utilisisation of different biochemical, biophysical and structural techniques. Structural studies combined with mutagenesis of the different ribosomal complexes comprising various RNAs and proteins enabled us to understand how this machine works inside a cell. Nowadays quite a number of ribosomal structures were published that confirmed biochemical studies on particular steps of protein synthesis by the ribosome. Four major steps were identified: initiation, elongation, termination and recycling. These steps lead us to the important question how the ribosome function can be regulated. Advances in technology for cryo electron microscopy: sample preparations, image recording, developments in algorithms for image analysis and processing significantly helped in revelation of structural details of the ribosome. We now have a library of ribosome structures from prokaryotes to eukaryotes that enable us to understand the complex mechanics of this nanomachine. As this structural library continues to grow, we gradually improve our understanding of this process and how it can be regulated and how the specific ribosomes can be stalled or activated, or completely disabled. This article provides a comprehensive overview of ribosomal structures that represent structural snapshots of the ribosome at its different functional states. Better understanding rises more particular questions that have to be addressed by determination structures of more complexes.

Synopsis: Structural biology of the ribosome.

Keywords Ribosome · Function · Structure · X-ray · cryoEM · Nascent chain

A. Javed · E. V. Orlova (✉)
Institute of Structural and Molecular Biology, Department of Biological Sciences, Birkbeck College, Malet Street, London WC1E 7HX, UK
e-mail: e.orlova@mail.cryst.bbk.ac.uk

A. Javed
e-mail: a.javed@mail.cryst.bbk.ac.uk

J. R. Harris and J. Marles-Wright (eds.), *Macromolecular Protein Complexes II: Structure and Function*, Subcellular Biochemistry 93,
https://doi.org/10.1007/978-3-030-28151-9_3

Why We Need to Study Ribosomes

Across all kingdoms of life, biological systems need proteins for their function. These proteins are encoded in their genomes. Specific nanomachines carry out translation of the genetic information into amino acid sequences of proteins. These bio-machines are named as ribosomes. Ribosomes decode the genetic information contained in a messenger RNA (mRNA) transcript and synthesise nascent polypeptide chains (NC), that will fold to become functional proteins.

In prokaryotes, ribosomes are dispersed in the cytoplasm, while in eukaryotes they can be both in the cytoplasm and can be bound to membranes of the nucleus and endoplasmic reticulum. Eukaryotes also have mitochondrial ribosomes, which translate mitochondrial proteins encoded in mitochondrial DNA. Depending on the cell type molecular mass of ribosomes varies from 2.5 to 4.5 MDa. Their sizes are characterised by Svedberg coefficients (S), which is a measure of a particle size based on its sedimentation rate in differential centrifugation. Prokaryotic ribosomes were named as 70S ribosomes, due to their lower S coefficient than eukaryotic ribosomes, which have the sedimentation coefficient of 80S (Taylor et al. 1967). In all organisms, ribosome consists of two ribonucleoproteins subunits (Tissieres and Watson 1958) (Fig. 3.1). The prokaryotic subunits are smaller and named as 30S and 50S whereas the eukaryotic ribosome subunits are termed as 40S and 60S respectively. Each subunit comprises of ribosomal RNA (16S for small and 23S for large subunits in bacteria) and 18S for small and 28S for large subunits (in eukaryotes) and ribosomal proteins (54 in bacteria and 80 in eukaryotes—Melnikov et al. 2012 and references herein) (Figs. 3.1 and 3.2).

The small subunit (SSU) of the ribosome binds the mRNA that contains the coded information of a protein to synthesise it. This step is followed by the formation of the ribosome/mRNA complex. The mRNA moves on the ribosome between the small and large subunit (LSU) where it is decoded (translated) with the assistance of different ribosomal factors into a sequence of amino acids. The SSU carries out one of the main ribosomal function such as decoding the mRNA and monitoring translation reliability. LSU operates a peptidyl transferase centre (PTC), where the polymerisation of amino acid into a new polypeptide chain takes place. Then the chain leaves the ribosome through the protein-exit tunnel. The data acquired by different studies revealed that the main steps of protein translation and formation of a polypeptide chain are profoundly conserved.

Due to the complexity of eukaryotic ribosomes and challenges in their purification from eukaryotic sources, bacterial ribosomes proved to be a good model

Fig. 3.1 Ribosome architecture. **a**—Bacterial ribosome (shown in grey) is composed of two ▶ subunits. The small subunit (SSU) is shown in gold, the large subunit (LSU) is shown in blue. Characteristic domains are labelled. **b**—the inter-subunit bridge points in SSU and LSU are shown (Liu and Frederick, 2016). **c**—components of ribosomal subunits located on the inner surface. **d**—components of the ribosome located on the outer surface of the ribosome. rRNA is shown in yellow (in SSU) or blue and dark blue (in LSU), and ribosomal proteins are shown in orange (SSU) or magenta (LSU)

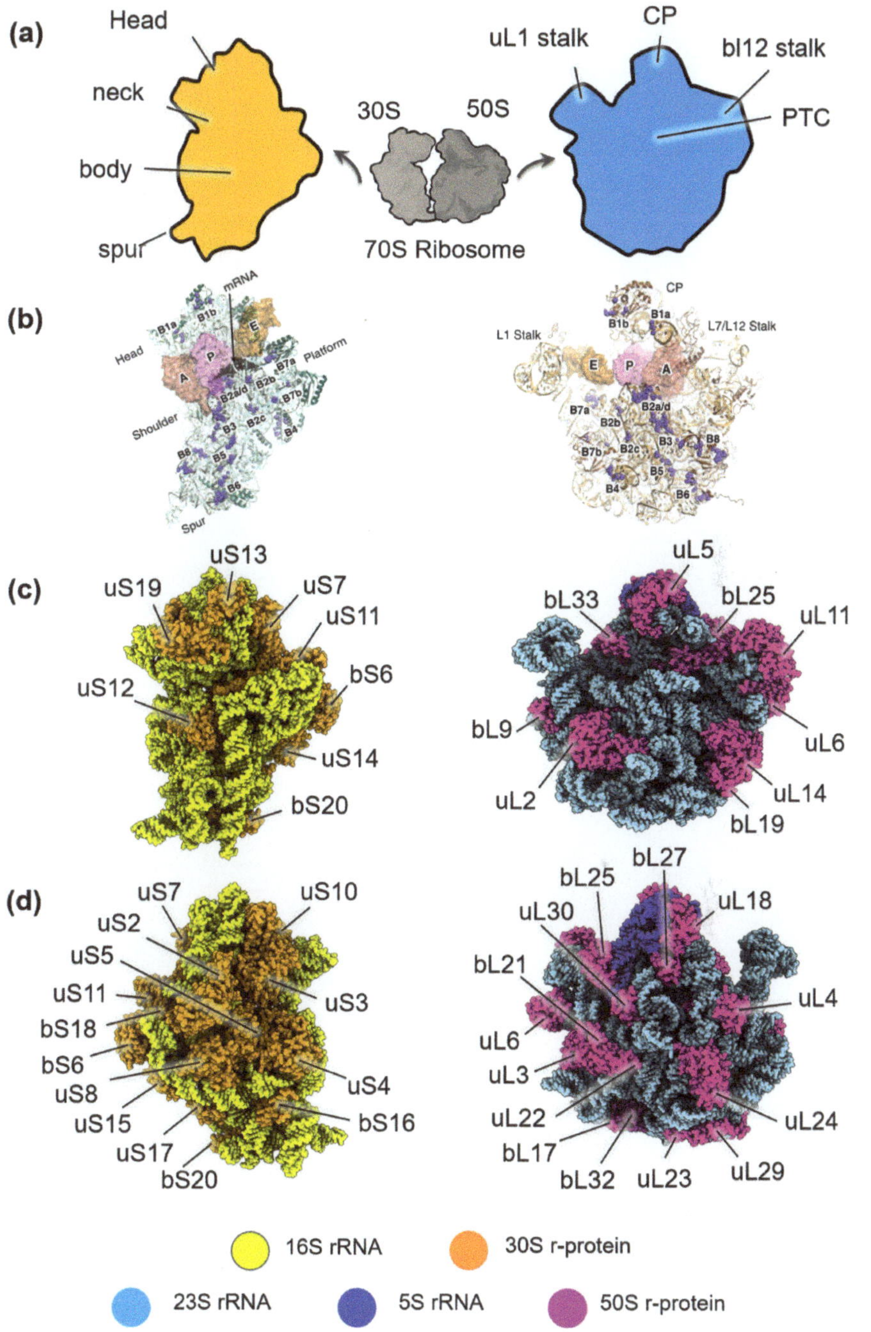
(a)
Head
neck
body
spur
30S
50S
70S Ribosome
CP
uL1 stalk
bl12 stalk
PTC
(b)
mRNA
Head
Platform
Shoulder
Spur
B1a
B1b
E
P
A
B7a
B2a/d
B2b
B2c
B7b
B3
B4
B5
B6
B8
CP
L1 Stalk
L7/L12 Stalk
(c)
uS13
uS19
uS7
uS11
bS6
uS12
uS14
bS20
uL5
bL33
bL25
uL11
bL9
uL6
uL2
uL14
bL19
(d)
uS7
uS10
uS2
uS5
uS11
uS3
bS18
bS6
uS8
uS4
uS15
bS16
uS17
bS20
bL27
bL25
uL18
uL30
bL21
uL4
uL6
uL3
uL22
uL24
bL17
bL32
uL23
uL29
16S rRNA
30S r-protein
23S rRNA
5S rRNA
50S r-protein

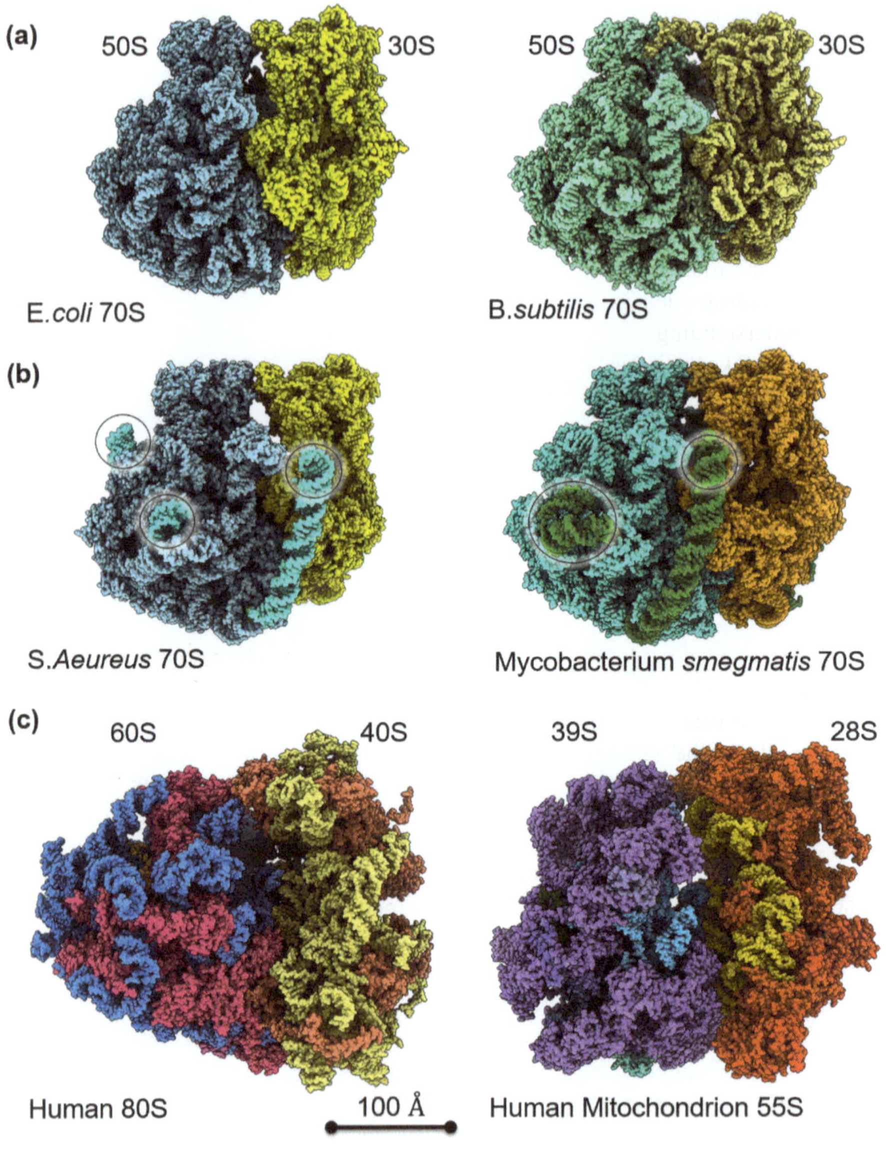

Fig. 3.2 Comparison of ribosomes from different organisms. **a**—*E. coli* 70S ribosome (PDB 4YBB) is shown with LSU in blue and SSU in yellow. The *B.subtilis* 70S ribosome (PDB 3J9M), very similar in structure to *E. coli* ribosome is shown with its LSU in cyan and SSU in pale yellow. **b**—The more divergent prokaryotic ribosomes are shown; ribosome structure from *S.aureus* (PDB 5NGM) and ribosome from *M.smegmatis* (PDB 5O61) shows rRNA extensions (circled and highlighted in blue or green). **c**—Eukaryotic ribosomes from human (PDB 4UG0) and mitochondrial 55S ribosome (PDB 3J9M). Human LSU rRNA is in dark blue and proteins in magenta whilst SSU rRNA is shown in pale yellow and proteins in orange. Mitochondrial LSU (39S) subunit rRNA is in blue and proteins in purple whilst SSU (28S) rRNA is shown in yellow and proteins in red. Scale bar is 100 Å

system to study ribosome structure and function and have contributed to a wealth of our existing knowledge of this complex nanomachine. Moreover, an extremely large number of diseases in mammals are caused by pathogenic bacteria. The leading tactic has been finding means to suppress these pathogenic microorganisms. Currently, usage of ribosome-specific antibiotics that will block translation in pathogenic ribosomes is the most efficient approach. The differences between prokaryotic and eukaryotic ribosome permits to use the antibiotics to block the activity of the bacterial ribosomes but allows mammalian ribosomes to function as normal. Furthermore, bacteria evolve much faster compared to the development of new specific drugs, leading to prevalent antibiotic resistance. Therefore, improvement in understanding structure-function relationship within ribosomes, particularly pathogenic prokaryotic ribosomes, became essential both in the sense of revealing crucial factors in the activity of these nanomachines and understanding specific mechanisms by which ribosome's activities are regulated.

Methods Used in Studies of Ribosome Structure/Function Relationship

The large size and complexity of the ribosome represented significant challenges for understanding their structural organisation, therefore a number of structural techniques were utilised to study the function of ribosomes. Ribosomes were discovered in the middle of the 1950s. Images of Palade particles were obtained by electron microscopy of stained thin sections of the rat pancreas where an unusual pattern of grains on the outer surface of the endoplasmic reticulum was reported (Palade 1955). Later these 'grains' were identified and renamed as ribosomes. Morphology and dimensions of these particles and individual subunits were assessed using negative stain EM (NS EM). The fully assembled ribosome particles have sizes ~250 Å for the 70S and ~250–300 Å for the 80S (Lake 1978; Bernabeu and Lake 1982). Then, the substantial impact for understanding their function came via extensive biochemical approaches that yielded illuminating insights into the essence of the ribosome function. Comparative DNA sequence analysis, sedimentation and SAXS methods have shown that the ribosome is a complex of ribosomal RNA (rRNA) with the well-ordered organisation and a number of globular proteins (r-proteins; Moore et al. 1968; Delius et al 1968; Noller and Herr 1974; Herr and Noller 1975; Brosius et al. 1978).

NS EM images of bacterial ribosomes revealed the basic structural features of the ribosome. These include the early description of ribosomal subunits. The SSU has four major structural domains: the "head" domain containing the major domain of rRNA, the "platform" with the RNA central domain, the "shoulder", and "foot" containing rRNA domain. Shoulder, foot and platform together form the "body" which is connected by the "neck" with the SSU head domain (Fig. 3.1a, left panel). The LSU of the prokaryotic ribosomes has been described as having a crown shape characterised by three protuberances: "ridge", "central" formed by 5S RNA and more elongated "stalks" on each side of the LSU: uL1, and bL12 stalks (Fig. 3.1a, right panel) (Frank and Agrawal 1998; Stark et al. 1997; Matadeen et al. 1999). Sample preparation by negative staining (NS) imaging, however, resulted in flattened molecular complexes (Orlova 2000; Frank 2006—and references herein), preventing to get reliable structural information on the ribosome, and the NS does not allow to see the details of the inner structure of the complex.

Studies on the individual components of the ribosome provided details of the overall organisation of this nano-machine. Neutron scattering studies of the individual subunits of bacterial ribosomes determined the relative positions of the ribosomal proteins (r-proteins) (Moore and Engelman 1975). Antibodies raised against the r-proteins located on the outer surface together with imaging using NS EM (immuno-electron microscopy) further localised some ribosomal proteins on both subunits on the surface (Tischendorf et al. 1974a, b). These observations were confirmed by chemical cross-linking studies on *E. coli* ribosome subunits, which provided details of r-protein-rRNA contacts (Moller and Brimacombe 1975).

Development of methods of bacterial ribosome purification commenced the quest for determination of a detailed architecture of this macromolecular machine by many research groups. X-ray crystallography has been known by that time as a routine technique for structural analysis of proteins at atomic resolution. Yet, ribosome represented an immense challenge for usage of this technique. A crucial prerequisite to have this method successful is the necessity of high-quality crystals and the size of the unit cell in these crystals. It took significant efforts and time until the first ribosome crystals were obtained for X-ray crystallography (Yonath et al. 1987). And, the process of obtaining a structure of the ribosome at an atomic resolution was not a well-paved road. It took nearly one third of the century to obtain atomic structures of ribosomes, with the help of other approaches.

Cryo-EM (cryo-electron microscopy at liquid nitrogen temperatures) is a fast-developing powerful tool used zealously in structural studies of macromolecular assemblies with molecular masses ranging from hundred KDa to MDa. Computational and technological advances extend boundaries of achievements by cryo-EM in structural studies of large bio-complexes and provide a wealth of information on the atomic level, assisting our understanding of functional mechanics and dynamics (Henderson 2015). Rapid freezing of thin layers of samples in liquid nitrogen enabled researchers to capture ribosomes in a near-native environment (Dubochet et al. 1988; Frank et al. 1991; Orlova 2000). The first structure of the bacterial 70S ribosome at nearly native conditions was obtained by cryo-EM in 1991 by Frank and colleagues. Electron crystallography of crystals of the chick embryo ribosome, and later low-resolution X-ray studies of *Bacillus stearothermophilus* 50S subunit indicated that there is a putative opening within the LSU for the exit of newly synthesised nascent polypeptide chains (NC) (Bernabeu and Lake 1982; Milligan and Unwin 1986; Yonath et al. 1987). The position of IgG antibody bound to the β–galactosidase NC in the NS EM images of the 80S translating ribosomes suggested that there is a channel through which the NC emerges from the ribosome (Bernabeu and Lake 1982). Further studies confirmed the existence of the exit channel for the newly synthesized polypeptide chain (Frank et al. 1995; Beckmann et al. 1997; Ban et al. 2000; Gabashvili et al. 2000, 2001). The next step in functional studies of the ribosomes was provided by the revelation of structural conformations in ribosomes using computational methods in the analysis of cryo-EM images. This helped to analyse the intrinsic dynamics of the ribosome complexes (Gabashvili et al. 2000; Elad et al. 2008). Due to the versatility of cryo-EM, it became the major tool for structural analysis of ribosome complexes allowing to understand their structural dynamics.

Combined studies of the ribosome by X-ray crystallography and cryo-EM represented an amazing example where this complementation provided one of the most significant breakthroughs in understanding the function of the ribosome bio-machine. The first 9 Å structure of the LSU ribosome (H50S) was obtained for the ribosome of *Haloarcula marismortui*, where the X-ray data were phased using the intermediate resolution EM map of the ribosome (Frank et al. 1995; Ban et al. 1998). Later, the refined structure at 2.4 Å of the LSU provided atomic details of 23S and 5S ribosomal RNA, revealed locations of ribosomal proteins and suggested the structural basis behind the catalytic peptide bond synthesis at the PTC (Ban et al. 2000). The structure of the SSU from the eubacteria *Thermus thermophilus* subsequently demonstrated the loci of the mRNA and tRNA binding sites, previously indicated by low-resolution cryo-EM maps, that allowed to propose a mechanism of mRNA decoding (Gabashvili et al. 2000; Wimberly et al. 2000). At the same time, studies of the complete *Thermus thermophilus* 70S ribosomes at high-resolution of 5.5 Å provided insights on the mRNA-tRNA binding interface between subunits, elucidating a key role for the inter-subunit RNA bridges in keeping the LSU and SSU together during protein translation (Yusupov et al. 2001).

NMR spectroscopy is another method used to study the dynamic regions of the ribosome, enabling us to both undertake structural studies of complexes with a relatively small mass (less than ~40 KDa; Clore and Gronenborn 1998), and to study protein folding and dynamics of large biocomplexes. Using site-specific isotope labelling, this method helped to study the mobility of ribosome components such as ribosome stalk component bL12 (Fig. 3.1a, left panel) (Christodoulou et al. 2004) and ribosome-bound nascent polypeptide chains (Cabrita et al. 2016).

Novel approaches to decipher the ribosome structure and its dynamics are emerging and are used to expand our understanding of this complex molecular machinery (Fig. 3.1). Single-molecule fluorescence (smFRET) studies of the ribosome have provided details on movements of the ribosome elements during protein translation (Petrov et al. 2012) or spontaneous ribosome ratcheting as a result of higher temperatures (Cornish et al. 2008). Ribosome profiling is another powerful tool, proving successful to discern which proteins are translated when inside a cell (Ingolia et al., 2019). Combination of these methods is enabling us to reach deeper knowledge of the intricacies of ribosome structure and function.

General Organisation of Ribosomes in Prokaryotes

Genetic studies demonstrated that bacterial and eukaryotic ribosomes share a common structural core having two subunits SSU and LSU (Figs. 3.1 and 3.2). The SSU consists of 16S rRNA with 1458 nucleotides and 15 conserved proteins. The LSU has 19 conserved proteins, 5S rRNA and 23S rRNA, and ~4400 RNA bases related to the decoding site, PTC and tRNA-binding sites, which represent the major functional regions of the ribosomes. 16S and 23S rRNA form part of the mRNA channel on the small subunit and the exit channel on the large subunit, respectively (Schmeing and Ramakrishnan 2009; Melnikov et al. 2012). However, there are important differences in structural details, sizes and regulation of protein translation between ribosomes from prokaryotes and eukaryotes, and they vary between species. The differences between the ribosomes are defined by the presence of specific elements such as domain-specific proteins, insertions and extensions within the conserved proteins, and expansion segments of rRNAs that are defined by species and environment in which they exist (Melnikov et al. 2018). The 70S ribosome contains 20 bacteria-specific proteins (6 in the 30S subunit, 14 in the 50S subunit) and ribosomal RNA (Fig. 3.1). Most of these rRNA and proteins cover the main core from the solvent side and are accessible for interactions with translation factors and chaperones. The surface of the inter-subunit interface is composed predominantly of rRNA and, in the assembled ribosome, all functional sites are located close to this interface (Fig. 3.1).

Insights into the organisation of ribosomes from both Gram-negative and Gram-positive species have shown that the core of ribosome structure is highly conserved (Melnikov et al. 2018; Khusainov et al. 2016; Schmeing and Ramakrishnan 2009; Sohmen et al. 2015). At the same time, comparison of the pathogenic bacterial ribosomes demonstrated that there are varieties of the equivalent r-proteins, which apparently have evolved distinct functions due to specific adaptability of ribosomes (Fig. 3.2). For instance, r-protein bS1 on the SSU, which in *E. coli* is essential for translation initiation of canonical mRNAs comprises six RNA binding domains in *E. coli*; but its equivalent in *Sappylococcus. aureus* (*S. aureus*) (Fig. 3.2b, left panel) and other Gram-positive bacteria with low-GC nucleotide content have only four domains (Eyal et al. 2015; Khusainov et al. 2016). Moreover, there are indications that bacterial ribosomes may change their components according to the environmental alterations (Vesper et al. 2011). Across bacterial ribosomes, both 16S and 23S rRNA contain insertions (varying from *E. coli* ribosomes), which typically protrude from the ribosome to different extents (Fig. 3.2). For example, rRNA helices h6, h10, h26 and h44 in the SSU, and H28 and H68 in the LSU have different lengths or adopt different folds and orientations, as revealed when comparing the structures of 70S ribosomes from *E. coli and B. subtilis* (Fig. 3.2a, Noeske et al. 2015; Sohmen et al. 2015) and the structure of the 50S from *Staphylococcus aureus* (Eyal et al. 2015). Such variations in peripheral extensions suggest involvement in translation regulation at one or several stages of the process. Recently, a structure of the 70S ribosome of *Mycobacterium smegmatis*

(Fig. 3.2b, right panel), which is a close relative to the human pathogen Mycobacterium tuberculosis was solved by cryo-EM at 3.3 Å resolution (Hentschel et al. 2017). The structure has revealed two Mycobacterium specific ribosomal proteins in the vicinity of two drug-target sites in the catalytic centre (on the LSU) and the decoding site (on the SSU) (Hentschel et al. 2017). The bS22 protein is located on the 30S subunit between 16S and rRNA helices h27, h44, and h45 beneath the mRNA channel. The bL37 protein is sandwiched on the 50S between 23S and rRNA domain II (helices H39 and H40) and domain V (helices H72 and H89), which contains the universally conserved PTC and harbors the essential tRNA binding loops (A and P loops) (Ban et al. 2000; Hentschel et al. 2017). The structure of a pathogenic bacterial ribosome enabled differences to be discerned in the antibiotic binding sites on the ribosome that will eventually aid in designing mycobacterium specific antibiotics.

tRNAs are the non-ribosomal substrates that decode the genetic information and bring the amino acids that match to a codon in mRNA and should be incorporated in the growing protein. Ribosomal subunits have three binding sites for tRNA substrates: The A site where incoming aminoacyl-tRNA (A-tRNA) is bound, the P site holds the peptidyl tRNA (p-tRNA) corresponding to the codon and amino acid became attached to a nascent polypeptide chain during elongation, and the E site, at which the deacylated P-site tRNA, after peptide formation, became released from the (Schmeing and Ramakrishnan 2009). Analysis of the ribosome conformations through classification of EM structures helped to unveil key functional regions on the ribosome including the mRNA channel on the SSU, binding sites for A-, P- and E-tRNAs and their movement along the 70S ribosome during translation (Fig. 3.1b) (Frank et al. 1995; Agrawal et al. 1996, 2000). At each elongation cycle, both subunits participate dynamically in translocating the mRNA and the tRNA molecules by a single codon (Bashan and Yonath 2008). Consequently, cryo-EM has demonstrated one of the characteristic ribosome motions known as the "ratcheting" of the subunits, as such movement ensures shifts along the mRNA transcript during translation (Frank and Agrawal 2000).

Structural Basis of Protein Synthesis by the Ribosome

The main function of the ribosome is to decode mRNA and synthesise proteins. Four main steps were distinguished in the process of protein synthesis by the ribosome: initiation, elongation, termination and recycling. The SSU mediates base-pairing interactions between the mRNAs and tRNA that define the amino acid sequence of the nascent polypeptide chain while the LSU catalyses peptide bond formation at the peptidyl transferase centre (PTC) between the amino acids covalently attached to tRNA during elongation (Schmeing and Ramakrishnan 2009; Steitz 2008; Moore 2009). Structures of ribosome complexes captured at various stages of translation solved by cryo-EM and X-ray crystallography provided much of our understanding of how this translational machinery works.

Initiation

Translation initiation is an important cellular checkpoint that ensures the timely production of proteins. In bacteria, this step involves the consecutive formation of three intermediate complexes that differ both in composition and conformation. In essence, the SSU binds the mRNA close to the translation start site (Fig. 3.3). The initial fMet-tRNA binds to the first codon of the mRNA at the P-site and the next tRNA, which enters the ribosome at the dynamic bL12 stalk, attaches to the next codon at the A-site. As soon as a peptide bond is formed, the A-site tRNA is translocated to the P-site and the deacylated tRNA moves from the P-site to the E-site on its way out of the ribosome through the mobile uL1 stalk (Fig. 3.3). Several structures of translation initiation intermediates have been determined by X-ray crystallography and cryo-EM (Fig. 3.3—McCutecheon et al. 1999; Carter et al. 2000; Myasnikov et al. 2005; Marzi et al. 2007; Simonetti et al. 2008; Julian et al. 2011; Lopez-Alonso et al. 2017), providing key insights on how this important step that commences protein translation.

The first step is the formation of "30S pre-initiation complex" (30S PIC) that is composed of mRNA, fMet-tRNA and three initiation factors (Fig. 3.3a, b). X-ray structure of T. *thermophilus* 30S bound to mRNA mimic (containing Shine-Dalgarno (SD) sequence at the 5′ end of mRNA) provided details on how the SD sequence of mRNA docks at the 30S during initiation located between the 'head' and 'platform' domains of the SSU (Fig. 3.3b; Kaminishi et al. 2007). This interaction fixes the SD sequence of mRNA on SSU positioning the AUG start site at the P-tRNA binding site (Fig. 3.3c, d). Structural snapshots captured using single-particle cryo-EM, visualised mRNA structures in its folded and unfolded forms on the SSU and provided details on how SSU recruits the correct mRNA sequence during initiation (Marzi et al. 2007).

The 30S subunit has to bind mRNA and fMet tRNA at a correct 'match' that transforms the pre-initiation complex into the initiation complex (Fig. 3.3b). There are three non-ribosomal factors that participate in this process: initiation factor 1 (IF1), IF2, and IF3. IF3 and IF1 are involved in the mRNA adjustment process, ensuring the correct mRNA codon is selected before the ribosome enters the elongation phase (Fig. 3.3c). IF2 takes care of recruitment of fMet-tRNA to the ribosome in the "30S initiation complex" (30S-IC) (Fig. 3.3d), correct positioning of fMet-tRNA at the P-site in 70S-IC, and dissociation of IFs from 70S initiation complex (70S-IC) (Fig. 3.3e), using energy from GTP hydrolysis. A recent cryo-EM structure of an antibiotic (GE81112) stalled SSU pre-initiation complex with IF1, IF2 and IF3 and f-Met tRNA bound to 30S was reported (Lopez-Alonso et al. 2017). The study revealed conformational changes that occur on the SSU, binding of IF1, IF2, and IF3 on three distinct sites on the SSU and the changes in positions of fMet-tRNA, critical to start active 70S ribosome assembly and protein translation (Fig. 3.3d; Lopez-Alonso et al. 2017).

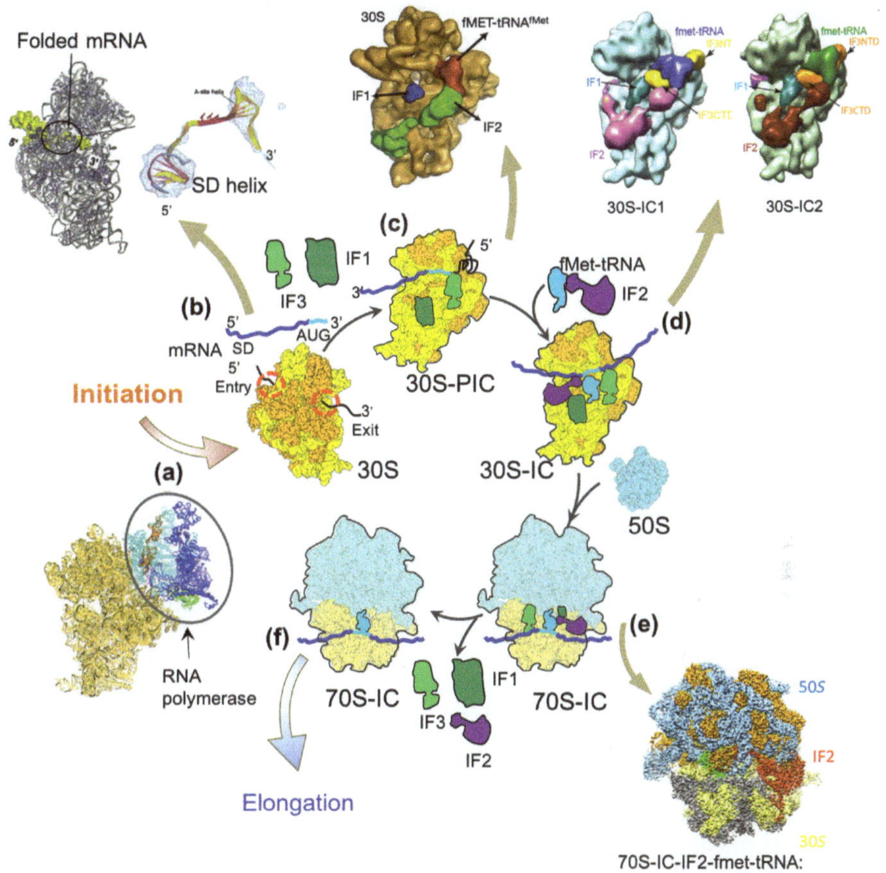

Fig. 3.3 Ribosome translation initiation. A schematic representation of bacterial translation initiation. **a**—The process begins with SSU (in yellow) coupled to mRNA synthesis by RNA polymerase (shown in blue) (Demo et al. 2017a, b). **b**—mRNA entry and exit on the SSU head are shown in red and the structure of mRNA with SSU shows the path mRNA takes on the SSU (Yusupov et al. 2001). Shine Dalgarno (SD) site on mRNA is indicated in the insert and AUG codon (start site) is shown in light blue on the mRNA. **c**—Initiation involves initiation factors: IF1 (in dark green), IF2 (purple) and IF3 (in light green) which bind the SSU to form pre-initiation complexes. **d**—The components that form these complexes are indicated: SSU in orange with fMet tRNA in red, and IF1 in blue and IF2 in green (Marzi et al. 2003); SSU in blue with IF1 in turquoise, IF2 in magenta and IF3 in yellow (Lopez-Alonso et al. 2017). **e**—Recently solved cryo-EM structure revealed how the last step of IF2 dissociation occurs (Sprink et al. 2016). **f**—Once LSU binds, the initiation factors dissociate to form the active 70S complex

Cryo-EM structure of 30S IC complex, comprising SSU, mRNA, fMet-tRNA and the three IF proteins shed light on how this conformational switch takes place, depending on the positioning of initiation substrates on the SSU (Julian et al. 2011). This was complemented by the structure of 30S-IC (lacking IF3), comprising 30S subunit, fMet-tRNA, IF1, IF2 and 27 nucleotide long mRNA. The structure demonstrated the organisation of 30S-IC, showing how the tRNA is stabilised by IF2 and provides details on the interaction between the SSU and initiation factors (Simonetti et al. 2008). The results reveal that the 30S IC complex has IF2 bound close to mRNA channel and blocking A tRNA binding site on SSU; IF2 has contacts with IF1 (bound closer to P tRNA binding site), the 30S subunit shoulder, and the CCA end of fMet-tRNA, which occupies an intermediate (P/I) tRNA binding position (Fig. 3.3d). The N-terminal domain of IF3 contacts the tRNA, whereas the C-terminal domain is bound to the platform of the 30S subunit (Julian et al. 2011). Interactions of IFs with SSU ensure the accurate position of the tRNA anticodon to AUG start site on the mRNA at the SSU P-site (Simonetti et al. 2008; Julian et al. 2011).

Consequently, the LSU (50S) is attached to the 30S-IC leading to the formation of the 70S-IC, which is ready to enter elongation phase (Schmeing and Ramakrishnan 2009) (Fig. 3.3e, f). IF2 plays a key role in this structural transition from 30S-IC to 70S-IC, leading to dissociation of IF1 and IF3. The cryo-EM structure of 70S-IC-IF2-GDP (Fig. 3.3e) has disclosed how IF2 induces IF dissociation from 70S-IC upon GTP hydrolysis (Myasnikov et al. 2005). This structure demonstrates that in the GTP-analog bound state, IF2 interacts with SSU and initiator tRNA at the P-site inducing structural changes in IFs and the ribosome, leading then to the dissociation and release of IF dissociation and formation of active 70S-IC (Myasnikov et al. 2005). IF3, which regulates the correct mRNA and tRNA match was also found to remain bound to 70S-IC, even after subunit joining. The high dissociation rates measured biochemically measured implicated the factors involvement in alternative (non-SD led) pathways that possibly exist in translation initiation (Goyal et al. 2017).

The interaction between the SSU and LSU takes place via 12 bridges between the subunits, made by RNA–RNA, RNA–protein, and protein–protein interactions (Yusupov et al. 2001; Selmer et al. 2006; Korostelev et al. 2006; Harms et al. 2001; Shaikh et al. 2014; Noeske et al. 2015) (Fig. 3.1b). Bridge B2a is particularly important because it connects the elements of the LSU and SSU at the decoding centre of the SSU and the region that forms the PTC on LSU. The B2a bridge has the ability to adopt several conformations, depending on the functional state of the ribosome (Ban et al. 2000; Bashan et al. 2003). At the end of the initiation process, an active 70S-IC is formed which can start peptide bond formation (Fig. 3.3f).

Elongation

After initiation, ribosome begins protein synthesis in a process called elongation that consists of well-defined cyclical steps (Fig. 3.4). The elongation cycle includes decoding, movement of tRNA to the P-site to form peptide-bond, amino acid polymerisation, detachment of the P-site tRNA from the growing polypeptide chain and release of the deacylated tRNA (Fig. 3.4). These actions are governed by the successive coordinated movements of the mRNA, associated tRNAs between the

SSU and LSUs from A-site to the P-site and then to the E-site, one codon at a time (in a 3′ to 5′ direction), and interactions between amino acids forming a nascent polypeptide chain (NC). Upon peptide bond formation, the ribosome fluctuates between two major conformations: from a non-rotated to a rotated state (the SSU rotates counterclockwise with respect to the LSU on ~ 10° between two positions). The non-rotated state has the two tRNAs bound to P and A sites on both SSU and LSU whereas the rotated state has tRNAs bound in hybrid P/E and A/P conformers (LSU/SSU).

Decoding

The ribosome nanomachine operates with high precision to ensure correct matching of the tRNA anticodon with the mRNA codon on the A-site. This process is assisted by Elongation Factor-Tu (EF-Tu) that delivers aminoacyl-tRNAs to the ribosome. EF-Tu is bound to GTP and has a high affinity for aminoacyl-tRNA; together these three components form the ternary complex (Fig. 3.4a). Binding of the ternary complex to the A-site tRNA and the ribosome involves the flexible bL12 stalk on the ribosome, tethering a charged tRNA close to the ribosomal A site. This step is codon-independent (Kothe et al. 2004; Diaconu et al. 2005). Then the aminoacyl-tRNA bound to EF-Tu and A-site at the ribosome undergoes verification via correspondence of its anticodon with the mRNA codon. Once the correct match is found between the mRNA and A-site tRNA, EF-Tu hydrolyses GTP, reducing the affinity to aminoacyl-tRNA and enabling EF-Tu to dissociate from the ribosome (Fig. 3.4b). Cryo-EM studies of EF-Tu on the ribosome show that it contacts the shoulder domain of the SSU. The closure of the shoulder domain of the SSU moves it towards the ternary complex stabilizing the transition state for GTP hydrolysis by EF-Tu and leading to activation of the GTPase if the matching with codon was correct (Ogle et al. 2002; Ogle and Ramakrishnan 2005; Stark et al. 2002; Valle et al. 2003). X-ray structure of the 70S ribosome-EF-Tu-A-tRNA complex further highlighted distortions introduced in A-site tRNA that permit aminoacyl-tRNA to interact with both the decoding centre of the SSU and EF-Tu at the factor-binding site (Schmeing et al. 2009). This distortion is necessary for the tRNA to be accurately positioned at the PTC, close to the P-site tRNA (Fig. 3.4b) (Schmeing et al. 2009). Latest developments in cryo-EM image analysis methods enabled researchers to distinguish micro-heterogeneity in ribosome complexes captured during decoding. In particular, intermediate structures of 70S-EF-Tu-Asite tRNA complex were solved by cryo-EM revealing the link between codon recognition and the activation of translational GTPases steps on ribosomes (Loveland et al. 2017). Classification of large datasets (between 500,000 and 1 million particles) has shown a series of short living states reflecting the differences during the selection of cognate and near-cognate tRNA matches at the A-site. These structures demonstrated that at the correct codon match, the SSU undergoes a conformational change and binds tightly EF-Tu. Structures of near-cognate complex fail to induce conformational changes in the SSU required for anchoring of the aminoacyl-tRNA on the ribosome (Loveland et al. 2017).

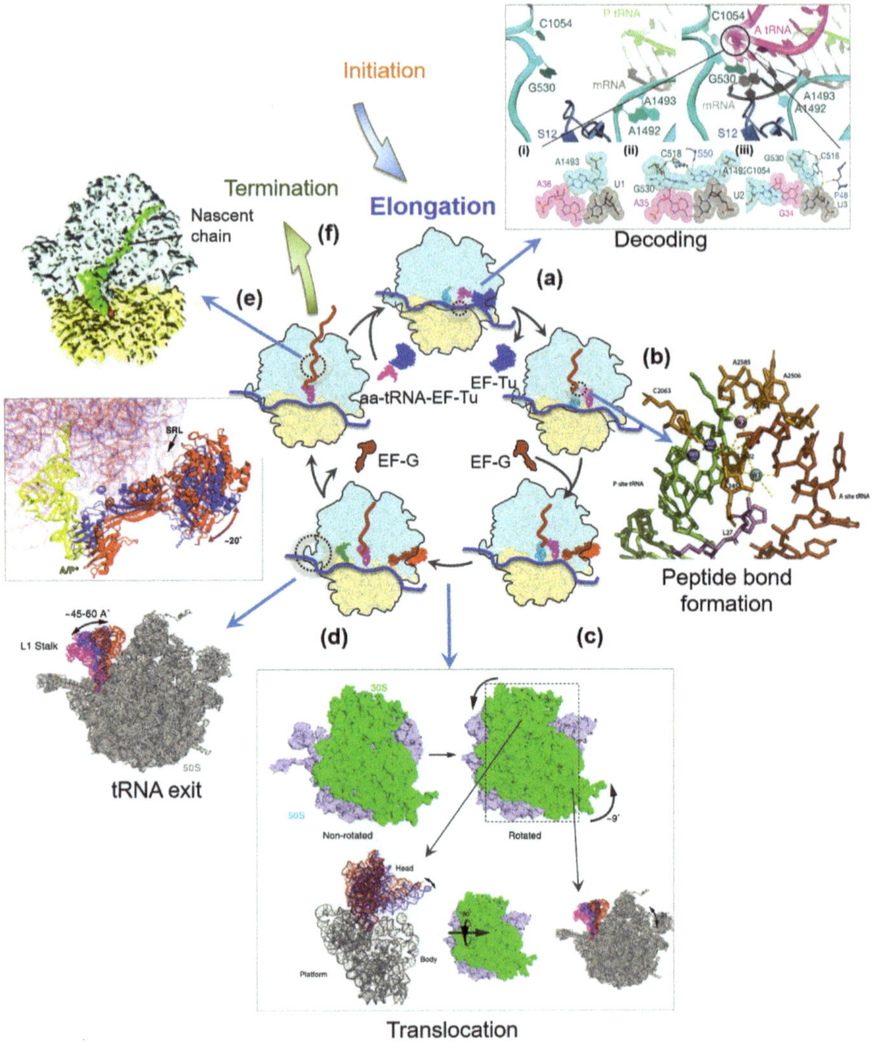

Fig. 3.4 Elongation cycle. The elongation starts as soon the 70S-fMet-tRNA complex is formed. **a**—The first step involves EF-Tu (in blue) bringing and binding the A-site amino-acid (magenta) for mRNA code recognition and selection—an insert shows structural details of the decoding and selection mechanism on the ribosome (adopted from Schmeing and Ramakrishnan 2009). **b**—Once the correct amino acid is selected, peptide bond formation occurs. The insert shows the P-site (green) and A-site tRNA (red) arrangement within the peptidyl transferase centre, surrounded by catalytic water molecules (W1-3) (Polikanov et al. 2014). **c**—EF-G (in red) then binds to catalyse the movement of tRNA and mRNA along the ribosome. **d**—The SSU physically moves ('ratcheting') with respect to the LSU. The insert shows non-rotated and rotated states of the 70S ribosome—the rotated SSU movement in head, shoulder and body domains (Ling and Ermolenko 2016). **e**—To allow E-site tRNA to exit, uL1 stalk (insert panel—Ling and Ermolenko 2016) moves inwards (highlighted as magenta to red). Upon GTP hydrolysis, EF-G dissociates (zoom in panel highlighting motions of EF-G in red and blue, after translocation and GTP hydrolysis), enabling the cycle to repeat and producing a nascent chain, emerging through the ribosome exit tunnel on the large subunit (zoom in panel—NC in green; Seidelt et al. 2009)

Peptide Bond Step

After the release of EF-Tu, the aminoacyl-tRNA moves form A-site to the active site of protein synthesis on the LSU known as the peptidyl transferase centre (PTC). The peptidyl tRNA (P-site tRNA) and aminoacyl-tRNA (A-site tRNA) react to form a peptide bond. The active site is located on nucleotides (domain V) from 23S rRNA. Several research groups during the early 90s undertook experiments to characterise the role of rRNA in the functioning of the PTC using biochemical and genetic screening. Mutational, footprinting, and crosslinking studies suggested the central loop of domain V 23S rRNA is involved in ribosome's peptidyl transferase activity (Polacek et al. 2001; Green and Noller 1997). These were subsequently verified by the high-resolution structures of the LSU.

Crystal structure of the archaea ribosome LSU confirmed that the PTC consists of nucleotides from domain V of 23S rRNA (Ban et al. 2000; Polacek and Mankin 2005). These structures of the ribosome indicated that the PTC active site is located at the bottom of a large cleft the LSU below the central protuberance (Ban et al. 2000; Harms et al. 2001; Noeske et al. 2015). The PTC cavity is formed by nucleotides of the central loop of domain V of 23S rRNA, and no ribosomal proteins were found in contact with the active site, confirming the suggestion that ribosomal enzyme activity is performed exclusively by rRNA. However, loops and long "tails" of ribosomal proteins uL2, uL3 and uL4 protrudes into the core of the ribosome (Nissen et al. 2000; Ban et al. 2000). It was proposed that ribosomal proteins located close to the PTC and tunnel entrance are essential for catalytic and regulatory rRNA activity. The structures also indicated that water molecules play an important role in interactions of the rRNA, tRNA and amino acid chemical groups within the active site during peptide bond formation (Polacek and Mankin 2005; Rodnina 2018).

During peptide synthesis at the PTC, carbonyl and amino groups from the amino acids attached on the ends of A- and P-site tRNA are positioned within the active site of the PTC at the LSU interface, where their universally conserved CCA ends are oriented and held in place by interactions with 23S rRNA (Fig. 3.4b; Yusupov et al. 2001). Peptide bond between the A-site amino acid and P-site amino acid is made when the α-amino group of the A-site aminoacyl-tRNA attacks the carbonyl group of P-site bound peptidyl tRNA (Fig. 3.4b). This chemical reaction at the PTC, producing the de-acylated tRNA at the P-site consisting of the de-acylated tRNA at the P-site and peptidyl-tRNA carrying an additional amino acid (+1 aa) at the A-site of the LSU (Fig. 3.4; Polacek and Mankin 2005).

mRNA and tRNA Translocation

Upon peptide bond formation and deacylation of P-site tRNA, ribosome participates in the translocation of tRNAs and mRNA between two subunits. This translocation requires the ribosome machinery to be adaptable enough to make the movement of tRNAs and mRNA smooth. Elongation factor G (EF-G) participates actively in this process providing energy for translocation. Several structures have revealed details on how mRNA, tRNA, and the ribosome interact to ensure re-adjustment of the components during translocation and preparing the whole complex for another round of elongation cycle (Schmeing and Ramakrishnan 2009;

Rodnina 2018). Translocation of the tRNAs and the mRNA on the ribosome is a multi-step process representing a great challenge to discern the intermediates states when tRNAs and mRNA move relative to the subunits of the ribosome. Current structural models complemented by results from an ensemble and single-molecule kinetic studies using a large variety of fluorescence reporters and fluorescence resonance energy transfer (FRET) pairs distinguished up to eight discrete steps (Guo and Noller 2012; Adio et al. 2015; Belardinelli et al. 2016; Wasserman et al. 2016). Contrary to the view that EF-G is essential for the movement of mRNA and tRNA, a recent crystal structure of ribosome in complex with mRNA and two tRNAs shows a spontaneous movement of mRNA and tRNAs from pre-translocation to post-translocation state, without the presence of EF-G (Zhou et al. 2019). This suggests that the movement of mRNA and tRNAs is facilitated by the ribosome itself.

During translocation, the SSU and LSU move relative to each other in the motion now known as ‘ratcheting’ where subunits rotate forward and reverse in a cyclical manner in a plane parallel to the inter-subunit interface. Furthermore, the ‘head’ domain of SSU carries out forward- and back-swivelling motions with respect to the body of the SSU around the axis is perpendicular to the inter-subunit plain (Fig. 3.4). These movements are essential to power movement of mRNA and tRNAs between the SSU and LSU. The ribosome ratchet motion was first visualised and described in the low-resolution cryo-EM structures of the ribosome translocation complex, providing the first evidence that interactions with the ribosome drive the substrate movements during elongation (Frank and Agrawal 2000). Subsequent high-resolution structures of the elongation pre-translocation and post-translocation complexes provided a visual picture of how elongation takes place (Rodnina 2018).

After EF-G binding to the ribosome, the SSU head and body domains move in the counter-clockwise direction relative to the large-subunit, which corresponds to the direction of translocation and accompanied by hydrolysis of GTP (Guo and Noller 2012; Belardinelli et al. 2016; Wasserman et al. 2016). Subsequently, the SSU body begins moving backwards in the clockwise direction, whereas the head remains in the forward-swivelled state (Guo and Noller 2012; Belardinelli et al. 2016; Wasserman et al. 2016) (Fig. 3.4c, d). This opens the decoding region allowing to uncouple the tRNAs from the interactions with the ribosome elements that hold the mRNA and the tRNA anticodons in the A and P site, respectively. Disconnection of the codon–anticodon complexes from the SSU moves 30S head domain to the non-rotated state (Guo and Noller 2012; Belardinelli et al. 2016; Wasserman et al. 2016). The tRNAs are then precisely positioned at the P and E sites and the EF-G is released (Fig. 3.4e; Savelsbergh et al. 2003).

Structures of post-translocation 70S ribosome demonstrate movements of the uL1 stalk, which acts as a sensor towards deacylated P-tRNA and moves by nearly 40 Å to help tRNA to exit from E-site and out from the ribosome (Schmeing and Ramakrishnan 2009; Rodnina 2018). Movements of both subunits, together with EF-G, display the mechanism of the ribosome that allows mRNA and tRNA to reposition and proceed to the next elongation cycle.

Termination

Once the ribosome reaches a stop codon on the mRNA, elongation (synthesis of the nascent chain) is completed. In bacteria, stop codons are recognized by the termination (or release) factors RF1 and RF2, which read the codons UAG/UAA and UGA/UAA, respectively (Fig. 3.5a). The third termination factor, RF3, regulates activities of RF1 and RF2 but is not required for peptidyl-tRNA hydrolysis. The process of termination consists of three main steps: recognition of the stop codon, hydrolysis of the ester bond of the peptidyl-tRNA (with the assistance of RF1 or RF2), and dissociation of RF1/RF2 with the help of RF3 (Fig. 3.5b). RF1 and RF2 select the respective stop codons by conserved recognition peptide motifs: "P-V-T" in RF1 or "S-P-F" in RF2 (Schmeing and Ramakrishnan 2009; Rodnina 2018).

Structures of ribosome termination complexes were determined by X-ray and cryo-EM that show that there are conformational changes in RF which trigger the termination of translation (Fig. 3.5; Schmeing and Ramakrishnan 2009). It was found that domain 1 of RF2 interacts with the uL11 stalk of the 70S ribosome (Korostelev et al. 2008; Weixlbaumer et al. 2008); however, this interaction is broken in complexes with RF1 bound (Fig. 3.5b, inserts) (Korostelev et al. 2008; Laurberg et al. 2008). This finding indicates that recognition of the stop codon is distinct for each RF. The sm-FRET studies also supported the fact that RF1 and RF2 stabilise the ribosome in the non-rotated state (Rawat et al. 2003, 2006; Korostelev et al. 2008, 2010; Laurberg et al. 2008; Weixlbaumer et al. 2008; Santos et al. 2013), whereas RF3 alone stabilizes the rotated state (Gao et al. 2007; Jin et al. 2011).

X-ray structure of a mutated RF1 bound to 70S suggested a mechanism of RF1 activation upon ribosome binding (Svidritskiy and Korostelev 2018). In the structure, RF1 catalytic domain was bound outside the PTC whilst the codon-recognition domain of RF1 was bound to the stop codon, accompanied with structural re-arrangements in the SSU decoding centre (Svidritskiy and Korostelev 2018). This suggests that RF1 recognises stop codon which induces conformational changes in SSU that cause the release of mRNA and tRNA.

In instances where ribosomes are stalled on a non-stop encoded mRNA (or truncated mRNA) caused by in-complete transcription or cellular stress, bacterial cells rescue these ribosomes by encoding ArfA factor (James et al. 2016; Huter et al. 2017a, b; Demo et al. 2017a, b). Cryo-EM studies of ribosomes in complex with ArfA, RF2, mRNA and tRNAs provided an insight on how ArfA, together with RF2, rescue stalled ribosomes. ArfA is able to identify ribosomes that have no mRNA in the 30S mRNA channel and recruits RF2 to mediate NC release from the ribosome, resulting in stalled ribosomes to terminate translation and to be recycled (Fig. 3.5a, left panel—Demo et al. 2017a, b).

Upon the release of NC, RF3 (with GTP hydrolysis) facilitates dissociation of remaining substrates: RF1 and RF2 from the ribosome, leaving tRNA in a hybrid P/E state and mRNA on the ribosome (Figs. 3.5b, c). Cryo-EM structures of 70S-RF1-RF3 complexes provided information on how RF3 activates release factor

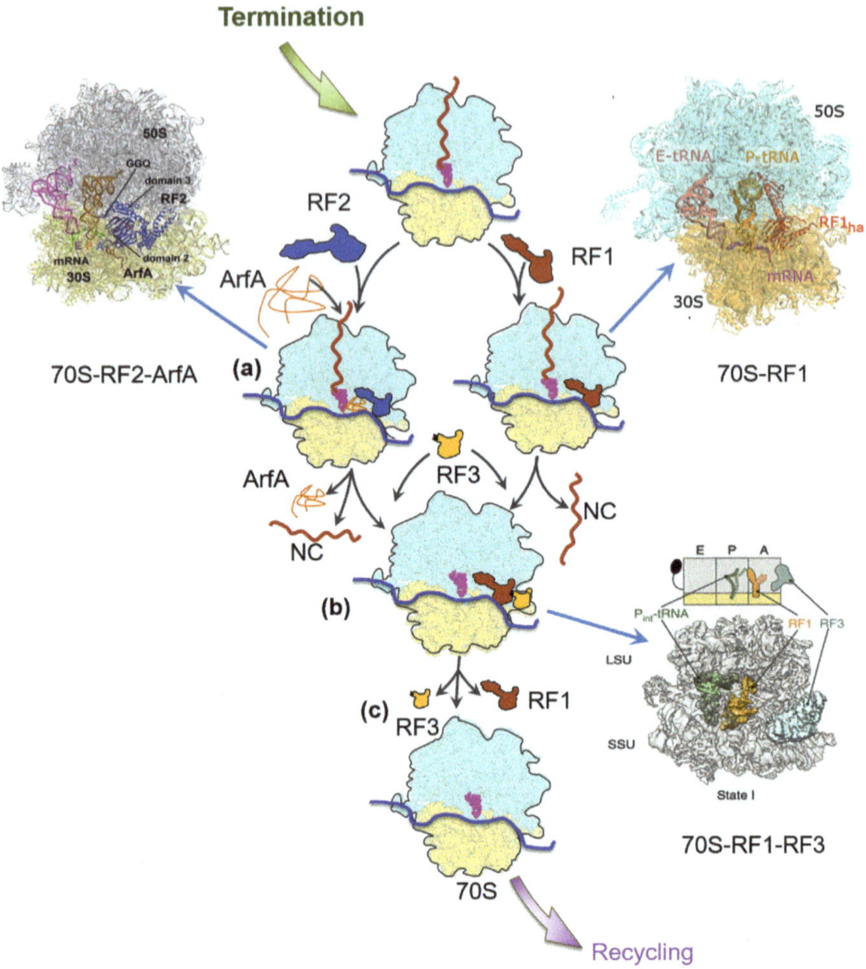

Fig. 3.5 Scheme of translation termination. Once a STOP codon is reached on the mRNA, the ribosome enters translation termination. **a**—Two factors: RF1 (in red) and RF2 (in blue) catalyse translation termination. For ribosomes that are stalled by non-stop codon mRNA (or truncated mRNA) for termination, an additional factor called ArfA (in orange) binds, together with RF2 to assist the ribosome termination. Left and right insert panels show recently solved cryo-EM strutures of 70S ribosomes with RF2 and ArfA (Demo et al. 2017a, b) and 70S with RF1 (Svidritskiy and Korostelev 2018). **b**—RF3 (in light orange) then binds the ribosome in either RF1 or RF2 bound states to catalyse the release of NC (red), as well as the RF1/RF2 bound to the ribosome. Cryo-EM structures of these complexes are shown in inserts: on the right is 70S bound with RF1-RF3 (Graf et al. 2018). **c**—Factors RF1 and RF3 dissociate from the ribosome, leaving hybrid P/E tRNA and mRNA on the ribosome followed by recycling

dissociation. The structures show that RF3 is recruited via bL12 stalk of the ribosome (Fig. 3.5c; Pallesen et al. 2013). Intermediate structures captured by cryo-EM show that binding of RF3 to the ribosome results in SSU head rotation and swivelling that shift the P-tRNA to hybrid P/E-site (Fig. 3.5c, right insert panel) (Graf et al. 2018). RF3 also induces rotation of the SSU that enables release of RF1 from its binding site. GTP hydrolysis of RF3 then enables RF3 to be released from the ribosome and for the subunits to be recycled (Fig. 3.5d) (Graf et al. 2018).

Recycling

In order for the ribosome to start a new round of protein synthesis, the post-termination 70S ribosome needs to be recycled: split into the individual subunits and be disassociated from mRNA and tRNA. In bacteria, the ribosome subunit separation is catalysed by the ribosome recycling factor (RRF) and EF-G. RRF binds to the A site of the ribosome (Gao et al. 2005) and stabilizes the SSU rotated with respect to the LSU in which the P-site tRNA in the hybrid P/E binding state (Dunkle et al. 2011). RRF consists of two domains (I and II), which interact with the LSU and SSU (Fig. 3.6; Fu et al. 2016).

Once EF-G and RRF dissociate the ribosome into subunits, binding of IF3 to the SSU prevents re-association of the SSU and LSUs (Peske et al. 2005; Zavialov et al. 2005). The mechanism by which both EF-G and RRF cooperate to induce ribosome recycling has remained less well-understood for a long time. Using time-resolved cryo-EM Fu et al. (2016) were able to obtain structural snapshots of short-lived intermediate states of ribosome recycling (Fig. 3.6a, b; Fu et al. 2016). The authors pre-incubated 70S post-termination ribosomes with RRF and then rapidly mixed them with EF-G, IF3, and GTP before spraying the sample onto a cryo-EM grid. Using classification of particle images several structures of ribosome recycling complexes were obtained at medium resolution, ranging between 7 and 20 Å. The 70S-EF-G-RRF complex provided information on how domain 4 of EF-G contacts domain 2 of RRF inducing separation of subunits (Fig. 3.6b). In the absence of EF-G, RRF moves closer to inter-subunit rRNA disrupting the bridge B2a (Fig. 3.6c). These experiments enabled authors to capture the post-termination complex of the LSU-EF-G-RRF complex, after dissociation from the SSU. The structure obtained shows RRF domain 1 in close proximity to H69 domain of 23S rRNA, in comparison to 70S-EF-G-RRF complex (Fu et al. 2016). Altogether, the structural snapshots illuminate a mechanism by which bacterial ribosomes undergo subunit splitting and recycling, ready until the next translation cycle begins.

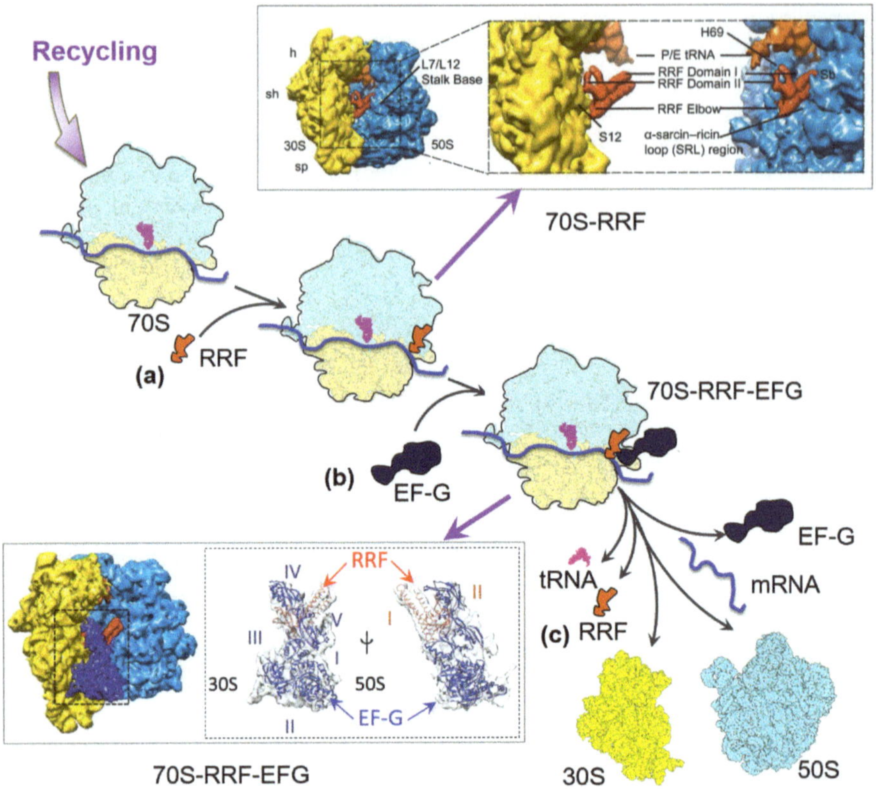

Fig. 3.6 Schematic diagram of ribosome recycling. **a**—Ribosome recycling factor (RRF—in red) recognise 70S ribosome with tRNA and mRNA followed by termination. **b**—Together with EF-G (blue) it catalyses the dissociation of the two subunits. **c**—Cryo-EM structures identifying structural intermediates during this process are shown in inserts: Top right panel shows structures of the 70S bound to RRF only (Fu et al. 2016) and the bottom panel shows the subunits dissociation in the last step of recycling, as the RRF, EFG, tRNA and mRNA leave the ribosome (Fu et al. 2016)

Regulatory Co-translational Events at the Ribosome Exit Tunnel

The ribosome has the ability to monitor the progress of translation and modulate nascent peptide chain (NC) folding by coupling with the speed of translation. During elongation, the NC moves through the exit tunnel of the ribosome. The exit tunnel is about 100 Å long with an average diameter of 15 Å that can accommodate a chain of ~30 amino acids before the NC emerges from the ribosome. The NC can start folding within the tunnel during elongation, accommodating secondary structure elements or even from small domains (Nilsson et al. 2015, 2017; Tian et al. 2018). Recent cryo-EM structures of the NC small domains on stalled ribosomes showed that folding of immunoglobulin domains can begin at the vestibule and close to tunnel exit (Nilsson et al. 2017; Tian et al. 2018; Javed et al. 2019 and references herein). NMR results confirm the fact that large domains fold primarily close to or outside the ribosome tunnel, during translation (Cabrita et al. 2016). The rate of the NC folding depends on interactions between the peptide emerging from the exit tunnel and the surface of the ribosome (Deckert et al. 2016). There is a link between sequence, number of rare codons in the mRNA, the rate of translation and the NC folding. These factors affect translation, alters the kinetics of co-translational folding, and changes the distribution of protein conformations in the resulting mature protein pool (Clarke and Clark 2008; Tsai et al. 2008; Zhang et al. 2009; Siller et al. 2010; Spencer et al. 2012; Yu et al. 2015; Buhr et al. 2016). Computational experiments modelling a process of folding on the ribosome suggest that the translation rate affects local folding rates and may induce or prevent mis-folding (O'Brien et al. 2014). The ribosome also acts as a hub during elongation coordinating co-translational events such as the NC folding with the recruitment of chaperones, and NC-modifying enzymes (Javed et al. 2017 and references herein).

To analyse interactions between the ribosome and NC during translation, one has to have the ribosome in a fixed, "semi" translational, state. Such a state is named as a stalled state of the ribosome. Cells have devised ways to achieve this by having non-stop mRNA sequence (or truncated mRNA) or by encoding short peptide sequences that interact with the exit tunnel and stall ribosomes (Javed et al. 2017; Wilson et al. 2016). One of the structurally characterised stalling peptides is the 17 residue SecM from bacterial secretion monitor protein, which regulates the expression of SecA translocon protein. Cryo-EM structures of SecM stalled ribosomes show specific points of interaction between the nascent peptide and the tunnel elements (rRNA and proteins) that define the strength of stalling (Bhushan et al. 2011). Small molecule drugs and antibiotics can stall the ribosome. EM structures of such stalled complexes demonstrated that these small molecules bound to the ribosome at the beginning of the tunnel close to PTC are triggering conformational changes blocking the channel (Li et al. 2018; Arenz et al. 2016). The speed of translation can also be slowed down by short repeat sequences of proline residues, translation of which stall the ribosome (Doerfel et al. 2015). To relieve poly-proline stalled ribosomes, bacterial cells express a specialised elongation factor (EF-P; Doerfel et al. 2015). Recent cryo-EM structures of EF-P bound ribosomes show that EF-P enters the E site of the ribosome and acts by bringing the P- and A-site substrates closer towards their catalytically productive orientation in the peptidyl transferase centre (Huter et al. 2017a, b). Altogether, the exit tunnel plays an important role during protein translation in monitoring events related to ribosome function in order to perform smooth production of the nascent polypeptide chain.

Future Prospects

Elucidation of the function of the ribosomes has come a long way due to the complexity of the ribosome and multitude of the factors involved into its function. The major accomplishments were made by amalgamation of biochemical and structural methods: X-ray crystallography, cryo-EM and NMR spectroscopy (Fig. 3.7). During the last decade, structural studies of bio-complexes progressed tremendously due to advances in computational technology and software development. That enabled us to gain a massive bulk of information on small details and essential conformational changes within the ribosome itself and in its complexes with assistant factors. The library of ribosomal structures determined using cryo-EM and X-ray crystallography allowed us to derive functional/structural relationship of this biological nanomachine (Fig. 3.7). Advances in automation of data collection in cryo-EM and software in statistical analysis particle images helped researchers to distinguish multiple states of ribosomes and extend characterisation of these nanomachines from gram negative bacterial ribosomes such as *E. coli* to gram positive and pathogenic bacterial, followed by analysis of eukaryotic ribosomes (Eyal et al. 2015; Hentschel et al. 2017). Obtaining high-resolution structures of non-characterised pathogenic bacterial ribosomes will help us map regions within the ribosome where existing antibiotics can be assessed for efficacy as well as the design of new effective antibiotics that can effectively block protein translation more widely.

The progress in studies of prokaryotic ribosomes paved the road to more systematic and efficient studies of much larger ribosome complexes and eukaryotic ribosomes, previously elusive or hypothesised based on biochemical studies. New areas in the understanding of the protein synthesis machine include analysis of properties, activity, and folding of the translation product, nascent polypeptide chain. It was long known that bacteria couple mRNA synthesis (i.e. the activity of RNA polymerase) to protein synthesis (i.e. ribosome protein translation). Yet the details of how RNA polymerase interacts with ribosome remains to be fully understood. Therefore, complexes of the RNA polymerase and the ribosome represent a special point of interest; how the RNA polymerase cooperates with ribosomes to couple bacterial transcription-translation and where is the link that makes such complex efficient in bacteria. Recent cryo-EM structures shed some light on this partnership between RNA polymerase and the ribosome (Demo et al. 2017a, b; Kohler et al. 2018). Nevertheless, how this coupling functions inside the cell remains to be seen. Particularly when ribosomes translate proteins inside a cell in polysomes (multiple ribosomes on a single mRNA) rather than as single ribosomes, many aspects of ribosome activity will require analysis of ribosomes and ribosome complexes in a cellular context.

Latest cryo-EM structures highlight the role of the ribosome also during co-translational folding of polypeptide chains as they emerge from the exit tunnel. Further structural analysis of ribosome-nascent chain complexes within cells will illuminate aspects related to the tuning of co-translational protein folding to

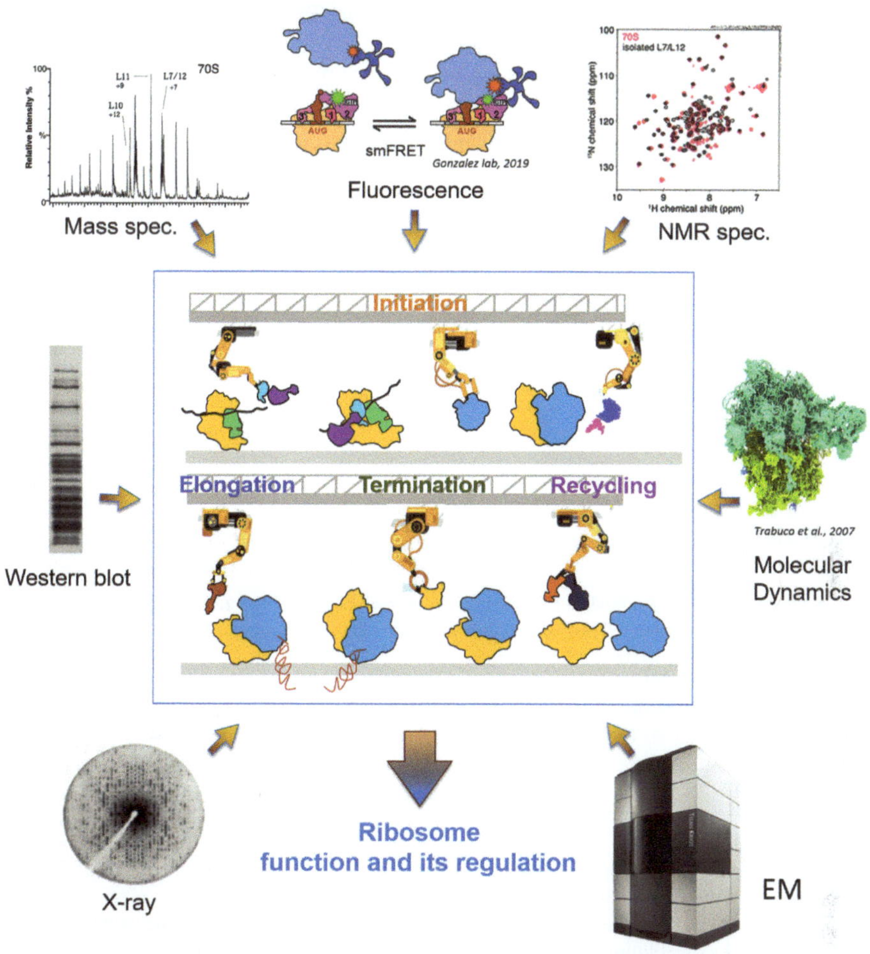

Fig. 3.7 Analysis of structure-function of the ribosome. Central panel (outlined in blue) shows the four major steps of translational (ribosomal subunits shown in yellow for 30S and blue for 50S, together with a variety of translation factors). Usage of different techniques (outlined as panels) illuminate the ribosome function as a nanomachine

activities of the ribosome. Both X-ray crystallography and cryo-EM can provide structures of intermediates of ribosomes, revealing many aspects of ribosome function. These intermediates however have to be artificially trapped, either by using antibiotics or peptides (Wilson et al. 2016). Undertaking structural studies in a time-resolved fashion without the use of antibiotics or non-hydrolysable GTP analogues will enable us to follow ribosome motions as they happen in real-time. A forthcoming method in this respect is time-resolved cryo-EM (Frank 2017), that is able to capture transient intermediates of ribosome complexes. For instance, analysis of molecular dynamics of ribosome machinery by time-resolved cryo-EM captured intermediate states of elongating ribosome (Fischer et al. 2010; Chen et al. 2015), particularly short-lived intermediates that give rise to ribosome subunit dissociation during ribosome recycling (Fu et al. 2016). An improvement of the time resolution of single-molecule techniques may yield additional, yet uncharacterized transient intermediates, as recently demonstrated by (Fu et al. 2016). We envisage the combination of these methods mentioned above will come into force revealing many important intermediate states of the ribosome complexes during protein translation.

Acknowledgements A. J. is supported by BBSRC grant BB/R002622/1. We would like to thank Dr. D. Houldershaw for the computational assistance in structural analysis and software used.

Competing Financial Interests
The authors declare no competing financial interests.

References

Adio S, Senyushkina T, Peske F, Fischer N et al (2015) Fluctuations between multiple EF-G-induced chimeric tRNA states during translocation on the ribosome. Nat Commun 6:7442. https://doi.org/10.1038/ncomms8442

Agrawal RK, Spahn CM, Penczek P et al (2000) Visualization of tRNA movements on the *Escherichia coli* 70S ribosome during the elongation cycle. J Cell Biol 150(3):447–460

Agrawal RK, Penczek P, Grassucci RA et al (1996) Direct visualization of A-, P-, and E-site transfer RNAs in the *Escherichia coli* ribosome. Science 271(5251):1000–1002

Arenz S, Bock LV, Graf M et al (2016) A combined cryo-EM and molecular dynamics approach reveals the mechanism of ErmBL-mediated translation arrest. Nat Commun 7. https://doi.org/10.1038/ncomms12026

Ban N, Freeborn B, Nissen P et al (1998) A 9 A resolution X-ray crystallographic map of the large ribosomal subunit. Cell 93(7):1105–1115

Ban N, Nissen P, Hansen J et al (2000) The complete atomic structure of the large ribosomal subunit at 2.4 Å resolution. Science 289:905–920

Bashan A, Agmon I, Zarivach R et al (2003) Structural basis of the ribosomal machinery for peptide bond formation, translocation, and nascent chain progression. Mol Cell 1:91–102

Bashan A, Yonath A (2008) Correlating ribosome function with high-resolution structures. Trends Microbiol 16(7):326–335

Beckmann R, Bubeck D, Grassucci RA, et al (1997) Alignment of conduits for the nascent polypeptide chain in the ribosome-Sec61 complex. Science 278 (5346)

Belardinelli R, Sharma H, Caliskan N et al (2016) Choreography of molecular movements during ribosome progression along mRNA. Nat Struct Mol Biol 23:342–348

Bernabeu C, Lake JA (1982) Nascent polypeptide chains emerge from the exit domain of the large ribosomal subunit: immune mapping of the nascent chain. Proc Natl Acad Sci USA 79:3111–3115

Bhushan S, Hoffmann T, Seidelt B et al (2011) SecM-stalled ribosomes adopt an altered geometry at the peptidyl transferase center. PLoS Biol 9 (1)

Brosius J, Palmer ML, Kennedy PJ et al (1978) Complete nucleotide sequence of 16S ribosomal RNA gene from *Escherichia coli*. Proc Nat Acad Sci USA 75:4801–4805

Buhr F, Jha S, Thommen M et al (2016) Synonymous codons direct cotranslational folding toward different protein conformations. Mol Cell 61:341–351

Cabrita LD, Cassaignau AM, Launay HM et al (2016) A structural ensemble of a ribosome-nascent chain complex during cotranslational protein folding. Nat Struct Mol Biol 23(4):278–285
Carter AP, Clemens WM, Brodersen DE et al (2000) Functional insights from the structure of the 30S ribosomal subunit and its interactions with antibiotics. Nature 407(6802):340–348
Chen B, Kaledhonkar S, Sun M et al (2015) Structural dynamics of ribosome subunit association studied by mixing-spraying time-resolved cryogenic electron microscopy. Structure 23 (6):1097–1105
Christodoulou J, Larsson G, Fucini P et al (2004) Heteronuclear NMR investigations of dynamic regions of intact *Escherichia coli* ribosomes. Proc Natl Acad Sci USA 101(30):10949–10954
Clarke TF, Clark PL (2008) Rare codons cluster. PLoS ONE 3:e3412
Clore GM, Gronenborn AM (1998) New methods of structure refinement for macromolecular structure determination by NMR. Proc Natl Acad Sci USA 95(11):5891–5898
Cornish PV, Ermolenko DN, Noller HF et al (2008) Spontaneous intersubunit rotation in single ribosomes. Mol Cell 30(5):578–588
Deckert A, Waudby CA, Wlodarski T et al (2016) Structural characterization of the interaction of α-synuclein nascent chains with the ribosomal surface and trigger factor. Proc Natl Acad Sci USA 113(18):5012–5017
Delius H, Traut RR, Moore PB et al (1968) Studies on purified *E.coli* ribosomal proteins. Molecular Genetics, Springer-Verlag, Berlin, pp 26–45
Demo G, Rasouly A, Vasilyev N et al (2017a) Structure of RNA polymerase bound to ribosomal 30S subunit. Elife. 6:e28560
Demo G, Svidritskiy E, Madireddy R et al (2017b) Mechanism of ribosome rescue by ArfA and RF2. Elife. 6
Diaconu M, Kothe U, Schluenzen F et al (2005) Structural basis for the function of the ribosomal L7/L12 stalk in factor binding and GTPase activation. Cell 121(7):991–1004
Doerfel LK, Wohlgemuth I, Kubyshkin V et al (2015) Entropic contribution of elongation factor P to proline positioning at the catalytic center of the ribosome. J Am Chem Soc 137:12997–13006
Dubochet J, Adrian M, Chang JJ et al (1988) Cryo-electron microscopy of vitrified specimens. Q Rev Biophys 21(2):129–228
Dunkle JA, Wang L, Feldman MB et al (2011) Structures of the bacterial ribosome in classical and hybrid states of tRNA binding. Science 332(6032):981–984
Elad N, Clare D, Saibil HR et al (2008) Detection and separation of heterogeneity in molecular complexes by statistical analysis of their two-dimensional projections. J Struct Biol 162:108–120
Eyal Z, Matzov D, Krupkin M et al (2015) Structural insights into species-specific features of the ribosome from the pathogen Staphylococcus aureus. Proc Natl Acad Sci 112(43):5805–5814
Fischer N, Konevega AL, Wintermeyer W et al (2010) Ribosome dynamics and tRNA movement by time-resolved electron microscopy. Nature 466(7304):329–33
Frank J, Penczek P, Grassucci RA et al (1991) Three-dimensional reconstruction of the 70S *Escherichia coli* ribosome in ice: the distribution of ribosomal RNA. J Cell Biol 115(3):597–605
Frank J (2017) Time-resolved cryo-electron microscopy: recent progress. J Struct Biol 3:303–306
Frank J (2006) Three-dimensional electron microscopy of macromolecular assemblies: visualization of biological molecules in their native state. Oxford University Press: Chapter 2:20–40
Frank J, Agrawal RK (1998) The movement of tRNA through the ribosome. Biophys J 74(1):589–594
Frank J, Agrawal RK (2000) A ratchet-like inter-subunit reorganization of the ribosome during translocation. Nature 406(6793):318–322
Frank J, Zhu J, Penczek P et al (1995) A model of protein synthesis based on cryo-electron microscopy of the *E. coli* ribosome. Nature 376(6539):441–444
Fu Z, Kaledhonkar S, Borg A et al (2016) Key intermediates in ribosome recycling visualized by time-resolved cryoelectron microscopy. Structure 24(12):2092–2101

Gabashvili IS, Agrawal RK, Spahn CM et al (2000) Solution structure of the *E. coli* 70S ribosome at 11.5 A resolution. Cell 100(5):537–549

Gabashvili IS, Gregory ST, Valle M et al (2001) The polypeptide tunnel system in the ribosome and its gating in erythromycin resistance mutants of L4 and L22. Mol Cell 8(1):181–188

Gao H, Zhou Z, Rawat U et al (2007) RF3 induces ribosomal conformational changes responsible for dissociation of class I release factors. Cell 129(5):929–941

Gao N, Zavialov AV, Li W et al (2005) Mechanism for the disassembly for the posttermination complex inferred from cryo-EM studies. Mol Cell 18(6):663–674

Goyal A, Belardinelli R, Rodnina MV (2017) Non-canonical binding site for bacterial initiation factor 3 on the large ribosomal subunit. Cell Rep. 20(13):3113–3122

Graf M, Huter P, Maracci C et al (2018) Visualization of translation termination intermediates trapped by the Apidaecin 137 peptide during RF3-mediated recycling RF1. Nat Comm. 9 (1):3053

Green R, Noller HF (1997) Ribosomes and translation. Annu Rev Biochem 66:679–716

Guo Z, Noller HF (2012) Rotation of the head of the 30S ribosomal subunit during mRNA translocation. Proc Natl Acad Sci 109:20391–20394

Hope H, Frolow F, Bohlen K et al (1989) Cryocrystallography of ribosomal particles. Acta Crystallogr Sect B 45:190–199

Harms J, Schluenzen F, Zarivach R et al (2001) High resolution structure of the large ribosomal subunit from a mesophilic eubacterium. Cell 107(5):679–688

Henderson R (2015) Overview and future of single particle electron cryomicroscopy. Arch Biochem Biophys 581:19–24

Hentschel J, Burnside C, Mignot I et al (2017) The complete structure of the *Mycobacterium smegmatis* 70S Ribosome. Cell. Rep. 20(1):149–160

Herr W, Noller HF (1975) A fragment of 23S RNA containing a nucleotide sequence complementary to a region of 5S RNA. FEBS Lett 53:248–252

Huter P, Arenz S, Bock LV et al (2017a) Structural basis for polyproline-mediated ribosome stalling and rescue by the translation elongation factor EF-P. Mol Cell 68:515–527.e516

Huter P, Muller C, Beckert B et al (2017b) Structural basis for ArfA-RF2-mediated translation termination on mRNAs lacking stop codons. Nature 541(7638):546–549

Ingolia NT, Hussmann JA, Weissman JS (2019) Ribosome Profiling: Global views of Translation. Cold Spring Harb Perspect Biol 11(5) pii:a032698

James NR, Brown A, Gordiyenko Y et al (2016) Translational termination without a stop codon. Science 354(6318):1437–1440

Javed A, Christodoulou J, Cabrita et al (2017) The ribosome and its role in protein folding: looking through a magnifying glass. Acta Cryst D Struc Biol 73(Pt6):509–521

Javed A, Cabrita D. Lisa, Cassaignau ME A, Wlodarski T, Christodoulou J, Orlova EV (2019) Visualising nascent chain dynamics at the ribosome exit tunnel by cryo-electron microscopy. BioRxiv. https://doi.org/10.1101/722611

Jin H, Kelley AC, Ramakrishnan V (2011) Crystal structure of the hybrid state of ribosome in complex with the guanosine triphosphatase release factor 3. Proc Natl Acad Sci 108 (38):15798–15803

Julian P, Milon P, Agirrezabala X et al (2011) The Cryo-EM structure of a complete 30S translation initiation complex from Escherichia coli. PLoS Biol 9(7):e1001095

Kaminishi T, Wilson DN, Takemoto C et al (2007) A snapshot of the 30S ribosomal subunit capturing mRNA via the Shine-Dalgarno interaction. Structure 15(3):289–297

Khusainov I, Vicens Q, Bochler A et al (2016) Structure of the 70S ribosome from human pathogen *Staphylococcus aureus*. Nuc Acid Res. 44(21):10491–10504

Kohler R, Mooney RA, Mills DJ et al (2018) Architecture of a transcribing-translating expressome. Science 356(6334):194–197

Korostelev A, Asahara H, Lancaster L et al (2008) Crystal structure of a translation termination complex formed with release factor RF2. Proc Natl Acad Sci 105(50):19684–19690

Korostelev A, Trakhanov S, Laurberg M et al (2006) Crystal structure of a 70S ribosome-tRNA complex reveals functional interactions and rearrangements. Cell 126(6):1065–1077

Korostelev A, Zhu J, Asahara H et al (2010) Recognition of the amber UAG stop codon by release factor RF1. EMBO J 29:2577–2585

Kothe U, Widen HJ, Mohr D et al (2004) Interaction of helix D of elongation factor Tu with helices 4 and 5 of protein L7/L12 on the ribosome. J Mol Biol 336(5):1011–21

Lake JA (1978) Protein synthesis. Science 200(4339):305–306

Laurberg M, Asahara H, Korostelev A et al (2008) Structural basis for translation termination on the 70S ribosome. Nature 454:852–857

Li W. McClure K, Montabana E et al (2018) Structural basis for selective stalling of human ribosome nascent chain complexes by a drug-like molecule. bioRxiv. https://doi.org/10.1101/315325

Ling C, Ermolenko DN (2016) Structural insights into ribosome translocation. Wiley Interdiscip Rev RNA. 5:620–36

Liu Q, Frederick K (2016) Intersubunit Bridges of the Bacterial Ribosome. J Mol Biol 428(10 Pt B): 2146–2164

Lopez-Alonso JP, Fabbretti A, Kaminishi T et al (2017) Structure of a 30S pre-initiation complex stalled by GE81112 reveals structural parallels in bacterial and eukaryotic protein synthesis initiation pathways. Nucleic Acids Res 45(4):2179–2187

Loveland AB, Demo G, Grigorieff N et al (2017) Ensemble cryo-EM elucidates the mechanism of translation fidelity. Nature 546:113–117

Marzi S, Knight W, Brandi L, Caserta E, Soboleva N, Hill WE, Gualerzi CO, Lodmell JS (2003) Ribosomal localization of translation initiation factor IF2. RNA 9(8):958–969

Marzi S, Myasnikov AG, Serganov A et al (2007) Structured mRNAs regulate translation initiation by binding to the platform of the ribosome. Cell 130(6):1019–1031

Matadeen R, Patwardhan A, Gowen B et al (1999) The *Escherichia coli* large ribosomal subunit at 7.5 A resolution. Structure 7 (12):1575–83

McCutcheon JP, Agrawal RK, Philips SM et al (1999) Location of translational initiation factor IF3 on the small ribosomal subunit. Proc Natl Acad Sci 96:4301–4306

Melnikov S, Ben-Shem A, Gareau de Loubresse N et al (2012) One core, two shells: bacterial and eukaryotic ribosomes. Nat Struct Mol Biol 19(6):560–567

Melnikov S, Manakongtreecheep K, Soll D (2018) Revising the structural diversity of ribosomal proteins across the three domains of life. Mol Biol and Evol. 35(7):1588–1598

Midgley JEM (1965) Effects of different extraction procedures on the molecular characteristics of bacterial ribosome ribonucleic acid. Biochem Biophys Acta 95:232–243

Milligan RA, Unwin PN (1986) Location of exit channel for nascent protein in 80S ribosome. Nature 319:693–695

Moller K, Brimacombe R (1975) Specific cross-linking of proteins S7 and L4 to ribosomal RNA, by UV irradiation of *Escherichia coli* ribosomal subunits. Mol Gen Genet 141(4):343–345

Moore PB (2009) The ribosome returned. J Biol. 8(1):1–8

Moore PB, Engelman DM (1975) A neutron scattering study of the distribution of protein and RNA in the 30S ribosomal subunit of *Escherichia coli*. J Mol Biol 91:101–120

Moore PB, Traut RR, Noller HF et al (1968) Ribosomal proteins of *Escherichia coli*. II. Proteins from the 30S subunit. J Mol Biol 31:441–461

Myasnikov AG, Marzi S, Simonetti A et al (2005) Conformational transition of initiation factor 2 from the GTP—to-GDP-bound state visualised on the ribosome. Nat Struct Mol Biol 12 (12):1145–1149

Nilsson OB, Nickson AA, Hollins JJ et al (2017) Cotranslational folding of spectrin domains via partially structured states. Nat Struct Mol Biol 24(3):221–225. https://doi.org/10.1038/nsmb.3355

Nilsson OB, Hedman R, Marino J et al (2015) Cotranslational protein folding inside the ribosome exit tunnel. Cell Rep. 12(10):1533–1540

Nissen P, Hansen J, Ban N et al (2000) The structural basis of ribosome activity in peptide bond synthesis. Science 289(5481):920–930

Noller, H.F. and Herr, W. (1974). Accessibility of 5S rRNA in 50S ribosomal subunits. J Mol Biol 90:181–184

Noeske J, Wasserman MR, Terry DS et al (2015) High-resolution structure of the *Escherichia coli* ribosome. Nat Struct Mol Biol 22(4):336–341

O'Brien EP, Ciryam P, Vendruscolo M et al (2014) Understanding the influence of codon translation rates on cotranslational protein folding. Acc Chem Res 47:1536–1544

Ogle JM, Murphy FV, Tarry MJ et al (2002) Selection of tRNA by the ribosome requires a transition from an open to a closed form. Cell 111:721–732

Ogle JM, Ramakrishnan V (2005) Structural insights into translational fidelity. Annu Rev Biochem 74:129–177

Orlova EV (2000) Structural analysis of non-crystalline macromolecules: the ribosome. Acta Crystallogr D Biol Crystallogr 56:1253–8

Palade GE (1955) A small particulate component of the cytoplasm. J Biophys Biochem Cytol 1:59–68

Pallesen J, Hashem Y, Korkmaz G et al (2013) Cryo-EM visualization of ribosome in termination complex with apo-RF3 and RF1. Elife 2

Peske F, Rodnina MV, Wintermeyer W (2005) Sequence of steps in ribosome recycling as defined by kinetic analysis. Mol Cell 18(4):403–412

Petrov A, Chen J, O'Leary S et al (2012) Single-molecule analysis of translational dynamics. Cold Spring Harb Perspective Biol. 4(9):a011551. https://doi.org/10.1101/cshperspect.a011551

Polacek N, Mankin AS (2005) The ribosomal peptidyl transferase center: structure, function, evolution, inhibition. Crit Rev Biochem Mol Biol 40(5):285–311

Polacek N, Gaynor M, Yassin A et al (2001) Ribosomal peptidyl transferase can withstand mutations at the putative catalytic nucleotide. Nature 411:498–501

Polikanov YS, Steitz TA, Innis CA (2014) A proton wire to couple aminoacyl-tRNA accommodation and peptide-bond formation on the ribosome. Nat Struct Mol Biol 21 (9):787–93

Rawat U, Gao H, Zavialov A et al (2006) Interactions of the release factor RF1 with the ribosome as revealed by cryo-EM. J Mol Biol 357:1144–1153

Rawat UBS, Zavialov AV, Sengupta J et al (2003) A cryo-electron microscopic study of ribosome-bound termination factor RF2. Nature 421:87–90

Rodnina MV (2018) Translation in Prokaryotes. Cold Spring Harb Perspect Biol 10(9):1–21

Santos N, Zhu J, Donohue PJ et al (2013) Crystal structure of the 70S ribosome bound with Q253P mutant form of release factor RF2. Structure 21(7):1258–1263

Savelsbergh A, Katunin VI, Mohr D et al (2003) An elongation factor G-induced ribosome rearrangement precedes tRNA-mRNA translocation. Mol Cell 11:1517–1523

Schmeing TM, Ramakrishnan V (2009) What recent ribosome structures have revealed about the mechanism of translation. Nature 461(7268):1234–42

Schmeing TM, Voorhees RM, Kelley AC et al (2009) The crystal structure of the ribosome bound to EF-Tu and aminoacyl-tRNA. Science 326(5953):688–694

Seidelt B, Innis CA, Wilson DN et al (2009) Structural insight into nascent polypeptide chain-mediated translational stalling. Science 326(5958):1412–1415

Selmer M, Dunham CM, Murphy FV 4th et al (2006) Structure of the 70S ribosome complexed with mRNA and tRNA. Science 313(5795):1935–1942

Shaikh TR, Yassin AS, Lu Z et al (2014) Initial bridges between two ribosomal subunits are formed within 9.4 milliseconds, as studied by time-resolved cryo-EM. Proc Natl Acad Sci. 111 (27):9822–9827

Siller E, DeZwaan DC, Anderson JF et al (2010) Slowing bacterial translation speed enhances eukaryotic protein folding efficiency. J Mol Biol 396:1310–1318

Simonetti A, Marzi S, Myasnikov AG et al (2008) Structure of the 30S initiation complex. Nature 455(7211):416–420

Sohmen D, Chiba S, Shimokawa-Chiba N et al (2015) Structure of the Bacillus subtilis 70S ribosome reveals the basis for species-specific stalling. Nat Commun 6:6941. https://doi.org/10.1038/ncomms7941

Spencer PS, Siller E, Anderson JF et al (2012) Silent substitutions predictably alter translation elongation rates and protein folding efficiencies. J Mol Biol 422:328–335

Sprink T, Ramrath D, Yamamoto H et al (2016) Structures of ribosome-bound initiation factor 2 reveal the mechanism of subunit association. Sci Adv. 2(3):e1501502

Stark H, Orlova EV, Rinke-Appel J et al (1997) Arrangement of tRNAs in pre- and posttranslational ribosomes revealed by electron cryomicroscopy. Cell 88:19–28

Stark H, Rodnina MV, Wieden HJ et al (2002) Ribosome interactions of aminoacyl-tRNA and elongation factor Tu in the codon-recognition complex. Nat Struct Biol 9:849–854

Steitz TA (2008) A structural understanding of the dynamic ribosome machine. Nat Rev Mol Cell Biol 9(3):242–253

Svidritskiy E, Korostelev A (2018) Conformational control of translation termination on the 70S ribosome. Structure. 26(6):821–828

Taylor MM, Glasgow JE, Storck R (1967) Sedimentation coefficients of RNA from 70S and 80S ribosomes. Proc Natl Acad Sci U S A. 57(1):164–9

Tian P, Steward A, Kudva R et al (2018) Folding pathway of an Ig domain is conserved on and off the ribosome. Proc Natl Acad Sci 115(48):E11284–E11293

Tischendorf GW, Zeichhardt H, Stoffler G (1974a) Determination of the location of proteins L14, L17, L18, L19, L22, L23 on the surface of 50S ribosomal subunit of *Escherichia coli* by immune electron microscopy. Mol Gen Genet 134:187–208

Tischendorf GW, Zeichhardt H, Stoffler G (1974b) Location of proteins, S5, S13 and S14 on the surface of the 30S ribosomal subunit from *Escherichia coli* as determined by immune electron microscopy. Mol Gen Genet 134:209–223

Tissieres A, Watson JD (1958) Ribonucleoprotein particles from the *Escherichia coli*. Nature 182 (4638):778–780

Tsai CJ, Sauna ZE, Kimchi-Sarfaty C et al (2008) Synonymous mutations and ribosome stalling can lead to altered folding pathways and distinct minima. J Mol Biol 383:281–291

Valle M, Zavialov A, Li W et al (2003) Incorporation of aminoacyl-tRNA into the ribosome as seen by cryo-electron microscopy. Nat Struct Biol 10:899–906

Vesper O, Amitai S, Belitsky M et al (2011) Selective translation of leaderless mRNAs by specialised ribosomes generated by MazF in *Escherichia coli*. Cell 147(1):147–157

Wasserman MR, Alejo JL, Altman RB et al (2016) Multiperspective smFRET reveals rate-determining late intermediates of ribosomal translocation. Nat Struct Mol Biol 23:333–341

Weixlbaumer A, Jin H, Neubauer C et al (2008) Insights into translational termination from the structure of RF2 bound to the ribosome. Science 322:953–956

Wilson DN, Arenz S, Beckmann R (2016) Translation regulation via nascent polypeptide-mediated ribosome stalling. Curr Opin Struct Biol 37:123–33

Wimberly BT, Brodersen DE, Clemons WM Jr (2000) Structure of the 30S ribosomal subunit. Nature 407(6802):327–339

Yonath A, Leonard RK, Wittmann GH (1987) A tunnel in the large ribosomal subunit revealed by three-dimensional image reconstruction. Science 236:813–6

Yu CH, Dang Y, Zhou Z et al (2015) Codon usage influences the local rate of translation elongation to regulate co-translational protein folding. Mol Cell 59:744–754

Yusupov M, Yusupova G, Baucom A et al (2001) Crystal structure of the ribosome at 5.5 A resolution. Science. 292(5518):883–96

Zavialov AV, Hauryliuk VV, Ehrenberg M (2005) Splitting of the posttermination ribosome into subunits by the concerted action of RRF and EF-G. Mol Cell 18(6):675–686

Zhang G, Hubalewska M, Ignatova Z (2009) Transient ribosomal attenuation coordinates protein synthesis and co-translational folding. Nat Struc Mol Biol. 16(3):274–80

Zhou J, Lancaster L, Donohue JP et al (2019) Spontaneous ribosomal translocation of mRNA and tRNAs into a chimeric hybrid state. Proc Natl Acad Sci.. https://doi.org/10.1073/pnas.1901310116

Chapter 4
Functions and Mechanisms of the Human Ribosome-Translocon Complex

Sven Lang, Duy Nguyen, Stefan Pfeffer, Friedrich Förster, Volkhard Helms and Richard Zimmermann

Abstract The membrane of the endoplasmic reticulum (ER) in human cells harbors the protein translocon, which facilitates membrane insertion and translocation of almost every newly synthesized polypeptide targeted to organelles of the secretory pathway. The translocon comprises the polypeptide-conducting Sec61 channel and several additional proteins, which are associated with the hetero-trimeric Sec61 complex. This ensemble of proteins facilitates ER targeting of precursor polypeptides, Sec61 channel opening and closing, and modification of precursor polypeptides in transit through the Sec61 complex. Recently, cryoelectron tomography of translocons in native ER membranes has given unprecedented insights into the architecture and dynamics of the native, ribosome-associated translocon and the Sec61 channel. These structural data are discussed in light of different Sec61 channel activities including ribosome receptor function, membrane

S. Lang (✉) · R. Zimmermann
Competence Center for Molecular Medicine, Saarland University Medical School, Building 44, 66421 Homburg, Germany
e-mail: sven.lang@uni-saarland.de

R. Zimmermann
e-mail: richard.zimmermann@uni-saarland.de

D. Nguyen · V. Helms
Center for Bioinformatics, Saarland University, 66041 Saarbrücken, Germany
e-mail: duy.nguyen@bioinformatik.uni-saarland.de

V. Helms
e-mail: volkhard.helms@bioinformatik.uni-saarland.de

S. Pfeffer · F. Förster
Department of Molecular Structural Biology, Max-Planck Institute of Biochemistry, 82152 Martinsried, Germany
e-mail: pfeffer@biochem.mpg.de

S. Pfeffer
ZMBH, 69120 Heidelberg, Germany

F. Förster
Center for Biomolecular Research, Utrecht University, 3584 CH Utrecht, The Netherlands
e-mail: f.g.forster@uu.nl

J. R. Harris and J. Marles-Wright (eds.), *Macromolecular Protein Complexes II: Structure and Function*, Subcellular Biochemistry 93,
https://doi.org/10.1007/978-3-030-28151-9_4

insertion or translocation of newly synthesized polypeptides as well as the possible roles of the Sec61 channel as a passive ER calcium leak channel and regulator of ATP/ADP exchange between cytosol and ER.

Keywords Endoplasmic reticulum · Membrane protein biogenesis · Protein secretion · Protein targeting · Protein translocation · Sec61 channel

Introduction: Structure, Function, Dynamics and Connectivity of the Mammalian Endoplasmic Reticulum (ER)

A fascinating hallmark of nucleated human cells is their complex compartmentalization, separating the cellular interior into different organelles. While some organelles like mitochondria occur in a multicopy fashion others such as the endoplasmic reticulum (ER) are usually present in one copy under steady-state conditions. Like other organelles the ER fulfills a plethora of functions many of which are interwoven with its morphological heterogeneity. Despite the fact that the ER represents a continuous single-membrane network within nucleated cells, different structural variations are known (Baumann and Walz 2001; Schwarz and Blower 2016; Voeltz et al. 2002). From the perspective of localization, the ER radiates as the outer membrane of the nuclear envelope to the perinuclear space and peripheral regions all the way to the plasma membrane, where it is considered as cortical ER (Westrate et al. 2015). Not strictly correlated to this spatial distribution, the ER can morph between different shapes often referred to as sheets (or cisternae), tubules and tubular-matrices (Nixon-Abell et al. 2016). In addition, near the nucleus, where the height of a cell is usually much greater than in the periphery, another structural peculiarity of the ER can be formed, Terasaki ramps (Terasaki et al. 2013). This structure is based on helicoidal ramps connecting adjacent stacks of ER sheets in a "parking garage" like fashion (Guven et al. 2014). From the 6000 μm^3 (6 pl) total cell volume of a COS-7 cell the ER occupies 1500 μm^3 (1.5 pl), nine times the volume occupied by mitochondria (Valm et al. 2017). The dynamics of the mesh-like shaped ER are further underscored by its intracellular mobility allowing the ER to scan and explore the majority of the cytosolic volume within minutes. With such mobility rates it is not surprising that the ER is the organelle with the highest contact rate to other organelles (Valm et al. 2017; Shim 2017). This interconnectivity between different organelles and extended cellular structures such as the cytoskeleton usually relying on proteinaceous tethers were elegantly reviewed previously and are not further discussed here (Csordás et al. 2018; Phillips and Voeltz 2016; Zhang and Hu 2016; Gatta and Levine 2017). In their landmark papers sixty years ago, Palade and colleagues also distinguished different ER morphologies. From their electron microscopic images they concluded, spot on, that the ER represents a "continuous, tridimensional reticulum" consisting of "cisternae [the term is used to designate a flat element of large size and irregular outline] which appear to communicate freely with the tubules". Furthermore, they wrote "although such cisternae may assume considerable breadth they seem to retain, in general, a depth of ~50 μm" and "the surface of the latter appears to be dotted with small, dense granules that cover them in part or in entirety" (Palade and Porter 1954). Nowadays, those original observations are coined by the key phrase rough sheets and smooth tubules, where rough and smooth refers to the presence or absence of the dense granules observed by Palade et al., i.e. ribosomes or polysomes attached to the cytosolic surface of the ER (Friedman and Voeltz 2011; Shibata et al. 2006; Pfeffer et al. 2012). Besides other factors, ribosome binding to the ER membrane is considered a major driving force for sheet formation (Shibata et al. 2010; Puhka et al. 2007). However, cells differ widely in the fraction of ER-bound ribosomes, from secretory cells in which almost all ribosomes are found at the ER to mature leukocytes in which the ER is barely detectable (Reid et al. 2014; Palade 1956). On average, half of all ribosomes and a third of all messenger RNAs are associated with the ER membrane of a typical mammalian cell (Reid et al. 2014). Important to note, the ratio of ER sheets to tubules is actively regulated by a cell and varies for example with cell type, cellular demands, and cell cycle stage (Puhka et al. 2007, 2012).

Similarly complex and versatile as the structural design of the mammalian ER is its function. With ribosomes bound to the membrane ER sheets are usually considered the primary domain for processes related to protein maturation including protein synthesis, membrane translocation or insertion, post-translational modification, folding, assembly as well as quality control and degradation. On the other hand, ER tubules with their higher surface-to-lumen ratio might be better suited for membrane-surface related ER functions such as lipid and steroid synthesis or inter-organelle signaling (Westrate et al. 2015). In addition, the ER represents the major intracellular calcium reservoir of mammalian cells (Brostrom and Brostrom 2003; Sammels et al. 2010). Under resting conditions the free calcium concentration of the ER lumen (>100 μM) exceeds the cytosolic counterpart (~50 nM) by several orders of magnitude, thus, generating a massive calcium gradient as prerequisite for efficient signaling purposes (Clapham 2007; Mogami et al. 1998; Suzuki et al. 2016). Hence, the ER is intimately linked to calcium signaling and related aspects such as muscle contraction, neuronal excitability, mitochondrial respiration or apoptosis (Berridge 2002; Berridge et al. 2003). Also, prominent, i.e. abundant, chaperones of the ER including BiP and calreticulin serve a dual function as calcium buffering protein on the one hand and folding chaperone on the other hand (Coe and Michalak 2009; Michalak et al. 2002). Common for ubiquitous calcium buffering chaperones is (i) a low affinity (k_d in mM range) paired with a high capacity (up to 50 calcium binding sites per molecule) for calcium binding and (ii) their folding activity relying on the ER calcium content (Lievremont et al. 1997; Meldolesi and Pozzan 1998; Ashby and Tepikin 2001). Given the continuous nature of both the membrane and the lumen between ER cisternae and tubules it is unclear to what extend different functions of the ER are spatially restricted or domain specific. Upcoming high-resolution live cell imaging approaches will certainly find an answer to this question.

At the crossroad of ER structure and function appears the protein translocase where the tasks of ribosome binding, protein transport, and calcium signaling coalesce. This heteromultimeric protein complex of the ER membrane has gained much attention over the past decades starting from the biochemical identification to evolutionary conservation and functional characterization all the way to its structural organization. From a biochemist's point of view, the protein translocase could be considered as an enzyme catalyzing the membrane passage of otherwise impermeable substrates such as the roughly 3000 presecretory proteins (Rychkova and Warshel 2013), which are encoded by the human genome (https://www.proteinatlas.org/humanproteome/tissue/secretome#plasma). In order to allow membrane passage, the precursors of secretory proteins are characterized by a cleavable N-terminal signal peptide with its tripartite structure (a positively charged N-terminal region, termed N-region, a central region containing hydrophobic residues, termed H-region, and a slightly polar C-terminal region, C-region). In the case of membrane proteins without a cleavable signal peptide, the most N-terminal transmembrane helix typically serves as an ER targeting and membrane insertion signal. To handle this wide range of different soluble and membrane protein substrates (Fig. 4.1), the active center of the protein translocase is designed with a lack of substrate specificity. Notably, the translocon of higher eukaryotes is even more promiscuous than the translocons of lower eukaryotes, archaea, and bacteria (Gonsberg et al. 2017). Therefore, multiple accessory cofactors support the active center to solve the issue of substrate specificity (Table 4.1). Following the idea of Koshland's induced-fit theory of specificity, insufficient compatibility between the substrate and the active center of the translocase might also explain the imperfect sealing of the translocase observed for small molecules including calcium ions (Koshland 1958; Harsman et al. 2011c).

In the following sections we will summarize our current knowledge and concepts about the functions and mechanisms of the eukaryotic protein translocon starting with the active center, the Sec61 complex (Fig. 4.2), followed by different cofactors and how these components affect the enzyme's kinetic for membrane permeability of proteins or small molecules.

Structures and Functions of Isolated and Native Sec61 Complexes

Structural Esthetics of the Sec61 Complex

As stated earlier, the protein translocon of the ER represents a complex machinery with a variable structural architecture and dynamic stoichiometry. In the past, major emphasis was usually given to the Sec61 complex, which is considered the pivotal

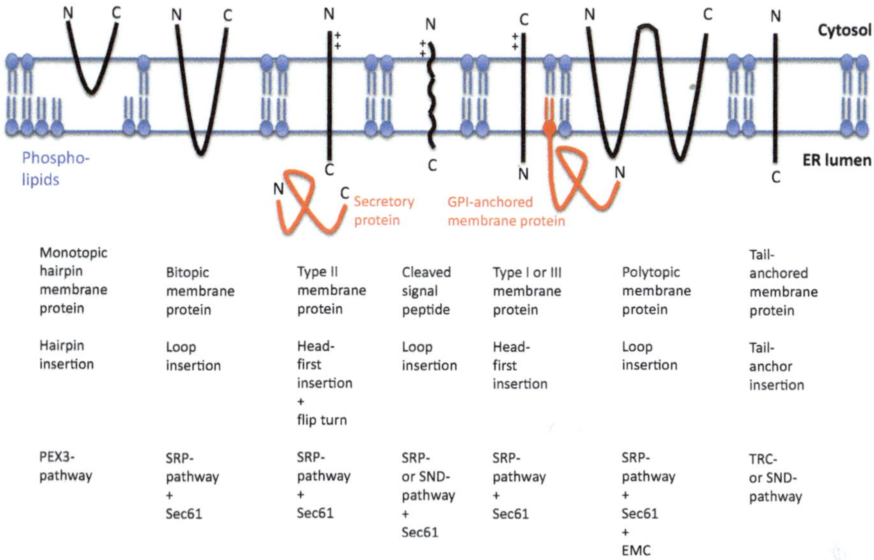

Fig. 4.1 Topologies of membrane proteins in the ER membrane. The cartoon depicts the membrane proteins of the ER membrane, together with their type, mechanism of membrane insertion, and targeting and insertion pathway. See text for details. We note that (i) bitopic and polytopic proteins can also have the opposite orientation, (ii) the shown bitopic protein is alternatively named double-spanning membrane protein, (iii) the shown polytopic protein is alternatively named tetra-spanning membrane protein, (iv) type I membrane proteins as well as bitopic and polytopic proteins with their N-terminus facing the ER lumen can be targeted to and inserted into the membrane via N-terminal signal peptides that are subsequently cleaved by ER luminal signal peptidase, (v) in case the shown type I membrane was not targeted by a cleavable signal peptide it is also defined as signal anchor protein, (vi) positively charged amino acid residues (+) play an important role in membrane protein orientation, i.e. typically, follow the positive inside rule. In the case of membrane proteins without N-terminal signal peptides, membrane insertion appears to involve the same components and mechanisms, which deliver secretory proteins and glycosylphosphatidylinositol (GPI)-anchored membrane proteins to the ER lumen. Subsequent to ER import, GPI-anchored membrane proteins become membrane anchored via their C-termini by GPI-attachment. N, N-terminus; C, C-terminus

subunit of the heteromultimeric translocon both structurally as its core subunit and enzymatically as its active center lowering the activation barrier for the membrane transport of polypeptides (Pfeffer et al. 2016; White and von Heijne 2008; Görlich and Rapoport 1993). Moreover, importance of the Sec61 complex as central component of a protein translocon is probably fortified best by its evolutionary conservation from bacteria and archaea to lower and higher eukaryotes (Calo and Eichler 2011; du Plessis et al. 2011; Park and Rapoport 2012; Dalal and Duong 2009). In all three domains of life the corresponding Sec61 complex is usually organized as a heterotrimeric protein ensemble consisting of a pore-forming α subunit accompanied by two smaller subunits, called β and γ (Fig. 4.3). Yet, nomenclature of the three Sec61 subunits is somewhat inconsistent. While we will

Table 4.1 Protein transport components/*complexes/networks* and associated proteins in HeLa cells

Component/subunit	Abundance	Location	Linked diseases
Calmodulin	9428	C	
Cytosolic chaperone network			
– Hsc70 (HSPA8)	3559		
– Hdj2 (DNAJA1)	660		
– Bag1 (HAP, RAP46)	46		
#NAC		C	
– NACα	1412		
– NACβ			
#SRP		C	
– SRP72	355		Aplasia, Myelodysplasia
– SRP68	197		
– SRP54	228		
– SRP19	33		
– SRP14	4295		
– SRP9	3436		
– 7SL RNA			
SRP receptor		ERM	
– SRα (docking protein)	249		
– SRβ	173		
hSnd1	?		
Snd receptor			
– hSnd2 (TMEM208)	81	ERM	
– hSnd3	?		
#Bag6 complex		C	
– TRC35 (Get4)	171		
– Ubl4A	177		
– Bag6 (Bat3)	133		
SGTA	549	C	
TRC40 (Asna1, Get3)	381	C	
TA receptor		ERM	
– CAML (CAMLG, Get2)	5		
– WRB (CHD5, Get1)	4		Congenital heart disease, down syndrome
ERM protein complex		ERM	
– EMC1	124		
– EMC2	300		
– EMC3	270		
– EMC4	70		
– EMC5 (MMGT1)	35		

(continued)

Table 4.1 (continued)

Component/subunit	Abundance	Location	Linked diseases
– EMC6 (TMEM93)	5		
– EMC7	247		
– EMC8	209		
– EMC9	1		
– EMC10	3		
#TMCO1	2013	ERM	Glaucoma, cerebrofaciothoracic dysplasia
PEX19	80	C	Zellweger syndrome
PEX3	103	ERM	Zellweger syndrome
#Sec62 (TLOC1)	26	ERM	Prostate cancer, lung cancer
#Sec61 complex		ERM	
– Sec61α1	139		Diabetes**, CVID, TKD
– Sec61β	456		Polycystic liver disease (PLD)
– Sec61γ	400		Glioblastoma
Alternative Sec61 complex			
– Sec61α2	?		
– Sec61β	456		
– Sec61γ	400		
ER chaperone network			
– Sec63 (ERj2)	168	ERM	Polycystic liver disease (PLD)
– #ERj1 (DNAJC1)	8	ERM	
– ERj3 (DNAJB11)	1001	ERL	Polycystic kidney disease (PKD)
– ERj4 (DNAJB9)	12	ERL	
– ERj5 (DNAJC10)	43	ERL	
– ERj6 (DNAJC3, $p58^{IPK}$)	237	ERL	Diabetes
– ERj7 (DNAJC25)	10	ERM	
– ERj8 (DNAJC16)	24	ERM	
– BiP (Grp78, HSPA5)	8253	ERL	Hemolytic uremic syndrome (HUS)
– Grp170 (HYOU1)	923	ERL	
– Sil1 (BAP)	149	ERL	Marinesco-Sjögren-syndrome (MSS)
#$Calnexin_{palmitoylated}$	7278	ERM	
#TRAM1	26	ERM	
TRAM2	40	ERM	
PAT-10	?		
#TRAP complex		ERM	
– TRAPα (SSR1)	568		
– TRAPβ (SSR2)			
– TRAPγ (SSR3)	1701		Congenital disorder of glycosylation (CDG), hepatocellular carcinoma

(continued)

Table 4.1 (continued)

Component/subunit	Abundance	Location	Linked diseases
– TRAPδ (SSR4)	3212		Congenital disorder of glycosylation (CDG)
#RAMP4 (SERP1)		ERM	
#Oligosaccharyltransferase		ERM	
– RibophorinI (Rpn1)	1956		
– RibophorinII (Rpn2)	527		
– OST48	273		Congenital disorder of glycosylation (CDG)
– OST4			
– TMEM258			
– DAD1	464		
– Stt3A*	430		Congenital disorder of glycosylation (CDG)
– Stt3B*	150		Congenital disorder of glycosylation (CDG)
– Kcp2			
– DC2			
– TUSC3			Congenital disorder of glycosylation (CDG)
– MagT1	33		
Signal peptidase (SPC)		ERM	
– SPC12	2733		
– SPC18*			
– SPC21*			
– SPC22/23	334		
– SPC25	94		
GPI transamidase (GPI-T)		ERM	
– GPAA1	9		
– PIG-K	38		
– PIG-S	86		
– PIG-T	20		
– PIG-U	42		
Signal peptide peptidase	424	ERM	
#p34 (LRC59, LRRC59)	2480	ERM	
#p180 (RRBP1)	135	ERM	
Kinectin 1 (KTN1)	263	ERM	

Alternative names of components/subunits are given in parentheses. We note that oligosaccharyltransferase exists as two paralogs, comprising Stt3a or Stt3b. Abundance is given in nM (Hein et al. 2015); 1 nM corresponds to roughly 1000 molecules/cell (Moran et al. 2010). C, cytosol; CVID, common variable immune deficiency; ERL, ER lumen; ERM, ER membrane; TKD, tubulo-interstitial kidney disease; *, catalytically active subunit; **, in mice; #, ribosome associated; ?, uncharacterized

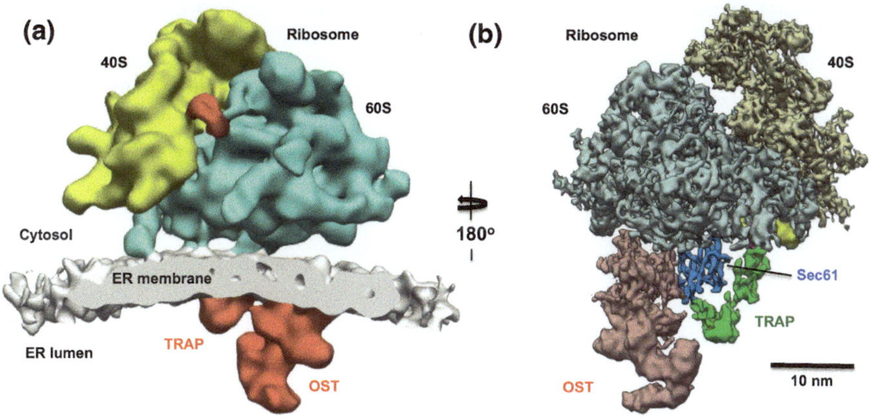

Fig. 4.2 Architecture of the ribosome associated protein translocon of the mammalian ER membrane. **a** In situ structure of the ER-associated mammalian ribosome after subtomogram-averaging at a resolution of about 20 Angström. The ribosome is present in rough microsomes and was imaged by cryo-electron tomography, using a FEI Titan Krios TEM and a FEI Falcon direct electron detector. The ER membrane was cut for better visibility of ER luminal electron densities (shown in red), the Sec61 complex is hidden by the phospholipid bilayer and the ribosome. An additional electron density (shown in red) was observed between 40S and 60S ribosomal subunits and indicates that a translating ribosome was imaged. **b** Further developments of cryo-EM instrumentation and computational algorithms allowed improvement of the ribosome-translocon structure to about 10 Angström resolution, which enables resolving separate transmembrane helices. Here, the membrane density was removed to highlight membrane integral parts of the translocon. Transmembrane helices for Sec61 complex, TRAP (translocon-associated protein), and OST (oligosaccharyl transferase) can be clearly distinguished under these conditions. Helix 51 of an rRNA expansion segment (shown in yellow) and ribosomal protein eL38 (shown in magenta) represent the contact site to the TRAP γ-subunit, but are partially hidden by other ribosomal densities in this view

adhere to the mammalian Sec61α-β-γ subunit terminology, these are referred to SecY-G-E in bacteria, SecY-β-E in archaea and Sec61p-Sbh1p-Sss1p in yeast, respectively (Auer et al. 1991; Hartmann et al. 1994; Kinch et al. 2002; Cao and Saier 2003; Görlich et al. 1992). Genetic studies in yeast and bacteria showed that the two subunits with highest sequence conservation across kingdoms, Sec61α and Sec61γ, are essential for protein translocation and cell viability, whereas the β subunit with lower sequence homology seems dispensable, hence some bacteria assemble only a dimeric SecYE complex (Nishiyama et al. 1994; Matlack et al. 1998; Tsukazaki et al. 2008). First structural insights for the arrangement of a trimeric Sec61 complex came from X-ray crystallography depicting the archaean SecYβE isolated from *Methanocaldococcus jannaschii* (Van den Berg et al. 2004). Subsequent crystal structures and cryo-electron microscopic (EM) studies using isolated pro- or eukaryotic Sec61 complexes confirmed the evolutionary conservation of its architecture (Egea and Stroud 2010; Becker et al. 2009; Gogala et al. 2014; Voorhees et al. 2014; Zimmer et al. 2008; Tanaka et al. 2015). Congruent for

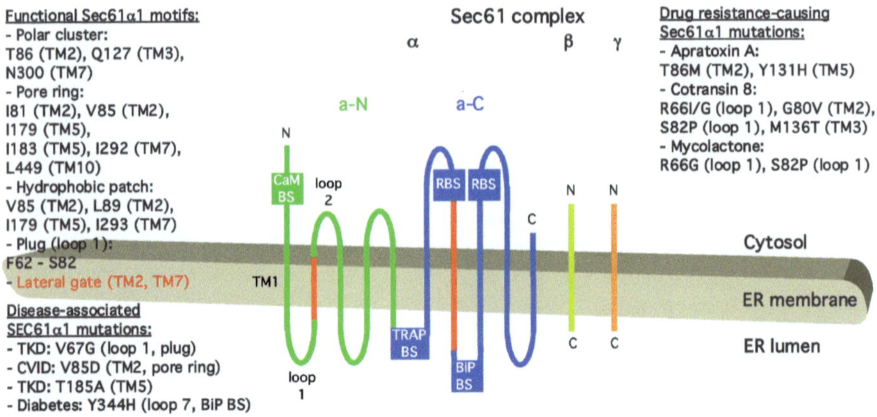

Fig. 4.3 Topology and functionally relevant domains of the heterotrimeric Sec61 complex. The membrane topology of the three subunits of the mammalian Sec61 complex is shown, highlighting binding sites (BS) of Ca^{2+}-calmodulin (CaM), the ribosome (R), the translocon-associated protein complex (TRAP), and immunoglobulin heavy-chain binding protein (BiP). Furthermore, functional motifs, disease associated mutations, and drug resistance causing mutations of the α-subunit are indicated, as are different targeting pathways (purple) to the Sec61 complex. The N- and C-terminal halves of the Sec61 α-subunit are shown in green and blue, respectively (a-N, a-C). Amino acid residues are given in single letter code; C, C-terminus; CVID, Common Variable Immune Deficiency; N, N-terminus; TKD, Tubulo-Interstitial Kidney Disease; TM, transmembrane helix. Notably, recent 3D reconstructions after single particle cryo-electron microscopic analysis of the yeast SEC complex, i.e. the Sec61 complex together with the Sec62/63/71/72 complex, suggested that in the post-translationally acting Sec61 complex, the Sec62/63 sub-complex interacts with the cytsoslic loops 6 and 8 on the cytosolic face of the Sec61 complex (i.e. clashes with ribosome binding), and that the ER luminal domain of Sec63 interacts with ER luminal loop 5 (i.e. clashes with TRAP binding)

all structures is the central, hourglass-like shaped Sec61α subunit forming the actual polypeptide-conducting channel. Its ten transmembrane helices are organized in a pseudo-symmetrical fashion generating an N-terminal half encompassing the cytosolic N-terminus and transmembrane helices 1–5 as well as a C-terminal half encompassing transmembrane helices 6–10 and the cytosolic C-terminus (Figs. 4.3 and 4.4). Loops connecting the ten transmembrane helices are numbered consecutively from 1 to 9 with the odd numbered ones located in the luminal (eukaryotic Sec61 complexes) or periplasmic (prokaryotic SecY complexes) space and even numbered ones in the cytosol. Interestingly, the cytosolic loops 6 and 8 share a functional conservation among all Sec61 homologs serving as universal docking port for different interaction partners including the polypeptide-delivering ribosome or SecA. Hence, loop 6 and 8 project in a lighthouse-like fashion into the cytosol directing incoming shipments. In its three-dimensional fold the Sec61α subunit forms in the plane of the membrane a central constriction called the pore ring. This structural element consists of six bulky, hydrophobic residues facing inwards to form a flexible gasket avoiding excessive membrane permeability of small molecules during the transport of polypeptides (Figs. 4.4 and 4.5). The six pore ring residues are dispersed across the primary structure of Sec61α, yet, localized in critical transmembrane helices 2, 5, 7 and 10 as will be discussed below. Fitting to the hourglass analogy, the pore ring separates two opposing funnels. In the closed conformation of the Sec61 channel the cytosolic funnel is "empty" (water-filled), whereas the luminal/periplasmic facing funnel is occupied by a short helical domain of loop 1, aptly named the plug domain (Rapoport et al. 2017; Junne et al. 2006). However, it appears that plug domains of orthologous Sec61 complexes have different structures, consistent with the facts that this region is the least conserved in the amino acid sequence of different homologs and has a subordinate role for function and cell viability (Li et al. 2007; Junne et al. 2006; Li et al. 2016). In addition, to act as a molecular switching device allowing membrane passage or insertion of incoming precursor polypeptides the Sec61α subunit has a lateral gate, formed between the two sterically adjacent transmembrane helices 2 and 7. In the closed state the lateral gate is stabilized by a polar cluster consisting of three conserved polar residues residing in the helices 2, 3 and 7 (Voorhees and Hegde 2016a). Structural and molecular dynamics studies demonstrated that transition from a closed to an open state spreads the N- and C-terminal halves of Sec61α apart laterally between helices 2 and 7, termed rigid body movement, thereby opening up the lateral gate and permitting access to the lipid phase (Fig. 4.5) (Denks et al. 2014; Egea and Stroud 2010; Park et al. 2014; Hizlan et al. 2012). Vis-à-vis to the lateral gate the Sec61α subunit is associated with the tail-anchored Sec61γ protein (Fig. 4.4a). It wraps transversely around the N- and C-terminal half like a U-shaped clip, potentially restricting excessive mobility of the complex. The third subunit Sec61β, another tail-anchored protein in eukaryotes (the bacterial SecG ortholog has two transmembrane segments), contacts the N-terminal half of Sec61α in vicinity to transmembrane helices 1 and 4, whereas its cytosolic domain might serve a regulatory function for the transport process, as described below (Figs. 4.4 and 4.5).

Structural Dynamics of the Sec61 Complex

Based on the multitude of structural data of Sec61 complexes gathered with X-ray crystallography and cryo-EM three common themes emerge. First, opening of the Sec61 complex requires some kind of ligand binding, ligands being the substrates as well as allosteric effectors, which bind to other parts of the Sec61 complex as compared to the substrates. Visualized ligands contributing in the transition from the idle to an open Sec61 complex include the bacterial ATPase SecA, the translating ribosome, the ER membrane protein Sec63, or even pseudo-ligands like a heterologous anti-Sec61α F_{ab} fragment or an autologous copy of a second Sec61 molecule arising from crystal packing (Li et al. 2016; Zimmer et al. 2008; Gogala et al. 2014; Voorhees et al. 2014; Voorhees and Hegde 2016a; Itskanov and Park 2018; Wu et al. 2018; Egea and Stroud 2010; Braunger et al. 2018). Second, all of those ligands, native or pseudo, interact with the cytosolic docking port (loops 6 and 8) of the

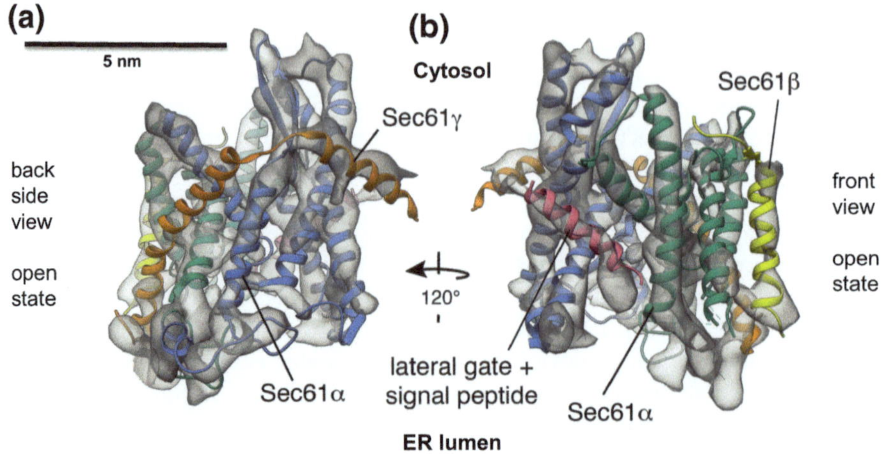

Fig. 4.4 Atomic model PDB 3jc2 for the heterotrimeric Sec61 complex, fitted into the ribosome-translocon structure, shown in Fig. 4.2b. **a, b** The laterally opened and translocating Sec61 channel is shown in two orientations, as seen form the plane of the membrane. b, In b the Sec61 complex is rotated counter clockwise in the plane of the membrane by 120° as compared to a. The N-terminal signal peptide of the translocating polypeptide is shown at the open lateral gate in magenta. The N- and C-terminal halves of the Sec61 α-subunit are shown in green and blue, respectively, the ß-subunit in yellow, and the γ-subunit in orange

Sec61 complex to initiate a conformational change (Figs. 4.2 and 4.5). Coming back to the enzymatic concept of the translocase such ligands probably act as catalysts that lower the activation energy (energy barrier) for the Sec61 channel gating reaction (Fig. 4.6). And third, the pivotal structural elements of the Sec61 complex mentioned before such as the universal docking port, the pore ring residues, the plug domain, the lateral gate helices, as well as the polar cluster stabilizing the latter seem to work hand in hand during the cycle of opening and closing the channel. Recent

Fig. 4.5 Atomic model for the structural dynamics of gating of the heterotrimeric Sec61 complex, as seen from the cytsosol. **a** Atomic model for the laterally closed Sec61 complex (PDB 3j7q). **b** Atomic model for the laterally opened Sec61 channel (PDB 3jc2). **a, b** N- and C-terminal halves of the Sec61 α-subunit, are shown in green and blue, respectively, lateral gate helices 2 and 7 are shown in red, and cytosolic loops are not shown for clarity. The ß-subunit is depicted in yellow and the γ-subunit in orange. At least three conformations of the Sec61 complex can be distinguished, (i) the closed state (closed even to calcium ions), (ii) a structurally ill-characterized primed state that is induced by interaction with either the ribosome or the Sec62/Sec63 complex, and (iii) the open state, which is induced by interaction with a strong signal peptide or N-terminal transmembrane helix of a precursor polypeptide or a weak signal peptide or N-terminal transmembrane helix plus allosteric effectors, such as TRAP or Sec62/Sec63 ± BiP (characterized by an open lateral gate and permeable to calcium ions). During protein translocation, the lateral gate is typically occupied by a signal peptide (Fig. 4.4b) and the central aqueous pore by the polypeptide chain in transit. We note that efficient closing of the Sec61 channel can also involve allosteric effectors, such as BiP with its ER luminal Hsp40-type co-chaperones ERj3 plus ERj6 or calcium-bound Sec62 plus calcium-bound calmodulin (CaM)

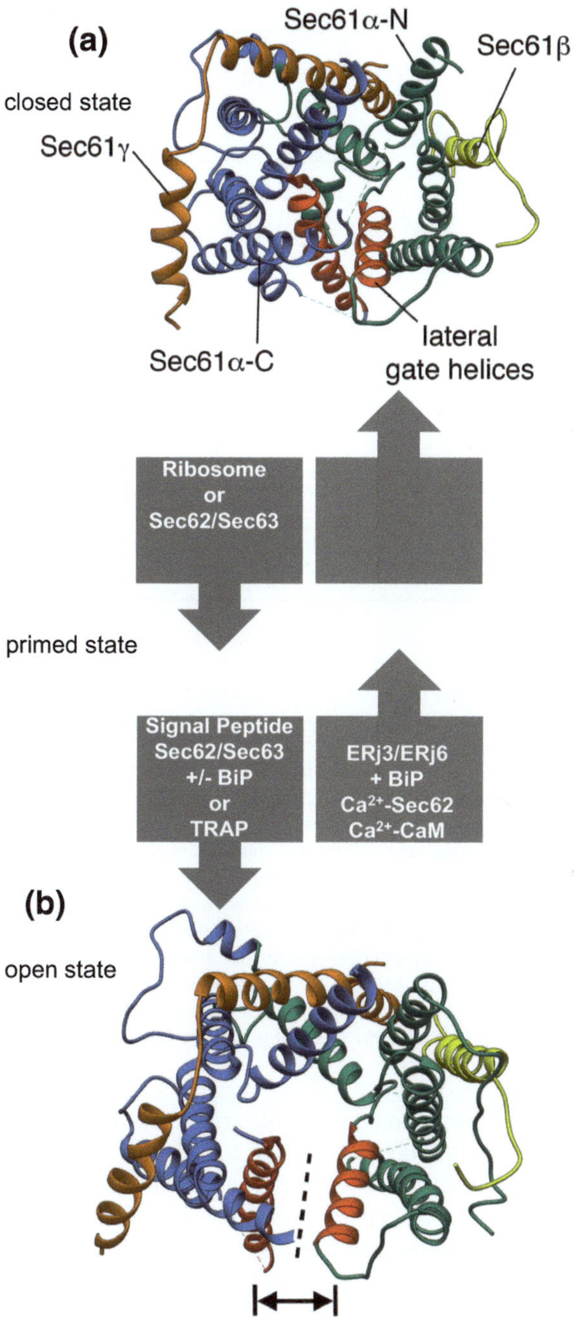
(a)
closed state
Sec61α-N
Sec61β
Sec61γ
Sec61α-C
lateral gate helices
Ribosome or Sec62/Sec63
primed state
Signal Peptide Sec62/Sec63 +/- BiP or TRAP
ERj3/ERj6 + BiP Ca2+-Sec62 Ca2+-CaM
(b)
open state

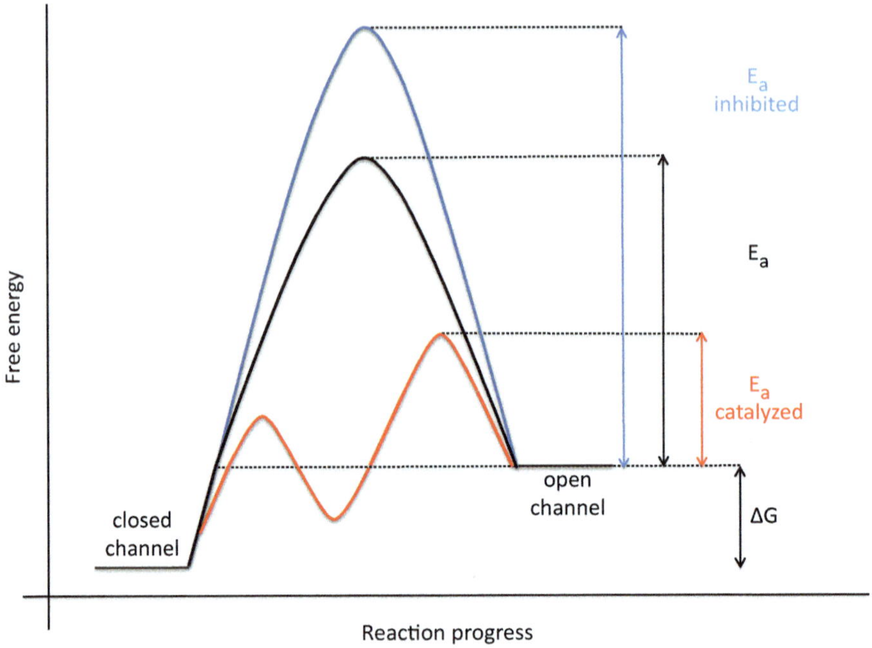

Fig. 4.6 Energetics and kinetics of Sec61 channel gating. The TRAP- or Sec63 ± BiP-mediated Sec61-channel gating is probably best considered in analogy to an enzyme-catalysed reaction. Accordingly, TRAP, Sec63 or BiP reduce the energetic barrier for full channel opening, which can apparently be reinforced by Sec61 channel inhibitors, such as cyclic heptadepsipeptides (e.g. CAM741) or certain eeyarestatins (e.g. ES1, ES24). At least in the case of ES1 and ES24, binding of the inhibitor within the channel pore arrests the channel in a partially open state (termed 'foot in the door'), which maybe identical with the primed state and is compatible with Ca^{2+}-efflux but not with full channel opening for protein translocation. TRAP and BiP contribute to full channel opening by direct interaction with ER luminal loops 5 or 7 of Sec61α (Fig. 4.6). SEC61A1 mutations can increase the energy barrier (E_a) for channel opening per se (V85D or V67G mutation) or indirectly, such as by interfering with BiP binding (Y344H mutation). Notably, all these effects are precursor specific because the N-terminal signal peptides are either efficient or inefficient in driving Sec61 channel opening. Typical for an enzyme-catalysed reaction, BiP can also support efficient gating of the Sec61 channel to the closed state, i.e. the reverse reaction

cryo-EM analyses of programmed and detergent extracted mammalian Sec61 complexes shed light on the orchestration of the opening process and its intermediary steps (Voorhees et al. 2014; Voorhees and Hegde 2016a, b). The closed, i.e. ligand-free, Sec61 complex is unable to conduct substrate transport due to a narrow pore ring and stabilized lateral gate topology (Fig. 4.5). Following the induced-fit model binding of a translating 80S ribosome to the cytosolic docking port of Sec61α induces a conformational change in the loops 6 and 8 reminiscent of a transitional state, also called the primed Sec61 complex. Specific interactions occur between the ribosomal components uL23, eL39, and 28S rRNA contacting conserved, basic residues in loop 6, loop 8, as well as the N-terminal helix of Sec61γ. As a result, the ribosomal exit tunnel with the nascent polypeptide aligns right on top of the cytosolic funnel of the primed Sec61 complex. Conformational changes in the docking port upon ligand binding propagate through associated transmembrane helices and the rest of the Sec61 complex with two ramifications. Binding of a translating ribosome causes destabilization of the polar cluster generating (i) a crack in the cytosolic half of the lateral gate and (ii) exposing a hydrophobic patch in the cytosolic funnel of the Sec61 complex. This single hydrophobic patch is ideally positioned to attract the hydrophobic stretch encoded in the targeting signal (cleavable signal peptide or transmembrane helix) of incoming precursor polypeptides and intercalates the targeting signal in the lateral gate (Fig. 4.4b). This intercalation displaces lateral gate helix 2 in a way that the targeting signal takes over the space occupied by helix 2 in the closed state. The intercalation step supports a rigid body movement of the N- and C-terminal halves of Sec61α and the channel is fully open with the pore ring widened and the plug displaced (Fig. 4.5b). The open Sec61 complex than allows the substrate access axially across or laterally into the membrane. Still, the pore ring residues surround placidly the polypeptide in transit to preserve the permeability barrier for other small molecules and ions during the transport process (Li et al. 2016; Park and Rapoport 2011). Strikingly, this two-stage model of activation from a closed to a primed to an opened Sec61 complex was observed for both prokaryotic and eukaryotic Sec61 complexes activated by a substrate-engaged SecA or translating 80S ribosome using X-ray crystallography or cryo-EM, respectively. Therefore, regardless of the organism or mode of substrate delivery the fundamental principle of protein transport is conserved at both levels component-wise (the Sec61 complex) and mechanistically (temporary intercalation of a targeting signal at the lateral gate).

Functions of the Mammalian Sec61 Complex

The Sec61 complex is a great example how structural, biochemical, biophysical, and cell biological methodologies can complement and guide one another to unravel its structure-function relationship. Historically, the abbreviation "Sec" was introduced to the eukaryotic system by Randy Schekman and coworkers to define mutants in a yeast screen for hampered secretion of secretory enzymes (Spang 2015; Novick et al. 1980). A refined version of the screen searching for mutant yeast cells that fail to translocate a secretory precursor protein into the lumen of the ER identified Sec61p (Deshaies and Schekman 1987; Schekman 2002). Subsequently, Tom Rapoport's research group cloned the mammalian ortholog of Sec61p and demonstrated in a couple of landmark papers its association with ribosomes and nascent chains, corroborating its central role in protein transport as polypeptide-conducting channel (Görlich et al. 1992; Görlich and Rapoport 1993; Hartmann et al. 1994; Kalies et al. 1994). Further crosslinking studies verified that the Sec61 complex also handles transmembrane helices and that targeting signals intercalate between the lateral gate helices 2 and 7, whose flexibility is a prerequisite for efficient translocation (High et al. 1993; Plath et al. 1998; du Plessis et al. 2009; Spiess 2014). Also, the high affinity of the Sec61 complex for ribosome binding was demonstrated under more physiological conditions using a fluorescence resonance energy transfer (FRET) based assay or directly after siRNA-mediated depletion of the Sec61 complex in conjunction with electron microscopy (Benedix et al. 2010; Lang et al. 2012). These data are in agreement with structural biology showing ribosome binding and transport of incoming polypeptides by the Sec61 complex. Actually, the phrase "incoming polypeptides" covers two themes, the arrival of the polypeptide relative to its synthesis and the topology of the polypeptide. Polypeptides arrive at the Sec61 complex either co-translationally, i.e. as nascent chain emerging from the ribosomal exit tunnel during their synthesis, or post-translationally, i.e. after completion of synthesis and release from the ribosome. Both options are conserved across all organisms and the polypeptides, fully synthesized or nascent, are transported by the Sec61 complex before substantial folding occurs (Dudek et al. 2015). While the ribosome acts as activating ligand for the Sec61 complex under co-translational conditions, structural studies mentioned above as well as in vitro reconstitutions highlight the importance of the Sec62/63 protein complex or the SecA ATPase for the post-translational transport in eukaryotic and prokaryotic systems, respectively (Haßdenteufel et al. 2018; Lakkaraju et al. 2012b; Schlenstedt et al. 1990; Johnson et al. 2012; Panzner et al. 1995; Akimaru et al. 1991; Brundage et al. 1990; Driessen and Nouwen 2008). Differentiating features driving the co- or post-translational transport mode are manifold and encoded by the primary structure of the precursor polypeptide, encompassing both the actual targeting signal as well as downstream located stretches (Chatzi et al. 2017). For the second theme, topology, precursor polypeptides can be classified as follows. Besides the complete translocation across the membrane in case of secretory proteins and glycosylphosphatidylinositol (GPI)-anchored proteins including cleavage of their N-terminally located signal peptide multiple variations of membrane topology are reported and summarized together with the mode of membrane insertion in Fig. 4.1. To facilitate transport, polypeptides insert either in a horseshoe bend coordination, called loop insertion (N-terminus of the targeting signal faces the cytosol), or head-first (N-terminus of the targeting signal faces away from the cytosol) into the Sec61 complex. The loop insertion corresponds nicely with structural data and can be envisioned as a result of the targeting signal intercalation at the lateral gate with the downstream mature part creating the horseshoe shaped loop (Li et al. 2016; Park et al. 2014; Voorhees and Hegde 2016b). Loop insertion is considered the more productive mode for cleavable signal peptides and every other transmembrane helix of multi-spanning membrane proteins whose N-terminus faces the cytosol. On the other hand, the transmembrane helices of multi-spanning membrane proteins whose N-termini face away from the cytosol are inserted head-first (Gogala et al. 2014). Interestingly, head-first insertion seems to be the preferred mode of transport for the two groups of inversely oriented type I and type II single-spanning membrane proteins (Fig. 4.1). The transmembrane helix of type II membrane protein starts out with a head-first, or a type I, orientation (the N-terminus translocates across the membrane) followed by an energetically unfavorable reaction of a 180° flip turn reversing orientation of the transmembrane helix to the final type II topology with the N-terminus now facing the cytosol (Devaraneni et al. 2011). This phenomenon of delayed topology determination allowing reorientation of transmembrane helices was also observed for an engineered poly-leucine model protein as well as a polytopic membrane protein (Seppälä et al. 2010; Goder and Spiess 2003). Two special cases of topology are represented by (i) monotopic hairpin proteins, whose membrane domains only dip into a membrane leaflet without traversing it and (ii) tail-anchored (also called type IV) membrane proteins, whose single transmembrane helix is located at the C-terminal end (Pataki et al. 2018; Borgese et al. 2009). Both classes of proteins seem to insert into the ER membrane Sec61-independently (Schrul and Kopito 2016; Yamamoto and Sakisaka 2012; Wang et al. 2014).

In addition to the snapshots of the transport process, live cell imaging and biophysical single-channel recordings from planar lipid bilayer experiments can address dynamic properties of the Sec61 complex over an extended period of time. As such, they can also shed light on the events during termination of polypeptide transport and provide insights into how quickly structural elements of the Sec61 channel "re-shape" to make the transition from the open back to the closed state (Figs. 4.5 and 4.6). The latter point is of particular interest given that the ER is considered the major intracellular calcium store in nucleated mammalian cells and the permeability of a powerful second messenger such as calcium across the ER membrane has to be precisely controlled (Clapham 2007). Indeed, studies with non-physiological molecules larger than a hydrated calcium ion show their permeation into the ER, likely via the Sec61 complex, and demonstrate the imperfect sealing of the mammalian channel for small molecules (Heritage and Wonderlin 2001; Roy and Wonderlin 2003; Le Gall et al. 2004). More recently in yeast, permeability of the physiological glutathione molecule was also shown to involve the Sec61 complex and two ER luminal proteins, Kar2 and Ero1 (Ponsero et al. 2017). Planar lipid bilayer experiments addressing the ion conductance of purified Sec61 complexes directly demonstrated its permeability for calcium with a main and sub-conductance state for calcium of 165 ± 10 pS and 733 ± 16 pS, which correlate to opening diameters of the pore from 5–7 to 12–14 Å, respectively (Lang et al. 2011a; Harsman et al. 2011a). Additional work with presecretory polypeptides and an inhibitor of protein synthesis showed the ion conductance of the Sec61 complex occurs at the end of protein translocation and the channel is fully closed only after washing off non-translating ribosomes (Wirth et al. 2003; Simon and Blobel 1991). The simultaneous use of ratiometric calcium sensitive dyes localized in the cytosol (such as Fura-2) and ratiometric biosensors for calcium in the ER lumen (such as D1ER) in combination with RNAi mediated gene silencing also demonstrated the calcium permeability of the vertebrate and invertebrate Sec61 complex in a cellular setting, under more physiological conditions (Lang et al. 2011a, b; Zhang et al. 2006; Gamayun et al. 2019).

Furthermore, work with dendritic cells showed that the Sec61 complex functions as polypeptide dislocase from endosomes during cross-presentation of extracellular antigens via MHC-I molecules, extending the portfolio of possible Sec61 complex functions and locations (Zehner et al. 2015). However, to restrict mobility of the Sec61 complex (none of the subunits harbors a known ER retention signal) the cytosolic N-terminus of the β-subunit interacts with microtubules allowing stable ER-cytoskeleton interaction. The loss of Sec61β in *Caenorhabditis elegans* induced ER stress, enhanced Sec61 complex mobility and reduced the amount of membrane-attached ribosomes (Zhu et al. 2018). It becomes apparent that the Sec61 complex is a multi-functional player able to (i) bind ribosomes and other ligands, (ii) transport structurally very different substrates, and (iii) represents a calcium permeable channel of the mammalian ER membrane. How these functions are concerted in situ and the different allosteric regulators supporting the Sec61 complex will be discussed in the next sections.

Architecture of the Native Sec61 Complex, the Translocon

While the aforementioned structures of purified Sec61 complexes are informative the situation in vivo within the native membrane might be more challenging. For example, competing forces within a living cell influencing biochemical reactions and transport processes arise from macromolecular crowding taking account of specific and nonspecific interactions between macromolecules (Minton 2006; Ellis 2001; Zhou et al. 2008). The high protein and solute concentrations in both the cytosol and the ER lumen as well as proteins and lipids of the membrane in the immediate vicinity of the Sec61 complex might influence the channel and details of the transport mechanism. Similarly, the plethora of transported clients (Fig. 4.1) with varying amino acid sequences of their targeting signal or mature part requires substrate-specific adjustments of the transport reaction. Advancements in the field of cryo-electron tomography (CET) in conjunction with subtomogram analysis enable this technology to address the heterogeneity of local, native environments and address the structure of Sec61 complexes in situ (Koning et al. 2018; Lučić et al. 2013). So far, CET studies were conducted for eukaryotic Sec61 complexes from intact HeLa cells or ER membrane vesicles derived from canine pancreatic cells, human cell lines, primary fibroblasts, or green algae (Mahamid et al. 2016; Pfeffer et al. 2012, 2014, 2015, 2017; Braunger et al. 2018). While CET structures of the mammalian ribosome-associated Sec61 complex are in agreement with the main conclusions drawn from cryo-EM and crystal structures, tomography adds an important aspect. The native Sec61 complex is a team player and associates (at least) with two membrane protein complexes to form the native ER protein translocon (Fig. 4.2). The translocon-associated protein (TRAP) complex localizes in a stoichiometric manner next to the C-terminal half of the ribosome-engaged Sec61 complex and the oligosaccharyl-transferase (OST) complex approaches in a substoichiometric manner (present only in 40–70% of complexes) the N-terminal half (Pfeffer et al. 2016). Therefore, TRAP and OST are not occluding the lateral gate and targeting signal intercalation. These data are further supported by cryo-EM studies of solubilized ribosome-bound translocon complexes (Ménétret et al. 2005, 2008; Braunger et al. 2018). Thus, the protein translocase seems dynamic by nature (Fig. 4.5). Not only the active center is subject to structural flexibility, also stoichiometry of subunits and partner components changes with different substrates or cellular cues. The functional implication of structurally visualized and other biochemically verified translocon components will be discussed next.

The Role of Allosteric Effectors of the Eukaryotic Sec61 Complex Previously Visualized by Structural Biology

In the living cell the Sec61 complex is continuously contacted and supported by an alternating repertoire of associated proteins. Up to now, the ribosome and three protein complexes were unequivocally identified by structural data and shown to affect gating and, therefore, functionality of the Sec61 complex (Fig. 4.5).

The Ribosome

As stated before, the ribosome contacts the Sec61 complex via charged residues in loop 6 and 8, the evolutionarily conserved docking port of the Sec61α subunit, inducing a conformational change from the closed to the primed state (Figs. 4.4 and 4.5). Mutational studies in yeast showed a phenotypic differentiation between mutants of Sec61α loop 6 and 8. While loop 8 mutants had a reduced binding affinity for 80S ribosomes, mutations in loop 6 inhibited co-translational transport without significantly affecting ribosome binding activity (Cheng et al. 2005). Similar studies in yeast and bacteria also highlighted the importance of the cytosolic C-terminus of Sec61α for both ribosome binding and viability. Positive charges in the C-terminus may interact with ribosomal rRNA to support positioning of the ribosome and protein translocation (Egea and Stroud 2010; Mandon et al. 2018). The ribosome also affects calcium permeability of the Sec61 channel. Protein synthesis inhibitors that arrest (e.g. emetine) or release (e.g. puromycin) the ribosome from the Sec61 complex can block or increase calcium efflux from the ER, respectively (Lang et al. 2011a; Klein et al. 2018; Gamayun et al. 2019). Thus, the ribosome acts as prominent modulator of the calcium leak from the ER opening an interesting connection between Sec61-mediated calcium efflux and protein synthesis by the ribosome. Furthermore, cryo-EM and CET studies have shown the interaction between the Sec61 complex and translating ribosome is not hermetically sealed. Instead, a considerable gap is visible between the N-terminal half of the Sec61 complex and the ribosome exit tunnel (Ménétret et al. 2007; Park et al. 2014; Voorhees et al. 2014; Pfeffer et al. 2015). This partially shielded micro-compartment could provide a space for the release and folding of cytosolic domains of membrane proteins, or a location for quality control, repair and de-clogging factors probing the transport process (Malsburg et al. 2015; Kayatekin et al. 2018; Ast et al. 2016). A cryo-EM structure of the bacterial ribosome-translocon complex (RTC) during synthesis of a polytopic membrane protein showed electrostatic interactions between positively charged residues in the cytosolic loop connecting the two transmembrane helices of the model precursor and negative charges of the ribosomal rRNA helix 59 (H59). Thus, the ribosome could be another player in decoding the positive-inside rule acting in concert with the Sec61/SecY complex and the decoding being orchestrated in the gap volume (von Heijne 1989; Bischoff et al. 2014). Alternatively, the calcium sensor calmodulin can occupy this micro-compartment to monitor or minimize the calcium flux associated with the transport process (Erdmann et al. 2011). However, elusion of the nascent polypeptide into the gap demonstrates that the GTP-driven elongation process is not necessarily a driving force for translocation and opens up a space and time-window for the recruitment of regulatory factors (Conti et al. 2015).

Aside from docking to the Sec61 complex, the ribosomal surface is an enormous hub for the recruitment of other ligands. Ligands, cytosolic as well as membrane-bound, allow the fine-tuning of protein transport in response to various stimuli. A recent example addressing the plethora of ribosomal ligands, dubbed the mammalian “ribo-interactome”, identified in addition to the 100 proteins constituting the canonical translation machinery 330 interactors with diverse functions. For example, the combination of high-throughput sequencing after UV crosslinking (iCLIP) and proximity-specific ribosome-profiling demonstrated that isoform 2 of the pyruvate kinase of the muscle, Pkm2, is enriched on ER-bound ribosomes near the A-site and acts as translational activator of ER destined mRNAs. A SILAC (stable isotope labeling by amino acids in cell culture) approach additionally verified that Pkm2-enriched ribosomes are contacting the Sec61 and OST complex, whereas the gamma subunit of the TRAP complex was found as general ribosomal interactor (Simsek et al. 2017). The cytosolic domain of the γ-subunit of the TRAP complex is in close proximity to the RNA expansion segments ES20L/ES26L and ribosomal protein eL38 of the 60S subunit (Pfeffer et al. 2016, 2017). Based on the structure of the bacterial RTC, the tip of a nascent targeting signal interacting with H59 and uL24 might come in close contact with the neighboring eL38 (Jomaa et al. 2016, 2017; Nguyen et al. 2018). Although not in conjunction with the translocon or the targeting signal, eL38 was shown to be a regulatory ribosomal protein that can support translation of subset of mRNAs carrying a specialized regulon motif in the 5’ untranslated region (Xue et al. 2015). Analogous to a regulon motif, TRAP could trigger the regulatory function of eL38 to support translation in vicinity to the Sec61 complex.

Recent structural data of the OST complex also provided striking insights into its interaction with the ribosome (Bai et al. 2018; Wild et al. 2018; Braunger et al. 2018). The cytosolic C-terminus of Rpn1, one of the core subunits shared amongst the two paralogous OST complexes containing either Stt3A or Stt3B as catalytic subunit, interacts with the ribosome. Rpn1 forms a quadruple-helix bundle aligning in a cavity made from rRNA helices H19/H20, rRNA expansion segment ES7a and ribosomal protein eL28 (Braunger et al. 2018). Importance of this interaction for efficient RTC formation was demonstrated by antibodies against the cytosolic Rpn1 segment which prevented ribosome targeting to and efficient protein translocation by the translocon (Yu et al. 1990). Though, the Rpn1 interaction with the ribosome is sterically hindered in case of the Stt3B-containging OST complex due to an additional sequence extension in a cytosolic loop and the presence of a paralog specific subunit (Braunger et al. 2018).

Besides the Sec61, TRAP, and OST complex the ribosome was shown to interact with additional membrane proteins of the ER, including palmitoylated Calnexin, ERj1, Sec62, or p180 (Table 4.1). These candidates support ribosome anchoring at the ER or serve as transient, regulatory proteins of the RTC. Calnexin, a type I membrane protein, is a lectin-like chaperone and assists maturation, folding and oligomerization of glycoproteins (Hebert et al. 1996). Upon modification by the ER palmitoyltransferase Dhhc6, palmitoylated calnexin associates with the RTC (next to the TRAP complex) to catch transported client proteins as they emerge from the Sec61 complex (Lakkaraju et al. 2012a; Wada et al. 1991; Chevet et al. 1999). Another single-pass type I membrane protein of the ER carrying a characteristic luminal J-domain, ERj1, can directly associate with ribosomes in the intact ER as was demonstrated by a FRET based assay employing antibody accessibility as a readout or from analysis of ribosome-associated ER membrane proteins in detergent extracts of canine pancreatic microsomes (Blau et al. 2005; Dudek et al. 2002; Benedix et al. 2010; Dudek et al. 2005). Using the same experimental strategies as well as surface plasmon resonance spectroscopy the double-spanning membrane protein Sec62 was shown to interact via two positively charged clusters encoded in the cytosolic N-terminus of Sec62 with the ribosome close to the exit tunnel (Müller et al. 2010). In case of p180, a single-pass type I membrane protein with a gigantic cytosolic coiled-coil domain, data are somewhat ambiguous with regard to what entity it actually attracts to the ER membrane. Besides a direct association of p180 to the ribosome as part of the RTC, some data highlight the possibility of p180 acting as direct mRNA anchor (Savitz and Meyer 1990, 1993; Cui et al. 2012, 2013; Dejgaard et al. 2010; Morrow and Brodsky 2001; Ueno et al. 2010, 2011). Regardless, both options would attract polysomes to the translocon to enhance protein translocation, substrate-specific or not. Further, predominantly cytosolic, ribosomal interactors supporting the substrate specific targeting of precursor polypeptides to the translocon will be discussed in the section "precursor protein targeting factors".

The TRAP Complex

The human TRAP complex consists of four consecutively named subunits TRAPα to TRAPδ (Hartmann et al. 1993). With the exception of TRAPγ carrying four transmembrane helices and negligible luminal mass, the other TRAP subunits are single spanning type I membrane proteins with a cleavable targeting signal and a luminal domain of over 100 amino acids (Bañó-Polo et al. 2017). A reasonable cytosolic mass comprised of roughly 60 amino acids is found only in TRAPα and TRAPγ. Despite the often discussed compositional heterogeneity of the native ribosome-associated translocon, TRAP seems to be a stoichiometric and permanent component of it (Braunger et al. 2018; Pfeffer et al. 2015). Functionally, TRAP supports protein translocation by the Sec61 complex in a substrate-specific manner. TRAP-mediated assistance was observed for precursor proteins with cleavable signal peptides or N-terminal transmembrane helices. Experiments based on biochemical reconstitution demonstrated for a small subset of substrates that only signal sequences with a strong translocon interaction and quick gating potential are able to be transported independently of TRAP (Fons et al. 2003). Similarly, in vitro studies testing mutations in the flanking charges of a type II signal anchor obscuring the positive-inside rule demonstrated the importance of TRAP for proper topogenesis of such a transmembrane helix handled by the Sec61 complex (Sommer et al. 2013; Baker et al. 2017; von Heijne 2006; von Heijne and Gavel 1988; Goder et al. 2004). Using quantitative proteomics to analyze changes of cellular protein abundance upon TRAP depletion revealed that signal peptides of TRAP-dependent clients exhibit a glycine-plus-proline content above and hydrophobicity below average (Nguyen et al. 2018). As "helix-breaking" residues the pronounced glycine-plus-proline content of TRAP-dependent signal peptides reduces their propensity of helix formation and likely their ability to (i) intercalate at the lateral gate of the Sec61 complex and (ii) displace the lateral gate helix 2 in order to advance the channel from the primed to the open state (Figs. 4.5 and 4.6). The reduced hydrophobicity of TRAP-dependent targeting signals probably reduces the likelihood for efficient binding/interaction with the single hydrophobic patch that opens up upon priming of the Sec61 complex by ribosome binding (Voorhees and Hegde 2016a). In the end, both parameters of TRAP clients reduce their ability to overcome the activation energy necessary to convert the primed state of the Sec61 complex into the open state in a reasonable dwell time (Figs. 4.5 and 4.6). CET provided an interesting view on how the TRAP complex could help to overcome this energetic deficit. To do so, a set of difference densities comparing the native canine TRAP complex with algal (lacking TRAPγ/δ) and TRAPδ-deficient human TRAP complexes demonstrated proximity of the luminal mass (likely the TRAPα/β subunits) to loop 5, the hinge region connecting the N- and C-terminal halves of Sec61α and permitting the rigid body movement (Pfeffer et al. 2017) (Figs. 4.5 and 4.7a). Different algorithms predicted for the luminal domains of both TRAPα and TRAPβ a beta-sandwich fold. This is a classical domain structure often found in immunoglobulins or lectins, protein classes specialized in binding to other polypeptide or carbohydrate moieties. Considering both proximity and domain fold, we assume that TRAP acts as allosteric effector of the Sec61 complex in a chaperone-equivalent fashion and catalyzes a reduction of the energy barrier enabling gating deficient signal peptides of TRAP-dependent substrates to open the Sec61 complex (Figs. 4.4, 4.5, 4.6 and 4.7). Alternatively, or additionally, TRAP could work as a ratchet on the nascent precursor polypeptides in transit into the ER lumen directly. Moreover, taking into consideration the vicinity between the cytosolic

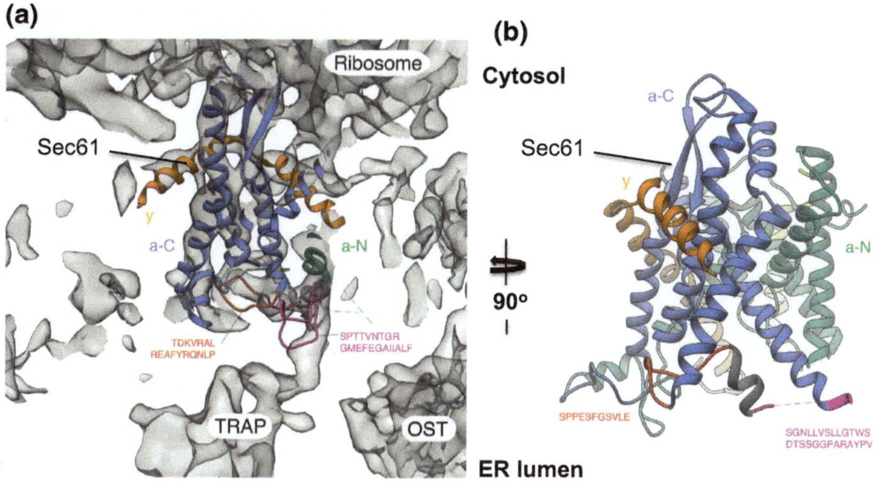

Fig. 4.7 Atomic model for the heterotrimeric Sec61 complex, fitted into the ribosome-translocon structure, shown in Fig. 4.2b and highlighting the ER luminal contact sites of TRAP and BiP, respectively. **a**, **b** The Sec61 channel is shown in two orientations, as seen from the plane of the membrane (PDB 3jc2, EMD 3069). The N- and C-terminal halves of the Sec61 α-subunit are shown in green and blue, respectively, and the γ-subunit in orange. **a** The same view of the Sec61 complex is shown as in Fig. 4.4b. However, most of the N-terminal half of the α-subunit and the complete ß-subunit were clipped for better visibility of the TRAP interaction site. The ER luminal domains of the TRAP α- and β-subunits interact with ER luminal loop 5 of the Sec61 α-subunit (connecting transmembrane helices 5 and 6), which is shown with the surrounding electron densities of the ribosome, TRAP, and OST. The primary structure of the TRAP binding site within loop 5 is N-terminal to the so-called hinge helix (connecting the N- and C-terminal halves of the Sec61 α-subunit; shown in grey) and is shown in magenta; the amino acid sequence C-terminal to the hinge region is shown in red. There is no atomic structure of TRAP, but secondary structure predictions for the ER luminal domains of the TRAP α- and β-subunits are consistent with a beta sandwich fold. **b** In (**b**) the Sec61 complex is rotated counter clockwise in the plane of the membrane by 90° as compared to (**a**). The substrate binding domain (SBD) of ER luminal Hsp70-type molecular chaperone BiP interacts with ER luminal loop 7 of the Sec61 α-subunit (connecting transmembrane helices 7 and 8), and is recruited to the Sec61 complex by ER membrane resident Hsp40-type co-chaperone Sec63 via the J domain of the latter. The BiP binding site within loop 7 includes the so-called minihelix (shown in grey) between the up-stream lying oligopeptide (shown in purple) and the down-stream lying oligopeptide (shown in red)

domain of TRAPγ and the ribosome (eL38, ES20L/ES26L) mentioned before, the TRAP complex acts as relay bridging incoming precursor information from the cytosol across the ER membrane to the luminal side to support the conformational switch of the Sec61complex necessary to accommodate TRAP-dependent substrates. This mechanism of action portraying TRAP as allosteric effector that supports opening of the Sec61 complex was further substantiated by two lines of evidence. One, live cell calcium imaging measurements showed that depletion of TRAP in human cells reduced the Sec61-mediated calcium efflux from the ER (Nguyen et al. 2018). Two, comparing the evolutionary conservation of the

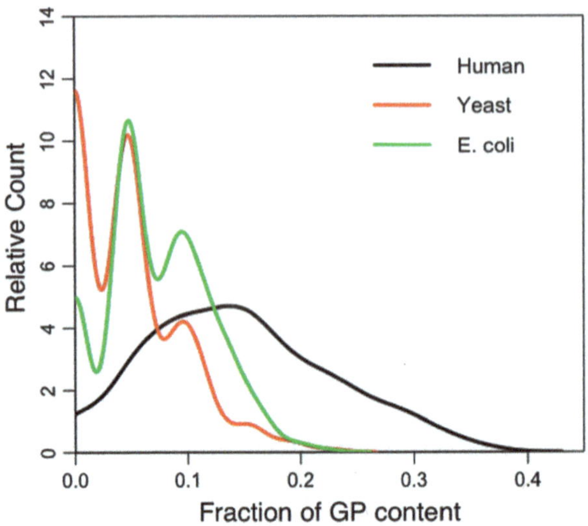

Fig. 4.8 Due to the presence of TRAP, the mammalian Sec61 complex can handle signal peptides with relatively high content of glycines and prolines, in contrast to the homologous complexes in yeast and bacteria. Client specificity of human TRAP was revealed by a combination of siRNA-mediated TRAP depletion in HeLa cells and quantitative proteomics plus differential protein abundance analysis. The combination of siRNA-mediated gene silencing, using two different siRNAs for each target and one non-targeting (control) siRNA, respectively with three replicates for each siRNA and label-free quantitative proteomic analysis plus differential protein abundance analysis was used to identify negatively affected proteins (i.e. TRAP clients). Subsequently, we used custom scripts to compute the glycine/proline (GP) content of signal peptide sequences of TRAP clients as the total fraction of glycine and proline in the respective signal peptide sequences (data not shown). We also used custom scripts to extract protein annotations for all human, *E. coli* and *S. cerevisiae* signal peptides from UniProtKB entries

glycine-plus-proline content of cleavable signal peptides encountered in TRAP-carrying humans and TRAP-deficient organisms such as yeast and *E. coli* showed a much higher glycine-plus-proline content in the former (Fig. 4.8). Thus, enabled by TRAP, the mammalian Sec61 complex can manage signal peptides with a higher content of glycines and prolines compared to its homologous ancestors in yeast and bacteria.

The OST Complex

In contrast to the monomeric oligosaccharyl-transferase in bacteria (PglB) and archaea (AglB) the human OST complex is represented by two multimeric paralogs, named after the catalytic core subunits Stt3A and Stt3B (Table 4.1). The latter two, which show highest sequence homology amongst the OST subunits to the prokaryotic monomers, catalyze the transfer of the lipid-linked glycan

($Glc_3Man_9GlcNAc_2$) onto the asparagine residue of a specific trimeric sequon motif (Asn-X-Ser/Thr, where X $\neq$ Pro) in the polypeptide chain (Larkin and Imperiali 2011). N-linked glycosylation in the mammalian ER can occur co-translationally, while the polypeptide is in transit, or post-translocationally. While the former is usually catalyzed by the Stt3A-containing OST complex skipped sequons are complemented by the Stt3B paralog which can also act post-translocationally to N-glycosylate C-terminally located sites within the polypeptide chain (Shrimal et al. 2015; Ruiz-Canada et al. 2009; Sato et al. 2012). Despite two paralogous OST complexes complementing each other one-third of Asn in glycosylation sequons are not modified (Petrescu et al. 2004; Kelleher and Gilmore 2006). Important for glycosylation is a conserved sequon binding motif (Trp-Trp-Asp) in the substrate binding groove of various Stt3 homologs, representing the active center of OST complexes (Bai et al. 2018). Structural data and early in vitro based assays demonstrated a distance of roughly 40–60 Å (equivalent to a distance spanned by 15–20 amino acids) from the translocon exit to the active center of Stt3 (Nilsson et al. 2003; Nilsson and von Heijne 1993; Wild et al. 2018; Bañó-Polo et al. 2011; Kowarik et al. 2002). The paralogous OST complexes in humans share a set of six subunits Rpn1, Rpn2, DAD1, OST4, OST48, and TMEM258. In the Stt3B-containing OST this set is supplemented by the catalytic Stt3B, and MagT1 or TUSC3 subunit. In case of the Stt3A paralog the complex-specific subunits are Stt3A, DC2 and Kcp2 (Cherepanova et al. 2016). The latter two, DC2 and Kcp2, were shown biochemically and by structural analysis to connect the Stt3A-containing OST to the Sec61 complex (Shrimal et al. 2017; Braunger et al. 2018). While the Stt3B-containing OST complex is a stand-alone unit, the Stt3A-containing OST can associate with the RTC, including TRAP, to form a co-translational super-complex as observed even in the native membrane (Wild et al. 2018; Pfeffer et al. 2014; Ruiz-Canada et al. 2009). Tethering of the Stt3A-containing OST to the RTC is ensured by two interfaces. The first interface refers to the specific association between DC2 and Stt3A. This interaction is mediated by cytosolic, membrane and luminal portions of both proteins including transmembrane helices 10–13 and the last cytosolic loop of Stt3A. Due to sequence variations between Stt3A and Stt3B in this contact area required for the DC2 association, DC2 binds selectively to the Stt3A paralog. Next, DC2 with the rest of the Stt3A-containing OST in tow binds to the Sec61 complex. The luminal loop of DC2 likely interacts with the N-terminal half of Sec61α as well as the C-termini of Sec61β and Sec61γ (Braunger et al. 2018). Put simply, the membrane-embedded part of the Stt3A-containing OST complex flanks the hinge region of Sec61α (Pfeffer et al. 2016). The second interface was mentioned earlier and refers to the quadruple-helix bundle of the cytosolic Rpn1 domain binding to the ribosome. Thus, the dissimilar subunit composition between the two alternative OST complexes explains both the paralog-specific association with the RTC and, by extrapolation, the different modes of action (co-translational Stt3A versus post-translocational Stt3B). The DC2 subunit is the bridging element that ties exclusively the Stt3A-containing OST to the N-terminal half of the Sec61 complex (Fig. 4.2). Further evidence showing the paralog-specific integration of the

Stt3A-containing OST- complex into the RTC comes from CET of Stt3 knockout cells. While the ratios of RTC populations carrying only TRAP or TRAP + OST densities were identical for wildtype and Stt3B knockout cells, the RTC population carrying TRAP + OST could not be observed at all in Stt3A knockout cells (Braunger et al. 2018). Yet, it is currently unclear if and how the OST complex might affect the Sec61-mediated calcium efflux. Although inhibitors of glycosylation such as tunicamycin trigger an elevated Sec61-mediated calcium efflux from the ER, this effect cannot be assigned to the inactivity of OST complexes rather than being a consequence of inaccurate protein maturation and subsequent BiP sequestration (Schäuble et al. 2012).

The Sec62/63 Complex

As discussed before, the Sec61 complex can handle incoming polypeptides either co-translationally or post-translationally. However, reasonable structural data depicting the organization of the post-translational eukaryotic Sec61 complex are scarce. So far, three studies highlighted the assembly of the detergent extracted, unoccupied, post-translational translocon complex from *S. cerevisiae* by cryo-EM (Itskanov and Park 2018; Wu et al. 2018; Harada et al. 2011). In yeast, the fully assembled post-translational translocon represents a heptameric protein ensemble referred to as the SEC complex (Deshaies et al. 1991; Panzner et al. 1995; Jermy et al. 2006). In the SEC assembly the trimeric Sec61 complex is accompanied by the tetrameric Sec62/63 complex. The latter consist of two essential, evolutionarily conserved membrane proteins, Sec62 and Sec63, and two dispensable subunits, Sec71 and Sec72. All seven subunits of the SEC complex are bundled in a 1:1 stoichiometry (Harada et al. 2011). Additionally, in vitro reconstitutions demonstrated that the functional SEC complex needs support from an ATP consuming Hsp70 chaperone of the ER lumen, Kar2p in yeast or BiP in mammals, for efficient post-translational transport (Panzner et al. 1995; Matlack et al. 1999; Brodsky et al. 1995; Brodsky and Scheckman 1993). Although the different cryo-EM structures did not include any BiP density, the data provided first insights into how the SEC complex is arranged to allow gating of the Sec61 complex and support transport of post-translational substrates. Most striking was the extensive interaction between Sec63 and the Sec61 complex including contacts in their cytosolic, membrane and luminal domains. Similar to the ribosome or the bacterial SecA ATPase, the cytosolic Brl domain of Sec63 interacts with loops 6 and 8 of Sec61α, the universal docking port, thereby "reserving" the docking port and blocking ribosome binding. Interestingly, as assumed for the interaction of the TRAPα/β subunits with the Sec61 complex, the Brl domain of Sec63 shows a canonical beta-sandwich fold for an antigen-antibody-like binding to loop 6. In the membrane, Sec63 (transmembrane helix 3) contacts all three subunits of the Sec61 complex in the hinge region opposite to the lateral gate including transmembrane helices 5 and 1 of Sec61α as well the membrane anchors of Sec61β and Sec61γ. In addition, the short luminal

N-terminus of Sec63 appears to intercalate on the luminal side of the channel between the hinge loop (Sec61α loop 5) and Sec61γ (Itskanov and Park 2018; Wu et al. 2018). Yeast viability assays with single point mutations introduced in the cytosolic or membrane contact area of Sec63 resulted in lethality, highlighting the importance of those contacts. On the other hand, mutations in the luminal interaction site did not affect cell viability (Wu et al. 2018). The two nonessential subunits, Sec71 and Sec72, sit on top of Sec63's Brl domain. Structurally, binding of the Sec62/63 complex to the Sec61 channel triggered opening of the lateral gate much wider than observed in any previous cryo-EM structure (Van den Berg et al. 2004; Voorhees and Hegde 2016a; Li et al. 2016; Zimmer et al. 2008; Egea and Stroud 2010). The functional implications for the translocon resulting from gating by the Sec62/63 complex are exquisite. First, targeting signals of many post-translational substrates are often less hydrophobic and therefore would have a lower chance to intercalate at the lateral gate in the primed Sec61 complex and drive further opening of the channel. However, in the post-translational SEC complex binding of the Sec62/63 complex seems to induce a fully opened channel that readily accommodates even "weak" or otherwise inefficient intercalating targeting signals (Trueman et al. 2011, 2012; Ng et al. 1996). Fitting to the concept of the Sec62/63 complex inducing wide opening of the lateral gate yeast Sec62 was found to mediate topology of moderately hydrophobic signal anchor proteins, in particular type II membrane proteins that undergo the energetically unfavorable 180° flip turn for reversing the initial type I orientation (Reithinger et al. 2013; Jung et al. 2014). Second, association of the Sec62/63 complex opposite to the lateral gate of the Sec61 complex might perturb binding of the co-translational acting Stt3A-containing OST complex. Thus, Sec63 might block both ribosomal binding and coordination of the Stt3A-containing OST complex found in most multicellular plants and metazoans (Cherepanova et al. 2016). This could imply that post-translational substrates are exclusively glycosylated post-translocationally and that the appearance of the Sec62/63 complex, which is absent in bacteria, occurred hand-in-hand with the gene duplications of specific OST complex subunits. Similar to metazoans, also yeast has two paralogous OST complexes containing either Ost3 or Ost6, the yeast homologs of TUSC3 and MagT1, respectively. Both variations of the OST complexes in yeast share the catalytic Stt3 subunit, which is more similar to the mammalian Stt3B subunit and explains why the OST complexes in yeast are considered stand-alone units (Wild et al. 2018). Interestingly, recent CET data from Stt3A depleted HEK cell microsomes observed a previously unidentified translocon population devoid of TRAP and OST (Braunger et al. 2018). The unknown density could represent the mammalian equivalent of the SEC complex with or without BiP. Third, binding of the Sec62/63 complex to the luminal end of the Sec61 channel might also interfere with functionality of the TRAP complex. Both accessory complexes Sec62/63 and TRAP appear on the luminal site of the channel in proximity to loop 5, which connects the N- and C-terminal halves of Sec61α. Interaction of the accessory complexes with loop 5 might support the rigid body movement during opening of the Sec61 complex, eventually in a substrate specific manner. Fourth, the extremely wide opening of the lateral gate triggered by binding of the Sec62/63 complex might also impact the Sec61-mediated calcium efflux from the ER. While this issue can be tolerated in yeast, whose major intracellular calcium store is the vacuole rather than the ER, in mammalian cells the excessive leakage of calcium would need to be compensated by other factors. Work from our own group suggests that Sec62 and BiP are efficient regulators of the ER calcium leakage (Schäuble et al. 2012; Greiner et al. 2011; Linxweiler et al. 2013).

Similar to the situation in yeast, studies of protein transport in mammalian cells also show the substrate-specific involvement of Sec62 in ER import (Fig. 4.5). Mammalian Sec62 is required for the efficient transport of small precursor proteins (such as preproapelin), which are, due to their short length, transported post-translationally (Lang et al. 2012; Haßdenteufel et al. 2018; Johnson et al. 2013; Lakkaraju et al. 2012b). However, in contrast to yeast, the mammalian Sec62 protein experienced a gain of function as it is able to interact with the ribosome near the ribosomal exit tunnel and also supports the co-translational transport of certain substrates, such as the precursors of ERj3 and prion protein (Müller et al. 2010; Fumagalli et al. 2016; Ziska et al. 2019). Accordingly, cross-linking experiments using different stalled precursor polypeptides in transit through the mammalian translocon demonstrated the dynamic recruitment of allosteric regulators including Sec62. The model precursor preprolactin recruited accessory factors like TRAP, OST, and the translocating chain-associating membrane (TRAM) protein to the Sec61 complex. Yet, when ERj3 or prion protein were used as precursor the Sec62/63 complex instead of TRAP and TRAM was recruited to the channel to support translocation of those substrates with a rather slowly or inefficiently gating targeting signal (Conti et al. 2015; Fumagalli et al. 2016; Ziska et al. 2019). The gap between the Sec61 complex and the ribosome exit tunnel providing a space for the elusion of nascent polypeptides could be critical for the substrate-specific recruitment of regulatory factors. Another dynamic transition of the translocon during specific translocation events was observed for Sec62 and the SRP receptor, a protein targeting complex that, as will be discussed in more detail later, is required for co-translational targeting of precursor proteins to the Sec61 complex. To allow co-translational targeting the SRP receptor apparently displaces Sec62 from the SEC complex switching the Sec61 channel from Sec62- to SRP-dependent translocation (Jadhav et al. 2015). However, according to above mentioned experiments, SRP receptor and Sec62 can also act sequentially, i.e. after SRP-dependent targeting of precursors of ERj3 and prion protein, Sec62 can displace SRP receptor from the Sec61 channel and together with Sec63 support channel gating. Furthermore, in this scenario Sec63 has to be expected to take over loops 6 and 8 of Sec61α from the ribosome. Lastly, depletion of Sec63 from mammalian cells, which is neither accompanied by a loss of Sec62 nor compensated by increased levels of other major translocon components (Fig. 4.9), causes a substrate specific defect in protein translocation of membrane and secretory proteins, but is without effect on the Sec61-mediated calcium leakage (Lang et al. 2012; Mades et al. 2012; Haßdenteufel et al. 2018; Schorr et al. 2015; Fedeles et al. 2011). To a certain extent even the cryo-EM structures of the post-translational SEC complex reflect the idea of a dynamic transition and flexibility of the Sec62/63

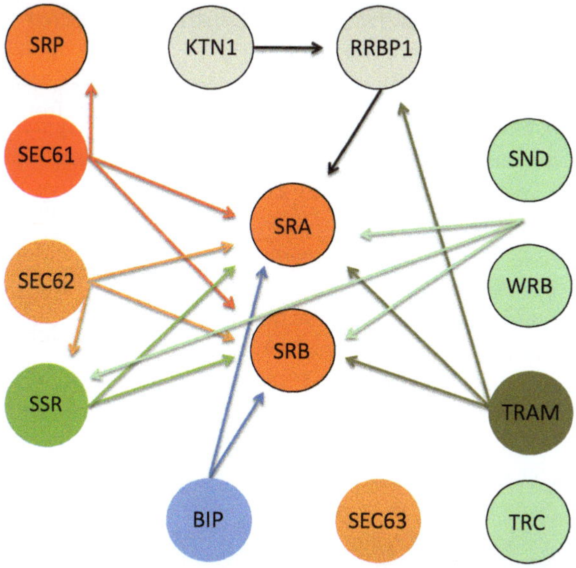

Fig. 4.9 Genetic interactions between ER targeting- or translocation-components, as revealed by a combination of siRNA-mediated component depletion in HeLa cells and quantitative proteomics plus differential protein abundance analysis. The combination of siRNA-mediated gene silencing, using two different siRNAs for each of the indicated targets and one non-targeting (control) siRNA, respectively with three replicates for each siRNA and label-free quantitative proteomic analysis plus differential protein abundance analysis was also used to identify positively affected proteins (i.e. compensatory mechanisms or genetic interactions, indicated as arrows)

complex. Both Sec62 and the characteristic luminal J-domain of Sec63 could not be sufficiently resolved in the particle analysis which might be due to their structural flexibility and dynamic integration into the translocon.

Additional Transport Components and Allosteric Effectors of the Sec61 Channel

In addition to the multimeric complexes from the previous section many more auxiliary transport components and allosteric effectors of the Sec61 channel, respectively, have been experimentally detected to transiently interact with the Sec61 complex. So far, these auxiliary components eluded structural visualization in combination with the polypeptide-conducting channel. Yet, a few examples will be named and summarized.

BiP, an Additional Allosteric Sec61 Channel Effector

One of the most abundant and versatile proteins within the ER lumen is the Hsp70-type molecular chaperone named BiP (Fig. 4.5). Originally identified as protein that binds non-covalently to free immunoglobulin heavy-chains its repertoire of functions was steadily extended as was reviewed before (Haas and Wabl 1983; Zimmermann 2016; Otero et al. 2010; Dudek et al. 2009; Ni and Lee 2007; Ma and Hendershot 2004). Consistent with its domain organization BiP reversibly binds to hydrophobic oligopeptides of loosely folded polypeptides in an ATP-regulated manner (Flynn et al. 1991). To do so, a flexible inter-domain linker region connects the N-terminal nucleotide binding domain (NBD) and C-terminal substrate-binding domain (SBD) of BiP (Kumar et al. 2011). Thus, binding and release of a substrate are coupled to an ATPase cycle that triggers conformational transition of BiP's different sub-domains and supports productive folding of a substrate (Smock et al. 2010; Marcinowski et al. 2011). For a complete and productive ATPase cycle BiP is part of a tripartite system. Aside from BiP, this system encompasses a J-domain carrying Hsp40-type co-chaperone and a nucleotide exchange factor (Table 4.1). ATP-bound BiP has a low affinity for substrates given that a sub-domain within the SBD called the lid is in the open conformation. Association with the characteristic J-domain of one of the multiple ER luminal Hsp40-type co-chaperones (e.g. Sec63) stimulates the ATPase activity of BiP and coincides with both a closure of the lid in the SBD and a drastic increase in substrate affinity. Reversal of this reaction is mediated by a nucleotide exchange factor that helps to replace ADP by ATP and convert BiP back into the low-affinity state with an open lid (Melnyk et al. 2015). The currently known repertoire of at least eight Hsp40 co-chaperones and two nucleotide exchange factors in human cells allows fine-tuning of the ATPase cycle of BiP and its function can be tailored for various substrates and/or occasions. The combinatorial assembly of the tripartite system with BiP at its center probably allows BiP to integrate its many known functions. Those functions relate to the import, folding/assembly, export, and degradation of polypeptides as well as regulation of folded proteins such as the UPR sensors IRE1, PERK, and ATF6 or the translocon component Sec61α (Dudek et al. 2009; Hennessy et al. 2000, 2005; Wang and Kaufman 2012; Walter and Ron 2011; Zhao and Ackerman 2006). For the latter, BiP was shown to support protein transport into the ER in two different modi operandi. One, as molecular ratchet BiP directly works on the incoming, unfolded precursor polypeptide in transit through the Sec61 complex (Tyedmers et al. 2003). The ratcheting supports both co- as well as post-translational translocating substrates (Liebermeister et al. 2001; Panzner et al. 1995; Brodsky and Scheckman 1993; Brodsky et al. 1995). Two, BiP acts as direct allosteric effector of the Sec61 complex for channel opening by binding to the luminal loop 7 of Sec61α (Fig. 4.7b). Prohibiting the binding of BiP to luminal loop 7 achieved either by BiP depletion or the introduction of a BiP repelling mutation in loop 7 caused a substrate-specific defect in transport for the precursors of ERj3 and prion protein as well as for the short, post-translationally transported preproapelin (Schäuble et al. 2012; Lloyd et al. 2010; Haßdenteufel et al. 2018). Again, those are the precursors with a rather inefficiently gating targeting signal plus inhibitory features in the mature part that require support for opening the channel (Fig. 4.5) (Haßdenteufel et al. 2018). Given that both ERj3 and prion protein were demonstrated to recruit the Sec62/63 complex to the translocon, suggests that BiP together with the luminal J-domain of the Hsp40 co-chaperone Sec63 forms a functional unit acting as allosteric effector of the Sec61 complex for substrate-specific channel opening (Conti et al. 2015). Simultaneously, the activity of BiP at the Sec61 complex is required for proper sealing of the Sec61 channel to prevent excessive calcium leakage (Schäuble et al. 2012). Interestingly, studies of the calcium permeability across the ER membrane provide an explanation of how the different activities of BiP at the Sec61 complex (opening for protein substrates; closing for calcium) might be connected to different co-chaperones. On the one hand, depletion of the Sec63 co-chaperone, i.e. a membrane-bound J-domain, caused a substrate-specific defect in protein transport without detectable impact on the Sec61-mediated calcium leakage (Lang et al. 2012; Schorr et al. 2015). On the other hand, depletion of the ER luminal, i.e. soluble, J-domain containing protein ERj3 or ERj6 caused a calcium specific phenotype which was attributed to inefficient sealing of the Sec61 channel (Schorr et al. 2015). This idea of BiP closing the Sec61 channel by direct interaction was further substantiated by single channel recordings from planar lipid bilayer experiments. Addition of a loop 7 antibody, binding as pseudo-ligand to loop 7, prevented ion currents through the translocon by shifting the equilibrium of the Sec61 channel to the closed state (Schorr et al. 2015). Taken together, functionality of the abundant calcium buffering chaperone BiP is linked to two metabolites, ATP and calcium. Therefore, BiP might be able to supervise and orchestrate ER homeostasis by integrating proteostasis, calcium balance, and, as will be discussed later, energy homeostasis of the ER (Lang et al. 2017).

Auxiliary Transport Components of the ER Membrane

Other auxiliary translocon components have been detected that reside in the ER membrane and modify the translocating polypeptide or foster its transport. As such, cleavable signal peptides from precursor proteins that intercalate at the lateral gate can be cleaved off by an intramembrane protease called the signal peptidase complex (Chen et al. 2001; Evans et al. 1986; Dalbey and von Heijne 1992). The existence of two enzymatically active subunits in this complex may hint at the possibility, in analogy to oligosaccharyltransferase, that there may actually be two signal peptidase paralogs with either SPC18 or SPC21 (Table 4.1). Chemical crosslinking suggested that the signal peptidase complex is transiently recruited to the translocon via an interaction with the Sec61β subunit and recruitment depended on the presence of membrane-bound ribosomes (Kalies et al. 1998). Signal peptides can be further processed by the signal peptide peptidase and translocating nascent chains can be modified by the GPI transamidase attaching a mixed lipid/sugar moiety called glycosylphosphatidylinositol (GPI) anchor to the C-terminus (Kapp et al. 2009; Lemberg and Martoglio 2002; Weihofen et al. 2002; Kamariah et al. 2011). Recently, attention was drawn to another ER membrane protein complex (EMC) that was first identified in yeast and later in human cells as heteromultimeric protein complex with 6 and 10 subunits, respectively (Jonikas et al. 2009; Christianson et al. 2011). Biochemical and cellular characterization of EMC characterized it as both stand-alone insertase for tail-anchored membrane proteins with a moderately hydrophobic transmembrane helix and as helper translocase likely in synergy with the Sec61 complex for the insertion of critical transmembrane helices of polytopic membrane proteins (Guna et al. 2017; Chitwood et al. 2018; Shurtleff et al. 2018). Besides those multimeric protein complexes other monomeric proteins have been shown to transiently contact or be in vicinity of the Sec61 complex (Table 4.1). Interestingly, it was deduced from sequence comparisons that TMCO1 is one of these proteins and, together with Get1 and EMC3, represents a remote Oxa1/Alg3/YidC homolog in the ER membrane (Anghel et al. 2017). In partial analogy with YidC, TMCO1 was found in association with both ribosomes and the Sec61 complex. It addition, it was found to be able to reversibly tetramerize and to restore calcium homeostasis upon calcium-overfilling of ER calcium stores (Wang et al. 2016). Using reconstituted proteoliposomes, TRAM was one of the first proteins found to provide substrate-specific assistance for the translocase (Görlich and Rapoport 1993). Follow-up studies demonstrated that precursor proteins with short charged N-terminal domains in their signal peptide require TRAM for efficient insertion into the lateral gate and that TRAM could regulate the cytosolic elusion of nascent chain domains into the gap between ribosome and translocon (Voigt et al. 1996; Hegde et al. 1998). Similar to Sec63 another J-domain containing membrane protein named ERj1 was demonstrated to associate with the Sec61 complex. ERj1 binds to ribosomes close to the exit tunnel and recruits BiP via the J-domain to both the Sec61 channel and incoming polypeptides (Dudek et al. 2005; Blau et al. 2005). However, the precise role of ERj1 in protein transport remains enigmatic. The dynamic recruitment of yet another auxiliary translocon component, RAMP4, can be triggered by transmembrane helices still buried within the ribosomal exit tunnel. Recruitment of RAMP4, also called stress-associated endoplasmic reticulum protein 1 (SERP1), is mediated by ribosomal protein uL22 which spans from the tunnel wall to the ribosomal surface. Thus, uL22 senses a transmembrane helix inside the ribosome and signals recruitment of RAMP4 to the translocon to prime it for subsequent transmembrane helix integration (Pool 2009; Yamaguchi et al. 1999). Also, bundles of transmembrane helices of certain polytopic membrane proteins were shown to reside in vicinity of the Sec61 complex for their collective release into the membrane during biogenesis (Cross and High 2009a, b; Sadlish et al. 2005; Ismail et al. 2008). In some cases, PAT-10 (protein associated with the ER translocon of 10 kDa) was found to interact in vicinity of the Sec61 complex and chaperone specific transmembrane helices of a polytopic membrane protein throughout its synthesis (Meacock et al. 2002; Ismail et al. 2008). To monitor arising problems with protein folding or transport early and right at the translocon the UPR sensor Ire1 directly interacts with the Sec61 complex generating a rendezvous point for surveillance, signaling and processing endeavors (Plumb et al. 2015). Ire1 is a membrane-anchored endonuclease that can cleave ribosome-engaged mRNAs after it is activated by misfolded proteins in the ER. To initiate UPR signaling Ire1 cleaves its key substrate XBP1u mRNA, which was targeted as ribosome-nascent chain complex to the translocon. And, Ire1 cleaves other ER destined mRNAs in a process identified as regulated Ire1-dependent decay (RIDD) to reduce the folding and synthesis burden of the ER under stress conditions (Sundaram et al. 2017; Yanagitani et al. 2011; Hollien et al. 2009; Hollien and Weissman 2006). In addition to the signaling and processing mode of Ire1, its close interaction with the RTC allows it to act in a surveillance mode to cleave mRNAs whose translation products show signs of mis-folding in the ER (Acosta-Alvear et al. 2018). Connection between the Sec61 complex and the Ire1 branch of the UPR was also demonstrated on a genetic level using a multiplexed, genome scale CRISPR screening showing a reciprocal feedback loop between the two. During ER stress, subunits of the Sec61 complex were exclusively transcribed in response to activation of the Ire1 branch of the UPR. In turn, the loss of Sec61 complex subunits was compensated by selective activation of Ire1 signaling (Adamson et al. 2016).

Additional Allosteric Effectors in the Cytosol Interacting with the Sec61 Complex

Multiple cytosolically located proteins have been identified to interact with the Sec61 complex. Many of those are part of coexisting targeting machineries that are composed of cytosolic and membrane-bound components. These targeting pathways delivering precursor proteins to the translocon will be discussed later. Here, we will focus on calmodulin (Fig. 4.5), the ubiquitous calcium binding protein of the cytosol involved in different second messenger systems and the regulation of ion channels and other pivotal proteins (Chin and Means 2000). Some binding sites for calmodulin are called IQ-motifs to which calmodulin binds either in its apo- (calcium-free) or holo-form (Bähler and Rhoads 2002; Tidow and Nissen 2013). Such an IQ-motif was identified in the cytosolic N-terminus of mammalian Sec61α (Erdmann et al. 2011). A series of protein-protein interaction studies, planar lipid bilayer recordings, molecular modeling and live cell calcium imaging demonstrated the calcium-dependent binding of calmodulin to this IQ-motif to limit the Sec61-mediated calcium efflux during protein translocation. The introduction of charge deleting mutations masking the IQ-motif or the use of calmodulin inhibitors caused dysregulated calcium permeability of the Sec61 complex (Erdmann et al. 2011; Harsman et al. 2011b; Lang et al. 2011b). Interestingly, binding of calcium-calmodulin to the translocon seemed not to interfere with the protein transport activity of the Sec61 complex and goes in line with the observed targeting function of calmodulin directing small precursors to the Sec61 complex (Shao and Hegde 2011; Erdmann et al. 2011). Thus, the N-terminus of Se61α serves as calmodulin docking site for the transport of certain precursors and regulation of calcium permeability without interference for the ribosome binding and formation of the RTC.

Small Molecules Directly Interfering with the Sec61 Complex

As discussed above, the efficient gating of the Sec61 channel is of crucial importance for its role in ER protein import and its potentially harmful role for calcium-homeostasis. In recent years a growing number of Sec61 channel inhibitors was identified, which to us are best discussed in light of the energetics and kinetics of Sec61 channel gating (Fig. 4.6). According to this point of view, inhibitor selectivity is based on the distinct efficiencies of different signal peptides in reducing the activation energy for Sec61 channel opening and the common principle that the bound inhibitors or ions may increase the energy barrier for opening of the Sec61 channel (Fig. 4.6). This view is supported by the observation that the ER import of the BiP- and Sec63-dependent preproapelin is sensitive to CAM741 (Haßdenteufel et al. 2018).

This common principle is e.g. demonstrated by the effect of Lanthanum ions on the channel. Binding of several Lanthanum ions to the Sec61 complex arrests the channel in the open state, restricts the Sec61 channel dynamics, and inhibits translocation of polypeptides. Molecular modeling indicated that Lanthanum binding sites cluster at the lateral gate (Erdmann et al. 2009). In addition, several structurally unrelated small molecules have non-identical binding sites in the Sec61 complex and also inhibit the Sec61 channel with respect to ER protein import (Fig. 4.3). The first described class of Sec61 inhibitors were the cyclic heptadepsipeptides, i.e. CAM741 and cotransins (such as CT8), which inhibit translocation of polypeptides in a precursor-specific manner (Garrison et al. 2005). Next, the natural compounds Apratoxin A and Mycolactone were characterized as Sec61 inhibitors and shown to have selective (Mycolactone) or non-selective (Apratoxin A) effects on ER protein import (Paatero et al. 2016; Baron et al. 2016). The binding sites of these Sec61 channel inhibitors were identified by clever strategies, which selected inhibitor resistant human cell lines. According to the analysis by MacKinnon et al., who obtained five mutations that showed resistance to CT8, four of which were in the plug and downstream region (R66I, R66G, G80V, and S82P), CT8 interacts with the short plug helix in loop 1 and transmembrane domains 2 and 3 of Sec61α (Fig. 4.3) (MacKinnon et al. 2014). In case of Mycolactone, the mutagenesis studies also identified the plug residues R66 and S82 in loop 1 as interaction site (Fig. 4.3) (Baron et al. 2016; McKenna et al. 2017). In contrast, the respective studies on Apratoxin A identified T86 and Y131 as binding site and indicated a distinct binding mode (Fig. 4.3) (Paatero et al. 2016). This is consistent with the effects of yet another class of Sec61 channel inhibitors, i.e. the Eeyarestatins (ES1, ES24), where there are no mutagenesis studies available as of yet, on calcium permeability of the channel. Here, binding of the inhibitor within the channel pore arrests the channel in a partially open state (termed 'foot in the door'), which may be identical with the primed state and is compatible with calcium efflux but not with full channel opening for protein translocation (Gamayun et al. 2019). We note, however, that the Sec61 channel is also affected by a bacterial protein toxin, *Pseudomonas aeruginosa* Exotoxin A, which enters human cells by retrograde transport and inhibits ER export of immunogenic peptides. Exotoxin A binds near the calcium-calmodulin binding site to the N-terminal tail of Sec61α and arrests the channel in the closed state (Schäuble et al. 2014).

Modalities of Precursor Targeting Factors Delivering Substrates to the Translocon

As outlined above, membrane proteins of the secretory pathway are integrated into the ER membrane by either the Sec61-channel, the tail-anchored (TA) receptor, or by PEX3, and, possibly also by TMCO1 or the SND receptor (Table 4.1). Before a precursor polypeptide can be membrane integrated or fully translocated by the polypeptide conducting Sec61 channel, however, its respective mRNA or the precursor itself has to be specifically delivered, i.e. targeted to the Sec61 complex in the ER membrane (Fig. 4.3). Current knowledge about mRNA targeting to the ER membrane is scarce (see below). In contrast, the following detailed concept emerged for ER protein targeting: Apparently, a molecular triage operates for ER-destined precursor polypeptides during their synthesis on ribosomes in the cytosol, which determines the fates of nascent or fully synthesized but not yet folded polypeptides by the complex network of targeting signals in nascent chains or completed polypeptides and a whole variety of cytosolic factors (SND1, SRP, TRC40) (Table 4.1), which recognize these signals and have overlapping specificities. In addition, these factors chaperone the precursors for staying in solution and competent for ER targeting as well as subsequent membrane insertion into or translocation across the ER membrane. The common principle seems to be that the cytosolic factors in complex with their clients interact with heterodimeric receptors on the ER surface, which are associated with or in the neighborhood of Sec61 complexes. The respective receptors are termed Snd receptor (comprising Snd2 and Snd3), SRP receptor (comprising SRα and SRβ), or TA receptor (comprising WRB and CAML) (Table 4.1). In addition, there may be direct targeting of fully synthesized precursor polypeptides to the Sec62 protein in the ER membrane.

Targeting of Precursor Polypeptides to the Sec61 Complex in the ER Membrane

The original concept for targeting of precursor polypeptides to the ER was formulated in 1971 by G. Blobel and colleagues in the signal hypothesis (Blobel 1980). Accordingly, the N-terminal signal peptide of a nascent presecretory protein is recognized and bound by SRP at the ribosomal tunnel exit and mediates a translational attenuation. Next, the respective ribosome-nascent chain-SRP complex associates with the ER membrane via the heterodimeric SRP receptor (SR), which is membrane-anchored via the β-subunit (Blobel and Dobberstein 1975a, b; Gilmore et al. 1982a, b). Contact between SRP and SR drives the mutual hydrolysis of bound GTP and leads to hand-over of the ribosome-nascent chain complex to the Sec61 complex (Voorhees and Hegde 2015; Halic and Beckmann 2005; Egea et al. 2004; Halic et al. 2004; Jomaa et al. 2017). Thus, by definition, SRP also represents a, albeit ribosome-dependent, mRNA targeting device. This latter concept of SRP acting eventually as precursor-mRNA dependent particle was addressed in yeast by a combination of ribosome profiling and biochemical fractionation of membrane-attached and soluble ribosome populations. The genome-wide analysis of ribosome footprints (mRNA snippets protected by the ribosome during nuclease treatment) showed that non-coding mRNA elements of the 3′ UTR promote recruitment of SRP even before the encoded targeting signal is synthesized. However, such SRP loading motifs of the 3′ UTR alone were insufficient to direct translocation of a substrate into the ER (Chartron et al. 2016; Ingolia 2016). Two more studies employing comparative ribosome profiling strategies addressed functionality of the bacterial and yeast SRP in vivo. They highlighted the strong preference of SRP for transmembrane helices as “SRP recognon” regardless of their position relative to the N-terminus and, most surprisingly, the efficient targeting of precursors with just cleavable signal peptides in absence of SRP (Costa et al. 2018; Schibich et al. 2016). Rather than being at odds with the current SRP targeting dogma, in our eyes, those studies stretch the versatility of SRP and reconcile two important considerations. First, the lower abundance of the SRP compared to translating mono- or polysomes can be overcome by an mRNA dependent pre-recruitment step, probably stretching the time-window for the target recognition by SRP. Second, the crowded environment at the ribosomal tunnel exit sieged by many factors with competing function (reviewed in Pfeffer et al. 2016) can be eased by multiple iterations for SRP recognition not limited to recognition of the first transmembrane helix. In the late 1980s, identification of precursor proteins with the ability for SRP-independent ER targeting, such as small presecretory proteins in mammalian cells (many of which act as hormones in intercellular signaling or as antibacterial proteins in the immune sytem), TA-membrane proteins in mammalian and yeasts cells, and, GPI-anchored membrane proteins in yeast, suggested alternative ER targeting machineries (Yabal et al. 2003; Shao and Hegde 2011; Schlenstedt et al. 1990; Ast et al. 2013; Hann and Walter 1991). TA proteins are defined as single spanning membrane proteins with a characteristic C-terminally located transmembrane helix (Fig. 4.1) (Kutay et al. 1993). Approximately 1% of the human genome code for TA proteins. However, not all of these have their functional locations in the secretory pathway (Kalbfleisch et al. 2007; Borgese and Righi 2010). TA proteins of the secretory pathway, such as the γ- and β-subunits of the Sec61 complex, the redox protein Cytochrome b_5, many apoptosis-associated proteins (such as the Bcl family members) and many vesicular transport components (such as Syntaxins and VAMPs), have to be targeted and inserted into the ER membrane (Borgese and Fasana 2011). Similar to SRP-mediated targeting, TA proteins are directed to the ER membrane via an ER membrane resident receptor complex. The minimal targeting machinery for TA proteins was termed transmembrane domain recognition complex (TRC) in mammalian cells (Table 4.1). The cytosolic ATPase TRC40 (also termed Asna1) with its hydrophobic binding pocket binds the TA protein and the heterodimeric receptor complex facilitates efficient ER targeting. The receptor may also facilitate the actual membrane insertion (Stefanovic and Hegde 2007; Schuldiner et al. 2008; Vilardi et al. 2011; Yamamoto and Sakisaka 2012). In addition, the TA targeting machinery involves a ribosome binding heterotrimeric complex (comprising Bag6, Ubl4A, and TRC35), which appears to act upstream of TRC40 (Mariappan et al. 2010).

Although about one dozen genes coding for yeast TA proteins were characterized as essential, knockout strains in the TA targeting pathway are viable, suggesting at least one additional route (Schuldiner et al. 2008). Indeed, in 2016 a high-throughput screening approach performed by the lab of Maya Schuldiner identified a novel targeting pathway in yeast, termed SRP-independent (SND) (Aviram et al. 2016). Three novel components were identified, characterized and termed Snd1, Snd2, and Snd3 (Table 4.1). Two hallmarks of the SND targeting pathway were described. First, similar to the SRP and TA targeting pathways, precursor polypeptides were targeted via the interplay of a cytosolic factor (termed Snd1) and a heterodimeric receptor located at the ER membrane (termed Snd2 and Snd3). Interestingly, Snd1 had previously been described as a ribosome-binding protein. Second, the SND pathway showed a preference for clients with a central transmembrane domain. In addition, the SND route was able to provide an alternative targeting pathway for substrates with a transmembrane helix at their extreme N- or C-terminus (Aviram et al. 2016). Sequence comparisons identified the previously described ER membrane protein TMEM208 as putative human Snd2 ortholog, which was termed hSnd2 (Zhao et al. 2013; Aviram et al. 2016). According to experiments, combining siRNA-mediated gene silencing and protein transport into the ER of semi-permeabilized human cells in cell-free transport assays, hSnd2 appears to have the same or at least a similar function as its yeast ortholog (Haßdenteufel et al. 2017; Casson et al. 2017). The TA membrane protein Cytochrome b_5 as well as some small presecretory proteins can be targeted to the ER or even the Sec61 complex in the mammalian cell-free assay. In brief, the human hormone precursor proteins preproapelin and prestatherin can use Sec62 as well as SR for ER targeting in the cell-free assay, which does not necessarily mean they actually do so in living cells (see below). Although smaller in overall size, prestatherin preferred SRα over Sec62-mediated targeting, whereas preproapelin did the opposite, which may be related to the higher hydrophobicity of the pre-statherin signal peptide (ΔG^{pred} −0.91 vs. −0.19). Taken together with our observation that C-terminal extension of preproapelin or prestatherin by the dihydrofolate reductase (i.e. by 187 amino acid residues) leads to Sec62 independence, our data support the hypothesis that small presecretory proteins use the SRP pathway for Sec61 targeting in human cells less efficiently, simply because the corresponding nascent chains are more likely released from ribosomes before SRP can efficiently interact (Haßdenteufel et al. 2018; Lakkaraju et al. 2012b; Schlenstedt et al. 1990). Therefore, these precursors have to use alternative targeting pathways. Notably, in yeast, low hydrophobicity of signal peptides and C-terminal signals for the attachment of GPI-anchors preclude effective use of SRP and, therefore, cause Sec62p- and TA-dependence (Aviram and Schuldiner 2014). In addition to SR and Sec62, co- and post-translational targeting of preproapelin and prestatherin can also involve both the TRC system and the recently identified SND pathway, albeit to different degrees for the two different precursors (Fig. 4.3) (Haßdenteufel et al. 2018). However, orthologs of Snd1 and Snd3 have not yet been characterized in the mammalian system and are subject of our current research.

Furthermore, some small model presecretory proteins were shown to be targeted to the mammalian ER membrane in an SRP-independent fashion by their interaction with the cytosolic protein calcium-calmodulin and its putative association with the calcium-calmodulin-binding site in the cytosolic N-terminus of the Sec61α protein (Fig. 4.3) (Shao and Hegde 2011). In terms of interconnections between pathways, it is interesting to note that calmodulin was found to inhibit rather than stimulate targeting of TA proteins to the ER membrane (Haßdenteufel et al. 2011).

Targeting of mRNAs to the ER Membrane

Apparently, the synthesis of many polypeptides can be initiated on ribosomes or large ribosomal subunits that are continuously attached to the ER-membrane (Potter et al. 2001). In these cases, the above discussed targeting pathways for precursor polypeptides may not be required for membrane insertion or translocation of the translation products by the Sec61 channel. Instead, mRNA targeting was suggested as an alternative ER targeting mechanism and the proteins RRBP1 (also termed p180) and kinectin 1 (KTN1) were suggested as possible mRNA receptors in the ER membrane (Table 4.1) (Savitz and Meyer 1990, 1993; Cui et al. 2012, 2013; Dejgaard et al. 2010; Morrow and Brodsky 2001; Ueno et al. 2010, 2011). So far, however, there is no clue about the possible specificity of this targeting reaction and to the best of our knowledge there is only a single example of a precursor polypeptide (Sec61ß), where mRNA targeting was found to be involved in subsequent Sec61-independent membrane insertion (Voigt et al. 2017). In contrast, polypeptides that lack a signal peptide for ER-targeting and whose synthesis was initiated on ER-bound ribosomes or large ribosomal subunits were shown to be recognized by the nascent chain associated complex (NAC). Apparently, this interaction leads to release of the respective ribosomes from the ER membrane and completion of protein synthesis proceeds in the cytosol (Möller et al. 1998; Gamerdinger et al. 2015).

Additional Putative Functions of the Human Sec61 Channel

The mammalian Sec61 complex forms a dynamic and precursor gated channel, which can provide an aqueous path for polypeptides into the ER lumen and is regulated by various allosteric effectors (Fig. 4.5). When the aqueous path is open, it can apparently also provide a pore for efflux of calcium from the ER lumen into the cytosol. We suggest that this Sec61 feature is physiologically linked to the regulation of ATP import into the ER and the initiation of the intrinsic pathway to apoptosis, respectively. Furthermore, it is pathophysiologically linked to various human diseases, which we termed Sec61-channelopathies (Linxweiler et al. 2017; Haßdenteufel et al. 2014).

As outlined above, the ER of nucleated mammalian cells depends on an Hsp70-type molecular chaperone, termed BiP. BiP is present in the ER lumen in millimolar concentration and requires ATP for its action. Moreover, ATP hydrolysis by BiP results in ADP and, therefore, necessitates ADP removal from the ER. Until recently, mammalian proteins catalyzing the ATP uptake and the concomitant ADP release remained unknown on the molecular level. Screening databases for solute carriers (SLCs) that are located in the ER membrane, drew our attention to SLC35B1, which is predicted to have ten transmembrane domains (Fig. 4.10). Heterologously expressed SLC35B1 was found to be highly specific for ATP and ADP and to operate in antiport mode, to name just two of four characteristics it shares with the ATP transport activity, present in rough ER membranes. In addition, siRNA-mediated depletion of SLC35B1 from HeLa cells was found to reduce ER ATP levels and, therefore, BiP activity. Together these findings implied that SLC35B1 mediates ATP uptake into the ER plus ADP release from the ER in living cells. Therefore, SLC35B1 was named AXER, ATP/ADP exchanger of the ER membrane (Klein et al. 2018). According to a hypothetical structural model, human AXER can be expected to catalyze the equimolar exchange of adenosine di- and triphosphates by an alternating access mechanism, in which a single substrate binding site is made available either to the cytosolic ER surface or the ER lumen through conformational changes (Fig. 4.10). In human cells, AXER appears to be part of a regulatory circuit and a calcium-dependent signaling pathway, termed low energy response (lowER), acting in the vicinity of the ER and supplying sufficient ATP to the ER (Fig. 4.10). We suggest the following scenario for lowER: High ATP/ADP ratio in the ER allows BiP to limit calcium leakage from the ER via the Sec61 channel. Low ATP/ADP ratio due to increased protein import and folding or due to protein misfolding, leads to BiP dissociation from the Sec61 channel and, therefore, induces calcium leakage from the ER. In the cytosol, calcium binds to calmodulin (CaM) near the ER surface and activates AMP-activated protein kinase

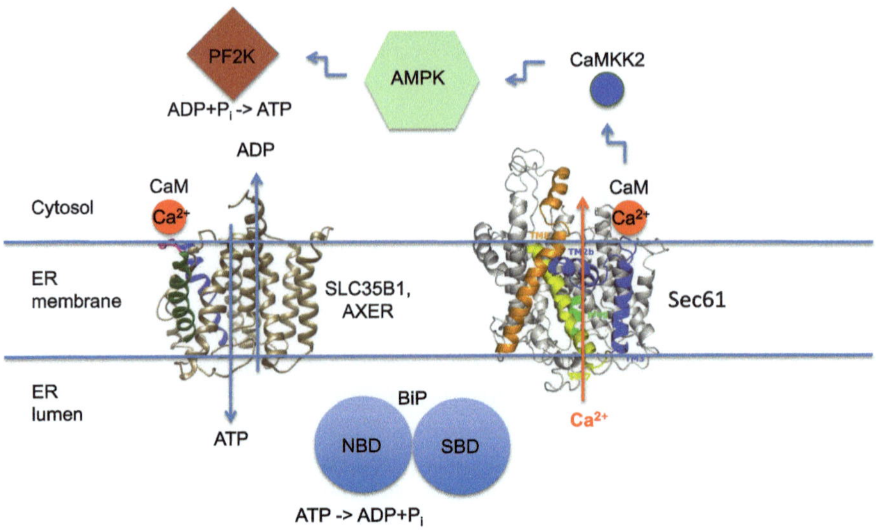

Fig. 4.10 ER low energy response (lowER) ensures a sufficient ATP supply of the mammalian ER. The given atomic structures of AXER (systematically termed SLC35B1) and Sec61 complex were derived from structure predictions. AXER, ATP/ADP exchanger in the ER membrane; AMPK, AMP-activated protein kinase; CaM, Calmodulin; CaMKK2, CaM kinase kinase 2; NBD, nucleotide binding domain of BiP; PF2K, 6-phospho-fructo-2-kinase; P_i, inorganic phosphate; SBD, substrate binding domain of BiP. See text for details

(AMPK), which in turn activates calcium-CaM kinase kinase 2 (CAMKK2) to eventually activate 6-phospho-fructo-2-kinase (PF2K). Activated PF2K causes increased ADP phosphorylation in glycolysis. The latter leads to ATP import into the ER via AXER, which is also activated by calcium efflux from the ER. Interestingly, mammalian AXER comprises an IQ motif in the cytosolic loop between transmembrane domains 2 and 3 (Fig. 4.10) and, thus, may also be activated by calcium-CaM. Normalization of the ER ATP/ADP ratio causes BiP to limit the calcium leakage and thus inactivates the signal transduction pathway. Sarcoplasmic/endoplasmic reticulum calcium ATPase (SERCA), which pumps calcium back into the ER lumen, balances the passive calcium efflux and protein phosphatase 2 (PP2) dephosphorylates AMPK. We note that activated AMPK was shown previously to lead to reduced cap-dependent translation and therefore ties the lowER to the unfolded protein response (UPR). While ADP is exported via AXER, phosphate may leave the ER via the Sec61 channel. Thus, under non-physiological or patho-physiological conditions, lowER can be expected to represent the first line of defense of a cell against ER stress. However, the details of this novel calcium-dependent signaling cascade from ER to cytosol remain to be worked out. Furthermore, future work will have to address the question if the mammalian ER membrane harbors additional ATP carriers. Additional open questions are discussed next.

Open Question

In all eukaryotic cells, nascent proteins, which are destined to membranes or the lumen of organelles of the endo- and exocytotic pathways or even the extracellular space, must be translocated across or integrated into the ER membrane. In mammals, most proteins are translocated concomitantly with their synthesis by cytosolic ribosomes (co-translationally), whereas many small presecretory proteins, which are essential for intercellular communication or pathogen defense, are post-translationally imported. A consensus on the major components of the ER-translocation machinery, almost 100 different proteins, is established, but their precise functions, as well as their spatial and temporal organization still remains largely elusive (Table 4.1).

Therefore the authors of this review, are trying to characterize the mammalian machinery for the co- and post-translational translocation of polypeptides into the ER and their accompanying covalent modifications in terms of composition, structure, as well as component functions and mechanisms. In addition, the obtained structural and mechanistic insights into the ER-translocation machinery are expected to provide a detailed understanding of the etiology of several human diseases and may even guide us to novel therapeutic strategies.

Systematic Knock Down of ER-Protein Translocation Machinery Components in Human Cells Combined with Characterization of Substrate Precursor Proteins and Compensatory Mechanisms by Quantitative Proteomic Analysis

Traditionally, the substrate specificities of mammalian protein transport components have been investigated in cell-free translation reactions in which a small set of model precursor proteins is synthesized one-by-one in the presence of reconstituted- or HeLa cell derived-ER membranes or in pulse/chase experiments in human cells that overproduce the model precursor of interest (Sharma et al. 2010; Dudek et al. 2015). These traditional approaches are suitable for addressing whether a certain component can stimulate ER import of a given precursor polypeptide. However, due to the bias of these experimental strategies, they fail to clearly define the characteristics of precursor polypeptides that lead to a certain dependence under physiological conditions. Therefore, we established a novel unbiased approach, which involves treatment of HeLa cells with either one of two different targeting siRNAs or a non-targeting siRNA, label-free quantitative proteomic analysis, and differential protein abundance analysis. As a proof of concept, HeLa cells were depleted of the Sec61-complex using two different *SEC61A1*-targeting siRNAs. We assessed the proteomic consequences of this knock-down via label-free quantitative proteomics and differential protein abundance analysis relative to cells treated with non-targeting siRNA. Roughly, 50% of the HeLa cell proteome was quantified in these experiments. Our experimental strategy was successfully used to analyze the client spectrum of the Sec61-complex, an essential transport component. These results set the stage for subsequent analysis of precursor-specific auxiliary transport components, such as the TRAP-complex. As discussed above, signal peptide analysis of the TRAP clients revealed above-average glycine-plus-proline content as the distinguishing feature for TRAP dependence and, thus, suggested a hitherto undetected signal peptide feature and heterogeneity (Nguyen et al. 2018). Therefore, we proposed that this signal peptide heterogeneity may provide an opportunity for regulation of transport of a subset of precursor polypeptides and may be linked to both TRAP mechanism and the etiology of TRAP-linked congenital disorder of glycosylation in human patients. Since TRAPα was found to be subject to phosphorylation and calcium-binding, this phenomenon may also provide a potential regulatory mechanism for the TRAP-dependent subset of precursors. At present, we are analyzing the results from similar experiments on another auxiliary transport component, i.e. the Sec62/Sec63-complex. Notably, this complex is also affected by phosphorylation (the Sec63 subunit) and binding of calcium (the Sec62 subunit) and, therefore, may provide a similar opportunity for transport regulation for another set of precursor polypeptides. As noted above, our novel approach also identified several genetic interactions between targeting- and Sec61 channel gating-pathways (Fig. 4.9), which may eventually pave the way towards understanding yet another layer of regulatory phenomena.

Integrative Determination of the Molecular Architecture of the Native ER Translocon Core Complexes

Cryo-electron tomography (CET) in conjunction with subtomogram averaging provided a three-dimensional map firstly of the core of the native co-translational translocation machinery. Next, we want to explore Sec62/Sec63-dependent import structurally in situ using CET. Here, our approach is based on the recent publication by the Bill Skach lab (Conti et al. 2015). The respective publication had followed our work on the import requirements of the precursors of ERj3 and prion protein (Lang et al. 2012; Schäuble et al. 2012) and defined in a sophisticated combination of nascent precursor polypeptide chains, cross-linking, and 2D gel electrophoresis, when these precursor polypeptides recruit Sec62 and Sec63 to the Sec61-complex. Apparently, they do so at a comparatively late stage of chain growth, possibly when a polybasic motif within the mature part of the precursor enters the Sec61-channel. We will use these particular nascent polypeptide chains for CET under the established conditions.

References

Acosta-Alvear D, Karagöz GE, Fröhlich F, Li H, Walther TC, Walter P (2018) The unfolded protein response and endoplasmic reticulum protein targeting machineries converge on the stress sensor IRE1. eLife 7:e43036. https://doi.org/10.7554/elife.43036

Adamson B, Norman TM, Jost M, Cho MY, Nuñez JK, Chen Y, Villalta JE, Gilbert LA, Horlbeck MA, Hein MY, Pak RA, Gray AN, Gross CA, Dixit A, Parnas O, Regev A, Weissman JS (2016) A multiplexed single-cell CRISPR screening platform enables systematic dissection of the unfolded protein response. Cell 167(7):1867–1882.e1821. https://doi.org/10.1016/j.cell.2016.11.048

Akimaru J, Matsuyama S-i, Tokuda H, Mizushima S (1991) Reconstitution of a Protein translocation system containing purified SecY, SecE, and SecA from *Escherichia coli*. Proc Natl Acad Sci USA 88:6545–6549

Anghel SA, McGilvray PT, Hegde RS, Keenan RJ (2017) Identification of Oxa1 homologs operating in the eukaryotic endoplasmic reticulum. Cell Reports 21(13):3708–3716. https://doi.org/10.1016/j.celrep.2017.12.006

Ashby MC, Tepikin AV (2001) ER calcium and the functions of intracellular organelles. Semin Cell Dev Biol 12(1):11–17

Ast T, Cohen G, Schuldiner M (2013) A network of cytosolic factors targets SRP-independent proteins to the endoplasmic reticulum. Cell 152(5):1134–1145. https://doi.org/10.1016/j.cell.2013.02.003

Ast T, Michaelis S, Schuldiner M (2016) The protease Ste24 clears clogged translocons. Cell 164 (1):103–114. https://doi.org/10.1016/j.cell.2015.11.053

Auer J, Spicker G, Böck A (1991) Presence of a gene in the archaebacterium *Methanococcus vannielii* homologous to secY of eubacteria. Biochimie 73(6):683–688

Aviram N, Schuldiner M (2014) Embracing the void—how much do we really know about targeting and translocation to the endoplasmic reticulum? Curr Opin Cell Biol 29:8–17. https://doi.org/10.1016/j.ceb.2014.02.004

Aviram N, Ast T, Costa EA, Arakel EC, Chuartzman SG, Jan CH, Haßdenteufel S, Dudek J, Jung M, Schorr S, Zimmermann R, Schwappach B, Weissman JS, Schuldiner M (2016) The SND proteins constitute an alternative targeting route to the endoplasmic reticulum. Nature 540(7631):134–138. https://doi.org/10.1038/nature20169. http://www.nature.com/nature/journal/v540/n7631/abs/nature20169.html#supplementary-information

Bähler M, Rhoads A (2002) Calmodulin signaling via the IQ motif. FEBS Lett 513(1):107–113. https://doi.org/10.1016/S0014-5793(01)03239-2

Bai L, Wang T, Zhao G, Kovach A, Li H (2018) The atomic structure of a eukaryotic oligosaccharyltransferase complex. Nature. https://doi.org/10.1038/nature25755. https://www.nature.com/articles/nature25755#supplementary-information

Baker JA, Wong W-C, Eisenhaber B, Warwicker J, Eisenhaber F (2017) Charged residues next to transmembrane regions revisited: "positive-inside rule" is complemented by the "negative inside depletion/outside enrichment rule". BMC Biol 15(1):66. https://doi.org/10.1186/s12915-017-0404-4

Bañó-Polo M, Baldin F, Tamborero S, Marti-Renom MA, Mingarro I (2011) N-glycosylation efficiency is determined by the distance to the C-terminus and the amino acid preceding an Asn-Ser-Thr sequon. Protein Sci: Publ Protein Soc 20(1):179–186. https://doi.org/10.1002/pro.551

Bañó-Polo M, Martínez-Garay CA, Grau B, Martínez-Gil L, Mingarro I (2017) Membrane insertion and topology of the translocon-associated protein (TRAP) gamma subunit. Biochim Biophys Acta (BBA)-Biomembr 1859(5):903–909. http://dx.doi.org/10.1016/j.bbamem.2017.01.027

Baron L, Paatero AO, Morel J-D, Impens F, Guenin-Macé L, Saint-Auret S, Blanchard N, Dillmann R, Niang F, Pellegrini S, Taunton J, Paavilainen VO, Demangel C (2016)

Mycolactone subverts immunity by selectively blocking the Sec61 translocon. J Exp Med 213 (13):2885–2896. https://doi.org/10.1084/jem.20160662
Baumann O, Walz B (2001) Endoplasmic reticulum of animal cells and its organization into structural and functional domains. In: International review of cytology, vol 205. Academic Press, pp 149–214
Becker T, Bhushan S, Jarasch A, Armache JP, Funes S, Jossinet F, Gumbart J, Mielke T, Berninghausen O, Schulten K, Westhof E, Gilmore R, Mandon EC, Beckmann R (2009) Structure of monomeric yeast and mammalian Sec61 complexes interacting with the translating ribosome. Science 326(5958):1369–1373
Benedix J, Lajoie P, Jaiswal H, Burgard C, Greiner M, Zimmermann R, Rospert S, Snapp EL, Dudek J (2010) BiP modulates the affinity of its co-chaperone ERj1 for ribosomes. J Biol Chem 285(47):36427–36433
Berridge MJ (2002) The endoplasmic reticulum: a multifunctional signaling organelle. Cell Calcium 32(5–6):235–249
Berridge MJ, Bootman MD, Roderick HL (2003) Calcium signalling: dynamics, homeostasis and remodelling. Nat Rev Mol Cell Biol 4(7):517–529
Bischoff L, Wickles S, Berninghausen O, van der Sluis EO, Beckmann R (2014) Visualization of a polytopic membrane protein during SecY-mediated membrane insertion. Nat Commun 5:4103. https://doi.org/10.1038/ncomms5103. https://www.nature.com/articles/ncomms5103#supplementary-information
Blau M, Mullapudi S, Becker T, Dudek J, Zimmermann R, Penczek PA, Beckmann R (2005) ERj1p uses a universal ribosomal adaptor site to coordinate the 80S ribosome at the membrane. Nat Struct Mol Biol 12(11):1015–1016
Blobel G (1980) Intracellular protein topogenesis. Proc Natl Acad Sci USA 77(3):1496–1500
Blobel G, Dobberstein B (1975a) Transfer of proteins across membranes. I. Presence of proteolytically processed and unprocessed nascent immunoglobulin light chains on membrane-bound ribosomes of murine myeloma. J Cell Biol 67(3):835–851
Blobel G, Dobberstein B (1975b) Transfer of proteins across membranes. II. Reconstitution of functional rough microsomes from heterologous components. J Cell Biol 67(3):852–862
Borgese N, Fasana E (2011) Targeting pathways of C-tail-anchored proteins. Biochem Biophys Acta 1808(3):937–946
Borgese N, Righi M (2010) Remote origins of tail-anchored proteins. Traffic 11(7):877–885
Borgese N, Brambillasca S, Righi M, Colombo S (2009) Membrane insertion of tail-anchored proteins. In: Zimmermann R (ed) Protein transport into the endoplasmic reticulum. Landes Bioscience, pp 91–101
Braunger K, Pfeffer S, Shrimal S, Gilmore R, Berninghausen O, Mandon EC, Becker T, Förster F, Beckmann R (2018) Structural basis for coupling protein transport and N-glycosylation at the mammalian endoplasmic reticulum. Science 360(6385):215–219. https://doi.org/10.1126/science.aar7899
Brodsky JL, Scheckman R (1993) A Sec63-BiP complex is required for protein translocation in a reconstituted proteoliposome. J Cell Biol 123:1355–1263
Brodsky JL, Goeckeler J, Schekman R (1995) BiP and Sec63p are required for both co- and posttranslational protein translocation into the endoplasmic reticulum. Proc Natl Acad Sci 92 (21):9643–9646
Brostrom MA, Brostrom CO (2003) Calcium dynamics and endoplasmic reticular function in the regulation of protein synthesis: implications for cell growth and adaptability. Cell Calcium 34 (4–5):345–363
Brundage L, Hendrick JP, Schiebel E, Driessen AJM, Wickner W (1990) The purified *E. coli* integral membrane protein SecY/E is sufficient for reconstitution of SecA-dependent precursor protein translocation. Cell 62:649–657
Calo D, Eichler J (2011) Crossing the membrane in Archaea, the third domain of life. Biochem Biophys Acta 1808(3):885–891
Cao TB, Saier MH Jr (2003) The general protein secretory pathway: phylogenetic analyses leading to evolutionary conclusions. Biochem Biophys Acta 1609(1):115–125

Casson J, McKenna M, Haßdenteufel S, Aviram N, Zimmerman R, High S (2017) Multiple pathways facilitate the biogenesis of mammalian tail-anchored proteins. J Cell Sci 130 (22):3851–3861. https://doi.org/10.1242/jcs.207829

Chartron JW, Hunt KCL, Frydman J (2016) Cotranslational signal-independent SRP preloading during membrane targeting. Nature 536(7615):224–228. https://doi.org/10.1038/nature19309

Chatzi KE, Sardis MF, Tsirigotaki A, Koukaki M, Šoštarić N, Konijnenberg A, Sobott F, Kalodimos CG, Karamanou S, Economou A (2017) Preprotein mature domains contain translocase targeting signals that are essential for secretion. J Cell Biol. https://doi.org/10.1083/jcb.201609022

Chen X, VanValkenburgh C, Liang H, Fang H, Green N (2001) Signal peptidase and oligosaccharyltransferase interact in a sequential and dependent manner within the endoplasmic reticulum. J Biol Chem 276(4):2411–2416. https://doi.org/10.1074/jbc.M007723200

Cheng Z, Jiang Y, Mandon EC, Gilmore R (2005) Identification of cytoplasmic residues of Sec61p involved in ribosome binding and cotranslational translocation. J Cell Biol 168(1):67–77

Cherepanova N, Shrimal S, Gilmore R (2016) N-linked glycosylation and homeostasis of the endoplasmic reticulum. Curr Opin Cell Biol 41:57–65. https://doi.org/10.1016/j.ceb.2016.03.021

Chevet E, Wong HN, Gerber D, Cochet C, Fazel A, Cameron PH, Gushue JN, Thomas DY, Bergeron JJM (1999) Phosphorylation by CK2 and MAPK enhances calnexin association with ribosomes. EMBO J 18(13):3655–3666

Chin D, Means AR (2000) Calmodulin: a prototypical calcium sensor. Trends Cell Biol 10 (8):322–328. https://doi.org/10.1016/s0962-8924(00)01800-6

Chitwood PJ, Juszkiewicz S, Guna A, Shao S, Hegde RS (2018) EMC is required to initiate accurate membrane protein topogenesis. Cell 175(6):1507–1519.e1516. https://doi.org/10.1016/j.cell.2018.10.009

Christianson JC, Olzmann JA, Shaler TA, Sowa ME, Bennett EJ, Richter CM, Tyler RE, Greenblatt EJ, Harper JW, Kopito RR (2011) Defining human ERAD networks through an integrative mapping strategy. Nat Cell Biol 14(1):93–105. https://doi.org/10.1038/ncb2383

Clapham DE (2007) Calcium signaling. Cell 131(6):1047–1058

Coe H, Michalak M (2009) Calcium binding chaperones of the endoplasmic reticulum. Gen Physiol Biophys 28(Focus Issue):F96–F103

Conti BJ, Devaraneni PK, Yang Z, David LL, Skach WR (2015) Cotranslational stabilization of Sec62/63 within the ER Sec61 translocon is controlled by distinct substrate-driven translocation events. Mol Cell 58(2):269–283. https://doi.org/10.1016/j.molcel.2015.02.018

Costa EA, Subramanian K, Nunnari J, Weissman JS (2018) Defining the physiological role of SRP in protein-targeting efficiency and specificity. Science 359(6376):689–692. https://doi.org/10.1126/science.aar3607

Cross BCS, High S (2009a) Dissecting the physiological role of selective transmembrane-segment retention at the ER translocon. J Cell Sci 122(11):1768–1777. https://doi.org/10.1242/jcs.046094

Cross S, High S (2009b) Membrane protein biosynthesis at the endoplasmic reticulum. In: Zimmermann R (ed) Protein transport into the endoplasmic reticulum. Landes Bioscience, pp 77–89

Csordás G, Weaver D, Hajnóczky G (2018) Endoplasmic reticulum-mitochondrial contactology: structure and signaling functions. Trends Cell Biol 28(7):523–540. https://doi.org/10.1016/j.tcb.2018.02.009

Cui XA, Zhang H, Palazzo AF (2012) p180 promotes the ribosome-independent localization of a subset of mRNA to the endoplasmic reticulum. PLoS Biol 10(5):e1001336. https://doi.org/10.1371/journal.pbio.1001336

Cui XA, Zhang Y, Hong SJ, Palazzo AF (2013) Identification of a region within the placental alkaline phosphatase mRNA that mediates p180-dependent targeting to the endoplasmic reticulum. J Biol Chem 288(41):29633–29641. https://doi.org/10.1074/jbc.M113.482505

Dalal K, Duong F (2009) The SecY complex: conducting the orchestra of protein translocation. Trends Cell Biol 21(9):506–514

Dalbey RE, von Heijne G (1992) Signal peptidases in prokaryotes and eukaryotes—a new protease family. Trends Biochem Sci 17(11):474–478. https://doi.org/10.1016/0968-0004(92)90492-R

Dejgaard K, Theberge J-F, Heath-Engel H, Chevet E, Tremblay ML, Thomas DY (2010) Organization of the Sec61 translocon, studied by high resolution native electrophoresis. J Proteome Res 9(4):1763–1771. https://doi.org/10.1021/pr900900x

Denks K, Vogt A, Sachelaru I, Petriman N-A, Kudva R, Koch H-G (2014) The Sec translocon mediated protein transport in prokaryotes and eukaryotes. Mol Membr Biol 31(2–3):58–84. https://doi.org/10.3109/09687688.2014.907455

Deshaies RJ, Schekman R (1987) A yeast mutant defective at an early stage in import of secretory protein precursors into the endoplasmic reticulum. J Cell Biol 105(2):633–645. https://doi.org/10.1083/jcb.105.2.633

Deshaies RJ, Sanders SL, Feldheim DA, Schekman R (1991) Assembly of yeast Sec proteins involved in translocation into the endoplasmic reticulum into a membrane-bound multisubunit complex. Nature 349(6312):806–808

Devaraneni PK, Conti B, Matsumura Y, Yang Z, Johnson AE, Skach WR (2011) Stepwise insertion and inversion of a type II signal anchor sequence in the ribosome-Sec61 translocon complex. Cell 146(1):134–147

Driessen AJM, Nouwen N (2008) Protein translocation across the bacterial cytoplasmic membrane. Annu Rev Biochem 77(1):643–667. https://doi.org/10.1146/annurev.biochem.77.061606.160747

du Plessis DJF, Berrelkamp G, Nouwen N, Driessen AJM (2009) The lateral gate of SecYEG opens during protein translocation. J Biol Chem 284(23):15805–15814. https://doi.org/10.1074/jbc.M901855200

du Plessis DJF, Nouwen N, Driessen AJM (2011) The Sec translocase. Biochem Biophys Acta 1808(3):851–865

Dudek J, Volkmer J, Bies C, Guth S, Müller A, Lerner M, Feick P, Schäfer K-H, Morgenstern E, Hennessy F, Blatch GL, Janoscheck K, Heim N, Scholtes P, Frien M, Nastainczyk W, Zimmermann R (2002) A novel type of co-chaperone mediates transmembrane recruitment of DnaK-like chaperones to ribosomes. EMBO J 21(12):2958–2967

Dudek J, Greiner M, Müller A, Hendershot LM, Kopsch K, Nastainczyk W, Zimmermann R (2005) ERj1p has a basic role in protein biogenesis at the endoplasmic reticulum. Nat Struct Mol Biol 12:1008. https://doi.org/10.1038/nsmb1007. https://www.nature.com/articles/nsmb1007#supplementary-information

Dudek J, Benedix J, Cappel S, Greiner M, Jalal C, Müller L, Zimmermann R (2009) Functions and pathologies of BiP and its interaction partners. Cell Mol Life Sci 66(9):1556–1569

Dudek J, Pfeffer S, Lee P-H, Jung M, Cavalié A, Helms V, Förster F, Zimmermann R (2015) Protein transport into the human endoplasmic reticulum. J Mol Biol 427(6, Part A):1159–1175. http://doi.org/10.1016/j.jmb.2014.06.011

Egea PF, Stroud RM (2010) Lateral opening of a translocon upon entry of protein suggests the mechanism of insertion into membranes. Proc Natl Acad Sci 107(40):17182–17187. https://doi.org/10.1073/pnas.1012556107

Egea PF, Shan SO, Napetschnig J, Savage DF, Walter P, Stroud RM (2004) Substrate twinning activates the signal recognition particle and its receptor. Nature 427(6971):215–221

Ellis RJ (2001) Macromolecular crowding: obvious but underappreciated. Trends Biochem Sci 26 (10):597–604. https://doi.org/10.1016/S0968-0004(01)01938-7

Erdmann F, Jung M, Eyrisch S, Lang S, Helms V, Wagner R, Zimmermann R (2009) Lanthanum ions inhibit the mammalian Sec61 complex in its channel dynamics and protein transport activity. FEBS Lett 583(14):2359–2364

Erdmann F, Schäuble N, Lang S, Jung M, Honigmann A, Ahmad M, Dudek J, Benedix J, Harsman A, Kopp A, Helms V, Cavalié A, Wagner R, Zimmermann R (2011) Interaction of calmodulin with Sec61alpha limits Ca^{2+} leakage from the endoplasmic reticulum. EMBO J 30 (1):17–31

Evans EA, Gilmore R, Blobel G (1986) Purification of microsomal signal peptidase as a complex. Proc Natl Acad Sci USA 83:581–585

Fedeles SV, Tian X, Gallagher A-R, Mitobe M, Nishio S, Lee SH, Cai Y, Geng L, Crews CM, Somlo S (2011) A genetic interaction network of five genes for human polycystic kidney and liver diseases defines polycystin-1 as the central determinant of cyst formation. Nat Genet 43 (7):639–647

Flynn GC, Pohl J, Flocco MT, Rothman JE (1991) Peptide-binding specificity of the molecular chaperone BiP. Nature 353:726–730

Fons RD, Bogert BA, Hegde RS (2003) Substrate-specific function of the translocon-associated protein complex during translocation across the ER membrane. J Cell Biol 160(4):529–539

Friedman JR, Voeltz GK (2011) The ER in 3D: a multifunctional dynamic membrane network. Trends Cell Biol 21(12):709–717

Fumagalli F, Noack J, Bergmann Timothy J, Cebollero E, Pisoni Giorgia B, Fasana E, Fregno I, Galli C, Loi M, Soldà T, D'Antuono R, Raimondi A, Jung M, Melnyk A, Schorr S, Schreiber A, Simonelli L, Varani L, Wilson-Zbinden C, Zerbe O, Hofmann K, Peter M, Quadroni M, Zimmermann R, Molinari M (2016) Translocon component Sec62 acts in endoplasmic reticulum turnover during stress recovery. Nat Cell Biol 18:1173. https://doi.org/10.1038/ncb3423. https://www.nature.com/articles/ncb3423#supplementary-information

Gamayun I, O'Keefe S, Pick T, Klein M-C, Nguyen D, McKibbin C, Piacenti M, Williams HM, Flitsch SL, Whitehead RC, Swanton E, Helms V, High S, Zimmermann R, Cavalié A (2019) Eeyarestatin compounds selectively enhance Sec61-mediated Ca^{2+} leakage from the endoplasmic reticulum. Cell Chem Biol. https://doi.org/10.1016/j.chembiol.2019.01.010

Gamerdinger M, Hanebuth MA, Frickey T, Deuerling E (2015) The principle of antagonism ensures protein targeting specificity at the endoplasmic reticulum. Science 348(6231):201–207. https://doi.org/10.1126/science.aaa5335

Garrison JL, Kunkel EJ, Hegde RS, Taunton J (2005) A substrate-specific inhibitor of protein translocation into the endoplasmic reticulum. Nature 436:285. https://doi.org/10.1038/nature03821. https://www.nature.com/articles/nature03821#supplementary-information

Gatta AT, Levine TP (2017) Piecing together the patchwork of contact sites. Trends Cell Biol 27 (3):214–229. https://doi.org/10.1016/j.tcb.2016.08.010

Gilmore R, Blobel G, Walter P (1982a) Protein translocation across the endoplasmic reticulum. I. Detection in the microsomal membrane of a receptor for the signal recognition particle. J Cell Biol 95(2):463–469

Gilmore R, Walter P, Blobel G (1982b) Protein translocation across the endoplasmic reticulum. II. Isolation and characterization of the signal recognition particle receptor. J Cell Biol 95(2):470–477

Goder V, Spiess M (2003) Molecular mechanism of signal sequence orientation in the endoplasmic reticulum. EMBO J 22(14):3645–3653

Goder V, Junne T, Spiess M (2004) Sec61p contributes to signal sequence orientation according to the positive-inside rule. Mol Biol Cell 15(3):1470–1478. https://doi.org/10.1091/mbc.e03-08-0599

Gogala M, Becker T, Beatrix B, Armache J-P, Barrio-Garcia C, Berninghausen O, Beckmann R (2014) Structures of the Sec61 complex engaged in nascent peptide translocation or membrane insertion. Nature 506(7486):107–110. https://doi.org/10.1038/nature12950

Gonsberg A, Jung S, Ulbrich S, Origi A, Ziska A, Baier M, Koch H-G, Zimmermann R, Winklhofer KF, Tatzelt J (2017) The Sec61/SecY complex is inherently deficient in translocating intrinsically disordered proteins. J Biol Chem. https://doi.org/10.1074/jbc.m117.788067

Görlich D, Rapoport TA (1993) Protein translocation into proteoliposomes reconstituted from purified components of the endoplasmic reticulum membrane. Cell 75(4):615–630

Görlich D, Prehn S, Hartmann E, Kalies KU, Rapoport TA (1992) A mammalian homolog of SEC61p and SECYp is associated with ribosomes and nascent polypeptides during translocation. Cell 71(3):489–503

Greiner M, Kreutzer B, Lang S, Jung V, Cavalié A, Unteregger G, Zimmermann R, Wullich B (2011) Sec62 protein level is crucial for the ER stress tolerance of prostate cancer. Prostate 71 (10):1074–1083. https://doi.org/10.1002/pros.21324

Guna A, Volkmar N, Christianson JC, Hegde RS (2017) The ER membrane protein complex is a transmembrane domain insertase. Science. https://doi.org/10.1126/science.aao3099
Guven J, Huber G, Valencia DM (2014) Terasaki spiral ramps in the rough endoplasmic reticulum. Phys Rev Lett 113(18):188101
Haas IG, Wabl M (1983) Immunoglobulin heavy chain binding protein. Nature 306(5941):387–389. https://doi.org/10.1038/306387a0
Halic M, Beckmann R (2005) The signal recognition particle and its interactions during protein targeting. Curr Opin Struct Biol 15(1):116–125
Halic M, Becker T, Pool MR, Spahn CM, Grassucci RA, Frank J, Beckmann R (2004) Structure of the signal recognition particle interacting with the elongation-arrested ribosome. Nature 427 (6977):808–814
Hann BC, Walter P (1991) The signal recognition particle in *S. cerevisiae*. Cell 67(1):131–144
Harada Y, Li H, Wall JS, Li H, Lennarz WJ (2011) Structural studies and the assembly of the heptameric post-translational translocon complex. J Biol Chem 286(4):2956–2965. https://doi.org/10.1074/jbc.M110.159517
Harsman A, Bartsch P, Hemmis B, Wagner R (2011a) Exploring protein import pores of cellular organelles at the single molecule level using the planar lipid bilayer technique. Eur J Cell Biol 90(9):721–730
Harsman A, Kopp A, Wagner R, Zimmermann R, Jung M (2011b) Calmodulin regulation of the calcium-leak channel Sec61 is unique to vertebrates. Channels 5(4):293–298
Harsman A, Krüger V, Bartsch P, Honigmann A, Schmidt O, Rao S, Meisinger C, Wagner R (2011c) Protein conducting nanopores. J Phys: Condens Matter 22(45):454102
Hartmann E, Görlich D, Kostka S, Otto A, Kraft R, Knespel S, Bürger E, Rapoport TA, Prehn S (1993) A tetrameric complex of membrane proteins in the endoplasmic reticulum. Eur J Biochem 214(2):375–381. https://doi.org/10.1111/j.1432-1033.1993.tb17933.x
Hartmann E, Sommer T, Prehn S, Gorlich D, Jentsch S, Rapoport TA (1994) Evolutionary conservation of components of the protein translocation complex. Nature 367(6464):654–657
Haßdenteufel S, Schäuble N, Cassella P, Leznicki P, Müller A, High S, Jung M, Zimmermann R (2011) Ca^{2+}-calmodulin inhibits tail-anchored protein insertion into the mammalian endoplasmic reticulum membrane. FEBS Lett 585(21):3485–3490
Haßdenteufel S, Klein M-C, Melnyk A, Zimmermann R (2014) Protein transport into the human ER and related diseases, Sec61-channelopathies. Biochem Cell Biol 92(6):499–509. https://doi.org/10.1139/bcb-2014-0043
Haßdenteufel S, Sicking M, Schorr S, Aviram N, Fecher-Trost C, Schuldiner M, Jung M, Zimmermann R, Lang S (2017) hSnd2 protein represents an alternative targeting factor to the endoplasmic reticulum in human cells. FEBS Lett 591(20):3211–3224. https://doi.org/10.1002/1873-3468.12831
Haßdenteufel S, Johnson N, Paton AW, Paton JC, High S, Zimmermann R (2018) Chaperone-mediated Sec61 channel gating during ER import of small precursor proteins overcomes Sec61 inhibitor-reinforced energy barrier. Cell Rep 23(5):1373–1386. https://doi.org/10.1016/j.celrep.2018.03.122
Hebert DN, Foellmer B, Helenius A (1996) Calnexin and calreticulin promote folding, delay oligomerization and suppress degradation of influenza hemagglutinin in microsomes. EMBO J 15(12):2961–2968. https://doi.org/10.1002/j.1460-2075.1996.tb00659.x
Hegde RS, Voigt S, Rapoport TA, Lingappa VR (1998) TRAM regulates the exposure of nascent secretory proteins to the cytosol during translocation into the endoplasmic reticulum. Cell 92 (5):621–631. https://doi.org/10.1016/S0092-8674(00)81130-7
Hein MY, Hubner NC, Poser I, Cox J, Nagaraj N, Toyoda Y, Gak IA, Weisswange I, Mansfeld J, Buchholz F, Hyman AA, Mann M (2015) A human interactome in three quantitative dimensions organized by stoichiometries and abundances. Cell 163(3):712–723. https://doi.org/10.1016/j.cell.2015.09.053
Hennessy F, Cheetham ME, Dirr HW, Blatch GL (2000) Analysis of the levels of conservation of the J domain among the various types of DnaJ-like proteins. Cell Stress Chaperones 5(4):347–358

Hennessy F, Nicoll WS, Zimmermann R, Cheetham ME, Blatch GL (2005) Not all J domains are created equal: implications for the specificity of Hsp40-Hsp70 interactions. Protein Sci 14 (7):1697–1709

Heritage D, Wonderlin WF (2001) Translocon pores in the endoplasmic reticulum are permeable to a neutral, polar molecule. J Biol Chem 276(25):22655–22662

High S, Andersen SSL, Görlich D, Hartmann E, Prehn S, Rapoport TA, Dobberstein B (1993) Sec61p is adjacent to nascent type I and type II signal-anchor proteins during their membrane insertion. J Cell Biol 121(4):743–750

Hizlan D, Robson A, Whitehouse S, Gold V, Vonck J, Mills D, Kühlbrandt W, Collinson I (2012) Structure of the SecY complex unlocked by a preprotein mimic. Cell Reports 1(1):21–28

Hollien J, Weissman JS (2006) Decay of endoplasmic reticulum-localized mRNAs during the unfolded protein response. Science 313(5783):104–107. https://doi.org/10.1126/science.1129631

Hollien J, Lin JH, Li H, Stevens N, Walter P, Weissman JS (2009) Regulated Ire1-dependent decay of messenger RNAs in mammalian cells. J Cell Biol 186(3):323–331. https://doi.org/10.1083/jcb.200903014

Ingolia NT (2016) Ribosome Footprint Profiling of Translation throughout the Genome. Cell 165 (1):22–33. https://doi.org/10.1016/j.cell.2016.02.066

Ismail N, Crawshaw Samuel G, Cross Benedict CS, Haagsma Anna C, High S (2008) Specific transmembrane segments are selectively delayed at the ER translocon during opsin biogenesis. Biochem J 411(3):495–506. https://doi.org/10.1042/bj20071597

Itskanov S, Park E (2018) Structure of the posttranslational Sec protein-translocation channel complex from yeast. Science eaav6740. https://doi.org/10.1126/science.aav6740

Jadhav B, McKenna M, Johnson N, High S, Sinning I, Pool MR (2015) Mammalian SRP receptor switches the Sec61 translocase from Sec62 to SRP-dependent translocation. Nat Commun 6. https://doi.org/10.1038/ncomms10133

Jermy AJ, Willer M, Davis E, Wilkinson BM, Stirling CJ (2006) The Brl domain in Sec63p is required for assembly of functional endoplasmic reticulum translocons. J Biol Chem 281 (12):7899–7906

Johnson N, Vilardi F, Lang S, Leznicki P, Zimmermann R, High S (2012) TRC40 can deliver short secretory proteins to the Sec61 translocon. J Cell Sci. https://doi.org/10.1242/jcs.102608

Johnson N, Haßdenteufel S, Theis M, Paton AW, Paton JC, Zimmermann R, High S (2013) The signal sequence influences post-translational ER translocation at distinct stages. PLoS ONE 8 (10):e75394. https://doi.org/10.1371/journal.pone.0075394

Jomaa A, Boehringer D, Leibundgut M, Ban N (2016) Structures of the *E. coli* translating ribosome with SRP and its receptor and with the translocon. Nat Commun 7. https://doi.org/10.1038/ncomms10471

Jomaa A, Fu Y-HH, Boehringer D, Leibundgut M, Shan S-O, Ban N (2017) Structure of the quaternary complex between SRP, SR, and translocon bound to the translating ribosome. Nat Commun 8:15470. https://doi.org/10.1038/ncomms15470. https://www.nature.com/articles/ncomms15470#supplementary-information

Jonikas MC, Collins SR, Denic V, Oh E, Quan EM, Schmid V, Weibezahn J, Schwappach B, Walter P, Weissman JS, Schuldiner M (2009) Comprehensive characterization of genes required for protein folding in the endoplasmic reticulum. Science 323(5922):1693–1697

Jung S-j, Kim JEH, Reithinger JH, Kim H (2014) The Sec62–Sec63 translocon facilitates translocation of the C-terminus of membrane proteins. J Cell Sci 127(19):4270–4278. https://doi.org/10.1242/jcs.153650

Junne T, Schwede T, Goder V, Spiess M (2006) The plug domain of yeast Sec61p is important for efficient protein translocation, but is not essential for cell viability. Mol Biol Cell 17(9):4063–4068. https://doi.org/10.1091/mbc.E06-03-0200

Kalbfleisch T, Cambon A, Wattenberg BW (2007) A bioinformatics approach to identifying tail-anchored proteins in the human genome. Traffic 8(12):1687–1694. https://doi.org/10.1111/j.1600-0854.2007.00661.x

Kalies KU, Görlich D, Rapoport TA (1994) Binding of ribosomes to the rough endoplasmic reticulum mediated by the Sec61p-complex. J Cell Biol 126:925–934

Kalies K-U, Rapoport TA, Hartmann E (1998) The beta-subunit of the Sec61 complex facilitates cotranslational protein transport and interacts with the signal peptidase during translocation. J Cell Biol 141(4):887–894. https://doi.org/10.1083/jcb.141.4.887

Kamariah N, Eisenhaber F, Adhikari S, Eisenhaber B, Gruber G (2011) Purification and crystallization of yeast glycosylphosphatidylinositol transamidase subunit PIG-S (PIG-S71-467). Acta Crystallogr Sect F 67(8):896–899. https://doi.org/10.1107/S1744309111024080

Kapp K, Schrempf S, Lemberg M, Dobberstein B (2009) Post-targeting functions of signal peptides. In: Zimmermann R (ed) Protein transport into the endoplasmic reticulum. Landes Bioscience, pp 1–16

Kayatekin C, Amasino A, Gaglia G, Flannick J, Bonner JM, Fanning S, Narayan P, Barrasa MI, Pincus D, Landgraf D, Nelson J, Hesse WR, Costanzo M, Myers CL, Boone C, Florez JC, Lindquist S (2018) Translocon declogger Ste24 protects against IAPP oligomer-induced proteotoxicity. Cell 173(1):62–73.e69. https://doi.org/10.1016/j.cell.2018.02.026

Kelleher DJ, Gilmore R (2006) An evolving view of the eukaryotic oligosaccharyltransferase. Glycobiology 16(4):47R–62R. https://doi.org/10.1093/glycob/cwj066

Kinch LN, Saier JMH, Grishin NV (2002) Sec61beta—a component of the archaeal protein secretory system. Trends Biochem Sci 27(4):170–171

Klein M-C, Zimmermann K, Schorr S, Landini M, Klemens PAW, Altensell J, Jung M, Krause E, Nguyen D, Helms V, Rettig J, Fecher-Trost C, Cavalié A, Hoth M, Bogeski I, Neuhaus HE, Zimmermann R, Lang S, Haferkamp I (2018) AXER is an ATP/ADP exchanger in the membrane of the endoplasmic reticulum. Nat Commun 9(1):3489. https://doi.org/10.1038/s41467-018-06003-9

Koning RI, Koster AJ, Sharp TH (2018) Advances in cryo-electron tomography for biology and medicine. Ann Anat-Anat Anz 217:82–96. https://doi.org/10.1016/j.aanat.2018.02.004

Koshland DE (1958) Application of a theory of enzyme specificity to protein synthesis. Proc Natl Acad Sci 44(2):98–104. https://doi.org/10.1073/pnas.44.2.98

Kowarik M, Küng S, Martoglio B, Helenius A (2002) Protein folding during cotranslational translocation in the endoplasmic reticulum. Mol Cell 10(4):769–778. https://doi.org/10.1016/S1097-2765(02)00685-8

Kumar DP, Vorvis C, Sarbeng EB, Cabra Ledesma VC, Willis JE, Liu Q (2011) The four hydrophobic residues on the Hsp70 inter-domain linker have two distinct roles. J Mol Biol 411(5):1099–1113

Kutay U, Hartmann E, Rapoport TA (1993) A class of membrane proteins with a C-terminal anchor. Trends Cell Biol 3(3):72–75

Lakkaraju AK, Abrami L, Lemmin T, Blaskovic S, Kunz B, Kihara A, Dal Peraro M, van der Goot FG (2012a) Palmitoylated calnexin is a key component of the ribosome–translocon complex. EMBO J 31(7):1823–1835. https://doi.org/10.1038/emboj.2012.15

Lakkaraju AKK, Thankappan R, Mary C, Garrison JL, Taunton J, Strub K (2012b) Efficient secretion of small proteins in mammalian cells relies on Sec62-dependent posttranslational translocation. Mol Biol Cell 23(14):2712–2722. https://doi.org/10.1091/mbc.E12-03-0228

Lang S, Erdmann F, Jung M, Wagner R, Cavalié A, Zimmermann R (2011a) Sec61 complexes form ubiquitous ER Ca^{2+} leak channels. Channels 5(4):228–235

Lang S, Schäuble N, Cavalié A, Zimmermann R (2011b) Live cell calcium imaging combined with siRNA mediated gene silencing identifies Ca^{2+} leak channels in the ER membrane and their regulatory mechanisms. J Vis Exp 53:e2730. https://doi.org/10.3791/2730

Lang S, Benedix J, Fedeles SV, Schorr S, Schirra C, Schäuble N, Jalal C, Greiner M, Haßdenteufel S, Tatzelt J, Kreutzer B, Edelmann L, Krause E, Rettig J, Somlo S, Zimmermann R, Dudek J (2012) Different effects of Sec61α, Sec62 and Sec63 depletion on transport of polypeptides into the endoplasmic reticulum of mammalian cells. J Cell Sci 125(8):1958–1969. https://doi.org/10.1242/jcs.096727

Lang S, Pfeffer S, Lee P-H, Cavalié A, Helms V, Förster F, Zimmermann R (2017) An update on Sec61 channel functions, mechanisms, and related diseases. Front Physiol 8(887). https://doi.org/10.3389/fphys.2017.00887

Larkin A, Imperiali B (2011) The expanding horizons of asparagine-linked glycosylation. Biochemistry 50(21):4411–4426. https://doi.org/10.1021/bi200346n

Le Gall S, Neuhof A, Rapoport T (2004) The endoplasmic reticulum membrane is permeable to small molecules. Mol Biol Cell 15(2):447–455

Lemberg MK, Martoglio B (2002) Requirements for signal peptide peptidase-catalyzed intramembrane proteolysis. Mol Cell 10(4):735–744

Li W, Schulman S, Boyd D, Erlandson K, Beckwith J, Rapoport TA (2007) The plug domain of the SecY protein stabilizes the closed state of the translocation channel and maintains a membrane seal. Mol Cell 26(4):511–521

Li L, Park E, Ling J, Ingram J, Ploegh H, Rapoport TA (2016) Crystal structure of a substrate-engaged SecY protein-translocation channel. Nature 531(7594):395–399. https://doi.org/10.1038/nature17163. http://www.nature.com/nature/journal/v531/n7594/abs/nature17163.html#supplementary-information

Liebermeister W, Rapoport TA, Heinrich R (2001) Ratcheting in post-translational protein translocation: a mathematical model. J Mol Biol 305(3):643–656

Lievremont JP, Rizzuto R, Hendershot L, Meldolesi J (1997) BiP, a major chaperone protein of the endoplasmic reticulum lumen, plays a direct and important role in the storage of the rapidly exchanging pool of Ca^{2+}. J Biol Chem 272(49):30873–30879

Linxweiler M, Schorr S, Schäuble N, Jung M, Linxweiler J, Langer F, Schäfers H-J, Cavalié A, Zimmermann R, Greiner M (2013) Targeting cell migration and the endoplasmic reticulum stress response with calmodulin antagonists: a clinically tested small molecule phenocopy of SEC62 gene silencing in human tumor cells. BMC Cancer 13(1):574. https://doi.org/10.1186/1471-2407-13-574

Linxweiler M, Schick B, Zimmermann R (2017) Let's talk about Secs: Sec61, Sec62 and Sec63 in signal transduction, oncology and personalized medicine. 2:17002. https://doi.org/10.1038/sigtrans.2017.2

Lloyd DJ, Wheeler MC, Gekakis N (2010) A point mutation in Sec61alpha1 leads to diabetes and hepatosteatosis in mice. Diabetes 59(2):460–470

Lučić V, Rigort A, Baumeister W (2013) Cryo-electron tomography: the challenge of doing structural biology in situ. J Cell Biol 202(3):407–419. https://doi.org/10.1083/jcb.201304193

Ma Y, Hendershot LM (2004) ER chaperone functions during normal and stress conditions. J Chem Neuroanat 28(1):51–65. https://doi.org/10.1016/j.jchemneu.2003.08.007

MacKinnon AL, Paavilainen VO, Sharma A, Hegde RS, Taunton J (2014) An allosteric Sec61 inhibitor traps nascent transmembrane helices at the lateral gate. eLife 3:e01483. https://doi.org/10.7554/elife.01483

Mades A, Gotthardt K, Awe K, Stieler J, Döring T, Füser S, Prange R (2012) Role of human Sec63 in modulating the steady-state levels of multi-spanning membrane proteins. PLoS ONE 7(11):e49243. https://doi.org/10.1371/journal.pone.0049243

Mahamid J, Pfeffer S, Schaffer M, Villa E, Danev R, Kuhn Cuellar L, Förster F, Hyman AA, Plitzko JM, Baumeister W (2016) Visualizing the molecular sociology at the HeLa cell nuclear periphery. Science 351(6276):969–972. https://doi.org/10.1126/science.aad8857

Malsburg KVD, Shao S, Hegde RS (2015) The ribosome quality control pathway can access nascent polypeptides stalled at the Sec61 translocon. Mol Biol Cell 26(12):2168–2180. https://doi.org/10.1091/mbc.E15-01-0040

Mandon EC, Butova C, Lachapelle A, Gilmore R (2018) Conserved motifs on the cytoplasmic face of the protein translocation channel are critical for the transition between resting and active conformations. J Biol Chem. https://doi.org/10.1074/jbc.ra118.004123

Marcinowski M, Höller M, Feige MJ, Baerend D, Lamb DC, Buchner J (2011) Substrate discrimination of the chaperone BiP by autonomous and cochaperone-regulated conformational transitions. Nat Struct Mol Biol 18(2):150–158

Mariappan M, Li X, Stefanovic S, Sharma A, Mateja A, Keenan RJ, Hegde RS (2010) A ribosome-associating factor chaperones tail-anchored membrane proteins. Nature 466 (7310):1120–1124

Matlack KES, Mothes W, Rapoport TA (1998) Protein translocation: tunnel vision. Cell 92 (3):381–390

Matlack KES, Misselwitz B, Plath K, Rapoport TA (1999) BiP acts as a molecular ratchet during posttranslational transport of prepro-a factor across the ER membrane. Cell 97(5):553–564

McKenna M, Simmonds RE, High S (2017) Mycolactone reveals the substrate-driven complexity of Sec61-dependent transmembrane protein biogenesis. J Cell Sci 130(7):1307–1320. https://doi.org/10.1242/jcs.198655

Meacock SL, Lecomte FJL, Crawshaw SG, High S (2002) Different transmembrane domains associate with distinct endoplasmic reticulum components during membrane integration of a polytopic protein. Mol Biol Cell 13(12):4114–4129. https://doi.org/10.1091/mbc.E02-04-0198

Meldolesi J, Pozzan T (1998) The endoplasmic reticulum Ca^{2+} store: a view from the lumen. Trends Biochem Sci 23(1):10–14

Melnyk A, Rieger H, Zimmermann R (2015) Co-chaperones of the mammalian endoplasmic reticulum. In: Blatch GL, Edkins AL (eds) The networking of chaperones by co-chaperones: control of cellular protein homeostasis. Springer International Publishing, Cham, pp 179–200. https://doi.org/10.1007/978-3-319-11731-7_9

Ménétret JF, Hegde RS, Heinrich SU, Chandramouli P, Ludtke SJ, Rapoport TA, Akey CW (2005) Architecture of the ribosome-channel complex derived from native membranes. J Mol Biol 348(2):445–457

Ménétret JF, Schaletzky J, Clemons WM Jr, Osborne AR, Skanland SS, Denison C, Gygi SP, Kirkpatrick DS, Park E, Ludtke SJ, Rapoport TA, Akey CW (2007) Ribosome binding of a single copy of the SecY complex: implications for protein translocation. Mol Cell 28(6):1083–1092

Ménétret JF, Hegde RS, Aguiar M, Gygi SP, Park E, Rapoport TA, Akey CW (2008) Single copies of Sec61 and TRAP associate with a nontranslating mammalian ribosome. Structure 16 (7):1126–1137

Michalak M, Robert Parker JM, Opas M (2002) Ca^{2+} signaling and calcium binding chaperones of the endoplasmic reticulum. Cell Calcium 32(5–6):269–278

Minton AP (2006) How can biochemical reactions within cells differ from those in test tubes? J Cell Sci 119(14):2863–2869. https://doi.org/10.1242/jcs.03063

Mogami H, Tepikin AV, Petersen OH (1998) Termination of cytosolic Ca^{2+} signals: Ca^{2+} reuptake into intracellular stores is regulated by the free Ca^{2+} concentration in the store lumen. EMBO J 17(2):435–442

Möller I, Jung M, Beatrix B, Levy R, Kreibich G, Zimmermann R, Wiedmann M, Lauring B (1998) A general mechanism for regulation of access to the translocon: competition for a membrane attachment site on ribosomes. Proc Natl Acad Sci 95(23):13425–13430. https://doi.org/10.1073/pnas.95.23.13425

Moran U, Phillips R, Milo R (2010) SnapShot: key numbers in biology. Cell 141(7):1262

Morrow MW, Brodsky JL (2001) Yeast ribosomes bind to highly purified reconstituted Sec61p complex and to mammalian p180. Traffic 2(10):705–716

Müller L, Diaz de Escauriaza M, Lajoie P, Theis M, Jung M, Müller A, Burgard C, Greiner M, Snapp EL, Dudek J, Zimmermann R (2010) Evolutionary gain of function for the ER membrane protein Sec62 from yeast to humans. Mol Biol Cell 21(5):691–703

Ng DT, Brown JD, Walter P (1996) Signal sequences specify the targeting route to the endoplasmic reticulum membrane. J Cell Biol 134(2):269–278. https://doi.org/10.1083/jcb.134.2.269

Nguyen D, Stutz R, Schorr S, Lang S, Pfeffer S, Freeze HH, Förster F, Helms V, Dudek J, Zimmermann R (2018) Proteomics reveals signal peptide features determining the client specificity in human TRAP-dependent ER protein import. Nat Commun 9(1):3765. https://doi.org/10.1038/s41467-018-06188-z

Ni M, Lee AS (2007) ER chaperones in mammalian development and human diseases. FEBS Lett 581(19):3641–3651

Nilsson IM, von Heijne G (1993) Determination of the distance between the oligosaccharyltransferase active site and the endoplasmic reticulum membrane. J Biol Chem 268(8):5798–5801

Nilsson I, Kelleher DJ, Miao Y, Shao Y, Kreibich G, Gilmore R, von Heijne G, Johnson AE (2003) Photocross-linking of nascent chains to the STT3 subunit of the oligosaccharyltransferase complex. J Cell Biol 161(4):715–725

Nishiyama K-i, Hanada M, Tokuda H (1994) Disruption of the gene encoding p12 (SecG) reveals the direct involvement and important function of SecG in the protein translocation of *Escherichia coli* at low temperature. EMBO J 13(14):3272–3277

Nixon-Abell J, Obara CJ, Weigel AV, Li D, Legant WR, Xu CS, Pasolli HA, Harvey K, Hess HF, Betzig E, Blackstone C, Lippincott-Schwartz J (2016) Increased spatiotemporal resolution reveals highly dynamic dense tubular matrices in the peripheral ER. Science 354(6311). https://doi.org/10.1126/science.aaf3928

Novick P, Field C, Schekman R (1980) Identification of 23 complementation groups required for post-translational events in the yeast secretory pathway. Cell 21(1):205–215

Otero JH, Lizak B, Hendershot LM (2010) Life and death of a BiP substrate. Semin Cell Dev Biol 21(5):472–478

Paatero AO, Kellosalo J, Dunyak BM, Almaliti J, Gestwicki JE, Gerwick WH, Taunton J, Paavilainen VO (2016) Apratoxin kills cells by direct blockade of the Sec61 protein translocation channel. Cell Chem Biol 23(5):561–566. https://doi.org/10.1016/j.chembiol.2016.04.008

Palade GE (1956) The endoplasmic reticulum. J Biophys Biochem Cytol 2(4 Suppl):85–98

Palade GE, Porter KR (1954) Studies on the endoplasmic reticulum. J Exp Med 100(6):641–656. https://doi.org/10.1084/jem.100.6.641

Panzner S, Dreier L, Hartmann E, Kostka S, Rapoport TA (1995) Posttranslational protein transport in yeast reconstituted with a purified complex of Sec proteins and Kar2p. Cell 81 (4):561–570

Park E, Rapoport TA (2011) Preserving the membrane barrier for small molecules during bacterial protein translocation. Nature 473(7346):239–242

Park E, Rapoport TA (2012) Mechanisms of Sec61/SecY-mediated protein translocation across membranes. Annu Rev Biophys 41(1):21–40. https://doi.org/10.1146/annurev-biophys-050511-102312

Park E, Menetret J-F, Gumbart JC, Ludtke SJ, Li W, Whynot A, Rapoport TA, Akey CW (2014) Structure of the SecY channel during initiation of protein translocation. Nature 506(7486):102–106. https://doi.org/10.1038/nature12720. http://www.nature.com/nature/journal/v506/n7486/abs/nature12720.html#supplementary-information

Pataki CI, Rodrigues J, Zhang L, Qian J, Efron B, Hastie T, Elias JE, Levitt M, Kopito RR (2018) Proteomic analysis of monolayer-integrated proteins on lipid droplets identifies amphipathic interfacial α-helical membrane anchors. Proc Natl Acad Sci 115(35):E8172–E8180. https://doi.org/10.1073/pnas.1807981115

Petrescu A-J, Milac A-L, Petrescu SM, Dwek RA, Wormald MR (2004) Statistical analysis of the protein environment of N-glycosylation sites: implications for occupancy, structure, and folding. Glycobiology 14(2):103–114. https://doi.org/10.1093/glycob/cwh008

Pfeffer S, Brandt F, Hrabe T, Lang S, Eibauer M, Zimmermann R, Förster F (2012) Structure and 3D arrangement of endoplasmic reticulum membrane-associated ribosomes. Structure 20 (9):1508–1518. https://doi.org/10.1016/j.str.2012.06.010

Pfeffer S, Dudek J, Gogala M, Schorr S, Linxweiler J, Lang S, Becker T, Beckmann R, Zimmermann R, Förster F (2014) Structure of the mammalian oligosaccharyl-transferase complex in the native ER protein translocon. Nat Commun 5. https://doi.org/10.1038/ncomms4072

Pfeffer S, Burbaum L, Unverdorben P, Pech M, Chen Y, Zimmermann R, Beckmann R, Förster F (2015) Structure of the native Sec61 protein-conducting channel. Nat Commun 6. https://doi.org/10.1038/ncomms9403

Pfeffer S, Dudek J, Zimmermann R, Förster F (2016) Organization of the native ribosome–translocon complex at the mammalian endoplasmic reticulum membrane. Biochim Biophys Acta (BBA)-Gen Subj 1860(10):2122–2129. http://dx.doi.org/10.1016/j.bbagen.2016.06.024

Pfeffer S, Dudek J, Schaffer M, Ng BG, Albert S, Plitzko JM, Baumeister W, Zimmermann R, Freeze HH, Engel BD, Förster F (2017) Dissecting the molecular organization of the translocon-associated protein complex. Nat Commun 8:14516. https://doi.org/10.1038/ncomms14516. http://www.nature.com/articles/ncomms14516#supplementary-information

Phillips MJ, Voeltz GK (2016) Structure and function of ER membrane contact sites with other organelles. Nat Rev Mol Cell Biol 17(2):69–82. https://doi.org/10.1038/nrm.2015.8

Plath K, Mothes W, Wilkinson BM, Stirling CJ, Rapoport TA (1998) Signal sequence recognition in posttranslational protein transport across the yeast ER membrane. Cell 94(6):795–807

Plumb R, Zhang Z-R, Appathurai S, Mariappan M (2015) A functional link between the co-translational protein translocation pathway and the UPR. eLife 4:e07426. https://doi.org/10.7554/elife.07426

Ponsero AJ, Igbaria A, Darch MA, Miled S, Outten CE, Winther JR, Palais G, D'Autréaux B, Delaunay-Moisan A, Toledano MB (2017) Endoplasmic reticulum transport of glutathione by Sec61 is regulated by Ero1 and Bip. Mol Cell 67(6):962–973.e965. https://doi.org/10.1016/j.molcel.2017.08.012

Pool MR (2009) A trans-membrane segment inside the ribosome exit tunnel triggers RAMP4 recruitment to the Sec61p translocase. J Cell Biol 185(5):889–902. https://doi.org/10.1083/jcb.200807066

Potter MD, Seiser RM, Nicchitta CV (2001) Ribosome exchange revisited: a mechanism for translation-coupled ribosome detachment from the ER membrane. Trends Cell Biol 11(3):112–115

Puhka M, Vihinen H, Joensuu M, Jokitalo E (2007) Endoplasmic reticulum remains continuous and undergoes sheet-to-tubule transformation during cell division in mammalian cells. J Cell Biol 179(5):895–909. https://doi.org/10.1083/jcb.200705112

Puhka M, Joensuu M, Vihinen H, Belevich I, Jokitalo E (2012) Progressive sheet-to-tubule transformation is a general mechanism for endoplasmic reticulum partitioning in dividing mammalian cells. Mol Biol Cell 23(13):2424–2432. https://doi.org/10.1091/mbc.E10-12-0950

Rapoport TA, Li L, Park E (2017) Structural and mechanistic insights into protein translocation. Annu Rev Cell Dev Biol 33(1):369–390. https://doi.org/10.1146/annurev-cellbio-100616-060439

Reid DW, Chen Q, Tay ASL, Shenolikar S, Nicchitta CV (2014) The unfolded protein response triggers selective mRNA release from the endoplasmic reticulum. Cell 158(6):1362–1374. https://doi.org/10.1016/j.cell.2014.08.012

Reithinger JH, Kim JEH, Kim H (2013) Sec62 protein mediates membrane insertion and orientation of moderately hydrophobic signal anchor proteins in the endoplasmic reticulum (ER). J Biol Chem 288(25):18058–18067. https://doi.org/10.1074/jbc.M113.473009

Roy A, Wonderlin WF (2003) The permeability of the endoplasmic reticulum is dynamically coupled to protein synthesis. J Biol Chem 278(7):4397–4403

Ruiz-Canada C, Kelleher DJ, Gilmore R (2009) Cotranslational and posttranslational N-glycosylation of polypeptides by distinct mammalian OST isoforms. Cell 136(2):272–283. https://doi.org/10.1016/j.cell.2008.11.047

Rychkova A, Warshel A (2013) Exploring the nature of the translocon-assisted protein insertion. Proc Natl Acad Sci 110(2):495–500. https://doi.org/10.1073/pnas.1220361110

Sadlish H, Pitonzo D, Johnson AE, Skach WR (2005) Sequential triage of transmembrane segments by Sec61alpha during biogenesis of a native multispanning membrane protein. Nat Struct Mol Biol 12(10):870–878

Sammels E, Parys JB, Missiaen L, De Smedt H, Bultynck G (2010) Intracellular Ca^{2+} storage in health and disease: a dynamic equilibrium. Cell Calcium 47(4):297–314

Sato T, Sako Y, Sho M, Momohara M, Suico Mary A, Shuto T, Nishitoh H, Okiyoneda T, Kokame K, Kaneko M, Taura M, Miyata M, Chosa K, Koga T, Morino-Koga S, Wada I, Kai H

(2012) STT3B-dependent posttranslational N-glycosylation as a surveillance system for secretory protein. Mol Cell 47(1):99–110. https://doi.org/10.1016/j.molcel.2012.04.015

Savitz AJ, Meyer DI (1990) Identification of a ribosome receptor in the rough endoplasmic reticulum. Nature 346(6284):540–544

Savitz AJ, Meyer DI (1993) 180-kD ribosome receptor is essential for both ribosome binding and protein translocation. J Cell Biol 120(4):853–863

Schäuble N, Lang S, Jung M, Cappel S, Schorr S, Ulucan O, Linxweiler J, Dudek J, Blum R, Helms V, Paton AW, Paton JC, Cavalie A, Zimmermann R (2012) BiP-mediated closing of the Sec61 channel limits Ca^{2+} leakage from the ER. EMBO J 31(15):3282–3296

Schäuble N, Cavalié A, Zimmermann R, Jung M (2014) Interaction of *Pseudomonas aeruginosa* Exotoxin A with the human Sec61 complex suppresses passive calcium efflux from the endoplasmic reticulum. Channels (Austin, Tex) 8(1):76–83. https://doi.org/10.4161/chan.26526

Schekman R (2002) SEC mutants and the secretory apparatus. Nat Med 8(10):1055–1058

Schibich D, Gloge F, Pöhner I, Björkholm P, Wade RC, von Heijne G, Bukau B, Kramer G (2016) Global profiling of SRP interaction with nascent polypeptides. Nature 536(7615):219–223. https://doi.org/10.1038/nature19070

Schlenstedt G, Gudmundsson GH, Boman HG, Zimmermann R (1990) A large presecretory protein translocates both cotranslationally, using signal recognition particle and ribosome, and posttranslationally, without these ribonucleoparticles, when synthesized in the presence of mammalian microsomes. J Biol Chem 265(23):13960–13968

Schorr S, Klein M-C, Gamayun I, Melnyk A, Jung M, Schäuble N, Wang Q, Hemmis B, Bochen F, Greiner M, Lampel P, Urban SK, Hassdenteufel S, Dudek J, Chen X-Z, Wagner R, Cavalié A, Zimmermann R (2015) Co-chaperone specificity in gating of the polypeptide conducting channel in the membrane of the human endoplasmic reticulum. J Biol Chem 290 (30):18621–18635. https://doi.org/10.1074/jbc.M115.636639

Schrul B, Kopito RR (2016) Peroxin-dependent targeting of a lipid-droplet-destined membrane protein to ER subdomains. Nat Cell Biol 18:740. https://doi.org/10.1038/ncb3373. https://www.nature.com/articles/ncb3373#supplementary-information

Schuldiner M, Metz J, Schmid V, Denic V, Rakwalska M, Schmitt HD, Schwappach B, Weissman JS (2008) The GET complex mediates insertion of tail-anchored proteins into the ER membrane. Cell 134(4):634–645

Schwarz DS, Blower MD (2016) The endoplasmic reticulum: structure, function and response to cellular signaling. Cell Mol Life Sci 73:79–94. https://doi.org/10.1007/s00018-015-2052-6

Seppälä S, Slusky JS, Lloris-Garcerá P, Rapp M, von Heijne G (2010) Control of membrane protein topology by a single C-terminal residue. Science 328(5986):1698–1700. https://doi.org/10.1126/science.1188950

Shao S, Hegde RS (2011) A calmodulin-dependent translocation pathway for small secretory proteins. Cell 147(7):1576–1588

Sharma A, Mariappan M, Appathurai S, Hegde RS (2010) In vitro dissection of protein translocation into the mammalian endoplasmic reticulum. Methods Mol Biol (Clifton, NJ) 619:339–363. https://doi.org/10.1007/978-1-60327-412-8_20

Shibata Y, Voeltz GK, Rapoport TA (2006) Rough sheets and smooth tubules. Cell 126(3):435–439

Shibata Y, Shemesh T, Prinz WA, Palazzo AF, Kozlov MM, Rapoport TA (2010) Mechanisms determining the morphology of the peripheral ER. Cell 143(5):774–788

Shim S-H (2017) Cell imaging: an intracellular dance visualized. Nature advance online publication. https://doi.org/10.1038/nature22500

Shrimal S, Cherepanova NA, Gilmore R (2015) Cotranslational and posttranslocational N-glycosylation of proteins in the endoplasmic reticulum. Semin Cell Dev Biol 41:71–78. https://doi.org/10.1016/j.semcdb.2014.11.005

Shrimal S, Cherepanova NA, Gilmore R (2017) DC2 and KCP2 mediate the interaction between the oligosaccharyltransferase and the ER translocon. J Cell Biol. https://doi.org/10.1083/jcb.201702159

Shurtleff MJ, Itzhak DN, Hussmann JA, Schirle Oakdale NT, Costa EA, Jonikas M, Weibezahn J, Popova KD, Jan CH, Sinitcyn P, Vembar SS, Hernandez H, Cox J, Burlingame AL, Brodsky J, Frost A, Borner GHH, Weissman JS (2018) The ER membrane protein complex interacts cotranslationally to enable biogenesis of multipass membrane proteins. eLife 7:e37018. https://doi.org/10.7554/elife.37018

Simon SM, Blobel G (1991) A protein-conducting channel in the endoplasmic reticulum. Cell 65:371–380

Simsek D, Tiu GC, Flynn RA, Byeon GW, Leppek K, Xu AF, Chang HY, Barna M (2017) The mammalian ribo-interactome reveals ribosome functional diversity and heterogeneity. Cell 169 (6):1051–1065.e1018. https://doi.org/10.1016/j.cell.2017.05.022

Smock RG, Rivoire O, Russ WP, Swain JF, Leibler S, Ranganathan R, Gierasch LM (2010) An interdomain sector mediating allostery in Hsp70 molecular chaperones. Mol Syst Biol 6(1)

Sommer N, Junne T, Kalies K-U, Spiess M, Hartmann E (2013) TRAP assists membrane protein topogenesis at the mammalian ER membrane. Biochim Biophys Acta (BBA)-Mol Cell Res 1833(12):3104–3111. http://dx.doi.org/10.1016/j.bbamcr.2013.08.018

Spang A (2015) Anniversary of the discovery of sec mutants by Novick and Schekman. Mol Biol Cell 26(10):1783–1785. https://doi.org/10.1091/mbc.E14-11-1511

Spiess M (2014) Protein translocation: the Sec61/SecYEG translocon caught in the act. Curr Biol 24(8):R317–R319. https://doi.org/10.1016/j.cub.2014.02.051

Stefanovic S, Hegde RS (2007) Identification of a targeting factor for posttranslational membrane protein insertion into the ER. Cell 128(6):1147–1159

Sundaram A, Plumb R, Appathurai S, Mariappan M (2017) The Sec61 translocon limits IRE1α signaling during the unfolded protein response. eLife 6:e27187. https://doi.org/10.7554/elife.27187

Suzuki J, Kanemaru K, Iino M (2016) Genetically encoded fluorescent indicators for organellar calcium imaging. Biophys J 111(6):1119–1131. https://doi.org/10.1016/j.bpj.2016.04.054

Tanaka Y, Sugano Y, Takemoto M, Mori T, Furukawa A, Kusakizako T, Kumazaki K, Kashima A, Ishitani R, Sugita Y, Nureki O, Tsukazaki T (2015) Crystal structures of SecYEG in lipidic cubic phase elucidate a precise resting and a peptide-bound state. Cell Rep 13 (8):1561–1568. https://doi.org/10.1016/j.celrep.2015.10.025

Terasaki M, Shemesh T, Kasthuri N, Klemm RW, Schalek R, Hayworth KJ, Hand AR, Yankova M, Huber G, Lichtman JW, Rapoport TA, Kozlov MM (2013) Stacked endoplasmic reticulum sheets are connected by helicoidal membrane motifs. Cell 154(2):285–296. https://doi.org/10.1016/j.cell.2013.06.031

Tidow H, Nissen P (2013) Structural diversity of calmodulin binding to its target sites. FEBS J 280 (21):5551–5565. https://doi.org/10.1111/febs.12296

Trueman SF, Mandon EC, Gilmore R (2011) Translocation channel gating kinetics balances protein translocation efficiency with signal sequence recognition fidelity. Mol Biol Cell 22 (17):2983–2993. https://doi.org/10.1091/mbc.E11-01-0070

Trueman SF, Mandon EC, Gilmore R (2012) A gating motif in the translocation channel sets the hydrophobicity threshold for signal sequence function. J Cell Biol 199(6):907–918. https://doi.org/10.1083/jcb.201207163

Tsukazaki T, Mori H, Fukai S, Ishitani R, Mori T, Dohmae N, Perederina A, Sugita Y, Vassylyev DG, Ito K, Nureki O (2008) Conformational transition of Sec machinery inferred from bacterial SecYE structures. Nature 455:988. https://doi.org/10.1038/nature07421. https://www.nature.com/articles/nature07421#supplementary-information

Tyedmers J, Lerner M, Wiedmann M, Volkmer J, Zimmermann R (2003) Polypeptide-binding proteins mediate completion of co-translational protein translocation into the mammalian endoplasmic reticulum. EMBO Rep 4(5):505–510

Ueno T, Tanaka K, Kaneko K, Taga Y, Sata T, Irie S, Hattori S, Ogawa-Goto K (2010) Enhancement of procollagen biosynthesis by p180 through augmented ribosome association on the endoplasmic reticulum in response to stimulated secretion. J Biol Chem 285(39):29941–29950. https://doi.org/10.1074/jbc.M109.094607

Ueno T, Kaneko K, Sata T, Hattori S, Ogawa-Goto K (2011) Regulation of polysome assembly on the endoplasmic reticulum by a coiled-coil protein, p180. Nucl Acids Res. Advance online publication. https://doi.org/10.1093/nar/gkr1197

Valm AM, Cohen S, Legant WR, Melunis J, Hershberg U, Wait E, Cohen AR, Davidson MW, Betzig E, Lippincott-Schwartz J (2017) Applying systems-level spectral imaging and analysis to reveal the organelle interactome. Nature 546:162–167. https://doi.org/10.1038/nature22369. https://www.nature.com/articles/nature22369#supplementary-information

Van den Berg B, Clemons WM Jr, Collinson I, Modis Y, Hartmann E, Harrison SC, Rapoport TA (2004) X-ray structure of a protein-conducting channel. Nature 427(6969):36–44

Vilardi F, Lorenz H, Dobberstein B (2011) WRB is the receptor for TRC40/Asna1-mediated insertion of tail-anchored proteins into the ER membrane. J Cell Sci 124(8):1301–1307. https://doi.org/10.1242/jcs.084277

Voeltz G, Rolls M, Rapoport T (2002) Structural organization of the endoplasmic reticulum. EMBO Rep 3(10):944–950. https://doi.org/10.1093/embo-reports/kvf202

Voigt S, Jungnickel B, Hartmann E, Rapoport TA (1996) Signal sequence-dependent function of the TRAM protein during early phases of protein transport across the endoplasmic reticulum membrane. J Cell Biol 134(1):25–35. https://doi.org/10.1083/jcb.134.1.25

Voigt F, Zhang H, Cui XA, Triebold D, Liu AX, Eglinger J, Lee ES, Chao JA, Palazzo AF (2017) Single-molecule quantification of translation-dependent association of mRNAs with the endoplasmic reticulum. Cell Rep 21(13):3740–3753. https://doi.org/10.1016/j.celrep.2017.12.008

von Heijne G (2006) Membrane-protein topology. Nat Rev Mol Cell Biol 7:909. https://doi.org/10.1038/nrm2063. https://www.nature.com/articles/nrm2063#supplementary-information

von Heijne G, Gavel Y (1988) Topogenic signals in integral membrane proteins. Eur J Biochem 174(4):671–678. https://doi.org/10.1111/j.1432-1033.1988.tb14150.x

von Heijne G (1989) Control of topology and mode of assembly of a polytopic membrane protein by positively charged residues. Nature 341(6241):456–458. https://doi.org/10.1038/341456a0

Voorhees RM, Hegde RS (2015) Structures of the scanning and engaged states of the mammalian SRP-ribosome complex. eLife 4:e07975. https://doi.org/10.7554/elife.07975

Voorhees RM, Hegde RS (2016a) Structure of the Sec61 channel opened by a signal sequence. Science 351(6268):88–91. https://doi.org/10.1126/science.aad4992

Voorhees RM, Hegde RS (2016b) Toward a structural understanding of co-translational protein translocation. Curr Opin Cell Biol 41:91–99. https://doi.org/10.1016/j.ceb.2016.04.009

Voorhees RM, Fernández IS, Scheres SHW, Hegde RS (2014) Structure of the mammalian ribosome-Sec61 complex to 3.4 Å resolution. Cell 157(7):1632–1643. https://doi.org/10.1016/j.cell.2014.05.024

Wada I, Rindress D, Cameron PH, Ou W-J, Doherty JJ, Louvard D, Bell AW, Dignard D, Thomas DY, Bergeron JJM (1991) SSRa and associated calnexin are major calcium binding proteins of the endoplasmic reticulum membrane. J Biol Chem 266(29):19599–19610

Walter P, Ron D (2011) The unfolded protein response: from stress pathway to homeostatic regulation. Science 334(6059):1081–1086. https://doi.org/10.1126/science.1209038

Wang S, Kaufman RJ (2012) The impact of the unfolded protein response on human disease. J Cell Biol 197(7):857–867. https://doi.org/10.1083/jcb.201110131

Wang F, Chan C, Weir NR, Denic V (2014) The Get1/2 transmembrane complex is an endoplasmic-reticulum membrane protein insertase. Nature 512(7515):441–444. https://doi.org/10.1038/nature13471. http://www.nature.com/nature/journal/v512/n7515/abs/nature13471.html#supplementary-information

Wang Q-C, Zheng Q, Tan H, Zhang B, Li X, Yang Y, Yu J, Liu Y, Chai H, Wang X, Sun Z, Wang J-Q, Zhu S, Wang F, Yang M, Guo C, Wang H, Zheng Q, Li Y, Chen Q, Zhou A, Tang T-S (2016) TMCO1 is an ER Ca^{2+} load-activated Ca^{2+} channel. Cell 165(6):1454–1466. https://doi.org/10.1016/j.cell.2016.04.051

Weihofen A, Binns K, Lemberg MK, Ashman K, Martoglio B (2002) Identification of signal peptide peptidase, a presenilin-type aspartic protease. Science 296(5576):2215–2218

Westrate LM, Lee JE, Prinz WA, Voeltz GK (2015) Form follows function: the importance of endoplasmic reticulum shape. Annu Rev Biochem 84(1):791–811. https://doi.org/10.1146/annurev-biochem-072711-163501

White SH, von Heijne G (2008) How translocons select transmembrane helices. Annu Rev Biophys 37(1):23–42. https://doi.org/10.1146/annurev.biophys.37.032807.125904

Wild R, Kowal J, Eyring J, Ngwa EM, Aebi M, Locher KP (2018) Structure of the yeast oligosaccharyltransferase complex gives insight into eukaryotic N-glycosylation. Science 359 (6375):545–550. https://doi.org/10.1126/science.aar5140

Wirth A, Jung M, Bies C, Frien M, Tyedmers J, Zimmermann R, Wagner R (2003) The Sec61p complex is a dynamic precursor activated channel. Mol Cell 12(1):261–268

Wu X, Cabanos C, Rapoport TA (2018) Structure of the post-translational protein translocation machinery of the ER membrane. Nature. https://doi.org/10.1038/s41586-018-0856-x

Xue S, Tian S, Fujii K, Kladwang W, Das R, Barna M (2015) RNA regulons in Hox 5[prime] UTRs confer ribosome specificity to gene regulation. Nature 517(7532):33–38. https://doi.org/10.1038/nature14010. http://www.nature.com/nature/journal/v517/n7532/abs/nature14010.html#supplementary-information

Yabal M, Brambillasca S, Soffientini P, Pedrazzini E, Borgese N, Makarow M (2003) Translocation of the C terminus of a tail-anchored protein across the endoplasmic reticulum membrane in yeast mutants defective in signal peptide-driven translocation. J Biol Chem 278 (5):3489–3496. https://doi.org/10.1074/jbc.M210253200

Yamaguchi A, Hori O, Stern DM, Hartmann E, Ogawa S, Tohyama M (1999) Stress-associated endoplasmic reticulum protein 1 (Serp1)/ribosome-associated membrane protein 4 (Ramp4) stabilizes membrane proteins during stress and facilitates subsequent glycosylation. J Cell Biol 147(6):1195–1204. https://doi.org/10.1083/jcb.147.6.1195

Yamamoto Y, Sakisaka T (2012) Molecular machinery for insertion of tail-anchored membrane proteins into the endoplasmic reticulum membrane in mammalian cells. Mol Cell 48(3):387–397. https://doi.org/10.1016/j.molcel.2012.08.028

Yanagitani K, Kimata Y, Kadokura H, Kohno K (2011) Translational pausing ensures membrane targeting and cytoplasmic splicing of XBP1u mRNA. Science 331(6017):586–589. https://doi.org/10.1126/science.1197142

Yu YH, Sabatini DD, Kreibich G (1990) Antiribophorin antibodies inhibit the targeting to the ER membrane of ribosomes containing nascent secretory polypeptides. J Cell Biol 111(4):1335–1342. https://doi.org/10.1083/jcb.111.4.1335

Zehner M, Marschall Andrea L, Bos E, Schloetel J-G, Kreer C, Fehrenschild D, Limmer A, Ossendorp F, Lang T, Koster Abraham J, Dübel S, Burgdorf S (2015) The translocon protein Sec61 mediates antigen transport from endosomes in the cytosol for cross-presentation to $CD8^+$ T cells. Immunity 42(5):850–863. https://doi.org/10.1016/j.immuni.2015.04.008

Zhang H, Hu J (2016) Shaping the endoplasmic reticulum into a social network. Trends Cell Biol. http://dx.doi.org/10.1016/j.tcb.2016.06.002

Zhang SL, Yeromin AV, Zhang XH, Yu Y, Safrina O, Penna A, Roos J, Stauderman KA, Cahalan MD (2006) Genome-wide RNAi screen of Ca^{2+} influx identifies genes that regulate Ca^{2+} release-activated Ca^{2+} channel activity. Proc Natl Acad Sci 103(24):9357–9362

Zhao L, Ackerman SL (2006) Endoplasmic reticulum stress in health and disease. Curr Opin Cell Biol 18(4):444–452. https://doi.org/10.1016/j.ceb.2006.06.005

Zhao Y, Hu J, Miao G, Qu L, Wang Z, Li G, Lv P, Ma D, Chen Y (2013) Transmembrane protein 208: a novel ER-localized protein that regulates autophagy and ER stress. PLoS ONE 8(5): e64228. https://doi.org/10.1371/journal.pone.0064228

Zhou H-X, Rivas G, Minton AP (2008) Macromolecular crowding and confinement: biochemical, biophysical, and potential physiological consequences. Annu Rev Biophys 37:375–397. https://doi.org/10.1146/annurev.biophys.37.032807.125817

Zhu Y, Zhang G, Lin S, Shi J, Zhang H, Hu J (2018) Sec61β facilitates the maintenance of endoplasmic reticulum homeostasis by associating microtubules. Protein Cell 9(7):616–628. https://doi.org/10.1007/s13238-017-0492-5

Zimmer J, Nam Y, Rapoport TA (2008) Structure of a complex of the ATPase SecA and the protein-translocation channel. Nature 455(7215):936–943

Zimmermann R (2016) Components and mechanisms of import, modification, folding, and assembly of immunoglobulins in the endoplasmic reticulum. J Clin Immunol 36(1):5–11. https://doi.org/10.1007/s10875-016-0250-0

Ziska A, Tatzelt J, Dudek J, Paton AW, Paton JC, Zimmermann R, Haßdenteufel S (2019) The signal peptide plus a cluster of positive charges in prion protein dictate chaperone-mediated Sec61 channel gating. Biol Open bio.040691. https://doi.org/10.1242/bio.040691

Chapter 5
The Structures of Eukaryotic Transcription Pre-initiation Complexes and Their Functional Implications

Basil J. Greber and Eva Nogales

Abstract Transcription is a highly regulated process that supplies living cells with coding and non-coding RNA molecules. Failure to properly regulate transcription is associated with human pathologies, including cancers. RNA polymerase II is the enzyme complex that synthesizes messenger RNAs that are then translated into proteins. In spite of its complexity, RNA polymerase requires a plethora of general transcription factors to be recruited to the transcription start site as part of a large transcription pre-initiation complex, and to help it gain access to the transcribed strand of the DNA. This chapter reviews the structure and function of these eukaryotic transcription pre-initiation complexes, with a particular emphasis on two of its constituents, the multisubunit complexes TFIID and TFIIH. We also compare the overall architecture of the RNA polymerase II pre-initiation complex with those of RNA polymerases I and III, involved in transcription of ribosomal RNA and non-coding RNAs such as tRNAs and snRNAs, and discuss the general, conserved features that are applicable to all eukaryotic RNA polymerase systems.

Keywords Transcription · Initiation · General transcription factors · TFIIH · TFIID · RNA polymerase · Gene expression · Structural biology · Cryo-electron microscopy

---

B. J. Greber (✉) · E. Nogales
California Institute for Quantitative Biosciences (QB3), University of California, Berkeley, CA 94720, USA
e-mail: basilgreber@berkeley.edu

B. J. Greber · E. Nogales
Molecular Biophysics and Integrative Bio-Imaging Division, Lawrence Berkeley National Laboratory, Berkeley, CA 94720, USA

E. Nogales
Howard Hughes Medical Institute, University of California, Berkeley, CA 94720, USA

E. Nogales
Department of Molecular and Cell Biology, University of California, Berkeley, CA 94720, USA

J. R. Harris and J. Marles-Wright (eds.), *Macromolecular Protein Complexes II: Structure and Function*, Subcellular Biochemistry 93,
https://doi.org/10.1007/978-3-030-28151-9_5

Introduction to Transcription in Eukaryotes

Transcription is the synthesis of RNA based on a DNA template. While phage, viral, and organellar gene expression systems generally use simple single-subunit RNA polymerase enzymes, bacterial, archaeal, and eukaryotic RNA polymerases are increasingly complex multisubunit enzymes. A 5-subunit core is conserved from bacteria to humans, though archaeal and eukaryotic RNA polymerases harbor up to 12 additional subunits (Fig. 5.1a–e). Eukaryotes use three structurally and functionally distinct RNA polymerases (Roeder and Rutter 1969, 1970; Sklar et al. 1975), abbreviated Pol I, Pol II, and Pol III, each of them specialized for the synthesis of certain classes of RNAs. Pol I synthesizes long ribosomal RNA (rRNA) precursors in the nucleolus that are later processed into 25S/28S, 18S, and 5.8S rRNAs, Pol II synthesizes mostly messenger RNAs (mRNAs) that are later translated into proteins, and Pol III synthesizes tRNAs, 5S rRNA, and other small non-coding RNAs (Roeder and Rutter 1970; Weinmann and Roeder 1974).

The DNA core promoters that control expression of their corresponding genes also differ substantially between the three eukaryotic polymerase systems. In the Pol II system, several core promoter elements have been identified (Burke and Kadonaga 1996, 1997; Chen and Struhl 1985; Deng and Roberts 2005; Gannon et al. 1979; Juven-Gershon and Kadonaga 2010; Lagrange et al. 1998; Lenhard et al. 2012; Lim et al. 2004; Roy and Singer 2015), and a synthetic promoter combining a TATA-box, initiator (Inr), downstream promoter element (DPE), and motif ten element (MTE), which provides a highly specific and high-affinity binding site for the transcription machinery (Juven-Gershon et al. 2006) has been used in structural analysis of promoter-bound human Pol II pre-initiation complexes. However, not all of these elements are typically present in any natural promoter. For example, most Pol II promoters lack a TATA box (Carninci et al. 2006; Sandelin et al. 2007), the canonical binding site of TATA-box binding protein (TBP), which is a subunit of the general transcription factor IID (TFIID) and also ubiquitously used in the Pol I and Pol III systems (see below). Nevertheless, TBP still participates in assembly of the transcription machinery on Pol II promoters lacking TATA boxes because the recognition of the other core promoter motifs by TFIID allows TBP loading on the DNA (Burke and Kadonaga 1997; Kutach and Kadonaga 2000; Lim et al. 2004; Pugh and Tjian 1991). Panels f–h of Fig. 5.1 provide examples of the promoter elements typically found in eukaryotic genes.

Transcription is commonly subdivided into three phases (Fig. 5.1i). After the step-wise assembly of a pre-initiation complex (PIC) on the DNA (Buratowski et al. 1989), whereby general transcription factors and RNA polymerase are recruited to the promoter, the transcription bubble opens, allowing the synthesis of the first phosphodiester bond in the active site of the RNA polymerase. This step is referred to as *initiation* of transcription. Subsequently, RNA polymerase needs to clear the promoter and, in the case of Pol II, overcome a promoter-proximal pause, allowing

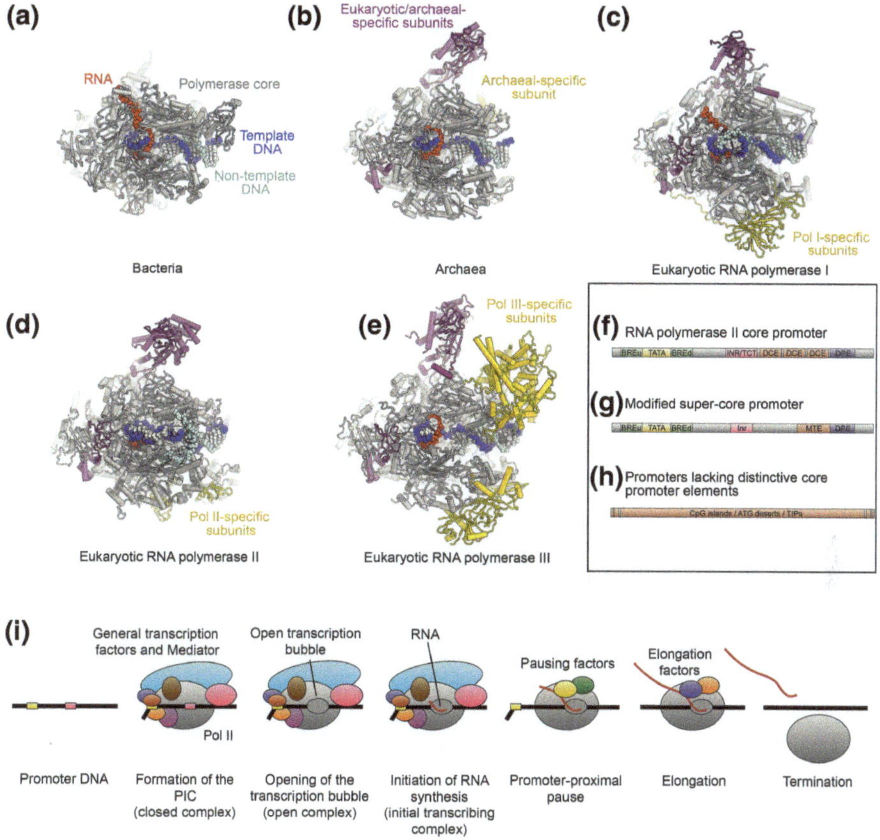

Fig. 5.1 Architecture of multisubunit RNA polymerases across kingdoms and organization of eukaryotic Pol II promoters. **a–e** Depictions of transcribing multisubunit RNA polymerases. Subunits occurring only in eukaryotes or archaea are colored purple, subunits specific to polymerase type are colored yellow. **a** Elongating bacterial RNA polymerase (Vassylyev et al. 2007). **b** Archaeal RNA polymerase (Korkhin et al. 2009), DNA superposed from (Vassylyev et al. 2007). **c** Elongating yeast Pol I (Tafur et al. 2016). **d** Elongating fungal Pol II (Ehara et al. 2017). **d** Elongating yeast Pol III (Hoffmann et al. 2015). **f–h** Organization of eukaryotic Pol II core promoters. **f** Idealized Pol II core promoter elements. Abbreviations: TATA, TATA-box; INR, initiator; TCT, initiator like element present in ribosomal protein coding genes; DCE, downstream core element; DPE, downstream promoter element (Lenhard et al. 2012; Roy and Singer 2015). **g** Modified super core promoter used for structural studies of the human Pol II-PIC and TFIID. The upstream and downstream TFIIB-response elements BREu (Lagrange et al. 1998) and BREd (Deng and Roberts 2005) were added to the super core promoter (Juven-Gershon et al. 2006), which comprises TATA-box (Gannon et al. 1979), Inr (Chen and Struhl 1985), MTE (Lim et al. 2004), and DPE (Burke and Kadonaga 1996, 1997). **h** Many eukaryotic Pol II promoters lack these distinct promoter elements and contain longer DNA segments harbouring CpG islands, ATG deserts, or transcription initiation platforms (TIPs) instead (Roy and Singer 2015). **i** Simplified schematic of the Pol II transcription cycle from PIC assembly to transcription termination

it to enter the *elongation* phase. After synthesis of the product RNA, transcription needs to be terminated during the *termination* phase of the transcription reaction.

The initiation of transcription is a highly regulated process. Together with the subsequent regulation of RNA processing, RNA stability, and regulated translation of mRNAs into proteins, transcription plays a major role in the control of gene expression as a whole. Deregulation of transcription is implicated in human pathologies, and elevated transcription levels by each of the three eukaryotic RNA polymerases have been associated with cancer (Cabarcas and Schramm 2011; Johnson et al. 2008; Lockwood et al. 2010).

The Pol II Pre-initiation Complex

Pol II and the General Transcription Factors

The enzymatic core of Pol II shows extensive structural homology to the 5-subunit $\alpha_2\beta\beta'\omega$ core of bacterial RNA polymerases (Cramer et al. 2000; Zhang et al. 1999) (Fig. 5.1a, d) and the fundamental mechanism of phosphodiester bond synthesis is highly conserved (Cramer et al. 2001; Gnatt et al. 2001; Kaplan 2013; Svetlov and Nudler 2013; Wang et al. 2006; Westover et al. 2004). However, Pol II includes 7 additional subunits to form a 12-subunit complex (Figs. 5.1a, d and 5.2a, b). Structural landmarks of Pol II (Fig. 5.2c, d) include the stalk, two jaws that interact with the incoming DNA substrate, the flexible clamp domain that closes over the bound nucleic acids in initiation and elongating complexes, and two pores that allow substrates to enter and product mRNAs to exit the catalytically active complex (Fig. 5.2c, d) (Cramer et al. 2000, 2001; Gnatt et al. 2001). Behind the unwound transcription bubble in the active site, a protein wall forces the DNA to adopt a bend before exiting the polymerase (Fig. 5.2c). In the transcribing complex (Fig. 5.2e, f), two fork loops from the Rpb2 subunit maintain the open transcription bubble (Cramer et al. 2001; Gnatt et al. 2001). Motions of the bridge helix that contacts the DNA-RNA hybrid in the active site are coupled to substrate translocation, while opening and closing of the trigger loop is associated with catalysis, which is mediated by two metal ions bound in the active site (Fig. 5.2f) (Gnatt et al. 2001; Wang et al. 2006; Westover et al. 2004).

Fig. 5.2 Architecture of eukaryotic RNA polymerase II. **a, b** Depiction of transcribing Pol II ▶ extracted from the structure of an elongation complex (Ehara et al. 2017), shown from the front and back and colored by protein subunits. **c, d** Structural domains of the Pol II subunits Rpb1 and Rpb2 are highlighted in color according to (Cramer et al. 2001). Additional structural landmarks mentioned in the text are highlighted as well (protein names are indicated). **e** View of the transcription bubble at the Pol II active site (Barnes et al. 2015). Key structural and functional elements near the active site that are mentioned in the text are indicated. **f** Detailed view of the Pol II active site (Barnes et al. 2015; Wang et al. 2006). Two conformations of the trigger loop are shown in orange and yellow

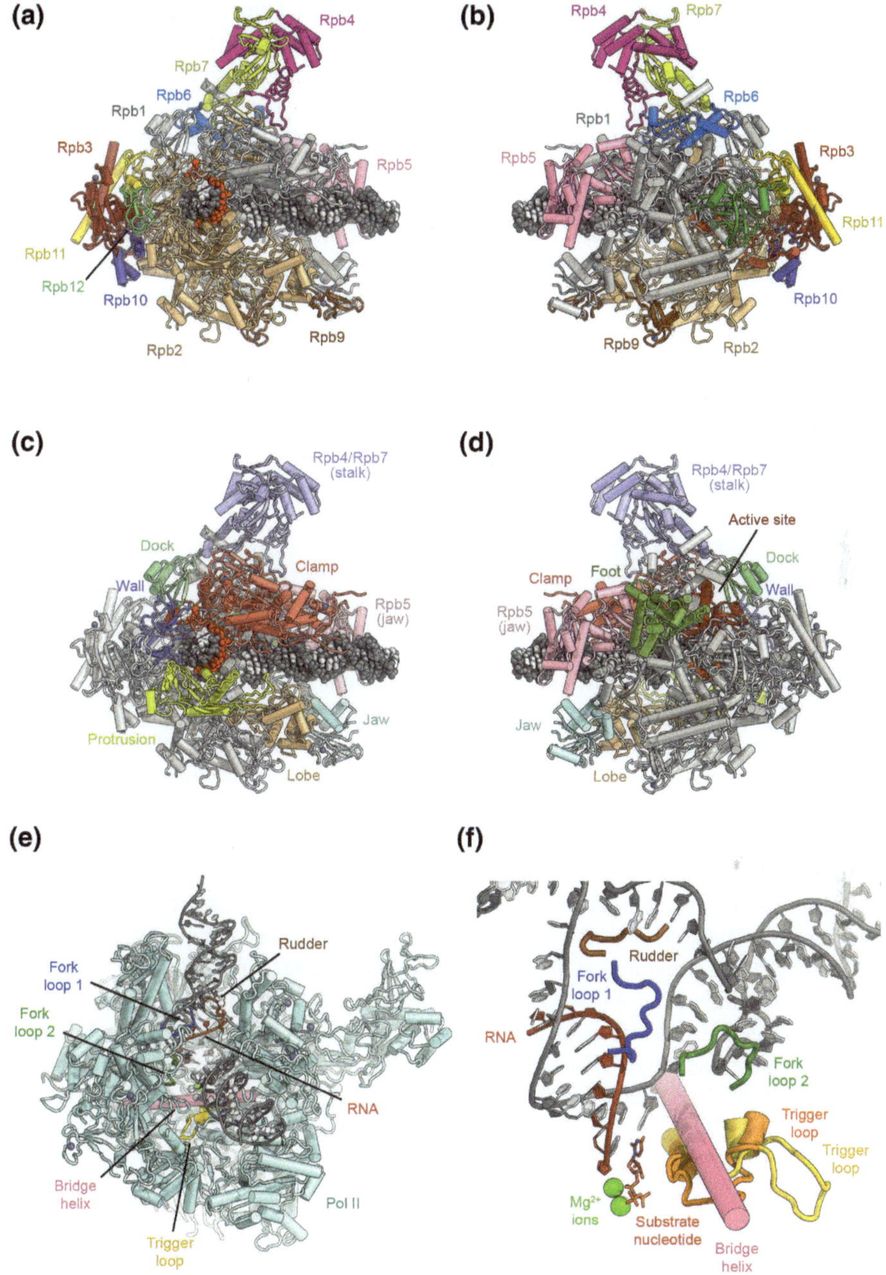

(a)
Rpb4
Rpb7
Rpb6
Rpb1
Rpb3
Rpb5
Rpb11
Rpb12
Rpb10
Rpb2
Rpb9
(b)
Rpb4
Rpb7
Rpb6
Rpb1
Rpb5
Rpb3
Rpb11
Rpb10
Rpb9
Rpb2
(c)
Rpb4/Rpb7 (stalk)
Dock
Clamp
Wall
Rpb5 (jaw)
Protrusion
Jaw
Lobe
(d)
Rpb4/Rpb7 (stalk)
Active site
Clamp
Foot
Dock
Wall
Rpb5 (jaw)
Jaw
Lobe
(e)
Rudder
Fork loop 1
Fork loop 2
RNA
Bridge helix
Trigger loop
Pol II
(f)
Rudder
Fork loop 1
RNA
Fork loop 2
Trigger loop
Trigger loop
Mg2+ ions
Substrate nucleotide
Bridge helix

The initiation of mRNA synthesis requires Pol II and the general transcription factors TFIIA, TFIIB, TFIID, TFIIE, TFIIF, and TFIIH (Sainsbury et al. 2015) to form a pre-initiation complex (PIC) on the promoter. Activated transcription additionally depends on Mediator, a 1.2 MDa multiprotein complex, and the activity of a number of cofactors and gene-specific activators and repressors. In contrast to the smaller, oligomeric general transcription factors TFIIA, TFIIB, TFIIE, and TFIIF, two general transcription factors are large multiprotein assemblies: TFIIH is a heterodecameric protein complex with a molecular weight of 0.5 MDa, and TFIID is a 1.3 MDa multiprotein assembly of TATA-box binding protein (TBP) and 13 different TBP-associated factors (TAFs) in humans. We will discuss the structures and functional mechanisms of these two general transcription factors in more detail in the following sections, while the remaining general transcription factors will be discussed in the context of the Pol II-PIC.

TFIID

Function of TFIID

The assembly of the PIC begins with the recognition of the core promoter DNA by TFIID, with the help of TFIIA (Burke and Kadonaga 1997; Chalkley and Verrijzer 1999; Lee et al. 2005; Verrijzer et al. 1995). TBP then interacts with TFIIB, which in turn interacts with Pol II, thereby nucleating the assembly of the full PIC (Buratowski et al. 1989). In addition to TBP, direct interactions between TFIID TAFs and other components of the general transcription machinery have also been proposed (Dubrovskaya et al. 1996; Ruppert and Tjian 1995). Furthermore, TFIID, together with Mediator and the multifunctional SAGA complex, integrates signals from transcriptional activators and has been shown to be required for activated transcription (Baek et al. 2002; Chen et al. 1993, 1994; Horikoshi et al. 1988; Pugh and Tjian 1990; Stringer et al. 1990; Wu and Chiang 2001). Additionally, a number of domains within the complex recognize specific histone modifications associated with active genes (Jacobson et al. 2000; Vermeulen et al. 2007). Thus, TFIID is important for nucleating PIC assembly at the promoter and also serves as a hub for the integration of regulatory cues that are important for regulated transcription.

Conformational Complexity of TFIID

Due to the size and complexity of TFIID, structural efforts to characterize the full complex have involved electron microscopy (EM) of natively purified samples. Early studies of negatively stained human and budding yeast TFIID showed it to have an overall horseshoe-shaped architecture of three major lobes, termed lobes A, B and C (Andel et al. 1999; Brand et al. 1999; Leurent et al. 2002). Improving the resolution by cryo-EM proved to be extremely challenging due to conformational heterogeneity (Elmlund et al. 2009; Grob et al. 2006; Papai et al. 2009). It was ultimately shown that lobes B and C form a relatively stable core (the BC core), while lobe A is flexibly attached and can undergo conformational rearrangements on the scale of 100 Å, from interacting with lobe C to interacting with lobe B (Fig. 5.3a) (Cianfrocco et al. 2013). Free TFIID was found to occupy two preferred

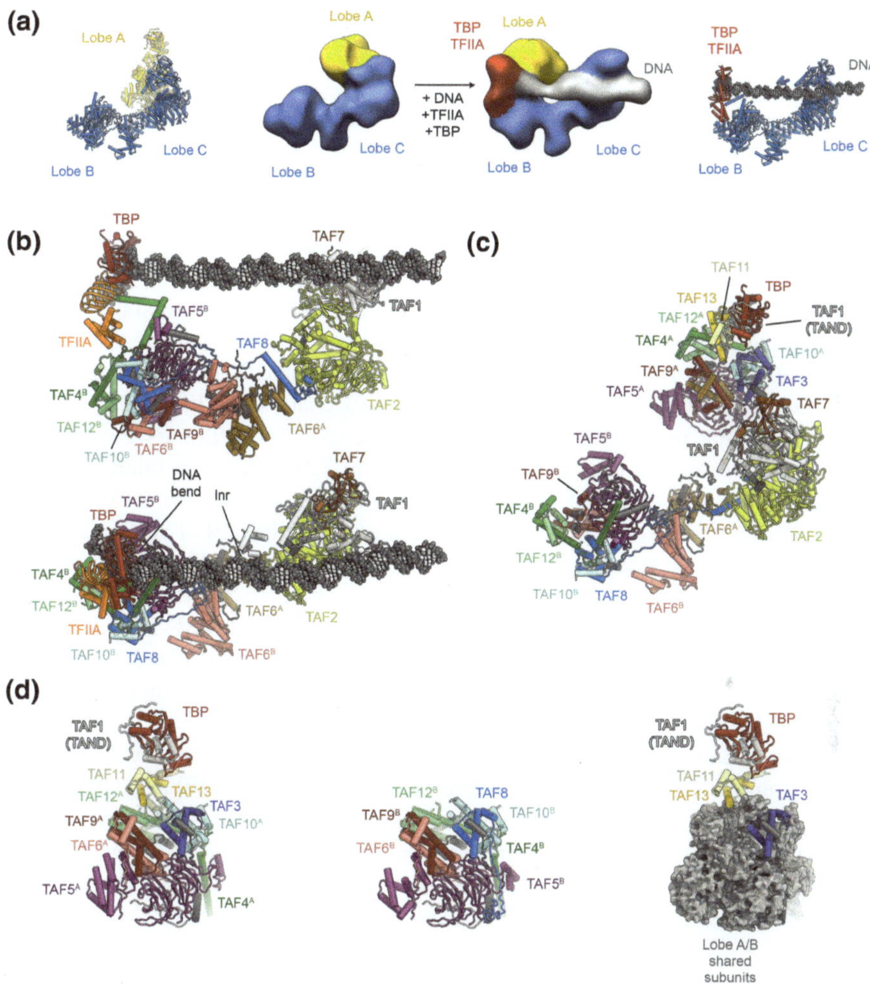

Fig. 5.3 The structure of human TFIID. **a** Organization of human TFIID into three lobes. Early cryo-EM maps (Cianfrocco et al. 2013) that revealed the tri-lobal overall structure of TFIID are shown in the center, with recent coordinate models (Patel et al. 2018) shown next to them for comparison. **b** Structure of TFIID bound to DNA and TFIIA (Patel et al. 2018) in side view (top) and top view (bottom). Protein subunits are shown in colour and labelled; the TBP-induced DNA bend and the location of the transcription start site (Inr) are labeled. **c** Structure of TFIID in the canonical state (Patel et al. 2018) shown in the same orientation as bottom panel in (**b**). **d** Composition of lobes A and B. Left: lobe A. Center: lobe B. Right: Subunits shared between lobes A and B are shown as grey surface to highlight the subunits in lobe A that distinguish between lobes A and B

conformational states, termed the canonical and rearranged states, with addition of TFIIA and promoter DNA shifting this equilibrium towards the rearranged state (Fig. 5.3a). In this conformation, lobe A interacts with lobe B and is positioned near the TATA box, suggesting the presence of TBP within lobe A (Cianfrocco et al. 2013) (Fig. 5.3a).

The Structure of TFIID

Subsequent structural studies of human TFIID aimed to provide higher-resolution analysis of both free and DNA-bound TFIID (Louder et al. 2016; Patel et al. 2018) (Fig. 5.3b, c). While we will focus on the structure of human TFIID below, similar work in yeast has also been reported recently, although promoter DNA could not be clearly visualized in the cryo-EM structure of promoter-bound yeast TFIID (Kolesnikova et al. 2018).

The structure of stably DNA-bound human TFIID (Fig. 5.3b), using the super core promoter, and in complex with TFIIA, showed that lobe C includes TAFs 1, 2, and 7, with a TAF1-TAF7 module binding the DPE and MTE downstream promoter elements, while a small domain within TAF1 contacts the Inr motif (indicated in Fig. 5.3b) that includes the transcription start site (TSS) (Louder et al. 2016). Lobe C is connected to lobe B via a long helix in TAF8 that also interacts with a centrally positioned dimer of HEAT repeats from the two copies of TAF6. The TAF6 dimer connects to lobe B, which in turn contacts the TBP-TFIIA module, as well as promoter sequences near the TATA box (Louder et al. 2016; Nogales et al. 2017a; Nogales et al. 2017c). The TATA box DNA in this structure is bent (Fig. 5.3b, bottom), mirroring the structure of the isolated TBP-DNA complex (Nikolov et al. 1996). By virtue of this arrangement, lobes B and C act as a molecular ruler around the TSS, providing a platform for transcription pre-initiation complex assembly (Louder et al. 2016; Nogales et al. 2017b).

The complete molecular architecture of all three lobes of human TFIID was derived from the cryo-EM analysis of free TFIID (Patel et al. 2018) (Fig. 5.3c). Due to the scarcity, structural heterogeneity, fragility, and preferred orientation of this specimen on the cryo-EM grid, determination of this structure was a tour-de-force. A striking structural feature of TFIID is that lobes A and B share a common core architecture (Fig. 5.3d), with six TAFs being present in two copies, one in each lobe (Bieniossek et al. 2013; Leurent et al. 2002; Patel et al. 2018; Sanders et al. 2002). In this common module, TAF5, which contains a helical N-terminal domain and a WD40 domain, binds to three dimers of TAFs that contain histone-fold domains (HFDs) and themselves form a hexameric assembly that resembles the overall architecture of the nucleosome core particle (Patel et al. 2018), in agreement with earlier biochemical and structural data (Hoffmann et al. 1996; Xie et al. 1996). TAF5 and five of the histone fold-containing TAFs, TAF10, and the pairs TAF4-TAF12 and TAF6-TAF9, are present in both lobes A and B (Fig. 5.3d). The TAF6 HFDs present in both lobes, A and B, are flexibly connected to their corresponding C-terminal HEAT repeat domains that exist as a dimer at the center of the complex, within lobe C (Fig. 5.3b). In spite of their shared core architecture, lobes A and B are functionally and structurally unique because their common TAF10 subunit associates with the HFD of TAF3 in lobe A but with the HFD of TAF8 in lobe B (Fig. 5.3d). TAF8 contributes to the lobe B-lobe C interaction, while TAF3 recruits the TAF11-TAF13 HFD dimer to lobe A. Together with N-terminal regions of TAF1 (Anandapadamanaban et al. 2013; Kokubo et al. 1993, 1994), the TAF11-TAF13 pair binds TBP, which consequently hangs peripherally on this mobile lobe A within TFIID.

Mechanism of TBP Loading onto the Promoter

The extreme mobility of lobe A, along with its ability to carry TBP, suggests a role for the dynamics of this lobe in the deposition of TBP onto the promoter (Fig. 5.4a). An exhaustive classification of a cryo-EM dataset where TFIID, TFIIA and core promoter DNA were mixed together revealed five distinct states, including previously observed as well as new conformational and functional states of lobe A. These states were termed canonical, extended, scanning, rearranged and engaged states (Fig. 5.4a), with the first corresponding to lobe A positioned proximal to lobe C, and consequent states being further from lobe C and ultimately contacting lobe B (Patel et al. 2018). The model derived from the co-existence of these states proposes that in the canonical state, where lobe A is far removed from the DNA, TBP is bound by the TAF11-TAF13 HFD dimer, as well as the TAF1 TAND1 and TAND2 motifs (Figs. 5.3d, 5.4b), which have been implicated in inhibition of DNA and TFIIA binding by TBP (Fig. 5.4c) (Anandapadamanaban et al. 2013; Kokubo et al. 1993, 1994). These inhibitory interactions are sequentially broken in the scanning and rearranged states as they are replaced by interactions of TBP with DNA and TFIIA (Fig. 5.4d), facilitated by the scaffolding role of TFIID and by lobe A-lobe B interactions. Specifically, we have proposed that in the scanning state TAND1 is displaced as TBP loosely associates with the DNA, while in the rearranged state TAND2 is displaced as TBP binds to TFIIA that also interacts with lobe B. In the final, engaged state, TBP binds tightly and bends the DNA, and it releases from TAF11-TAF13. Displacement of TAF11-TAF13 opens up the binding site for TFIIB, enabling PIC assembly (Patel et al. 2018).

Notably, current data indicate that the overall architecture of TFIID is conserved between promoters that do and do not contain TATA boxes (Cianfrocco et al. 2013; Patel et al. 2018), in agreement with the observation that TBP also participates in initiation on TATA-less promoters (Pugh and Tjian 1991). The intricate mechanism of TBP loading onto the upstream promoter may ensure that even in the absence of a TATA box, TBP engages with DNA in a defined, controlled, and productive fashion, rather than binding DNA nonspecifically. Completion of TBP loading onto the core promoter in such cases (reaching the engaged state) may depend on chromatin marks, transcriptional activators, or co-activator complexes, because both structural and biochemical analysis shows that TBP only weakly binds TATA-free promoters

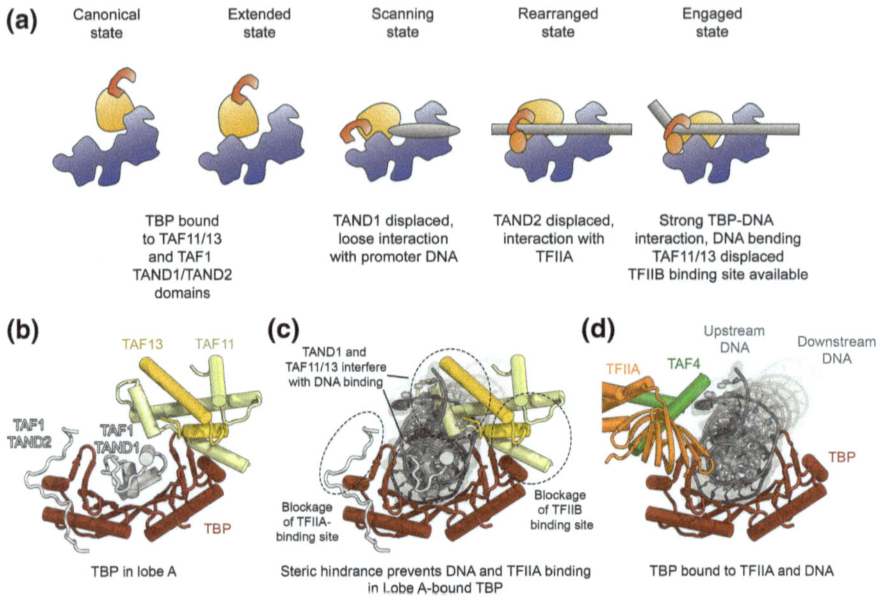

Fig. 5.4 Proposed conformational transitions of TFIID during DNA binding and mechanism of TBP deposition. **a** Five different conformational states (canonical, extended, scanning, rearranged, and engaged state) of TFIID as identified by cryo-EM analysis of a mixed dataset (Patel et al. 2018). For each state, a schematic depiction and the interaction environment of TBP, which is initially bound to TFIID lobe A and eventually deposited onto upstream promoter DNA in a stably bound manner, is provided. **b** TBP in lobe A is contacted by the TAF11-TAF13 dimer and the TAND1 and TAND2 domains of TAF1 (Anandapadamanaban et al. 2013; Patel et al. 2018). **c** TAF11-TAF13 and the TAF1 TAND1 and TAND2-domains block DNA- and transcription factor-binding sites on TBP. **d** TBP bound to and bending the DNA in the context of the engaged state of DNA-bound TFIID (Patel et al. 2018)

in reconstituted systems that contain TFIID and TFIIA (Cianfrocco et al. 2013) even though transcription assays show transcription from such promoters (Juven-Gershon et al. 2006; Lim et al. 2004).

TFIIH

Functional Roles of TFIIH in Transcription

TFIIH is a protein complex of ten different subunits organized into a core and a CdK activating kinase (CAK) subcomplex (Fig. 5.5a). Both the core and the CAK are required for TFIIH to function in transcription initiation, while only the TFIIH core complex is required in DNA repair (Svejstrup et al. 1995). The TFIIH core complex includes seven protein subunits: XPB, XPD, p62, p52, p44, p34, and p8

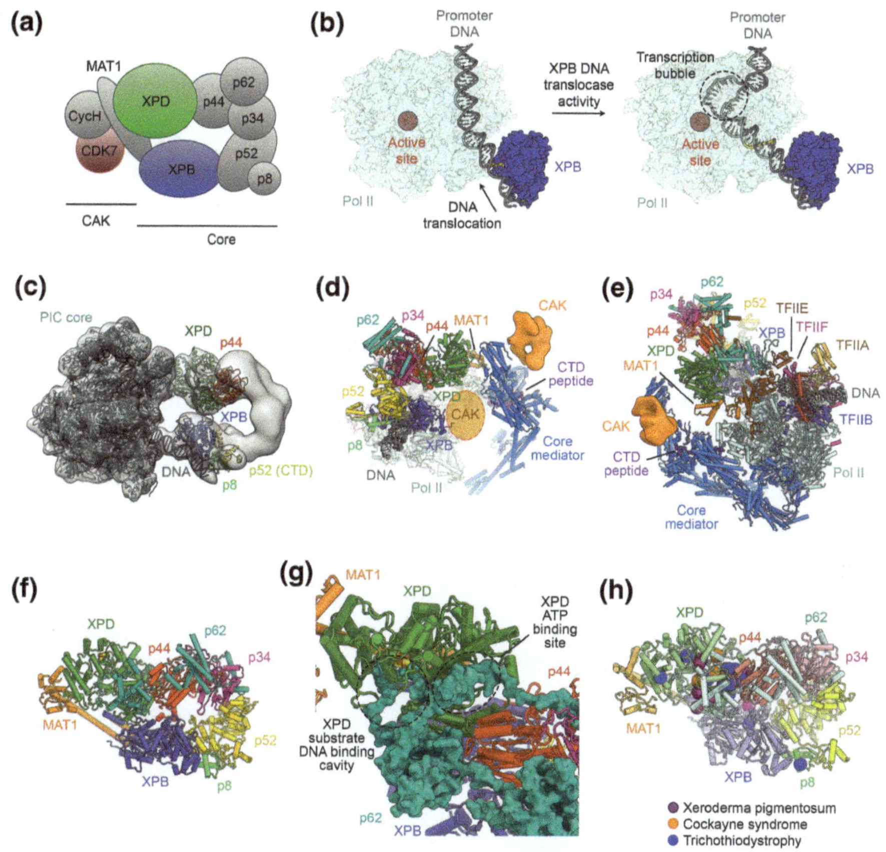

Fig. 5.5 Structure and function of TFIIH. **a** Schematic of TFIIH architecture. Main enzymatic subunits are shown in color (XPD and XPB, SF2-family DNA-dependent ATPases; CDK7, Cyclin-dependent kinase). Core and CAK subcomplexes are indicated. **b** Model for the activity of the TFIIH subunit XPB during promoter opening; depiction based on the structures in (He et al. 2016). **c** Cryo-EM reconstruction of the human Pol II-PIC (He et al. 2016) with fitted coordinate models. TFIIH subunits are highlighted in color. **d**, **e** Relocalization of the CAK kinase-cyclin module in the Mediator-bound Pol II-PIC (Schilbach et al. 2017): The cryo-EM volume corresponding to the CAK in the Mediator-containing PIC is shown in orange. The presumed location of the CAK in the absence of Mediator is indicated by an orange ellipse. **f** Structure of the TFIIH core complex with bound MAT1 (Greber et al. 2019). **g** Interactions of p62 with the ATP- and DNA-binding sites of XPD. **h** Mapping of human disease mutations onto the structure of TFIIH (Cleaver et al. 1999; Greber et al. 2019)

(Fig. 5.5a). Two of these, XPB and XPD, are SF2-family DNA-dependent ATPases/helicases (Guzder et al. 1994; Sung et al. 1993; Weeda et al. 1990). While the enzymatic function of XPD is dedicated to DNA repair (Kuper et al. 2014), only XPB activity is required to help promoter opening during transcription initiation (Fishburn et al. 2015; Grünberg et al. 2012; Kim et al. 2000). Originally thought to act like a wrench to introduce torsion into downstream promoter DNA to enable opening of the transcription bubble (Kim et al. 2000), it was later proposed that XPB acts as a DNA translocase that promotes the movement of downstream promoter DNA towards the Pol II active site, thereby facilitating bubble opening (Fig. 5.5b) (Fishburn et al. 2015; Grünberg et al. 2012; He et al. 2013).

The CAK subcomplex includes the CDK7, Cyclin H, and MAT1 subunits (Devault et al. 1995; Fisher et al. 1995; Fisher and Morgan 1994; Shiekhattar et al. 1995) (Fig. 5.5a). CDK7 is a cyclin dependent kinase that requires the presence of Cyclin H for full activity, while MAT1 aids formation of the CDK-cyclin complex (Devault et al. 1995; Fisher et al. 1995) and anchors the CAK subcomplex to the core complex via interactions with XPD and XPB (Abdulrahman et al. 2013; Busso et al. 2000; Greber et al. 2017; Rossignol et al. 1997). The role of XPD in recruitment of the CAK subcomplex (which is required only for transcription) explains why its presence in TFIIH is important for transcription initiation, while its enzymatic activity is not (Dubaele et al. 2003; Kuper et al. 2014). The CAK is able to phosphorylate a number of targets, which include the C-terminal low-complexity region of the Pol II subunit Rpb1, referred to as the Pol II CTD here, as well as transcriptional activators, general transcription factors, and, in humans, cell-cycle CDKs (Fisher and Morgan 1994; Shiekhattar et al. 1995).

TFIIH as a DNA Repair Complex

In addition to its role as a Pol II general transcription factor, TFIIH also serves as a helicase complex in the nucleotide excision DNA repair pathway. This pathway is responsible for removing a large number of bulky base alterations that result in distortion of the DNA backbone, including 6-4 photoproducts, cyclobutane pyrimidine dimers, and other large base adducts (Koch et al. 2016). In these pathways, TFIIH serves to open a DNA bubble around damaged nucleotides, which can then be excised by the structure-specific endonucleases XPF-ERCC1 and XPG (Marteijn et al. 2014). Initial recognition of the damage occurs independently of TFIIH, either by a global genome surveillance pathway that depends on the DDB1-DDB2 and XPC complexes, or via transcription-coupled repair (TCR), in which Pol II stalled by DNA lesions acts as the initial damage sensor, leading to the recruitment of specialized TCR factors such as CSA and CSB, and ultimaltely, TFIIH (Compe and Egly 2012). Unlike in transcription initiation, the enzymatic activity of both XPD and XPB is required during NER. XPD acts as a bona-fide helicase to unwind the DNA double helix and scan for the DNA lesion (Kuper et al. 2014; Li et al. 2015), while the role of XPB has been described either as an ATP-dependent conformational switch that anchors TFIIH to DNA (Coin et al. 2007), or as a weak helicase and DNA translocase that supports scanning of the DNA for the lesion that originally triggered TFIIH recruitment (Li et al. 2015; Sugasawa et al. 2009). Upon lesion recognition in a binding pocket in XPD (Buechner et al. 2014; Mathieu et al. 2013), both helicase motors stall (Li et al. 2015;

Naegeli et al. 1993), enabling assembly of the repair bubble, double incision, and ultimately repair of the DNA lesion by DNA polymerases Pol δ and Pol κ (Compe and Egly 2012; Marteijn et al. 2014).

The Structure of TFIIH

Due to its importance for two critical cellular processes, the structure of TFIIH has been intensively investigated for almost two decades. Initial studies resulted in low-resolution EM reconstructions that showed a ring- or horseshoe-shaped complex (Chang and Kornberg 2000; Schultz et al. 2000) and allowed the approximate assignment of the XPB and XPD ATPase subunits to the open ends of a horseshoe (Gibbons et al. 2012). An integrative modelling approach provided a tentative overall architecture for the complex, which was, however, complicated by the horseshoe-shaped overall architecture of the complex that puts many subunits within crosslinking distance of each other, and the high flexibility of the CAK subcomplex (Luo et al. 2015). Sub-nanometer resolution structures of TFIIH in the context of both human and yeast PICs (Fig. 5.5c) confirmed the assignment of the ATPase subunits to density at the open end of the horseshoe-shaped TFIIH complex and revealed the locations of p44 near XPD and of the p52-p8 dimer near XPB (Fig. 5.5c) (He et al. 2013, 2016; Murakami et al. 2015), in agreement with regulatory roles of these subunits on the two ATPase subunits (Coin et al. 2007; Dubaele et al. 2003; Jawhari et al. 2002; Kim et al. 2015). The interpretation of these cryo-EM maps relied heavily on docking of crystal structures or homology models. Structures of bacterial and archaeal homologs of XPB (Fan et al. 2006) and XPD (Bienstock et al. 2003; Fan et al. 2008; Liu et al. 2008; Wolski et al. 2008) provided structural templates for interpretation of the low-resolution cryo-EM maps, and for modelling of human disease mutations.

The TFIIH-containing PIC structures (He et al. 2013, 2016; Tsai et al. 2017) also suggested that in the absence of Mediator, the CAK subcomplex localizes close to XPD and the C-terminal region of Rpb1, thus in proximity to its major substrate (Greber et al. 2017; He et al. 2013; Tsai et al. 2017). However, later Mediator-bound structures showed that in this context, which likely better resembles the in vivo situation, CAK is found in a different location (Robinson et al. 2016; Schilbach et al. 2017) (Fig. 5.5d, e). Both in the absence and in the presence of Mediator, the CAK subcomplex is poorly resolved, and in spite of the availability of high-resolution crystal structures for both CDK7 and Cyclin H (Andersen et al. 1997; Kim et al. 1996; Lolli et al. 2004), the interaction with the third CAK subunit, MAT1, and the mechanism of CAK regulation remain unresolved.

A more detailed and complete molecular model of the TFIIH core complex resulted from the improved cryo-EM reconstructions of apo-TFIIH (Greber et al. 2017) and of the yeast Pol II-PIC (Schilbach et al. 2017). These studies revealed the overall structure of additional TFIIH subunits, including p62, and the positioning of the zinc-binding domains of p34, p44, and the CAK subunit MAT1. The latest breakthrough came from the 3.7 Å-resolution cryo-EM structure of human apo-TFIIH (Greber et al. 2019). This structure revealed the complete network of interactions that governs the assembly of the TFIIH core complex (Fig. 5.5f, g), allowed the mapping of numerous disease mutations in XPD and XPB (Fig. 5.5h and see below), and showed that XPD is probably tightly regulated, not only by MAT1 and p44, as demonstrated by previous biochemical and mutagenesis experiments (Abdulrahman et al. 2013; Dubaele et al. 2003; Kim et al. 2015; Sandrock and Egly 2001), but also by XPB and p62, which block access to known XPD DNA- and ATP-binding sites (Fig. 5.5g) (Greber et al. 2019).

TFIIH in Health and Disease

Defects in TFIIH subunits are associated with xeroderma pigmentosum, Cockayne syndrome, and trichothiodystrophy (Fig. 5.5h), human disease syndromes characterized by high incidence of cancers, premature ageing, or transcriptional defects (Cleaver et al. 1999; Rapin 2013). Mutations that cause xeroderma pigmentosum, characterized by extreme sensitivity to sunlight and a high incidence of cancers, interfere with the global genome NER pathway by impairing (i) detection of lesions by the specialized damage recognition sensors (e.g. XPC), (ii) repair bubble opening by interfering with the function of the TFIIH helicases XPB and XPD, or (iii) excision of the damaged nucleotide by the XPF-ERCC1 or XPG endonucleases. Cockayne syndrome is a premature ageing condition, which in the case of mutations in the XPD subunit of TFIIH, occurs in combination with xeroderma pigmentosum (Cleaver et al. 1999; Rapin 2013). Cockayne syndrome is thought to be a disease induced by defects in TCR. However, recent data suggest that CS mutations in XPD may lead to stalling of the helicase after initiation of the repair process, which may interfere with completion of the repair reaction and with the resumption of transcription (Herrera-Moyano et al. 2014; Moriel-Carretero et al. 2015; Theron et al. 2005). Mapping of XP and XP-CS mutations onto the structure of human XPD (Greber et al. 2017, 2019) (Fig. 5.5h) or its archaeal and bacterial homologs (Bienstock et al. 2003; Fan et al. 2008; Liu et al. 2008; Wolski et al. 2008) shows that they localize primarily to the enzymatic core of the helicase, affecting either nucleotide binding and hydrolysis, or DNA binding and translocation, in agreement with the enzymatic function of XPD in NER (Kuper et al. 2014), and the observation that these mutations primarily lead to NER defects.

In contrast, trichothiodystrophy (TTD) mutations in TFIIH mostly affect peripheral regions of XPD (Fig. 5.5h) (Bienstock et al. 2003; Fan et al. 2008; Greber et al. 2017; Liu et al. 2008; Wolski et al. 2008), where they structurally destabilize the protein or its interactions within TFIIH, thereby disrupting TFIIH assembly and impairing proper placement of the CAK during transcription initiation (Coin et al. 1998; Dubaele et al. 2003). Additional TTD mutations in XPB and p8 have been shown to destabilize these proteins and reduce the overall levels of properly assembled TFIIH (Botta et al. 2002; Dubaele et al. 2003; Giglia-Mari et al. 2004; Kainov et al. 2008). This lack of structurally intact TFIIH, either due to reduced protein levels or disrupted protein-protein interfaces, is what leads to a general transcription defect in addition to DNA repair defects.

In addition to its association with these inherited disorders, TFIIH has also been implicated in promoting cancer cell growth due to the transcription-promoting activity of its CAK module and the requirement for elevated transcription in cancer cells. Therefore, TFIIH is a possible drug target in cancer chemotherapy (Berico and Coin 2017; Fisher 2018; Gervais et al. 2018).

Structural Insight into the Pol II-PIC

Visualization of the Step-Wise Assembly of the Human Pol II-PIC

Early crystallographic studies revealed the structures of promoter-bound TBP, alone and in complex with TFIIA and TFIIB (Fig. 5.6a, b), as well as TFIIB bound to Pol II (Pol II-ITC with bound TFIIB shown in Fig. 5.6c). These studies showed that TBP binding to the TATA-box induces an almost 90° bend in the overall trajectory of the upstream promoter DNA (Fig. 5.6a), and how TFIIB interacts with TBP and with the BRE promoter elements near the TATA-box (Fig. 5.6b) (Bleichenbacher et al. 2003; Nikolov et al. 1992, 1995, 1996; Tan et al. 1996; Tsai and Sigler 2000). Structures of TFIIB bound to Pol II (Bushnell et al. 2004; Kostrewa et al. 2009; Liu et al. 2010; Sainsbury et al. 2013) showed how the cyclin domains of TFIIB bind above the Pol II wall, while the N-terminal B-ribbon binds to the Pol II dock (Fig. 5.6c). The segment between the B-ribbon and the cyclin folds is seen inserted into the Pol II cleft, approaching the active site, which led to the suggestion that this B-reader aids in start site selection (Kostrewa et al. 2009; Liu et al. 2010). A combination of these structures with the previously mentioned structure of the TPB-TFIIB-DNA complex (Nikolov et al. 1995; Tsai and Sigler 2000) and the structure of the Pol II elongation complex (Gnatt et al. 2001) suggested approximate models for the closed and open Pol II-PIC, and the subsequent structure of the TFIIB-bound Pol II-ITC (Fig. 5.6c), complete with substrate DNA and an RNA product mimic, revealed that the TFIIB B-reader contacts the single stranded DNA in the polymerase active site, leads to binding of one of the catalytic metals, and stimulates RNA synthesis (Sainsbury et al. 2013).

The architecture of the complete Pol II-PIC and a structural characterization of its assembly pathway have remained refractory to crystallographic approaches and were finally obtained using cryo-EM analysis of in vitro reconstituted complexes (Fig. 5.6d). In order to arrive at a more complete understanding of the complete Pol II-PIC, and to map the location of all the individual general transcription factors, the human Pol II-PIC was assembled in a step-wise fashion, following the generally accepted model for PIC assembly (Buratowski et al. 1989), and analyzed by cryo-EM (He et al. 2013). The system was simplified by substituting TFIID with just TBP (a consensus TATA box was included in the promoter DNA), and Mediator was not included in the reconstitution. Additionally, although TFIIF is

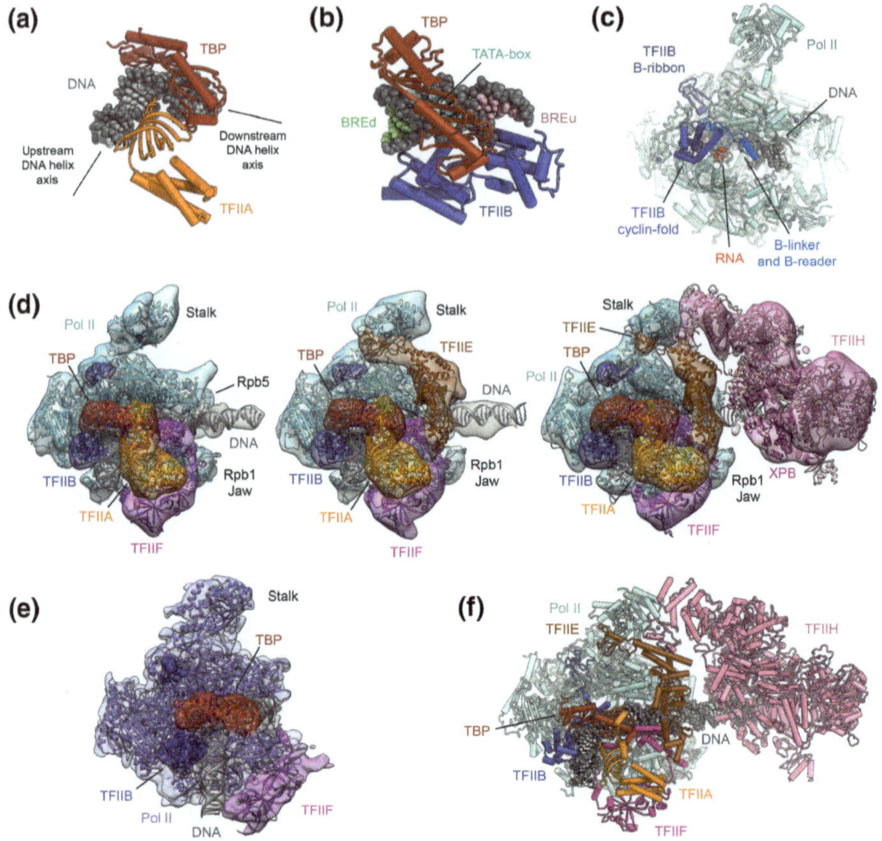

Fig. 5.6 The structure of the Pol II-PIC and its components. **a** Structure of promoter DNA bound to TBP and TFIIA (Bleichenbacher et al. 2003). TBP introduces a pronounced bend into the TATA box (indicated by dotted black lines representing the upstream and downstream DNA helix axes). **b** Structure of promoter DNA bound to TBP and TFIIB (Tsai and Sigler 2000). The TATA box and regions corresponding to the BREd and BREu regions recognized by TFIIB are indicated by light blue, green, and pink shading of the DNA bases. **c** Structure of initially transcribing Pol II bound to TFIIB (Sainsbury et al. 2013). **d** Stepwise assembly of the human Pol II-PIC: Pol II-DNA-TBP-TFIIA-TFIIB-TFIIF (left), Pol II-DNA-TBP-TFIIA-TFIIB-TFIIF-TFIIE (middle), and Pol II-DNA-TBP-TFIIA-TFIIB-TFIIF-TFIIE-TFIIH (right), the latter showing a bilobal architecture. Depiction based on (He et al. 2013) with fitted coordinates from (Greber et al. 2019; He et al. 2016). **e** The structure of the yeast Pol II-ITC lacking TFIIH, TFIIA, and TFIIE (Plaschka et al. 2015) is in good agreement with the human Pol II-PIC core (compare e.g. to leftmost complex in panel d). **f** Complete molecular structure of the human Pol II-PIC with TFIIH, assembled from coordinates for the human core Pol II-PIC and the structure of human TFIIH (Greber et al. 2019; He et al. 2016)

thought to join the growing PIC pre-bound to Pol II (Killeen and Greenblatt 1992; Rani et al. 2004; Roeder 1996), the structure was determined for complexes with and without TFIIF to allow mapping of the density corresponding to this transcription factor.

The cryo-EM density of the TBP-TFIIA-TFIIB-Pol II complex resulting from these studies (He et al. 2013) could be fully accounted for by the crystallographic structures of isolated sub-complexes (see above) and showed the general transcription factors clustered on the upstream promoter DNA near the TATA box, with density for the DNA approaching the Pol II cleft. The otherwise poorly ordered downstream promoter DNA became stabilized upon addition of TFIIF, which binds to the cluster of general transcription factors near the TATA box (Fig. 5.6d, left), Pol II, and the upstream promoter DNA itself (He et al. 2013). The stabilization of the DNA was induced by interactions of TFIIF with DNA near the BREd element and the opening of the Pol II clamp, allowing DNA to access the Pol II cleft and form interactions with the clamp head and Rpb5. These interactions are consistent with roles of TFIIF in start site selection and transcription bubble opening (Ghazy et al. 2004; Yan et al. 1999).

Addition of TFIIE, which spans from TFIIF to the Pol II stalk (Fig. 5.6d, middle), led to further stabilization of the core PIC (He et al. 2013). Winged-helix domains in TFIIE connect the Pol II stalk region to the TFIIF subunit Rap30, topologically trapping the DNA on the Pol II-PIC. The DNA in this complex is slightly bent and contacted at either side of the initiator element by Pol II elements near the clamp head and at the Rpb5 jaw (He et al. 2013, 2016).

Finally, TFIIH was seen to bind to this core-PIC (Fig. 5.6d, right) such that its XPB subunit is in contact with the downstream promoter DNA near Rpb5 (Figs. 5.5c, 5.6d, right) (He et al. 2013, 2016), in support of its proposed function as a double-stranded DNA translocase during promoter opening (Fishburn et al. 2015; Grünberg et al. 2012; Kim et al. 2000). During the closed-to-open transition, mimicked in the studies of the human system by synthetic constructs that provide single-stranded DNA at the side of the transcription bubble, the downstream promoter DNA is bent around the Rpb5 contact for insertion of the template strand into the active site, while the Pol II clamp domain closes and assumes a conformation similar to that observed in elongation complexes (He et al. 2013, 2016).

The TFIIH-containing Pol II-PIC exhibits a bi-lobed overall appearance (Figs. 5.5c, 5.6d), with TFIIH forming one lobe and Pol II and the remaining GTFs forming the other lobe (He et al. 2013, 2016). In addition to the interaction between XPB and the downstream promoter DNA, TFIIH and the core PIC are connected by two main protein-protein interfaces: (i) Mobile regions in the C-terminal half of TFIIEα interact with the N-terminal PH domain of p62 (He et al. 2016; Okuda et al. 2008; Schilbach et al. 2017) and form additional contacts with the C-terminal region of p62, and with XPB (Schilbach et al. 2017), with the latter interaction explaining previous biochemical data that suggest a role of TFIIE in XPB regulation (Drapkin et al. 1994; Ohkuma and Roeder 1994; Maxon et al. 1994); and (ii) the N-terminal RING domain of MAT1 interacts with the Pol II stalk and TFIIE (Schilbach et al. 2017).

In spite of initial claims to the contrary (Murakami et al. 2013), the model derived from the sequential assembly of human proteins on promoter DNA (He et al. 2013) is in good agreement with subsequent structures from both the human (He et al. 2016) and the yeast Pol II system (Fig. 5.6d, e) (Murakami et al. 2015;

Plaschka et al. 2015, 2016; Schilbach et al. 2017). However, slight differences between the human and yeast systems were visualized in the higher-resolution structures. For example, the promoter DNA in the human closed complex is contacted by the Rpb1 clamp and Rbp5, while the DNA is suspended further away from Pol II in the yeast complexes, where such contacts are not present (Nogales et al. 2017b) (for further discussion see next section).

High-Resolution Analysis of the Pol II Core-PIC and Mechanism of Open Complex Formation

High-resolution cryo-EM analysis of human and yeast Pol II core-PICs (the latter excluding TFIIH) (Figs. 5.6f, 5.7a–c) have revealed the intricate architecture of closed, open, and initially transcribing complexes in near-atomic detail (He et al. 2016; Plaschka et al. 2016). In the human open Pol II-PIC, the B-reader and B-linker elements of TFIIB contact the single stranded DNA of the open transcription bubble (Fig. 5.7d–f), and all of these elements are resolved in the map (He et al. 2016). In the yeast Pol II-PIC reconstructions (Plaschka et al. 2016), both the single-stranded DNA and the interacting TFIIB elements are not visualized or poorly resolved. The better definition of the open DNA bubble in the human system might be linked to the scanning mode of start site selection in yeast (Nogales et al. 2017b).

The reverse is true for interactions of TFIIE (Fig. 5.7g–i), where the human PIC lacks density for part of the TFIIE E-wing domain near the upstream end of the open bubble, while an interaction between this domain and the DNA is visible in the yeast structure (He et al. 2016; Plaschka et al. 2016). This might account for the propensity of the yeast Pol II-PIC system to spontaneously open the transcription bubble (Plaschka et al. 2016), while this has not been observed in the reconstituted human complexes. Nearby, the WH1 domain of TFIIEβ (Tfa2 in yeast) that interacts with the TFIIFβ (Rap30 in humans, Tfg2 in yeast) WH domain and in particular with the additional helix of this domain in yeast is, accordingly, also in a different position between the human and yeast systems. The WH2 domain of TFIIEβ (Tfa2) also has a slightly different position in the two systems.

The structure of the human Pol II-ITC is very similar to the open complex (Fig. 5.7a, b, e, f), with the exception of an enlarged DNA bubble and the presence of RNA, which had been introduced with the substrate DNA template (He et al. 2016).

The Mediator-Bound Pol II PIC

Mediator is a co-activator complex composed of more than 20 subunits and with a molecular weight of approximately 1.2 MDa. Mediator integrates the signals from transcriptional activators and repressors for regulated transcription initiation (Borggrefe and Yue 2011; Brzovic et al. 2011; Flanagan et al. 1991; Jeronimo et al.

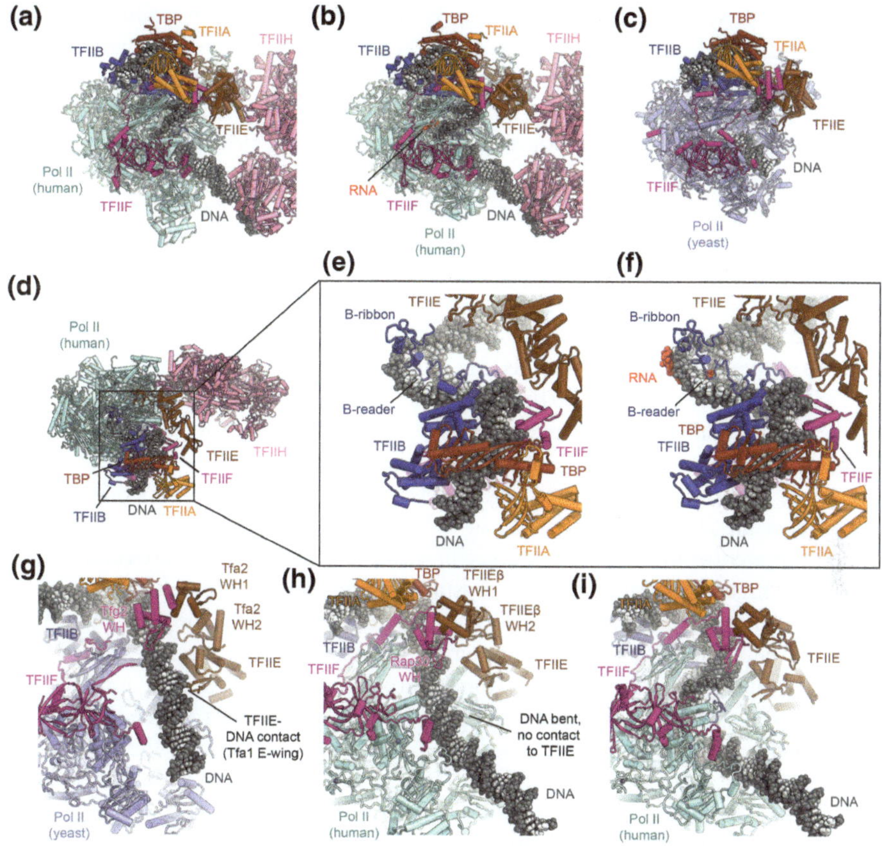

Fig. 5.7 Structure of the open and initially transcribing human Pol II-PICs and comparison with their yeast counterparts. **a** Structure of the open human Pol-II PIC (He et al. 2016). **b** Structure of the human Pol II-ITC (He et al. 2016). RNA near the Pol II active site is shown in red. **c** Comparison to the structure of the yeast open complex (Plaschka et al. 2016) reveals overall conservation of PIC architecture. **d–f** DNA interactions of TFIIB in the human Pol II open complex (**e**) and ITC (**f**) (He et al. 2016). **g** Interaction between the TFIIE E-wing domain and bound promoter DNA in the yeast closed PIC (Plaschka et al. 2016). **h** In the human Pol II-PIC (He et al. 2016), the closed promoter DNA is inserted more deeply into the Pol II cleft, and the TFIIE-DNA interaction visualized in yeast is not present. **i** Human Pol II open PIC (He et al. 2016)

2016; Lee et al. 1999; Natarajan et al. 1999; Soutourina 2018), a prerequisite for cellular development and differentiation. Additionally, Mediator stimulates the activity of CDK7 within the TFIIH CAK module towards the C-terminal, low-complexity region of the Pol II subunit Rpb1 (Kim et al. 1994). Mediator itself includes the CDK8-Cyclin C pair, which together with the Med12 and Med13 subunits form a dissociable module (Borggrefe et al. 2002; Tsai et al. 2013) that phosphorylates a range of targets, with broad implications for the mechanism of transcription regulation and for cellular development and disease (e.g. (Bancerek et al. 2013; Donner et al. 2010; Firestein et al. 2008; Knuesel et al. 2009)).

The overall structure of Mediator can be subdivided into four modules, termed head, middle, tail, and arm (Cai et al. 2009), as well as the dissociable kinase module (Borggrefe et al. 2002; Tsai et al. 2013). Early low-resolution electron microscopy analysis indicated that an open conformation of Mediator wraps around one side of Pol II to form the holoenzyme (Asturias et al. 1999; Davis et al. 2002; Dotson et al. 2000). Because of its size, complexity, conformational variability, and low abundance in cells, structure determination of Mediator has been pursued by X-ray crystal structures of subcomplexes, such as structures of its head and kinase modules (Imasaki et al. 2011; Lariviere et al. 2012a; Robinson et al. 2012; Schneider et al. 2011). These structures were then combined with chemical crosslinking-mass spectrometry distance restraints and low-resolution cryo-EM maps to arrive at structural models for the full complex (Lariviere et al. 2012b, 2013; Robinson et al. 2015, 2016; Tsai et al. 2014; Verger et al. 2019).

The most complete atomic model for any Mediator complex to date is that of the *Schizosaccharomyces pombe* Mediator head and middle modules (or core Mediator), determined by X-ray crystallography at 3.4 Å resolution (Fig. 5.8a) (Nozawa et al. 2017) and cryo-EM at 4.4 Å resolution (Tsai et al. 2017). Even though the architecture of the complex in these two structures is in overall agreement, the higher-resolution X-ray crystallographic analysis resulted in a more accurate and more complete model. The structure shows that the Med14 subunit spans the entire middle module and serves as a scaffold for the assembly of the remaining subunits of this module (Fig. 5.8a). The head and middle modules are bridged by the Med17 and Med6 subunits (Fig. 5.8a), each of which contributes folded domains to both modules, with flexible linkers connecting them (Nozawa et al. 2017). The head module had been visualized previously in isolation (Imasaki et al. 2011; Lariviere et al. 2012a; Robinson et al. 2012). Interestingly, one of these structures had been solved in complex with a CTD peptide bound in a groove formed by the Med6, Med8, and Med17 subunits (Fig. 5.8b) (Robinson et al. 2012). The Mediator tail module is critical for interactions with transcriptional activators, but in spite of the availability of partial X-ray crystal structures (e.g. for the human Med23 and Med25 subunits (Monté et al. 2018; Vojnic et al. 2011)), it has not been visualized in its entirety at high resolution. However, an approximate subunit mapping has been obtained from an integrative modeling approach (Robinson et al. 2015).

Detailed insight into the architecture of the Mediator-bound Pol II-PIC came from cryo-EM analysis of reconstituted complexes of yeast core-PIC (without TFIIH) or ITC complexes (Plaschka et al. 2015; Tsai et al. 2017) bound to a core-Mediator complex (Fig. 5.8c). These cryo-EM reconstructions revealed extensive density for the Mediator middle and head modules, and initially allowed fitting of several sub-modules of the head module, aided by distance restraints from chemical crosslinking-mass spectrometry (Plaschka et al. 2015). Subsequently, with higher resolution cryo-EM structures and with the determination of the X-ray crystal structure of the head and middle modules (Nozawa et al. 2017), the entire core Mediator complex could be placed in the density of a TFIIH-containing and Mediator-bound Pol II-PIC (Schilbach et al. 2017) (Fig. 5.8d). Core Mediator binds

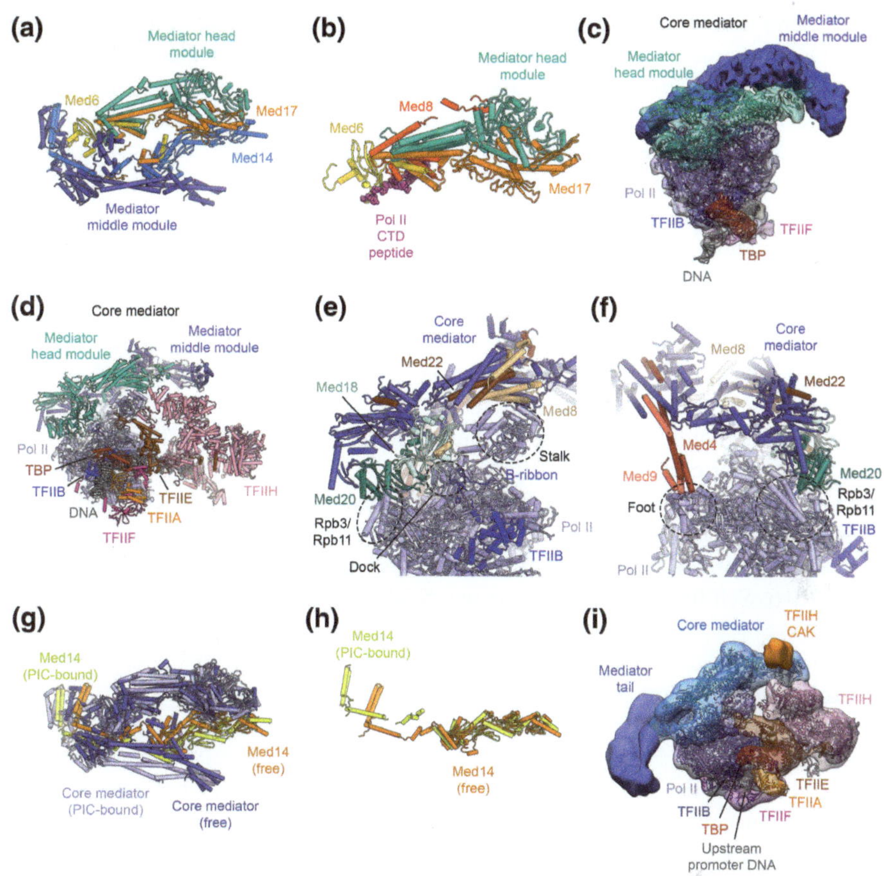

Fig. 5.8 The Mediator-bound Pol II-PIC. **a** Crystal structure of the Mediator head and middle modules (core Mediator) (Nozawa et al. 2017). **b** Crystal structure of the Mediator head module bound to a Pol II CTD peptide (Robinson et al. 2012). **c** Low-resolution cryo-EM structure of the core-Pol II-ITC bound to core Mediator (Plaschka et al. 2015). **d** Cryo-EM structure of the yeast Pol II-PIC, including TFIIH, bound to core Mediator (Schilbach et al. 2017). **e, f** Contacts sites between core Mediator and Pol II, according to the structure shown in (**d**). **g, h** Conformational changes in Mediator upon binding to the PIC (**g**) and associated with conformational changes in Med14 (Nozawa et al. 2017; Schilbach et al. 2017). **h** Low-resolution cryo-EM map of the yeast Pol-II PIC with Mediator (Robinson et al. 2016), including the Mediator tail module, fitted with the coordinates of the most recent yeast Pol II-core Mediator structure (Schilbach et al. 2017)

to the PIC at three contact sites that are arranged near the Pol II Rpb4-Rpb7 stalk (Fig. 5.8e, f). First, the so-called movable jaw (Med18-Med20) forms contacts with the TFIIB B-ribbon, the Pol II dock, and the Pol II subunits Rpb3 and Rpb11 (Fig. 5.8e). Second, the Mediator head module subunits Med8 and Med22 directly contact the Rpb4-Rpb7 stalk (Fig. 5.8e). Finally, the plank domain of the middle module forms transient contacts to the Pol II foot (Fig. 5.8f), as deduced from different conformations observed in cryo-EM particle sub-classes (Plaschka et al.

2015; Schilbach et al. 2017). While the Mediator head and middle modules themselves assume a relatively constant conformation when associating with Pol II —which also does not undergo any major conformational rearrangements upon complex formation—the relative orientation of the two Mediator modules is changed upon core-Mediator-Pol II complex formation due to conformational rearrangements in the Med14 subunit (Schilbach et al. 2017; Tsai et al. 2017) (Fig. 5.8g, h).

This structural model just described, together with a low-resolution cryo-EM reconstruction of a complete Pol II-PIC, including TFIIH and the Mediator tail module (Robinson et al. 2016), showed the Mediator tail module pointing towards the upstream promoter DNA, where it may interact with bound transcriptional activators, consistent with their function in Mediator recruitment (Fig. 5.8i) (Borggrefe and Yue 2011; Brzovic et al. 2011; Jeronimo et al. 2016; Lee et al. 1999; Natarajan et al. 1999). The structures of TFIIH-containing PICs in complex with Mediator or core Mediator suggest that the hook motif of the Mediator middle module aids in positioning of the TFIIH CAK subcomplex, such that it is optimally placed to phosphorylate the Pol II CTD, which runs inside a cradle formed by the inner surface of the remainder of the Mediator middle module (Robinson et al. 2012, 2016; Schilbach et al. 2017) (Figs. 5.5d, e, 5.8b).

Possible Organization of a Complete Pol II-PIC Containing TFIID and Mediator

The structure of the Pol II-PIC in the presence of full TFIID has not yet been obtained. An approximate architecture can be deduced by superimposing the components that are shared between the high-resolution Pol II-PIC reconstructions and the structure of promoter-bound TFIID, specifically TBP, TFIIA, and the upstream promoter DNA (Louder et al. 2016; Patel et al. 2018). Similarly, by superposition of Pol II in the Mediator-containing Pol II-PIC reconstructions (Plaschka et al. 2015; Robinson et al. 2016; Schilbach et al. 2017; Tsai et al. 2017) allows approximate positioning of Mediator in this architectural model (Fig. 5.9).

This analysis shows that Mediator and TFIID would occupy opposite sides of Pol II (Fig. 5.9), and thus would not clash with each other. However, structural rearrangements may be necessary concerning the position of TAF contacts with the DNA during PIC assembly. The region of TAF4 that, within lobe B, approaches TFIIA and the upstream promoter DNA partially overlaps with the Rap30 winged-helix domain within TFIIF in the model (Louder et al. 2016; Patel et al. 2018). Furthermore, the downstream promoter contacts formed by TFIID lobe C partially overlap with regions of Pol II and TFIIH that contact the DNA in the PIC, and would, in any case, need to be released for Pol II to transcribe through the downstream promoter region. Indeed, it has been proposed that an isomerization of these contacts and release of TAF7, localized in the immediate vicinity of the downstream promoter DNA, is required for transcription initiation (Gegonne et al. 2006; Zhang et al. 2015). Because this isomerization is apparently required only for

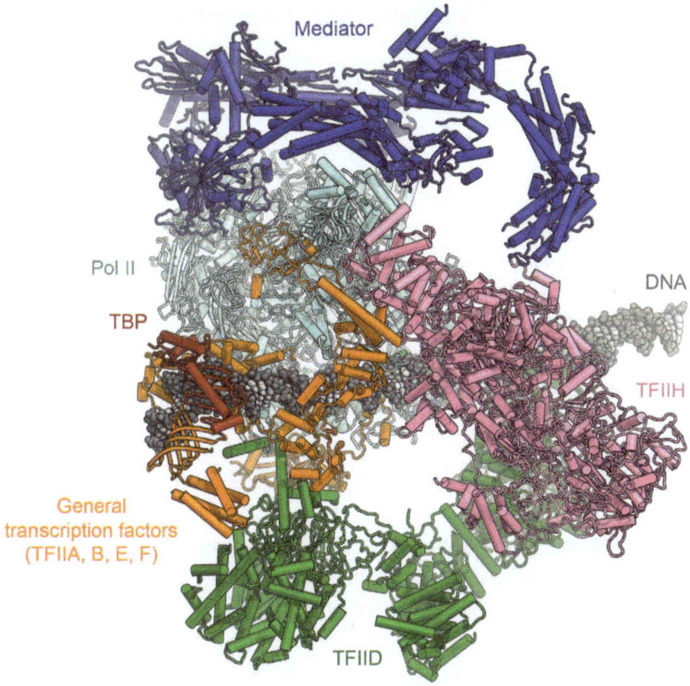

Fig. 5.9 Architectural model of the holo-Pol II-PIC with TFIID and Mediator. Proposed model of an entire Pol II-PIC, including Mediator and TFIID, generated by superimposing component structures (Greber et al. 2019; He et al. 2016; Patel et al. 2018; Schilbach et al. 2017)

the first round of initiation, but not for subsequent re-initiation events (Zhang et al. 2015), it is possible that TFIID does not re-form these downstream promoter contacts after it has functioned as a molecular ruler to load TBP for PIC recruitment.

In spite of the release of downstream promoter contacts, TFIID may remain bound to the promoter if the upstream TFIID architecture can rearrange such that the sterical hindrance between the TFIID TAFs and the GTFs of the core Pol II-PIC can be mitigated. It has been shown that transcription re-initiation can be up to fourfold faster than the first round of initiation (Jiang and Gralla 1993) and activator-independent (Joo et al. 2017), suggesting that a set of general transcription factors remains bound at the promoter and allows incoming Pol II complexes to re-initiate rapidly. Transcription re-initiation depends on ATP hydrolysis, implicating the XPB or CDK7 subunits of TFIIH in the process (Yudkovsky et al. 2000). Even though much remains to be elucidated to gain a full understanding of the molecular details that distinguish the processes of initiation and re-initiation in Pol II transcription, the recent detailed structures of the Pol II-PIC and its components provide an architectural framework for future studies.

The Pol I and Pol III Pre-initiation Complexes

Structure of the Pol I-PIC and Mechanism of Initiation of rRNA Transcription

Pol I and Its General Transcription Factors

Pol I is responsible for synthesis of long rRNA precursors in the nucleolus, contributing up to 60% of total RNA synthesis in yeast (Warner 1999). The X-ray crystal structure of the 14-subunit Pol I was that of a transcription-incompetent dimer (Fig. 5.10a) with a widened cleft that was blocked by inserted protein segments (Engel et al. 2013; Fernández-Tornero et al. 2013). In solution, Pol I exists in an equilibrium of inactive dimers and functional monomers (Milkereit et al. 1997), a situation that may parallel the physiological response to nutrient starvation (Torreira et al. 2017). Binding of Rrn3, one of the Pol I transcription initiation factors, renders Pol I monomeric (Fig. 5.10b) and with a widened cleft that can be accessed by substrate DNA (Engel et al. 2016; Pilsl et al. 2016).

The Pol I initiation system differs substantially from that of Pol II. In yeast, the Pol I-PIC comprises the general transcription factors TBP, Rrn3, upstream activating factor (UAF), and the heterotrimeric core factor (CF) (Fig. 5.10c) (Bedwell et al. 2012; Hontz et al. 2008; Keener et al. 1998; Keys et al. 1996; Schneider 2012; Steffan et al. 1998). Rrn3 and CF support basal Pol I initiation, while binding of UAF and TBP precedes CF recruitment in vivo and is required for full activity (Bordi et al. 2001; Hontz et al. 2008; Keys et al. 1996; Steffan et al. 1998). Yeast CF is formed by Rrn6, Rrn7, and Rrn11 (Keys et al. 1994; Lalo et al. 1996; Lin et al. 1996). Its mammalian counterpart, selectivity factor 1 (SL1), harbors three subunits homologous to the yeast system and two additional mammalian-specific factors (Comai et al. 1992; Denissov et al. 2007; Friedrich et al. 2005; Gorski et al. 2007; Learned et al. 1985; Naidu et al. 2011), suggesting a conserved core architecture with certain mammalian-specific features. The structure of CF (Fig. 5.10d), determined independently by both X-ray crystallography and cryo-EM, shows a bi-lobal assembly, with Rrn11 and Rrn7 each forming the core of one lobe, and with Rrn6 spanning across both lobes (Engel et al. 2017; Han et al. 2017; Sadian et al. 2017). In spite of sequence and structural homology between Rrn7 and TFIIB (Knutson and Hahn 2011), these two initiation factors are not functionally equivalent because, due to its position in the Pol I-PIC, Rrn7 cannot bind to DNA in the way TFIIB does (Fig. 5.10e) (Engel et al. 2017; Han et al. 2017; Sadian et al. 2017).

Structural Insight into RNA Polymerase I Pre-initiation Complexes

Three independent cryo-EM studies have provided detailed insights into the structures of the Pol I-PIC and the promoter-free Pol I-Rrn3-CF complex to

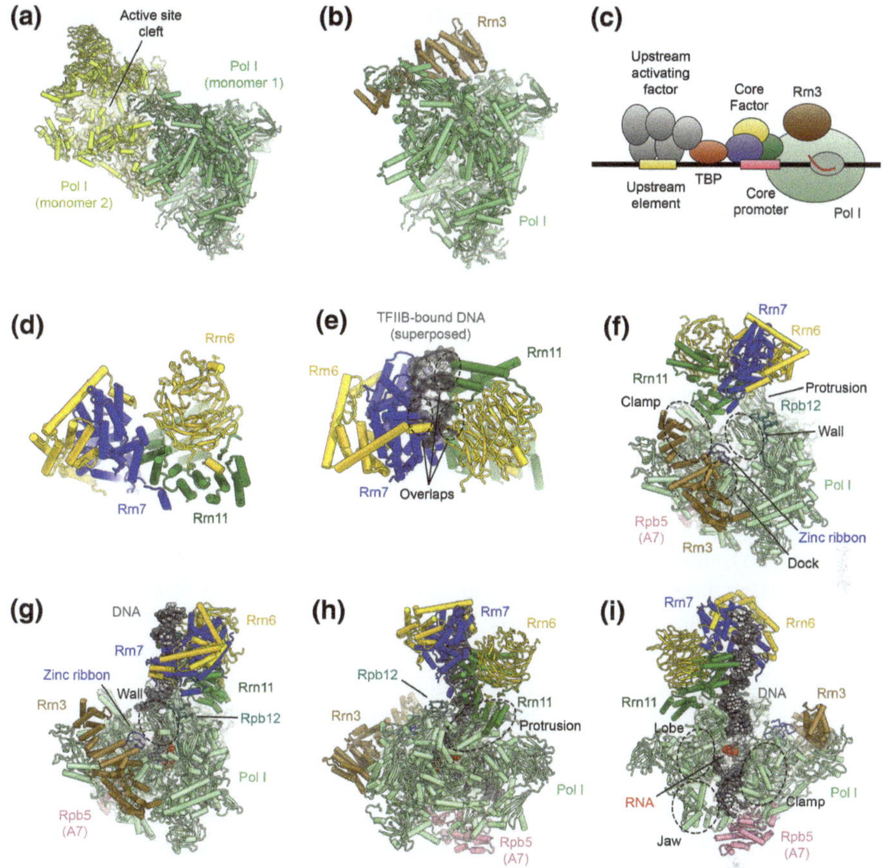

Fig. 5.10 The architecture of the RNA polymerase I PIC and ITC. **a** Structure of the Pol I-dimer (Engel et al. 2013), with a widened active site cleft, as observed in Pol I crystals. **b** Structure of Pol I bound with Rrn3 (Engel et al. 2016), which blocks the Pol I dimerization interface. **c** Schematic of the Pol I core promoter and general transcription factors in yeast. **d** Structure of yeast core factor (CF) (Engel et al. 2017). **e** Structure of yeast core factor shown with DNA superposed from a TFIIB-DNA structure (Tsai and Sigler 2000). The superposed DNA overlaps with CF, indicating that Rrn7 in CF cannot bind to DNA by the same mechanism as TFIIB (also see Fig. 5.15). **f** Structure of the Pol I-Rrn3-CF complex (Engel et al. 2017). Regions of Pol I that are in contact with CF are indicated. **g–i** Depiction of the Pol I-ITC, with bound DNA, Rrn3, and CF (Engel et al. 2017; Han et al. 2017). Contacts differ from the DNA-free complex (shown in **f**) and corresponding regions on Pol I are indicated. Downstream promoter DNA is stabilized by contacts from the Pol I clamp, lobe, and jaw domains and Rpb5

elucidate the mechanism of transcription initiation by Pol I (Engel et al. 2017; Han et al. 2017; Sadian et al. 2017). In the Pol I-Rrn3-CF complex, lacking promoter DNA (Fig. 5.10f), CF binds to the upstream end of the Pol I cleft, which is partially widened (Engel et al. 2017). There are three contact sites between CF and the PIC core (Fig. 5.10f): (i) The N-terminal zinc-ribbon domain of Rrn7 contacts the Pol I dock domain and a loop in Rrn3; (ii) the Rrn7 insertion domain contacts the Pol I wall; and (iii) Rrn11 contacts the Pol I clamp and protrusion via its TPR domain (Engel et al. 2017). Notably, the N-terminal cyclin domain of Rrn7 is embedded in CF and therefore unable to form contacts equivalent to the TFIIB-Pol II wall interactions present in the Pol II system (Engel et al. 2017; Han et al. 2017; Sadian et al. 2017).

Upon binding of DNA and formation of an ITC, mimicked experimentally by introduction of a mismatch bubble with bound RNA, the Pol I active site cleft assumes a contracted conformation (Engel et al. 2017; Han et al. 2017; Sadian et al. 2017), similar to that observed for the elongation complex structures (Neyer et al. 2016; Tafur et al. 2016). The promoter DNA in the active site cleft of Pol I is stabilized by the clamp head, cleft, and jaw of Pol I subunit A190, the lobe domain of A135, and Rpb5 (Engel et al. 2017; Han et al. 2017; Sadian et al. 2017). Upon DNA binding, CF is relocated to a different binding site on Pol I (Fig. 5.10g) (Engel et al. 2017). Though CF uses the same binding motifs, its footprint on Pol I moves to the protrusion, Rpb12, and a Pol I-specific structural element of the wall (Engel et al. 2017) (Fig. 5.10g–i). Density for Rrn3 was absent in one of the Pol I-ITC cryo-EM maps, indicating that it may dissociate under conditions that might mimic a late initiation intermediate (Han et al. 2017), consistent with Rrn3 dissociation after Pol I initiation (Bier et al. 2004; Milkereit and Tschochner 1998).

In all cryo-EM reconstructions of Pol I-ITCs, the upstream promoter DNA appears bent (Engel et al. 2017; Han et al. 2017; Sadian et al. 2017). In one case, two kinks, by 35° and 45°, are observed near nucleotide −16 (Han et al. 2017); in the second, the DNA bends by approximately 30° between the CF binding site, where it runs along Rrn11 (DNA nucleotides −35 to −25), and the entry point into the Pol I cleft between the protrusion and the wall (DNA nucleotides −20 to −12) (Engel et al. 2017); and in the third, the bend localizes to approximately nucleotide −30 (Sadian et al. 2017). The ability of the DNA to bend and form these contacts is likely important for promoter recognition, along with the ability of the region around the transcription start site to melt and form the transcription bubble.

Interestingly, the DNA is suspended above the cleft in the closed Pol II-PIC (He et al. 2013, 2016; Louder et al. 2016; Murakami et al. 2015; Plaschka et al. 2016), while it is already sandwiched between the wall and the protrusion in the CF-bound PIC within the Pol I system (Fig. 5.11a, b) (Engel et al. 2017; Han et al. 2017; Sadian et al. 2017). This difference indicates that Pol I-DNA interactions in the Pol I-PIC already assume an elongation complex-like state, while the Pol II-PIC requires the DNA to shift by approx. 20 Å between the PIC and the elongation complex to reach this state (Fig. 5.11c).

Model for Initiation by RNA Polymerase I

The model for Pol I initiation that emerges from the structural studies just described is that Rrn3 prepares Pol I for initiation by stabilizing it in a monomeric, open-cleft conformation, while CF binds the upstream promoter DNA, and, after docking to

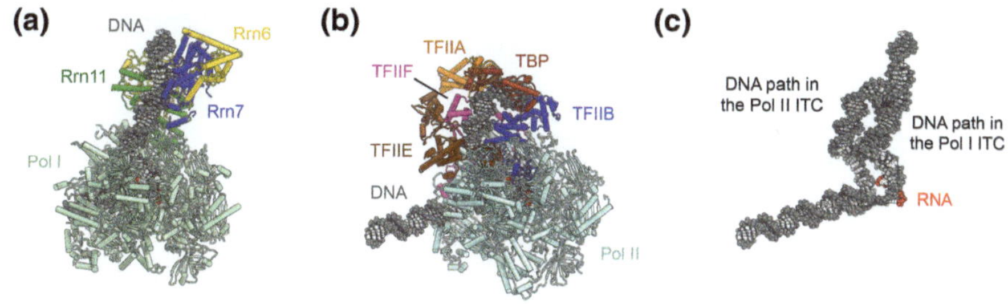

Fig. 5.11 Comparison of DNA paths in the Pol I and Pol II ITCs. **a** Structure of the Pol I-ITC (Han et al. 2017). **b** Structure of the Pol II-ITC (He et al. 2016). **c** The path of the upstream promoter DNA is different between the Pol I and Pol II ITCs because of differences in the mode of binding of Rrn7 and TFIIB to the DNA. See also Fig. 5.15

Pol I, loads the DNA into the Pol I cleft. Contacts between proximal promoter regions and Pol I specific structural elements likely contribute to promoter recognition at this stage. DNA unwinding and open complex formation leads to initiation of RNA synthesis and eventual displacement of Rrn7 and CF once the RNA reaches a suitable length (Engel et al. 2017; Han et al. 2017; Sadian et al. 2017).

Structure of the Pol III-PIC and Similarities to the Pol II System

Pol III and Its Redox-Sensing General Transcription Factor TFIIIB

Pol III contains 17 subunits, 10 of which form the conserved core shared across multisubunit RNA polymerases, whereas 2 form a stalk reminiscent of that in Pol II (Fig. 5.1d, e). 5 Pol III-specific subunits, the C82-C34-C31 trimer and the C53-C37 dimer, appear to be homologous to general transcription factors in the Pol II system (Fig. 5.12), but have become stably incorporated into Pol III (Khatter et al. 2017; Vannini and Cramer 2012). Specifically, there is strong structural homology between TFIIF and the Pol III C53-C37 dimer, which is involved in both transcription initiation and termination (Arimbasseri and Maraia 2015; Kassavetis et al. 2010; Rijal and Maraia 2013; Wu et al. 2011) while the C82-C34-C31 trimer, located on the Pol III C160 clamp, shows homology to TFIIE (Vannini and Cramer 2012) and appears to be a functional fusion of TFIIE and TFIIF (Vorländer et al. 2018; Wu et al. 2012). C82 and C34 contain multiple winged-helix domains that participate in interactions with the Pol III general transcription factors and facilitate Pol III-PIC formation (Khoo et al. 2014, 2018).

Transcription initiation by Pol III requires the trimeric general transcription factor TFIIIB (Fig. 5.13a–c), which is sufficient for Pol III transcription initiation in vitro (Kassavetis et al. 1990, 1999). TFIIIB is formed by TBP, TFIIB-related

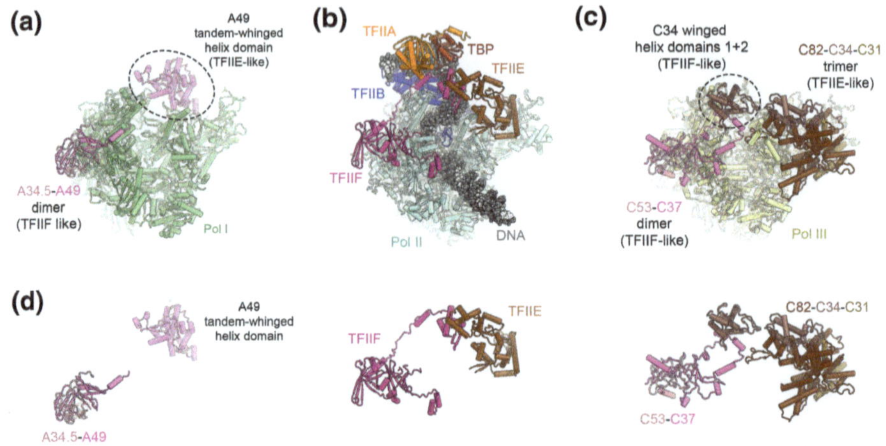

Fig. 5.12 Transcription factor-like subunits of Pol III. **a** Structure of Pol I extracted from the structure of a Pol I elongation complex in which the tandem winged helix domain of A49 is resolved (Tafur et al. 2016). **b** Structure of the Pol II-PIC shown in the same orientation for comparison (He et al. 2016). **c** Structure of Pol III extracted from the structure of an open Pol III-PIC (Abascal-Palacios et al. 2018) showing the locations of the C82-C34-C31 trimer and the C53-C37 dimer. **d** Comparison of the overall architecture of the A34.5-A49 dimer (Pol I system, left), TFIIE and TFIIF (Pol II system, middle), and the C82-C34-C31 trimer and C53-C37 dimer (Pol III system, right). The Pol I A34.5-A49 and Pol III C53-C37 dimers are positioned similarly to TFIIF, except that the A49 tandem-winged helix domain shows similarities with TFIIE, though the binding site on the polymerase is not identical. Most of the Pol III C82-C34-C31 trimer is positioned similarly to TFIIE, with the C82 cleft loop and TFIIE E-wing domain occupying similar positions, juxtaposed to bound promoter DNA in closed PICs. The C34 winged-helix 1 domain is positioned similarly to the winged-helix domain of the TFIIF subunit Rap30

factor 1 (Brf1), and the SANT-domain containing protein B-double prime 1 (Bdp1) (Colbert and Hahn 1992; Ishiguro et al. 2002; Kassavetis et al. 1995; Lobo et al. 1992; López-De-León et al. 1992; Schramm and Hernandez 2002; Schramm et al. 2000; Wang and Roeder 1995). In addition to Brf1, vertebrates encode a second TFIIB-related subunit, termed Brf2 (Cabart and Murphy 2001; Mital et al. 1996; Schramm et al. 2000; Teichmann et al. 2000), which acts at certain promoters, including the U6 small nuclear RNA (snRNA) promoter, characterized by a strong TATA-box and an upstream proximal sequence element (PSE) that is recognized by the SNAPc complex (Henry et al. 1995; Sadowski et al. 1993). TFIIIB not only functions in recruitment of Pol III to the promoter, but also supports initial promoter melting and extension of the open bubble via its Bdp1 and Brf1 subunits, respectively (Kassavetis et al. 2001).

The N-terminal halves of Brf1 and Brf2 are structurally homologous to TFIIB (Fig. 5.13a), contributing to the conservation of the overall architecture of the Pol II and Pol III-PICs (López-De-León et al. 1992; Vannini and Cramer 2012; Wang and Roeder 1995), while the C-terminal half includes a Brf-specific domain that forms tight and specific interactions with TBP (Juo et al. 2003; Khoo et al. 1994).

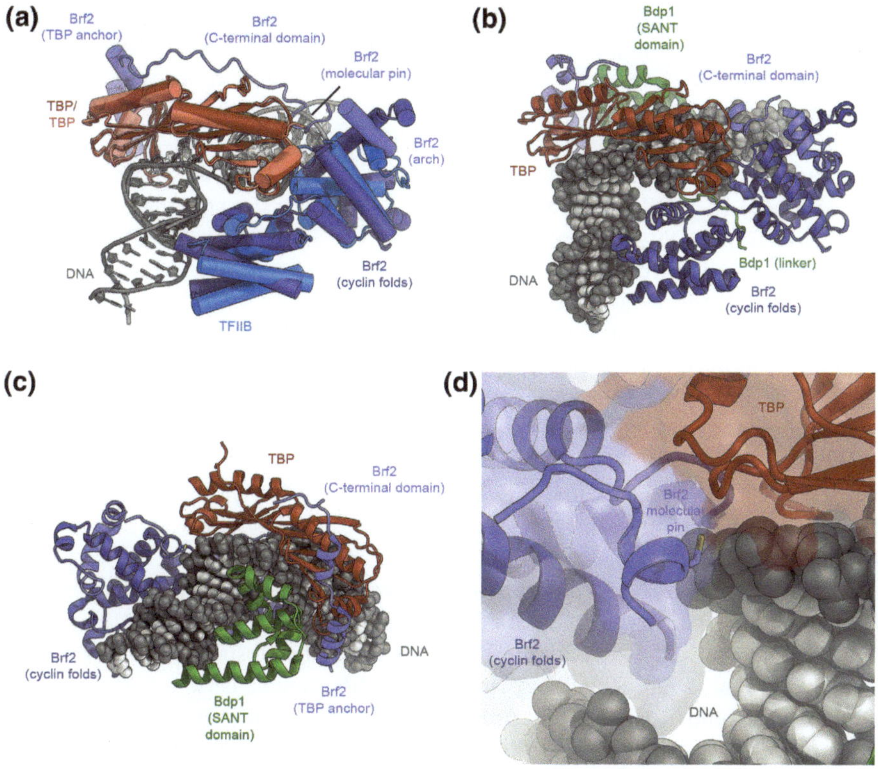

Fig. 5.13 The structure of TFIIIB. **a** Superposition of the structures of the DNA-bound TBP-Brf2 complex (Gouge et al. 2015) and the TBP-TFIIB-DNA complex (Tsai and Sigler 2000). TBP and the Brf2 cyclin folds in TFIIIB occupy very similar positions as their Pol II-system counterparts (TBP and TFIIB, respectively). The C-terminal domain of Brf2 comprises arch, molecular pin, and TBP anchor regions (as indicated). **b**, **c** Structure of DNA-bound TFIIIB (Gouge et al. 2017). The Bdp1 SANT domain binds to DNA at the side opposite to the Brf2 cyclin folds. **d** Structure of the redox-sensitive molecular pin in the context of bound DNA and TBP (Gouge et al. 2017)

The crystal structure of a Brf2-TBP-DNA complex (Gouge et al. 2015) shows that TBP engages DNA in the same way as in the TFIIB-TBP-DNA complex (Nikolov et al. 1995; Tsai and Sigler 2000) and induces the same strong bend at the TATA box (Gouge et al. 2015; Nikolov et al. 1996) (Fig. 5.13a). Even though details differ, the interactions between DNA and the regions conserved between Brf2 and TFIIB are similar overall, including contacts of the cyclin domains with the promoter DNA immediately upstream and downstream of the TATA box (Gouge et al. 2015; Nikolov et al. 1995; Tsai and Sigler 2000). The Brf2 C-terminal region, however, is unique, and contains three architectural regions, termed arch, anchor, and molecular pin (Gouge et al. 2015) (Fig. 5.13a). The molecular pin harbors a conserved LPPC-motif and binds to a ternary interface between the Brf2-cyclin domain, TBP, and the DNA (Fig. 5.13d). Interestingly, the cysteine in the molecular pin allows Brf2 to serve as a redox-sensing transcription factor that links oxidative stress to cellular responses, including apoptosis, because oxidation of the cysteine impairs complex assembly and leads to down-regulation of survival-promoting genes (Gouge et al. 2015).

The third component of TFIIIB, Bdp1, is specific to the Pol III system and contributes to an extremely tight interaction of TFIIIB with promoter DNA (Kassavetis et al. 1990, 2005; Shah et al. 1999). The crystal structure of a TFIIIB (TBP-Brf2-Bdp1)-DNA complex (Fig. 5.13b, c) shows that the Bdp1 SANT domain plays a key role in association of the protein with the remainder of the complex and binds to a location similar to that of TFIIA in the Pol II system (Bleichenbacher et al. 2003; Gouge et al. 2017; Tan et al. 1996).

Once assembled on the promoter, TFIIIB binds very stably (Kassavetis et al. 1990) and remains bound even after Pol III starts elongating, allowing for efficient Pol III recycling and high rates of transcription initiation on Pol III promoters (Dieci et al. 2013; Dieci and Sentenac 1996). Indeed, the TFIIIB association is so stable that it serves as a roadblock that impedes pervasive Pol II transcription and leads to dissociation of lagging strand replicative polymerases (Roy et al. 2016; Smith and Whitehouse 2012).

As detailed above, TFIIIB is the key transcription factor in the Pol III system. Two additional transcription factors, TFIIIA and TFIIIC are assembly and specificity factors that aid in positioning of TFIIIB on the upstream promoter DNA (Kassavetis et al. 1990; Roberts et al. 1995). TFIIIC a hexameric 0.5 MDa protein complex that binds to intragenic promoter elements (Conesa et al. 1993; Ducrot et al. 2006; Male et al. 2015; Stillman and Geiduschek 1984), but is dispensable after TFIIIB binding and is displaced by transcribing Pol III (Bardeleben et al. 1994).

The Structure of the RNA Polymerase III Pre-initiation Complex

Three independent cryo-EM studies (Abascal-Palacios et al. 2018; Han et al. 2018; Vorländer et al. 2018) have revealed the mechanism of Pol III-PIC assembly and promoter opening at high resolution. Two of these structures were determined on a U6 snRNA promoter (Abascal-Palacios et al. 2018; Vorländer et al. 2018), one of them on the asparagine-tRNA promoter (Han et al. 2018). These studies used promoter DNA substrates mimicking closed and open Pol III-PICs as well as Pol III-ITCs (Abascal-Palacios et al. 2018; Han et al. 2018; Vorländer et al. 2018). Interestingly, it was found that the open Pol III-PIC spontaneously forms even from fully base-paired templates (Abascal-Palacios et al. 2018; Han et al. 2018; Vorländer et al. 2018) and even in presence of mutants deficient in promoter melting (Abascal-Palacios et al. 2018; Han et al. 2018). The RNA substrate was lost from the Pol III-ITC under some conditions, either due to washing or nucleolytic cleavage aided by the Pol III subunit C11 (Han et al. 2018; Vorländer et al. 2018), and the Pol III-OC and ITC complexes exhibited similar architecture (Han et al. 2018).

The cryo-EM reconstructions comprise Pol III, TFIIIB, and promoter DNA, and show TFIIIB tightly bound to and wrapped around the upstream DNA (Fig. 5.14a). Similarly to the Brf2-containing structure (Gouge et al. 2015, 2017), the cyclin folds of Brf1, present in the cryo-EM sample as a Brf1-TBP fusion protein (Kassavetis et al. 2005), bind to TBP and promoter DNA upstream and downstream of TBP (Figs. 5.13a, 5.14b). The SANT domain of Bdp1 binds TBP, DNA, and Brf1, leading to a highly stable assembly of TFIIIB on the upstream promoter DNA, suitable for supporting multiple rounds of initiation by Pol III (Dieci et al. 2013; Dieci and Sentenac 1996).

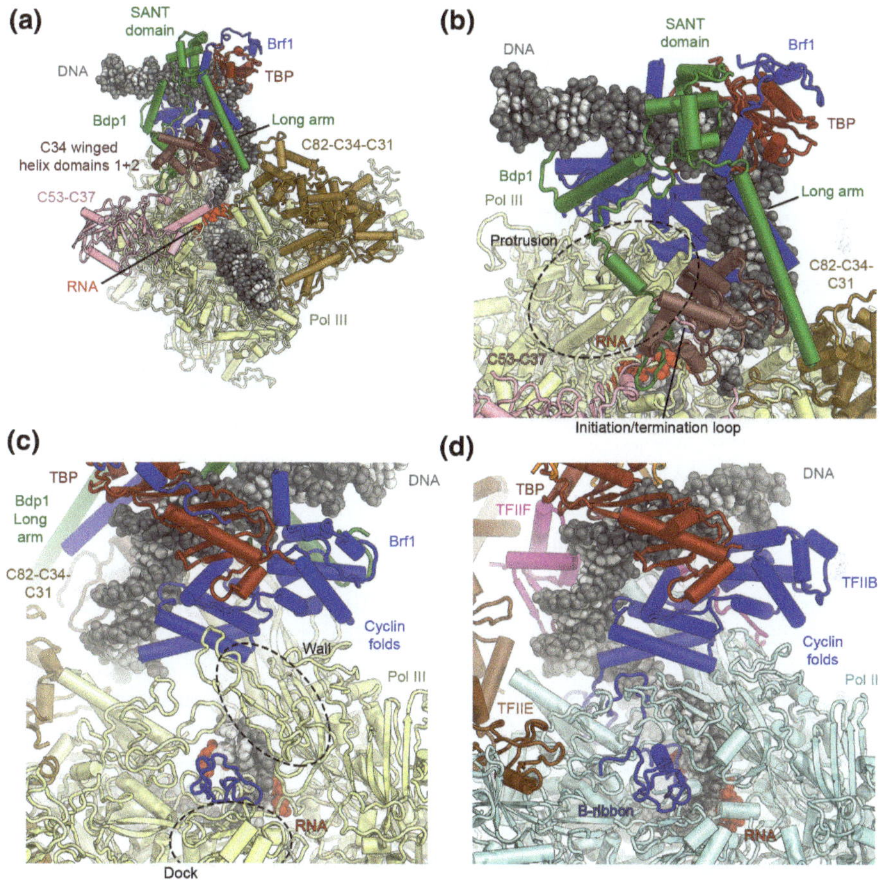

Fig. 5.14 The structure of the Pol III-ITC. **a** Overview of the structure of the yeast Pol III-ITC (Vorländer et al. 2018). Compared to the structure of isolated TFIIIB, additional elements reaching towards the polymerase are visualized in the complete PIC and ITC structures. **b** TFIIIB encloses the upstream promoter. Structural elements of Bdp1 (green) contact the transcription-factor like subunits of Pol III (pink and brown), thereby contributing to a stable platform for ordering of some of these factors. **c**, **d** The B-ribbon domains of Brf1 (**c**) and TFIIB (**d**) occupy similar positions and interact with nucleic acids

In the open Pol III-PIC, Pol III itself exhibits a closed clamp conformation and tightly stabilizes the open DNA bubble (Abascal-Palacios et al. 2018; Han et al. 2018; Vorländer et al. 2018). Pol III subunits C37, C34 and C31, all of them part of the "built-in transcription factors" of Pol III are flexible in the structure of transcribing Pol III (Hoffmann et al. 2015), but ordered in the open PIC (Fig. 5.14a). Bound to the upstream promoter DNA, TFIIIB sits above the Pol III cleft, where its Bdp1 and Brf1 subunits form several contacts with Pol III. Due to these contacts, parts of the general transcription factor subunits that are not visualized in the crystal structures (Gouge et al. 2015, 2017) are visible in the cryo-EM reconstructions (Abascal-Palacios et al. 2018; Han et al. 2018; Vorländer et al. 2018).

In Brf1, the N-terminal zinc-ribbon and the cyclin folds contact the Pol III dock, wall, and protrusion, similar to the TFIIB contacts observed in the Pol II system (Fig. 5.14c, d) (He et al. 2016; Murakami et al. 2015; Nikolov et al. 1995; Plaschka et al. 2016; Tsai and Sigler 2000). The zinc-ribbon is inserted through the active site cleft and contacts the Pol III dock domain (Fig. 5.14c), and an adjacent linker region interacts with the template DNA strand (Abascal-Palacios et al. 2018; Han et al. 2018; Vorländer et al. 2018). The zinc-ribbon interaction is also found in the Pol I system (Engel et al. 2017; Han et al. 2017; Sadian et al. 2017), while the interaction of the N-terminal cyclin fold with the wall and protrusion (Fig. 5.14c, d) is restricted to Pol II and Pol III.

Notably, Brf1 includes a structural element termed the helical pin, structurally homologous to the Brf2 molecular pin and occupying the same site at the interface between TBP, DNA, and the Brf1 cyclin folds (Gouge et al. 2017), but without the redox-sensing activity of its Brf2 counterpart (Abascal-Palacios et al. 2018; Han et al. 2018).

The extended SANT domain and a linker of Bdp1 bind to the major and minor grooves of the DNA, as visualized previously (Gouge et al. 2017). Additionally, several long helices of Bdp1 are seen in the cryo-EM maps (Fig. 5.14a), reaching across the Pol III C34 subunit (part of the TFIIE/TFIIF-like Pol III subunits), and together with C37 (part of the TFIIF-like Pol III subunits) form a platform that stabilizes C34 (Abascal-Palacios et al. 2018; Han et al. 2018; Vorländer et al. 2018). Notably, these interactions between Bdp1 and C34-C37 also lead to the stabilization of the initiation/termination loop of C37 (C37 residues 211–224) (Kassavetis et al. 2010; Rijal and Maraia 2013; Wu et al. 2011) by interactions with part of a Bdp1 region termed the "tether" (Bdp1 residues 360–398). After stabilization by the C37-Bdp1 platform, winged-helix domain 2 in C34 contacts the open transcription bubble near its upstream edge (Fig. 5.14b), implicating C34 in promoter melting or stabilization of the open bubble (Abascal-Palacios et al. 2018; Brun et al. 1997; Han et al. 2018; Vorländer et al. 2018). Bdp1 mutants that are defective in open complex formation (Kassavetis et al. 2001) map to this platform region, underscoring the importance of the molecular arrangement in this region for promoter opening (Abascal-Palacios et al. 2018).

The cryo-EM maps visualized the interactions that contribute to melting and stabilization of the transcription bubble at the active site. In the Pol III-PIC, the active site assumes a conformation that is reminiscent of elongating Pol III

(Hoffmann et al. 2015), with a disordered trigger loop and rudder, and a bent bridge helix. Upon transition to the ITC, the upstream bubble edge is stabilized mostly by contacts from Pol III subunits, including the C34 and C82 winged helix domains and the C160 clamp, as well as the N-terminal Brf1 cyclin fold (Abascal-Palacios et al. 2018; Han et al. 2018; Vorländer et al. 2018). The arrangement of TFIIIB elements, with the Brf1 N-terminal zinc-ribbon, the adjacent linker, and the N-terminal cyclin fold close to the transcription bubble, and with Bdp1 stabilizing C34 (Abascal-Palacios et al. 2018; Han et al. 2018; Vorländer et al. 2018), explains the dual function of TFIIIB in promoter melting (Kassavetis et al. 2001).

Model for Transcription Initiation by RNA Polymerase III and Transition to Elongation

The structural results on the Pol III-PIC described in the previous section, combined with structures of apo-Pol III and transcribing Pol III (Hoffmann et al. 2015), suggest a model for Pol III transcription initiation. Initially, Pol III has a closed clamp, as visualized in the closed Pol III-PIC complex, with disordered downstream DNA projecting away from the polymerase (Vorländer et al. 2018). This is in contrast to e.g. the closed Pol II-PIC, where the DNA runs along the length of the Pol II active site cleft and interacts with the jaws at the downstream end (He et al. 2013, 2016; Murakami et al. 2015). The transcription factor-like modules of Pol III that later become stabilized, as well as TFIIIB elements interacting with them, are initially partially disordered in closed Pol III-PIC (Vorländer et al. 2018). Upon insertion of promoter DNA into the cleft by TFIIIB and the consequent Pol III cleft closure, TFIIIB stabilizes the winged-helix domains of C34 above the cleft, leading to entrapment of the promoter DNA and initial promoter melting. Subsequently, the Brf1 zinc-ribbon and an adjacent linker region become stabilized, with the linker region contacting the opening bubble and facilitating bubble extension, leading to an ITC and eventually an elongating state (Abascal-Palacios et al. 2018; Han et al. 2018; Vorländer et al. 2018).

As detailed above, TFIIIB interacts with promoter DNA very stably and may participate in the Pol III transcription re-initiation pathway (Dieci et al. 2013; Dieci and Sentenac 1996; Kassavetis et al. 1990). An interesting and unexpected finding was the visualization of a transcribing Pol III complex that is still attached to promoter-bound TFIIIB, with up to 33 bp of melted DNA accommodated by Pol III (Han et al. 2018). Based on these findings, and the fact that most Pol III-transcribed genes are short, it was proposed that Pol III may stay associated with TFIIIB throughout the entire transcription cycle (Han et al. 2018).

Comparison of PIC Architectures

Conserved and Divergent Features of Eukaryotic Pre-initiation Complex Architectures

Now that detailed structures of pre-initiation complexes of Pol I, II, and III have been determined, it is possibly to compare the shared and unique features of these three polymerase systems on an architectural level.

Universality of TFIIB-like Factors and Positioning of the Promoter DNA in the PIC

TFIIB-like factors play a critical role in all three eukaryotic RNA polymerase systems discussed in this chapter (see e.g. Figures 5.6, 5.7, 5.10, 5.11, 5.13, 5.14). While Pol II employs TFIIB itself, the Pol I core factor harbours the TFIIB homolog Rrn7, and the Pol III initiation factor TFIIIB contains the TFIIB-related factors Brf1 or Brf2 (Fig. 5.15a–c). The structures of the Pol II and Pol III-PICs (Abascal-Palacios et al. 2018; Han et al. 2018; He et al. 2016; Plaschka et al. 2016; Vorländer et al. 2018) have revealed that in spite of the presence of additional domains in the Pol III initiation factors Brf1/Brf2, the general architecture and role of TFIIB and Brf1/Brf2 are conserved, with all of them forming similar interactions with the upstream promoter DNA and the polymerase, including interactions of the cyclin domains with the polymerase wall and the positioning of the zinc-ribbon domain at the polymerase dock and the adjacent linker in proximity of the transcription bubble (Figs. 5.14c, d, 5.15b, c). The structure of the Pol I-PIC revealed that the zinc-ribbon of Rrn7 assumes a similar position as its TFIIB and Brf1/Brf2 counterparts. However, in spite of the conservation of the two cyclin folds that lie at the core of all TFIIB-related initiation factors, the Rrn7 cyclin folds interact differently with both the upstream promoter DNA (Fig. 5.15a–c) and the RNA polymerase, as compared to TFIIB and Brf1/Brf2 (Fig. 5.15d–f).

As a consequence of this difference in PIC architecture, the path of the DNA differs between the Pol I-PIC on the one hand and the Pol II- and Pol III-PICs on the other (Engel et al. 2017, 2018; Han et al. 2017; Sadian et al. 2017). In the Pol I-ITC, the promoter DNA is accommodated in the active site cleft very similarly as in the Pol I elongation complex, likely assuming such a conformation as early as during closed complex formation. In contrast, in both the Pol II- and Pol III-PICs, the upstream promoter DNA in the closed complex is suspended high above the active site cleft, remains in this conformation during ITC formation, and rearranges to its final trajectory only in the elongation complex (Fig. 5.11c) (Abascal-Palacios et al. 2018; Han et al. 2018; He et al. 2016; Plaschka et al. 2016; Vorländer et al. 2018).

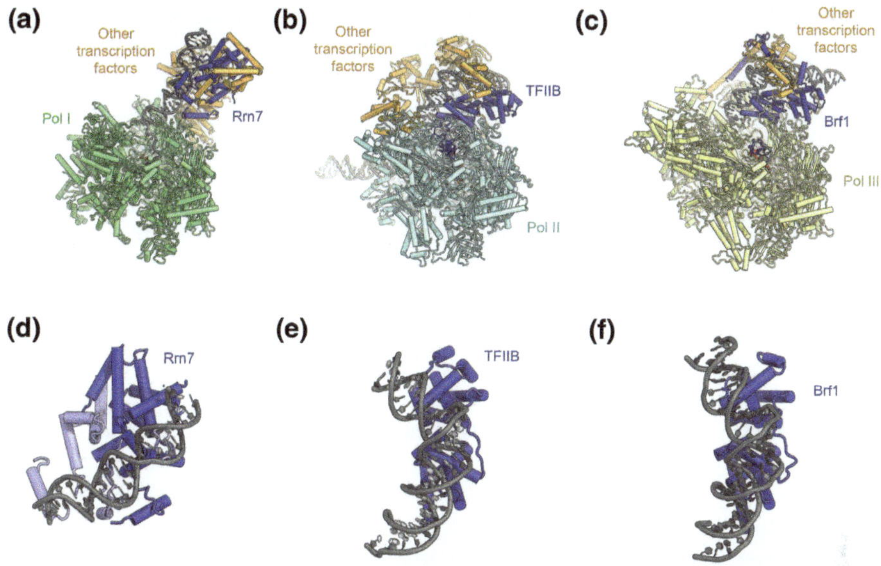

Fig. 5.15 Comparison of DNA-bound structures of TFIIB-like initiation factors. **a–c** Structure of the Pol I (**a**), Pol II (**b**), and Pol III (**c**) ITCs (Han et al. 2017; He et al. 2016; Vorländer et al. 2018). TFIIB-related factors are colored dark blue, other transcription factors orange. **d–f** Conformation of upstream promoter DNA bound to the cyclin folds of Rrn7 (**d**), TFIIB (**e**), and Brf1 (**f**)

Built-in General Transcription Factor-like Subunits in Pol I and Pol III

Pol II is a 12-subunit enzyme, while Pol I has 14 subunits, and Pol III comprises 17 subunits (see Fig. 5.1). As noted above, the supernumerary subunits in Pol I and Pol III share structural and functional similarity to general transcription factors of the Pol II system. Specifically, the location of the Pol I A49-A34.5 and Pol III C53-C37 heterodimers is highly similar to the binding site of TFIIF on the Rpb2 lobe in the Pol II-PIC (Fig. 5.12). This suggests structural and functional overlap with TFIIF, even though A49 in the Pol I system also contains a tandem winged helix domain reminiscent of TFIIE in terms of its DNA interactions (Khatter et al. 2017; Tafur et al. 2016; Vannini and Cramer 2012). Notably, TFIIF is thought to arrive to the Pol II-PIC pre-bound to Pol II (Killeen and Greenblatt 1992; Rani et al. 2004; Roeder 1996), suggesting a relatively straightforward evolutionary path to how its Pol I and Pol III equivalents became incorporated into the multisubunit polymerase (Khatter et al. 2017). In spite of these similarities, there are also differences. The C53-C37 dimer in Pol III is involved not only in initiation, but also in termination, enabling Pol III to terminate autonomously upon encountering the poly-thymidine stretch that serves as the termination signal in this system (Khatter et al. 2017; Landrieux et al. 2006; Rijal and Maraia 2013).

The second polymerase subassembly that shows similarity to Pol II transcription factors is the Pol III C82-C34-C31 trimer. Similarly to TFIIE, the C82-C34-C31 trimer binds to the clamp head (Fig. 5.12b–d) and may contact and stabilize the transcription bubble (Vorländer et al. 2018). At the same time, the C34 winged helix domains 1 and 2 bind in a location that is highly similar to the TFIIFβ (yeast Tfg2) winged helix domain (Fig. 5.7g, h), indicating that the C82-C34-C31 may be a functional fusion of TFIIE and TFIIF (Vorländer et al. 2018).

Pausing of elongating Pol II can lead to backtracking, at which stage the 3′-end of the nascent mRNA leaves the active site and is extruded through a tunnel in the enzyme (Cramer et al. 2000; Gnatt et al. 2001). At this point, critical elements of the Pol II active centre are trapped in inactive conformations and the nascent RNA can no longer be elongated (Cheung and Cramer 2011; Kettenberger et al. 2003), unless the polymerase either advances to place the 3′-end in the active site again, or the transcript is cleaved. In the Pol II-system, TFIIS aids in transcript cleavage and permits resumption of elongation (Izban and Luse 1992; Reines 1992; Wang and Hawley 1993). While the intrinsic Pol II nuclease activity is weak and requires enhancement by TFIIS, the Pol I and Pol III systems harbour strong intrinsic nuclease activity, residing in a TFIIS-like extension domain of their A12.2 and C11 subunits (Ruan et al. 2011; Vannini and Cramer 2012), which for the Pol III system has also been implicated in the Pol III re-initiation pathway (Iben et al. 2011; Khatter et al. 2017). Lacking or unexpectedly weak density for the product RNA in some cryo-EM reconstructions of Pol I and Pol III complexes that were formed with a synthetic DNA bubble scaffold and bound RNA may be due to cleavage of initially bound RNA by intrinsic nuclease activity (Han et al. 2018; Sadian et al. 2017; Vorländer et al. 2018).

Opening of the Transcription Bubble

The Pol II system is unique in its ability to exploit the activity of a DNA-dependent ATPase, XPB within TFIIH, for opening of the transcription bubble. Bacterial RNA polymerases and eukaryotic Pol I and Pol III are able to independently melt the promoter to form open PICs, as strikingly illustrated by structural studies in which attempted structure determination of closed complexes resulted in the formation of open complexes (Abascal-Palacios et al. 2018; Engel et al. 2017; Han et al. 2018; Vorländer et al. 2018). Recent structural, biochemical, and computational analyses (Alekseev et al. 2017; Dienemann et al. 2018; Plaschka et al. 2016) suggest that at least under some circumstances, Pol II is also able to spontaneously melt the promoter, suggesting a universally conserved mechanism of promoter opening in multi-subunit RNA polymerases.

Open Complex Formation in Yeast

For the yeast Pol II transcription initiation system, a reconstitution of the yeast Pol II core-PIC using double stranded DNA for cryo-EM analysis resulted in an open complex (Plaschka et al. 2016). A subsequent study showed that in yeast, there are two distinct promoter classes concerning the requirement for TFIIH; one promoter class can be melted spontaneously without contribution from TFIIH, while the second class benefits more strongly from the presence of TFIIH (Dienemann et al. 2018). The two promoter classes seem to be distinguished by subtle differences in their free energy of melting. Interestingly, the HIS4 promoter used for the structural analysis that resulted in spontaneous open-complex formation (Plaschka et al. 2016) belongs to the first, spontaneously melting class. A follow-up experiment using the more highly TFIIH-responsive GAT1 promoter resulted in structure determination of both open and closed core-PICs, indicating that indeed, promoter melting is less energetically favorable compared to HIS4 (Dienemann et al. 2018). The DNA in one of these closed complexes assumes a different, more highly distorted and more deeply inserted conformation compared to previously determined closed PICs (Dienemann et al. 2018) and may represent an initiation intermediate. This DNA conformation is observed both in the presence and absence of TFIIH.

The Mammalian RNA Polymerase System

Whether these findings can be applied to the human Pol II system is not fully established. Critically, the closed Pol II-PIC could be successfully reconstituted and structurally characterized using human proteins (He et al. 2016). However, experiments in human cells where XPB was depleted using spironolactone indicate that XPB depletion predominantly affects DNA repair, while global transcription is less affected. In contrast, when XPB is present, but its enzymatic activity is inhibited by triptolide, global transcription is reduced (Alekseev et al. 2017). A hypothesis that is consistent with these observation is that XPB may, at least in some cases, act in a checkpoint-like manner. According to this model, inactive XPB binds to the downstream promoter DNA and prevents its insertion into the Pol II active site. Activation of the enzymatic DNA translocase activity of XPB—or alternatively, its depletion—would allow the downstream promoter DNA to reach the active site of Pol II and promoter melting would occur driven by binding energy alone (Alekseev et al. 2017).

Further analysis will be required to establish whether spontaneous melting of Pol II promoters is a general feature and occurs in vivo, consistent with a universally applicable mechanism for melting of promoters by multisubunit polymerases, or if higher eukaryotes have indeed evolved a unique, TFIIH-dependent solution, possibly to allow for more precise regulation of gene expression.

Conclusion

Recent breakthroughs in the structure determination of transcriptional assemblies have revealed the detailed structures of the PICs of Pol I, Pol II, and Pol III. These structures uncovered the intricate architecture of these molecular machines in atomic detail, and have provided valuable mechanistic insight into the process of PIC assembly, transcription bubble opening, and initiation of transcription. This progress has been critically enabled by technical advances in cryo-EM, which allowed the study of highly complex but low-abundance and conformationally heterogeneous molecular assemblies at ever-higher resolution.

Future progress will likely involve the reconstitution and visualization of even larger complexes, in the presence of gene-specific transcription factors, and in the context of chromatin.

Acknowledgements Molecular depictions were created using UCSF Chimera (Pettersen et al. 2004) and PyMol (The PyMOL Molecular Graphics System, Version 1.8, Schrödinger, LLC.). This work was funded through NIGMS grant R35-GM127018 to E. N; B. J. G. was supported by the Swiss National Science Foundation (projects P300PA-160983 and P300PA-174355). E. N. is a Howard Hughes Medical Investigator.

References

Abascal-Palacios G, Ramsay EP, Beuron F, Morris E, Vannini A (2018) Structural basis of RNA polymerase III transcription initiation. Nature 553:301–306

Abdulrahman W et al (2013) ARCH domain of XPD, an anchoring platform for CAK that conditions TFIIH DNA repair and transcription activities. Proc Natl Acad Sci USA 110:E633–642

Alekseev S, Nagy Z, Sandoz J, Weiss A, Egly J-M, Le May N, Coin F (2017) Transcription without XPB establishes a unified helicase-independent mechanism of promoter opening in eukaryotic gene expression. Mol Cell 65:504–513.e505

Anandapadamanaban M et al (2013) High-resolution structure of TBP with TAF1 reveals anchoring patterns in transcriptional regulation. Nat Struct Mol Biol 20:1008–1014

Andel F, Ladurner AG, Inouye C, Tjian R, Nogales E (1999) Three-dimensional structure of the human TFIID-IIA-IIB complex. Science 286:2153–2156

Andersen G et al (1997) The structure of cyclin H: common mode of kinase activation and specific features. EMBO J 16:958–967

Arimbasseri AG, Maraia RJ (2015) Mechanism of transcription termination by RNA polymerase III utilizes a non-template strand sequence-specific signal element. Mol Cell 58:1124–1132

Asturias FJ, Jiang YW, Myers LC, Gustafsson CM, Kornberg RD (1999) Conserved structures of mediator and RNA polymerase II holoenzyme. Science 283:985–987

Baek HJ, Malik S, Qin J, Roeder RG (2002) Requirement of TRAP/mediator for both activator-independent and activator-dependent transcription in conjunction with TFIID-associated TAF(II)s. Mol Cell Biol 22:2842–2852

Bancerek J et al (2013) CDK8 kinase phosphorylates transcription factor STAT1 to selectively regulate the interferon response. Immunity 38:250–262

Bardeleben C, Kassavetis GA, Geiduschek EP (1994) Encounters of *Saccharomyces cerevisiae* RNA polymerase III with its transcription factors during RNA chain elongation. J Mol Biol 235:1193–1205

Barnes CO et al (2015) Crystal structure of a transcribing RNA polymerase II complex reveals a complete transcription bubble. Mol Cell 59:258–269

Bedwell GJ, Appling FD, Anderson SJ, Schneider DA (2012) Efficient transcription by RNA polymerase I using recombinant core factor. Gene 492:94–99

Berico P, Coin F (2017) Is TFIIH the new Achilles heel of cancer cells? Transcription 9:47–51

Bieniossek C et al (2013) The architecture of human general transcription factor TFIID core complex. Nature 493:699–702

Bienstock RJ, Skorvaga M, Mandavilli BS, Van Houten B (2003) Structural and functional characterization of the human DNA repair helicase XPD by comparative molecular modeling and site-directed mutagenesis of the bacterial repair protein UvrB. J Biol Chem 278:5309–5316

Bier M, Fath S, Tschochner H (2004) The composition of the RNA polymerase I transcription machinery switches from initiation to elongation mode. FEBS Lett 564:41–46

Bleichenbacher M, Tan S, Richmond TJ (2003) Novel interactions between the components of human and yeast TFIIA/TBP/DNA complexes. J Mol Biol 332:783–793

Bordi L, Cioci F, Camilloni G (2001) In vivo binding and hierarchy of assembly of the yeast RNA polymerase I transcription factors. Mol Biol Cell 12:753–760

Borggrefe T, Yue X (2011) Interactions between subunits of the Mediator complex with gene-specific transcription factors. Semin Cell Dev Biol 22:759–768

Borggrefe T, Davis R, Erdjument-Bromage H, Tempst P, Kornberg RD (2002) A complex of the Srb8, -9, -10, and -11 transcriptional regulatory proteins from yeast. J Biol Chem 277:44202–44207

Botta E, Nardo T, Lehmann AR, Egly J-M, Pedrini AM, Stefanini M (2002) Reduced level of the repair/transcription factor TFIIH in trichothiodystrophy. Hum Mol Genet 11:2919–2928

Brand M, Leurent C, Mallouh V, Tora L, Schultz P (1999) Three-dimensional structures of the TAFII-containing complexes TFIID and TFTC. Science 286:2151–2153

Brun I, Sentenac A, Werner M (1997) Dual role of the C34 subunit of RNA polymerase III in transcription initiation. EMBO J 16:5730–5741

Brzovic PS et al (2011) The acidic transcription activator Gcn4 binds the mediator subunit Gal11/Med15 using a simple protein interface forming a fuzzy complex. Mol Cell 44:942–953

Buechner CN, Heil K, Michels G, Carell T, Kisker C, Tessmer I (2014) Strand-specific recognition of DNA damages by XPD provides insights into nucleotide excision repair substrate versatility. J Biol Chem 289:3613–3624

Buratowski S, Hahn S, Guarente L, Sharp PA (1989) Five intermediate complexes in transcription initiation by RNA polymerase II. Cell 56:549–561

Burke TW, Kadonaga JT (1996) Drosophila TFIID binds to a conserved downstream basal promoter element that is present in many TATA-box-deficient promoters. Genes Dev 10:711–724

Burke TW, Kadonaga JT (1997) The downstream core promoter element, DPE, is conserved from Drosophila to humans and is recognized by TAFII60 of Drosophila. Genes Dev 11:3020–3031

Bushnell DA, Westover KD, Davis RE, Kornberg RD (2004) Structural basis of transcription: an RNA polymerase II-TFIIB cocrystal at 4.5 Angstroms. Science 303:983–988

Busso D, Keriel A, Sandrock B, Poterszman A, Gileadi O, Egly JM (2000) Distinct regions of MAT1 regulate cdk7 kinase and TFIIH transcription activities. J Biol Chem 275:22815–22823

Cabarcas S, Schramm L (2011) RNA polymerase III transcription in cancer: the BRF2 connection. Mol Cancer 10:47

Cabart P, Murphy S (2001) BRFU, a TFIIB-like factor, is directly recruited to the TATA-box of polymerase III small nuclear RNA gene promoters through its interaction with TATA-binding protein. J Biol Chem 276:43056–43064

Cai G, Imasaki T, Takagi Y, Asturias FJ (2009) Mediator structural conservation and implications for the regulation mechanism. Structure 17:559–567

Carninci P et al (2006) Genome-wide analysis of mammalian promoter architecture and evolution. Nat Genet 38:626–635

Chalkley GE, Verrijzer CP (1999) DNA binding site selection by RNA polymerase II TAFs: a TAF(II)250-TAF(II)150 complex recognizes the initiator. EMBO J 18:4835–4845

Chang WH, Kornberg RD (2000) Electron crystal structure of the transcription factor and DNA repair complex, core TFIIH. Cell 102:609–613

Chen W, Struhl K (1985) Yeast mRNA initiation sites are determined primarily by specific sequences, not by the distance from the TATA element. EMBO J 4:3273–3280

Chen X, Farmer G, Zhu H, Prywes R, Prives C (1993) Cooperative DNA binding of p53 with TFIID (TBP): a possible mechanism for transcriptional activation. Genes Dev 7:1837–1849

Chen JL, Attardi LD, Verrijzer CP, Yokomori K, Tjian R (1994) Assembly of recombinant TFIID reveals differential coactivator requirements for distinct transcriptional activators. Cell 79:93–105

Cheung ACM, Cramer P (2011) Structural basis of RNA polymerase II backtracking, arrest and reactivation. Nature 471:249–253

Cianfrocco MA, Kassavetis GA, Grob P, Fang J, Juven-Gershon T, Kadonaga JT, Nogales E (2013) Human TFIID binds to core promoter DNA in a reorganized structural state. Cell 152:120–131

Cleaver JE, Thompson LH, Richardson AS, States JC (1999) A summary of mutations in the UV-sensitive disorders: xeroderma pigmentosum, Cockayne syndrome, and trichothiodystrophy. Hum Mutat 14:9–22

Coin F, Marinoni JC, Rodolfo C, Fribourg S, Pedrini AM, Egly JM (1998) Mutations in the XPD helicase gene result in XP and TTD phenotypes, preventing interaction between XPD and the p44 subunit of TFIIH. Nat Genet 20:184–188

Coin F, Oksenych V, Egly J-M (2007) Distinct roles for the XPB/p52 and XPD/p44 subcomplexes of TFIIH in damaged DNA opening during nucleotide excision repair. Mol Cell 26:245–256

Colbert T, Hahn S (1992) A yeast TFIIB-related factor involved in RNA polymerase III transcription. Genes Dev 6:1940–1949

Comai L, Tanese N, Tjian R (1992) The TATA-binding protein and associated factors are integral components of the RNA polymerase I transcription factor, SL1. Cell 68:965–976

Compe E, Egly J-M (2012) TFIIH: when transcription met DNA repair. Nat Rev Mol Cell Biol 13:343–354

Conesa C, Swanson RN, Schultz P, Oudet P, Sentenac A (1993) On the subunit composition, stoichiometry, and phosphorylation of the yeast transcription factor TFIIIC/tau. J Biol Chem 268:18047–18052

Cramer P et al (2000) Architecture of RNA polymerase II and implications for the transcription mechanism. Science 288:640–649

Cramer P, Bushnell DA, Kornberg RD (2001) Structural basis of transcription: RNA polymerase II at 2.8 angstrom resolution. Science 292:1863–1876

Davis JA, Takagi Y, Kornberg RD, Asturias FA (2002) Structure of the yeast RNA polymerase II holoenzyme: mediator conformation and polymerase interaction. Mol Cell 10:409–415

Deng W, Roberts SGE (2005) A core promoter element downstream of the TATA box that is recognized by TFIIB. Genes Dev 19:2418–2423

Denissov S et al (2007) Identification of novel functional TBP-binding sites and general factor repertoires. EMBO J 26:944–954

Devault A et al (1995) MAT1 ('menage à trois') a new RING finger protein subunit stabilizing cyclin H-cdk7 complexes in starfish and Xenopus CAK. EMBO J 14:5027–5036

Dieci G, Sentenac A (1996) Facilitated recycling pathway for RNA polymerase III. Cell 84:245–252

Dieci G, Bosio MC, Fermi B, Ferrari R (2013) Transcription reinitiation by RNA polymerase III. Biochim Biophys Acta 1829:331–341

Dienemann C, Schwalb B, Schilbach S, Cramer P (2018) Promoter distortion and opening in the RNA polymerase II cleft. Mol Cell 73:97–106.e4

Donner AJ, Ebmeier CC, Taatjes DJ, Espinosa JM (2010) CDK8 is a positive regulator of transcriptional elongation within the serum response network. Nat Struct Mol Biol 17:194–201
Dotson MR et al (2000) Structural organization of yeast and mammalian mediator complexes. Proc Natl Acad Sci USA 97:14307–14310
Drapkin R, Reardon JT, Ansari A, Huang J-C, Zawel L, Ahn KJ, Sancar A, Reinberg D (1994) Dual role of TFIIH in DNA excision repair and in transcription by RNA polymerase II. Nature 368:769–772
Dubaele S, Proietti De Santis L, Bienstock RJ, Keriel A, Stefanini M, Van Houten B, Egly J-M (2003) Basal transcription defect discriminates between xeroderma pigmentosum and trichothiodystrophy in XPD patients. Mol Cell 11:1635–1646
Dubrovskaya V, Lavigne AC, Davidson I, Acker J, Staub A, Tora L (1996) Distinct domains of hTAFII100 are required for functional interaction with transcription factor TFIIF beta (RAP30) and incorporation into the TFIID complex. EMBO J 15:3702–3712
Ducrot C, Lefebvre O, Landrieux E, Guirouilh-Barbat J, Sentenac A, Acker J (2006) Reconstitution of the yeast RNA polymerase III transcription system with all recombinant factors. J Biol Chem 281:11685–11692
Ehara H, Yokoyama T, Shigematsu H, Yokoyama S, Shirouzu M, Sekine S-I (2017) Structure of the complete elongation complex of RNA polymerase II with basal factors. Science 357:921–924
Elmlund H et al (2009) Cryo-EM reveals promoter DNA binding and conformational flexibility of the general transcription factor TFIID. Structure 17:1442–1452
Engel C, Sainsbury S, Cheung AC, Kostrewa D, Cramer P (2013) RNA polymerase I structure and transcription regulation. Nature 502:650–655
Engel C, Plitzko J, Cramer P (2016) RNA polymerase I-Rrn3 complex at 4.8 Å resolution. Nat Commun 7:12129
Engel C et al (2017) Structural basis of RNA polymerase I transcription initiation. Cell 169:120–131.e22
Engel C, Neyer S, Cramer P (2018) Distinct mechanisms of transcription initiation by RNA polymerases I and II. Annu Rev Biophys 47:425–446
Fan L, Arvai AS, Cooper PK, Iwai S, Hanaoka F, Tainer JA (2006) Conserved XPB core structure and motifs for DNA unwinding: implications for pathway selection of transcription or excision repair. Mol Cell 22:27–37
Fan L et al (2008) XPD helicase structures and activities: insights into the cancer and aging phenotypes from XPD mutations. Cell 133:789–800
Fernández-Tornero C et al (2013) Crystal structure of the 14-subunit RNA polymerase I. Nature 502:644–649
Firestein R et al (2008) CDK8 is a colorectal cancer oncogene that regulates beta-catenin activity. Nature 455:547–551
Fishburn J, Tomko E, Galburt E, Hahn S (2015) Double-stranded DNA translocase activity of transcription factor TFIIH and the mechanism of RNA polymerase II open complex formation. Proc Natl Acad Sci USA 112:3961–3966
Fisher RP (2018) Cdk7: a kinase at the core of transcription and in the crosshairs of cancer drug discovery. Transcription 10:47–56
Fisher RP, Morgan DO (1994) A novel cyclin associates with MO15/CDK7 to form the CDK-activating kinase. Cell 78:713–724
Fisher RP, Jin P, Chamberlin HM, Morgan DO (1995) Alternative mechanisms of CAK assembly require an assembly factor or an activating kinase. Cell 83:47–57
Flanagan PM, Kelleher RJ, Sayre MH, Tschochner H, Kornberg RD (1991) A mediator required for activation of RNA polymerase II transcription in vitro. Nature 350:436–438
Friedrich JK, Panov KI, Cabart P, Russell J, Zomerdijk JCBM (2005) TBP-TAF complex SL1 directs RNA polymerase I pre-initiation complex formation and stabilizes upstream binding factor at the rDNA promoter. J Biol Chem 280:29551–29558
Gannon F et al (1979) Organisation and sequences at the 5′ end of a cloned complete ovalbumin gene. Nature 278:428–434

Gegonne A, Weissman JD, Zhou M, Brady JN, Singer DS (2006) TAF7: a possible transcription initiation check-point regulator. Proc Natl Acad Sci USA 103:602–607

Gervais V et al (2018) Small molecule-based targeting of TTD-A dimerization to control TFIIH transcriptional activity represents a potential strategy for anticancer therapy. J Biol Chem 293:14974–14988

Ghazy MA, Brodie SA, Ammerman ML, Ziegler LM, Ponticelli AS (2004) Amino acid substitutions in yeast TFIIF confer upstream shifts in transcription initiation and altered interaction with RNA polymerase II. Mol Cell Biol 24:10975–10985

Gibbons BJ et al (2012) Subunit architecture of general transcription factor TFIIH. Proc Natl Acad Sci USA 109:1949–1954

Giglia-Mari G et al (2004) A new, tenth subunit of TFIIH is responsible for the DNA repair syndrome trichothiodystrophy group A. Nat Genet 36:714–719

Gnatt AL, Cramer P, Fu J, Bushnell DA, Kornberg RD (2001) Structural basis of transcription: an RNA polymerase II elongation complex at 3.3 A resolution. Science 292:1876–1882

Gorski JJ, Pathak S, Panov K, Kasciukovic T, Panova T, Russell J, Zomerdijk JCBM (2007) A novel TBP-associated factor of SL1 functions in RNA polymerase I transcription. EMBO J 26:1560–1568

Gouge J et al (2015) Redox signaling by the RNA polymerase III TFIIB-related factor Brf2. Cell 163:1375–1387

Gouge J et al (2017) Molecular mechanisms of Bdp1 in TFIIIB assembly and RNA polymerase III transcription initiation. Nat Commun 8:130

Greber BJ, Nguyen THD, Fang J, Afonine PV, Adams PD, Nogales E (2017) The cryo-electron microscopy structure of human transcription factor IIH. Nature 549:414–417

Greber BJ, Toso D, Fang J, Nogales E (2019) The complete structure of the human TFIIH core complex. eLife 8:e44771

Grob P, Cruse MJ, Inouye C, Peris M, Penczek PA, Tjian R, Nogales E (2006) Cryo-electron microscopy studies of human TFIID: conformational breathing in the integration of gene regulatory cues. Structure 14:511–520

Grünberg S, Warfield L, Hahn S (2012) Architecture of the RNA polymerase II preinitiation complex and mechanism of ATP-dependent promoter opening. Nat Struct Mol Biol 19:788–796

Guzder SN, Sung P, Bailly V, Prakash L, Prakash S (1994) RAD25 is a DNA helicase required for DNA repair and RNA polymerase II transcription. Nature 369:578–581

Han Y, Yan C, Nguyen THD, Jackobel AJ, Ivanov I, Knutson BA, He Y (2017) Structural mechanism of ATP-independent transcription initiation by RNA polymerase I. Elife 6:e27414

Han Y, Yan C, Fishbain S, Ivanov I, He Y (2018) Structural visualization of RNA polymerase III transcription machineries. Cell Discov 4:40

He Y, Fang J, Taatjes DJ, Nogales E (2013) Structural visualization of key steps in human transcription initiation. Nature 495:481–486

He Y, Yan C, Fang J, Inouye C, Tjian R, Ivanov I, Nogales E (2016) Near-atomic resolution visualization of human transcription promoter opening. Nature 533:359–365

Henry RW, Sadowski CL, Kobayashi R, Hernandez N (1995) A TBP-TAF complex required for transcription of human snRNA genes by RNA polymerase II and III. Nature 374:653–656

Herrera-Moyano E, Moriel-Carretero M, Montelone BA, Aguilera A (2014) The rem mutations in the ATP-binding groove of the Rad3/XPD helicase lead to *Xeroderma pigmentosum*-Cockayne syndrome-like phenotypes. PLoS Genet 10:e1004859

Hoffmann A, Chiang CM, Oelgeschläger T, Xie X, Burley SK, Nakatani Y, Roeder RG (1996) A histone octamer-like structure within TFIID. Nature 380:356–359

Hoffmann NA et al (2015) Molecular structures of unbound and transcribing RNA polymerase III. Nature 528:231–236

Hontz RD, French SL, Oakes ML, Tongaonkar P, Nomura M, Beyer AL, Smith JS (2008) Transcription of multiple yeast ribosomal DNA genes requires targeting of UAF to the promoter by Uaf30. Mol Cell Biol 28:6709–6719

Horikoshi M, Carey MF, Kakidani H, Roeder RG (1988) Mechanism of action of a yeast activator: direct effect of GAL4 derivatives on mammalian TFIID-promoter interactions. Cell 54:665–669

Iben JR, Mazeika JK, Hasson S, Rijal K, Arimbasseri AG, Russo AN, Maraia RJ (2011) Point mutations in the Rpb9-homologous domain of Rpc11 that impair transcription termination by RNA polymerase III. Nucleic Acids Res 39:6100–6113

Imasaki T et al (2011) Architecture of the Mediator head module. Nature 475:240–243

Ishiguro A, Kassavetis GA, Geiduschek EP (2002) Essential roles of Bdp1, a subunit of RNA polymerase III initiation factor TFIIIB, in transcription and tRNA processing. Mol Cell Biol 22:3264–3275

Izban MG, Luse DS (1992) The RNA polymerase II ternary complex cleaves the nascent transcript in a 3′–5′ direction in the presence of elongation factor SII. Genes Dev 6:1342–1356

Jacobson RH, Ladurner AG, King DS, Tjian R (2000) Structure and function of a human TAFII250 double bromodomain module. Science 288:1422–1425

Jawhari A et al (2002) p52 Mediates XPB function within the transcription/repair factor TFIIH. J Biol Chem 277:31761–31767

Jeronimo C, Langelier M-F, Bataille AR, Pascal JM, Pugh BF, Robert F (2016) Tail and kinase modules differently regulate core mediator recruitment and function in vivo. Mol Cell 64:455–466

Jiang Y, Gralla JD (1993) Uncoupling of initiation and reinitiation rates during HeLa RNA polymerase II transcription in vitro. Mol Cell Biol 13:4572–4577

Johnson SAS, Dubeau L, Johnson DL (2008) Enhanced RNA polymerase III-dependent transcription is required for oncogenic transformation. J Biol Chem 283:19184–19191

Joo YJ, Ficarro SB, Soares LM, Chun Y, Marto JA, Buratowski S (2017) Downstream promoter interactions of TFIID TAFs facilitate transcription reinitiation. Genes Dev 31:2162–2174

Juo ZS, Kassavetis GA, Wang J, Geiduschek EP, Sigler PB (2003) Crystal structure of a transcription factor IIIB core interface ternary complex. Nature 422:534–539

Juven-Gershon T, Kadonaga JT (2010) Regulation of gene expression via the core promoter and the basal transcriptional machinery. Dev Biol 339:225–229

Juven-Gershon T, Cheng S, Kadonaga JT (2006) Rational design of a super core promoter that enhances gene expression. Nat Meth 3:917–922

Kainov DE, Vitorino M, Cavarelli J, Poterszman A, Egly J-M (2008) Structural basis for group A trichothiodystrophy. Nat Struct Mol Biol 15:980–984

Kaplan CD (2013) Basic mechanisms of RNA polymerase II activity and alteration of gene expression in *Saccharomyces cerevisiae*. Biochim Biophys Acta 1829:39–54

Kassavetis GA, Braun BR, Nguyen LH, Geiduschek EP (1990) *S. cerevisiae* TFIIIB is the transcription initiation factor proper of RNA polymerase III, while TFIIIA and TFIIIC are assembly factors. Cell 60:235–245

Kassavetis GA, Nguyen ST, Kobayashi R, Kumar A, Geiduschek EP, Pisano M (1995) Cloning, expression, and function of TFC5, the gene encoding the B″ component of the *Saccharomyces cerevisiae* RNA polymerase III transcription factor TFIIIB. Proc Natl Acad Sci USA 92:9786–9790

Kassavetis GA, Letts GA, Geiduschek EP (1999) A minimal RNA polymerase III transcription system. EMBO J 18:5042–5051

Kassavetis GA, Letts GA, Geiduschek EP (2001) The RNA polymerase III transcription initiation factor TFIIIB participates in two steps of promoter opening. EMBO J 20:2823–2834

Kassavetis GA, Soragni E, Driscoll R, Geiduschek EP (2005) Reconfiguring the connectivity of a multiprotein complex: fusions of yeast TATA-binding protein with Brf1, and the function of transcription factor IIIB. Proc Natl Acad Sci USA 102:15406–15411

Kassavetis GA, Prakash P, Shim E (2010) The C53/C37 subcomplex of RNA polymerase III lies near the active site and participates in promoter opening. J Biol Chem 285:2695–2706

Keener J, Josaitis CA, Dodd JA, Nomura M (1998) Reconstitution of yeast RNA polymerase I transcription in vitro from purified components. TATA-binding protein is not required for basal transcription. J Biol Chem 273:33795–33802

Kettenberger H, Armache K-J, Cramer P (2003) Architecture of the RNA polymerase II-TFIIS complex and implications for mRNA cleavage. Cell 114:347–357

Keys DA, Vu L, Steffan JS, Dodd JA, Yamamoto RT, Nogi Y, Nomura M (1994) RRN6 and RRN7 encode subunits of a multiprotein complex essential for the initiation of rDNA transcription by RNA polymerase I in *Saccharomyces cerevisiae*. Genes Dev 8:2349–2362

Keys DA et al (1996) Multiprotein transcription factor UAF interacts with the upstream element of the yeast RNA polymerase I promoter and forms a stable preinitiation complex. Genes Dev 10:887–903

Khatter H, Vorländer MK, Müller CW (2017) RNA polymerase I and III: similar yet unique. Cur Opin Struct Biol 47:88–94

Khoo B, Brophy B, Jackson SP (1994) Conserved functional domains of the RNA polymerase III general transcription factor BRF. Genes Dev 8:2879–2890

Khoo S-K, Wu C-C, Lin Y-C, Lee J-C, Chen H-T (2014) Mapping the protein interaction network for TFIIB-related factor Brf1 in the RNA polymerase III preinitiation complex. Mol Cell Biol 34:551–559

Khoo S-K, Wu C-C, Lin Y-C, Chen H-T (2018) The TFIIE-related Rpc82 subunit of RNA polymerase III interacts with the TFIIB-related transcription factor Brf1 and the polymerase cleft for transcription initiation. Nucleic Acids Res 46:1157–1166

Killeen MT, Greenblatt JF (1992) The general transcription factor RAP30 binds to RNA polymerase II and prevents it from binding nonspecifically to DNA. Mol Cell Biol 12:30–37

Kim YJ, Björklund S, Li Y, Sayre MH, Kornberg RD (1994) A multiprotein mediator of transcriptional activation and its interaction with the C-terminal repeat domain of RNA polymerase II. Cell 77:599–608

Kim KK, Chamberlin HM, Morgan DO, Kim SH (1996) Three-dimensional structure of human cyclin H, a positive regulator of the CDK-activating kinase. Nat Struct Biol 3:849–855

Kim TK, Ebright RH, Reinberg D (2000) Mechanism of ATP-dependent promoter melting by transcription factor IIH. Science 288:1418–1422

Kim JS, Saint-André C, Lim HS, Hwang C-S, Egly J-M, Cho Y (2015) Crystal structure of the Rad3/XPD regulatory domain of Ssl1/p44. J Biol Chem 290:8321–8330

Knuesel MT, Meyer KD, Bernecky C, Taatjes DJ (2009) The human CDK8 subcomplex is a molecular switch that controls Mediator coactivator function. Genes Dev 23:439–451

Knutson BA, Hahn S (2011) Yeast Rrn7 and human TAF1B are TFIIB-related RNA polymerase I general transcription factors. Science 333:1637–1640

Koch SC, Simon N, Ebert C, Carell T (2016) Molecular mechanisms of xeroderma pigmentosum (XP) proteins. Q Rev Biophys 49:e5

Kokubo T, Gong DW, Yamashita S, Horikoshi M, Roeder RG, Nakatani Y (1993) Drosophila 230-kD TFIID subunit, a functional homolog of the human cell cycle gene product, negatively regulates DNA binding of the TATA box-binding subunit of TFIID. Genes Dev 7:1033–1046

Kokubo T, Yamashita S, Horikoshi M, Roeder RG, Nakatani Y (1994) Interaction between the N-terminal domain of the 230-kDa subunit and the TATA box-binding subunit of TFIID negatively regulates TATA-box binding. Proc Natl Acad Sci USA 91:3520–3524

Kolesnikova O, Ben-Shem A, Luo J, Ranish J, Schultz P, Papai G (2018) Molecular structure of promoter-bound yeast TFIID. Nat Commun 9:4666

Korkhin Y et al (2009) Evolution of complex RNA polymerases: the complete archaeal RNA polymerase structure. PLoS Biol 7:e1000102–1000110

Kostrewa D, Zeller ME, Armache K-J, Seizl M, Leike K, Thomm M, Cramer P (2009) RNA polymerase II-TFIIB structure and mechanism of transcription initiation. Nature 462:323–330

Kuper J et al (2014) In TFIIH, XPD helicase is exclusively devoted to DNA repair. PLoS Biol 12: e1001954

Kutach AK, Kadonaga JT (2000) The downstream promoter element DPE appears to be as widely used as the TATA box in Drosophila core promoters. Mol Cell Biol 20:4754–4764

Lagrange T, Kapanidis AN, Tang H, Reinberg D, Ebright RH (1998) New core promoter element in RNA polymerase II-dependent transcription: sequence-specific DNA binding by transcription factor IIB. Genes Dev 12:34–44

Lalo D, Steffan JS, Dodd JA, Nomura M (1996) RRN11 encodes the third subunit of the complex containing Rrn6p and Rrn7p that is essential for the initiation of rDNA transcription by yeast RNA polymerase I. J Biol Chem 271:21062–21067

Landrieux E, Alic N, Ducrot C, Acker J, Riva M, Carles C (2006) A subcomplex of RNA polymerase III subunits involved in transcription termination and reinitiation. EMBO J 25:118–128

Lariviere L, Plaschka C, Seizl M, Wenzeck L, Kurth F, Cramer P (2012a) Structure of the Mediator head module. Nature 492:448–451

Lariviere L, Seizl M, Cramer P (2012b) A structural perspective on Mediator function. Curr Opin Cell Biol 24:305–313

Lariviere L, Plaschka C, Seizl M, Petrotchenko EV, Wenzeck L, Borchers CH, Cramer P (2013) Model of the Mediator middle module based on protein cross-linking. Nucleic Acids Res 41:9266–9273

Learned RM, Cordes S, Tjian R (1985) Purification and characterization of a transcription factor that confers promoter specificity to human RNA polymerase I. Mol Cell Biol 5:1358–1369

Lee YC, Park JM, Min S, Han SJ, Kim YJ (1999) An activator binding module of yeast RNA polymerase II holoenzyme. Mol Cell Biol 19:2967–2976

Lee D-H, Gershenzon N, Gupta M, Ioshikhes IP, Reinberg D, Lewis BA (2005) Functional characterization of core promoter elements: the downstream core element is recognized by TAF1. Mol Cell Biol 25:9674–9686

Lenhard B, Sandelin A, Carninci P (2012) Metazoan promoters: emerging characteristics and insights into transcriptional regulation. Nat Rev Genet 13:233–245

Leurent C et al (2002) Mapping histone fold TAFs within yeast TFIID. EMBO J 21:3424–3433

Li C-L, Golebiowski FM, Onishi Y, Samara NL, Sugasawa K, Yang W (2015) Tripartite DNA lesion recognition and verification by XPC, TFIIH, and XPA in nucleotide excision repair. Mol Cell 59:1025–1034

Lim CY, Santoso B, Boulay T, Dong E, Ohler U, Kadonaga JT (2004) The MTE, a new core promoter element for transcription by RNA polymerase II. Genes Dev 18:1606–1617

Lin CW, Moorefield B, Payne J, Aprikian P, Mitomo K, Reeder RH (1996) A novel 66-kilodalton protein complexes with Rrn6, Rrn7, and TATA-binding protein to promote polymerase I transcription initiation in *Saccharomyces cerevisiae*. Mol Cell Biol 16:6436–6443

Liu H et al (2008) Structure of the DNA repair helicase XPD. Cell 133:801–812

Liu X, Bushnell DA, Wang D, Calero G, Kornberg RD (2010) Structure of an RNA polymerase II-TFIIB complex and the transcription initiation mechanism. Science 327:206–209

Lobo SM, Tanaka M, Sullivan ML, Hernandez N (1992) A TBP complex essential for transcription from TATA-less but not TATA-containing RNA polymerase III promoters is part of the TFIIIB fraction. Cell 71:1029–1040

Lockwood WW et al (2010) Integrative genomic analyses identify BRF2 as a novel lineage-specific oncogene in lung squamous cell carcinoma. PLoS Med 7:e1000315

Lolli G, Lowe ED, Brown NR, Johnson LN (2004) The crystal structure of human CDK7 and its protein recognition properties. Structure 12:2067–2079

López-De-León A, Librizzi M, Puglia K, Willis IM (1992) PCF4 encodes an RNA polymerase III transcription factor with homology to TFIIB. Cell 71:211–220

Louder RK, He Y, López-Blanco JR, Fang J, Chacón P, Nogales E (2016) Structure of promoter-bound TFIID and model of human pre-initiation complex assembly. Nature 531:604–609

Luo J et al (2015) Architecture of the human and yeast general transcription and DNA repair factor TFIIH. Mol Cell 59:794–806

Male G et al (2015) Architecture of TFIIIC and its role in RNA polymerase III pre-initiation complex assembly. Nat Commun 6:7387

Marteijn JA, Lans H, Vermeulen W, Hoeijmakers JHJ (2014) Understanding nucleotide excision repair and its roles in cancer and ageing. Nat Rev Mol Cell Biol 15:465–481

Mathieu N, Kaczmarek N, Rüthemann P, Luch A, Naegeli H (2013) DNA quality control by a lesion sensor pocket of the xeroderma pigmentosum group D helicase subunit of TFIIH. Curr Biol 23:204–212

Maxon ME, Goodrich JA, Tjian R (1994) Transcription factor IIE binds preferentially to RNA polymerase IIa and recruits TFIIH: a model for promoter clearance. Genes Dev 8:515–524

Milkereit P, Tschochner H (1998) A specialized form of RNA polymerase I, essential for initiation and growth-dependent regulation of rRNA synthesis, is disrupted during transcription. EMBO J 17:3692–3703

Milkereit P, Schultz P, Tschochner H (1997) Resolution of RNA polymerase I into dimers and monomers and their function in transcription. Biol Chem 378:1433–1443

Mital R, Kobayashi R, Hernandez N (1996) RNA polymerase III transcription from the human U6 and adenovirus type 2 VAI promoters has different requirements for human BRF, a subunit of human TFIIIB. Mol Cell Biol 16:7031–7042

Monté D et al (2018) Crystal structure of human Mediator subunit MED23. Nat Commun 9:3389

Moriel-Carretero M, Herrera-Moyano E, Aguilera A (2015) A unified model for the molecular basis of *Xeroderma pigmentosum*-Cockayne Syndrome. Rare Dis 3:e1079362

Murakami K et al (2013) Architecture of an RNA polymerase II transcription pre-initiation complex. Science 342:1238724

Murakami K, Tsai K-L, Kalisman N, Bushnell DA, Asturias FJ, Kornberg RD (2015) Structure of an RNA polymerase II preinitiation complex. Proc Natl Acad Sci USA 112:13543–13548

Naegeli H, Bardwell L, Friedberg EC (1993) Inhibition of Rad3 DNA helicase activity by DNA adducts and abasic sites: implications for the role of a DNA helicase in damage-specific incision of DNA. Biochemistry 32:613–621

Naidu S, Friedrich JK, Russell J, Zomerdijk JCBM (2011) TAF1B is a TFIIB-like component of the basal transcription machinery for RNA polymerase I. Science 333:1640–1642

Natarajan K, Jackson BM, Zhou H, Winston F, Hinnebusch AG (1999) Transcriptional activation by Gcn4p involves independent interactions with the SWI/SNF complex and the SRB/mediator. Mol Cell 4:657–664

Neyer S et al (2016) Structure of RNA polymerase I transcribing ribosomal DNA genes. Nature 540:607–610

Nikolov DB et al (1992) Crystal structure of TFIID TATA-box binding protein. Nature 360:40–46

Nikolov DB et al (1995) Crystal structure of a TFIIB-TBP-TATA-element ternary complex. Nature 377:119–128

Nikolov DB, Chen H, Halay ED, Hoffman A, Roeder RG, Burley SK (1996) Crystal structure of a human TATA box-binding protein/TATA element complex. Proc Natl Acad Sci USA 93:4862–4867

Nogales E, Fang J, Louder RK (2017a) Structural dynamics and DNA interaction of human TFIID. Transcription 8:55–60

Nogales E, Louder RK, He Y (2017b) Structural insights into the eukaryotic transcription initiation machinery. Annu Rev Biophys 46:59–83

Nogales E, Patel AB, Louder RK (2017c) Towards a mechanistic understanding of core promoter recognition from cryo-EM studies of human TFIID. Curr Opin Struct Biol 47:60–66

Nozawa K, Schneider TR, Cramer P (2017) Core Mediator structure at 3.4 Å extends model of transcription initiation complex. Nature 545:248–251

Ohkuma Y, Roeder RG (1994) Regulation of TFIIH ATPase and kinase activities by TFIIE during active initiation complex formation. Nature 368:160–163

Okuda M, Tanaka A, Satoh M, Mizuta S, Takazawa M, Ohkuma Y, Nishimura Y (2008) Structural insight into the TFIIE-TFIIH interaction: TFIIE and p53 share the binding region on TFIIH. EMBO J 27:1161–1171

Papai G, Tripathi MK, Ruhlmann C, Werten S, Crucifix C, Weil PA, Schultz P (2009) Mapping the initiator binding Taf2 subunit in the structure of hydrated yeast TFIID. Structure 17:363–373

Patel AB et al (2018) Structure of human TFIID and mechanism of TBP loading onto promoter DNA. Science 362(6421):eaau8872

Pettersen EF, Goddard TD, Huang CC, Couch GS, Greenblatt DM, Meng EC, Ferrin TE (2004) UCSF Chimera–a visualization system for exploratory research and analysis. J Comput Chem 25:1605–1612
Pilsl M et al (2016) Structure of the initiation-competent RNA polymerase I and its implication for transcription. Nat Commun 7:12126
Plaschka C et al (2015) Architecture of the RNA polymerase II-Mediator core initiation complex. Nature 518:376–380
Plaschka C, Hantsche M, Dienemann C, Burzinski C, Plitzko J, Cramer P (2016) Transcription initiation complex structures elucidate DNA opening. Nature 533:353–358
Pugh BF, Tjian R (1990) Mechanism of transcriptional activation by Sp1: evidence for coactivators. Cell 61:1187–1197
Pugh BF, Tjian R (1991) Transcription from a TATA-less promoter requires a multisubunit TFIID complex. Genes Dev 5:1935–1945
Rani PG, Ranish JA, Hahn S (2004) RNA polymerase II (Pol II)-TFIIF and Pol II-mediator complexes: the major stable Pol II complexes and their activity in transcription initiation and reinitiation. Mol Cell Biol 24:1709–1720
Rapin I (2013) Disorders of nucleotide excision repair. Handb Clin Neurol 113:1637–1650
Reines D (1992) Elongation factor-dependent transcript shortening by template-engaged RNA polymerase II. J Biol Chem 267:3795–3800
Rijal K, Maraia RJ (2013) RNA polymerase III mutants in TFIIFα-like C37 that cause terminator readthrough with no decrease in transcription output. Nucleic Acids Res 41:139–155
Roberts S, Colbert T, Hahn S (1995) TFIIIC determines RNA polymerase III specificity at the TATA-containing yeast U6 promoter. Genes Dev 9:832–842
Robinson PJJ, Bushnell DA, Trnka MJ, Burlingame AL, Kornberg RD (2012) Structure of the mediator head module bound to the carboxy-terminal domain of RNA polymerase II. Proc Natl Acad Sci USA 109:17931–17935
Robinson PJ et al (2015) Molecular architecture of the yeast Mediator complex. Elife 4:e08719
Robinson PJ, Trnka MJ, Bushnell DA, Davis RE, Mattei P-J, Burlingame AL, Kornberg RD (2016) Structure of a complete mediator-RNA polymerase II pre-initiation complex. Cell 166:1411–1422.e1416
Roeder RG (1996) The role of general initiation factors in transcription by RNA polymerase II. Trends Biochem Sci 21:327–335
Roeder RG, Rutter WJ (1969) Multiple forms of DNA-dependent RNA polymerase in eukaryotic organisms. Nature 224:234–237
Roeder RG, Rutter WJ (1970) Specific nucleolar and nucleoplasmic RNA polymerases. Proc Natl Acad Sci USA 65:675–682
Rossignol M, Kolb-Cheynel I, Egly JM (1997) Substrate specificity of the cdk-activating kinase (CAK) is altered upon association with TFIIH. EMBO J 16:1628–1637
Roy AL, Singer DS (2015) Core promoters in transcription: old problem, new insights. Trends Biochem Sci 40:165–171
Roy K, Gabunilas J, Gillespie A, Ngo D, Chanfreau GF (2016) Common genomic elements promote transcriptional and DNA replication roadblocks. Genome Res 26:1363–1375
Ruan W, Lehmann E, Thomm M, Kostrewa D, Cramer P (2011) Evolution of two modes of intrinsic RNA polymerase transcript cleavage. J Biol Chem 286:18701–18707
Ruppert S, Tjian R (1995) Human TAFII250 interacts with RAP74: implications for RNA polymerase II initiation. Genes Dev 9:2747–2755
Sadian Y et al (2017) Structural insights into transcription initiation by yeast RNA polymerase I. EMBO J 36(18):2698–2709
Sadowski CL, Henry RW, Lobo SM, Hernandez N (1993) Targeting TBP to a non-TATA box cis-regulatory element: a TBP-containing complex activates transcription from snRNA promoters through the PSE. Genes Dev 7:1535–1548
Sainsbury S, Niesser J, Cramer P (2013) Structure and function of the initially transcribing RNA polymerase II-TFIIB complex. Nature 493:437–440

Sainsbury S, Bernecky C, Cramer P (2015) Structural basis of transcription initiation by RNA polymerase II. Nat Rev Mol Cell Biol 16:129–143

Sandelin A, Carninci P, Lenhard B, Ponjavic J, Hayashizaki Y, Hume DA (2007) Mammalian RNA polymerase II core promoters: insights from genome-wide studies. Nat Rev Genet 8:424–436

Sanders SL, Garbett KA, Weil PA (2002) Molecular characterization of *Saccharomyces cerevisiae* TFIID. Mol Cell Biol 22:6000–6013

Sandrock B, Egly JM (2001) A yeast four-hybrid system identifies Cdk-activating kinase as a regulator of the XPD helicase, a subunit of transcription factor IIH. J Biol Chem 276:35328–35333

Schilbach S, Hantsche M, Tegunov D, Dienemann C, Wigge C, Urlaub H, Cramer P (2017) Structures of transcription pre-initiation complex with TFIIH and Mediator. Nature 551:204–209

Schneider DA (2012) RNA polymerase I activity is regulated at multiple steps in the transcription cycle: recent insights into factors that influence transcription elongation. Gene 493:176–184

Schneider EV, Böttcher J, Blaesse M, Neumann L, Huber R, Maskos K (2011) The structure of CDK8/CycC implicates specificity in the CDK/cyclin family and reveals interaction with a deep pocket binder. J Mol Biol 412:251–266

Schramm L, Hernandez N (2002) Recruitment of RNA polymerase III to its target promoters. Genes Dev 16:2593–2620

Schramm L, Pendergrast PS, Sun Y, Hernandez N (2000) Different human TFIIIB activities direct RNA polymerase III transcription from TATA-containing and TATA-less promoters. Genes Dev 14:2650–2663

Schultz P, Fribourg S, Poterszman A, Mallouh V, Moras D, Egly JM (2000) Molecular structure of human TFIIH. Cell 102:599–607

Shah SM, Kumar A, Geiduschek EP, Kassavetis GA (1999) Alignment of the B″ subunit of RNA polymerase III transcription factor IIIB in its promoter complex. J Biol Chem 274:28736–28744

Shiekhattar R et al (1995) Cdk-activating kinase complex is a component of human transcription factor TFIIH. Nature 374:283–287

Sklar VE, Schwartz LB, Roeder RG (1975) Distinct molecular structures of nuclear class I, II, and III DNA-dependent RNA polymerases. Proc Natl Acad Sci USA 72:348–352

Smith DJ, Whitehouse I (2012) Intrinsic coupling of lagging-strand synthesis to chromatin assembly. Nature 483:434–438

Soutourina J (2018) Transcription regulation by the Mediator complex. Nat Rev Mol Cell Biol 19:262–274

Steffan JS, Keys DA, Vu L, Nomura M (1998) Interaction of TATA-binding protein with upstream activation factor is required for activated transcription of ribosomal DNA by RNA polymerase I in *Saccharomyces cerevisiae* in vivo. Mol Cell Biol 18:3752–3761

Stillman DJ, Geiduschek EP (1984) Differential binding of a *S. cerevisiae* RNA polymerase III transcription factor to two promoter segments of a tRNA gene. EMBO J 3:847–853

Stringer KF, Ingles CJ, Greenblatt J (1990) Direct and selective binding of an acidic transcriptional activation domain to the TATA-box factor TFIID. Nature 345:783–786

Sugasawa K, Akagi J-i, Nishi R, Iwai S, Hanaoka F (2009) Two-step recognition of DNA damage for mammalian nucleotide excision repair: Directional binding of the XPC complex and DNA strand scanning. Mol Cell 36:642–653

Sung P, Bailly V, Weber C, Thompson LH, Prakash L, Prakash S (1993) Human xeroderma pigmentosum group D gene encodes a DNA helicase. Nature 365:852–855

Svejstrup JQ et al (1995) Different forms of TFIIH for transcription and DNA repair: holo-TFIIH and a nucleotide excision repairosome. Cell 80:21–28

Svetlov V, Nudler E (2013) Basic mechanism of transcription by RNA polymerase II. Biochim Biophys Acta 1829:20–28

Tafur L et al (2016) Molecular structures of transcribing RNA polymerase I. Mol Cell 64:1135–1143

Tan S, Hunziker Y, Sargent DF, Richmond TJ (1996) Crystal structure of a yeast TFIIA/TBP/DNA complex. Nature 381:127–151

Teichmann M, Wang Z, Roeder RG (2000) A stable complex of a novel transcription factor IIB-related factor, human TFIIIB50, and associated proteins mediate selective transcription by RNA polymerase III of genes with upstream promoter elements. Proc Natl Acad Sci USA 97:14200–14205

Theron T et al (2005) Transcription-associated breaks in xeroderma pigmentosum group D cells from patients with combined features of xeroderma pigmentosum and Cockayne syndrome. Mol Cell Biol 25:8368–8378

Torreira E et al (2017) The dynamic assembly of distinct RNA polymerase I complexes modulates rDNA transcription. Elife 6:e20832

Tsai K-L et al (2017) Mediator structure and rearrangements required for holoenzyme formation. Nature 544:196–201

Tsai FT, Sigler PB (2000) Structural basis of preinitiation complex assembly on human pol II promoters. EMBO J 19:25–36

Tsai K-L, Sato S, Tomomori-Sato C, Conaway RC, Conaway JW, Asturias FJ (2013) A conserved Mediator-CDK8 kinase module association regulates Mediator-RNA polymerase II interaction. Nat Struct Mol Biol 20:611–619

Tsai K-L, Tomomori-Sato C, Sato S, Conaway RC, Conaway JW, Asturias FJ (2014) Subunit architecture and functional modular rearrangements of the transcriptional mediator complex. Cell 157:1430–1444

Vannini A, Cramer P (2012) Conservation between the RNA polymerase I, II, and III transcription initiation machineries. Mol Cell 45:439–446

Vassylyev DG, Vassylyeva MN, Perederina A, Tahirov TH, Artsimovitch I (2007) Structural basis for transcription elongation by bacterial RNA polymerase. Nature 448:157–162

Verger A, Monté D, Villeret V (2019) Twenty years of Mediator complex structural studies. Biochem Soc Trans 47:399–410

Vermeulen M et al (2007) Selective anchoring of TFIID to nucleosomes by trimethylation of histone H3 lysine 4. Cell 131:58–69

Verrijzer CP, Chen JL, Yokomori K, Tjian R (1995) Binding of TAFs to core elements directs promoter selectivity by RNA polymerase II. Cell 81:1115–1125

Vojnic E et al (2011) Structure and VP16 binding of the Mediator Med25 activator interaction domain. Nat Struct Mol Biol 18:404–409

Vorländer MK, Khatter H, Wetzel R, Hagen WJH, Müller CW (2018) Molecular mechanism of promoter opening by RNA polymerase III. Nature 553:295–300

Wang D, Hawley DK (1993) Identification of a 3′–5′ exonuclease activity associated with human RNA polymerase II. Proc Natl Acad Sci USA 90:843–847

Wang Z, Roeder RG (1995) Structure and function of a human transcription factor TFIIIB subunit that is evolutionarily conserved and contains both TFIIB- and high-mobility-group protein 2-related domains. Proc Natl Acad Sci USA 92:7026–7030

Wang D, Bushnell DA, Westover KD, Kaplan CD, Kornberg RD (2006) Structural basis of transcription: role of the trigger loop in substrate specificity and catalysis. Cell 127:941–954

Warner JR (1999) The economics of ribosome biosynthesis in yeast. Trends Biochem Sci 24:437–440

Weeda G, van Ham RC, Vermeulen W, Bootsma D, van der Eb AJ, Hoeijmakers JH (1990) A presumed DNA helicase encoded by ERCC-3 is involved in the human repair disorders xeroderma pigmentosum and Cockayne's syndrome. Cell 62:777–791

Weinmann R, Roeder RG (1974) Role of DNA-dependent RNA polymerase 3 in the transcription of the tRNA and 5S RNA genes. Proc Natl Acad Sci USA 71:1790–1794

Westover KD, Bushnell DA, Kornberg RD (2004) Structural basis of transcription: nucleotide selection by rotation in the RNA polymerase II active center. Cell 119:481–489

Wolski SC, Kuper J, Hänzelmann P, Truglio JJ, Croteau DL, Van Houten B, Kisker C (2008) Crystal structure of the FeS cluster-containing nucleotide excision repair helicase XPD. PLoS Biol 6:e149

Wu SY, Chiang CM (2001) TATA-binding protein-associated factors enhance the recruitment of RNA polymerase II by transcriptional activators. J Biol Chem 276:34235–34243

Wu C-C, Lin Y-C, Chen H-T (2011) The TFIIF-like Rpc37/53 dimer lies at the center of a protein network to connect TFIIIC, Bdp1, and the RNA polymerase III active center. Mol Cell Biol 31:2715–2728

Wu C-C et al (2012) RNA polymerase III subunit architecture and implications for open promoter complex formation. Proc Natl Acad Sci USA 109:19232–19237

Xie X et al (1996) Structural similarity between TAFs and the heterotetrameric core of the histone octamer. Nature 380:316–322

Yan Q, Moreland RJ, Conaway JW, Conaway RC (1999) Dual roles for transcription factor IIF in promoter escape by RNA polymerase II. J Biol Chem 274:35668–35675

Yudkovsky N, Ranish JA, Hahn S (2000) A transcription reinitiation intermediate that is stabilized by activator. Nature 408:225–229

Zhang G, Campbell EA, Minakhin L, Richter C, Severinov K, Darst SA (1999) Crystal structure of Thermus aquaticus core RNA polymerase at 3.3 A resolution. Cell 98:811–824

Zhang Z et al (2015) Chemical perturbation of an intrinsically disordered region of TFIID distinguishes two modes of transcription initiation. Elife 4:e07777

Chapter 6
Regulation of Antiviral Innate Immunity Through APOBEC Ribonucleoprotein Complexes

Jason D. Salter, Bogdan Polevoda, Ryan P. Bennett and Harold C. Smith

Abstract The DNA mutagenic enzyme known as APOBEC3G (A3G) plays a critical role in innate immunity to Human Immunodeficiency Virus-1 (HIV-1). A3G is a zinc-dependent enzyme that mutates select deoxycytidines (dC) to deoxyuridine (dU) through deamination within nascent single stranded DNA (ssDNA) during HIV reverse transcription. This activity requires that the enzyme be delivered to viral replication complexes by redistributing from the cytoplasm of infected cells to budding virions through what appears to be an RNA-dependent process. Once inside infected cells, A3G must bind to nascent ssDNA reverse transcripts for dC to dU base modification gene editing. In this chapter we will discuss data indicating that ssDNA deaminase activity of A3G is regulated by RNA binding to A3G and ribonucleoprotein complex formation along with evidence suggesting that RNA-selective interactions with A3G are temporally and mechanistically important in this process.

Keywords A3G · AID · Antiviral · APOBEC · Crosslinking · Crystal structure · Cure · Cytidine deaminase · DNA · Gene editing · HIV · Hypermutation · Innate immunity · Mass spectrometry · Noncoding RNA · Protein RNA interactions · Ribonucleoprotein particles · RNA · RNA binding domains

---

J. D. Salter · R. P. Bennett · H. C. Smith
OyaGen, Inc, 77 Ridgeland Road, Rochester, NY 14623, USA
e-mail: jason.salter@oyageninc.com

R. P. Bennett
e-mail: rbennett@oyageninc.com

B. Polevoda · H. C. Smith (✉)
Department of Biochemistry and Biophysics, School of Medicine and Dentistry, University of Rochester, 601 Elmwood Ave, Rochester, NY 14642, USA
e-mail: harold.smith@rochester.edu

B. Polevoda
e-mail: bogdan_polevoda@urmc.rochester.edu

J. R. Harris and J. Marles-Wright (eds.), *Macromolecular Protein Complexes II: Structure and Function*, Subcellular Biochemistry 93,
https://doi.org/10.1007/978-3-030-28151-9_6

Abbreviations

A3	Collectively all APOBEC3 proteins, A3A, A3B, A3C, A3D, A3F, A3G, A3H
AID	Activation induced deaminase
APOBEC	Apolipoprotein B editing catalytic unit
CD1	N-terminal domain of dual domain APOBEC proteins
CD2	C-terminal domain of dual domain APOBEC proteins
CLIP-Seq	Crosslinking immunoprecipitation and sequencing of RNA bound to proteins
gRNA	HIV genomic RNA
HIV	Human immunodeficiency virus
hY	Human Y ncRNA
NC	Nucleocapsid portion of HIV Gag
ncRNA	Noncoding RNA
RNP	Ribonucleoprotein particle
RT	Reverse transcriptase
RNA-Seq	RNA sequencing
ssDNA	Single stranded DNA
ZBD	Zinc binding domain

The APOBEC3 Proteins in Innate Antiviral Immunity

There are eleven APOBEC/AID proteins encoded in the human genome (Salter et al. 2016). With the exception of APOBEC2 and APOBEC4, the other nine APOBEC proteins have cytidine or deoxycytidine deaminase activity on nucleic acids. Cytidine to uridine transition involves hydrolytic deamination of cytidine within a nucleic acid substrate with demonstrable nearest neighbor dinucleotide preferences for each APOBEC (Salter et al. 2016). There are four APOBEC3 (A3) deaminases (A3D, A3F, A3G and A3H) that can restrict HIV infection in T cells, macrophages and dendritic cells (Salter and Smith 2018). The present-day antiviral requirement for A3 proteins may have evolved from their role in restricting retroviral and endogenous retroviral-like elements (Sawyer et al. 2004; Zhang and Webb 2004; Esnault et al. 2005; Bulliard et al. 2009; Munk et al. 2012; Krupp et al. 2013). Evolutionary models predict that A3 proteins have served a significant role as antiviral vanguards for many species for millions of years by mutating retroviral genomes and impairing their productive replication. These predictions are based on paleo DNA database analyses identifying deoxyguanine (dG) to deoxyadenosine (dA) changes that arose with nearest neighbor preferences of the A3 family, foremost and notably within integrated endogenous retrovirus-like sequences (Sawyer et al. 2004; Conticello et al. 2005; Munk et al. 2012).

The antiviral mechanism of action of A3 proteins has been demonstrated through in vitro studies (Harris and Dudley 2015; Pery et al. 2015; Bennett et al. 2016), in humanized mice infected with HIV (Krisko et al. 2013; Sato et al. 2010) and suggested in long-term, nonprogressing, HIV infected patients who have enhanced A3G expression (Jin et al. 2005; Refsland et al. 2010; De Pasquale et al. 2013; Kikuchi et al. 2015). A3G, the focus of this review, is the most abundant and active A3 protein in lymphoid tissues (Refsland et al. 2010; Aydin et al. 2014) responsible for dG to dA hypermutations within the coding strand of HIV.

APOBEC3 Antiviral Mechanism

A3 antiviral deaminase activity occurs on ssDNA during reverse transcription of the viral RNA genome (Harris and Dudley 2015; Salter et al. 2016). This requires that A3 proteins are packaged and delivered into cells as cargo within virions during infection as A3 proteins preexisting in cells do not access HIV replication complexes (Svarovskaia et al. 2004; Soros et al. 2007; Xu et al. 2007; Thangavelu et al. 2014). The ssDNA deaminase activity of A3 proteins therefore appears to be limited temporally to when nascent reverse transcripts are being made and within the spatial confines of virions that are 'unpacking' within the cell's cytoplasm (pre-integration complexes).

The amount of A3 protein already present in a cell being infected however can be an important determinant of the number of A3 proteins that will be packaged in virions during the late viral life cycle (Soros et al. 2007; Xu et al. 2007; Thangavelu et al. 2014). As a case supporting this point, two to five percent of the HIV positive, treatment-naïve individuals remain asymptomatic for many years while maintaining high CD4+T cell counts and low plasma HIV RNA levels (Poropatich et al. 2011). Although the scientific basis for low replicative virus in these elite controller patients may be due to multiple factors, in several of these patients there was a direct correlation between increased A3G mRNA and protein expression and lower viremia/higher CD4+T cell counts (Jin et al. 2005; Peng et al. 2007; Poropatich et al. 2011; Kourteva et al. 2012; De Pasquale et al. 2013; Kikuchi et al. 2015). It is worth noting that the HIV accessory protein known as Viral Infectivity Factor or Vif has a primary function, to reduce the abundance of A3 proteins in infected cells by binding to them and mediating their polyubiquitination and proteosomal degradation (Sova and Volsky 1993; von Schwedler et al. 1993; Sheehy et al. 2002; Guo et al. 2014; Salter et al. 2014b; Kouno et al. 2015). However, in vitro studies in which A3G was over expressed or knocked down in virus producing cells suggested that the abundance of A3G in a cell was unlikely to be the primary determinant of resistance to HIV in elite controller patients (Sato et al. 2010; MacMillan et al. 2013; Stavrou et al. 2014).

Several reports suggested that A3G can restrict HIV by a deaminase-independent mechanism that most probably involves inhibition of reverse transcriptase (Iwatani et al. 2007; Adolf et al. 2013; Belanger et al. 2013). It was proposed that A3G binds to HIV ssDNA as a monomer and subsequently forms catalytically inactive dimers that block reverse transcriptase from elongating viral ssDNA (Chaurasiya et al. 2014; Morse et al. 2017).

HIV Vif-Dependent Proteosomal Degradation of A3 Reduces Innate Antiviral Immunity

The significance of the interaction between HIV Vif and antiviral A3 proteins in controlling A3 antiviral activity was first discovered by Anne Sheehy et al. (2002) and confirmed in studies using mouse models infected with virus expressing a wild type or lacking a functional Vif gene (Okeoma et al. 2007; Sato et al. 2010; Krisko et al. 2013; Stavrou et al. 2014; Cadena et al. 2016). In the absence of Vif, A3 antiviral activity was sufficient to neutralize a viral infection but virus expressing Vif was resistant to A3 proteins. Effective A3 antiviral activity has been seen in clinical data from patients whose virus expressed Vif polymorphisms with reduced efficacy in promoting A3 degradation (Simon et al. 2005; Ooms et al. 2013; Aydin et al. 2014; Reddy et al. 2016). Both animal and clinical data suggest that A3 hypermutations can lead to catastrophic failure of viral genome functions (Zhang and Webb 2004; Simon et al. 2005; Okeoma et al. 2007; Rangel et al. 2009; Kourteva et al. 2012; Stavrou et al. 2014; Cadena et al. 2016; Reddy et al. 2016). Studies such as these have led to a consensus that Vif protects HIV from accumulating A3-dependent dG to dA hypermutations.

Until Vif antagonists can be tested in clinical trials (Bennett et al. 2016), we can only speculate what the threshold for a catastrophic level of mutation in HIV is and the conditions under which A3 proteins will achieve this level of activity. The in vivo mutation frequency of reverse transcriptase (RT) is 1.8×10^{-6}/bp/cycle (Mansky et al. 1995). The current estimate is that the mutation frequency need only to increase 3- to 4-fold for mutational events to reach a distribution in proviral DNA to neutralize HIV (Drake et al. 1999). Each viral particle may contain 6–8 A3G proteins associated with HIV RNAs in the virion core (Burnett and Spearman 2007; Xu et al. 2007). However it has been estimated that only 1–2 A3G per virion may be sufficient for hypermutation activity and HIV restriction (Thangavelu et al. 2014). Taken together these data suggest that even small changes in the availability of catalytically active A3 deaminases may have a significant impact on A3G antiviral efficacy.

RNA Binding to A3G Can Inhibit Deaminase Activity

Other than enhanced A3 gene expression or alternative mRNA splicing to produce more active A3 protein haplotypes (Ooms et al. 2013; Reddy et al. 2016), regulatory mechanisms for A3 antiviral capacity in cells are not understood. There is no known protein co-factor required for A3 binding to ssDNA and dC deaminase activity. Most proteins identified in complex with A3 proteins in co-immunoprecipitates are bridged to A3 proteins via RNase-sensitive linkers (Kozak et al. 2006; Gallois-Montbrun et al. 2007; Gallois-Montbrun et al. 2008; Friew et al. 2009). While some APOBECs can edit RNA (Salter et al. 2016), the function of RNA binding to A3 proteins is less well understood (Smith 2016). Both ssDNA and RNA binding induce homo oligomerization of A3G (Chelico et al. 2008; Feng and Chelico 2011; McDougall et al. 2011; McDougall and Smith 2011; Polevoda et al. 2015). A3G binds with high affinity to both RNA and ssDNA oligonucleotides (McDougall et al. 2011; McDougall and Smith 2011; Polevoda et al. 2015); DNA can be displaced from A3G in vitro and deaminase activity on ssDNA can be inhibited when RNA oligos bind to A3G (McDougall and Smith 2011; McDougall et al. 2011; Polevoda et al. 2015). In this regard, A3G deaminase activity in vitro (as well as that of other antiviral homologs A3F, A3D and A3H) requires activation by RNase digestion when they are isolated from mammalian cells (Kreisberg et al. 2006; Wichroski et al. 2006; Gallois-Montbrun et al. 2007; Khan et al. 2007; Shaban et al. 2018) or HIV virions (Soros et al. 2007), and when expressed as recombinant protein in *E. coli* (Shaban et al. 2018) or insect cells (Wedekind et al. 2006).

RNase digestion of A3H isolated from mammalian cells is also required for its in vitro deaminase activity. However, structural and functional studies of purified A3H reveal that a small RNase-insensitve RNA duplex remains bound to A3H after RNase treatment. The presence of this RNA duplex not only maintains an RNA-dependent A3H dimer but also supports A3H deaminase activity on DNA (Shaban et al. 2018; Ito et al. 2018; Feng et al. 2018); A3H RNA-mediated dimers from pig-tailed macaque (Bohn et al. 2017) and chimpanzee (Matsuoka et al. 2018) displayed similar characteristics.

A variety of RNA sequences and secondary structure and viral genomic RNA (gRNA) bind to A3G to form megadalton sized RNP complexes evident by size exclusion chromatography, electrophoretic mobility gel shift assay or glycerol gradient sedimentation (Opi et al. 2006; McDougall et al. 2011; McDougall and Smith 2011). A3 RNP may become sequestered within cellular P-bodies (Wichroski et al. 2006; Gallois-Montbrun et al. 2007 and reviewed in Smith 2011) which is particularly apparent in proliferating cells (Kreisberg et al. 2006). P-body-localized A3 does not associate with HIV virions nor contribute to antiviral activity. A3 that becomes localized in viral particle assembly centers is bound to HIV-1 gRNA and cellular RNAs (Khan et al. 2005; Wichroski et al. 2006; Gallois-Montbrun et al. 2007; Bach et al. 2008; Wang et al. 2007, 2008a, b; Zhang et al. 2010; Keene and Telesnitsky 2012). The regulatory mechanism governing this functional dichotomy of A3 RNP localization is not understood. Although A3H forms large cytoplasmic catalytically inactive RNPs, the catalytically active A3H RNA-duplex dimer form may hint at a mechanism for selecting viral encapsidation over cellular localization. It is notable that the most frequent HIV gRNA sites bound by A3H by CLIP-Seq are predicted to form RNA duplexes of 7 nt or greater

(Shaban et al. 2018). Whether A3H is an anomaly among A3s or if other A3s are capable of selecting structured RNAs for viral encapsidation remains an open question.

The combined effect of Vif-dependent degradation of A3 proteins and RNA-dependent inactivation of A3 deaminase activity may contribute to why the level of A3 expressed in most people is not robust enough to protect them against an HIV infection. The mechanisms that regulate suppression of A3 antiviral activity are not understood but are believed to determine HIV infectivity and may influence the success of antiviral therapies that seek to enable A3 antiviral activity. Suppression of A3 antiviral activity also raises concern that suboptimal A3 mutagenic activity may serve in facilitating viral evolution and mechanistically contribute to the emergence of drug resistance (Fourati et al. 2010; Kim et al. 2010; Sadler et al. 2010; Smith 2011 and reviewed in Smith 2011; Simon and Landau 2015; Venkatesan et al. 2018).

Biochemical Analysis of A3 Interactions with Nucleic Acids

The cytidine deaminase domains within the N- and C-termini of A3G (CD1 and CD2, respectively) contain zinc binding domains (ZBD) (reviewed in (Aydin et al. 2014; Salter and Smith 2018) (Fig. 6.2a). The ZBD in CD2 of A3G is indispensable in carrying out deamination of dC to dU (Fig. 6.2b). In isolation, CD2 has low or no catalytic activity and requires CD1 of A3G for robust catalytic activity (Navarro et al. 2005). The ZBD in CD1 is catalytically inactive and the mechanistic basis for the requirement of full length A3G for robust deaminase activity has not been determined but may be as structural support of CD2 and its binding to nucleic acids (Salter et al. 2014a; Salter and Smith 2018) and protein dimerization (Huthoff et al. 2009).

Historically all RNA binding to A3G and RNA-dependent inhibition of A3G deaminase activity was presumed to be through CD1 [(Navarro et al. 2005) and reviewed in (Strebel and Khan 2008; Smith 2011, 2016)]. This 'one domain' perspective of RNA binding led to a focused search of CD1 residues that are involved in binding RNA. Only a few of the residues predicted to have hydrophobic and hydrophilic R-groups capable of binding to the bases or phosphate backbone of RNA, respectively, actually were required for A3G binding to RNA and RNP formation (Opi et al. 2006; Huthoff and Malim 2007; Wang et al. 2007; Zhang et al. 2007; Friew et al. 2009; Huthoff et al. 2009; Chelico et al. 2010; Bélanger and Langlois 2015). From these studies, residues C97, R122, F126, W127 and D128 (Opi et al. 2006; Huthoff and Malim 2007; Wang et al. 2007; Friew et al. 2009; Huthoff et al. 2009; Khan et al. 2009; Chelico et al. 2010; Bélanger and Langlois 2015) and W94, Y124 and W127 (Bulliard et al. 2009; Bélanger et al. 2013) were implicated in RNA-dependent A3G homo oligomerization and assembly with virions. Also, R24 and R30 may be indirectly involved in RNA binding by stabilizing A3G (Bulliard et al. 2009). Based on the previously published NMR solution structure snapshots of a soluble mutant A3G-CD1 (1-200 aa) Fukuda et al. (2019) generated *in silico* docking models to stimulate the RNA-binding propensity of A3G-CD1 and found a novel amino acid residue, I26 that may contact RNA.

Mutational analyses also suggested that W94, C97, R122, F126, W127 and D128 may be required to bind cellular 7SL1, hY1,3-5 and HIV gRNA (Huthoff and Malim 2007; Wang et al. 2007; Huthoff et al. 2009; Chelico et al. 2010; Bélanger and Langlois 2015). Mutation of A3F at amino acid residues highly conserved among A3 proteins (W126A, A3G W127A and A3H W115A) significantly decreased A3 affinity for RNA, reduced the amount of A3 packaged in virions and reduced antiviral activity (Wang et al. 2008a, b; Bach et al. 2008; Zheng et al. 2012; Bélanger and Langlois 2015). A3H W126A mutation specifically affected protein interaction with 7SL1 ncRNA (and partially with tRNALys3) but not with HIV-1 gRNA or 5S RNA (Wang et al. 2008a, b). A similar reduction in A3G and A3H antiviral activity was observed when 7SL1 ncRNA was depleted by overexpressing the 7SL1 ncRNA interacting partner SRP19 invovled in forming the signal recognition particle required for endoplasmic reticulum-bound protein translation (Wang et al. 2008a, b). While A3G mutants W94A and W127A were unable to bind to RNA, S28E and Y124A mutants retained their ability to bind to 7SL1 and *Alu* RNAs though they had significantly diminished binding to hY1,Y3 RNAs (Bulliard et al. 2009). In contrast, an F126L mutant was defective in binding 7SL1, *Alu* and hY1 RNAs but only slightly defective in hY3 RNA binding. It is possible that the A3G motif YYFWD at positions 124–128, in addition to its suggested roles in protein oligomerization and assembly with virions, serves as an RNA-specificity box, similar to the proposed DNA-nucleotide specificity box in homologous C-terminal domains of A3 proteins (A3F, YYFWD and A3G, YDDQG, positions 307–311 and 315–319, respectively) (Siu et al. 2013).

EMSA and fluorescence anisotropy analyses (McDougall et al. 2011; Polevoda et al. 2015) and atomic force microscopy (AFM) (Pan et al. 2016) suggested that titration of A3G with increasing amounts of RNAs 15–100 nt in primary sequence length achieved saturation of multiple binding sites. The aggregate size of RNP assembled with A3G was dependent on the stoichiometry of A3G to RNA (or ssDNA) in the reaction (Opi et al. 2006; McDougall et al. 2011; Polevoda et al. 2015; Polevoda et al. 2016). Interestingly, RNAs $\geq$ 25 nt were sufficient to inhibit 25 nt to 99 nt ssDNA substrate binding to A3G and inhibit dC to dU deaminase activity in an RNA concentration-dependent manner (McDougall and Smith 2011; McDougall et al. 2011; Polevoda et al. 2015).

Emergence of the Dual RNA-Binding Domain Hypothesis

A major challenge to the 'one domain nucleic acid binding model' was made through mass spectrometry studies of A3G crosslinked to bound RNA which showed for the first time that RNA can bind to distinct amino acid residues of the catalytically active CD2 of A3G as well as to residues within CD1 (Polevoda et al. 2016, 2017). The data suggested that with multiple binding sites, there may be multiple RNA interactions along the length of an A3 protein and that more than one RNA molecule may be bound to A3G simultaneously (Smith 2016). Multiple RNA binding domains with A3G CD1 has been corroborated through in silco RNA docking studies Fukuda et al. (2019). Support for the two RNA binding domain hypotheses was provided by AFM experiments which showed that A3G could bind two short RNA oligos (one each within CD1 and CD2) or a ssDNA oligo and an RNA oligo at opposite ends (Pan et al. 2016). Taken together the studies suggested that the requirement of CD1 for robust CD2 deaminase activity may be to facilitate or stabilize nucleic acid binding. Mass spectrometry analyses of native A3G that was UV crosslinked to RNAs identified multiple tryptic peptides covalently bound to nucleotides and provided evidence that tyrosine 181, 182 and 315 were in direct contact with RNA and ssDNA (Polevoda et al. 2017).

These data led to the dual domain hypothesis that RNA-binding A3G amino acids are dispersed along the surface of both domains and that different RNA binding preferences of these residues contribute to the underlying structural diversity of A3G nucleic acid complexes (Fig. 6.1 and Table 6.1). A critical concept in the identification of RNA binding to CD2 was the identification of tyrosine 315 as critical for both high molecular mass RNP formation and ssDNA binding and deaminase activity (Polevoda et al. 2017). The data provided a mechanistic explanation for how RNA binding to A3G can competitively inhibit the antiviral deaminase activity both in P-bodies and within viral particles. Conversely, the RNA duplex that bound and mediated dimerization of the A3H catalytic domain does not inhibit DNA binding or catalytic deamination, supporting a model wherein A3s may simultaneously bind multiple nucleic acids at unique locations with different functional outcomes.

The Functional Significance of the RNA Binding Partner for A3G

An open question for nearly two decades has been which A3G-binding RNA(s) has the most significant impact on A3G antiviral function and A3G/A3F/A3D/A3H packaging into virions (Strebel and Khan 2008; Smith 2011, 2016). Numerous reports suggested that 7SL1 ncRNA and not HIV gRNA may play a more significant role in helping A3 proteins to assemble with virions (Wang et al. 2008a, b; Bach et al. 2008; Zheng et al. 2012). A3G appears to be unique among A3 proteins in binding to the *Alu* region of 7SL1 RNA (Zhang et al. 2010; York et al. 2016). A3G also bound preferentially to the conserved stem-loop structures 1 and 3 within HIV gRNA, and A3G interaction with SL1 enhanced the recovery of A3G with virions (Khan et al. 2005). A3G has greater than an order of magnitude higher affinity for the *Alu* domain of 7SL1 RNA in vitro as compared to bulk cellular RNAs binding

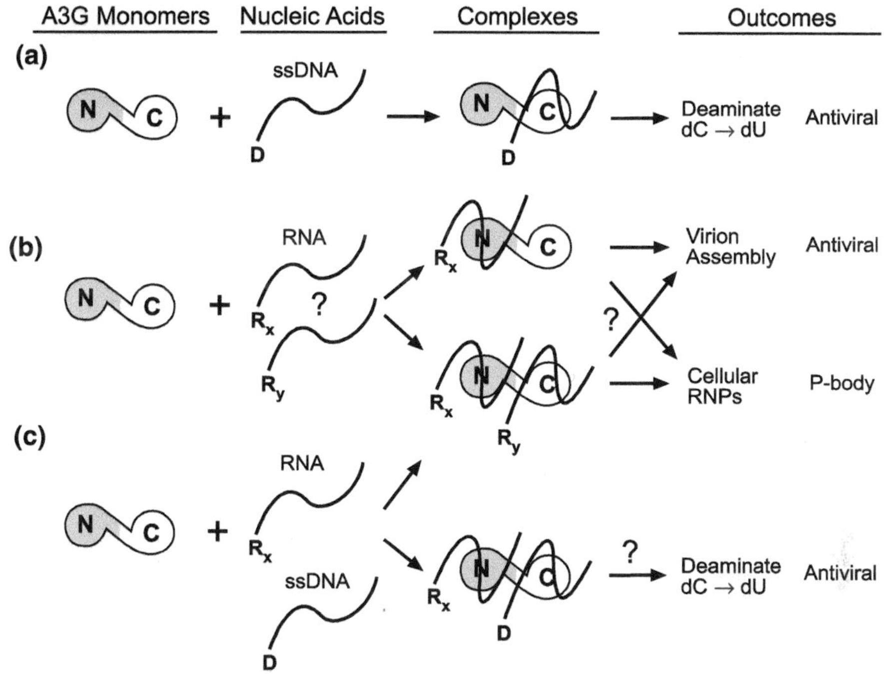

Fig. 6.1 Models Depicting the Single-Domain and Two-Domain RNA Binding Hypotheses. The cartoon depicts ssDNA or RNA binding to A3G (shown as a monomer for clarity of discussion) in the single RNA-binding domain model (**a** and top half of **b**) and the two domain RNA-binding model (bottom of **b** and **c**). For N-terminal half of the A3G monomer is shaded. The general locations of the zinc-dependent cytidine deaminase domains are shown in the N-terminus 'N' and C-terminus 'C' corresponding to CD1 and CD2 respectively. (**a**), The single-domain RNA binding model depicts ssDNA only binds to CD2 (c) which deaminates dC to dU in HIV reverse transcripts with an anticipated antiviral outcome. (**b**, **top**) All RNAs, regardless of sequence of secondary structure, only bind to CD1 which enabled A3G to assemble with newly forming HIV virions but also may sequester A3G within cellular RNP for localization within cytoplasmic P-bodies. The outcome of N-terminal RNA binding in this model is uncertain and could either have antiviral activity or prevent antiviral activity. (**b**, **bottom**) The two-domain RNA-binding model is depicted wherein different RNA sequences or secondary structures within one or multiple RNA molecules may bind selectively or nonspecifically to residues with the N- and C-termini of A3G. The anticipated outcome is that nonspecific RNA interactions such as with mRNAs would lead to RNP that largely sequester in P-bodies and therefore do not promote antiviral activity. RNAs that bind selectively to A3G within either domain may promote assembly of A3G with HIV virions. (**c**) The two-domain RNA binding model predicts that post infection, A3G in preintegration replication complexes will displace HIV genomic RNAs from CD2, enable nascent ssDNA HIV reverse transcripts to bind and become deaminated. RNAs bound to CD1 may be retained and promote A3G oligomerization which may augment the catalytic activity and/or processivity of A3G hypermutation on proviral DNA

regardless of whether they contain *Alu* repeats (Wang et al. 2007; Chelico et al. 2010). However, comparative studies of cellular RNP complexes associated with A3F and A3G found that in contrast to A3G, A3F complexes did not contain 7SL1

Table 6.1 Summary of RNAs Crosslinked to A3G Tryptic Peptides Determine by Mass Spectrometry

	Peptide, aa	ApoB	HIV	7SL	tRNA^{Lys3}
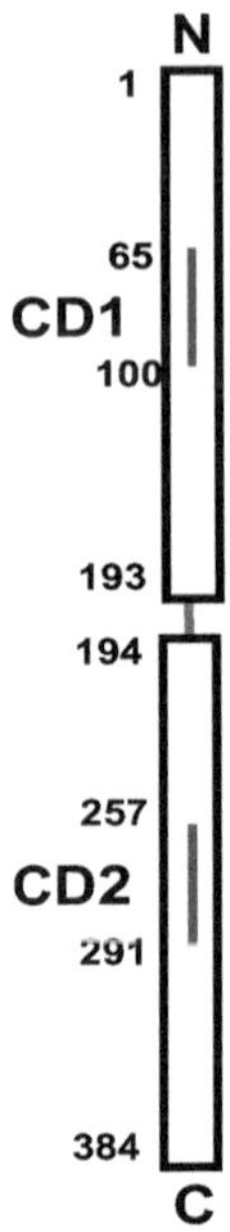	15–29	Y	N	N	N
	70–76	N	Y	Y	N
	100–113	Y	N	N	N
	123–136	N	N	N	Y
	151–163	Y	N	N	N
	169–180	Y	Y	N	N
	181–194	Y	Y	Y	Y
	195–213	N	N	N	Y
	257–270	Y	Y	Y	Y
	279–297	Y	Y	Y	Y
	314–320	Y	Y	N	N
	321–326	Y	Y	N	N
	327–334	Y	Y	Y	N
	345–374	Y	Y	Y	Y

Affinity purified recombinant A3G produced in insect cells (Wedekind et al. 2006) was incubated with the RNAs listed across the top of the table, then irradiated with short wavelength ultraviolet light to crosslink protein RNA complexes which were SDS-PAGE purified, digested with RNase A+T1, followed by digestion with trypsin and analyzed by mass spectrometry (Polevoda et al. 2016, 2017). Tryptic peptides identified as bound to ribonucleotides from the various RNAs tested are listed in the left column. Cartoon of A3G (right) shows CD1 and CD2 domains with the deaminase domains marked with a red line.

and Y1-4 RNAs and contained only minor proportion of *Alu* RNA (Gallois-Montbrun et al. 2008). Further research is needed to determine whether the other antiviral A3 have binding preferences for HIV gRNA and ncRNAs.

The RNA Binding Preference of A3G Changes Following HIV Infection

A3G RNA-Seq of RNAs crosslinked to A3G in uninfected and infected cells, or recovered with virions, suggested that A3G RNA binding preferences changed upon infection from predominantly random bulk mRNAs with a weak complement of ncRNAs in uninfected cells to largely 7SL1 ncRNA and HIV gRNA in HIV virions (York et al. 2016; Kutluay et al. 2014) (Fig. 6.2a–c). During viral maturation, we and others have observed that A3 binding specificity for 7SL1 RNA may increase 10-fold on a molar basis relative to that for HIV gRNA (Eckwahl et al. 2015, 2016; York et al. 2016) (Fig. 6.1c). A3G bound preferentially to the *Alu* secondary structure domain of 7SL1 RNA in vitro but not to the linker RNA or adjoining secondary structure known as the S domain (Bach et al. 2008). Other ncRNAs and a variety of incompletely processed cytoplasmic RNAs also were enriched in the viral particle relative to cytoplasmic mRNAs (Eckwahl et al. 2016; York et al. 2016) (Fig. 6.2c).

A critical future milestone will be to establish which RNA binding residues affect RNA binding preferences and to determine their roles in A3G oligomerization and RNP formation. It is intriguing to consider that either post-translation modifications of A3 proteins or post-transcriptional modifications of gRNA and 7SL1 ncRNA could be involved in determining the alterations in A3 RNA binding preferences. These data highlight the need to understand the mechanism whereby A3G alters its preference for mRNAs in uninfected cells to ncRNAs and gRNA in virus producing cells and virions.

Despite these details, the functional role that various RNAs might have when they bind to A3 proteins remains under determined (Smith 2016). There is conflicting evidence in that several studies suggested that A3G sites required to bind 7SL1 and HIV gRNA during viral particle assembly were also involved in A3G binding to bulk cellular RNAs and inhibition of A3G antiviral activity through sequestration of A3G RNP within cytoplasmic P-bodies and stress granules (Kozak et al. 2006; Wichroski et al. 2006; Gallois-Montbrun et al. 2007; Soros et al. 2007; Gallois-Montbrun et al. 2008; Wang et al. 2008a, b; Friew et al. 2009).

RNA Binding of A3G Paradoxically May Both Inhibit and Promote Antiviral Deaminase Activity

For A3G to be positioned proximal to nascent cDNA emanating from HIV reverse transcription complexes, A3G must minimally bind to viral genomic RNA (gRNA). As described above, A3G recovered from virions is also bound to cellular RNAs, such as 7SL1 ncRNA (Khan et al. 2007; Wang et al. 2007, 2008a, b; Refsland et al. 2010; Smith 2011; Keene and Telesnitsky 2012). An important open question is whether 7SL1 ncRNA or HIV gRNA differentially regulate A3G deaminase activity during viral reverse transcription. A3G must disengage from HIV gRNA and bind to nascent ssDNA reverse transcripts (Chelico et al. 2008; McDougall et al. 2011) as substrates for cytidine deamination. RNase H-dependent degradation of the gRNA template during reverse transcription, exposes viral ssDNA for A3G binding. This necessitates gRNA removal from CD2 by RNase H. However, binding of RNAs to both the N- and C-terminal domains of A3G (Fig. 6.1c) may regulate A3G deaminase activity through RNA binding sites that interact with cellular RNAs. When bound to these RNAs certain residues may facilitate A3G assembly with

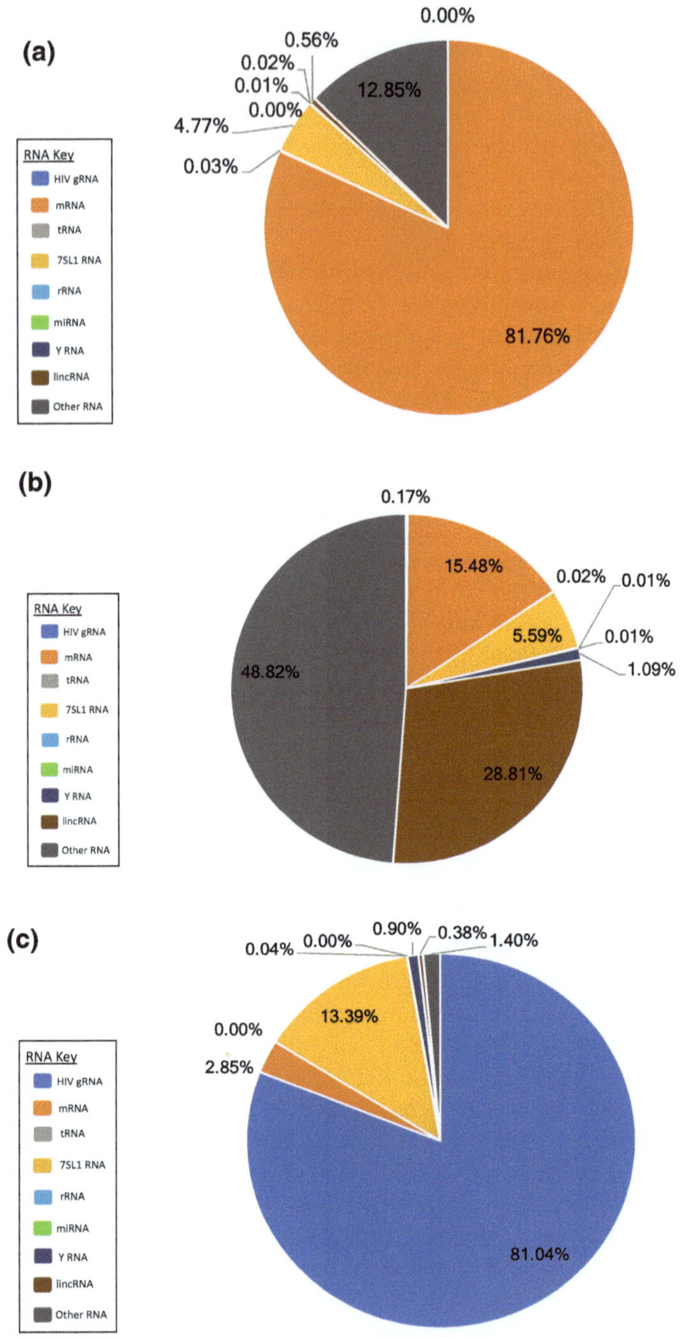

(a)
RNA Key
HIV gRNA
mRNA
tRNA
7SL1 RNA
rRNA
miRNA
Y RNA
lincRNA
Other RNA
0.00%
0.56%
0.02%
0.01%
0.00%
4.77%
0.03%
12.85%
81.76%
(b)
RNA Key
HIV gRNA
mRNA
tRNA
7SL1 RNA
rRNA
miRNA
Y RNA
lincRNA
Other RNA
0.17%
15.48%
0.02%
0.01%
5.59%
0.01%
1.09%
48.82%
28.81%
(c)
RNA Key
HIV gRNA
mRNA
tRNA
7SL1 RNA
rRNA
miRNA
Y RNA
lincRNA
Other RNA
0.04%
0.00%
0.90%
0.38%
1.40%
13.39%
0.00%
2.85%
81.04%

◂**Fig. 6.2** RNA binding Preference of A3G is Altered by HIV Infection and Localization within Virions. Classes of RNA species associated with A3G in uninfected cells (**a**), HIV-infected cells (**b**) and within HIV virions (**c**). 6XHis-A3G proteins and associated RNA were purified from cells or virions with use of Ni-NTA resin, RNA was isolated by using RNeasy Plus Mini and RNeasy MinElute columns (Qiagen) and combined for each sample RNA was subjected for RNA sequencing procedure based on Illumina technology. Illumina compatible library construction was performed using the TruSeq Total Stranded RNA Sample Preparation Kit. Single end reads were mapped to the human and HIV-1 genomes and proportions of each RNA class (in colors, as indicated on the figure) associated with A3G in each sample were presented in pie charts

virions but not inhibit A3G ssDNA binding and deamination during reverse transcription. Which RNA might have this regulatory role is not understood, but given the enrichment of 7SL1 ncRNA bound to A3G in virions compared with A3G recovered from the cytoplasm of uninfected cells, one might speculate that A3G binding to 7SL1 ncRNA has a unique purpose.

For example, 7SL1 ncRNA does not form duplexes with nascent viral cDNA and is not degraded by RNase H. Is the catalytic domain (CD2) of A3G active on proviral ssDNA when 7SL1 ncRNA is bound to CD1? Interestingly, RNA bridging of A3G through CD1 to form oligomers may serve in promoting robust and processive ssDNA deaminase activity (Huthoff et al. 2009; Feng and Chelico 2011). This suggests a paradox in that while RNA binding to A3G is essential for assembly of A3G with virions, antiviral DNA mutagenic deaminase activity is inhibited by RNA binding to A3G. The structural basis for this dicotomy of RNA interactions with A3, their functional consequence(s) and importantly the regulatory mechanisms responsible for A3G changing RNA binding preferences are poorly understood.

Noteworthy, Itano et al. (2018) demonstrated that 7SL1 RNA is recruited to the plasma membrane (the site of virion formation) during the time of assembly of the Gag HIV polyprotein but separately from HIV gRNA. 7SL1 RNA interaction with Gag likely occurs before Gag protein oligomerization at the plasma membrane and could be initiated in the cytoplasm since Gag-7SL1 complexes were immunoprecipitated from the fractions of both of these cell compartments (Itano et al. 2018). A3G binding to 7SL1, *Alu* and gRNA also requires the nucleocapsid (NC) portion of the HIV Gag but this interaction may be independent from Gag gRNA interaction during which NC provides a critical RNA chaperoning functions and gRNA dimerization for viral particle assembly (Muriaux and Darlix 2010). Immunoprecipitation data demonstrated that Gag and NC either associated directly with A3G or interacted via an RNA bridge (Cen et al. 2004; Friew et al. 2009). Both NC and A3G exhibited very similar RNA binding patterns to HIV gRNA and 7SL1 with a strong preference for A- or G-rich sequences in both immature and mature virions (Kultuay et al. 2014; York et al. 2016). Due to the large molar excess of Gag over A3G that assemble with virions ($\sim$2000:7) (Xu et al. 2007) A3G must face vigorous competition for gRNA and 7SL1 binding. A possibility is that A3G binding to those RNAs may be facilitated by Gag remodeling of RNAs within HIV SL1-SL3 and 7SL1 *Alu*-regions. As described above, in HIV infected cells

A3G showed selective preference for binding to 7SL1 RNA(York et al. 2016). Therefore, it is reasonable to suggest that 7SL1 ncRNA plays a major role in the virion assembly pathway that A3G utilizes or shares with NC-Gag.

It is of interest that in the preparation of RNA-depleted full-length A3G for small angle X-ray scattering analysis, we monitored the removal of RNA from affinity purified A3G by silver staining of the RNA retained bound to A3G and resolved by denaturing PAGE (Fig. 6.3). We noted that while increasing amounts of RNase A or micrococcal nuclease (MNase) digestion markedly diminished the recovery of A260 absorbing material and activated in vitro deaminase activity (Wedekind et al. 2006), an RNase-resistant fragment of RNA of ~35 nt appeared to be protected by A3G (Fig. 6.3) (Bennett 2007). RNase A resistant RNAs were significantly shorter, suggesting greater nuclease access to RNAs bound to A3G (Fig. 6.3). Despite the

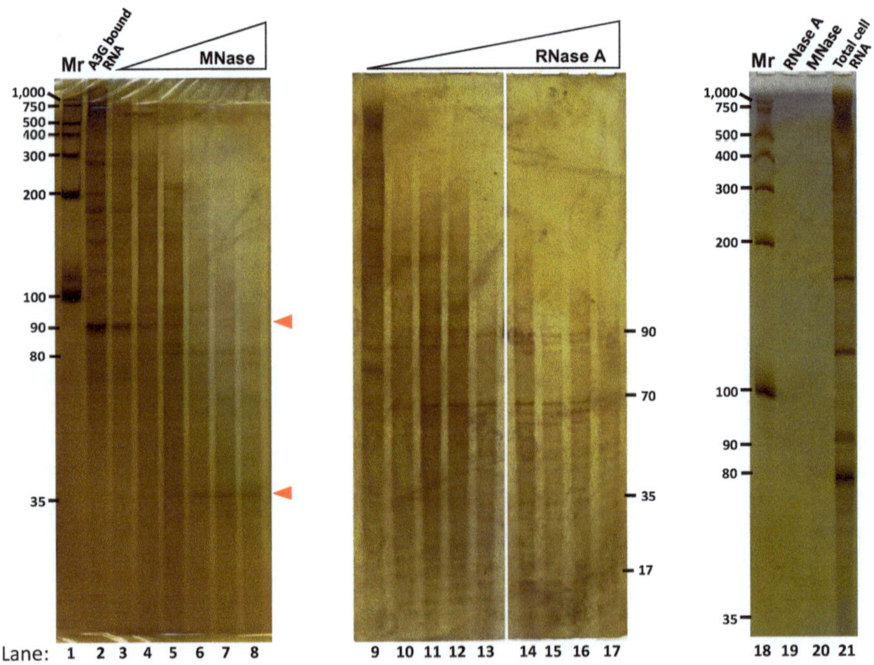

Fig. 6.3 RNA of specific size protected from RNase degradation by A3GRNA was analyzed by silver stained denaturing RNA PAGE. The length of RNA standards (lane 1 and 18) are identified by number of bases. Following extensively washing with 1M NaCl, 1M urea and 125 mM imidazole RNA that has been co-purified with A3G by Ni-NTA IMAC (lane 2 from the left) was digested at 4 °C with increasing amounts of MNase (lanes 3–8) and RNase A (lanes 9–17). Total cellular RNA (lane 21) purified from Sf9 insect cells without A3G overexpression was not protected from digestion when incubated with either RNase A (lane 19) or MNase (lane 20). Red arrowheads point to silver stained RNAs whose MNase-dependent reduction in abundance (~90 nt) and resistance to MNase digestion (~35 nt) suggest A3G protection of bound RNA. Additional RNA fragments that appeared to be relatively refractory to RNase A digest are visible with average sizes of ~90, ~70, ~35 and ~17 nt

presence of residual fragments of RNA, the specific activity of RNase A digested A3G was 1.2 fmol C to U/μg A3G/min compared to 0.25 fmol C to U/μg A3G/min of the pre-digested A3G (Wedekind et al. 2006; Polevoda et al. 2015). An RNase-protected fragment of RNA has also been demonstrated for A3H RNP (discussed further below). The detection of RNase-protected fragments of RNA bound to A3G and A3H suggested that a conformational pocket may sequester RNA through intimate contacts between amino acids and ribonucleotides but this does not rule out A3 protein surface exposed interactions with RNA that could also have important functions. The discoveries to date underscore the need to determine the characteristics of RNA bound to A3G that may promote deaminase activity, enable A3G assembly with viral particles or inhibit A3G antiviral activities.

Structural Models of APOBECs Suggest Mechanisms for Nucleic Acid Binding

Recent co-crystal structural models of APOBECs with nucleic acid (Xiao et al. 2016; Bohn et al. 2017; Kouno et al. 2017; Shi et al. 2017b; Fang et al. 2018; Matsuoka et al. 2018; Shaban et al. 2018) or nucleotides (Shi et al. 2015; Qiao et al. 2017; Ziegler et al. 2018) have begun to shed light on how APOBECs select and bind substrate and regulatory nucleic acids alike. There is likely a great diversity of macromolecular complexes that APOBECs form with nucleic acid. X-ray crystallographic structures of the A3G C-terminal domain (Maiti et al. 2018), A3A (Kouno et al. 2017; Shi et al. 2017a) and of an A3A-Loop 1 substituted A3B C-terminal domain (Shi et al. 2017b) have been solved with short single-stranded deoxynucleic acids bound in catalytically relevant manners at the respective active sites. A structure of Activation Induced Deaminase (AID) with deoxynucleotides bound at the active site revealed unique positively charged surface grooves proximal to the catalytic site (Qiao et al. 2017). This structure supported a model wherein AID specifically selects G4-structured DNA targets for deamination; in contrast, A3A and A3G select for linear ssDNA. Together, these models suggest mechanisms by which APOBECs select deamination targets with specific di- or tri-nucleotide nearest neighbor sequences or within a specific structural context.

In addition to those of APOBEC-substrate complexes, structures of APOBECs bound with non-substrate nucleic acid revealed potential mechanisms of action for regulatory DNA and RNA binding. In particular, a structure of the A3F catalytic C-terminal domain solved with ssDNA bound outside of the active site supported a model wherein non-specific nucleic acid binding helps guide substrate to the active site (Fang et al. 2018). Likewise, co-crystal structures of the single domain A3H from a variety of primates (Bohn et al. 2017; Matsuoka et al. 2018; Shaban et al. 2018) with endogenous RNA suggested how an RNA-mediated dimer of an antiretroviral APOBEC may promote packaging within nascent virions and inhibit

DNA editing. Although A3H is the only APOBEC to be co-crystallized with RNA, these structures may provide insight into how other APOBECs form ribonucleoprotein complexes and the mechanism by which RNA binding can regulate enzymatic activity on DNA.

Nearest Neighbor Base Sequence Determines APOBEC Target Specificity

The co-crystal structures of catalytically inactivated mutants of A3A and the C-terminal domain of A3G with short ssDNA comprising either a $dT_{-1}dC_0$ or $dC_{-2}dC_{-1}dC_0$ hotspot motif for A3A and A3G catalytic activity respectively, revealed that DNA binds within a deep surface channels formed between loops L1, L3, and L7 that coalesce adjacent to the Zn-centered active site (Fig. 6.4a). In each case, DNA binding was centered around a conserved L1 histidine residue (H29 and H216 for A3A and A3G C-term respectively) which underwent extensive rearrangement upon binding. Several other differences in conformation of L1 and L7 residues observed between structures of DNA-bound and -unbound A3A and A3G CD2 suggested an induced fit mechanism controls APOBEC substrate selection. The targeted dC_0 bases were flipped into the deep active site pockets where the base was stabilized by stacking interactions with aromatic sidechains and through several hydrogen bonds (Fig. 6.4a). A3A and A3G have strong preferences for substrate with specific base identity at the −1 position (dT for A3A and dC for A3G). This specificity was born out through multiple hydrogen bonds between specific L7 residues of either A3A or A3G CD2 and the WC face of the preferred base at position −1 (Fig. 6.4a). A3G formed additional water-mediated hydrogen bonds with the WC face of the base at the −2 position which supported a mechanism for its preference of dC at the −2 position. Non-specific base stacking interactions were also observed between aromatic sidechains of either A3A or the CD2 of A3G and the base at the +1 position. For A3G CD2, base stacking occurred at the −3 position as well. The lack of base specific interactions with substrate other than with the −1 (or −2 for A3G) through +1 positions provided a rationale for the relative lack of sequence specificity other than the two nearest neighbors of the target dC_0 nucleotide. Whether other APOBECs form enzyme-substrate complexes comparable to that observed for A3A and A3G remains to be determined.

AID Selects Structured DNA Substrates

In contrast to the linear DNA binding preference of A3A and A3G, AID has been shown to preferentially bind G4 quadruplex (G4) structured DNA (Qiao et al. 2017). AID targets dC at the third position in a 5′ overhang adjacent to the G4

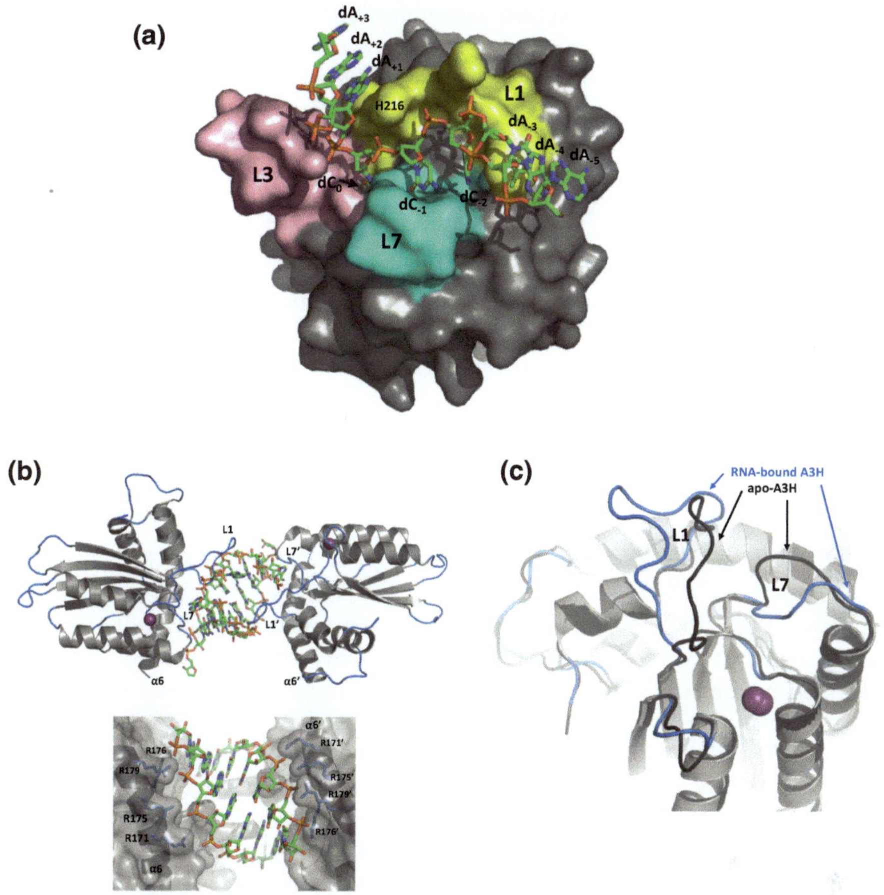

Fig. 6.4 Structural models illustrate the diversity of APOBEC-nucleic acid interactions. (a) The co-crystal structure of A3G CD2 with the 9 nt ssDNA substrate (5′-$dA_{-5}dA_{-4}dT_{-3}dC_{-2}dC_{-1}\underline{dC}_0dA_{+1}dA_{+2}dA_{+3}$-3′) where $\underline{dC}_0$ is the substrate cytidine bound within the active site of this catalytically inactive variant. The A3G CD2 molecular surface is colored gray with residues of L1, L3 and L7 colored yellow, pink and blue respectively. DNA is shown as sticks and atoms of carbon, phosphate, oxygen, and nitrogen are colored green, orange, red, and blue respectively binds. (**b**) The co-crystal structure of dimeric A3H in complex with an 8-mer RNA duplex (PDB 6BOB) (top panel). A3H molecules are depicted as cartoons with gray α-helices and blueβ-strands while RNA molecules are depicted as sticks and colored as in (**a**). A3H molecules form significant interactions with each RNA molecule but protein-protein interactions are absent. Extensive interactions between arginine residues (sidechains depicted as blue sticks) of α6-helices and the RNA backbone are shown in the bottom panel. (**c**) The same cocrystal structure from (**b**) is depicted as a light gray cartoon with loops colored blue and the RNA has been removed for clarity. The crystal structure of human apoA3H (PDB 5W45) (dark gray loops) is superposed to illustrate changes in A3H loop conformation upon RNA binding. The conformation of L1 and L7 are shifted apart to accommodate the RNA duplex interaction compared with those from apoA3H

structured core during class-switch recombination (CSR). The X-ray crystallographic structure of AID revealed a positively charged surface groove that is bifurcated to form two channels and may explain how AID selects dC specifically associated with G4-structured DNA for deamination. One channel, termed the substrate binding channel, was formed by L1, L3, and L7 and centers on the active site. The second channel, or 'assistant patch', was located between the L7 and α6. These channels deviated from one another at a negatively charged region of L7 called the 'separation wedge'. The use of such positively charged grooves bifurcated by a separation wedge for binding branched DNA structures has been observed for macromolecular complexes involving Cas9 (Jiang et al. 2016) and T4 RNase H (Devos et al. 2007), but this substrate binding mode may be unique to AID among the APOBECs as it is the only family member hypothesized to bind structured DNA substrate.

DNA Binding Sites Distal to the APOBEC Active Site May Assist Substrate Selection

Single-stranded DNA binding to the catalytic domains of A3A and A3G has been clearly demonstrated by the previously described crystal structures. Although APOBEC binding of DNA at sites distal to the active site or at the non-catalytically active domains of the dual-domain A3 s has been demonstrated (Polevoda et al. 2015, 2017), little is known how binding at such sites regulates deamination of substrate at the active site. A co-crystal structure of the A3F C-terminal domain with a poly-dT_{10} nucleic acid revealed how a large positively charged surface patch binds a deoxyribonucleic acid in a non-sequence specific manner (Fang et al. 2018). This positively charged surface patch is unique among the A3 family and comprised five surface exposed lysine residues within α-5 and α-6 helices and L10 (K334, K337, K352, K355, and K358) that formed the bulk of an extensive electrostatic interaction network with the negatively charged poly-dT backbone. Two conserved tyrosine residues (Y333 and Y359) within the path contributed additional pi-stacking interactions with two bases. A mechanism for substrate selection has been proposed in which nucleic acids were first bound non-specifically by the positive surface patch. Once bound, nucleic acids of specific sequence could be selected and guided towards the active site for deamination. However, the mechanism for sequence selection is unknown and it is unknown if other APOBECs may utilize a similar mechanism for capture of nucleic acid substrates.

RNP Formation with APOBECs Regulate Enzymatic Activity

RNA mediated regulation of DNA binding (McDougall and Smith 2011; McDougall et al. 2011), enzymatic activity (Iwatani et al. 2006; McDougall et al. 2011) and localization of APOBEC proteins has been extensively documented (Salter and Smith 2018). RNA binds A3G CD1 and CD2 at several sites and RNA competition with ssDNA, specifically at tyrosine 315, has been shown to diminish binding and editing of DNA (Polevoda et al. 2015, 2017). Likewise, A3B CD1 non-catalytic domain attenuates the DNA editing rate at the CD2 catalytic domain (Caval et al. 2015) through RNA binding by a positively charged surface patch (Xiao et al. 2017). Localization of the antiretroviral A3s (A3D, F, G and H) to budding virions is partially dependent upon interaction with cellular and viral RNAs (Khan et al. 2005; Wang et al. 2007). The structure of human A3H haplotype II (Shaban et al. 2018) and those from both a pigtailed macaque polymorphic variant (Bohn et al. 2017) and chimpanzee (Matsuoka et al. 2018) show A3H forms a dimer, mediated entirely by a cell-derived RNA duplex (Fig. 6.4b). The A3H molecules within each co-crystal structure form extensive electrostatic interactions with the backbone of both RNA strands through a positively charged surface patch localized to α-6 helix arginine residues (Shaban et al. 2018). Several tyrosine residues found in A3H-specific motifs of L1 and L7 add non-sequence specific base-stacking interactions (Shaban et al. 2018). The arginine residues within the patch are conserved in A3H across species and have been shown to be required for A3H localization to virions (Shaban et al. 2018).

What is most intriguing about this structure is that it reveals two potential mechanisms for RNA mediated regulation of A3H. First, a mutational analysis of the residues that interact with the duplex RNA were shown to be required for virion packaging and antiretroviral restriction suggesting that an RNA-mediated dimer may be a necessary step for antiretroviral activity (Shaban et al. 2018). Second, the RNA does not bind within the active site (Fig. 6.4b) and the RNA mediated dimer of A3H is catalytically active. It is thus implicit that the A3H dimer is capable of binding DNA at the active site while also bound due to RNA. Notably, monomeric A3H without RNA bound is also catalytically active (Ito et al. 2018). A comparision of monomeric and RNA-bound structures of A3H reveals the catalytic site loops L1 and L7 are further displaced from one another in the RNA-bound structure (Fig. 6.4c) suggesting that the RNA duplex may allosterically modulate the catalytic rate and that structural rearrangement may be necessary for DNA binding (Feng et al. 2018). While the RNA binding motifs of L1 and L7 that stabilize the RNA duplex interaction are specific to A3H it is unlikely that RNA specifically regulates other APOBECs in this manner. However, positively charged patches are common among APOBECs and may be directly involved in RNA-regulation of A3B and A3G. In particular, it has been demonstrated that A3G dimers are bridged by RNA-bound CD1 domains which led to a model wherein RNA-linked dimers of A3G are required for the processivity of A3G antiretroviral catalytic activity on HIV during reverse transcription (Feng and Chelico 2011).

Concluding Remarks

In this review the role of RNA binding to A3G has been emphasized and used as a platform for introducing the concept of regulatory RNAs and the diversity of protein RNA interactions that may take place A3 family members. The discussion has embraced a historical perspective of discoveries in the APOBEC field over the past 30+ years; pointing out leading hypotheses and open questions when appropriate.

The APOBEC field has at its origin the discovery of psoriasis-upregulated and phorbol ester-induced group of proteins known as the phorbolins which are currently understood to be in the APOBEC3 class of proteins (Gene ID 9582), but have no known functional or enzymatic activities such as those described for A3 proteins. However, the functional role of RNP assembly and RNA binding to APOBECs did not come into focus until APOBEC1 (A1) was discovered as the enzyme responsible for C to U editing of cytidine 6666 in apolipoprotein B mRNA (Salter et al. 2016). In the following years the role of RNA binding to APOBECs became more difficult to understand as the discovery of AID and the A3 family ushered in the understanding that although these proteins bind RNA, the natural substrate for their deaminase activities was deoxycytidines within ssDNA stretches of genomic DNA.

Structural studies of both apo and nucleic acid bound APOBECs are beginning to reveal there are a variety of mechanisms at play for both substrate selection and RNA-mediated regulation of activity. Models of substrate-bound APOBECs have revealed structural details for di- or trinucleotide sequence motifs. Protein surface features such as grooves, and hydrophobic and charged patches for binding nucleic acids both at and away from the active site suggest mechanisms for differentiating between linear and structured nucleic acid substrate. Likewise, an unexpected role for a structured RNA in the regulation of A3H antiretroviral activity has been proposed based upon an RNP crystallographic model comprising an A3H dimer and a short RNA duplex. The implicit ability of the A3H RNP to simultaneously bind and deaminate DNA may have a tremendous impact on the direction of future investigations into how RNA regulates the other antiretroviral APOBECs, in particular the dual domain A3G which comprises numerous predicted RNA binding sites on both domains.

The field has advanced in its focus not only on the RNAs that bind to APOBEC proteins, but the APOBEC protein residues required for RNA binding and the effect that select APOBEC RNP formation has on DNA deaminase and antiviral activities. Specifically, the open question in the field is whether there are unique RNA recognition motifs in APOBEC that determine selective binding to RNAs of different sequence or secondary structures. Moreover, can RNP formation functionally differentiate APOBEC structures for substrate specific binding and gene editing acivity?

The enzymatic activities of APOBECs on RNA and in gene editing suggest that they may be attractive targets of therapeutics. Key areas for future therapeutic development are (i) determining if Vif inhibition will enable sufficient A3 antiviral activity for a catastrophic level of HIV genomic hypermutation, (ii) identifying selective inhibitors of AID and A3 proteins as potential interventions in the progression of lymphatic and solid tumors and (iii) harnessing the APOBEC proteins as Cas3/9 chimeras for CRISPR targeting of base modification gene editing to either restore gene function or disrupt disease causing alleles. We anticipate that RNA binding to APOBECs and RNA-dependent regulation of their deoxycytidine deaminase activity will modulate gene editing activity within the microenvironment of the cells and tissues and determine the success of the therapeutic approaches.

References

Adolf MB, Webb J, Chelico L (2013) Retroviral Restriction Factor APOBEC3G Delays the Initiation of DNA Snythesis by HIV-1 ReverseTranscriptase. PLoS ONE 8:e64196

Aydin H, Taylor MW, Lee JE (2014) Structure-guided analysis of the human APOBEC3-HIV restrictome. Structure 22:668–684

Bach D, Peddi S, Mangeat B, Lakkaraju A, Strub K, Trono D (2008) Characterization of APOBEC3G binding to 7SL1 RNA. Retrovirology 5:54

Bennett RP, Stewart RA, Hogan PA, Ptak RG, Mankowski MK, Hartman TL, Buckheit RW Jr, Snyder BA, Salter JD, Morales GA et al (2016) An analog of camptothecin inactive against Topoisomerase I is broadly neutralizing of HIV-1 through inhibition of Vif-dependent APOBEC3G degradation. Antiviral Res 136:51–59

Bennett RP (2007) A functional analysis of the domains of APOBEC3G involved in dimerization, Antiviral activity and cellular localization. University of Rochester School of Medicine and Dentistry, Department of Biochemistry and Biophysics. Ph.D. thesis Chapter 2, pp 48–90. Proquest ID # 304818534

Bohn JA, Thummar K, York A, Raymond A, Brown WC, Bieniasz PD, Hatziioannou T, Smith JL (2017) APOBEC3H structure reveals an unusual mechanism of interaction with duplex RNA. Nat Commun 8:1021

Bulliard Y, Turelli P, Rohrig UF, Zoete V, Mangeat B, Michielin O, Trono D (2009) Functional analysis and structural modeling of human APOBEC3G reveal the role of evolutionarily conserved elements in the inhibition of human immunodeficiency virus type 1 infection and Alu transposition. J Virology 83:12611–12621

Burnett A, Spearman P (2007) APOBEC3G multimers are recruited to the plasma membrane for packaging into human immunodeficiency virus type 1 virus-like particles in an RNA-dependent process requiring the NC basic linker. J Virology 81:5000–5013

Bélanger K, Langlois MA (2015) RNA-binding residues in the N-terminus of APOBEC3G influence its DNA sequence specificity and retrovirus restriction efficiency. Virology 483:141–148

Bélanger K, Savoie M, Rosales Gerpe MC, Couture JF, Langlois MA (2013) Binding of RNA by APOBEC3G controls deamination-independent restriction of retroviruses. Nucleic Acids Res 41:7438–52

Cadena C, Stavrou S, Manzoni T, Iyer SS, Bibollet-Ruche F, Zhang, W, Hahn BH, Browne EP, Ross SR (2016) The effect of HIV-1 Vif polymorphisms on A3G anti-1 viral activity in an in vivo mouse model. Retrovirology. (In Press)

Caval V, Bouzidi MS, Suspène R, Laude H, Dumargne MC, Bashamboo A, Krey T, Vartanian JP, Wain-Hobson S (2015) Molecular basis of the attenuated phenotype of human APOBEC3B DNA mutator enzyme. Nucleic Acids Res 43:9340–9349

Cen S, Guo F, Niu M, Saadatmand J, Deflassieux J, Kleiman L (2004) The interaction between HIV-1 Gag and APOBEC3G. J Biol Chem 279:33177–33184

Chaurasiya KR, McCauley MJ, Wang W, Qualley DF, Wu T, Kitamura S, Geertsema H, Chan DS, Hertz A, Iwatani Y, Levin JG, Musier-Forsyth K, Rouzina I, Williams MC (2014) Oligomerization transforms human APOBEC3G from an efficient enzyme to a slowly dissociating nucleic acid-binding protein. Nat Chem 6:28–33

Chelico L, Prochnow C, Erie DA, Chen XS, Goodman MF (2010) Structural model for deoxycytidine deamination mechanisms of the HIV-1 inactivation enzyme APOBEC3G. J Biol Chem 285:16195–16205

Chelico L, Sacho EJ, Erie DA, Goodman MF (2008) A model for oligomeric regulation of APOBEC3G cytosine deaminase-dependent restriction of HIV. J Biol Chem 283:13780–13791

Conticello SG, Thomas CJ, Petersen-Mahrt SK, Neuberger MS (2005) Evolution of the AID/ APOBEC family of polynucleotide (deoxy) cytidine deaminases. Molecular Biol Evolution 22:367–377

Devos JM, Tomanicek SJ, Jones CE, Nossal NG, Mueser TC (2007) Crystal structure of bacteriophage T4 5′ nuclease in complex with a branched DNA reveals how flap endonuclease-1 family nucleases bind their substrates. J Biol Chem 282:31713–31724

Drake JWaH, JJ (1999) Mutation rates among RNA viruses. Proc Natl Acad Sci USA 96:13910–13913

Eckwahl MJ, Arnion H, Kharytonchyk S, Zang T, Bieniasz PD, Telesnitsky A, Wolin SL (2016) Analysis of the human immunodeficiency virus-1 RNA packageome. RNA 22:1228–1238

Eckwahl MJ, Sim S, Smith D, Telesnitsky A, Wolin SL (2015) A retrovirus packages nascent host noncoding RNAs from a novel surveillance pathway. Genes Dev 29:646–657

Esnault C, Heidmann O, Delebecque F, Dewannieux M, Ribet D, Hance AJ, Heidmann T, Schwartz O (2005) APOBEC3G cytidine deaminase inhibits retrotransposition of endogenous retroviruses. Nature 433:430–433

Fang Y, Xiao X, Li SX, Wolfe A, Chen XS (2018) Molecular interactions of a dna modifying enzyme APOBEC3F catalytic domain with a single-stranded DNA. J Mol Biol 430:87–101

Feng Y, Chelico L (2011) Intensity of deoxycytidine deamination of HIV-1 proviral DNA by the retroviral restriction factor APOBEC3G is mediated by the noncatalytic domain. J Biol Chem 286:11415–11426

Feng Y, Wong L, Morse M, Rouzina I, Williams MC, Chelico L (2018) RNA-mediated dimerization of the human deoxycytidine deaminase APOBEC3H influences enzyme activity and interaction with nucleic acids. J Mol Biol 430:4891–4907

Fourati S, Malet I, Binka M, Boukobza S, Wirden M, Sayon S, Simon A, Katlama C, Simon V, Calvez V et al (2010) Partially active HIV-1 Vif alleles facilitate viral escape from specific antiretrovirals. Aids 24:2313–2321

Friew YN, Boyko V, Hu WS, Pathak VK (2009) Intracellular interactions between APOBEC3G, RNA, and HIV-1 Gag: APOBEC3G multimerization is dependent on its association with RNA. Retrovirology 6:56

Fukuda H, Songling L, Sardo, L, Smith JL, Yamashita K, Sarca AD, Shirakawa K, Standley DM, Takaori-Kondo A, Izumi T (2019) Structural Determinants of the APOBEC3G N-Terminal Domain for HIV-1 RNA Association. Frontiers in Cellular and Infection Microbiology 9

Gallois-Montbrun S, Holmes RK, Swanson CM, Fernandez-Ocana M, Byers HL, Ward MA, Malim MH (2008) Comparison of cellular ribonucleoprotein complexes associated with the APOBEC3F and APOBEC3G antiviral proteins. J Virol 82:5636–5642

Gallois-Montbrun S, Kramer B, Swanson CM, Byers H, Lynham S, Ward M, Malim MH (2007) Antiviral protein APOBEC3G localizes to ribonucleoprotein complexes found in P bodies and stress granules. J Virology 81:2165–2178

Guo Y, Dong L, Qiu X, Wang Y, Zhang B, Liu H, Yu Y, Zang Y, Yang M, Huang Z (2014) Structural basis for hijacking CBF-beta and CUL5 E3 ligase complex by HIV-1 Vif. Nature 505:229–233

Harris RS, Dudley JP (2015) APOBECs and virus restriction. Virology (479–480):131–145

Huthoff H, Autore F, Gallois-Montbrun S, Fraternali F, Malim MH (2009) RNA-dependent oligomerization of APOBEC3G is required for restriction of HIV-1. PLoS Pathog 5:e1000330
Huthoff H, Malim MH (2007) Identification of amino acid residues in APOBEC3G required for regulation by human immunodeficiency virus type 1 Vif and Virion encapsidation. J Virology 81:3807–3815
Itano MS, Arnion H, Wolin SL, Simon SM (2018) Recruitment of 7SL1 RNA to assembling HIV-1 virus-like particles. Traffic. 19:36–43
Ito F, Yang H, Xiao X, Li SX, Wolfe A, Zirkle B, Arutiunian V, Chen XS (2018) Understanding the structure, multimerization, subcellular localization and mC selectivity of a genomic mutator and anti-HIV factor APOBEC3H. Sci Rep 8:3763
Iwatani Y, Chan DS, Wang F, Maynard KS, Sugiura W, Gronenborn AM, Rouzina I, Williams MC, Musier-Forsyth K, Levin JG (2007) Deaminase-independent inhibition of HIV-1 reverse transcription by APOBEC3G. Nucleic Acids Res 35:7096–108
Iwatani Y, Takeuchi H, Strebel K, Levin JG (2006) Biochemical activities of highly purified, catalytically active human APOBEC3G: correlation with antiviral effect. J Virology 80:5992–6002
Jiang F, Taylor DW, Chen JS, Kornfeld JE, Zhou K, Thompson AJ, Nogales E, Doudna JA (2016) Structures of a CRISPR-Cas9 R-loop complex primed for DNA cleavage. Sci (New York, NY) 351:867–871
Jin X, Brooks A, Chen H, Bennett R, Reichman R, Smith H (2005) APOBEC3G/CEM15 (hA3G) mRNA levels associate inversely with human immunodeficiency virus viremia. J Virology 79:11513–11516
Keene SE, Telesnitsky A (2012) cis-Acting determinants of 7SL1 RNA packaging by HIV-1. J Virology 86:7934–7942
Khan MA, Goila-Gaur R, Kao S, Miyagi E, Walker RC Jr, Strebel K (2009) Encapsidation of APOBEC3G into HIV-1 virions involves lipid raft association and does not correlate with APOBEC3G oligomerization. Retrovirology 6:99
Khan MA, Goila-Gaur R, Opi S, Miyagi E, Takeuchi H, Kao S, Strebel K (2007) Analysis of the contribution of cellular and viral RNA to the packaging of APOBEC3G into HIV-1 virions. Retrovirology 4:48
Khan MA, Kao S, Miyagi E, Takeuchi H, Goila-Gaur R, Opi S, Gipson CL, Parslow TG, Ly H, Strebel K (2005) Viral RNA is required for the association of APOBEC3G with human immunodeficiency virus type 1 nucleoprotein complexes. J Virol 79:5870–5874
Kikuchi T, Iwabu Y, Tada T, Kawana-Tachikawa A, Koga M, Hosoya N, Nomura S, Brumme ZL, Jessen H, Pereyra F et al (2015) Anti-APOBEC3G activity of HIV-1 Vif protein is attenuated in elite controllers. J Virol 89:4992–5001
Kim EY, Bhattacharya T, Kunstman K, Swantek P, Koning FA, Malim MH, Wolinsky SM (2010) Human APOBEC3G-mediated editing can promote HIV-1 sequence diversification and accelerate adaptation to selective pressure. J Virology 84:10402–10405
Kouno T, Luengas EM, Shigematsu M, Shandilya SM, Zhang J, Chen L, Hara M, Schiffer CA, Harris RS, Matsuo H (2015) Structure of the Vif-binding domain of the antiviral enzyme APOBEC3G. Nature Struct Molecular Biol 22:485–491
Kouno T, Silvas TV, Hilbert BJ, Shandilya SMD, Bohn MF, Kelch BA, Royer WE, Somasundaran M, Kurt Yilmaz N, Matsuo H, Schiffer CA (2017) Crystal structure of APOBEC3A bound to single-stranded DNA reveals structural basis for cytidine deamination and specificity. Nat Commun 8:15024
Kourteva Y, De Pasquale M, Allos T, McMunn C, D'Aquila RT (2012) APOBEC3G expression and hypermutation are inversely associated with human immunodeficiency virus type 1 (HIV-1) burden in vivo. Virology 430:1–9
Kozak SL, Marin M, Rose KM, Bystrom C, Kabat D (2006) The anti-HIV-1 editing enzyme APOBEC3G binds HIV-1 RNA and messenger RNAs that shuttle between polysomes and stress granules. J Biol Chem 281:29105–29119

Kreisberg JF, Yonemoto W, Greene WC. (2006) Endogenous factors enhance HIV infection of tissue naive CD4 T cells by stimulating high molecular mass APOBEC3G complex formation. J Exp Med 203:865–870

Krisko JF, Martinez-Torres F, Foster JL, Garcia JV (2013) HIV restriction by APOBEC3 in humanized mice. PLoS Pathog 9:e1003242

Krupp A, McCarthy KR, Ooms M, Letko M, Morgan JS, Simon V, Johnson WE (2013) APOBEC3G polymorphism as a selective barrier to cross-species transmission and emergence of pathogenic SIV and AIDS in a primate host. PLoS Pathog 9:e1003641

Kutluay SB, Zang T, Blanco-Melo D, Powell C, Jannain D, Errando M, Bieniasz PD (2014) Global changes in the RNA binding specificity of HIV-1 gag regulate virion genesis. Cell 159:1096–1109

MacMillan AL, Kohli RM, Ross SR (2013) APOBEC3 inhibition of mouse mammary tumor virus infection: the role of cytidine deamination versus inhibition of reverse transcription. J Virology 87:4808–4817

Maiti A, Myint W, Kanai T, Delviks-Frankenberry K, Sierra Rodriguez C, Pathak VK, Schiffer CA, Matsuo H (2018) Crystal structure of the catalytic domain of HIV-1 restriction factor APOBEC3G in complex with ssDNA. Nat Commun 9:2460

Mansky LMaT HM (1995) Lower in vivo mutation rate of human immunodeficiency virus type 1 than that predicted from the fidelity of purified reverse transcriptase. J Virol 69:5087–5094

Matsuoka T, Nagae T, Ode H, Awazu H, Kurosawa T, Hamano A, Matsuoka K, Hachiya A, Imahashi M, Yokomaku Y et al (2018) Structural basis of chimpanzee APOBEC3H dimerization stabilized by double-stranded RNA. Nucleic Acids Res 46:10368–10379

McDougall WM, Okany C, Smith HC (2011) Deaminase activity on single-stranded DNA (ssDNA) occurs in vitro when APOBEC3G cytidine deaminase forms homotetramers and higher-order complexes. J Biol Chem 286:30655–30661

McDougall WM, Smith HC (2011) RNA-Dependent inhibition of APOBEC3G ssDNA cytidine deaminase activity. Biochem Biophys Res Commun 412:612–7

Morse M, Huo R, Feng Y, Rouzina I, Chelico L, Williams MC (2017) Dimerization regulates both deaminase-dependent and deaminase-independent HIV-1 restriction by APOBEC3G. Nat Commun. 8:597

Munk C, Willemsen A, Bravo IG (2012) An ancient history of gene duplications, fusions and losses in the evolution of APOBEC3 mutators in mammals. BMC Evol Biol 12:71

Muriaux D, Darlix JL (2010) Properties and functions of the nucleocapsid protein in virus assembly. RNA Biol 7:744–53

Navarro F, Bollman B, Chen H, Konig R, Yu Q, Chiles K, Landau NR (2005) Complementary function of the two catalytic domains of APOBEC3G. Virology 333:374–386

Okeoma CM, Lovsin N, Peterlin BM, Ross SR (2007) APOBEC3 inhibits mouse mammary tumour virus replication in vivo. Nature 445:927–930

Ooms MBB, Letko M, Maio SM, Pilcher CD, Hecht FM, Barbour JD, Simon V (2013) HIV-1 Vif adaptation to human APOBEC3H haplotypes. Cell Host Microbe 14:411–421

Opi S, Takeuchi H, Kao S, Khan MA, Miyagi E, Goila-Gaur R, Iwatani Y, Levin JG, Strebel K (2006) Monomeric APOBEC3G is catalytically active and has antiviral activity. J Virol 80:4673–4682

Pan Y, Sun Z, Maiti A, Kanai T, Matsuo H, Li M, Harris RS, Shlyakhtenko LS, Lyubchenko YL (2016) Nanoscale characterization of interaction of APOBEC3G with RNA. Biochemistry

De Pasquale M, Kourteva Y, Allos T, D'Aquila RT (2013) Lower HIV provirus levels are associated with more APOBEC3G protein in blood resting memory CD4+T lymphocytes of controllers in vivo. PLoS ONE 8:e76002

Peng G, Greenwell-Wild T, Nares S, Jin W, Lei KJ, Rangel ZG, Munson PJ, Wahl SM (2007) Myeloid differentiation and susceptibility to HIV-1 are linked to APOBEC3 expression. Blood 110:393–400

Pery E, Sheehy A, Nebane NM, Brazier AJ, Misra V, Rajendran KS, Buhrlage SJ, Mankowski MK, Rasmussen L, White EL et al (2015) Identification of a novel HIV-1

inhibitor targeting Vif-dependent degradation of human APOBEC3G protein. J Biol Chem 290:10504–10517
Polevoda B, Joseph R, Friedman AE, Bennett RP, Greiner R, De Zoysa T, Stewart RA, Smith HC (2017) DNA mutagenic activity and capacity for HIV-1 restriction of the cytidine deaminase APOBEC3G depend on whether DNA or RNA binds to tyrosine 315. J Biol Chem 292:8642–8656
Polevoda B, McDougall WM, Bennett RP, Salter JD, Smith HC (2016) Structural and functional assessment of APOBEC3G macromolecular complexes. Methods 107:10–22
Polevoda B, McDougall WM, Tun BN, Cheung M, Salter JD, Friedman AE, Smith HC (2015) RNA binding to APOBEC3G induces the disassembly of functional deaminase complexes by displacing single-stranded DNA substrates. Nucleic Acids Res 43:9434–9445
Poropatich KaS, DJ, Jr (2011) Human immunodeficiency virus type 1 long-term non-progressors: the viral, genetic and immunological basis for disease non-progression. J General Virology 92:247–268
Qiao Q, Wang L, Meng FL, Hwang JK, Alt FW, Wu H (2017) AID recognizes structured DNA for class switch recombination. Mol Cell 67:361–373.e364
Rangel HR, Garzaro D, Rodriguez AK, Ramirez AH, Ameli G, Del Rosario Gutierrez C, Pujol FH (2009) Deletion, insertion and stop codon mutations in vif genes of HIV-1 infecting slow progressor patients. J Infect Dev Ctries 3:531–538
Reddy K, Ooms M, Letko M, Garrett N, Simon V, Ndung'u T (2016) Functional characterization of Vif proteins from HIV-1 infected patients with different APOBEC3G haplotypes. Aids
Refsland EW, Stenglein MD, Shindo K, Albin JS, Brown WL, Harris RS (2010) Quantitative profiling of the full APOBEC3 mRNA repertoire in lymphocytes and tissues: implications for HIV-1 restriction. Nucleic Acids Res 38:4274–4284
Sadler HA, Stenglein MD, Harris RS, Mansky LM (2010) APOBEC3G contributes to HIV-1 variation through sublethal mutagenesis. J Virology 84:7396–7404
Salter et al (2014a) Structural insights for HIV-1therapeutic strategies targeting Vif. Trends Biochem Sci 39:373–380
Salter JD, Bennett RP, Smith HC (2016) The APOBEC protein family: united by structure, divergent in function. Trends Biochem Sci 41:578–594
Salter JD, Morales GA, Smith HC (2014b) Structural insights for HIV-1 therapeutic strategies targeting Vif. Trends Biochem Sci 39:373–380
Salter JD, Smith HC (2018) Modeling the embrace of a mutator: APOBEC selection of nucleic acid ligands. Trends in Biochem Sci 43:606–622
Sato K, Izumi T, Misawa N, Kobayashi T, Yamashita Y, Ohmichi M, Ito M, Takaori-Kondo A, Koyanagi Y (2010) Remarkable lethal G-to-A mutations in vif-proficient HIV-1 provirus by individual APOBEC3 proteins in humanized mice. J Virology 84:9546–9556
Sawyer SL, Emerman M, Malik HS (2004) Ancient adaptive evolution of the primate antiviral DNA-editing enzyme APOBEC3G. PLoS Biol 2:E275
von Schwedler U, Song J, Aiken C, Trono D (1993) Vif is crucial for human immunodeficiency virus type 1 proviral DNA synthesis in infected cells. J Virology 67:4945–4955
Shaban NM, Shi K, Lauer KV, Carpenter MA, Richards CM, Salamango D, Wang J, Lopresti MW, Banerjee S, Levin-Klein R, Brown WL, Aihara H, Harris RS (2018) The antiviral and cancer genomic DNA deaminase APOBEC3H is regulated by an RNA-Mediated dimerization mechanism. Mol Cell 69:75–86
Sheehy AM, Gaddis NC, Choi JD, Malim MH (2002) Isolation of a human gene that inhibits HIV-1 infection and is suppressed by the viral Vif protein. Nature 418:646–650
Shi K, Carpenter MA, Kurahashi K, Harris RS, Aihara H (2015) Crystal structure of the DNA deaminase APOBEC3B catalytic domain. J Biol Chem 290:28120–28130
Shi K, Demir Ö, Carpenter MA, Wagner J, Kurahashi K, Harris RS, Amaro RE, Aihara H (2017) Conformational switch regulates the DNA cytosine deaminase activity of human APOBEC3B. Sci Rep 7:17415
Shi K, Carpenter, MA, Banerjee, S, Shaban, NM, Kurahashi, K1, Salamango, DJ, McCann, JL, Starrett, GJ, Duffy, JV, Demir, Ö, Amaro, RE, Harki, DA, Harris, RS, Aihara, H (2017a)

Structural basis for targeted DNA cytosine deamination and mutagenesis by APOBEC3A and APOBEC3B. Nature Struct Molecular Biol 24:131–139
Simon V, Zennou V, Murray D, Huang Y, Ho DD, Bieniasz PD (2005) Natural variation in Vif: differential impact on APOBEC3G/3F and a potential role in HIV-1 diversification. PLoS Pathog 1:e6
Simon VBN, Landau NR (2015) Intrinsic host restrictions to HIV-1 and mechanisms of viral escape. Nat Immunol 16:546–553
Siu KK, Sultana A, Azimi FC, Lee JE (2013) Structural determinants of HIV-1 Vif susceptibility and DNA binding in APOBEC3F. Nat Commun 4:2593
Smith HC (2011) APOBEC3G: a double agent in defense. Trends Biochem Sci 36:239–244
Smith HC (2016) RNA binding to APOBEC deaminases; Not simply a substrate for C to U editing. RNA Biol 14:1153–1165
Soros VB, Yonemoto W, Greene WC (2007) Newly synthesized APOBEC3G is incorporated into HIV virions, inhibited by HIV RNA, and subsequently activated by RNase H. PLoS Pathog 3: e15
Sova P, Volsky DJ (1993) Efficiency of viral DNA synthesis during infection of permissive and nonpermissive cells with vif-negative human immunodeficiency virus type 1. J Virology 67:6322–6326
Stavrou S, Crawford D, Blouch K, Browne EP, Kohli RM, Ross SR (2014) Different modes of retrovirus restriction by human APOBEC3A and APOBEC3G in vivo. PLoS Pathog 10: e1004145
Strebel K, Khan MA (2008) APOBEC3G encapsidation into HIV-1 virions: which RNA is it? Retrovirology 5:55
Svarovskaia ES, Xu H, Mbisa JL, Barr R, Gorelick RJ, Ono A, Freed EO, Hu WS, Pathak VK (2004) Human apolipoprotein B mRNA-editing enzyme-catalytic polypeptide-like 3G (APOBEC3G) is incorporated into HIV-1 virions through interactions with viral and nonviral RNAs. J Biol Chem 279:35822–35828
Thangavelu PU, Gupta V, Dixit NM (2014) Estimating the fraction of progeny virions that must incorporate APOBEC3G for suppression of productive HIV-1 infection. Virology 449:224–228
Venkatesan SRR, Kanu N, McGranahan N, Bartek J, Quezada SA, Hare J, Harris RS8, Swanton C1 (2018) Perspective: APOBEC mutagenesis in drug resistance and immune escape in HIV and cancer evolution. Ann Oncology 29:563–572
Wang T, Tian C, Zhang W, Luo K, Sarkis PT, Yu L, Liu B, Yu Y, Yu XF (2007) 7SL1 RNA mediates virion packaging of the antiviral cytidine deaminase APOBEC3G. J Virology 81:13112–13124
Wang TTC, Zhang W, Sarkis PT, Yu XF (2008a) Interaction with 7SL1 RNA but not with HIV-1 genomic RNA or P bodies is required for APOBEC3F virion packaging. J Mol Biol 375:1098–1112
Wang T, Zhang W, Tian C, Liu B, Yu Y, Ding L, Spearman P, Yu XF (2008b) Distinct viral determinants for the packaging of human cytidine deaminases APOBEC3G and APOBEC3C. Virology 377:71–79
Wedekind JE, Gillilan R, Janda A, Krucinska J, Salter JD, Bennett RP, Raina J, Smith HC (2006) Nanostructures of APOBEC3G support a hierarchical assembly model of high molecular mass ribonucleoprotein particles from dimeric subunits. J Biological Chem 281:38122–38126
Wichroski MJ, Robb GB, Rana TM (2006) Human retroviral host restriction factors APOBEC3G and APOBEC3F localize to mRNA processing bodies. PLoS Pathog 2:e41
Xiao X, Li SX, Yang H, Chen XS (2016) Crystal structures of APOBEC3G N-domain alone and its complex with DNA. Nat Commun 7:12193
Xiao X, Yang H, Arutiunian V, Fang Y, Besse G, Morimoto C, Zirkle B, Chen XS (2017) Structural determinants of APOBEC3B non-catalytic domain for molecular assembly and catalytic regulation. Nucleic Acids Res 45:7494–7506
Xu H, Chertova E, Chen J, Ott DE, Roser JD, Hu WS, Pathak VK (2007) Stoichiometry of the antiviral protein APOBEC3G in HIV-1 virions. Virology 360:247–256

York A, Kutluay SB, Errando M, Bieniasz PD (2016) The RNA binding specificity of human APOBEC3 proteins resembles That of HIV-1 Nucleocapsid. PLoS Pathog 12:e1005833
Zhang W, Du J, Yu K, Wang T, Yong X, Yu XF (2010) Association of potent human antiviral cytidine deaminases with 7SL1 RNA and viral RNP in HIV-1 virions. J Virology 84:12903–12913
Zhang KL, Mangeat B, Ortiz M, Zoete V, Trono D, Telenti A, Michielin O (2007) Model structure of human APOBEC3G. PLoS ONE 2:e378
Zhang J, Webb DM (2004) Rapid evolution of primate antiviral enzyme APOBEC3G. Hum Mol Genet 13:1785–1791
Zheng Y-H, Jeang K-T, Tokunaga K (2012) Host restriction factors in retroviral infection: promises in virus-host interaction. Retrovirology 9:112
Ziegler SJ, Liu C, Landau M, Buzovetsky O, Desimmie BA, Zhao Q, Sasaki T, Burdick RC, Pathak VK, Anderson KS et al (2018) Insights into DNA substrate selection by APOBEC3G from structural, biochemical, and functional studies. PLoS ONE 13:e0195048

Chapter 7
Structure and Function of the AAA+ ATPase p97, a Key Player in Protein Homeostasis

Petra Hänzelmann, Carolina Galgenmüller and Hermann Schindelin

Abstract p97 belongs to the functional diverse superfamily of AAA+ (ATPases Associated with diverse cellular Activities) ATPases and is characterized by an N-terminal regulatory domain and two stacked hexameric ATPase domains forming a central protein conducting channel. p97 is highly versatile and has key functions in maintaining protein homeostasis including protein quality control mechanisms like the ubiquitin proteasome system (UPS) and autophagy to disassemble poly-ubiquitylated proteins from chromatin, membranes, macromolecular protein complexes and aggregates which are either degraded by the proteasome or recycled. p97 can use energy derived from ATP hydrolysis to catalyze substrate unfolding and threading through its central channel. The function of p97 in a large variety of different cellular contexts is reflected by its simultaneous association with different cofactors, which are involved in substrate recognition and processing, thus leading to the formation of transient multi-protein complexes. Dysregulation in protein homeostasis and proteotoxic stress are often involved in the development of cancer and neurological diseases and targeting the UPS including p97 in cancer is a well-established pharmacological strategy. In this chapter we will describe structural and functional aspects of the p97 interactome in regulating diverse cellular processes and will discuss the role of p97 in targeted cancer therapy.

Keywords AAA+ ATPase · p97 · Ubiquitin · Cancer therapy · Protein homeostasis · Protein quality control · Unfoldase · Protein disassembly

---

P. Hänzelmann (✉) · C. Galgenmüller · H. Schindelin
Rudolf Virchow Center for Experimental Biomedicine, University of Würzburg, Josef-Schneider-Str. 2, 97080 Würzburg, Germany
e-mail: petra.haenzelmann@virchow.uni-wuerzburg.de

C. Galgenmüller
e-mail: carolina.galgenmueller@virchow.uni-wuerzburg.de

H. Schindelin
e-mail: hermann.schindelin@virchow.uni-wuerzburg.de

J. R. Harris and J. Marles-Wright (eds.), *Macromolecular Protein Complexes II: Structure and Function*, Subcellular Biochemistry 93,
https://doi.org/10.1007/978-3-030-28151-9_7

Introduction

p97 (also referred to as VCP), which is called Cdc48 in yeast, archaea and *Caenorhabditis elegans* as well as TER94 in drosophila, is a member of the large AAA+ (ATPases Associated with diverse cellular Activities) superfamily of ATPases (for reviews see Erzberger and Berger 2006; Miller and Enemark 2016; Wendler et al. 2012). AAA+ proteins are found in all kingdoms of life and are functionally diverse. For example, in humans they are involved in protein degradation processes, in the biogenesis and maturation of organelles and multi-protein complexes, in replication, in transcriptional activation, in the dynamics of microtubules and in membrane fusion. In many different processes they serve as unfolding and/or proteolytic systems and they are involved in the remodeling of protein-DNA complexes, protein-protein complexes as well as of protein aggregates (for reviews see Abid Ali and Costa 2016; Mogk et al. 2015; Yedidi et al. 2017; Zhao and Brunger 2016). To fulfill their functions, they use ATP hydrolysis to generate mechanical forces to act on their respective substrates.

Typically, AAA+ proteins form hexameric ring structures with their ATPase modules as common features (for reviews see Miller and Enemark 2016; Wendler et al. 2012). Mg^{2+}-ATP is bound at the interface between adjacent monomers and is stabilized by residues from both monomers. Classical Walker A (P-loop, consensus Gx_4GKS/T; x, any residue) and Walker B motifs (consensus h_4DE; h, hydrophobic residue), involved in ATP-binding and catalysis as well as Mg^{2+} coordination, respectively, together with so-called sensors 1 (asparagine/threonine) and 2 (arginine) residues, which sense nucleotide binding/hydrolysis acting in cis, whereas a conserved so-called arginine-finger acts in trans. AAA+ proteins can be divided into two classes: (i) Type I with one hexameric ATPase module; (ii) Type II with two hexameric ATPase modules stacked on top of each other. Both classes may harbor additional domains or insertions, for example an N-terminal substrate-binding/recognition domain. Common to both families is a central cavity formed by the hexameric assembly, which features two substrate-binding loops. Especially the pore loop 1 with its conserved aromatic residues is involved in coupling ATP hydrolysis to substrate-binding. Typically, a specific AAA+ protein has a distinct function in a specific pathway like the motor protein dynein in transporting cellular cargo along microtubules (for a review see Kato et al. 2018) or the AAA+ ring of the regulatory particle of the proteasome (for a review see Bard et al. 2018), which is involved in substrate unfolding and translocation into the proteolytic chamber. However, p97 is involved in a large variety of functionally divergent cellular processes, which include both degradative and non-degradative pathways (for a review see e.g. Meyer and Weihl 2014).

p97 is an essential and highly abundant protein found in the cytoplasm and nucleus. p97 plays a key role in maintaining protein homeostasis and is involved in protein quality control (PQC) mechanisms such as the ubiquitin proteasome system (UPS) and autophagy by disassembling polyubiquitylated proteins from chromatin, membranes, macromolecular protein complexes and aggregates including endoplasmic reticulum associated degradation (ERAD), mitochondria associated degradation (MAD), ribosome quality control (RQC), lysophagy and mitophagy as well as the clearance of stress granules (SGs) (for reviews see Bug and Meyer 2012; Franz et al. 2016a; Meyer et al. 2012; Meyer and Weihl 2014; Stach and Freemont 2017; Xia et al. 2016; Ye et al. 2017). The fate of the segregated protein can be either proteasomal degradation or recycling. p97 further acts in many chromatin-associated processes like transcription, replication and DNA repair. Nondegradative functions include endosomal trafficking, Golgi and nuclear envelope reassembly after mitosis as well as NF-κB (nuclear factor kappa-light-chain-enhancer of activated B cells) activation. Targeting p97 to all these diverse cellular pathways is regulated through the assembly with a large variety of different so-called cofactors that are involved in substrate recognition and processing (for reviews see Buchberger et al. 2015; Hänzelmann and Schindelin 2017).

Due to the participation of p97 and its associated cofactors in many different cellular pathways p97 has been linked to several neurological diseases, aging and cancer (for reviews see Chapman et al. 2011; Fessart et al. 2013; Franz et al. 2014; Haines 2010). These diseases are mainly characterized by an imbalance of protein homeostasis causing the accumulation of aberrant proteins and the formation of protein aggregates under cellular stress conditions (for a review see Alberti et al. 2017), which need to be handled by PQC mechanisms. However, proteotoxic stress can also lead to the formation of SGs, which are highly dynamic cytoplasmic dense aggregates composed of RNA derived from stalled pre-initiation complexes and a large number of proteins (for a review see Protter and Parker 2016). SGs have been implicated in age-related neurodegenerative diseases including amyotrophic lateral sclerosis (ALS) and frontotemporal dementia (FTD) (for reviews see Alberti et al. 2017; Ramaswami et al. 2013; Taylor et al. 2016). Furthermore, several heterozygous missense mutations have been identified in p97 which are responsible for rare autosomal dominant disorders: Multisystem proteinopathy 1 (MSP1), also referred to as Inclusion Body Myopathy associated with Paget's disease of the bone and Frontotemporal Dementia (IBMPFD), Familial Amyotrophic Lateral Sclerosis (FALS) and Charcot-Marie-Tooth Disease Type 2Y (CMT2Y) (for a review see Tang and Xia 2016).

Structure and Conformational Changes of p97

As a type II AAA+ ATPase p97 harbors two stacked hexameric ATPase rings called the D1 and D2 domains (Davies et al. 2008; DeLaBarre and Brunger 2003; Huyton et al. 2003) (Fig. 7.1a). In addition, p97 contains an N-terminal domain as well as an unstructured C-terminal tail, which are both involved in cofactor-binding. In addition, the C-terminus participates in the regulation of the ATPase activity (Niwa et al. 2012) and phosphorylation of the penultimate tyrosine regulates cofactor interactions (for a review see Ewens et al. 2010) (Li et al. 2008; Zhao et al. 2007). The N, D1 and D2 domains are connected via flexible linkers. Although p97 harbors two ATPase rings, p97 only features substrate-binding loops in the central channel of the D2 domain (Fig. 7.1b). This central pore has a restriction formed by six histidine residues located in the D1 domain (the so-called His-gate). In the D2 domain the smaller pore loop 1 features the conserved AAA+ tripeptide ϕ-h-G (aromatic-hydrophobic-Gly) and the larger, less conserved pore loop 2 harbors positively and negatively charged residues. During the ATPase cycle p97 undergoes only minor conformational changes in its two ATPase rings (for a review see Xia et al. 2016) (Banerjee et al. 2016; Schuller et al. 2016). Structural studies revealed slight rotational movements of the ATPase rings and a widening of the central pore within the D2 domain. The most striking changes are associated with the N domains, which, in the ADP-bound structure, are co-planar with the D1 ring (down-conformation) and show an up-movement (up-conformation) upon ATP-binding (Fig. 7.1c). The down-movement is driven by ATP hydrolysis in the D1 domain. Movements of pore loops 1 and 2 leads to a more open D2 pore with flexible loops for substrate engagement. However, these rather small conformational changes where the ATPase rings stay in a co-planar conformation are in contrast with most other AAA+ proteins, which show an asymmetric spiral architecture of their ATPase domain including both open and closed spiral conformations (e.g. (Deville et al. 2017; Lander et al. 2012)) but are similar to the peroxisome biogenesis proteins PEX1/PEX6, the closest p97 homolog (for a review see Saffert et al. 2017). Conformational changes induced by ATP-binding and hydrolysis in AAA+ ATPases are commonly transmitted via a complex network of intra- and intersubunit signaling mechanisms (Hänzelmann and Schindelin 2016b; Huang et al. 2012; Li et al. 2012). In intrasubunit signaling changes are transmitted via the flexible linkers connecting N-D1-D2 (Fig. 7.1d). A so-called intersubunit signaling network (ISS) is responsible for signal transmission between two adjacent monomers and coordinates ATP hydrolysis (Hänzelmann and Schindelin 2016b) (Fig. 7.1e). Via the ISS, the ATP-binding site located at the interface of two monomers is coupled to the central cavity with its substrate-binding loops. ATP-binding to the cis-site is sensed by residues in trans, including the arginine finger. In contrast to other AAA+ ATPases, p97 features two arginine fingers as well as a third arginine located in the unstructured C-terminus which inserts into the nucleotide-binding pocket of the trans monomer and coordinates the γ-phosphate of ATP, thus also contributing to inter-domain communication (Hänzelmann and Schindelin 2016b) (Fig. 7.1e inset). Both inter- and intrasubunit signaling are interconnected via Gly480 to the D1D2 linker of the same monomer, which is involved in the coordination of the adenine and via Gly610 to the D1D2 linker of the adjacent monomer, which is in contact with the ISS motif. Interestingly, a low resolution cryo-EM study revealed that the small flexible N-terminal extension of 24 aa of the p97 N domain undergoes conformational changes upon ATP-binding and inserts into the D2 domain (Schuller et al. 2016).

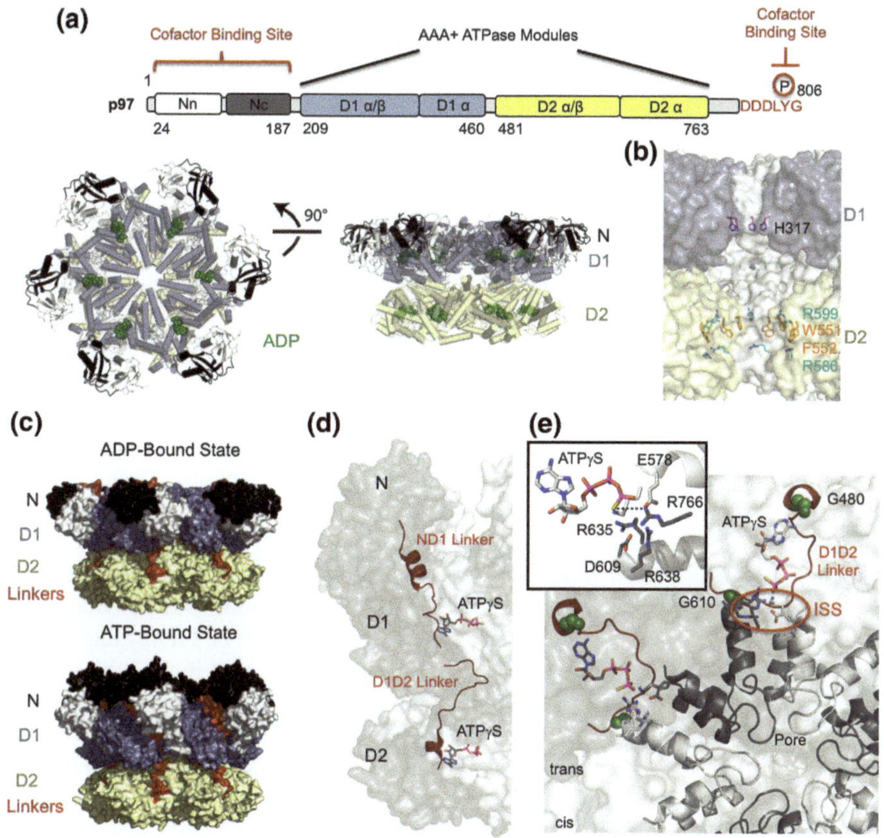

Fig. 7.1 Structure of p97. **a** Domain architecture and cartoon representation of p97 in the ADP-bound state viewed from the top and side colored according to its domains (pdb: 3cf3). ADP is shown as green spheres. **b** Surface representation of the central pore of p97 in the ADP-bound state (pdb: 5ftk). The restriction in the D1 domain formed by the His-gate (colored in purple) and residues in pore loops 1 and 2 (colored in orange and cyan, respectively) are shown as sticks. **c** Surface representation of p97 in the ADP- and ATP-bound state colored as in (**a**) (pdb: 5ftk and 5ftn). The ND1 and D1D2 linkers are shown in red. **d** Transparent surface representation of p97 with intersubunit signaling elements in the ATP-bound state in red (pdb: 5ftn). The signal traverses from the D2 ATP binding site via the D1D2 linker to the D1 ATP binding site and further via the ND1 linker to the N domain. ATPγS is shown in stick representation and the ND1 and D1D2 linkers in red. **e** Intrasubunit signaling in the ATP-bound state (pdb: 5ftn). Transparent surface representation of p97 at the height of the ATP-binding site in the D2 domain; the pore is indicated. Structural elements involved in signal transmission from the D2 ATP binding site to the central pore with monomers colored alternating in white (cis-subunit) and dark-gray (trans-subunit) are shown in cartoon representation while ATPγS and participating residues are shown in stick representation. The D1D2 linker is shown in red. Glycine residues involved in the interconnection of inter- and intrasubunit signaling are shown as green spheres. The inset shows a zoom in of ATPγS and residues involved in ISS in a stick representation

Macromolecular p97–Cofactor Complexes

p97 Cofactor Interaction Modes

p97 fulfills its many functions with the help of a large variety of cofactors (for recent reviews see Buchberger et al. 2015; Hänzelmann and Schindelin 2017). Cofactors are multi-domain proteins, which harbor, besides specific p97 interacting domains/motifs, further structural modules that can be involved in substrate recognition like ubiquitin-binding domains (UBD), substrate-processing domains involved in ubiquitylation, deubiquitylation or deglycosylation, protein-protein interactions and regulation. However, the specific functions of many cofactors and their associated domains are still poorly understood. p97 can interact with its cofactors either via its N domain or the extreme C-terminus of its C-terminal tail. The majority of cofactors have been shown to target the N domain while there are currently only two known cofactors that bind to the C-terminus. Cofactors that target the N domain can bind either via their ubiquitin-related UBX (ubiquitin regulatory X) (for a review see Buchberger et al. 2001) or UBX-like (UBXL) domains as well as via short peptide stretches called SHP (also referred to as BS1, binding segment 1) (Bruderer et al. 2004), VCP-binding motif (VBM) (Boeddrich et al. 2006) and VCP-interacting motif (VIM) (Stapf et al. 2011). Interestingly, most of these domains and motifs target the same binding site located in the hydrophobic cleft formed between both subdomains of the N domain and hence compete with each other (Fig. 7.2a). In contrast, cofactors harboring a SHP motif bind to a different hydrophobic groove in the C-terminal subdomain close to the ND1 linker (Fig. 7.2a).

The molecular details of p97 cofactor interactions derived from X-ray structures using small binding domains/motifs as well their regulation has been recently described in detail and will be summarized briefly (for a review see Hänzelmann and Schindelin 2017): (i) UBX and UBXL domains: In humans 12 different UBX proteins are known with five of them harboring an additional ubiquitin-associated (UBA) domain for binding to polyubiquitylated substrates. Structures of different UBX proteins revealed that they bind to p97 in a conserved manner via a signature R…FPR motif with the proline in a cis configuration (also called cis-Pro touch-turn motif (Kang and Yang 2011)) into the hydrophobic interdomain groove (Fig. 7.2a) (Arumughan et al. 2016; Dreveny et al. 2004; Hänzelmann et al. 2011; Kim et al. 2011; Lee et al. 2013; Li et al. 2017). UBXL domains, which have been identified in the two deubiquitylating enzymes (DUBs) YOD1 and VCIP135, bind in a similar manner via an FPP loop (Kim et al. 2014). In contrast, NPL4 the third member of UBXL proteins does not feature a loop, but also binds to this region (Hao et al. 2015; Isaacson et al. 2007). (ii) VIM/VBM motifs: Both motifs are short linear amino acid stretches containing several arginines, which are essential for the interaction with p97. The VIM motif (consensus motif $Rx_2h_3A_2x_2Rh$; x, any residue; h, hydrophobic residue) of the E3 ligase gp78 and the VBM motif (consensus motif EhR_4Lxh_2; h, hydrophobic residue; x, any residue) of the rhomboid

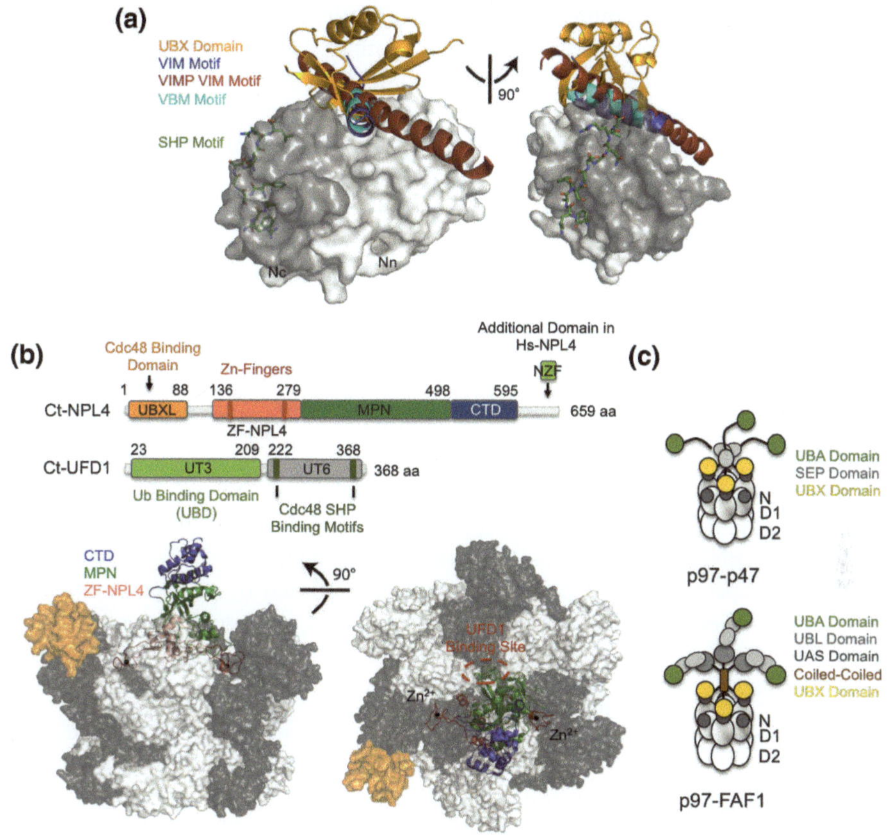

Fig. 7.2 p97 cofactor interactions. **a** Surface representation of the p97 N domain in complex with different cofactors in cartoon representation including the FAF1-UBX domain (pdb: 3qq8), the VBM motif of RHBDL4 (pdb: 5epp), the VIM motif of gp78 (pdb: 3tiw) and the extended VIM motif of VIMP (pdb: 5kiw), which all target the interdomain cleft of the p97 N domain, as well as the SHP motif of UFD1 (pdb: 5b6c), which binds into a hydrophobic pocket on the C-terminal subdomain. **b** Domain architecture of *C. thermophilum* NPL4 and UFD1 and the cryo-EM structure of the Cdc48–UFD1-NPL4 complex (pdb: 6chs). Side and top views of a surface representation of Cdc48 with monomers colored alternating in dark-gray and white. NPL4 is shown in cartoon representation and is colored according to its domains with the Zn^{2+}-ions as black spheres. The Cdc48 N domains (pdb: 5ftn) and the UBXL domain (pdb: 2pjh) of NPL4 are modelled based on other structures. The putative UFD1 binding site on NPL4 is indicated. **c** Cryo-EM based models for the p97–p47 (top) and p97–FAF1 (bottom) complexes

intermembrane protease RHBDL4 form an α-helix and insert via their arginines between the two subdomains of the N domain (Fig. 7.2a) (Hänzelmann and Schindelin 2011; Lim et al. 2016a). Interestingly, structural studies of the interaction of the ER membrane protein VIMP with the p97 N domain revealed that the VIM binding motif located in a rather long α-helix exhibits a different binding mode with only a partial overlap with the classical VIM/VBM α-helices (Fig. 7.2a)

(Tang et al. 2017). The minimal VIMP-binding motif, composed of about 40 amino acids, binds to a different area and covers a much larger surface. Although no structural data are available for the interaction of the VIMP motif of SVIP, it could be shown that it also uses a longer α-helix (Hänzelmann and Schindelin 2011). This indicates that the VIM interaction is not conserved and there are variations how this helical segment is oriented, thus also defining its specificity. (iv) SHP motif: The SHP-binding motif (consensus $h(x)_{1\text{-}2}F/W(x)_{0\text{-}1}GxGx_2L$; h, hydrophobic residue; x, any residue) was identified in UFD1 and other proteins but also in several UBX proteins, which contain this motif in addition to their UBX domain, thus enabling them to engage in a bipartite binding mode. The SHP-binding motif has been shown to insert into a hydrophobic binding crevice on the C-terminal subdomain of the N domain, thereby forming a kink via its GxG motif and extending the Nc mixed β-sheet by a short antiparallel β-strand (Fig. 7.2a) (Hänzelmann and Schindelin 2016a; Le et al. 2016; Lim et al. 2016b). Cofactor interactions with the extreme C-terminus of p97 with the sequence D_3LYG is accomplished by binding to two structurally unrelated domains, the PUB (PNGase/UBA or UBX containing proteins) and PUL domains (PLAP, Ufd3p, and Lub1p) found in peptide N-glycanase (PNGase) and PLAA (also called phospholipase A2-activating protein or PLAP), respectively. The main interaction is mediated by the penultimate tyrosine, which inserts into a hydrophobic binding pocket on these domains (Qiu et al. 2010; Zhao et al. 2007), hence phosphorylation of this residue abolishes interactions with these cofactors (Li et al. 2008).

Regulation of p97 Cofactor Assembly

Although p97 harbors six N domains and six C-termini for binding of cofactors, p97 typically associates with its cofactors in a sub-stoichiometric manner (for a review see Hänzelmann and Schindelin 2017). This could be due to steric reasons caused by these rather large multi-domain cofactors, but, at the same time, it may enable p97 to interact with additional cofactors in a coordinated assembly and, as a result, induce specific cellular responses. Known exceptions are the small protein SVIP (small VCP interacting protein), which binds with a 6:6 stoichiometry (Hänzelmann and Schindelin 2011), as well as the UBX protein ASPL (also called TUG and UBXD9), which disassembles the p97 hexamer forming a heterotetrameric $p97_2$-$ASPL_2$ complex (Arumughan et al. 2016). The two UBA-UBX cofactors p47 and FAF1 trimerize and bind only to every second N domain (Beuron et al. 2006; Ewens et al. 2014). The heterodimeric UFD1-NPL4 complex is a key protein complex involved in many cellular pathways where it recognizes polyubiquitylated proteins via specific UBDs. Only one UFD1-NPL4 heterodimer binds via the SHP motif of UFD1 and the UBXL domain of NPL4 to the p97 hexamer (Bebeacua et al. 2012; Pye et al. 2007). A 6:1 binding has also been shown for the DUB ATAXIN-3 (Rao et al. 2017). A bipartite binding mechanism by using two different binding modules like UFD1-NPL4 or the UBX proteins p47 and p37 with a UBX domain and a SHP-binding motif could increase the affinity but may also result in more conformational flexibility during catalysis due to different binding affinities of the individual binding motifs/domains. For example, for p37 it could be shown that the SHP motif provides the main interactions (Weith et al. 2018). In contrast to all other cofactors UBXD1 interacts via a bipartite interaction involving the N domain as well as the C-terminus (Hänzelmann and Schindelin 2011; Kern et al. 2009), specifically, UBXD1 associates via its PUB domain with the C-terminus and binds with its VIM motif to the N domain. It is currently unknown whether domains/motifs involved in bipartite binding associate with the same monomer or bind to adjacent monomers and whether they bind at the same time, which may also depend on conformational changes during ATP hydrolysis.

Earlier cryo-EM studies of the p97–UFD1-NPL4 complex revealed a high degree of nucleotide-dependent conformational flexibility (Bebeacua et al. 2012). Unfortunately, a recent cryo-EM structure of this complex from the thermophilic fungus *Chaetomium thermophilum* at a resolution of 4.3 Å did not resolve these open questions due to disorder of the N domains and UFD1 (Bodnar et al. 2018). Nevertheless, cryo-EM studies of the p97–p47, p97–FAF1 and p97–UFD1-NPL4 complexes revealed a common binding mode of these cofactors on top of the D1 ring (Fig. 7.2b, c) (Bebeacua et al. 2012; Beuron et al. 2006; Bodnar et al. 2018; Ewens et al. 2014). At present, only low-resolution models are available for the p97–p47 and p97–FAF1 complexes (Beuron et al. 2006; Ewens et al. 2014). Both cofactors trimerize, either via the p47 SEP domain or the coiled-coiled region of FAF1, and associate with their UBX domains with non-adjacent p97 N domains (Fig. 7.2c). The additional domains of both cofactors are flexible and reside on top of the assembly. The structure of *C. thermophilum* Cdc48–UFD1-NPL4 complex shows how NPL4 interacts with its UBXL domain with the N domain and assembles with its two Zn-finger containing ZF-NPL4 domain as well as its catalytically inactive MPN (Mpr1-Pad1 N-terminal) domain above the D1 ATPase ring (Fig. 7.2b) (Bodnar et al. 2018). Although both the Zn-fingers as well as the MPN domain are engaged in binding to the D1 domain, the interaction with the UBXL domain is the main contact site, and deletion of the UBXL domain completely abolishes binding. UFD1-NPL4 binds as a tower on top of Cdc48 with the NPL4 Zn-fingers and the MPN domain forming the base and the NPL4 C-terminal domain (CTD) on top. The central pore of p97 is barely contacted at all and would hence be available for interactions with substrates. UFD1 is composed of an N-terminal UBD (called UT3 domain) and a C-terminal UT6 domain, which harbors both NPL4 interaction segments as well as two SHP-binding motifs for interactions with Cdc48. The UT3 domain is disordered in the structure and based on some unassigned density and crosslinking experiments the UT6 domain is probably bound to the MPN domain. Although the last 36 amino acids of p97 are unstructured, in fact the C-terminal residues are highly flexible and not resolved in any p97 structure, the p97–cofactor structures described above could indicate that cofactors associated with the long flexible C-terminus could also act on substrates bound on top of the D1 domain.

Since only one UFD-NPL4 heterodimer per p97 hexamer is bound, further cofactors can bind to non-occupied N domains forming higher order complexes. Indeed, different UBX proteins are found together with UFD1-NPL4 (Alexandru et al. 2008), which have been shown to bind in a hierarchical manner to p97 with a 6:1:1 stoichiometry (Hänzelmann et al. 2011). The assembly and disassembly of p97 cofactors could be regulated by ATP hydrolysis, which is associated with the movement of the N domain from its up-position back to its position co-planar with the D1 domain. Altered cofactor interactions as well as cofactor-induced differential effects on the ATPase activity have been described for wild type p97 as well as variants involved in IBMPFD (Bulfer et al. 2016; Chia et al. 2012; Fernandez-Saiz and Buchberger 2010; Zhang et al. 2015). Most of the identified IBMPFD mutations are located in the interface between the N domain and the D1 domain. They are gain of function mutations leading to elevated ATPase levels and affect the usually tight control of N domain movement, thus leading to a more constitutively active up-conformation due a disturbed interdomain communication (for reviews see Schuetz and Kay 2016; Tang et al. 2010; Tang and Xia 2013). A further level of regulation are posttranslational modifications (PTMs) like ubiquitylation, SUMOylation and phosphorylation of p97 and its associated cofactors (for a review see Hänzelmann and Schindelin 2017).

p97–Cofactor Complexes Involved in Diverse Cellular Functions

p97s function in a large variety of different cellular functions is reflected in its association with different cofactors and the formation of transient multi-protein complexes with p97 acting as the platform for its regulatory cofactors. The benefits of such complexes are that through spatial proximity catalysis is faster, coordinated and selective. Furthermore, unstable intermediates are protected and not released. Dependent on the cellular pathway, p97 forms distinct complexes to perform different functions. Selected examples are described in the following part. Although the function of the p97–UFD1-NPL4 complex in different cellular pathways is well established, not much is known for most other cofactors including pathways they are involved in and the substrates being targeted.

p97 is a key player in maintaining protein hoemeostasis and plays multiple roles in PQC mechanisms like the UPS system and autophagy:

(i) p97–cofactor complexes in ERAD (Fig. 7.3a): Dependent on the misfolded protein to be degraded, the ERAD system can be divided into ERAD-L (lumenal proteins), ERAD-M (membrane proteins) and ERAD-C (cytosolic domain of membrane proteins) (for reviews see Christianson and Ye 2014; Wu and Rapoport 2018). For example, soluble secretory proteins that do not fold properly in the ER need to be retro-translocated through the membrane into the cytosol for degradation by the proteasome in the ERAD-L pathway.

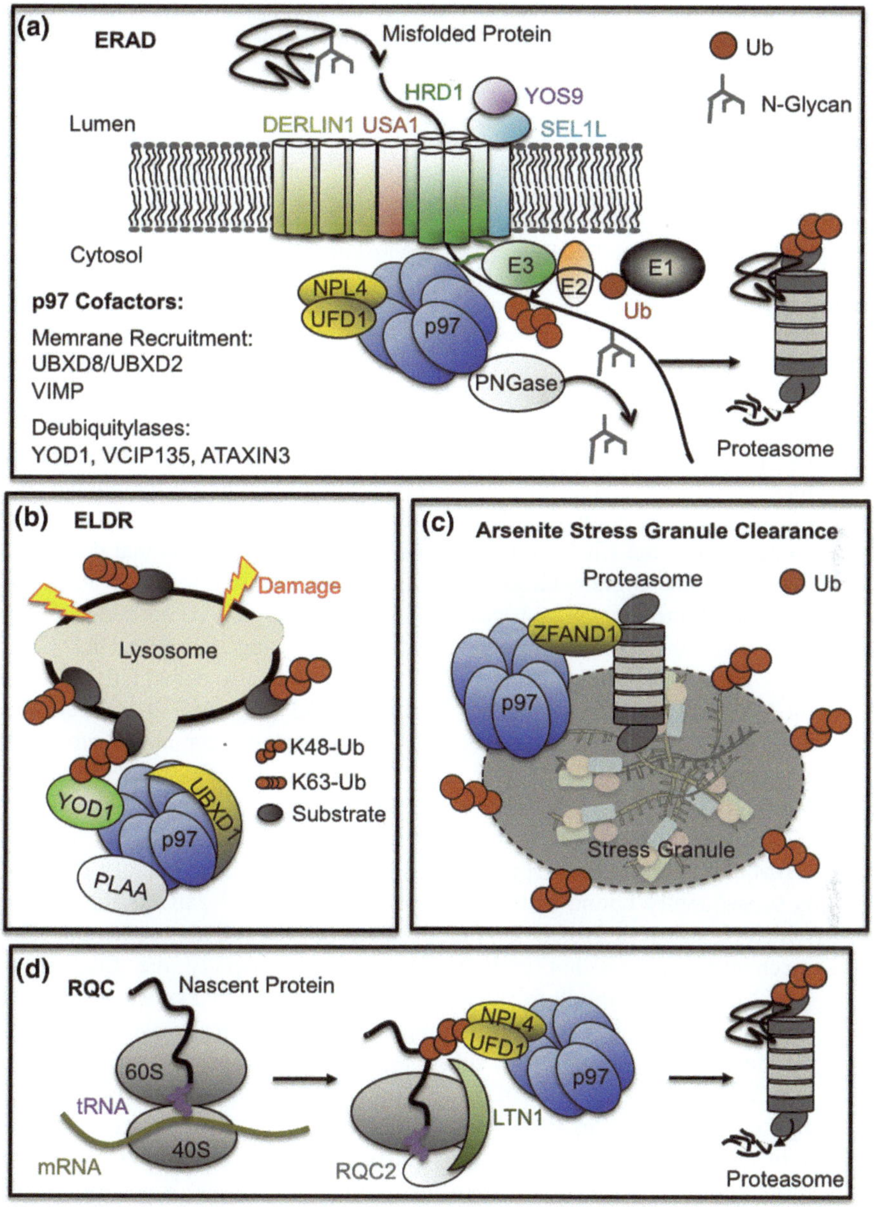
(a)
ERAD
Misfolded Protein
Ub
N-Glycan
HRD1
YOS9
Lumen
DERLIN1
USA1
SEL1L
Cytosol
E3
E1
E2
NPL4
UFD1
p97
Ub
p97 Cofactors:
Memrane Recruitment:
UBXD8/UBXD2
VIMP
Deubiquitylases:
YOD1, VCIP135, ATAXIN3
PNGase
Proteasome
(b)
ELDR
Damage
Lysosome
K48-Ub
K63-Ub
Substrate
YOD1
UBXD1
p97
PLAA
(c)
Arsenite Stress Granule Clearance
Proteasome
Ub
ZFAND1
p97
Stress Granule
(d)
RQC
Nascent Protein
60S
tRNA
40S
mRNA
NPL4
UFD1
p97
LTN1
RQC2
Proteasome

◀**Fig. 7.3** Schematic models for p97 function in PQC mechanisms. **a** Endoplasmic reticulum associated protein degradation (ERAD). Misfolded glycosylated luminal ER proteins are recognized by YOS9 and SEL1L and are inserted, with the help of DERLIN1, into the HRD1 protein conducting channel of the retro-translocation complex. At the cytosolic site proteins are polyubiquitylated by HRD1. The p97–UFD1-NPL4 complex is recruited to the membrane and pulls out the protein under ATP consumption. The polypeptide chain is subsequently deglycosylated by PNGase and delivered to the proteasome for degradation. **b** Endo-lysosomal damage response (ELDR). Upon lysosome damage lysosomal proteins are polyubiquitylated, either via K48 or K63 chains. At least some of the K48-linked ubiquitylated substrates are subsequently extracted by the ELDR complex composed of p97, UBXD1, YOD1 and PLAA. **c** Arsenite induced stress granule (SG) clearance. After arsenite stress ZFAND1 associates with SGs and both p97 and the proteasome are recruited for clearance of ubiquitylated SG proteins. **d** Ribosome quality control (RQC). Ribosomes stalled with a translocating polypeptide chain are recognized by RQC2 and polyubiquitylated by LTN1. Nascent chains are disassembled by the p97–UFD1-NPL4 complex and degraded

p97 plays an important role in the retro-translocation step by providing the mechanical force to pull the unfolded protein through the ER membrane once the polypeptide emerges on the cytoplasmic site. Misfolded proteins are recognized in the ER lumen by the lectin YOS9 and the luminal domain of the membrane protein HRD3 (also called SEL1L) (Carvalho et al. 2006). This multiprotein retro-translocation complex is composed of the multi-spanning membrane protein HRD1, which, together with HRD3, and other accessory proteins like DERLIN1 and USA1 (also called HERP) forms the protein-conducting channel (Baldridge and Rapoport 2016; Schoebel et al. 2017). Once the polypeptide chain reaches the cytoplasmic site it gets polyubiquitylated by the cytosolic RING-finger domain of the E3 ligase HRD1 and the p97–UFD1-NPL4 complex is targeted to the membrane where p97 can bind to the retro-translocon proteins HRD1 and DERLIN1 via their VBM and SHP motifs, respectively. The polyubiquitylated misfolded protein is subsequently bound to the UBD of UFD1 and p97 extracts the protein, which is then delivered to the proteasome via shuttle cofactors. This mechanism is still not entirely understood, especially the function of DERLIN1, which might facilitate the insertion of lumenal substrates into the channel and neither is the exact composition of the retro-translocon known. In addition, dependent on the substrate, further cofactors like UBX proteins (e.g. UBXD8 or UBXD2), DUBs (YOD1, VCIP135 and ATAXIN-3), the E3 ligase gp78 as well as PNGase, which is involved in the deglycosylation of substrates, might be involved. One further candidate is the ER membrane protein VIMP (SelS), which binds via its VIM motif to the N domain of p97, and has been shown to play a nucleotide-dependent essential role in p97 recruitment to the ER membrane (Tang et al. 2017). In summary, during ERAD p97 can associate via its N domain and its C-terminus with multiple cofactors, and it can be speculated that some of the cofactors will be bound simultaneously. Probably the whole process is highly dynamic with cofactors associating and dissociating in rapid succession. The p97–UFD1-NPL4 complex performs a similar function during mitochondria assisted degradation (MAD) where it extracts several K48-linked polyubiquitylated outer membrane proteins for proteasomal degradation (Heo et al. 2010; Xu et al. 2011).

(ii) p97–cofactor complexes in cytosolic PQC: The cytosolic hexameric chaperone BAG6 (BCL2-associated athanogene) displays a holdase activity, thus keeping proteins in an unfolded state to prevent their aggregation (Wang et al. 2011). It plays several roles together with p97 in PQC at the ER, either prior to translocation into the ER or during ERAD. For example, p97 and BAG6 are involved in the so-called ER stress induced pre-emptive quality control (ERpQC), a control mechanism, which attenuates the translocation of newly synthesized ER targeted proteins during ER stress to avoid an overload of proteins in the ER and reroutes them for proteasomal degradation, which also applies to mislocalized ER proteins. Newly synthesized and signal sequence-containing proteins emerging from the ribosome are captured at the translocon site by the C-terminal part of DERLIN1 and are ubiquitylated by HRD1. p97, which can associate with both DERLIN1 and HRD1, together with BAG6, which protects the unfolded protein from aggregation by interaction with hydrophobic segments, are then involved in the degradation step (Kadowaki et al. 2018). Several other p97 cofactors including the E3 ligase gp78, UBXD8 and the ubiquitin-binding membrane protein AIRAPL (arsenite-inducible RNA-associated protein-like) have been implicated in BAG6-mediated processes (Braunstein et al. 2015; Xu et al. 2013). Furthermore, p97 together with UFD1-NPL4 and its cofactor SAKS1 have been shown to be involved in the triage of ubiquitylated BAG6-bound cytosolic proteins (Ganji et al. 2018), indicating multiple PQC mechanisms in the cytosol.

(iii) The multi-protein endo-lysosomal damage response (ELDR) complex (Fig. 7.3b): p97 is involved in the recognition and the clearance of damaged lysosomes by selective macroautophagy, also called lysophagy, and, in particular, p97 is involved in the so-called endo-lysosomal damage response (ELDR) (for a review see Papadopoulos and Meyer 2017). Three different types of autophagy exist: Microautophagy, macroautophagy as well as chaperone-mediated autophagy. During autophagy, for example, aggregated or long-lived proteins and damaged organelles or cytosolic components, such as lipids, are engulfed by double membranous vesicles called autophagosomes, which subsequently fuse with the lysosome to form the autophagolysosome. After degradation of the engulfed components the products are recycled. Upon damage of lysosomes, triggered among other factors by pathogens, drugs, oxidative stress as well as proteases, and detection of the damage, autophagy is initiated by extensive K63-linked polyubiquitylation of lysosomal proteins and the subsequent recruitment of the autophagy receptor p62, which connects ubiquitylated proteins to LC3-decorated phagophores. However, to some extent also K48 ubiquitin conjugates were identified, which leads to the recruitment of p97 to damaged lysosomes for the extraction of K48-modified lysosomal proteins to promote lysophagy (Papadopoulos et al. 2017). To achieve this, p97 acts together with the DUB YOD1, UBXD1 and PLAA to form the so-called macromolecular ELDR complex. Depletion of these cofactors leads to the accumulation of K48-linked ubiquitylated damaged lysosomes. Similar to lysophagy, p97 also acts in the clearance of damaged mitochondria via mitophagy (for a review see Karbowski and Youle 2011). p97 is involved in the PINK1/PARKIN-mediated clearance of K48-linked polyubiquitylated mitofusins to prevent the fusion of fragments derived from damaged mitochondria, which would interfere with mitophagy (Tanaka et al. 2010). A similar function is conceivable for the action of p97 in lysophagy. Although not much is known about cofactors participating in mitophagy, recently, it could be shown that p97 is recruited via its UBXD1 cofactor to damaged mitochondria (Bento et al. 2018). In this process UBXD1 is translocated in a PARKIN-dependent manner to depolarized mitochondria via its UBX domain leading to the subsequent recruitment of p97.

(iv) p97 in association with the proteasome: The eukaryotic 26S proteasome is composed of a 20S core particle and a 19S regulatory particle (for a recent review see (Bard et al. 2018)). The 20S core particle consists of four stacked heptameric rings (two α- and two β-rings) forming a barrel. The proteolytic activity is located in the central cavity located in between the two β-rings. The two α-rings act as caps located on the top and bottom of the central β-rings and are involved in restricting access to the proteolytic chamber through a narrow channel. The 19S regulatory particle is composed of six different AAA+ ATPase proteins (RPT) forming a hexameric ring structure and 12 RPN proteins, which do not have ATPase activity and are involved in substrate binding/recognition and processing (e.g. ubiquitin binding, deubiquitylation). The AAA+ RPT-ring, which is attached to the α-ring via its C-terminal HbYX (Hb, hydrophobic; Y, tyrosine; X, any residue) motifs and regulates, via this interaction, the opening and closing of the α-ring channel, is involved in the unfolding of substrates and gates them to the α-ring. Interestingly, p97 also harbors a HbYX motif at its C-terminus and it has been suggested already some time ago that p97 could bind to the α-ring, thereby replacing the regulatory particle and forming an alternative proteasome (for a review see Esaki et al. 2018) (Barthelme and Sauer 2013). In archaea, which only have a 20S core particle and do not harbor a 19S regulatory particle, several proteasomal ATPases including VAT, the archaeal homolog of p97, can associate with the 20S proteasome, thereby increasing the capabilities of the archaeal proteasome in degradation processes (Forouzan et al. 2012). Although VAT can activate the proteolytic activity of the 20S proteasome, the interaction between Cdc48 and the 20S core particle is rather weak or transient and needs to be stabilized by cross-linking (Barthelme et al. 2014). In case of the 26S proteasome it is difficult to imagine that indeed the 19S regulatory particle dissociates to enable attachment of p97 to the 20S core particle. The direct association of p97 and the proteasome could be restricted to archaea, a function that might have been subsequently lost during evolution. How p97 substrates destined for degradation are delivered to the proteasome is still barely understood. In analogy to the yeast system, shuttling factors like yeast Rad23 and Dsk2, which feature UBA domains for binding to polyubiquitylated substrates and ubiquitin-like (UBL) domains for association with the proteasome, might be involved (for reviews see Finley 2009; Saeki 2017). In yeasts, both proteins transport more than 90% of substrates to the proteasome (Tsuchiya et al. 2017). However, a close contact of p97 and the proteasome for a direct delivery of substrates, possibly bridged by a cofactor, is also possible, at least for certain substrates. In line with this, the p97 cofactor ZFAND1 has been shown to regulate the turnover of arsenite-induced SGs (Turakhiya et al. 2018). After association with SGs both p97 and the proteasome are recruited for clearance of ubiquitylated SGs (Fig. 7.3c). ZFAND1 binds with its UBXL domain to p97 and associates via its Zn-finger domain to the proteasome. These data indicate that p97 and the proteasome, at least under certain conditions, can act together in a concerted manner to efficiently clear damaged proteins.

(v) p97–cofactor complexes in ribosome quality control (RQC) (Fig. 7.3d): Protein biosynthesis at the ribosome is error-prone and faulty RNA or an insufficient number of charged tRNAs causes ribosome stalling during translation (for review see Joazeiro 2017) leading to, for example, truncated proteins or damaged proteins, which aggregate and induce proteotoxic stress. To avoid this, cells use a so-called ribosomal rescue system to release the damaged mRNA followed by its degradation to recycle the ribosomes and to degrade the nascent polypeptide chain by the RQC machinery (for reviews see Brandman and Hegde 2016; Defenouillere and Fromont-Racine 2017). After the recognition of stalled ribosomes, the 80S ribosome is dissociated with the peptidyl-tRNA still bound to the 60S subunit. The tRNA is exposed at the 40S interface and is recognized by the RQC machinery protein RQC2, leading to the recruitment of the E3 ligase listerin 1 (LTN1), which polyubiquitylates the nascent polypeptide. After recruitment of the p97 - UFD1-NPL4 complex, and with the help of RQC1, the polyubiquitylated nascent chain is extracted and finally degraded by the proteasome (Brandman et al. 2012; Defenouillere et al. 2013; Verma et al. 2013).

p97 is a key regulator of chromatin-associated processes including transcriptional activation, DNA repair processes and replication (for a review see Franz et al. 2016a). Several studies describe the association of UFD1-NPL4 with UBX proteins in chromatin-related processes. UBA-UBX proteins can associate with a large range of E3 ligases including the multi-subunit culling ring ligases (CRL), the largest group of E3 ligases. In particular the p97–UFD1-NPL4–UBXD7 complex has been shown (Alexandru et al. 2008) to be involved in the degradation of the alpha subunit of the transcription factor hypoxia-inducible factor 1 (HIF1α). Under normoxic conditions HIF1α is polyubiquitylated by CUL2, which binds to UBXD7 and recruits p97–UFD1-NPL4 (den Besten et al. 2012). During DNA double strand break (DSB) repair the heterodimeric Ku70/80 complex senses DSBs and binds with its central cavity to the damaged DNA, thus forming a tightly bound and sterically trapped ring structure. This then initiates non-homologous end joining

(NHEJ) by the recruitment of additional NHEJ factors (for reviews see Schwertman et al. 2016; Torrecilla et al. 2017). Subsequently, trapped Ku70/80 is actively disassembled by the action of p97 (van den Boom et al. 2016). To achieve this an opening of the ring structure is necessary, in addition to K48 ubiquitylation. p97 also plays a role in UV-induced DNA damage repair (for reviews see Ruthemann et al. 2016; Zhang and Gong 2016). Bulky UV lesions are recognized by the proteins XPC and the DDB1-DDB2 complex, which initiate global nucleotide excision repair (GG-NER) (Fig. 7.4a). For the subsequent excision reaction, catalyzed by the TFIIH complex, both initiation factors need to be extracted. However, the fate of these two proteins is different. The DDB1-DDB2 complex recruits CUL4 forming the cullin-type $CRL4^{DDB2}$ ligase complex which modifies XPC with K48-linked ubiquitin chains resulting in enhanced binding to damaged DNA. XPC is then recognized by p97 and its cofactors UFD1-NPL4 as well as UBXD7 and is extracted from the DNA (Puumalainen et al. 2014). Although K48-linked chains are the typical degradation signal, XPC is not degraded by the proteasome, instead it is deubiquitylated and recycled (He et al. 2014). DDB2, which itself is a substrate of the $CRL4^{DDB2}$ ligase, also undergoes K48-linked polyubiquitylation, which reduces its affinity to DNA lesions leading to its extraction by p97. However, in contrast to XPC, DDB2 is targeted to the proteasome for degradation.

p97 also fulfills many different functions during replication where it is foremost involved in the regulation of replication licensing and fork progression as well as in the termination step (for reviews see Franz et al. 2016a; Ramadan et al. 2017). Replication is initiated by the origin replication complex (ORC)-mediated recruitment of the licensing factor CDT1, which binds to the ORC. Licensing of origins is achieved by the recruitment of the CDC45–MCM2-7–GINS (CMG) DNA helicase which unwinds the DNA and starts replication fork progression. To prevent relicensing CDT1 must be inactivated and is therefore polyubiquitylated by two CRLs. During this process p97, together with its cofactors UFD1-NPL4 and FAF1, is involved in the extraction of ubiquitylated CDT1 and its proteasomal degradation (Franz et al. 2011, 2016b) (Fig. 7.4b). In the termination step of replication, the replisome needs to be detached. This reaction is initiated by a CRL-dependent K48-linked polyubiquitylation of MCM7. p97, in complex with UFD1-NPL4, is involved in this key step of disassembling the ubiquitylated CDC45–MCM2-7–GINS (CMG) DNA helicase (Dewar et al. 2017; Gaggioli and Zegerman 2017; Maric et al. 2017; Moreno et al. 2014) (Fig. 7.4c). After deubiquitylation the CMG helicase complex dissociates.

p97 also functions in responses to replication stress, a cause of genomic instability, cancer and aging. A stalled replication machinery caused by bulky DNA lesions triggers, for example, translesion DNA synthesis (TLS) where, instead of the replicative DNA polymerase, a TLS polymerase bypasses this lesion. p97 together with its cofactor SPARTN is involved in this process called DNA polymerase switching (Davis et al. 2012; Ghosal et al. 2012; Mosbech et al. 2012). In addition to TLS p97 is involved in the Fanconi anaemia DNA repair pathway, which processes DNA interstrand crosslinks. p97 and SPARTN are also involved in the removal of ubiquitylated FANCI and FANCD2 from sites of damage (Gibbs-Seymour et al. 2015).

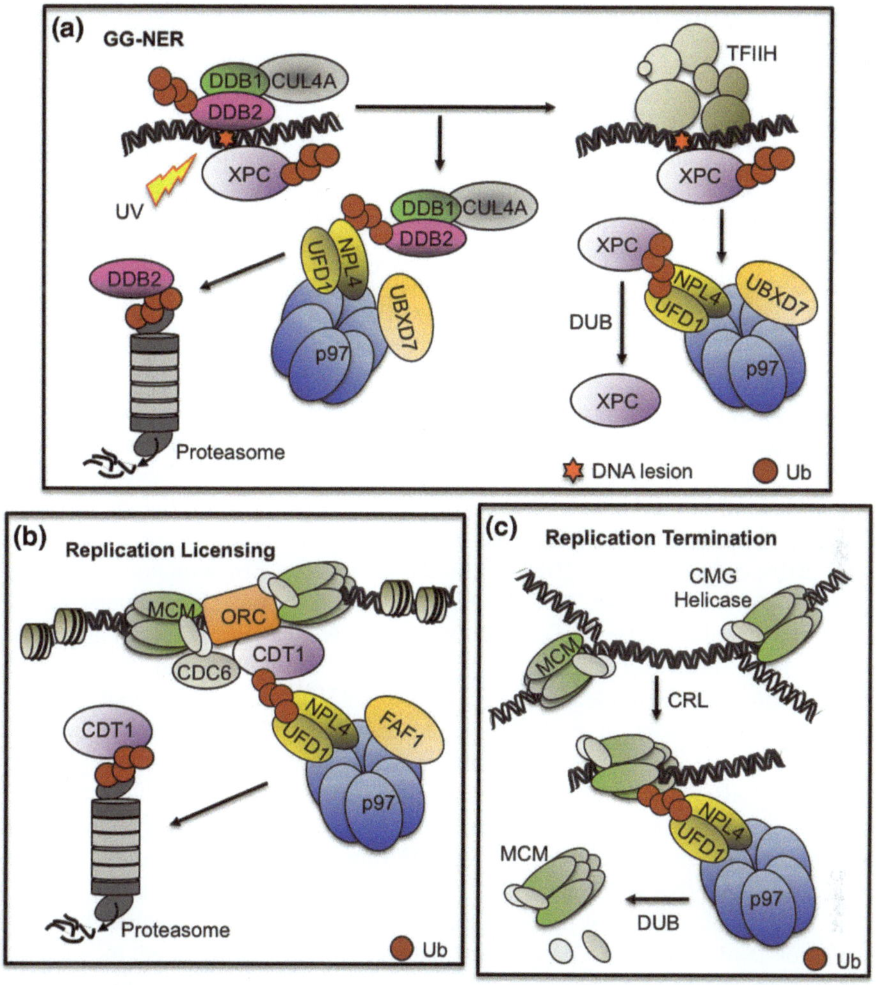

Fig. 7.4 Schematic models for p97 function in chromatin associated processes. **a** Global genome nucleotide excision repair (GG-NER). UV-induced DNA damage is recognized by the initiation factors XPC and the DDB1-DDB2 complex. For the subsequent excision reaction catalyzed by the TFIIH complex both initiation factors are K48-linked polyubiquitylated and extracted by the p97–UFD1-NPL4–UBXD7 complex. DDB2 is degraded by the proteasome whereas XPC is recycled after deubiquitylation. **b** Replication licensing. After recruitment of the CMG helicase the ORC-bound licensing factor CDT1 is polyubiquitylated and extracted by the p97–UFD1-NPL4–FAF1 complex for proteasomal degradation. **c** Replication termination. The p97–UFD1-NPL4 complex extracts K48-linked polyubiquitylated MCM7 form the DNA. After deubiquitylation the CMG helicase complex dissociates

Mechanistic Insights into p97 Function

How p97 in conjunction with its many different cofactors fulfills its function in such a large variety of cellular processes is currently under investigation. As p97 and its cofactors are not only involved in the degradation of polyubiquitylated substrates and their proteasomal targeting, but instead also play important roles in deubiquitylation and the functional release of proteins, this might indicate different mechanisms.

For a long time it had been suggested that, in analogy to other AAA+ ATPase and VAT, the archaeal homolog of p97 (for a review see Yedidi et al. 2017), that p97 also possesses an unfolding activity which involves a threading of the substrate through its central channel (Ye et al. 2003), however, how threading through the channel is achieved, has been hotly debated. p97 does not contain conserved aromatic residues at the entry site in the D1 domain, which could assist with the insertion of substrates. Based on structures of p97 in different nucleotide states the central channel is too narrow and, due to a restriction in the D1 domain by the His-gate, is not able to accommodate even a partly unfolded substrate (Banerjee et al. 2016; Schuller et al. 2016). In addition, in cryo-EM structures of p97 in complex with its cofactors UFD1-NPL4 (Fig. 7.5a) or p47 the diameter of this restriction is unaltered (Beuron et al. 2006; Bodnar et al. 2018). Different alternative models have been proposed (for reviews see Barthelme and Sauer 2016; Hänzelmann and Schindelin 2017). Since p97 harbors conserved putative substrate-binding loops in the D2 pore, in contrast to the D1 pore, it has been speculated that the substrate enters and exits the pore from the D2 site for processing (DeLaBarre et al. 2006). Alternatively, the conformational changes as well as the up-movement of the N domains during the catalytic cycle may result in the formation of a cleft between the D1 and D2 rings, which may represent an alternative access site (Na and Song 2016). However, concerning the threading model it could be possible that only in the presence of a ubiquitylated substrate bound to a cofactor conformational changes are induced which are required to open the access point in the D1 ring, thus enabling substrate entry. Typically, proteins destined for degradation are delivered to the proteasome where polyubiquitylated substrates are deubiquitylated and finally degraded, whereas p97 acts upstream of the proteasome. As mentioned above (see Section "p97–Cofactor Complexes Involved in Diverse Cellular Functions"), substrates destined for proteasomal degradation could be transported to the proteasome via shuttling cofactors and/or BAG6 or be directly delivered by p97, possible with the help of p97 cofactors, which raises the question whether p97 needs to act as an unfoldase. However, recent reports clearly show that p97 indeed can unfold substrates and is involved in ubiquitin-dependent substrate processing as well as ubiquitin-independent disassembly of macromolecular complexes as will be described below (Blythe et al. 2017; Bodnar and Rapoport 2017b; Weith et al. 2018).

Ubiquitin-Dependent Unfolding and Substrate Processing (see Fig. 7.5b): Two independent in vitro studies working either with yeast Cdc48 or human p97

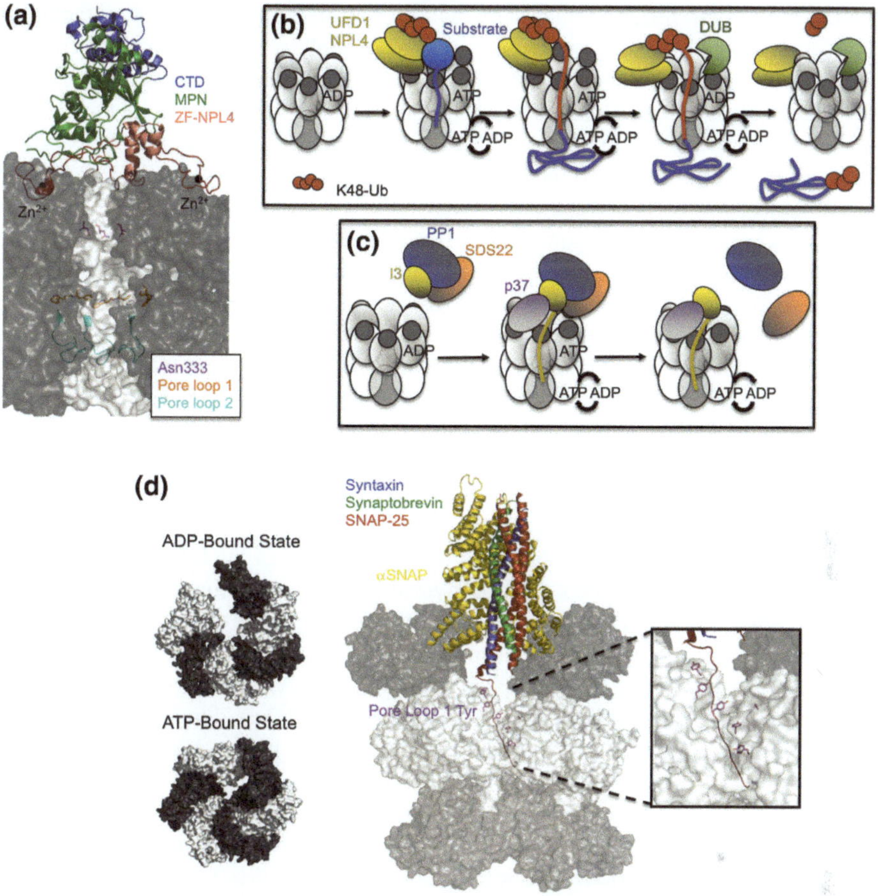

Fig. 7.5 Mechanistic insights into p97 action. **a** Molecular surface representation of the *C. thermophilum* Cdc48–UFD1-NPL4 complex (pdb: 6chs) viewed into the central channel. The restriction in the D1 ring formed by asparagine residues (replacing the His-gate in higher eukaryotes) and the pore loops 1 and 2 in the D2 domain are indicated. Binding of NPL4 on top of the D1 ring is shown in cartoon representation and Zn^{2+}-ions as black spheres. **b** Model for the ubiquitin-dependent unfoldase activity of p97. UFD1-NPL4 is recruited to p97 with the N domains in the up-position. The polyubiquitylated substrate bound to UFD1-NPL4 inserts into the central channel and is unfolded as well as translocated through the pore driven by ATP hydrolysis in the D2 domain. After ATP-hydrolysis in the D1 domain (down conformation of N domains) the polyubiquitin chain is cleaved partially by a DUB and the remaining ubiquitin molecules are pulled, together with the polypeptide, through the pore. **c** Model for the ubiquitin-independent disassembly activity of p97. p97 with the N domains in their up-conformation leads to the association of the PP1-SDS22-I3 complex via the p97 cofactor p37. I3 inserts into the p97 central pore and ATP hydrolysis generates the necessary force to disassemble I3 from PP1, thus leading to dissociation of SDS22. **d** Cryo-EM structures of NSF: Left, top views of a surface representation of NSF in the ADP- and ATP-bound states (pdb: 3j95 and 3j94) with monomers shown alternating in white and black. Right, NSF in complex with SNARE-SNAP (pdb: 6mdm): Surface representation of NSF in the ATP-bound state viewed into the central pore with a cartoon representation of SNARE-SNAP. Inset: The spiral arrangement of Tyr residues in pore loop 1 is shown as a zoom in

demonstrated that p97, together with its heterodimeric UFD1-NPL4 cofactor, is catalyzing the unfolding of polyubiquitylated model substrates and their translocation through the p97 channel (Blythe et al. 2017; Bodnar and Rapoport 2017b) (for recent reviews see Bodnar and Rapoport 2017a; van den Boom and Meyer 2018). The model substrates used in their studies were either a K48-linked polyubiquitylated 43 amino acid long degron sequence fused to GFP or the polyubiquitylated UFD pathway substrate UB^{G76V}-GFP containing either linear or branched chains. This reaction strictly depends on UFD1-NPL4 and K48-linked polyubiquitylated substrates with a minimum chain length of five ubiquitin molecules. UFD1-NPL4 cannot be replaced by p47 or UBXD7, cofactors with different cellular functions and the unfolding activity can be enhanced by using branched ubiquitin chains (Blythe et al. 2017). In the first step ATP-binding to the D1 domain triggers the movement of the N domains to their up-conformation leading to the recruitment of UFD1-NPL4 and binding of polyubiquitylated substrates to the p97-associated UFD1-NPL4 cofactor, in particular to their UBDs, which are either K48-linkage specific (UFD1) or linkage unspecific (NPL4) (Bodnar and Rapoport 2017b). Substrate-/polyubiquitin-binding leads to a stimulation of the ATPase activity in the D2 domain, accompanied by an inhibition of the D1 ATPase activity. Subsequently, repeated cycles of ATP hydrolysis in the D2 domain constitute the main driving force for threading substrates through the entire channel. The threading mechanism could be confirmed by photo-crosslinking studies using p97 variants modified with p-benzoyl-phenylalanine (pBPA) in the presence of ATP, which revealed that the substrate cross-links to both ATPase domains, the His-gate as well as the D2 pore loops. Furthermore, fusion of a protease domain to the D2 pore revealed substrate degradation in vitro, thus indicating that the substrate is translocated through the complete channel and is released at the D2 site. Probably during the translocation process ATP stays bound to the D1 domains with the N domains in the up-position. The release of the substrate is not spontaneous and in vitro the involvement of the DUB Otu1, which binds to the N domain and catalyzes the trimming of the polyubiquitin chain, could be demonstrated for yeast Cdc48 (Bodnar and Rapoport 2017b). Interestingly, not all ubiquitin molecules are cleaved off, and it could be shown by cross-linking that, after at least partial unfolding, the remaining ubiquitin molecules, are also translocated through the central pore. For this step ATP hydrolysis in the D1 domain is necessary leading to the down-conformation of the N domains, suggesting that the N domains in the up-position may hinder access of the DUB to the polyubiquitin attached to the substrate. In the human system the three DUBs YOD1 (ortholog of Otu1), ATAXIN-3 and VCIP135 have been identified to interact with the N domain of p97 and potentially are involved in this step. Ubiquitin chain trimming by DUBs is similar to the proteasome, where also DUBs act upstream of substrate translocation (for a review see Mevissen and Komander 2017). An analysis of gain-of function IBMPFD mutants with an increased ATPase activity and imbalanced cofactor interaction also showed an enhanced unfolding activity (Blythe et al. 2017). An uncontrolled unfolding would lead to damage and non-proper function of p97 and its cofactors.

Ubiquitin-Independent Disassembly Activity (see Fig. 7.5c): The p97 UBX cofactors p37 and p47 are involved in different cellular pathways than UFD1-NPL4 and play, for example, roles in membrane fusion and mitosis (Kondo et al. 1997; Uchiyama et al. 2006) (for a review see Stach and Freemont 2017). Both harbor, besides interaction modules for p97 (UBX domain and SHP binding motif), a central SEP domain. Mass spectrometry identified several SEP domain-containing proteins as interactors of protein phosphatase-1 (PP1) (Raman et al. 2015). The p97–p37 complex, in particular, has recently been shown to be involved in the biogenesis of PP1 (Weith et al. 2018), a protein involved in mitotic spindle orientation (Lee et al. 2018). PP1, as a multimeric protein, is composed of a catalytic and a regulatory subunit. To achieve its diverse functions PP1 acts on more than 200 different targeting or regulatory partners, which also confer substrate specificity. To fulfill its mitotic function PP1 forms, together with SDS22 and I3, a transient and inactive trimeric intermediate during its biogenesis. SDS22 and I3 are maturation factors and are involved in the stabilizing of folding intermediates or are required to keep PP1 in an inactive state until it is required for processing of its substrates (for a review see Verbinnen et al. 2017). To be active PP1 needs to be liberated from its complex with SDS22 and I3 to allow it to associate with other partners, an essential step for cell viability and cell cycle regulation.

Recently, it could be shown that the p97–p37 complex as well as p97 in complex with UBXD4 can catalyze this reaction, indicating that these cofactors are partially redundant (Weith et al. 2018). The other two SEP domain-containing cofactors p47 and UBXD5, which contain additional domains exhibited only weak or no interaction at all, indicating that these additional domains exhibit different substrate specificities, hence the major catalytic function is derived from the p97–p37 complex. The disassembly is ubiquitin-independent as p37 does not harbor UBDs and depends on the ATPase activity of p97 since the inactive complex is stabilized in a catalytically inactive variant of the D2 ATPase domain (E578Q) and can be blocked by p97 inhibitors. To disassemble the complex p37 associates mainly via its SHP binding motif and, to a lower extent, via its UBX domain with p97 as well as via its SEP domain with I3. The I3 interaction is p97-dependent and only occurs within the trimeric PP1–SDS22–I3 complex. To identify whether I3 is threaded through the central channel of p97 photo cross-linking studies with pBPA cross-linked to different positions of the central pore were performed. These studies revealed that I3 is crosslinked to both the D1 and D2 pore showing that I3 is indeed pulled into the central pore, which possibly leads to the dissociation of I3 from PP1. UFD1-NPL4, which functions in ubiquitin-dependent unfolding reactions (see above), cannot catalyze this reaction. Fusing I3 to a fluorescent reporter indicated that I3 is unfolded and threaded through the central channel. How SDS22 is finally disassembled is currently unknown, possibly the elimination of I3 destabilizes the interaction of SDS22 with PP1 or, after dissociation of I3, SDS22 is exchanged by other proteins.

The studies described above suggest that p97 uses a common threading mechanism to generate the force necessary for different functions. Besides ubiquitin-dependent unfolding of substrates with the help of UFD1-NPL4 (Blythe et al. 2017;

Bodnar and Rapoport 2017b), p97, together with other cofactors like p37, can catalyze ubiquitin-independent protein complex disassembly (Weith et al. 2018). These data clearly show that the diameter of the central pore needs to increase to accommodate both the substrates and the associated ubiquitin polypeptides. Maybe this indicates that p97, like other unfolding machines such as the proteasome regulatory particle or VAT (Lander et al. 2012; Ripstein et al. 2017), can adopt different conformations, e.g. a spiral staircase arrangement of its ATPase domains. This would lead to different heights of both substrate-processing loops resulting in a bigger substrate-processing chamber to achieve translocation. A threading mechanism will create a pulling force important for extraction of proteins from membranes as for example in ERAD. It could be also involved in the pre-processing of stably folded substrates for downstream proteasomal degradation. In contrast to the proteasome, which needs unstructured regions to be recognized, there is currently no evidence that p97 needs unstructured regions to bind to its substrates. Whether there is a global unfolding or only a partial unfolding is still being debated (for a review see van den Boom and Meyer 2018). Global unfolding of substrates destined for proteasomal degradation implies that refolding is required to further transport substrates via shuttle factors to the proteasome and that substrates need to be ubiquitylated again to be recognized by the proteasome. Partial unfolding would avoid refolding and would be especially beneficial for proteins acting downstream of p97, which would argue against a common threading mechanism. Dependent on the substrate only partial unfolding and insertion into the pore could occur, a so-called hybrid model (for a review see Barthelme and Sauer 2016). The decision of partial or global unfolding most likely depends on the cellular pathway and the particular cofactor/substrate associated with p97.

One of the major unanswered questions is whether there is indeed a common threading mechanism, irrespective of the substrate, or do substrate- and cofactor-dependent alternative mechanisms exist? Specifically, how does p97 catalyze the disassembly of macromolecular complexes? For p47, which has an additional UBA domain and therefore probably has a ubiquitin-dependent disassembly activity, it has been suggested that the force generated by ATP-dependent movement of the N domains is transferred to the substrate in analogy to NSF (Beuron et al. 2006; Rouiller et al. 2000). Recently, several structures of AAA+ proteins in complex with substrates became available including the disaggregases Hsp104 and ClpB (Deville et al. 2017; Gates et al. 2017), the archaeal VAT (Ripstein et al. 2017), Vps4 (vacuolar protein sorting-associated protein 4) (Han et al. 2017), which is involved in the disassembly of the ESCRT-III (endosomal sorting complex required for transport) complex in membrane fission, TRIP13 (pachytene checkpoint protein 2 homolog) (Alfieri et al. 2018), which is involved in the regulation of spindle assembly checkpoint as well as NSF (White et al. 2018). All these ATPases perform substrate-remodeling similar to p97. Interestingly, common to all structures is that during catalysis the ATPase rings undergo large conformational changes leading to an asymmetric helical architecture including both open and closed spiral conformations (Fig. 7.5d). All structures are consistent with the generalized hand-over-hand model in which substrates are translocated in an ATP-dependent and coordinated manner. These arrangements are typically already observed in the absence of substrates, whereas such changes have not been observed for other AAA+ ATPases including p97 and PEX1/PEX6, the latter being involved in the transport of folded proteins into peroxisomes during their biogenesis (for a review see Saffert et al. 2017), which both show a co-planar conformation of their ATPase rings.

NSF is a protein involved in membrane fusion events and disassembles stable four-helix bundle SNARE (soluble N-ethylmaleimide-sensitive factor attachment protein receptor) complexes. NSF binds, together with several molecules of SNAP (soluble N-ethylmaleimide-sensitive factor attachment protein), to the SNARE complex forming a 20S complex, which is then disassembled in an ATP-dependent manner that does not involve unfolding (for a review see Zhao and Brunger 2016). Recent cryo-EM studies of the neuronal SNARE complex (syntaxin, synaptobrevin, SNAP-25) in complex with two αSNAPs and NSF revealed that the N-terminal residues of SNAP-25 insert into the spiral shaped ATP-bound hexamer and interact with conserved aromatic residues including a Tyr of the pore loops in the D1 domain (White et al. 2018) (Fig. 7.5d). It should be noted that in NSF, in contrast to p97, substrate-processing loops are present in the D1 domain but absent in the D2 domain, which does not open upon nucleotide-binding. The two αSNAPs are bound to the N domains and position the SNARE complex, which is not in contact with the N domain, above the D1 ring. Binding of the SNARE-SNAP complex enhances the ATPase activity of NSF, and it has been suggested that the resulting conformational changes in the D1 ring translate to the N domain with the attached complex and that N domain movement could therefore provide an additional force during disassembly (Zhao et al. 2015). Such a scenario would also be conceivable for p97 and a translocation through the entire central pore would hence not be necessary. To clarify all these unresolved questions structures of p97 with different cofactors and their physiological substrates are necessary, which will help to decipher conformational changes associated with substrate threading.

p97 as a Potential Drug Target in Cancer Therapy

Targeting PQC Pathways in Cancer Therapy

The UPS is essential in protein homeostasis (for reviews see Dikic 2017; Pilla et al. 2017). Dysregulations of protein homeostasis and proteotoxic stress are often involved in the development of cancer as well as neurological diseases, and targeting the UPS in cancer therapy is a well-established approach (for a review see Bastola et al. 2018). An imbalance in protein homeostasis can be caused by many genetic and epigenetic alterations as well as changes in the cancer genome such as altered copy numbers, mutations, duplication of chromosomes and aneuploidy leading for example to unbalanced protein synthesis and the accumulation of misfolded proteins. Therefore, the survival and proliferation of cancer cells largely relies on PQC mechanisms like the UPS, autophagy and chaperones such as the HSP70/HSP90 system (Fig. 7.6), which is involved in the refolding of misfolded proteins, to reduce proteotoxic stress (for a review see Bastola et al. 2018). Consequently, proteins involved in the UPS are quite often upregulated in a range of cancer cells compared to normal cells. PQC mechanisms have a dual role in cancer. On the one hand they display a tumor-suppressing function to minimize genome damage while on the other hand they exhibit a tumor-promoting function in cancer cell survival and proliferation.

Ubiquitylation is a sequential multistep process involving the action of ubiquitin-activating enzymes (E1), ubiquitin-conjugating enzymes (E2) and ubiquitin ligases (E3) (Fig. 7.6) (for reviews see Cappadocia and Lima 2018; Hänzelmann et al. 2012; Kerscher et al. 2006). In the first step the C-terminus of ubiquitin is activated in an ATP-dependent manner by forming a thioester bond with the E1 enzyme. The activated ubiquitin is then transferred to the active site Cys of an E2 enzyme and, finally, in an E3 ligase-dependent reaction is conjugated to substrates. In humans two E1s, about 40 E2s and several hundred E3 ligases, which mediate substrate specificity, are involved. The monoubiquitylated protein then can run several times through this process getting multi-monoubiquitylated or poly-ubiquitylated,

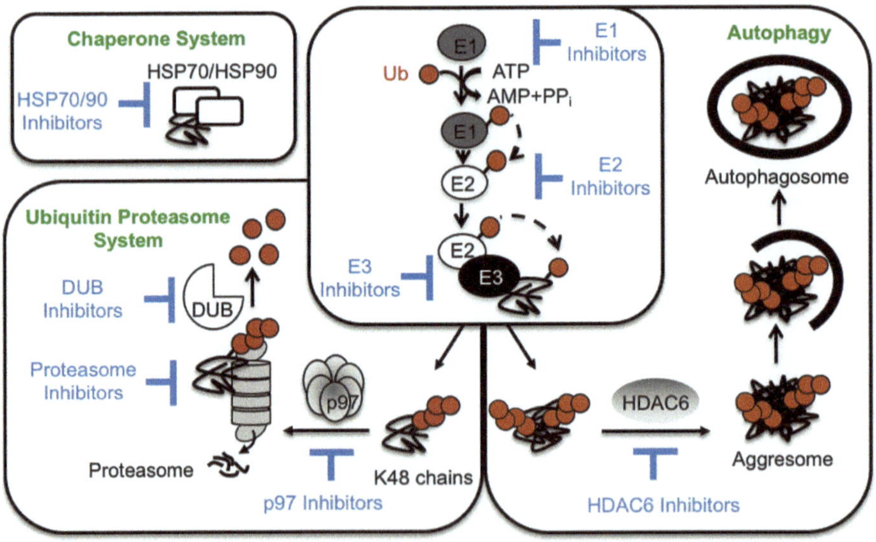

Fig. 7.6 Targeting PQC mechanisms for cancer therapy. The UPS and autophagy both depend on ubiquitylation, whereas the HSP70/HSP90 chaperone system is involved in the refolding of misfolded proteins. E1, E2 and E3 of the ubiquitylation cascade present independent sites for pharmacological intervention. In the UPS branch most commonly proteins with K48-linked chains are targeted, with the help of p97, for proteasomal degradation after deubiquitylation. In the autophagy branch larger protein aggregates are targeted, with the help of HDAC6, for subsequent clearance by autophagy and lysosomal degradation. Points of inhibition are indicated

resulting in the formation of linear, branched or forked chains, which can be either homotypic or heterotypic (for reviews see Akutsu et al. 2016; Yau and Rape 2016). Depending on the fate and function of the protein ubiquitin is conjugated via one of its seven lysine residues (K6, K11, K27, K29, K33, K48, and K63) or its N-terminal methionine. In the UPS K48-linked poly-ubiquitylation is the major signal for protein degradation via the proteasome. Before proteasomal degradation the ubiquitylation of substrates can be counteracted by DUBs (Fig. 7.6), which have essential functions in ubiquitin recycling, editing of ubiquitin chains and the regulation of activity, localization and stability (for reviews see Husnjak and Dikic 2012; Mevissen and Komander 2017; Sahtoe and Sixma 2015). E1s, E2s, E3s, DUBs as well as the proteasome have been shown to be druggable targets (for reviews see Ao et al. 2017; Bastola et al. 2018; Huang and Dixit 2016; Kumari et al. 2018; Wertz and Wang 2019) (Fig. 7.6).

Inhibition of the proteasome leads to the accumulation of ubiquitylated proteins destined for proteolytic degradation. Several proteasome inhibitors have been approved by the FDA (Food and Drug Administration) for the treatment of multiple myeloma (MM), which, however, non-specifically inhibit protein degradation. Due to their toxicity, these drugs cannot be used in the treatment of solid tumors, which require high drug concentrations (for a review see Manasanch and Orlowski 2017). Since the proteasome is also essential in normal cells, a more specific inhibition by targeting for example upstream components such as the E1s, E2s, and, in particular, E3s as well as DUBs is necessary, which potentially affect fewer proteins and would be probably less toxic. Over the last years also p97, as a key regulator of protein homeostasis, has been described as a potential target for cancer therapy (for a review see Deshaies 2014). Elevated expression levels of p97 in many types of cancer have been shown to correlate with poor prognosis in cancer as well as metastasis due to an anti-apoptotic function of p97 (for reviews see Fessart et al. 2013; Lan et al. 2017). Inhibition of p97 induces, through the accumulation of ubiquitylated proteins destined for degradation, proteotoxic ER stress and the unfolded protein response (UPR), which subsequently leads to apoptosis in cancer cells and decreases tumor cell proliferation and survival, thus indicating a potential neoplastic activity.

Beside the UPS, autophagy is the other major quality control system involved in protein homeostasis regulating tumor promoting as well as tumor suppressing pathways (for a review see Bastola et al. 2018) (Fig. 7.6). Both systems depend on ubiquitylation but use different mechanisms, however, a cross-talk and interplay between both systems has been shown (for a review see Kocaturk and Gozuacik 2018). Impairment of one of the systems leads to the upregulation of the other, and both systems complement each other. p97 and the cytoplasmic histone deacetylase 6 (HDAC6) have been suggested to be involved in this connection and are important in the decision whether degradation proceeds via the UPS or autophagy (for a review see Dargemont and Ossareh-Nazari 2012). HDAC6, a microtubule associated deacetylase with a Zn-finger ubiquitin-binding domain (ZnF-UBD), plays an important role in PQC when misfolded proteins cannot be eliminated by the chaperone refolding system or the UPS (Lee et al. 2010). HDAC6 deacetylates tubulin and mediates the retrograde transport of ubiquitylated protein aggregates to microtubules and the inclusion in aggresomes for subsequent clearance by autophagy and lysosomal degradation. p97 can use its disassembly activity to extract HDAC6 from polyubiquitylated proteins for proteasomal degradation. However, for larger protein aggregates the degradation pathway is shifted and HDAC6 is involved in aggresome formation and autophagy. HDAC6 is overexpressed in hematologic malignancies including MM and several inhibitors have been developed. Depending on whether the catalytic domain (deacetylation function) or the ZnF-UBD domain is targeted, HDAC inhibitors either antagonize the transport of aggregated proteins to the aggresome or to microtubules (Ferreira de Freitas et al. 2018; Mishima et al. 2015). Inhibition of HDAC6 in the aggresome-proteolysis pathway could be a good strategy to treat MM, especially to overcome resistance to proteasomal inhibitors used for the treatment of MM (for a review see Wallington-Beddoe et al. 2018). Accordingly, a combined treatment of proteasomal and HDAC6 inhibitors has been used in the treatment of MM (Mishima et al. 2015). A simultaneous inhibition of both the UPS and lysosomal degradation will increase cell stress and the accumulation of misfolded proteins followed by apoptosis.

p97 Inhibitors

Several potent and specific p97 inhibitors have been identified over the years (for reviews see Chapman et al. 2015; Vekaria et al. 2016) but only one inhibitor has reached the stage of clinical I trials so far. Two different classes of p97 inhibitors, which both target the D2 domain have been developed, ATP-competitive and, more recently, allosteric inhibitors. They will be described in more detail below, with a focus on high potency inhibitors (Fig. 7.7a).

ATP-Competitive Inhibitors Targeting the D2 ATP-Binding Site

High-throughput screening (HTS) efforts using an ATPase assay identified lead scaffolds like the quinazolines. DBeQ (Chou et al. 2011), the first reversible ATP-competitive inhibitor, which targets both ATPase domains with an IC_{50} of 2.6 μM, then served as starting point for inhibitors with improved specificity for the D2 ATPase domain and higher potency like ML240 and ML241 (IC_{50} = 0.11 μM) (Chou et al. 2013) and, finally, led to the potent inhibitor CB-5083 (Zhou et al. 2015), which progressed to early-phase clinical I trials where it has been tested for relapsed and refractory MM and advanced solid tumors (ClinicalTrials.gov Identifier: NCT02243917 and NCT02223598). CB-5083 (Cleave Biosciences, Inc.) is a reversible orally bioavailable highly potent inhibitor (IC_{50} = 0.011 μM) with high selectivity for p97 against other ATPases and kinases (Anderson et al. 2015). It specifically targets the ATPase site in the D2 domain. Based on molecular modeling and side-chain structure-affinity relationship (SAR) studies the putative binding site partially overlaps with the adenosine-binding site in the D2 domain, but points, compared to the phosphate groups of ADP/ATP, in the opposite direction into a

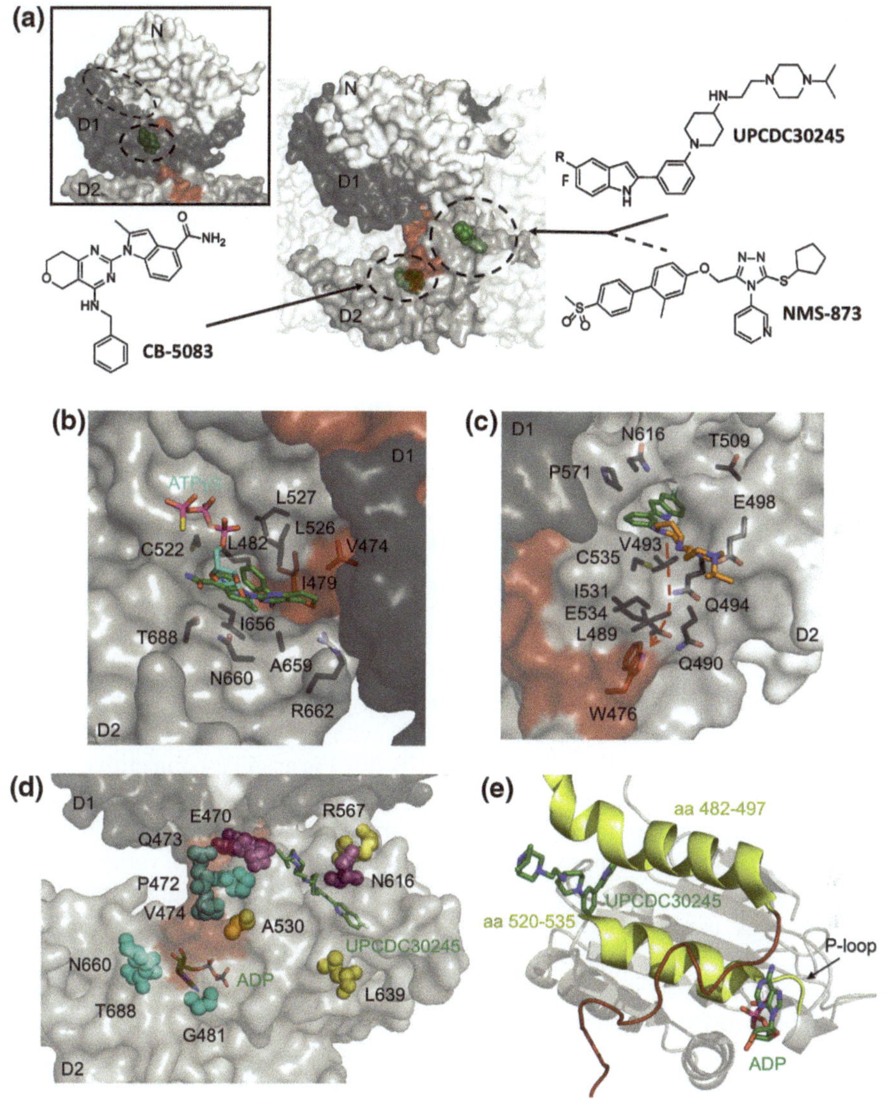
(a)
N
D1
D2
UPCDC30245
NMS-873
CB-5083
(b)
D1
ATPγS
L527
L526
V474
C522
L482
I479
I656
T688
N660
A659
R662
D2
(c)
D1
N616
T509
P571
E498
V493
C535
I531
E534
L489
Q494
Q490
W476
D2
(d)
D1
E470
R567
Q473
P472
N616
V474
A530
UPCDC30245
N660
ADP
T688
G481
L639
D2
(e)
aa 482-497
UPCDC30245
aa 520-535
P-loop
ADP

◂**Fig. 7.7** Inhibition of p97. **a** Target sites for ATP-competitive (CB-5083) and allosteric (UPCDC30245 and NMS-873) inhibitors in the D2 domain. Surface representation of p97 colored by domains (pdb: 5ftj). UPCDC30245 and ADP are shown as green spheres and the D1D2 linker in red. The inset shows potentialy druggable target sites in ND1: ATP binding site and allosteric binding site in the ND1 interface. ADP is shown as green spheres. **b** Binding site of CB-5083 (green C-atoms) with interacting residues (pdb: 6mck) in stick representation together with a surface representation of the D2-nucleotide binding site. For comparison ATPγS (pdb: 5ftn; C-atoms in cyan) has been superimposed onto the p97-CB-5803 complex. The D1D2 linker is highlighted in red. **c** Molecular details of UPCDC30245 binding to the allosteric binding site. Cryo EM structure of p97 in complex with UPCDC30245 (pdb: 5ftj). Carbon atoms of the disordered side chain of the inhibitor are colored in orange. The proposed alternative binding site, based on molecular modeling and side-chain SAR studies (LaPorte et al. 2018), is indicated with a dashed arrow. Residues identified to be important for binding are shown in stick representation together with a surface rendering of p97. The D1D2 linker is highlighted in red. **d** Identified mutations in resistant cancer cell lines. Surface representation of p97 in complex with UPCDC30245 and ADP in stick representation (pdb: 5ftj). Residues implicated in CB-5083 and NMS-873 resistance are shown as cyan and yellow spheres, respectively. Mutations at codons 470 and 616 are shown as purple spheres. The D1D2 linker is displayed in red. **e** Interconnection of the nucleotide and allosteric binding sites. Cartoon representation of the D2-ADP and UPCDC30245 binding sites (pdb: 5ftj). Structural elements that connect both binding sites are colored in yellow and the D1D2 linker is shown in red. UPCDC30245 and ADP are displayed in stick representation

different binding pocket located in close proximity to the D1D2 linker (Fig. 7.7b) (Zhou et al. 2015). This binding mode could be confirmed by a crystal structure (Tang et al. 2019), however, recent cellular and modeling studies suggest that CB-5083 additionally can bind into the allosteric binding site formed at the D1D2 interface next to the D1D2 linker (see below) with similar thermodynamic favorability (Bastola et al. 2017), indicating that CB-5083 is an ATP-competitive as well as an allosteric inhibitor. However, to fully understand its inhibitory properties further investigations are needed. The recent crystal structure does not rule out the possibility that CB-5083 binds to both sites, since an inactive p97 variant lacking the N domain and containing a nucleotide free D1 domain, which affects the D1D2 interface, was used. CB-5083 activates the UPR even stronger than the FDA approved proteasomal inhibitor bortezomib (Millennium Pharmaceuticals). Furthermore, a broad and strong anti-tumor activity in hematological and solid tumor models (MM, lymphoma, lung and colorectal cancer) could be demonstrated (Anderson et al. 2015; Gareau et al. 2018; Gugliotta et al. 2017; Le Moigne et al. 2017). CB-5083 displays antileukemic activity and synergetic effects with drugs that represent the backbone in the treatment of B-acute lymphoblastic leukemia (B-ALL) (Gugliotta et al. 2017). Preclinical studies with MM models showed that the p97 inhibitor CB-5083 inhibits tumor growth and can also be combined with MM standard-of-care agents (Le Moigne et al. 2017). However, phase I trials of this first-generation drug have been discontinued due to off-target inhibition of retinal PDE6 (cGMP-specific 3′,5′-cyclic phosphodiesterase), which is involved in phototransduction in the visual process. Based on this first-generation inhibitor a second-generation inhibitor (CB-5339) has already been developed, which displays a 40-fold reduced effect on PDE6 as well as improved physiochemical and in vivo drug metabolism and pharmacokinetic (DMPK) properties (Doroshow 2018).

Reversible Allosteric Inhibitors Targeting the Intersubunit Signaling Pathway

Dependent on the scaffold different types of allosteric inhibitors have been identified:

(i) Triazole-based inhibitors: Based on HTS the first identified series of allosteric inhibitors were based on alkylsulfanyl-1,2,4-triazoles scaffolds including the intensively studied and highly potent inhibitor NMS-873 (Magnaghi et al. 2013; Polucci et al. 2013). NMS-873 inhibits the ATPase activity with an IC_{50} of about 0.03 μM in an in vitro ATPase assay and exhibits a sub-micromolar antiproliferative activity in tumor cell lines. Based on biochemical studies (mutational studies, cross-linking and modeling) the binding site of NMS-873 was identified in a cavity between the D1 and D2 domain as well as the D1 domain of the adjacent monomer involving the D1D2 linker as well as Lys615 and Asn616 (Magnaghi et al. 2013). Unfortunately, NMS-873 is only poorly soluble and displays metabolic instability, thus hindering its usage in cancer therapy, nevertheless this compound has been used in many biochemical studies on p97 function.

(ii) Phenylindole-based inhibitors: In HTS approaches the indole amide-based scaffold which is characterized by excellent pharmacological properties, was identified (Alverez et al. 2016). It has been further optimized to 2-phenylindole-based inhibitors like UPCDC30245, which inhibits p97 with an IC_{50} of about 0.055 μM in an in vitro ATPase assay (Burnett et al. 2017). The p97-binding site could be identified in a cryo-EM structure of UPCDC30245 in complex with ADP-bound p97 (Banerjee et al. 2016) (Fig. 7.7c). Based on this structure the postulated allosteric binding site of NMS-873 could be roughly confirmed. The inhibitor is bound in the intramolecular cleft between the D1 and D2 domains in close proximity to the D1D2 linker, however, binding of the inhibitor does not lead to conformational changes including the nucleotide-binding site. The inhibitor-binding mode is conformation-selective, and the inhibitor is no longer able to bind to this cavity in the ATP-bound state of p97, suggesting that p97 needs to be in a D1-ADP locked state with the N domains in the down-position to allow inhibitor-binding. However, the cryo-EM structure of p97 in complex with UPCDC30245 most likely only reflects the correct interaction site for the phenyl indole component and not for the side chain, which is disordered (Banerjee et al. 2016). The phenyl-indole component binds into a deep pocket of the allosteric binding cleft while the side chain points away from the protein and stretches into the solvent. Molecular modeling based on SAR studies and placing the piperidine ring in a different conformation indicates that the side chain is oriented towards the protein and is engaged in several hydrogen bonds (LaPorte et al. 2018) (Fig. 7.7c); the piperazine ring is in contact with the side chains of Gln494 and Glu534 and the piperidine ring with the side chains of Cys535 and Glu498. Based on this model the phenyl indole amide scaffold has been further optimized (LaPorte et al. 2018). In this new series of inhibitors, the phenyl indole amide scaffold remains unchanged and the side chain is optimized by incorporation of small cyclic moieties like cyclobutane as kinks in the flexible side chain to obtain a more constrained conformation. These new inhibitors are highly potent with IC_{50} values of about 0.02 μM (LaPorte et al. 2018). Compared to NMS-873, UPCDC30245 and these optimized allosteric inhibitors exhibit higher solubilities and significantly improved metabolic stabilities. Interestingly, all phenyl indole amide-based inhibitors exhibit an uncompetitive inhibitory mechanism and bind to p97 in the ADP-bound state, whereas the triazole-based NMS-873 is a non-competitive inhibitor and binds in the absence of nucleotide. These properties indicate that, despite the fact that both inhibitor classes target the D1-D2 interface, different binding modes exist. Nevertheless, despite these positive features, the phenylindole-based inhibitors show only modest effects in cell-based assays measuring the accumulation of ubiquitylated proteins compared to NMS-873. An explanation of this discrepancy between the in vitro and in vivo activities could be that uncompetitive inhibitors are less efficient compared to ATP-competitive and non-competitive inhibitors. Another likely possibility, which is currently under investigation, is that in the Ub^{G76V}-GFP cellular assay being employed not all p97 relevant functions are recapitulated, for example, only autophagy is inhibited. Nevertheless, these inhibitors show an anti-proliferative activity in human tumor cell line screens similar to NMS-873.

Resistance to Inhibitors

Similar to proteasome inhibitors, several reports revealed that prolonged treatment of cancer cells with cytotoxic concentrations of p97 inhibitors leads to mutations resulting in resistance, thus suppressing cytotoxicity (Anderson et al. 2015; Bastola et al. 2017; Her et al. 2016; Wei et al. 2018). This has been reported for both types of inhibitors, ATP-competitive (CB-5083) as well as allosteric (NMS-873) compounds. The mechanism how resistances develop is still unknown, however, a correlation between resistance and high p97 copy numbers as well as expression levels are observed, thus indicating that p97 is involved in cancer progression (Anderson et al. 2015).

Analysis of resistant cell lines revealed that treatment with the ATP-competitive inhibitor CB-5083 leads to single homozygous mutations (P472L, Q473P, V474A, G481A) located in the D1D2 linker region as well as in the proposed CB-5083 binding pocket (N660 K, T688A) (Anderson et al. 2015; Wei et al. 2018) (Fig. 7.7b + d). Mutations in the D1D2 linker will affect interprotomer motion transmission, a mechanism where nucleotide-induced conformational changes in the D2 domain are transmitted via the D1D2 linker to the D1 domain of the adjacent monomer (Huang et al. 2012; Li et al. 2012). In addition, in ovarian cancer cells also heterozygous mutations with point mutations at codon 470 (E470K/D located in the D1D2 linker) on one allele leading to enhanced ATPase activity and inactivating mutations at codons 603/616 (nonsense mutation Q603* and frameshift stop mutation N616Mfs*63) on the other allele have been identified (Bastola and Chien 2018; Bastola et al. 2017). The latter mutations result in a reduced expression of p97 since one allele is not expressed. Modelling studies suggest that the linker residue Glu470 is in close proximity of the proposed allosteric binding sites of CB-5083 and NMS-873 (Bastola et al. 2017). Interestingly, analysis of cancer samples revealed that frameshift mutations at codon 616 are the most frequently observed hotspot mutations indicating that patients harboring this kind of mutation are not amenable to the treatment with these inhibitors.

Analysis of cells with resistance against the allosteric inhibitor NMS-873 identified the point mutations P472L, A530T, R567H, and L639F (Her et al. 2016; Wei et al. 2018) (Fig. 7.7d). The heterozygous mutation A530T does not directly affect the inhibitor-binding site but leads to a 9-fold higher catalytic efficiency (Her et al. 2016). It affects the sequential binding of nucleotides to the D1 and D2 domains, which is associated with cooperative conformational changes. Furthermore, it changes the sensitivity to cofactor-induced alterations during ATP hydrolysis. Ala530 is located in the α-helix following the P-loop involved in D2 ATP-binding, which leads into the allosteric binding site where Glu534 and Cys535 are proposed to be in hydrogen bonding distance of the UPCDC30245 inhibitor (LaPorte et al. 2018) (Fig. 7.7c + e). The nucleotide-binding site is additionally interconnected with the allosteric binding site via its D1D2 linker with the ensuing α-helix encompassing Gln494 and Glu498 being in hydrogen bonding distance of the inhibitor. Interestingly, the P472L mutation identified in CB-5083-resistant cell lines displayed cross-resistance to NMS-873 demonstrating a desensitization of p97 to both ATP-competitive as well as allosteric inhibitors (Wei et al. 2018). This mutation affects the intrasubunit signaling between the D2 and D1 domains due to an enhanced cooperative ATPase activity in the D2 domain.

Most of the identified mutations are located in the D1D2 linker and are hyperactive gain-of function variants resulting in an altered communication between the two ATPase domains, which is responsible for the desensitization of p97 to inhibitors. Understanding the determinants leading to the bypass of the inhibitory effect would be a good starting point for the rational development of second-generation inhibitors.

Conformation/Cofactor Specificity of Currently Studied Inhibitors

The function of p97 in different cellular pathways is regulated by the association of p97 with a large number of regulatory cofactors, which influence the ATPase activity differently (Meyer et al. 1998; Trusch et al. 2015; Zhang et al. 2015). In addition, cofactor assembly depends on nucleotide-dependent conformational changes of p97 (Bulfer et al. 2016; Chia et al. 2012). A general model suggests that conformational changes induced by cofactor-binding to the N domain regulate ADP and ATP binding to the D1 domain and thereby ATP hydrolysis in the D1 and D2 domains (Zhang et al. 2015). For example, p47 and p37 exhibit a tight binding to the ATP-bound state of p97 when the N domains are in the up-position and only weakly interact with the ADP-bound state (Bulfer et al. 2016). Furthermore, p37 up-regulates the ATPase activity, whereas p47 is an inhibitory cofactor and, only at high concentration, acts as an activating cofactor (Zhang et al. 2015). In contrast to p47 and p37, binding of UFD1-NPL4 does not regulate the ATPase activity of p97

(Blythe et al. 2017). Interestingly, inhibition of p97 by ATP-competitive as well as allosteric inhibitors typically does not result in a complete loss of p97 function (Anderson et al. 2015; LaPorte et al. 2018). For example, unlike other inhibitors the ATP-competitive inhibitor CB-5083 does not affect the important function of p97 in homotypic membrane fusion and in autophagy, suggesting that CB-5083 only binds to specific p97-cofactor complexes, thus suggesting conformational selectivity (Anderson et al. 2015). A similar behavior has been reported for the D2-selective and ATP-competitive inhibitor ML241 but not for DBeQ and ML240 (Chou et al. 2011, 2013).

In vitro studies revealed that the association of p97 with cofactors affects inhibitor potencies differentially (Chou et al. 2014; Fang et al. 2015; Gui et al. 2016). In the presence of p47 the inhibition of the ATPase activity by the D2-selective competitive inhibitors ML240 and ML241 decreases the potency of the inhibitor 50-fold, whereas the allosteric inhibitor NMS-873 only displays a two-fold effect (Chou et al. 2014; Fang et al. 2015). This indicates that the affinity for ATP is largely enhanced, consistent with a decreased K_m for ATP. Presumably, binding of p47 induces conformational changes in the D2 ATP-binding pocket, which prevent binding of D2-selective inhibitors. Similar to NMS-873, the ATP-competitive inhibitor CB-5083 also decreases the potency only 3-fold (Gui et al. 2016), indicating that both target different binding sites in comparison with the D2 selective ATP-competitive inhibitors ML240 and ML241. Modelling studies suggest that NMS-873 and CB-5083 are in close contact with the D1D2 linker residue Glu470 next to the allosteric binding site (Bastola et al. 2017). Both inhibitors display cross-resistance in cancer cell lines with mutations affecting codon 470, which is not observed with the inhibitors DBeQ, ML240, ML241 and UPCDC30245. Interestingly, the ATP-competitive inhibitor DBeQ, which targets both the D1 and D2 ATPase sites shows only a 4-6-fold decrease in the presence of p47, indicating that targeting the D1 ATPase site is more p97-p47 specific (Fang et al. 2015). Due to many important cellular functions in homotypic membrane fusion events, targeting the p97-p47 complex is of high interest. In contrast to p47, p37 and UFD1-NPL4 have almost no effect on the potencies of a large number of inhibitors (Gui et al. 2016). Both cofactors have no significant effect on the K_m value of ATP but result in an increase of k_{cat}. The effects of the p97-p47 complex described above might at least in part be an explanation why inhibitors display an effect on the ATPase activity in vitro but many have no effects in cell-based assays (LaPorte et al. 2018). The association with these cofactors could induce different conformational changes, which might contribute to the observed differences. The inhibition studies with p97 cofactors clearly need to be extended to further well-known cofactors to study effects on different cellular pathways. For example, it would be of interest to study the effect of the allosteric inhibitor UPCDC30245, which is conformation-selective and only binds to p97 in the ADP-bound state with the N domains in the down-conformation following association with cofactors. The approach of the Chou lab to screen about 200 different inhibitors on their effect on the D1 and D2 ATPase activity as well as on p47, p37 and UFD1-NPL4 represents an interesting possibility to identify p97 conformation-selective and cofactor-specific inhibitors (Fang et al. 2015; Gui et al. 2016).

Since inhibition of p97 with the ATP-competitive inhibitor CB-5083 does not lead to a complete functional loss and for example the autophagic pathway is not affected, a combined treatment with HDAC6 inhibitors was suggested and has already been demonstrated to be effective for the p97 inhibitor CB-5083 and the HDAC6 inhibitor ACY-125 (Vekaria et al. 2019). Combinatorial treatments are also good options to reduce side-effects as well as to offset or slow-down the development of resistance (see above) and cancer relapse.

p97 Cofactors as Potential Drug Targets

Depending on the p97 inhibitor a larger or smaller number of different cellular pathways will be affected, nevertheless, p97 inhibitors will always elicit broad cellular responses. A more specific inhibition could be achieved by directly targeting p97 cofactors to either decrease or increase their affinity to p97, as described below.

Targeting NPL4

The UFD1-NPL4 heterodimer plays multiple roles in protein degradation processes. The FDA-approved drug disulfiram (tetraethylthiuram disulfide, trade name Antabuse) (Fig. 7.8), which has been used for a long time in the treatment of alcoholism, has recently been identified via drug repurposing approaches to target the p97 cofactor NPL4 (Ding and Zhu 2018; Skrott et al. 2017). Drug repurposing is a promising and popular approach to identify new inhibitors of a given target. With this approach, well characterized but abandoned or even FDA-approved drugs are screened for other specific molecular targets. This strategy was also used to identify new p97 inhibitors targeting the allosteric binding site (Segura-Cabrera et al. 2017).

Disulfiram shows a tumor suppressing activity in diverse cancer types via disrupting protein homeostasis. Disulfiram itself does not react with p97–UFD1-NPL4, however, it rapidly forms, together with Cu^{2+}-ions, a bis(diethyldithiocarbamate)-copper complex (CuET)) (Fig. 7.8), which accumulates in tumors, thereby enhancing the anti-tumor activity of this compound. Binding of CuET to NPL4 interferes with the cellular functions of the p97–UFD1-NPL4 complex. Site directed mutagenesis studies suggested that the Zf-NPL4 domain of NPL4 is involved in binding. The recently solved cryo-EM structure of *C. thermophilum* Cdc48 in complex with the UFD1-NPL4 cofactor might explain the molecular mechanism of disulfiram inhibition (Bodnar et al. 2018). The MPN and Zf-NPL4 domains of NPL4 are positioned on top of p97 and both Zn-fingers form stalks that insert into grooves, harboring a conserved triple-phenylalanine (FFF) sequence, formed at the interface of adjacent D1 domains. Both Zn-fingers, which are of the CHCC type with three cysteine residues and one

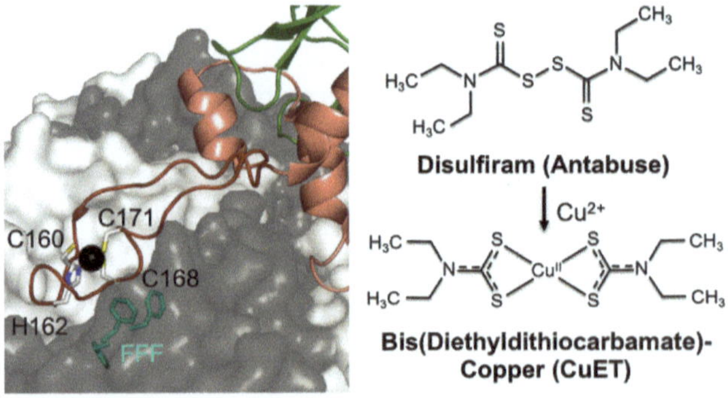

Fig. 7.8 NPL4 inhibition by disulfiram. Left, NPL4 Zn^{2+} binding site on p97. *C. thermophilum* Cdc48 in complex with NPL4 (pdb: 6chs) with the Cdc48 monomers in surface representation alternating in white and black as well as NPL4 in cartoon representation. The potential target sites of CuET are the Zn-fingers of NPL4 (colored in red). The coordination of Zn^{2+} (shown as black sphere) via three Cys and one His as well as the tri-Phe sequence formed at the interface of adjacent D1 domains is shown in stick representation. CuET might directly bind via its reactive Cu to the Cys residues of the Zn-fingers, thereby displacing the Zn^{2+}. Right, structure of disulfiram and its conversion to the bis(diethyldithiocarbamate)-copper complex (CuET) in the body

histidine coordinating the Zn^{2+}, are important for the ATP-dependent unfoldase activity of Cdc48 (Bodnar et al. 2018). The CuET might directly bind via its reactive Cu to the Cys residues of the Zn-fingers, thereby displacing the Zn^{2+} ions. Disruption of Zn-binding to NPL4 leads to aggregation, consistent with the observed immobilization of NPL4 and the subsequent prominent aggregation/clustering of NPL4 in the nucleus, which triggers the heat-shock response (HSR) as well as the UPR (Skrott et al. 2017). The interaction of NPL4 via its UBXL domain with the N domain of p97 is not affected in agreement with a co-localization of p97 with nuclear NPL4 clusters.

Targeting UBX Proteins

Several UBX proteins were shown to play critical regulatory roles in different types of cancer (reviewed in (Rezvani 2016)) and were suggested to have tumor suppressing functions. For example, SAKS1 has an inhibitory function on cellular inhibitors of apoptosis proteins (cIAPs), which are typically upregulated in cancer cells resulting in tumor cell proliferation and poor prognosis. Furthermore, SAKS1 as well as p47 and FAF1 have been identified to be negative regulators (suppressors) of the NF-κB pathway in cancer cells. FAF1 engages in a cross-talk with diverse tumor suppressor oncoproteins indicating important anti-tumor functions, e.g. an apoptotic function while FAF1 itself is a potential tumor suppressor. Recently it could be shown that phosphorylation of Ser582 of the UBX domain of FAF1 by the AKT kinase disrupts the interaction of FAF1 with p97, thus resulting in a reduced level of FAF1 at the plasma membrane and accumulation of the TCF-β type II (TβII) receptor (transforming growth factor), which then promotes breast cancer metastasis (Xie et al. 2017). This is in accordance with the observation that

FAF1 is highly expressed in normal epithelial cells but attenuated in several cancer cells. Under normal conditions, the FAF1-p97 complex is involved in the ubiquitylation and degradation of TβII via the lysosomal pathway, thus indicating a metastasis-suppressing role of FAF1.

Whether UBX proteins are indeed potentially useful targets in targeted cancer therapies needs to be investigated in pre-clinical studies. Nevertheless, targeting tumor suppressor proteins is known to be challenging. The role of UBX proteins and other p97 cofactors in specific cancers should be analyzed: Are they up- or downregulated, are there mutant proteins that drive cancer progression? Analysis of The Cancer Genome Atlas (TCGA) (Cancer Cancer Genome Atlas Research Network et al. 2013), which contains many data like genetic mutations identified in cancer genomes as well as gene expression profiling, revealed many mutations like missense and frameshift mutations in the gene encoding p97 including the described hotspot frameshift mutation at codon 616, which is located in the allosteric binding site (Bastola et al. 2017). A detailed integrated genomic analysis across cancer types of p97 and its associated cofactors should be performed, a successful approach recently described for a large range of E1, E2, E3 as well as DUBs (Ge et al. 2018). A systematic characterization of mutations that drive cancer, which will give answers on the frequency of mutations in certain cancer types, whether they are cancer type specific and whether they are loss of function mutations or hotspot mutations, will yield valuable information for potential therapeutic strategies. In addition, information regarding a mutually exclusive pattern of p97 and/or its associated cofactors with other genes that drive cancer would suggest possible combination strategies for treatment.

Proteomic Approaches to Analyze the p97 Interactome

Mass Spectrometry (MS) is used as an indispensable tool for the identification of PTMs as well as the analysis of protein-protein interaction networks, thus helping in understanding diverse global biological processes. During the last years with the help of MS techniques, a better understanding of the p97 interactome and physiological as well as pathophysiological functions could be gained. In the following part selected examples leading to new insights into the involvement of p97 in numerous cellular processes including the identification of new p97 substrates and the regulation of p97 by PTMs will be described.

To analyze the complexity of p97-cofactor and cofactor-substrate interactions different approaches using p97-specific inhibitors or p97 variants like the so-called substrate-trapping mutant Glu578Gln, a variant with impaired ATPase activity in the D2 domain, were used (Her et al. 2016; Hülsmann et al. 2018; Xue et al. 2016). Most of the studies involved affinity purification and MS (AP-MS), which represents only a static overview of the interactome. However, multiple MS-based proteomic studies by the Deshaies lab revealed that p97 interactions with its cofactors are highly dynamic (Xue et al. 2016). Both, inhibition of p97 or the proteasome, which lead to the accumulation of ubiquitylated proteins, slow down the rapid dynamics of p97 and its cofactor interactions. Furthermore, by using an in vivo cross-linking approach a much higher number of known p97 cofactors could be identified. To better understand the dynamic changes of the p97 interactome and to also analyze temporal changes AP-MS combined with SWATH (sequential window acquisition of all theoretical spectra) of the substrate-trapping variant compared with wt-p97 was used (Hülsmann et al. 2018). With this approach most of the known p97 cofactors could be trapped, yet also many proteins involved in the ERAD pathway as well as E3 ligases associated with the mitochondrial quality control system were identified. Furthermore, with the AP-SWATH approach it was possible to also detect individual substrate proteins like ubiquitylated CReP (constitutive repressor of elF2α phosphorylation), an elF2α (eucaryotic translation initiation factor 2 α)-specific PP1 regulator, or glutamine synthetase (Nguyen et al. 2017), which are both targeted to the proteasome via the p97–UFD1-NPL4 complex.

Another approach to investigate the p97 interactome and p97 functions in many different cellular processes involves p97 cofactor proteomics to identify substrates in particular pathways. Via proteomics approaches it could be shown that p97 binds to a large number of UBX proteins and that, in particular, UBA-UBX proteins can associate with many E3 ligases including CRLs, the largest group of E3 ligases (Alexandru et al. 2008). These studies also revealed for the first time that p97 together with its UFD1-NPL4 cofactor and UBA-UBX proteins form higher order complexes (see Section "Regulation of p97 Cofactor Assembly"). Systematic proteomics of the UFD1-NPL4 heterodimer and UBX proteins revealed a large number of interacting proteins with implications in many different cellular pathways (Raman et al. 2015). The p97-UBX cofactor interaction network identified a cross-talk between UBX proteins, thus indicating that they possibly assemble on the same p97 hexamer. Via this approach functions of so far poorly characterized UBX proteins could be identified, like for example the participation of UBXD3 in the regulation of ciliogenesis (Raman et al. 2015), the function of the p97–SAKS complex in the degradation of BAG6-bound proteins (Ganji et al. 2018) (see Section "p97–Cofactor Complexes Involved in Diverse Cellular Functions") and the function of p97–p37 in the activation of PP1 (Weith et al. 2018) (see Section "Mechanistic Insights into p97 Function"). Using SILAC (stable isotope labeling by amino acids in cell culture)-based proteomics of the ERAD-associated E3 ligase HRD1 a large number of physiological substrates were identified (Ye et al. 2018). Furthermore, an unbiased MS approach to identify cofactors and substrates involved in IBMPFD identified the involvement of UBXD1 and caveolin-1 (CAV1) in endolysosomal sorting (Ritz et al. 2011). In a different approach the membrane protein ERGIC-53 was shown to specifically interact with UBXD1, thus establishing a function of p97 in protein trafficking (Haines 2010). p97 has been directly linked to the clearance of SGs under oxidative stress, in particular arsenite stress as described above (see Section "p97–Cofactor Complexes Involved in Diverse Cellular Functions"). A MS approach identified that the p97 cofactor ZFAND1 can directly interact with the 26S proteasome (Turakhiya et al.

2018). ZFAND1 recruits both p97 and the proteasome to SGs to ensure their normal turnover by proteasomal degradation. Depletion of ZFAND1 results in the formation of abnormal SGs and the elimination of SGs via autophagy. MS approaches also unraveled the role of p97 in DSB repair. During DSB repair by NHEJ the p97–UFD1-NPL4 complex specifically targets Ku80 modified with K48-linked chains for releasing Ku rings from the DNA (van den Boom et al. 2016) (see Section "p97–Cofactor Complexes Involved in Diverse Cellular Functions"). Label-free quantitative MS using biotinylated single-strand DNA fragments isolated from egg extracts revealed Ku70/80 as the major substrate of p97 when p97 activity was impaired. In addition, the regulatory kinase DNA-PKcs (DNA dependent protein kinase, catalytic subunit) together with the p97 cofactors UFD1, NPL4 and FAF1 were identified.

PTMs are reversible covalent modification of proteins and can cause conformational changes in proteins leading to the masking or exposition of specific residues which affect protein-protein interactions in a diverse manner and regulate protein stability, activity and protein localization by enzymatic deactivation or activation (for reviews see Beltrao et al. 2013; Ryslava et al. 2013; Venne et al. 2014). Based on primarily high-throughput proteomic studies, p97 and its cofactors were shown to be modified by a large number of PTMs including phosphorylation, acetylation, ubiquitylation and SUMOylation (for a review see Hänzelmann and Schindelin 2017), but for most modifications the structural and functional consequences are not known. Examples for PTMs studied in more detail include the phosphorylation of the p97 penultimate residue Tyr805 by c-Src kinase, which impairs the interaction of p97 with PUB/PUL-domain containing cofactors (Li et al. 2008; Zhao et al. 2007), the phosphorylation of Ser582 of the UBX protein FAF1 by AKT, which disrupts the interaction of FAF1 with p97, thus leading to the accumulation of the TβII receptor (see Section "p97 Cofactors as Potential Drug Targets") (Xie et al. 2017) or SUMOylation of ATAXIN-3, which tightens the binding of ATAXIN-3 to p97 (Almeida et al. 2015).

The function of p97 in maintaining protein homeostasis is regulated by ubiquitylation, which modulates its downstream cellular activities. Some of the p97 cofactors possess a UBD to interact with ubiquitylated substrates, which accumulate upon p97 inhibition. Ubiquitin is linked via its C-terminal di-glycine motif to substrate lysines, hence di-Gly-Lys specific antibodies targeting ubiquitin remnant peptides left after tryptic digestion followed by subsequent enrichment can be used to study changes in the ubiquitin-modified proteome and to also identify specific substrates of individual proteins after their inhibition (e.g. Xu et al. 2010). The Beli lab (Heidelberger et al. 2018) demonstrated, by enrichment of ubiquitin remnant peptides after p97 as well as proteasomal inhibition and SILAC-based quantitative MS, that a subset of ubiquitylated proteins differs between these two inhibitors. With this approach around 450 new putative p97-substrates, which are characterized by disordered regions, could be identified. These proteins play crucial roles in gene expression, cell cycle regulation and DNA repair processes. One-third of the identified substrates only showed an increase in ubiquitylation after p97-inhibition but not after proteasome inhibition, indicating that these substrates are recycled and not degraded. Interestingly, besides an upregulation of K48- and K11-linked ubiquitylation, p97 inhibition also upregulated atypical K6-linked ubiquitylation, which depended on the HECT-type ubiquitin E3-ligase HUWE1. Furthermore, MS revealed that p97 inhibition leads to a HUWE1-dependent K48-linked ubiquitylation of the transcription factor c-Myc, which promotes p97 assisted proteasomal degradation, thereby regulating the activity of c-Myc. In a different approach, the so-called VCP inhibitor substrate trapping approach (VISTA), the poorly characterized spectrum of endogenous ERAD substrates was analyzed (Huang et al. 2018). This technique uses the ATP-competitive p97 inhibitor CB-5083 to trap ubiquitylated ERAD substrates still associated with the membrane-embedded ER retro-translocation complex, followed by ubiquitin remnant peptide enrichment from the membrane fraction. Upon inhibition of p97 and SILAC-labeling, about 500 ubiquitylated proteins could be identified. K48-, K11- and K33-linked ubiquitylations were upregulated consistent with proteasomal degradation.

SUMOylation has also been linked to cellular homeostasis, in particular to genome stability where it either acts together with ubiquitin in degradation processes or has a ubiquitin-independent regulatory role (for a review see Liebelt and Vertegaal 2016). p97 itself can be modified by SUMO and SUMOylated p97 was shown to be targeted to SGs and the nucleus upon oxidative and ER stress. This targeting is impaired in IBMPFD variants due to reduced SUMOylation (Wang et al. 2016). Furthermore, MS revealed that the UFD1 cofactor harbors several SUMO interaction sites (Hendriks et al. 2017), indicating an association with SUMOylated substrates. Accordingly, polyubiquitylation of SUMOylated proteins, which is catalyzed by SUMO-targeting ubiquitin ligases (STUbl) like RNF4 and leads to hybrid signals, has been associated with p97–UFD1-NPL4 mediated proteasomal degradation (for a review see Nie and Boddy 2016). A similar approach as employed by the Beli lab to study the ubiquitin-modified proteome upon p97 inhibition is also conceivable for investigating the SUMO-modified proteome by using SUMO1-3-specific antibodies.

In addition to ubiquitylation and SUMOylation, NEDDylation has been identified as a new regulator of protein homeostasis (Maghames et al. 2018), which might also influence the p97-cofactor and substrate composition and thus the function of p97 in PQC pathways. It could be shown that under proteotoxic stress NEDDylation, which depends on atypical NEDD8 activation by the ubiquitin specific E1 enzyme UBA1 and not by the typical NEDD8 activating enzyme, is involved in the nuclear PQC by promoting transient protein aggregation (Maghames et al. 2018). By utilizing SILAC-based proteomic approaches, besides NEDD8, also ubiquitin as well as SUMO1-2 were found to be upregulated upon heat shock, indicating a cross-talk between ubiquitin and ubiquitin-like modifiers. The E3 ligase HUWE1 is required for NEDDylation but also for ubiquitylation, resulting in the formation of possible hybrid NEDD8/ubiquitin chains on substrates, thereby protecting the nuclear UPS. p97 might be directly linked to this regulatory mechanism by its interaction with NUB1L (NEED8 ultimate buster-1 long), a protein responsible for downregulating NEDD8 levels and NEDDylation (Liu et al. 2013). p97 together with its cofactor UFD1-NPL4 has been shown to be involved in proteasomal degradation of NEDD8. Furthermore, a link between NEDDylated CRLs and p97 has been demonstrated (Bandau et al. 2012; den Besten et al. 2012). The p97 cofactor UBXD7 directly binds in a ubiquitin-independent manner via its ubiquitin-interacting motif (UIM) to NEDDylated CRLs, thereby probably controlling E3 ligase activity by locking the CRL in its NEDDylated state.

Proteins are quite often regulated by several PTMs which interplay or cross-talk with each other (for reviews see Beltrao et al. 2013; Csizmok and Forman-Kay 2018; Venne et al. 2014). An example for a PTM cross-talk involving p97 is the regulation of the cellular activities of the cytoplasmic protein tyrosine phosphatase PTP-PEST, which controls via phosphorylation-dependent ubiquitylation cell proliferation and invasion in glioblastoma (Chen et al. 2018). MS analysis of PTP-PEST interacting proteins revealed that p97 and the focal adhesion protein Cas (Crk-associated substrate) are substrates. This study demonstrated that action of PTP-PEST on Cas creates a phosphodegron, which, in turn, acts as a signal for downstream ubiquitylation and targeting of phosphorylated-ubiquitylated Cas (pY-Cas-Ub) to the proteasome in a p97-dependent manner. Low expression levels of PTP-PEST as well as phosphorylation of p97 Tyr805, which impairs interaction with PTP-PEST and pY-Cas-Ub, is involved in increased focal adhesion disassembly and tumor cell invasion.

Conclusions and Future Perspectives

p97 as a key player in the maintenance of protein homeostasis is clearly associated with cancer and neurodegenerative diseases, which are characterized by the accumulation of misfolded proteins and aggregation. Currently no therapies are available to reverse misfolding and aggregation. p97 missense mutations identified in IBMPFD are characterized by an enhanced ATPase activity as well as an increased unfoldase activity resulting from an impaired interdomain communication and are considered to be gain off function mutants. p97 inhibitors, which adjust the ATPase activity to wt-p97 levels could be attractive for treatment of the two autosomal dominant disorders FALS and IBMPFD (Blythe et al. 2017; Zhang et al. 2015). Since most of the IBMPFD mutations are located in the interface between the N domain and the D1 domain, targeting the ND1 interface may lead to a new class of allosteric inhibitors exhibiting tighter control of the ATPase activity. In contrast to age-related neurodegenerative diseases like Alzheimer's disease and Parkinson's disease, which are counteracted by functional PQR systems including p97 and therefore cannot be treated with p97 inhibitors, p97 as a target in cancer treatment holds promise and several inhibitors have been recently developed, which are effective in both hematological cancers as well as solid tumors, a clear benefit over proteasomal inhibitors. However, the important functions of p97 in PQC pathways in normal cells and inhibitor cytotoxicity overcame the inhibitory effect by the generation of gain of function mutations. Together with the fact that p97 is overexpressed in cancer cells and that there are several indications that induction of ER stress upon p97 inhibition is involved in the epithelial-mesenchymal transition (EMT) and subsequent tumor metastasis (for a review see Lan et al. 2017), renders this a difficult task. Future strategies for the development of p97 inhibitors for cancer treatment should also consider targeting the D1 domain and, in particular the N domain with its cofactor-binding sites, which would lead to inhibitors with higher specificity and reduced cytotoxicity. A screening approach to identify hotspots on the ND1 fragment was described (Chimenti et al. 2015). Analysis for good druggable sites revealed, besides the D1 ATP binding site also the large ND1 interface (Fig. 7.7a). In addition, an alternative approach would be to target specific p97–cofactor complexes to obtain pathway-specific inhibitors, which would also reduce the cytotoxicity of currently known inhibitors that, in general, affect the ATPase activity or inter- and intrasubunit signaling. However, targeting the N domain will be challenging since the hydrophobic interdomain cleft as well as the SHP-binding site are rather large and flat surface features. Irrespective of targeting p97 in cancer therapy, p97 inhibitors will certainly provide valuable insights into the physiological and pathophysiological functions of p97.

The p97 interactome in diverse cellular pathways is still barely understood. Proteomic studies of p97 and its cofactors, the benefit of specific p97-mutants and p97-inhibitors influencing its activity as well as in situ cross-linking facilitated the identification of so far unknown p97 substrates and new p97 functions. Future systematic proteomic studies to study dynamic changes of p97–cofactor as well as p97-substrate interactions should be performed in response to stress conditions known to affect particular pathways p97 is involved in, like DNA repair pathways, replication, autophagy and PQC. Posttranslational modifications of p97 by phosphorylation, ubiquitylation and SUMOylation were identified, and it would be of interest to elucidate their interplay with p97 cofactors/substrates in different cellular pathways. Proteomics of p97 and its cofactors as well as substrates to investigate the dynamic protein-protein interaction networks will lay a strong foundation in understanding p97 function and will help to uncover the molecular basis underlying certain disease states, like neurodegenerative disorders and cancer.

Although much progress has been made recently in elucidating the p97 unfoldase and disassembly activity, suggesting a general threading mechanism through the entire p97 channel, open questions still remain. Studies so far either involved ubiquitylated model substrates or rather small physiological proteins, but it is difficult to envisage that all ubiquitylated proteins destined for proteasomal degradation are unfolded and threaded through the p97 channel. For the delivery to the proteasome the substrates probably need to be refolded and ubiquitylated to be recognized by the proteasome. Of course, other possibilities could be that at least some small proteins could be targeted to the proteasome via the cytosolic chaperone BAG6 or even directly by p97 without the help of additional shuttle factors. However, disassembling entire macromolecular complexes like Ku70/80 in DSB repair or the CMG helicase during replication termination is unlikely to involve unfolding of these rather big proteins, thus suggesting that the substrate is only partly unfolded and inserted into the pore. Upon ATP hydrolysis conformational changes are transmitted to the N domains with the bound cofactor-substrate complex and the N domain movement could provide an additional force during disassembly. These data suggest that p97, depending on the respective substrate, uses different mechanisms. To clarify all these unresolved questions structures of p97 with different cofactors and their physiological substrates are necessary, which will help to decipher conformational changes associated with substrate threading.

Acknowledgements This work was supported by the Deutsche Forschungsgemeinschaft (Grant GRK2243/1) and the Rudolf Virchow Center for Experimental Biomedicine.

Note Added in Proof Two recent cryo-EM structures (Cooney et al. 2019; Twomey et al. 2019) of Cdc48–cofactor complexes bound to either a polyubiquitylated model substrate or a native substrate visualized substrate/ubiquitin-derived residues in the central channel of Cdc48. Furthermore, the Cdc48 subunits adopted a helical arrangement consistent with the generalized hand-over-hand mechanism of protein translocation by AAA+ ATPases. Processing of ubiquitylated substrates was found to be initiated by ubiquitin unfolding and the insertion of its N-terminal segment into the central Cdc48 channel.

References

Abid Ali F, Costa A (2016) The MCM helicase motor of the eukaryotic replisome. J Mol Biol 428:1822–1832. https://doi.org/10.1016/j.jmb.2016.01.024

Akutsu M, Dikic I, Bremm A (2016) Ubiquitin chain diversity at a glance. J Cell Sci 129:875–880. https://doi.org/10.1242/jcs.183954

Alberti S, Mateju D, Mediani L, Carra S (2017) Granulostasis: protein quality control of RNP granules. Front Mol Neurosci 10:84. https://doi.org/10.3389/fnmol.2017.00084

Alexandru G, Graumann J, Smith GT, Kolawa NJ, Fang R, Deshaies RJ (2008) UBXD7 binds multiple ubiquitin ligases and implicates p97 in HIF1alpha turnover. Cell 134:804–816. https://doi.org/10.1016/j.cell.2008.06.048

Alfieri C, Chang L, Barford D (2018) Mechanism for remodelling of the cell cycle checkpoint protein MAD2 by the ATPase TRIP13. Nature 559:274–278. https://doi.org/10.1038/s41586-018-0281-1

Almeida B et al (2015) SUMOylation of the brain-predominant Ataxin-3 isoform modulates its interaction with p97. Biochim Biophys Acta 1852:1950–1959. https://doi.org/10.1016/j.bbadis.2015.06.010

Alverez C et al (2016) Allosteric indole amide inhibitors of p97: identification of a novel probe of the ubiquitin pathway. ACS Med Chem Lett 7:182–187. https://doi.org/10.1021/acsmedchemlett.5b00396

Anderson DJ et al (2015) Targeting the AAA ATPase p97 as an approach to treat cancer through disruption of protein homeostasis cancer. Cell 28:653–665. https://doi.org/10.1016/j.ccell.2015.10.002

Ao N, Chen Q, Liu G (2017) The small molecules targeting ubiquitin-proteasome system for cancer therapy. Comb Chem High Throughput Screen 20:403–413. https://doi.org/10.2174/1386207320666170710124746

Arumughan A et al (2016) Quantitative interaction mapping reveals an extended UBX domain in ASPL that disrupts functional p97 hexamers. Nat Commun 7:13047. https://doi.org/10.1038/ncomms13047

Baldridge RD, Rapoport TA (2016) Autoubiquitination of the Hrd1 ligase triggers protein retrotranslocation in ERAD. Cell 166:394–407. https://doi.org/10.1016/j.cell.2016.05.048

Bandau S, Knebel A, Gage ZO, Wood NT, Alexandru G (2012) UBXN7 docks on neddylated cullin complexes using its UIM motif and causes HIF1alpha accumulation. BMC Biol 10:36. https://doi.org/10.1186/1741-7007-10-36

Banerjee S et al (2016) 2.3 A resolution cryo-EM structure of human p97 and mechanism of allosteric inhibition. Science 351:871–875. https://doi.org/10.1126/science.aad7974

Bard JAM, Goodall EA, Greene ER, Jonsson E, Dong KC, Martin A (2018) Structure and function of the 26S proteasome. Annu Rev Biochem 87:697–724. https://doi.org/10.1146/annurev-biochem-062917-011931

Barthelme D, Sauer RT (2013) Bipartite determinants mediate an evolutionarily conserved interaction between Cdc48 and the 20S peptidase. Proc Natl Acad Sci U S A 110:3327–3332. https://doi.org/10.1073/pnas.1300408110

Barthelme D, Sauer RT (2016) Origin and functional evolution of the Cdc48/p97/VCP AAA+ protein unfolding and remodeling machine. J Mol Biol 428:1861–1869. https://doi.org/10.1016/j.jmb.2015.11.015

Barthelme D, Chen JZ, Grabenstatter J, Baker TA, Sauer RT (2014) Architecture and assembly of the archaeal Cdc48*20S proteasome. Proc Natl Acad Sci U S A 111:1687–1694. https://doi.org/10.1073/pnas.1404823111

Bastola P, Chien J (2018) Co-selected mutations in VCP: a novel mechanism of resistance to VCP inhibitors. Cell Death Dis 9:35. https://doi.org/10.1038/s41419-017-0049-9

Bastola P, Wang F, Schaich MA, Gan T, Freudenthal BD, Chou TF, Chien J (2017) Specific mutations in the D1-D2 linker region of VCP/p97 enhance ATPase activity and confer resistance to VCP inhibitors. Cell Death Discov 3:17065. https://doi.org/10.1038/cddiscovery.2017.65

Bastola P, Oien DB, Cooley M, Chien J (2018) Emerging cancer therapeutic targets in protein homeostasis. AAPS J 20:94. https://doi.org/10.1208/s12248-018-0254-1

Bebeacua C, Forster A, McKeown C, Meyer HH, Zhang X, Freemont PS (2012) Distinct conformations of the protein complex p97-Ufd1-Npl4 revealed by electron cryomicroscopy. Proc Natl Acad Sci U S A 109:1098–1103. https://doi.org/10.1073/pnas.1114341109

Beltrao P, Bork P, Krogan NJ, van Noort V (2013) Evolution and functional cross-talk of protein post-translational modifications. Mol Syst Biol 9:714. https://doi.org/10.1002/msb.201304521

Bento AC, Bippes CC, Kohler C, Hemion C, Frank S, Neutzner A (2018) UBXD1 is a mitochondrial recruitment factor for p97/VCP and promotes mitophagy. Sci Rep 8:12415. https://doi.org/10.1038/s41598-018-30963-z

Beuron F et al (2006) Conformational changes in the AAA ATPase p97–p47 adaptor complex. EMBO J 25:1967–1976. https://doi.org/10.1038/sj.emboj.7601055

Blythe EE, Olson KC, Chau V, Deshaies RJ (2017) Ubiquitin- and ATP-dependent unfoldase activity of p97/VCP*NPLOC4*UFD1L is enhanced by a mutation that causes multisystem proteinopathy. Proc Natl Acad Sci U S A 114:E4380–E4388. https://doi.org/10.1073/pnas.1706205114

Bodnar NO et al (2018) Structure of the Cdc48 ATPase with its ubiquitin-binding cofactor Ufd1-Npl4. Nat Struct Mol Biol 25:616–622. https://doi.org/10.1038/s41594-018-0085-x

Bodnar N, Rapoport T (2017a) Toward an understanding of the Cdc48/p97 ATPase. F1000Res 6:1318 https://doi.org/10.12688/f1000research.11683.1

Bodnar NO, Rapoport TA (2017b) Molecular mechanism of substrate processing by the Cdc48 ATPase complex. Cell 169:722–735 e 729 https://doi.org/10.1016/j.cell.2017.04.020

Boeddrich A et al (2006) An arginine/lysine-rich motif is crucial for VCP/p97-mediated modulation of ataxin-3 fibrillogenesis. EMBO J 25:1547–1558. https://doi.org/10.1038/sj.emboj.7601043

Brandman O, Hegde RS (2016) Ribosome-associated protein quality control. Nat Struct Mol Biol 23:7–15. https://doi.org/10.1038/nsmb.3147

Brandman O et al (2012) A ribosome-bound quality control complex triggers degradation of nascent peptides and signals translation stress. Cell 151:1042–1054. https://doi.org/10.1016/j.cell.2012.10.044

Braunstein I, Zach L, Allan S, Kalies KU, Stanhill A (2015) Proteasomal degradation of preemptive quality control (pQC) substrates is mediated by an AIRAPL-p97 complex. Mol Biol Cell 26:3719–3727. https://doi.org/10.1091/mbc.e15-02-0085

Bruderer RM, Brasseur C, Meyer HH (2004) The AAA ATPase p97/VCP interacts with its alternative co-factors, Ufd1-Npl4 and p47, through a common bipartite binding mechanism. J Biol Chem 279:49609–49616. https://doi.org/10.1074/jbc.m408695200

Buchberger A, Howard MJ, Proctor M, Bycroft M (2001) The UBX domain: a widespread ubiquitin-like module. J Mol Biol 307:17–24. https://doi.org/10.1006/jmbi.2000.4462

Buchberger A, Schindelin H, Hänzelmann P (2015) Control of p97 function by cofactor binding. FEBS Lett 589:2578–2589. https://doi.org/10.1016/j.febslet.2015.08.028

Bug M, Meyer H (2012) Expanding into new markets–VCP/p97 in endocytosis and autophagy. J Struct Biol 179:78–82. https://doi.org/10.1016/j.jsb.2012.03.003

Bulfer SL, Chou TF, Arkin MR (2016) p97 disease mutations modulate nucleotide-induced conformation to alter protein–protein interactions. ACS Chem Biol 11:2112–2116. https://doi.org/10.1021/acschembio.6b00350

Burnett JC et al (2017) A threonine turnstile defines a dynamic amphiphilic binding motif in the AAA ATPase p97 allosteric binding site. Org Biomol Chem 15:4096–4114. https://doi.org/10.1039/c7ob00526a

Cancer Genome Atlas Research Network et al (2013) The Cancer Genome Atlas Pan-Cancer analysis project. Nat Genet 45:1113–1120. https://doi.org/10.1038/ng.2764

Cappadocia L, Lima CD (2018) Ubiquitin-like protein conjugation: structures, chemistry, and mechanism. Chem Rev 118:889–918. https://doi.org/10.1021/acs.chemrev.6b00737

Carvalho P, Goder V, Rapoport TA (2006) Distinct ubiquitin-ligase complexes define convergent pathways for the degradation of ER proteins. Cell 126:361–373. https://doi.org/10.1016/j.cell.2006.05.043

Chapman E, Fry AN, Kang M (2011) The complexities of p97 function in health and disease. Mol Biosyst 7:700–710. https://doi.org/10.1039/c0mb00176g

Chapman E, Maksim N, de la Cruz F, La Clair JJ (2015) Inhibitors of the AAA+ chaperone p97. Molecules 20:3027–3049. https://doi.org/10.3390/molecules20023027

Chen Z, Morales JE, Guerrero PA, Sun H, McCarty JH (2018) PTPN12/PTP-PEST regulates phosphorylation-dependent ubiquitination and stability of focal adhesion substrates in invasive glioblastoma. Cells Cancer Res 78:3809–3822. https://doi.org/10.1158/0008-5472.can-18-0085

Chia WS, Chia DX, Rao F, Bar Nun S, Geifman Shochat S (2012) ATP binding to p97/VCP D1 domain regulates selective recruitment of adaptors to its proximal N-domain. PLoS One 7:e50490. https://doi.org/10.1371/journal.pone.0050490

Chimenti MS et al (2015) A fragment-based ligand screen against part of a large protein machine: the ND1 domains of the AAA+ ATPase p97/VCP. J Biomol Screen 20:788–800. https://doi.org/10.1177/1087057115570550

Chou TF et al (2014) Specific inhibition of p97/VCP ATPase and kinetic analysis demonstrate interaction between D1 and D2 ATPase domains. J Mol Biol 426:2886–2899. https://doi.org/10.1016/j.jmb.2014.05.022

Chou TF et al (2011) Reversible inhibitor of p97, DBeQ, impairs both ubiquitin-dependent and autophagic protein clearance pathways. Proc Natl Acad Sci U S A 108:4834–4839. https://doi.org/10.1073/pnas.1015312108

Chou TF, Li K, Frankowski KJ, Schoenen FJ, Deshaies RJ (2013) Structure-activity relationship study reveals ML240 and ML241 as potent and selective inhibitors of p97 ATPase. ChemMedChem 8:297–312. https://doi.org/10.1002/cmdc.201200520

Christianson JC, Ye Y (2014) Cleaning up in the endoplasmic reticulum: ubiquitin in charge. Nat Struct Mol Biol 21:325–335. https://doi.org/10.1038/nsmb.2793

Cooney I et al (2019) Structure of the Cdc48 segregase in the act of unfolding an authentic substrate. Science. https://doi.org/10.1126/science.aax0486

Csizmok V, Forman-Kay JD (2018) Complex regulatory mechanisms mediated by the interplay of multiple post-translational modifications. Curr Opin Struct Biol 48:58–67. https://doi.org/10.1016/j.sbi.2017.10.013

Dargemont C, Ossareh-Nazari B (2012) Cdc48/p97, a key actor in the interplay between autophagy and ubiquitin/proteasome catabolic pathways. Biochim Biophys Acta 1823:138–144. https://doi.org/10.1016/j.bbamcr.2011.07.011

Davies JM, Brunger AT, Weis WI (2008) Improved structures of full-length p97, an AAA ATPase: implications for mechanisms of nucleotide-dependent conformational change. Structure 16:715–726. https://doi.org/10.1016/j.str.2008.02.010

Davis EJ, Lachaud C, Appleton P, Macartney TJ, Nathke I, Rouse J (2012) DVC1 (C1orf124) recruits the p97 protein segregase to sites of DNA damage. Nat Struct Mol Biol 19:1093–1100. https://doi.org/10.1038/nsmb.2394

Defenouillere Q et al (2013) Cdc48-associated complex bound to 60S particles is required for the clearance of aberrant translation products. Proc Natl Acad Sci U S A 110:5046–5051. https://doi.org/10.1073/pnas.1221724110

Defenouillere Q, Fromont-Racine M (2017) The ribosome-bound quality control complex: from aberrant peptide clearance to proteostasis maintenance. Curr Genet 63:997–1005. https://doi.org/10.1007/s00294-017-0708-5

DeLaBarre B, Brunger AT (2003) Complete structure of p97/valosin-containing protein reveals communication between nucleotide domains. Nat Struct Biol 10:856–863. https://doi.org/10.1038/nsb972

DeLaBarre B, Christianson JC, Kopito RR, Brunger AT (2006) Central pore residues mediate the p97/VCP activity required for ERAD. Mol Cell 22:451–462. https://doi.org/10.1016/j.molcel.2006.03.036

den Besten W, Verma R, Kleiger G, Oania RS, Deshaies RJ (2012) NEDD8 links cullin-RING ubiquitin ligase function to the p97 pathway. Nat Struct Mol Biol 19:511–516, S511. https://doi.org/10.1038/nsmb.2269

Deshaies RJ (2014) Proteotoxic crisis, the ubiquitin-proteasome system, and cancer therapy. BMC Biol 12:94. https://doi.org/10.1186/s12915-014-0094-0

Deville C, Carroni M, Franke KB, Topf M, Bukau B, Mogk A, Saibil HR (2017) Structural pathway of regulated substrate transfer and threading through an Hsp100 disaggregase. Sci Adv 3:e1701726. https://doi.org/10.1126/sciadv.1701726

Dewar JM, Low E, Mann M, Raschle M, Walter JC (2017) CRL2(Lrr1) promotes unloading of the vertebrate replisome from chromatin during replication termination. Genes Dev 31:275–290. https://doi.org/10.1101/gad.291799.116

Dikic I (2017) Proteasomal and autophagic degradation systems. Annu Rev Biochem 86:193–224. https://doi.org/10.1146/annurev-biochem-061516-044908

Ding N, Zhu Q (2018) Disulfiram combats cancer via crippling valosin-containing protein/p97 segregase adaptor NPL4. Transl Cancer Res 7:S495–S499. https://doi.org/10.21037/tcr.2018.03.33

Doroshow JH (2018) https://deainfo.nci.nih.gov/advisory/fac/0518/Doroshow.pdf

Dreveny I, Kondo H, Uchiyama K, Shaw A, Zhang X, Freemont PS (2004) Structural basis of the interaction between the AAA ATPase p97/VCP and its adaptor protein p47. EMBO J 23:1030–1039. https://doi.org/10.1038/sj.emboj.7600139

Erzberger JP, Berger JM (2006) Evolutionary relationships and structural mechanisms of AAA+ proteins. Annu Rev Biophys Biomol Struct 35:93–114. https://doi.org/10.1146/annurev.biophys.35.040405.101933

Esaki M, Johjima-Murata A, Islam MT, Ogura T (2018) Biological and pathological implications of an alternative ATP-powered proteasomal assembly with Cdc48 and the 20S peptidase. Front Mol Biosci 5:56. https://doi.org/10.3389/fmolb.2018.00056

Ewens CA et al (2014) The p97-FAF1 protein complex reveals a common mode of p97 adaptor binding. J Biol Chem 289:12077–12084. https://doi.org/10.1074/jbc.m114.559591
Ewens CA, Kloppsteck P, Forster A, Zhang X, Freemont PS (2010) Structural and functional implications of phosphorylation and acetylation in the regulation of the AAA+ protein p97. Biochem Cell Biol 88:41–48. https://doi.org/10.1139/o09-128
Fang CJ et al (2015) Evaluating p97 inhibitor analogues for their domain selectivity and potency against the p97–p47 complex. ChemMedChem 10:52–56. https://doi.org/10.1002/cmdc.201402420
Fernandez-Saiz V, Buchberger A (2010) Imbalances in p97 co-factor interactions in human proteinopathy. EMBO Rep 11:479–485. https://doi.org/10.1038/embor.2010.49
Ferreira de Freitas R et al (2018) Identification and structure-activity relationship of HDAC6 zinc-finger ubiquitin binding domain inhibitors. J Med Chem 61:4517–4527. https://doi.org/10.1021/acs.jmedchem.8b00258
Fessart D, Marza E, Taouji S, Delom F, Chevet E (2013) p97/CDC-48: proteostasis control in tumor cell biology. Cancer Lett 337:26–34. https://doi.org/10.1016/j.canlet.2013.05.030
Finley D (2009) Recognition and processing of ubiquitin-protein conjugates by the proteasome. Annu Rev Biochem 78:477–513. https://doi.org/10.1146/annurev.biochem.78.081507.101607
Forouzan D, Ammelburg M, Hobel CF, Stroh LJ, Sessler N, Martin J, Lupas AN (2012) The archaeal proteasome is regulated by a network of AAA ATPases. J Biol Chem 287:39254–39262. https://doi.org/10.1074/jbc.M112.386458
Franz A et al (2011) CDC-48/p97 coordinates CDT-1 degradation with GINS chromatin dissociation to ensure faithful DNA replication. Mol Cell 44:85–96. https://doi.org/10.1016/j.molcel.2011.08.028
Franz A et al (2016b) Chromatin-associated degradation is defined by UBXN-3/FAF1 to safeguard DNA replication fork progression. Nat Commun 7:10612. https://doi.org/10.1038/ncomms10612
Franz A, Ackermann L, Hoppe T (2014) Create and preserve: proteostasis in development and aging is governed by Cdc48/p97/VCP. Biochim Biophys Acta 1843:205–215. https://doi.org/10.1016/j.bbamcr.2013.03.031
Franz A, Ackermann L, Hoppe T (2016a) Ring of change: CDC48/p97 drives protein dynamics at chromatin. Front Genet 7:73. https://doi.org/10.3389/fgene.2016.00073
Gaggioli V, Zegerman P (2017) Terminating the replication helicase. Nat Cell Biol 19:410–412. https://doi.org/10.1038/ncb3519
Ganji R, Mukkavalli S, Somanji F, Raman M (2018) The VCP-UBXN1 complex mediates triage of ubiquitylated cytosolic proteins bound to the BAG6 complex. Mol Cell Biol. https://doi.org/10.1128/mcb.00154-18
Gareau A, Rico C, Boerboom D, Nadeau ME (2018) In vitro efficacy of a first-generation valosin-containing protein inhibitor (CB-5083) against canine lymphoma. Vet Comp Oncol 16:311–317. https://doi.org/10.1111/vco.12380
Gates SN et al (2017) Ratchet-like polypeptide translocation mechanism of the AAA+ disaggregase Hsp104. Science 357:273–279. https://doi.org/10.1126/science.aan1052
Ge Z et al (2018) Integrated genomic analysis of the ubiquitin pathway across cancer types. Cell Rep 23(213–226):e213. https://doi.org/10.1016/j.celrep.2018.03.047
Ghosal G, Leung JW, Nair BC, Fong KW, Chen J (2012) Proliferating cell nuclear antigen (PCNA)-binding protein C1orf124 is a regulator of translesion synthesis. J Biol Chem 287:34225–34233. https://doi.org/10.1074/jbc.m112.400135
Gibbs-Seymour I et al (2015) Ubiquitin-SUMO circuitry controls activated fanconi anemia ID complex dosage in response to DNA damage. Mol Cell 57:150–164. https://doi.org/10.1016/j.molcel.2014.12.001
Gugliotta G et al (2017) Valosin-containing protein/p97 as a novel therapeutic target in acute lymphoblastic leukemia. Neoplasia 19:750–761. https://doi.org/10.1016/j.neo.2017.08.001
Gui L et al (2016) Evaluating p97 inhibitor analogues for potency against p97–p37 and p97–Npl4–Ufd1 complexes. ChemMedChem 11:953–957. https://doi.org/10.1002/cmdc.201600036

Haines DS (2010) p97-containing complexes in proliferation control and cancer: emerging culprits or guilt by association? Genes Cancer 1:753–763. https://doi.org/10.1177/1947601910381381
Han H, Monroe N, Sundquist WI, Shen PS, Hill CP (2017) The AAA ATPase Vps4 binds ESCRT-III substrates through a repeating array of dipeptide-binding pockets. Elife 6. https://doi.org/10.7554/elife.31324
Hänzelmann P, Schindelin H (2011) The structural and functional basis of the p97/valosin-containing protein (VCP)-interacting motif (VIM): mutually exclusive binding of cofactors to the N-terminal domain of p97. J Biol Chem 286:38679–38690. https://doi.org/10.1074/jbc.m111.274506
Hänzelmann P, Schindelin H (2016a) Characterization of an additional binding surface on the p97 N-terminal domain involved in bipartite cofactor interactions. Structure 24:140–147. https://doi.org/10.1016/j.str.2015.10.027
Hänzelmann P, Schindelin H (2016b) Structural basis of ATP hydrolysis and intersubunit signaling in the AAA+ ATPase p97. Structure 24:127–139. https://doi.org/10.1016/j.str.2015.10.026
Hänzelmann P, Schindelin H (2017) The interplay of cofactor interactions and post-translational modifications in the regulation of the AAA+ ATPase p97. Front Mol Biosci 4:21. https://doi.org/10.3389/fmolb.2017.00021
Hänzelmann P, Buchberger A, Schindelin H (2011) Hierarchical binding of cofactors to the AAA ATPase p97. Structure 19:833–843 https://doi.org/10.1016/j.str.2011.03.018
Hänzelmann P, Schäfer A, Völler D, Schindelin H (2012) Structural insights into functional modes of proteins involved in ubiquitin family pathways. Methods Mol Biol 832:547–576. https://doi.org/10.1007/978-1-61779-474-2_39
Hao Q et al (2015) A non-canonical role of the p97 complex in RIG-I antiviral signaling. EMBO J 34:2903–2920. https://doi.org/10.15252/embj.201591888
He J et al (2014) Ubiquitin-specific protease 7 regulates nucleotide excision repair through deubiquitinating XPC protein and preventing XPC protein from undergoing ultraviolet light-induced and VCP/p97 protein-regulated proteolysis. J Biol Chem 289:27278–27289. https://doi.org/10.1074/jbc.m114.589812
Heidelberger JB, Voigt A, Borisova ME, Petrosino G, Ruf S, Wagner SA, Beli P (2018) Proteomic profiling of VCP substrates links VCP to K6-linked ubiquitylation and c-Myc function. EMBO Rep 19. https://doi.org/10.15252/embr.201744754
Hendriks IA, Lyon D, Young C, Jensen LJ, Vertegaal AC, Nielsen ML (2017) Site-specific mapping of the human SUMO proteome reveals co-modification with phosphorylation. Nat Struct Mol Biol 24:325–336. https://doi.org/10.1038/nsmb.3366
Heo JM et al (2010) A stress-responsive system for mitochondrial protein degradation. Mol Cell 40:465–480. https://doi.org/10.1016/j.molcel.2010.10.021
Her NG, Toth JI, Ma CT, Wei Y, Motamedchaboki K, Sergienko E, Petroski MD (2016) p97 composition changes caused by allosteric inhibition are suppressed by an on-target mechanism that increases the enzyme's ATPase activity cell. Chem Biol 23:517–528. https://doi.org/10.1016/j.chembiol.2016.03.012
Huang EY et al (2018) A VCP inhibitor substrate trapping approach (VISTA) enables proteomic profiling of endogenous ERAD substrates. Mol Biol Cell 29:1021–1030. https://doi.org/10.1091/mbc.e17-08-0514
Huang X, Dixit VM (2016) Drugging the undruggables: exploring the ubiquitin system for drug development. Cell Res 26:484–498. https://doi.org/10.1038/cr.2016.31
Huang C, Li G, Lennarz WJ (2012) Dynamic flexibility of the ATPase p97 is important for its interprotomer motion transmission. Proc Natl Acad Sci U S A 109:9792–9797. https://doi.org/10.1073/pnas.1205853109
Hülsmann J, Kravic B, Weith M, Gstaiger M, Aebersold R, Collins BC, Meyer H (2018) AP-SWATH reveals direct involvement of VCP/p97 in integrated stress response signaling through facilitating CReP/PPP1R15B degradation. Mol Cell Proteomics 17:1295–1307. https://doi.org/10.1074/mcp.ra117.000471

Husnjak K, Dikic I (2012) Ubiquitin-binding proteins: decoders of ubiquitin-mediated cellular functions. Annu Rev Biochem 81:291–322. https://doi.org/10.1146/annurev-biochem-051810-094654

Huyton T et al (2003) The crystal structure of murine p97/VCP at 3.6 Å. J Struct Biol 144:337–348. https://doi.org/10.1016/j.jsb.2003.10

Isaacson RL, Pye VE, Simpson P, Meyer HH, Zhang X, Freemont PS, Matthews S (2007) Detailed structural insights into the p97-Npl4-Ufd1 interface. J Biol Chem 282:21361–21369. https://doi.org/10.1074/jbc.m610069200

Joazeiro CAP (2017) Ribosomal stalling during translation: providing substrates for ribosome-associated protein quality control. Annu Rev Cell Dev Biol 33:343–368. https://doi.org/10.1146/annurev-cellbio-111315-125249

Kadowaki H, Satrimafitrah P, Takami Y, Nishitoh H (2018) Molecular mechanism of ER stress-induced pre-emptive quality control involving association of the translocon, Derlin-1, and HRD1. Sci Rep 8:7317. https://doi.org/10.1038/s41598-018-25724-x

Kang W, Yang JK (2011) Crystal structure of human FAF1 UBX domain reveals a novel FcisP touch-turn motif in p97/VCP-binding region. Biochem Biophys Res Commun 407:531–534. https://doi.org/10.1016/j.bbrc.2011.03.052

Karbowski M, Youle RJ (2011) Regulating mitochondrial outer membrane proteins by ubiquitination and proteasomal degradation. Curr Opin Cell Biol 23:476–482. https://doi.org/10.1016/j.ceb.2011.05.007

Kato Y, Miyakawa T, Tanokura M (2018) Overview of the mechanism of cytoskeletal motors based on structure. Biophys Rev 10:571–581. https://doi.org/10.1007/s12551-017-0368-1

Kern M, Fernandez-Saiz V, Schafer Z, Buchberger A (2009) UBXD1 binds p97 through two independent binding sites. Biochem Biophys Res Commun 380:303–307. https://doi.org/10.1016/j.bbrc.2009.01.076

Kerscher O, Felberbaum R, Hochstrasser M (2006) Modification of proteins by ubiquitin and ubiquitin-like proteins. Annu Rev Cell Dev Biol 22:159–180. https://doi.org/10.1146/annurev.cellbio.22.010605.093503

Kim SJ et al (2014) Structural basis for ovarian tumor domain-containing protein 1 (OTU1) binding to p97/valosin-containing protein (VCP). J Biol Chem 289:12264–12274. https://doi.org/10.1074/jbc.m113.523936

Kim KH, Kang W, Suh SW, Yang JK (2011) Crystal structure of FAF1 UBX domain in complex with p97/VCP N domain reveals a conformational change in the conserved FcisP touch-turn motif of UBX domain. Proteins 79:2583–2587. https://doi.org/10.1002/prot.23073

Kocaturk NM, Gozuacik D (2018) Crosstalk between mammalian autophagy and the ubiquitin-proteasome system. Front Cell Dev Biol 6:128. https://doi.org/10.3389/fcell.2018.00128

Kondo H, Rabouille C, Newman R, Levine TP, Pappin D, Freemont P, Warren G (1997) p47 is a cofactor for p97-mediated membrane fusion. Nature 388:75–78. https://doi.org/10.1038/40411

Kumari N, Lee KK, Jha S (2018) Targeting the ubiquitin proteasome system in cancer. IntechOpen. https://doi.org/10.5772/intechopen.76705

Lan B, Chai S, Wang P, Wang K (2017) VCP/p97/Cdc48, a linking of protein homeostasis and cancer therapy. Curr Mol Med 17:608–618. https://doi.org/10.2174/1566524018666180308111238

Lander GC, Estrin E, Matyskiela ME, Bashore C, Nogales E, Martin A (2012) Complete subunit architecture of the proteasome regulatory particle. Nature 482:186–191. https://doi.org/10.1038/nature10774

LaPorte MG et al (2018) Optimization of phenyl indole inhibitors of the AAA+ ATPase p97. ACS Med Chem Lett 9:1075–1081. https://doi.org/10.1021/acsmedchemlett.8b00372

Le Moigne R et al (2017) The p97 inhibitor CB-5083 is a unique disrupter of protein homeostasis in models of multiple myeloma. Mol Cancer Ther 16:2375–2386. https://doi.org/10.1158/1535-7163.mct-17-0233

Le LT, Kang W, Kim JY, Le OT, Lee SY, Yang JK (2016) Structural details of Ufd1 binding to p97 and their functional implications in ER-associated degradation. PLoS One 11:e0163394. https://doi.org/10.1371/journal.pone.0163394

Lee JY et al (2010) HDAC6 controls autophagosome maturation essential for ubiquitin-selective quality-control autophagy. EMBO J 29:969–980. https://doi.org/10.1038/emboj.2009.405

Lee JJ, Park JK, Jeong J, Jeon H, Yoon JB, Kim EE, Lee KJ (2013) Complex of Fas-associated factor 1 (FAF1) with valosin-containing protein (VCP)-Npl4-Ufd1 and polyubiquitinated proteins promotes endoplasmic reticulum-associated degradation (ERAD). J Biol Chem 288:6998–7011. https://doi.org/10.1074/jbc.m112.417576

Lee BH, Schwager F, Meraldi P, Gotta M (2018) p37/UBXN2B regulates spindle orientation by limiting cortical NuMA recruitment via PP1/Repo-Man. J Cell Biol 217:483–493. https://doi.org/10.1083/jcb.201707050

Li G, Zhao G, Schindelin H, Lennarz WJ (2008) Tyrosine phosphorylation of ATPase p97 regulates its activity during ERAD. Biochem Biophys Res Commun 375:247–251. https://doi.org/10.1016/j.bbrc.2008.08.018

Li G, Huang C, Zhao G, Lennarz WJ (2012) Interprotomer motion-transmission mechanism for the hexameric AAA ATPase p97. Proc Natl Acad Sci U S A 109:3737–3741. https://doi.org/10.1073/pnas.1200255109

Li ZH, Wang Y, Xu M, Jiang T (2017) Crystal structures of the UBX domain of human UBXD7 and its complex with p97 ATPase. Biochem Biophys Res Commun 486:94–100. https://doi.org/10.1016/j.bbrc.2017.03.005

Liebelt F, Vertegaal AC (2016) Ubiquitin-dependent and independent roles of SUMO in proteostasis. Am J Physiol Cell Physiol 311:284–296. https://doi.org/10.1152/ajpcell.00091.2016

Lim JJ et al (2016a) Structural insights into the interaction of p97 N-terminus domain and VBM in rhomboid protease, RHBDL4. Biochem J 473:2863–2880. https://doi.org/10.1042/bcj20160237

Lim JJ et al (2016b) Structural insights into the interaction of human p97 N-terminal domain and SHP motif in Derlin-1 rhomboid pseudoprotease. FEBS Lett 590:4402–4413. https://doi.org/10.1002/1873-3468.12447

Liu S, Yang H, Zhao J, Zhang YH, Song AX, Hu HY (2013) NEDD8 ultimate buster-1 long (NUB1L) protein promotes transfer of NEDD8 to proteasome for degradation through the P97UFD1/NPL4 complex. J Biol Chem 288:31339–31349. https://doi.org/10.1074/jbc.m113.484816

Maghames CM et al (2018) NEDDylation promotes nuclear protein aggregation and protects the Ubiquitin Proteasome System upon proteotoxic stress. Nat Commun 9:4376. https://doi.org/10.1038/s41467-018-06365-0

Magnaghi P et al (2013) Covalent and allosteric inhibitors of the ATPase VCP/p97 induce cancer cell death. Nat Chem Biol 9:548–556. https://doi.org/10.1038/nchembio.1313

Manasanch EE, Orlowski RZ (2017) Proteasome inhibitors in cancer therapy. Nat Rev Clin Oncol 14:417–433. https://doi.org/10.1038/nrclinonc.2016.206

Maric M, Mukherjee P, Tatham MH, Hay R, Labib K (2017) Ufd1-Npl4 recruit Cdc48 for disassembly of ubiquitylated CMG helicase at the end of chromosome replication. Cell Rep 18:3033–3042. https://doi.org/10.1016/j.celrep.2017.03.020

Mevissen TET, Komander D (2017) Mechanisms of deubiquitinase specificity and regulation. Annu Rev Biochem 86:159–192. https://doi.org/10.1146/annurev-biochem-061516-044916

Meyer H, Weihl CC (2014) The VCP/p97 system at a glance: connecting cellular function to disease pathogenesis. J Cell Sci 127:3877–3883. https://doi.org/10.1242/jcs.093831

Meyer HH, Kondo H, Warren G (1998) The p47 co-factor regulates the ATPase activity of the membrane fusion protein, p97. FEBS Lett 437:255–257. https://doi.org/10.1016/S0014-5793(98)01232-0

Meyer H, Bug M, Bremer S (2012) Emerging functions of the VCP/p97 AAA-ATPase in the ubiquitin system. Nat Cell Biol 14:117–123. https://doi.org/10.1038/ncb2407

Miller JM, Enemark EJ (2016) Fundamental characteristics of AAA+ protein family structure and function. Archaea 2016:9294307. https://doi.org/10.1155/2016/9294307

Mishima Y et al (2015) Ricolinostat (ACY-1215) induced inhibition of aggresome formation accelerates carfilzomib-induced multiple myeloma cell death. Br J Haematol 169:423–434. https://doi.org/10.1111/bjh.13315

Mogk A, Kummer E, Bukau B (2015) Cooperation of Hsp70 and Hsp100 chaperone machines in protein disaggregation. Front Mol Biosci 2:22. https://doi.org/10.3389/fmolb.2015.00022

Moreno SP, Bailey R, Campion N, Herron S, Gambus A (2014) Polyubiquitylation drives replisome disassembly at the termination of DNA replication. Science 346:477–481. https://doi.org/10.1126/science.1253585

Mosbech A et al (2012) DVC1 (C1orf124) is a DNA damage-targeting p97 adaptor that promotes ubiquitin-dependent responses to replication blocks. Nat Struct Mol Biol 19:1084–1092. https://doi.org/10.1038/nsmb.2395

Na H, Song G (2016) Predicting the functional motions of p97 using symmetric normal modes. Proteins 84:1823–1835. https://doi.org/10.1002/prot.25164

Nguyen TV, Li J, Lu CJ, Mamrosh JL, Lu G, Cathers BE, Deshaies RJ (2017) p97/VCP promotes degradation of CRBN substrate glutamine synthetase and neosubstrates. Proc Natl Acad Sci U S A 114:3565–3571. https://doi.org/10.1073/pnas.1700949114

Nie M, Boddy MN (2016) Cooperativity of the SUMO and ubiquitin pathways in genome stability. Biomolecules 6:14. https://doi.org/10.3390/biom6010014

Niwa H, Ewens CA, Tsang C, Yeung HO, Zhang X, Freemont PS (2012) The role of the N-domain in the ATPase activity of the mammalian AAA ATPase p97/VCP. J Biol Chem 287:8561–8570. https://doi.org/10.1074/jbc.m111.302778

Papadopoulos C et al (2017) VCP/p97 cooperates with YOD1, UBXD1 and PLAA to drive clearance of ruptured lysosomes by autophagy. EMBO J 36:135–150. https://doi.org/10.15252/embj.201695148

Papadopoulos C, Meyer H (2017) Detection and clearance of damaged lysosomes by the endo-lysosomal damage response and lysophagy. Curr Biol 27:R1330–R1341. https://doi.org/10.1016/j.cub.2017.11.012

Pilla E, Schneider K, Bertolotti A (2017) Coping with protein quality control failure. Annu Rev Cell Dev Biol 33:439–465. https://doi.org/10.1146/annurev-cellbio-111315-125334

Polucci P et al (2013) Alkylsulfanyl-1,2,4-triazoles, a new class of allosteric valosine containing protein inhibitors. Synthesis and structure-activity relationships. J Med Chem 56:437–450. https://doi.org/10.1021/jm3013213

Protter DSW, Parker R (2016) Principles and properties of stress granules. Trends Cell Biol 26:668–679. https://doi.org/10.1016/j.tcb.2016.05.004

Puumalainen MR, Lessel D, Ruthemann P, Kaczmarek N, Bachmann K, Ramadan K, Naegeli H (2014) Chromatin retention of DNA damage sensors DDB2 and XPC through loss of p97 segregase causes genotoxicity. Nat Commun 5:3695. https://doi.org/10.1038/ncomms4695

Pye VE et al (2007) Structural insights into the p97-Ufd1-Npl4 complex. Proc Natl Acad Sci U S A 104:467–472. https://doi.org/10.1073/pnas.0603408104

Qiu L et al (2010) Structure and function of the PLAA/Ufd3-p97/Cdc48 complex. J Biol Chem 285:365–372. https://doi.org/10.1074/jbc.m109.044685

Ramadan K, Halder S, Wiseman K, Vaz B (2017) Strategic role of the ubiquitin-dependent segregase p97 (VCP or Cdc48) in DNA replication. Chromosoma 126:17–32. https://doi.org/10.1007/s00412-016-0587-4

Raman M et al (2015) Systematic proteomics of the VCP-UBXD adaptor network identifies a role for UBXN10 in regulating ciliogenesis. Nat Cell Biol 17:1356–1369. https://doi.org/10.1038/ncb3238

Ramaswami M, Taylor JP, Parker R (2013) Altered ribostasis: RNA-protein granules in degenerative disorders. Cell 154:727–736. https://doi.org/10.1016/j.cell.2013.07.038

Rao MV, Williams DR, Cocklin S, Loll PJ (2017) Interaction between the AAA(+) ATPase p97 and its cofactor ataxin3 in health and disease: nucleotide-induced conformational changes regulate cofactor binding. J Biol Chem 292:18392–18407. https://doi.org/10.1074/jbc.m117.806281

Rezvani K (2016) UBXD Proteins: a family of proteins with diverse functions in cancer. Int J Mol Sci 17. https://doi.org/10.3390/ijms17101724

Ripstein ZA, Huang R, Augustyniak R, Kay LE, Rubinstein JL (2017) Structure of a AAA+ unfoldase in the process of unfolding substrate. Elife 6. https://doi.org/10.7554/elife.25754

Ritz D et al (2011) Endolysosomal sorting of ubiquitylated caveolin-1 is regulated by VCP and UBXD1 and impaired by VCP disease mutations. Nat Cell Biol 13:1116–1123. https://doi.org/10.1038/ncb2301

Rouiller I, Butel VM, Latterich M, Milligan RA, Wilson-Kubalek EM (2000) A major conformational change in p97 AAA ATPase upon ATP binding. Mol Cell 6:1485–1490. https://doi.org/10.1016/S1097-2765(00)00144-1

Ruthemann P, Balbo Pogliano C, Naegeli H (2016) Global-genome nucleotide excision repair controlled by ubiquitin/sumo modifiers. Front Genet 7:68. https://doi.org/10.3389/fgene.2016.00068

Ryslava H, Doubnerova V, Kavan D, Vanek O (2013) Effect of posttranslational modifications on enzyme function and assembly. J Proteomics 92:80–109. https://doi.org/10.1016/j.jprot.2013.03.025

Saeki Y (2017) Ubiquitin recognition by the proteasome. J Biochem 161:113–124. https://doi.org/10.1093/jb/mvw091

Saffert P, Enenkel C, Wendler P (2017) Structure and function of p97 and Pex1/6 Type II AAA+ complexes. Front Mol Biosci 4:33. https://doi.org/10.3389/fmolb.2017.00033

Sahtoe DD, Sixma TK (2015) Layers of DUB regulation. Trends Biochem Sci 40:456–467. https://doi.org/10.1016/j.tibs.2015.05.002

Schoebel S et al (2017) Cryo-EM structure of the protein-conducting ERAD channel Hrd1 in complex with Hrd3. Nature 548:352–355. https://doi.org/10.1038/nature23314

Schuetz AK, Kay LE (2016) A dynamic molecular basis for malfunction in disease mutants of p97/VCP. Elife 5. https://doi.org/10.7554/elife.20143

Schuller JM, Beck F, Lossl P, Heck AJ, Forster F (2016) Nucleotide-dependent conformational changes of the AAA+ ATPase p97 revisited. FEBS Lett 590:595–604. https://doi.org/10.1002/1873-3468.12091

Schwertman P, Bekker-Jensen S, Mailand N (2016) Regulation of DNA double-strand break repair by ubiquitin and ubiquitin-like modifiers. Nat Rev Mol Cell Biol 17:379–394. https://doi.org/10.1038/nrm.2016.58

Segura-Cabrera A, Tripathi R, Zhang X, Gui L, Chou TF, Komurov K (2017) A structure- and chemical genomics-based approach for repositioning of drugs against VCP/p97 ATPase. Sci Rep 7:44912. https://doi.org/10.1038/srep44912

Skrott Z et al (2017) Alcohol-abuse drug disulfiram targets cancer via p97 segregase adaptor NPL4. Nature 552:194–199. https://doi.org/10.1038/nature25016

Stach L, Freemont PS (2017) The AAA+ ATPase p97, a cellular multitool. Biochem J 474:2953–2976. https://doi.org/10.1042/bcj20160783

Stapf C, Cartwright E, Bycroft M, Hofmann K, Buchberger A (2011) The general definition of the p97/valosin-containing protein (VCP)-interacting motif (VIM) delineates a new family of p97 cofactors. J Biol Chem 286:38670–38678. https://doi.org/10.1074/jbc.m111.274472

Tanaka A, Cleland MM, Xu S, Narendra DP, Suen DF, Karbowski M, Youle RJ (2010) Proteasome and p97 mediate mitophagy and degradation of mitofusins induced by Parkin. J Cell Biol 191:1367–1380. https://doi.org/10.1083/jcb.201007013

Tang WK, Xia D (2013) Altered intersubunit communication is the molecular basis for functional defects of pathogenic p97 mutants. J Biol Chem 288:36624–36635. https://doi.org/10.1074/jbc.m113.488924

Tang WK, Xia D (2016) Mutations in the human AAA(+) chaperone p97 and related diseases. Front Mol Biosci 3:79. https://doi.org/10.3389/fmolb.2016.00079

Tang WK, Li D, Li CC, Esser L, Dai R, Guo L, Xia D (2010) A novel ATP-dependent conformation in p97 N-D1 fragment revealed by crystal structures of disease-related mutants. EMBO J 29:2217–2229. https://doi.org/10.1038/emboj.2010.104

Tang WK, Zhang T, Ye Y, Xia D (2017) Structural basis for nucleotide-modulated p97 association with the ER membrane. Cell Discov 3:17045. https://doi.org/10.1038/celldisc.2017.45

Tang WK, Odzorig T, Jin W, Xia D (2019) Structural basis of p97 inhibition by the site-selective anti-cancer compound CB-5083. Mol Pharmacol 95:286-293. https://doi.org/10.1124/mol.118.114256

Taylor JP, Brown RH Jr, Cleveland DW (2016) Decoding ALS: from genes to mechanism. Nature 539:197–206. https://doi.org/10.1038/nature20413

Torrecilla I, Oehler J, Ramadan K (2017) The role of ubiquitin-dependent segregase p97 (VCP or Cdc48) in chromatin dynamics after DNA double strand breaks. Philos Trans R Soc Lond B Biol Sci 372. https://doi.org/10.1098/rstb.2016.0282

Trusch F et al (2015) The N-terminal region of the ubiquitin regulatory X (UBX) domain-containing protein 1 (UBXD1) modulates interdomain communication within the valosin-containing protein p97. J Biol Chem 290:29414–29427. https://doi.org/10.1074/jbc.m115.680686

Tsuchiya H, Ohtake F, Arai N, Kaiho A, Yasuda S, Tanaka K, Saeki Y (2017) In vivo ubiquitin linkage-type analysis reveals that the Cdc48-Rad23/Dsk2 axis contributes to K48-linked chain specificity of the proteasome. Mol Cell 66:488–502 e 487. https://doi.org/10.1016/j.molcel.2017.04.024

Turakhiya A et al (2018) ZFAND1 recruits p97 and the 26S proteasome to promote the clearance of arsenite-induced stress granules. Mol Cell 70:906–919 e 907. https://doi.org/10.1016/j.molcel.2018.04.021

Twomey EC et al (2019) Substrate processing by the Cdc48 ATPase complex is initiated by ubiquitin unfolding. Science. https://doi.org/10.1126/science.aax1033

Uchiyama K et al (2006) p37 is a p97 adaptor required for Golgi and ER biogenesis in interphase and at the end of mitosis. Dev Cell 11:803–816. https://doi.org/10.1016/j.devcel.2006.10.016

van den Boom J et al (2016) VCP/p97 extracts sterically trapped Ku70/80 rings from DNA in double-strand break repair. Mol Cell 64:189–198. https://doi.org/10.1016/j.molcel.2016.08.037

van den Boom J, Meyer H (2018) VCP/p97-mediated unfolding as a principle in protein homeostasis and signaling. Mol Cell 69:182–194. https://doi.org/10.1016/j.molcel.2017.10.028

Vekaria PH, Home T, Weir S, Schoenen FJ, Rao R (2016) Targeting p97 to disrupt protein homeostasis in cancer. Front Oncol 6:181. https://doi.org/10.3389/fonc.2016.00181

Vekaria PH et al (2019) Functional cooperativity of p97 and histone deacetylase 6 in mediating DNA repair in mantle cell lymphoma cells. Leukemia 33:1675–1686. https://doi.org/10.1038/s41375-018-0355-y

Venne AS, Kollipara L, Zahedi RP (2014) The next level of complexity: crosstalk of posttranslational modifications. Proteomics 14:513–524. https://doi.org/10.1002/pmic.201300344

Verbinnen I, Ferreira M, Bollen M (2017) Biogenesis and activity regulation of protein phosphatase 1. Biochem Soc Trans 45:89–99. https://doi.org/10.1042/bst20160154

Verma R, Oania RS, Kolawa NJ, Deshaies RJ (2013) Cdc48/p97 promotes degradation of aberrant nascent polypeptides bound to the ribosome. Elife 2:e00308. https://doi.org/10.7554/elife.00308

Wallington-Beddoe CT, Sobieraj-Teague M, Kuss BJ, Pitson SM (2018) Resistance to proteasome inhibitors and other targeted therapies in myeloma. Br J Haematol 182:11–28. https://doi.org/10.1111/bjh.15210

Wang T et al (2016) Pathogenic mutations in the valosin-containing protein/p97(VCP) N-domain inhibit the SUMOylation of VCP and lead to impaired stress response. J Biol Chem 291:14373–14384. https://doi.org/10.1074/jbc.m116.729343

Wang Q, Liu Y, Soetandyo N, Baek K, Hegde R, Ye Y (2011) A ubiquitin ligase-associated chaperone holdase maintains polypeptides in soluble states for proteasome degradation. Mol Cell 42:758–770. https://doi.org/10.1016/j.molcel.2011.05.010

Wei Y, Toth JI, Blanco GA, Bobkov AA, Petroski MD (2018) Adapted ATPase domain communication overcomes the cytotoxicity of p97 inhibitors. J Biol Chem 293:20169–20180. https://doi.org/10.1074/jbc.ra118.004301

Weith M et al (2018) Ubiquitin-independent disassembly by a p97 AAA-ATPase complex drives PP1 holoenzyme formation. Mol Cell 72:766–777 e 766. https://doi.org/10.1016/j.molcel.2018.09.020

Wendler P, Ciniawsky S, Kock M, Kube S (2012) Structure and function of the AAA+ nucleotide binding pocket. Biochim Biophys Acta 1823:2–14. https://doi.org/10.1016/j.bbamcr.2011.06.014

Wertz IE, Wang X (2019) From discovery to bedside: targeting the ubiquitin system. Cell Chem Biol 26:156-177. https://doi.org/10.1016/j.chembiol.2018.10.022

White KI, Zhao M, Choi UB, Pfuetzner RA, Brunger AT (2018) Structural principles of SNARE complex recognition by the AAA+ protein NSF. Elife 7. https://doi.org/10.7554/elife.38888

Wu X, Rapoport TA (2018) Mechanistic insights into ER-associated protein degradation. Curr Opin Cell Biol 53:22–28. https://doi.org/10.1016/j.ceb.2018.04.004

Xia D, Tang WK, Ye Y (2016) Structure and function of the AAA+ ATPase p97/Cdc48p. Gene 583:64–77. https://doi.org/10.1016/j.gene.2016.02.042

Xie F et al (2017) FAF1 phosphorylation by AKT accumulates TGF-beta type II receptor and drives breast cancer metastasis. Nat Commun 8:15021. https://doi.org/10.1038/ncomms15021

Xu G, Paige JS, Jaffrey SR (2010) Global analysis of lysine ubiquitination by ubiquitin remnant immunoaffinity profiling. Nat Biotechnol 28:868–873. https://doi.org/10.1038/nbt.1654

Xu S, Peng G, Wang Y, Fang S, Karbowski M (2011) The AAA-ATPase p97 is essential for outer mitochondrial membrane protein turnover. Mol Biol Cell 22:291–300. https://doi.org/10.1091/mbc.e10-09-0748

Xu Y, Liu Y, Lee JG, Ye Y (2013) A ubiquitin-like domain recruits an oligomeric chaperone to a retrotranslocation complex in endoplasmic reticulum-associated degradation. J Biol Chem 288:18068–18076. https://doi.org/10.1074/jbc.M112.449199

Xue L et al (2016) Valosin-containing protein (VCP)-adaptor interactions are exceptionally dynamic and subject to differential modulation by a VCP inhibitor. Mol Cell Proteomics 15:2970–2986. https://doi.org/10.1074/mcp.m116.061036

Yau R, Rape M (2016) The increasing complexity of the ubiquitin code. Nat Cell Biol 18:579–586. https://doi.org/10.1038/ncb3358

Ye Y, Meyer HH, Rapoport TA (2003) Function of the p97-Ufd1-Npl4 complex in retrotranslocation from the ER to the cytosol: dual recognition of nonubiquitinated polypeptide segments and polyubiquitin chains. J Cell Biol 162:71–84. https://doi.org/10.1083/jcb.200302169

Ye Y, Tang WK, Zhang T, Xia D (2017) A mighty "protein extractor" of the cell: structure and function of the p97/CDC48 ATPase. Front Mol Biosci 4:39. https://doi.org/10.3389/fmolb.2017.00039

Ye Y, Baek SH, Ye Y, Zhang T (2018) Proteomic characterization of endogenous substrates of mammalian ubiquitin ligase Hrd1. Cell Biosci 8:46. https://doi.org/10.1186/s13578-018-0245-z

Yedidi RS, Wendler P, Enenkel C (2017) AAA-ATPases in protein degradation. Front Mol Biosci 4:42. https://doi.org/10.3389/fmolb.2017.00042

Zhang X et al (2015) Altered cofactor regulation with disease-associated p97/VCP mutations. Proc Natl Acad Sci U S A 112:1705–1714. https://doi.org/10.1073/pnas.1418820112

Zhang L, Gong F (2016) The emerging role of deubiquitination in nucleotide excision repair. DNA Repair (Amst) 44:118–122. https://doi.org/10.1016/j.dnarep.2016.05.035

Zhao M, Brunger AT (2016) Recent advances in deciphering the structure and molecular mechanism of the AAA+ ATPase N-ethylmaleimide-sensitive factor (NSF). J Mol Biol 428:1912–1926. https://doi.org/10.1016/j.jmb.2015.10.026

Zhao G, Zhou X, Wang L, Li G, Schindelin H, Lennarz WJ (2007) Studies on peptide: N-glycanase-p97 interaction suggest that p97 phosphorylation modulates endoplasmic reticulum-associated degradation. Proc Natl Acad Sci U S A 104:8785–8790. https://doi.org/10.1073/pnas.0702966104

Zhao M, Wu S, Zhou Q, Vivona S, Cipriano DJ, Cheng Y, Brunger AT (2015) Mechanistic insights into the recycling machine of the SNARE complex. Nature 518:61–67. https://doi.org/10.1038/nature14148

Zhou HJ et al (2015) Discovery of a first-in-class, potent, selective, and orally bioavailable inhibitor of the p97 AAA ATPase (CB-5083). J Med Chem 58:9480–9497. https://doi.org/10.1021/acs.jmedchem.5b01346

Chapter 8
Penicillin-Binding Proteins (PBPs) and Bacterial Cell Wall Elongation Complexes

Mayara M. Miyachiro, Carlos Contreras-Martel and Andréa Dessen

Abstract The bacterial cell wall is the validated target of mainstream antimicrobials such as penicillin and vancomycin. Penicillin and other β-lactams act by targeting Penicillin-Binding Proteins (PBPs), enzymes that play key roles in the biosynthesis of the main component of the cell wall, the peptidoglycan. Despite the spread of resistance towards these drugs, the bacterial cell wall continues to be a major Achilles' heel for microbial survival, and the exploration of the cell wall formation machinery is a vast field of work that can lead to the development of novel exciting therapies. The sheer complexity of the cell wall formation process, however, has created a significant challenge for the study of the macromolecular interactions that regulate peptidoglycan biosynthesis. New developments in genetic and biochemical screens, as well as different aspects of structural biology, have shed new light on the importance of complexes formed by PBPs, notably within the cell wall elongation machinery. This chapter summarizes structural and functional details of PBP complexes involved in the periplasmic and membrane steps of peptidoglycan biosynthesis with a focus on cell wall elongation. These assemblies could represent interesting new targets for the eventual development of original antibacterials.

Keywords Penicillin-Binding proteins (PBPs) · Peptidoglycan · Cell wall elongation · Protein complexes · Periplasm · Membrane proteins

---

M. M. Miyachiro · A. Dessen (✉)
Brazilian Biosciences National Laboratory (LNBio), CNPEM, Campinas, Brazil
e-mail: andrea.dessen@ibs.fr

M. M. Miyachiro
e-mail: mayara.miyachiro@lnbio.cnpem.br

C. Contreras-Martel · A. Dessen
Univ Grenoble Alpes, CNRS, CEA, Institut de Biologie Structurale (IBS), Bacterial Pathogenesis Group, Grenoble, France
e-mail: carlos.contreras-martel@ibs.fr

J. R. Harris and J. Marles-Wright (eds.), *Macromolecular Protein Complexes II: Structure and Function*, Subcellular Biochemistry 93,
https://doi.org/10.1007/978-3-030-28151-9_8

Introduction

Bacteria are unicellular organisms whose growth and morphology are strictly dependent on their cell wall. Despite the fact that bacteria exist in different shapes, such as rods, spirals or cocci, the cell wall ensures that their morphologies remain the same during growth, with progeny inheriting the same shape as the original cells (Egan et al. 2017). The cell wall is a complex structure that surrounds the entire cell, and also plays important roles in maintenance of osmotic pressure and association of surface-bound virulence factors and lipopolysaccharides (Matteï et al. 2010; Egan et al. 2017). Both the bacterial cell wall biosynthetic pathway and elements of the wall itself are major targets of antibiotics employed in clinics worldwide.

One of the major components of the bacterial cell wall is the peptidoglycan, a mesh-like structure that surrounds the inner membrane and is composed of polymerized disaccharide subunits, N-acetylglucosamine (GlcNAc) and N-acetylmuramic acid (MurNAc), cross-linked by stem peptides (Fig. 8.1). Cell growth and division are highly dynamic processes, requiring the biosynthesis of

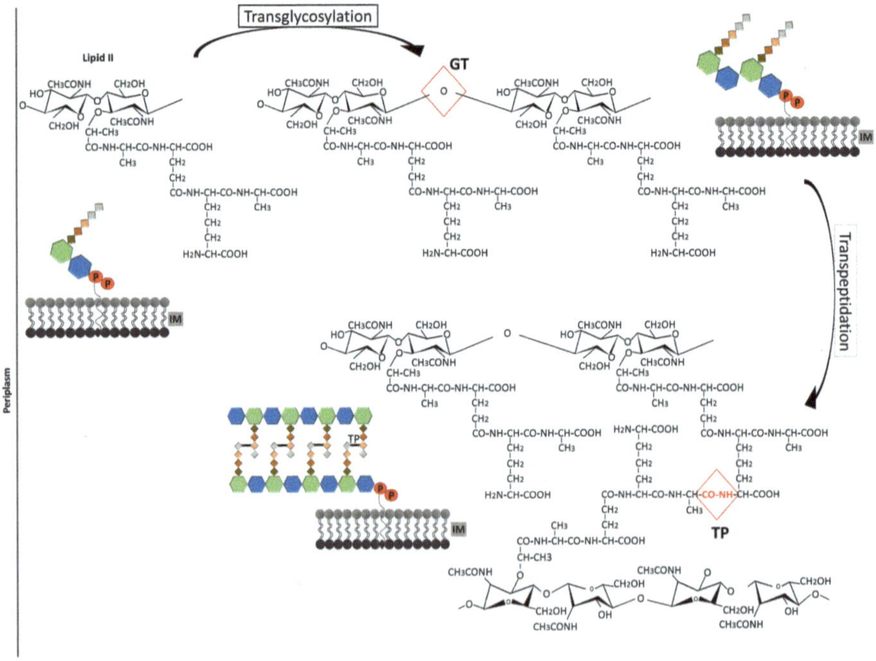

Fig. 8.1 Transglycosylation and transpeptidation reactions involved in peptidoglycan biosynthesis. After Lipid II is flipped to the periplasmic side, glycosyltransferases (GT) polymerize the sugar chains. Subsequently, transpeptidases (TP) catalyze the formation of cross-bridges between the peptides, removing the last amino acid (D-Ala) from the pentapeptide. GlcNAc and MurNAc are represented by blue and green hexagons, respectively. Amino acids in stem peptides are shown as small lozenges. Adapted from (Vollmer et al. 2008)

new peptidoglycan and the cleavage of the old material in order to allow the attachment of the new one to the membrane. Synthesis, modification, and recycling of peptidoglycan are all catalyzed by synthases and hydrolases that associate into different complexes and act in spatio-temporal fashion, in order to guarantee maintenance of cell shape during different phases of the cycle (Pazos et al. 2017).

Peptidoglycan biosynthesis in rod-shaped bacteria such as *E. coli* and *P. aeruginosa* occurs in two phases: cell division, where the septum and polar caps of the cell are synthesized, and elongation, where cellular growth occurs along the longitudinal axis of the cell (den Blaauwen et al. 2008). Three cellular compartments are involved in these processes. In the cytoplasm, the synthesis of UDP-N-acetylmuramyl-pentapeptide (UDP-MurNAc-pentapeptide) from UDP-N-acetyl-glucosamine (UDP-GlcNAc) occurs via the action of enzymes MurA through MurF (Pazos et al. 2017). The second stage involves the inner face of the cytoplasmic membrane, where the integral membrane protein MraY catalyzes the formation of Lipid I by transferring the phospho-MurNAc-pentapeptide motif of UDP-MurNAc-pentapeptide to a lipid carrier, undecaprenyl phosphate (Höltje 1998; Matteï et al. 2010; Barreteau et al. 2008). Association between Lipid I and a GlcNAc molecule by the glycosyltransferase MurG subsequently generates Lipid II (Liu and Breukink 2016) that is eventually translocated to the periplasm by flippases (Mohammadi et al. 2011; Meeske et al. 2015; Sham et al. 2014). In the periplasm, Penicillin Binding Proteins (PBPs) and Shape, Elongation, Division, and Sporulation (SEDS) proteins incorporate Lipid II into the growing peptidoglycan layer through glycosylation and transpeptidation reactions (Leclercq et al. 2017; Matteï et al. 2010; Cho et al. 2016; Meeske et al. 2016). Upon Lipid II polymerization, undecaprenyl pyrophosphate is released, and the polymerized glycan chains are inserted into the cell wall. In view of the complexity of these steps, it is noteworthy that bacteria must also 'switch' between the septal and cylindrical modes of peptidoglycan biosynthesis in a regulated fashion. It is of interest that rod-like cells elongate by inserting newly synthesized peptidoglycan into different sites along the wall, in a process called *dispersed elongation* (Typas et al. 2012).

Several of the proteins related to the above mentioned processes are involved in distinct complexes that comprise cytoskeletal elements. The 'divisome', which is essential for cell division, depends on tubulin-like polymers of FtsZ, while the 'elongasome', or Rod complex, plays a key role during the cell wall elongation phase and requires participation of the actin homolog MreB (Fig. 8.2) (Pazos et al. 2017). These multi-protein assemblies implicate all three compartments of the cell, with those located in the cytoplasm having been the most studied. The recent structural characterization of a complex between a PBP and an elongasome partner protein, as well as the description of interactions between PBPs and other essential periplasmic macromolecules, underline the timely nature of addressing protein complexes formed in the periplasm, which are the subject of this chapter.

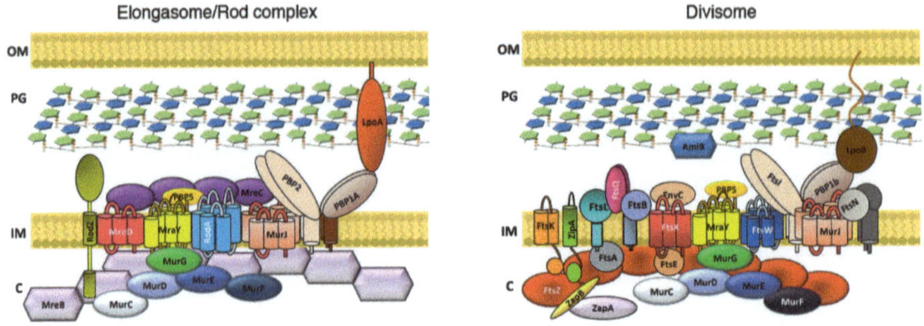

Fig. 8.2 Schematic diagrams of the elongasome and divisome complexes in Gram-negative bacteria. The peptidoglycan (PG) is indicated as a mesh of polymerized GlcNac (blue) and MurNAc (green) units, with the stem peptides indicated in red. Lpo regulating factors are associated to the outer membrane (OM), while numerous protein partners are embedded in the inner membrane (IM). The cytoskeletal factors MreB and FtsZ polymerize within the cytoplasm (C) and associate to the inner side of the IM. Figure adapted from (Hugonnet et al. 2016)

Fold and Functions of Penicillin-Binding Proteins (PBPs)

The cytoplasmic and membrane-bound steps of PG biosynthesis culminate with the synthesis of Lipid II and its 'flipping' to the outside of the inner membrane, and thus into the periplasm. The glycan chains of Lipid II are then acted upon by glycosyltransferases, producing the nascent, linear PG, whose stem peptides are then a substrate for the transpeptidase activity of PBPs. High molecular mass (HMM) PBPs were classified as class A, carrying both glycosyltransferase and transpeptidase activities, or class B, being able only to catalyze transpeptidation (Macheboeuf et al. 2006) (Fig. 8.3). However, proteins from the SEDS family have also been shown to possess the ability to catalyze glycan chain polymerization (Meeske et al. 2015), and will be further discussed below.

In what relates to transpeptidation, it is precisely this PBP-catalyzed reaction that is inhibited by β-lactams, whose use eventually leads to peptidoglycan weakening and cell lysis (Walsh 2003). Transpeptidation generally requires that the PBP recognize the terminal D-alanine, D-alanine moiety of the stem peptide, subsequently catalyzing the attack of the carbonyl group of the penultimate D-alanine by the lateral amino group at position (3) of an adjacent chain (in a classical '4–3' crosslink, see Fig. 8.1). The acceptor involved in enzyme deacylation depends on the organism; in *S. aureus*, it is a glycine residue, while it is a lysine or a diamino-pimelate group in *S. pneumoniae* or *E. coli*, respectively. If the acceptor is a water molecule, stem pentapeptides are hydrolyzed into tetrapeptides (with D-alanine being released), in a reaction catalyzed by Low molecular mass (LMM) PBPs, preventing further PG crosslinking (Goffin and Ghuysen 2002).

Blocking of either the transpeptidation or carboxypeptidation reactions by β-lactam antibiotics, which structurally resemble the D,D-stem peptide moiety, weakens the PG and may engender cell death (Walsh 2003). Being the targets of

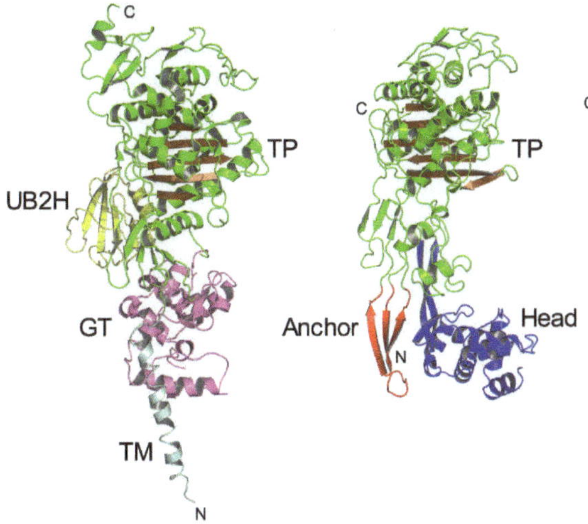

Fig. 8.3 Gallery of PBP structures and folds. (Left) PBP1b from *E. coli*, HMM Class A (PDB 3FWL); (Center) PBP2b from *S. pneumoniae*, HMM Class B (PDB 2WAD); (Right), PBP3 from *S. pneumoniae*, carboxypeptidase (PDB 1XP4)

β-lactam antibiotics, many PBPs are also involved in the development of resistance towards these drugs in a number of pathogens. Resistance mechanisms towards β-lactams occur through PBP active site modification, *pbp* gene overexpression, or, in the case of PBP-unrelated events, the production of β-lactamases that cleave the β-lactam ring, inactivating the drug (Nikolaidis et al. 2014; Alexander et al. 2018; Contreras-Martel et al. 2009; Contreras-Martel et al. 2006; Pernot et al. 2004; Moon et al. 2018; Lim and Strynadka 2002). Inhibition of the glycosyltransferase reaction by the natural product moenomycin also weakens the peptidoglycan and kills bacterial cells, but moenomycin's toxicity, a direct consequence of its long half-life, has precluded it from human use (Goldman and Gange 2000). However, the GT domain has been the target of multiple studies viewing the identification of novel, less toxic inhibitors (Wu et al. 2018; Wang et al. 2014; Ema et al. 2017).

Due to the complexity of these tasks and the importance of synthesizing PG in a timely and controlled manner, bacteria often present multiple PBPs. The rod-shaped model organism *E. coli*, for example, has 13 PBPs (5 HMM, 7 LMM, and one generated by proteolytic cleavage (Kocaoglu et al. 2015) while the pathogenic organism *S. pneumoniae* has 6 (3 HMM and 3 LMM). In most bacteria, at least one class A PBP is required for viability (Macheboeuf et al. 2006).

Most PBPs are associated to the outer leaflet of the cytoplasmic membrane by a single transmembrane (TM) region, which for HMM enzymes is located at the N-terminus of the molecule. Thus, they harbor large periplasmic domains consisting of up to 600–800 residues (Macheboeuf et al. 2006). Following the TM, HMM Class harbor the glycosyltransferase (GT) domain, which is separated into large and small lobes that form a crevice into which the glycan chain substrate potentially binds. PBP4 from *Listeria monocytogenes* and PBP2 from *S. aureus* are classical examples that were crystallized in the presence of moenomycin (Lovering et al. 2007; Jeong et al. 2013). Interestingly, in PBP1b from *E. coli*, the linker region folds into the UB2H domain, that presents a compact fold consisting mostly of four short strands (Sung et al. 2009; King et al. 2017) shown to play a key role in interactions with partner molecules (below). The latter structure also presents the TM domain, that has a small hydrophobic platform that interacts with the GT region (Fig. 8.3).

The transpeptidase (TP) domain is composed of a central, five-stranded β-sheet (β1–β5) surrounded by α-helices (green and brown in Fig. 8.3), and has been the target of many studies viewing the development of novel β-lactam-related or non-β-lactam inhibitors (Macheboeuf et al. 2005, 2007; Fedarovich et al. 2012; Contreras-Martel et al. 2011; Zervosen et al. 2011; van der Akker and Bonomo 2018; King et al. 2017). The fold of this domain is conserved within most enzymes that bind β-lactam antibiotics, including most β-lactamases (Nikolaidis et al. 2014). Interestingly, in order to catalyze the transpeptidation reaction, the TP domain must bind two stem peptides simultaneously, which indicates that the active site could exist in 'open' and 'closed' states. A recent model of allostery suggests that nascent peptidoglycan first binds to a location on the N-terminal region of certain class B PBPs that is 60 Å away from the active site, triggering its opening through small conformational modifications. These observations led the authors to suggest a mechanism for regulation of peptidoglycan biosynthesis that could involve interactions with partner molecules (Otero et al. 2013; Bernardo-García et al. 2018). However, the potential allosteric site is not present in many class B PBPs, and details regarding how and if PBP allostery plays a role in PG biosynthesis regulation are still under investigation.

PBP Interactions Within the Elongasome/Rod Complex

PBPs are key members of both the divisome and the elongasome (Egan et al. 2017), and their inhibition or deregulation can lead to defects in cell shape, impaired growth, and often cell wall lysis and death (den Blaauwen et al. 2008; Höltje 1998). PBPs have been reported to interact with various members of both complexes within the periplasm and the bacterial membrane; in this chapter, we will concentrate mostly on new knowledge regarding cell wall elongation. Within the elongasome of a variety of bacteria including *E. coli*, *Caulobacter crescentus*, *Bacillus subtilis*, and *Helicobacter pylori*, PBPs have been shown to interact with, and be recruited by, MreC, a bitopic membrane protein that is essential for cell shape (Kruse et al. 2005; Figge et al. 2004; Dye et al. 2005; El Ghachi et al. 2011; Leaver and Errington 2005; Lee et al. 2014; Divakaruni et al. 2005, 2007; Vats et al. 2009). MreC has thus been suggested as being a scaffold for elongasome formation, docking PBPs (Vats et al. 2009) and generating a core complex onto which other PG-biosynthesis partners can successively tether (El Ghachi et al. 2011; Contreras-Martel et al. 2017).

MreC variants from different species co-localize not only with HMM PBPs, as mentioned above, but also lytic transglycosylases, inner and outer membrane proteins (MreD, MipA), and MreB (which is cytoplasmic). These observations have led to the suggestion that MreC could stabilize elongasome-forming proteins across the three cellular compartments (Dye et al. 2005; van den Ent et al. 2006; Divakaruni et al. 2005, 2007; White et al. 2010; Leaver and Errington 2005; Slovak et al. 2006; El Ghachi et al. 2011).

The recent structure of a complex between the LMM PBP2 and MreC from *H. pylori*, as well as of PBP2 in its MreC-free form, started to shed light on structural details of elongasome formation (Contreras-Martel et al. 2017). In order to bind MreC, the N-terminal domain of PBP2, and notable its 'anchor' region, must sway away from the body of the protein, generating a V-shaped region that exposes a hydrophobic surface (Fig. 8.4). This 'open' form of PBP2 is thus apt to bind an apolar region of MreC, generating a 'hydrophobic zipper' whose stability was shown to be important not only for complex stability but also the maintenance of cell shape (Contreras-Martel et al. 2017). The employment of the N-terminal region of this class B PBP to interact with a partner molecule closes the argument regarding the function of this region in PBP biology, since it had been suggested

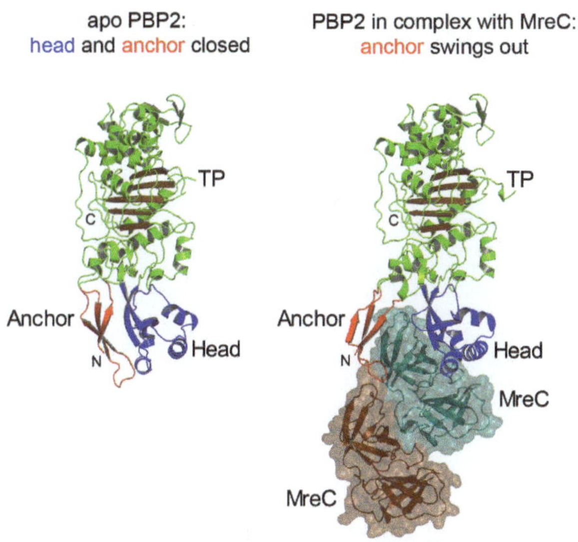

Fig. 8.4 Architecture and structural arrangement of PBP2 and the PBP2:MreC complex from *H. pylori* (left) PBP2 folds into anchor, head, linker, and transpeptidase domain. The anchor is 'clasped' against the helical head. (right) The PBP2 anchor region sways away from the head to expose a previously hidden hydrophobic region that allows binding of MreC, shown here with a transparent surface. In the structure, one PBP2 monomer binds to 2 MreC molecules (Contreras-Martel et al. 2017). N- and C-termini are indicated, as is the transpeptidase (TP) domain of PBP2

that it could either serve as a protein-binding region or merely as a pedestal, to separate the TP domain from the membrane (Macheboeuf et al. 2006). It is clear now that even if the role of pedestal is still a possibility, a much more important function lies in the interaction that the N-terminal region makes with other members of the elongasome.

If MreC is to serve as a scaffold for the elongasome, then it must be able to bind different proteins concomitantly. How this could be accomplished by such a small (28 kDa) protein is an interesting question, but this can be envisioned if it is taken into consideration that MreC could employ different surfaces for partner recognition. Thus, the hydrophobic patch would bind partner molecules such as PBP2, while the central body of the MreC fold could ensure a bidirectional interaction that could form short patches and allow for multiple binding partners, as suggested in the model below (Fig. 8.5). This exciting idea opens the path for understanding how multi-partite PG biosynthesis complexes are associated. As can be inferred from Fig. 8.2, the bacterial elongasome also includes other members whose structures have not yet been solved in complex with any other partners, such as class A PBPs, LpoA, hydrolases, Slt/Mlt lytic transglycosylases, or the inner membrane proteins MraY, MreD and MurJ.

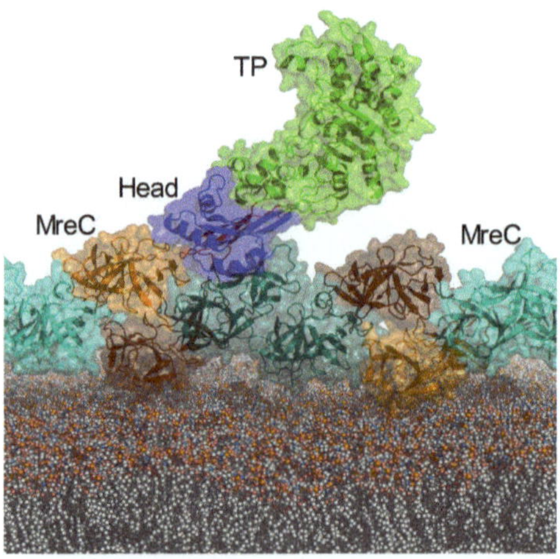

Fig. 8.5 Model of a core bacterial elongasome complex consisting of a class B PBP and MreC. The model was generated using the MreC pattern formed along the z-axis of the PBP2:MreC crystal structure (Contreras-Martel et al. 2017) and one of the PBP2 molecules associated to it (the 'Head' and 'TP' domain are shown in dark blue and green, respectively). A representation of a membrane was included for clarity purposes. MreC self-associates to form patches (cyan/blue and orange/brown) that provide ample docking sites for the recognition of PBP2 and/or different elongasome members. IM, inner membrane

Regulation of Cell Wall Elongation

As mentioned above, SEDS proteins such as RodA have recently been shown to possess GT activity (Meeske et al. 2016; Cho et al. 2016). SEDS are also known to associate to class B PBPs both in the cell and in vitro (Fay et al. 2010; Fraipont et al. 2011; Trip and Scheffers 2015), and often they are encoded by the same operon (Nikolaidis et al. 2012). However, recent fluorescence microscopy work shows that, in some species, class A enzymes may not co-localize with the rest of the elongation machinery (Meeske et al. 2016; Cho et al. 2016). That having been said, in *S. pneumoniae*, the integral membrane protein CozE was shown to direct the activity of the HMM PBP1a to the midcell plane, and there play an important role in cell elongation and the maintenance of bacterial shape (Fenton et al. 2016). These complex, species-specific observations bring up questions regarding the different roles that PBPs may play in cell wall formation, and bring to mind the possibility that their functions could vary slightly depending on the nature of the bacterium, the phase of the cell cycle, and the local environment of the cell.

The PBP2:MreC complex structure described above (Contreras-Martel et al. 2017), in addition to revealing details of the function of the N-terminal domain of class B PBPs, also highlighted the importance of another intriguing aspect of this association: conformational change. The opening of the N-terminal region of PBP2 to accommodate MreC is crucial for partner recognition, but recent work performed in *E. coli* suggests that such conformational modifications in class B PBPs can also be important for activation of peptidoglycan biosynthesis. By using a library of suppressor mutants as well as purified proteins, Rohs and co-workers identified that glycan chain polymerization was activated in cells carrying a class B PBP (PBP2) with a mutation within its N-terminal region that could maintain this domain in an 'open' conformation (Rohs et al. 2018). This PBP2 was able to activate the glycosyltransferase activitiy of RodA, generating the production of longer peptidoglycan fragments and the quicker depletion of Lipid II substrate (Rohs et al. 2018).

These results are also of interest in light of the recently solved structure of apo-RodA (Sjodt et al. 2018). RodA (from *Thermus thermophilus*) presents 10 transmembrane helices, most of which are perpendicular to the membrane, and surprisingly lacks a transmembrane channel that could bind the Lipid II substrate. However, the structure presented electron density suggestive of a bound lipid between TM2 and TM3, which could indicate the/a binding site for Lipid II. Notably, the potential PBP2 interaction region was mapped to be between TM8 and TM9 (Fig. 8.6), and the introduction of mutations within this interface engendered cell shape abnormalities (Sjodt et al. 2018). It is of interest that this region corresponds to the class B PBP-binding interface identified in FtsW, the RodA SEDS counterpart involved in cell division (Leclercq et al. 2017), lending further support to the hypothesis of the existence of homologous SEDS-PBP complexes within the elongasome and the divisome (Sjodt et al. 2018).

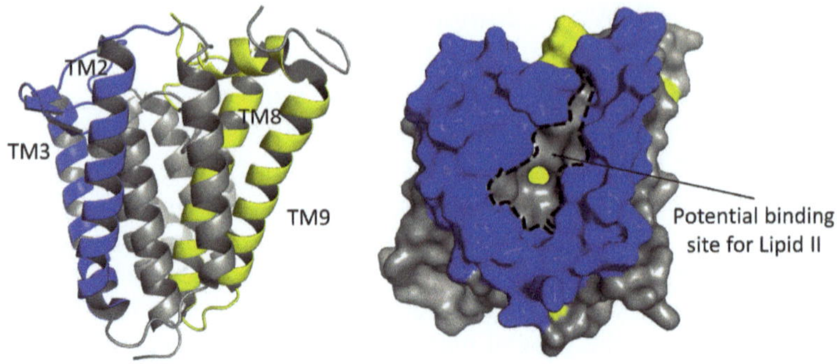

Fig. 8.6 RodA from *Thermus thermophilus* (PDB 6BAR) has 10 TMHs, and both N- and C-termini point into the cytoplasm. A central cavity between TM2 and TM3 forms a potential binding site for Lipid II (helices in blue). TM8 and TM9 could potentially interact with PBP2 (yellow helices)

Partnerships with Other Inner Membrane Proteins

Membrane proteins play key roles for both cell division and elongation processes. MraY is an integral membrane protein essential for bacterial growth, responsible for the catalysis of the first step in membrane-associated PG synthesis by transfer of the phosphoMurNAc-pentapeptide portion to the undecaprenyl phosphate (C55-P), producing C55-PPMurNAc pentapeptide (lipid I) (Chung et al. 2013; Bouhss et al. 2004). This enzyme is a member of the polyprenyl-phosphate N-acetyl hexosamine 1-phosphate transferase (PNPT) superfamily, which is involved in cell envelope polymer synthesis (Lovering et al. 2012; Chung et al. 2016). In addition, MraY is the target of different classes of natural inhibitors with antibacterial activity, such as capuramycin and tunicamycin, and has been studied by numerous laboratories in search of novel inhbitors (Hakulinen et al. 2017; Chung et al. 2013; Tanino et al. 2011; Zhang et al. 2011).

MraY from both *Aquifex aeolicus* and *Clostridium bolteae* were crystallized as dimers in the asymmetric unit, and both displayed N- and C-termini facing the cytoplasm. The MraY monomer consists of 10 transmembrane helices, an interfacial helix, a periplasmic β hairpin, a periplasmic helix, and five cytoplasmic loops (Chung et al. 2013; Hakulinen et al. 2017). At the center of the dimer is located an oval-shaped tunnel, surrounded by hydrophobic amino acids, possibly to accommodate lipids (Fig. 8.7). Mutational mapping indicated that the amino acids important for catalytic activity, Asp117, Asp118, Asp265, His234 and His235, which are conserved among enzymes of the PNPT family (Lloyd et al. 2004; Chung et al. 2013; Hakulinen et al. 2017) are clustered near H9 and H10. The loop between the two latter helices could also play a role in the recognition of the phospho-MurNAc pentapeptide (Chung et al. 2013, 2016; Hakulinen et al. 2017).

MraY has been found to interact with several other proteins involved in cell wall division and elongation. Interestingly, most of these partner proteins are

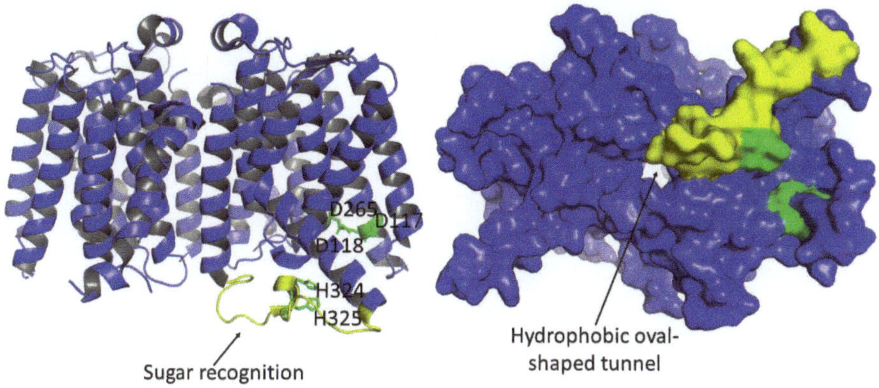

Fig. 8.7 Different views of the structure of MraY from *Aquifex aeolicus* (PDB 4J72). (Left) MraY harbors a dimer in the asymmetric unit. (Right) A hydrophobic oval-shaped tunnel between the monomers is wide enough to accommodate lipids. The cytoplasmic loop involved in sugar recognition is highlighted in yellow and involves TMs 9 and 10 (cyan), while the amino acids that participate in the catalytic activity are shown in green

cytoplasmic, and include MurG and MreB (White et al. 2010; Mohammadi et al. 2007). As indicated in Fig. 8.2, MraY could thus play a central bridging role between cytoplasmic and membrane-associated members of PG-biosynthesis complexes.

Upon formation of Lipid II, the next step in PG biosynthesis, which is common to both division and elongation, is 'flipping' of this major building block to the periplasm, where the sugar units will be polymerized and the stem peptides will be cross-linked. The nature of the Lipid II flippase has been a matter of discussion in the literature, and to date the SEDS proteins RodA and FtsW, as well as MurJ, which belongs to the MOP (multidrug/oligo-saccharidyl-lipid/polysaccharide) transporter superfamily, have been proposed to exert this role, either in division, elongation, or in both (Mohammadi et al. 2011, 2014; Sham et al. 2014; Kuk et al. 2017). FtsW has been shown to bind Lipid II and flip it and other lipids in vitro but not in cells; in the case of MurJ, elegant studies have shown that depletion of *murJ* in *E. coli* causes accumulation of PG precursors in the cytoplasm and blocks PG biosynthesis, resulting in irregularly shaped cell and eventual cell lysis (Meeske et al. 2015), linking the protein with Lipid II flipping in cells but to date not in vitro. In Gram-positives, unrelated proteins Amj and Wzk have recently been shown to also be able to translocate Lipid II (Elhenawy et al. 2016; Meeske et al. 2015).

Recent crystal structures of MurJ from *Thermosipho africanus* and *E. coli* (Zheng et al. 2018; Kuk et al. 2017) reveal 14 TMHs with both C- and N-termini in the cytoplasm. The central cavity involves residues from the amino-terminal/N-lobe (TMs 1–6) and Carboxy-terminal/C-lobe (TMs 7–14) arranged in inward-facing (open to the cytoplasm) and outward-facing (open to the periplasm) conformations

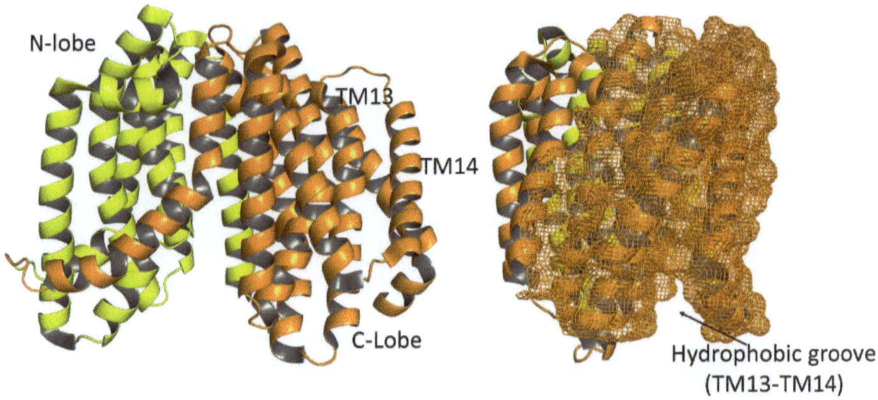

Fig. 8.8 Crystal structure of MurJ. MurJ from *Thermosipho africanus* (PDB 5T77) has 14 TMs divided into an N-lobe (TM 1–6, in yellow) and C-lobe (TM 7–14, in orange). A hydrophobic groove is formed by TMs 13 and 14 and is highlighted by an orange mesh

(Fig. 8.8), and both states have been proposed to be adopted by MurJ. In addition, TMs 13 and 14 form a hydrophobic groove which penetrates into the central transport cavity and could play a role in Lipid II flipping (Kumar et al. 2019; Kuk et al. 2017; Zheng et al. 2018); however, this functionality remains to be shown in atomic detail.

Recent work has suggested that MurJ depends on Lipid II for localization and flipping, and these activities could be dependent on its positioning on the vicinity of the elongasome, thus linking it with the presence of RodA (Liu et al. 2018). Moreover, MraY, MurG, and MurJ were shown to co-localize at midcell. Interestingly, MurJ requires a functional FtsW at midcell in the absence of which Lipid II flipping cannot be measured (Liu et al. 2018). These observations highlight the intricate web of interactions within PG-biosynthesis complexes, with several members being shared between divisome and elongasome with their activities being most probably regulated in independent fashion.

Interactions with the Outer Membrane

In addition to the challenge of coordinating the steps described above, Gram-negative bacteria must still regulate PG biosynthesis and outer membrane assembly. The outer membrane also harbors cofactors that are essential for PG-formation complexes, and that regulate the activity of PBPs. By using genetic screens, large-scale phenotyping, and proteomics in *E. coli*, two independent groups identified LpoA and LpoB, activators of the bifunctional enzymes PBP1a and PBP1b, respectively (Paradis-Bleau et al. 2010; Typas et al. 2010). Both Lpo proteins contain signal peptides with sequences indicative of outer membrane lipoproteins, and localization experiments indicted that they are broadly distributed throughout the bacterial envelope. LpoA and LpoB were shown not only to stimulate the activity of their respective PBPs, but also to be essential for their function in vivo. *E. coli* PBP1a and PBP1b are HMM class A enzymes, and thus harbor both GT and TP domains. In vitro experiments with purified Lipid II indicated that the glycosyltransferase activity of PBP1b was affected by LpoB (Paradis-Bleau et al. 2010), and both Lpo factors were shown to increase the transpeptidase activity of their cognate proteins (Typas et al. 2010). Interestingly, LpoB was shown by NMR to cause structural changes leading to destabilization of the UB2H domain of PBP1b (Fig. 8.3), which was suggested as representing a potential activation mechanism of this class A enzyme (Egan et al. 2018). Last but not least, one hypothesis that needs testing is that Lpo proteins may also bind to or control the activity of hydrolases, whose role is to release peptidoglycan fragments during cell growth (Typas et al. 2010). These observations indicate that PG-biosynthesis complexes such as the divisome and the elongasome may be regulated at several levels, including by Lpo factors at the level of the outer membrane and the cytoskeletal proteins FtsZ and MreB within the cytoplasm.

Concluding Remarks

PBPs have been studied with great interest for decades, not only due to the fact that they are the targets of β-lactam antibiotics, but also thanks to the roles they play in antibiotic resistance in many bacterial species. Despite the fact that more classical approaches had led to the study of PBPs as individual entities, in the last few years the advances in structural biology, fluorescence microscopy and high throughput genome/transposon screening techniques have expanded our understanding regarding PBPs, especially in what concerns complexes that they form in the cell. The next great challenges will involve the description of how, when, and by which protein partners PBPs are activated and regulated, and structural biology will certainly help shed light on these important issues.

Acknowledgements Work in the Dessen lab on Penicillin-Binding Proteins and cell wall elongation complexes is supported by grants from the Agence Nationale de la Recherche (ANR-18-CE11-0019), FAPESP (São Paulo Research Foundation) grant 2017/12,436-9, and the Laboratoire Intenational Associé (LIA) BACWALL (CNRS). M. M. M. was supported by grant 2013/02451-0 from FAPESP.

References

Alexander JAN, Chatterjee SS, Hamilton SM, Eltis LD, Chambers HF, Strynadka NC (2018) Structural and kinetic analyses of penicillin-binding protein 4 (PBP4)-mediated antibiotic resistance in *Staphylococcus aureus*. J Biol Chem 293:19854–19865

Barreteau H, Kovac A, Boniface A, Sova M, Gobec S, Blanot D (2008) Cytoplasmic steps of peptidoglycan biosynthesis. FEMS Microbiol Rev 32:168–207

Bernardo-García N, Mahasenan KV, Batuecas MT, Lee M, Hesek D, Petráčková D, Doubravová L, Branny P, Mobashery S, Hermoso JA (2018) Allostery, recognition of nascent peptidoglycan, and cross-linking of the cell Wall by the essential Penicillin-Binding Protein 2x of *Streptococcus pneumoniae*. ACS Chem Biol 13:694–702

Bouhss A, Crouvoisier M, Blanot D, Mengin-Lecreulx D (2004) Purification and characterization of the bacterial MraY translocase catalyzing the first membrane step of peptidoglycan biosynthesis. J Biol Chem 279:29974–29980

Cho H, Wivagg CN, Kapoor M, Barry Z, Rohs PD, Suh H, Marto JA, Garner EC, Bernhardt TG (2016) Bacterial cell wall biogenesis is mediated by SEDS and PBP polymerase families functioning semi-autonomously. Nat Microbiol 19:16172

Chung BC, Mashalidis EH, Tanino T, Kim M, Matsuda A, Hong J, Ichikawa S, Lee SY (2016) Structural insights into inhibition of Lipid I production in bacterial cell wall synthesis. Nature 533:557–560

Chung BC, Zhao J, Gillespie RA, Kwon D-Y, Guan Z, Hong J, Zhou P, Lee S-Y (2013) Crystal structure of MraY, an essential membrane enzyme for bacterial cell wall synthesis. Science 341:1012–1016

Contreras-Martel C, Amoroso A, Woon ECY, Zervosen A, Inglis S, Martins A, Verlaine O, Rydzik AM, Job V, Luxen A, Joris B, Schofield CJ, Dessen A (2011) Structure-guided design of cell wall biosynthesis inhibitors that overcome β-lactam resistance in *Staphylococcus aureus* (MRSA). ACS Chem Biol 6:943–951

Contreras-Martel C, Dahout-Gonzalez C, Dos Santos Martins A, Kotnik M, Dessen A (2009) PBP active site flexibility as the key mechanism for beta-lactam resistance in pneumococci. J Mol Biol 387:899–909

Contreras-Martel C, Job V, Di Guilmi AM, Vernet T, Dideberg O, Dessen A (2006) Crystal structure of Penicillin-Binding Protein 1a (PBP1a) reveals a mutational hotspot implicated in b-lactam resistance in *Streptococcus pneumoniae*. J Mol Biol 355:684–696

Contreras-Martel C, Martins A, Ecobichon C, Trindade DM, Mattei PJ, Hicham S, Hardouin P, Ghachi M, Boneca IG, Dessen A (2017) Molecular architecture of the PBP2-MreC core bacterial cell wall synthesis complex. Nature Commun 8:776

den Blaauwen T, de Pedro MA, Nguyen-Distèche M, Ayala JA (2008) Morphogenesis of rod-shaped sacculi. FEMS Microbiol Rev 32:321–344

Divakaruni AV, Baida C, White CL, Gober JW (2007) The cell shape proteins MreB and MreC control cell morphogenesis by positioning cell wall synthetic complexes. Mol Microbiol 66:174–188

Divakaruni AV, Ogorzalek Loo RR, Xie Y, Loo JA, Gober JW (2005) The cell-shape protein MreC interacts with extracytoplasmic proteins including cell wall assembly complexes in *Caulobacter crescentus*. Proc Natl Acad Sci USA 51:18602–18607

Dye NA, Pincus Z, Theriot JA, Shapiro L, Gitai Z (2005) Two independent spiral structures control cell shape in *Caulobacter*. Proc Natl Acad Sci USA 102:18608–18613

Egan AJ, Cleverley RM, Peters K, Lewis RJ, Vollmer W (2017) Regulation of bacterial cell wall growth. FEBS J 284:851–867

Egan AJF, Maya-Martinez R, Ayala I, Bougault CM, Banzhaf M, Breukink E, Vollmer W, Simorre JP (2018) Induced conformational changes activate the peptidoglycan synthase PBP1b. Mol Microbiol 110:335–356

El Ghachi M, Matteï PJ, Ecobichon C, Martins A, Hoos S, Schmitt C, Colland F, Ebel C, Prevost MC, Gabel F, Dessen A, Boneca IG (2011) Characterization of the elongasome core PBP2: MreC complex of *Helicobacter pylori*. Mol Microbiol 82:68–86

Elhenawy W, Davis RM, Fero J, Salama NR, Felman MF, Ruiz N (2016) The O-antigen flippase Wzk can substitute for MurJ in peptidoglycan synthesis in *Helicobacter pylori* and *Escherichia coli*. PLoS ONE 11:e0161587

Ema M, Xu Y, Gehrke S, Wagner GK (2017) Identification of non-substrate-like glycosyltransferase inhibitors from library screening: Pitfalls & hits. Med Chem Comm 9:131–137

Fay A, Meyer P, Dworkin J (2010) Interactions between late-acting proteins required for peptidoglycan synthesis during sporulation. J Mol Biol 399:547–561

Fedarovich A, Djordjevic KA, Swanson SM, Peterson YK, Nicholas RA, Davies C (2012) High-throughput screening for novel inhibitors of *Neisseria gonorrhoeae* penicillin-binding protein 2. PLoS ONE 7:e44918

Fenton AK, El Mortaji L, Lau DTC, Rudner DZ, Bernhardt TG (2016) CozE is a member of the MreCD complex that directs cell elongation in *Streptococcus pneumoniae*. Nat Microbiol 2:16237

Figge RM, Divakaruni AV, Gober JW (2004) MreB, the cell shape-determining bacterial actin homologue, coordinates cell wall morphogenesis in *Caulobacter crescentus*. Mol Microbiol 51:1321–1332

Fraipont C, Alexeeva S, Wolf B, van der Ploeg R, Schloesser M, den Blaauwen T, Nguyen-Distèche M (2011) The integral membrane FtsW protein and peptidoglycan synthase PBP3 form a subcomplex in *Escherichia coli*. Microbiology 157:251–259

Goffin C, Ghuysen JM (2002) Biochemistry and comparative genomics of SXXK superfamily acyltransferases offer a clue to the mycobacterial paradox. Microb Mol Biol Rev 66:702–738

Goldman RC, Gange D (2000) Inhibition of transglycosylation involved in bacterial peptidoglycan synthesis. Curr Med Chem 7:801–820

Hakulinen JK, Hering J, Brändén G, Chen H, Snijder A, Ek M, Johansson P (2017) MraY-antibiotic complex reveals details of tunicamycin mode of action. Nat Chem Biol 13:265–267

Höltje JV (1998) Growth of the stress-bearing and shape-maintaining murein sacculus of *Escherichia coli*. Microbiol Mol Biol Rev 62:181–203

Hugonnet JE, Mengin-Lecreulx D, Monton A, den Blaauwen T, Carbonnelle E, Veckerlé C, Brun YV, van Niewenhze M, Bouchier C, Tu K, Rice LB, Arthur M (2016) Factors essential for L, D-transpeptidase-mediated peptidoglycan cross-linking and β-lactam resistance in Escherichia coli. Elife 5:e19469

Jeong J-H, Kim Y-S, Rojviriya C, Ha S-C, Kang BS, Kim Y-G (2013) Crystal structures of bifunctional Penicillin-Binding Protein 4 from *Listeria monocytogenes*. Antimicrob Agents Chemother 57:3507–3512

King DT, Wasney GA, Nosella M, Fong A, Strynadka NC (2017) Structural insights into inhibition of *Escherichia coli* Penicillin-Binding Protein 1b. J Biol Chem 292:979–993

Kocaoglu O, Tsui HC, Winkler ME, Carlson EE (2015) Profiling of β-lactam selectivity for penicillin-binding proteins in *Streptococcus pneumoniae* D39. Antimicrob Agents Chemother 59:3548–3555

Kruse T, Bork-Jensen J, Gerdes K (2005) The morphogenetic MreBCD proteins of *Escherichia coli* form an essential membrane-bound complex. Mol Microbiol 55:78–89

Kuk AC, Mashalidis EH, Lee SY (2017) Crystal structure of the MOP flippase MurJ in an inward-facing conformation. Nat Struct Mol Biol 24:171–176

Kumar S, Rubino FA, Mendoza AG, Ruiz N (2019) The bacterial lipid II flippase MurJ functions by an alternating-access mechanism. J Biol Chem 294:981–990

Leaver M, Errington J (2005) Roles for MreC and MreD proteins in helical growth of the cylindrical cell wall in *Bacillus subtilis*. Mol Microbiol 57:1196–1209

Leclercq S, Derouaux A, Olatunji S, Fraipont C, Egan AJ, Vollmer W, Breukink E, Terrak M (2017) Interplay between Penicillin-Binding Proteins and SEDS proteins promotes bacterial cell wall synthesis. Sci Rep 7:43306

Lee TK, Tropini C, Hsin J, Desmarais SM, Ursell TS, Gong E, Gitai Z, Monds RD, Huang KC (2014) A dynamically assembled cell wall synthesis machinery buffers cell growth. Proc Natl Acad Sci USA 111:4554–4559

Lim D, Strynadka NC (2002) Structural basis for the beta lactam resistance of PBP2a from methicillin-resistant *Staphylococcus aureus*. Nat Struct Biol 9(11):870–876

Liu Y, Breukink E (2016) The membrane steps of bacterial cell wall synthesis as antibiotic targets. Antibiotics (Basel) 5:E28

Liu X, Meiresonne NY, Bouhss A, den Blaauwen T (2018) FtsW activity and Lipid II synthesis are required for recruitment of MurJ to midcell during cell division in *Escherichia coli*. Mol Microbiol 109:865–884

Lloyd AJ, Brandish PE, Gilbey AM, Bugg TD (2004) Phospho-N-acetyl-muramyl-pentapeptide translocase from *Escherichia coli*: catalytic role of conserved aspartic acid residues. J Bacteriol 186:1747–1757

Lovering AL, de Castro LH, Lim D, Strynadka NC (2007) Structural insight into the transglycosylation step of bacterial cell wall biosynthesis. Science 315:1402–1405

Lovering AL, Safadi SS, Strynadka NCJ (2012) Structural perspective of peptidoglycan biosynthesis and assembly. Annu Rev Biochem 81:451–478

Macheboeuf P, Contreras-Martel C, Job V, Dideberg O, Dessen A (2006) Penicillin Binding Proteins: key players in bacterial cell cycle and drug resistance processes. FEMS Microbiol Rev 30(5):673–691

Macheboeuf P, Di Guilmi AM, Job V, Vernet T, Dideberg O, Dessen A (2005) Active site restructuring regulates ligand recognition in class A penicillin-binding proteins. Proc Natl Acad Sci USA 102(3):577–582

Macheboeuf P, Fischer DS, Brown T Jr, Zervosen A, Luxen A, Joris B, Dessen A, Schofield CJ (2007) Structural and mechanistic basis of penicillin-binding protein inhibition by lactivicins. Nat Chem Biol 3:565–569

Matteï P-J, Neves D, Dessen A (2010) Bridging cell wall biosynthesis and bacterial morphogenesis. Curr Opin Struct Biol 20:749–766

Meeske AJ, Riley EP, Robins WP, Uehara T, Mekalanos JJ, Kahne D, Walker S, Kruse AC, Bernhardt TG, Rudner DZ (2016) SEDS proteins are a widespread family of bacterial cell wall polymerases. Nature 537:634–638

Meeske AJ, Sham LT, Kimsey H, Koo BM, Gross CA, Bernhardt TG, Rudner DZ (2015) MurJ and a novel lipid II flippase are required for cell wall biogenesis in *Bacillus subtilis*. Proc Natl Acad Sci USA 112:6437–6442

Mohammadi T, Karczmarek A, Crouvoisier M, Bouhss A, Mengin-Lecreulx D, den Blaauwen T (2007) The essential peptidoglycan glycosyltransferase MurG forms a complex with proteins involved in lateral envelope growth as well as with proteins involved in cell division in *Escherichia coli*. Mol Microbiol 65:1106–1121

Mohammadi T, Sijbrandi R, Lutters M, Verheul J, Martin NI, den Blaauwen T, de Kruijff B, Breukink E (2014) Specificity of the transport of lipid II by FtsW in *Escherichia coli*. J Biol Chem 289:14707–14718

Mohammadi T, van Dam V, Sijbrandi R, Vernet T, Zapun A, Bouhss A, Diepeveen-de Bruin M, Nguyen-Disteche M, de Kruijff B, Breukink E (2011) Identification of FtsW as a transporter of lipid-linked cell wall precursors across the membrane. EMBO J 30:1425–1432

Moon TM, D'Andréa ED, Lee CW, Soares A, Jakoncic J, Desbonnet C, Garcia-Solache M, Rice LB, Page R, Peti W (2018) The structures of penicillin-binding protein 4 (PBP4) and PBP5 from *Enterococci* provide structural insights into β-lactam resistance. J Biol Chem 293:18574–18584

Nikolaidis I, Favini-Stabile S, Dessen A (2014) Resistance to antibiotics targeted to the bacterial cell wall. Protein Sci 23:243–259

Nikolaidis I, Izoré T, Job V, Thielens N, Breukink E, Dessen A (2012) Calcium-dependent complex formation between PBP2 and lytic transglycosylate SltB1 of *Pseudomonas aeruginosa*. Microb Drug Resist 18:298–305

Otero LH, Rojas-Altuve A, Llarrull LI, Carrasco-López C, Kumarasiri M, Lastochkin E, Fishovitz J, Dawley M, Hesek D, Lee M, Johnson JW, Fisher JF, Chang M, Mobashery S, Hermoso JA (2013) How allosteric control of *Staphylococcus aureus* penicillin binding protein 2a enables methicillin resistance and physiological function. Proc Natl Acad Sci USA 110:16808–16813

Paradis-Bleau C, Markovski M, Uehara T, Lupoli TJ, Walker S, Kahne DE, Bernhardt TG (2010) Lipoprotein cofactors located in the outer membrane activate bacterial cell wall polymerases. Cell 143:1110–1120

Pazos M, Peters K, Vollmer W (2017) Robust peptidoglycan growth by dynamic and variable multi-protein complexes. Curr Opin Microbiol 36:55–61

Pernot L, Chesnel L, Le Gouellec A, Croize J, Vernet T, Dideberg O, Dessen A (2004) A PBP2x from a clinical isolate of *Streptococcus pneumoniae* exhibits an alternative mechanism for reduction of susceptibility to b-lactam antibiotics. J Biol Chem 279:16463–16470

Rohs PDA, Buss J, Sim SI, Squyres GR, Srisuknimit V, Smith M, Cho H, Sjodt M, Kruse AC, Garner EC, Walker S, Kahne DE, Bernhardt TG (2018) A central role for PBP2 in the activation of peptidoglycan polymerization by the bacterial cell elongation machinery. PLoS Genet 14:e1007726

Sham LT, Butler EK, Lebar MD, Kahne D, Bernhardt TG, Ruiz N (2014) MurJ is the flippase of lipid-linked precursors for peptidoglycan biogenesis. Science 345:220–222

Sjodt M, Brock K, Dobihal G, Rohs PDA, Green AG, Hopf TA, Meeske AJ, Srisuknimit V, Kahne D, Walker S, Marks DS, Bernhardt TG, Rudner DZ, Kruse AC (2018) Structure of the peptidoglycan polymerase RodA resolved by evolutionary coupling analysis. Nature 556:118–121

Slovak PM, Porter SL, Armitage JP (2006) Differential localization of Mre proteins with PBP2 in *Rhodobacter sphaeroides*. J Bacteriol 188:1691–1700

Sung M-T, Lai Y-T, Huang C-Y, Chou L-Y, Shih H-W, Cheng W-C, Wong C-H, Ma C (2009) Crystal structure of the membrane-bound bifunctional transglycosylase PBP1b from *Escherichia coli*. Proc Natl Acad Sci USA 106:8824–8829

Tanino T, Al-Dabbagh B, Mengin-Lecreulx D, Bouhss A, Oyama H, Ichikawa S, Matsuda A (2011) Mechanistic analysis of muraymycin analogues: a guide to the design of MraY inhibitors. J Med Chem 54:8421–8439

Trip E, Scheffers D-J (2015) A 1MDa protein complex containing critical components of the *Escherichia coli* divisome. Sci Rep 5:18190

Typas A, Banzhaf M, Gross CA, Vollmer W (2012) From the regulation of peptidoglycan synthesis to bacterial growth and morphology. Nat Rev Microbiol 10:123–136

Typas A, Banzhaf M, van den Berg van Saparoea B, Verheul J, Bilboy J, Nichols RJ, Zietek M, Beilharz K, Kannenberg K, von Rechenberg M, Breukink E, den Blaauwen T, Gross CA, Vollmer W (2010) Regulation of peptidoglycan synthesis by outer-membrane proteins. Cell 143:1097–1109

van den Ent F, Leaver M, Bendezú F, Errington J, de Boer P, Löwe J (2006) Dimeric structure of the cell shape protein MreC and its functional implications. Mol Microbiol 62:1631–1642

van der Akker F, Bonomo RA (2018) Exploring additional dimensions of complexity in inhibitor design for serine β-lactamases: mechanistic and intra- and inter-molecular chemistry approaches. Front Microbiol 9:622

Vats P, Shigh Y-L, Rothfield L (2009) Assembly of the MreB-associated cytoskeletal ring of *Escherichia coli*. Mol Microbiol 72:170–182

Vollmer W, Blanot D, de Pedro MA (2008) Peptidoglycan structure and architecture. FEMS Microbiol Rev 32:149–167

Walsh C (2003) Where will new antibiotics come from? Nat Rev Microbiol 1:65–70

Wang Y, Chan FY, Sun N, Lui HK, So PK, Yan SC, Chan KF, Chiou J, Chen S, Abagyan R, Leung YC, Wong KY (2014) Structure-based design, synthesis, and biological evaluation of isatin derivatives as potential glycosyltransferase inhibitors. Chem Biol Drug Des 84:685–696

White CL, Kitich A, Gober JW (2010) Positioning cell wall synthetic complexes by the bacterial morphogenetic proteins MreB and MreD. Mol Microbiol 76:616–633

Wu WS, Cheng WC, Cheng TR, Wong CH (2018) Affinity-based screen for inhibitors of bacterial transglycosylase. J Am Chem Soc 140:2752–2755

Zervosen A, Herman R, Kerff f, Herman A, Bouillez a, Prati F, Pratt RF, Frère J-M, Joris B, Luxen A, Charlier P, Sauvage E (2011) Unexpected tricovalent binding mode of boronic acids within the active site of a penicillin-binding protein. J Am Chem Soc 133:10839–10848

Zhang W, Ntai I, Bolla ML, Malcolmson SJ, Kahne D, Kelleher NL, Walsh CT (2011) Nine enzymes are required for assembly of the pacidamycin group of peptidyl nucleoside antibiotics. J Am Chem Soc 133:5240–5243

Zheng S, Sham LT, Rubino FA, Brock KP, Robins WP, Mekalanos JJ, Marks DS, Bernhardt TG, Kruse AC (2018) Structure and mutagenic analysis of the lipid II flippase MurJ from *Escherichia coli*. Proc Natl Acad Sci USA 115:6709–6714

Chapter 9
Structure and Function of Roundabout Receptors

Francesco Bisiak and Andrew A. McCarthy

Abstract The creation of complex neuronal networks relies on ligand-receptor interactions that mediate attraction or repulsion towards specific targets. Roundabouts comprise a family of single-pass transmembrane receptors facilitating this process upon interaction with the soluble extracellular ligand Slit protein family emanating from the midline. Due to the complexity and flexible nature of Robo receptors, their overall structure has remained elusive until now. Recent structural studies of the Robo1 and Robo2 ectodomains have provided the basis for a better understanding of their signalling mechanism. These structures reveal how Robo receptors adopt an auto-inhibited conformation on the cell surface that can be further stabilised by *cis* and/or *trans* oligmerisation arrays. Upon Slit-N binding Robo receptors must undergo a conformational change for Ig4 mediated dimerisation and signaling, probably via endocytosis. Furthermore, it's become clear that Robo receptors do not only act alone, but as large and more complex cell surface receptor assemblies to manifest directional and growth effects in a concerted fashion. These context dependent assemblies provide a mechanism to fine tune attractive and repulsive signals in a combinatorial manner required during neuronal development. While a mechanistic understanding of Slit mediated Robo signaling has advanced significantly further structural studies on larger assemblies are required for the design of new experiments to elucidate their role in cell surface receptor complexes. These will be necessary to understand the role of Slit-Robo signaling in neurogenesis, angiogenesis, organ development and cancer progression. In this chapter, we provide a review of the current knowledge in the field with a particular focus on the Roundabout receptor family.

Keywords Neurons · Guidance cues · Receptors · Signalling · Robo · Slit

F. Bisiak (✉) · A. A. McCarthy (✉)
European Molecular Biology Laboratory, Grenoble Outstation, 71 Avenue Des Martyrs, 38042 Grenoble, France
e-mail: fbisiak@embl.fr

A. A. McCarthy
e-mail: andrewmc@embl.fr

J. R. Harris and J. Marles-Wright (eds.), *Macromolecular Protein Complexes II: Structure and Function*, Subcellular Biochemistry 93,
https://doi.org/10.1007/978-3-030-28151-9_9

Introduction

Neurons are a highly differentiated cell type that can receive, process, and transmit nerve impulses over long distances. They are characterised by a cell body (soma) surrounded by short branched extensions (dendrites) and a long extension (axon). Neurons typically make multiple connections to form complex circuits such as the central nervous system (CNS). During development the coordinated extension of neurons to their target destination, essential for proper CNS wiring, is called axon guidance (Chédotal and Richards 2010). In Bilateria, the midline forms an important division between the two symmetric halves of the CNS, and therefore acts as an important guidance checkpoint during neuronal development. Critical for proper CNS function is a connection of the two bilateral sides, requiring a crossing of the midline in a strictly regulated manner (Placzek and Briscoe 2005). Here, a subset of commissural neurons cross the midline for connection, while another, ipsilateral neurons, remain on the same side (Fig. 9.1). In order to achieve this remarkable feat a navigational pathway is delineated by a series of attractive and repulsive cues.

Neuronal development cues bind their cognate cell surface receptors to trigger attractive or repulsive cellular responses. There are four major families of axon guidance cues that can interact with one (or more) receptors (Kolodkin and Tessier-Lavigne 2011; Seiradake et al. 2016): ephrins and Eph receptors (Kania and Klein 2016); semaphorins and plexin receptors, including neurophilin co-receptors

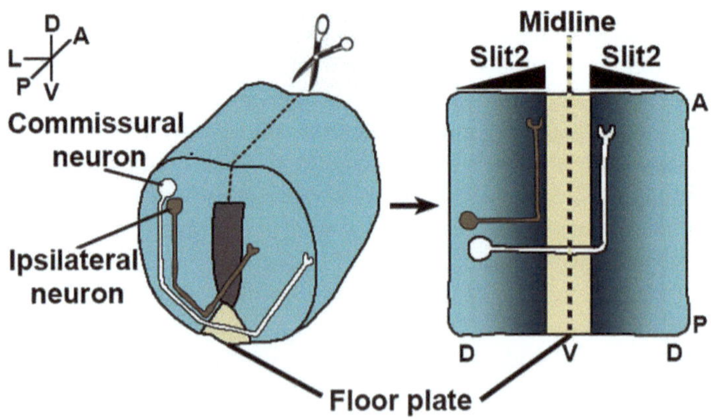

Fig. 9.1 Schematic of classical commissural and ipsilateral axon pathfinding in the spinal cord of mouse embryos. On the left is a spinal cord section from a mouse embryo showing commissural and ipsilateral axons, and midline floor plate. Commissural neurons express Robo receptors only upon crossing the midline to prevent their re-crossing, while ipsilateral maintain Robo receptor expression to always remain on the same side. The more classic 'open-book' flat format, as dissected along the spinal cord, is shown on the right to illustrate how a Slit gradient emanating from the midline is used for axon repulsion. Commissural and ipsilateral neurons are white and grey, respectively. A, anterior; P, posterior; D, dorsal; V, ventral; L, lateral

(Koropouli and Kolodkin 2014; Kolodkin and Tessier-Lavigne 2011); netrins with multiple receptors (DCC/neogenin, UNC5, Dscam and NGL1) (Seiradake et al. 2016); and Slits with Robo receptors (Blockus and Chedotal 2016). However, more recently Slit has been shown to interact with multiple partners, including Eva1C (Fujisawa et al. 2007), plexin A1 (Delloye-Bourgeois et al. 2015), dystroglycan (Wright et al. 2012), and Dscam (Dascenco et al. 2015; Alavi et al. 2016). Several other morphogen-receptor interactions are also known to contribute to axon guidance, including the Wnt and Sonic hedgehog (Shh) families that signal via interaction with multiple receptors, fibronectin leucine-rich repeat transmembrane proteins (FLRT) and some members of the Cadherin superfamily (Kolodkin and Tessier-Lavigne 2011; Seiradake et al. 2016).

The *roundabout* gene, *Robo*, was first identified in *Drosophila* genetic screens for commissural axons midline crossing defects (Kidd et al. 1998; Seeger et al. 1993). The 'Robo' name originates from stalled commissural axons at the midline creating a 'ROundaBOut' or Robo phenotype (Tear et al. 1993). Many homologs have since been identified in other species (Chédotal 2007). Following their discovery the Robo receptors were shown to act as the cognate receptors for the secreted guidance factor Slit (Brose et al. 1999; Kidd et al. 1999). Subsequently, the Slit-Robo signalling pathway has been shown to be important for many other developmental processes, including organogenesis of the kidney (Grieshammer et al. 2004), lungs (Domyan et al. 2013), and heart (Mommersteeg et al. 2013). Given their importance during development the Robo receptors have also been implicated in several types of cancer (Seth et al. 2005; Legg et al. 2008; Ballard and Hinck 2012; Huang et al. 2015), and chronic diseases such as kidney disease (Hwang et al. 2015) and liver fibrosis (Chang et al. 2015). As such they are now considered as attractive therapeutic targets. Here, we provide an overview of the Robo receptors from a structural perspective and how these insights have shed light on their activation mechanism.

The Robo Receptor Family

Robo receptors are evolutionarily conserved across bilateral anatomical species. Three *robo* genes have been identified in *Drosophila* (Seeger et al. 1993; Simpson et al. 2000; Rajagopalan et al. 2000), Zebrafish (Yuan et al. 1999b; Lee et al. 2001), and chick (Vargesson et al. 2001) while *C. elegans* contains a single *robo* ortholog, SAX-3 (Hao et al. 2001). In mammals four Robo receptors (Robo1, Robo2, Robo3 and Robo4) have been identified (Kidd et al. 1998; Yuan et al. 1999a; Huminiecki et al. 2002). Robo1, Robo3 and Robo4 are sometimes referred to as Deleted in U twenty twenty (DUTT1), Retinoblastoma-inhibiting gene 1 (Rig-1) and magic Robo, respectively, in older manuscripts. The *ROBO1* and *ROBO2* genes are located on human chromosome 3, while the *ROBO3* and *ROBO4* genes are located on chromosome 11, both in a similar head to head configuration. This closely linked chromosomal organization suggests the Robo receptors emerged from a single ancestral vertebrate *ROBO* gene by a tandem duplication that was followed by two whole genome duplications and losses before vertebrate radiation (Zelina et al. 2014). Two tandem duplication copies exist in most vertebrates' genomes today. To add further complexity alternative splicing of mammalian *ROBO* genes with varied expression patterns and distinct guidance responses have also been reported (Clark et al. 2002; Dalkic et al. 2006; Chen et al. 2008). This mechanism was shown to be particularly important for Robo3 function (Chen et al. 2008), as discussed later, but whether this is valid for Robo1 and Robo2 function remains to be shown.

The Robo receptors are single-pass type I membrane proteins that belong to the immunoglobulin (Ig) superfamily of cell adhesion molecules (CAMs). Robo receptors typically contain five C2-type Ig domains, three fibronectin (FN) type III domains, a transmembrane helix and a large unstructured intracellular region incorporating four conserved cytoplasmic (CC) motifs (Fig. 9.2). All vertebrate Robo receptors are expressed by CNS neurons, except Robo4, which is vascular specific (Huminiecki et al. 2002; Park et al. 2003). In addition, Robo4 is much smaller than the other family members, containing only two Ig and FN3 domains (Fig. 9.2). The structures of several individual and tandem Robo receptor domains, alone or in complex with interaction partners, have been determined. In particular hRobo1 Ig1–Ig2 (PDB 2V9R, 2V9Q) (Morlot et al. 2007), dRobo Ig1–2–heparin complex (PDB 2VRA) (Fukuhara et al. 2008), hRobo1 FN2-3 juxta-membrane (PDB 4HLJ) (Barak et al. 2014), and hRobo1 FN3 in complex with antigen-binding fragment (Fab) B2212A (PDB 3WIH) (Nakayama et al. 2015) (Fig. 9.3). These were further complemented by the hRobo2 Ig4-5 (PDB 5NOI) (Yom-Tov et al. 2017), hRobo1 Ig5 (5O5I) and hRobo1 Ig1–4 (PDB 5OPE, 5O5G) crystal structures (Fig. 9.3), as well as a 20 Å 3D negative stain electron microscopy (EM) reconstruction of the whole hRobo1 ectodomain (Aleksandrova et al. 2018). But the most detailed structural insights on Robo receptors to date comes from a crystal structure of the complete hRobo2 ectodomain at 3.6 Å resolution (PDB 6IAA) (Barak et al. 2019) (Fig. 9.3).

Robo1 and Robo2 can form homophilic adhesion interactions (Hivert et al. 2002) that are dependent on the whole ectodomain when coated on fluorescent beads (Liu et al. 2004), with no single domain being responsible, implying the entire intact Robo ectodomain is required for homophilic binding. Similar results were obtained at the cell surface using Fluorescence Resonance Energy Transfer (FRET) and Spatial Intensity Distribution Analysis (SpIDA) imaging techniques (Zakrys et al. 2014). Here, the whole ectodomain is again required for dimer formation, with the Ig domains essential for homophilic interactions. Interestingly, this predominant Robo1 dimeric assembly was not altered upon Slit binding (Zakrys et al. 2014). However, a recombinantly produced Robo1 ectodomain from baculovirus was also shown to be monomeric in solution (Barak et al. 2014). This may indicate that other factors, such as high local concentrations, may be required for dimerisation.

These results are all consistent with the low resolution reconstruction of hRobo1 (Aleksandrova et al. 2018). Here, tetrameric Robo1 assemblies from recombinant mammalian cells stabilised using the GraFix cross-linking technique before

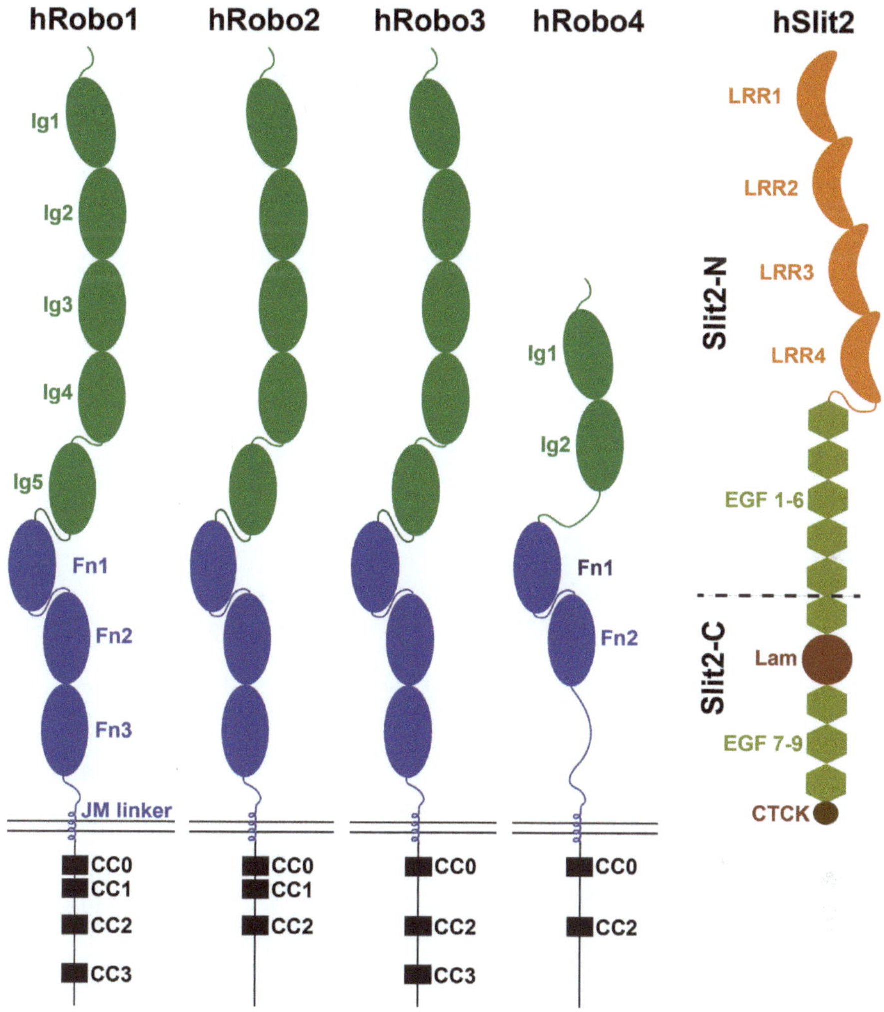

Fig. 9.2 Domain composition of human Robo receptors and Slit2 ligand. The Robo receptors domains are coloured as follows: Ig domains in green; FN in blue; JM linker in blue; and CC0-3 in black. Slit2 domains are coloured as follows: LRR in orange, EGF in lime, Lam in light brown, and CTCK in dark brown

negative staining were observed (Fig. 9.4). The available hRobo1 crystal structures (hRobo1 Ig1–4, Ig5 and FN2–3) were combined with homology and SAXS models to fit a Robo1 ectodomain 'dimer-of-dimers' tetrameric assembly. In this model the hRobo1 ectodomain monomers adopt a domain arrangement that folds back on itself, before turning and extending towards the membrane. This rather compact association is facilitated by the longer linker regions between the Ig4–5 (7 amino acids), Ig5-FN1 (11 amino acids), and FN1–2 (9 amino acids), whose length are

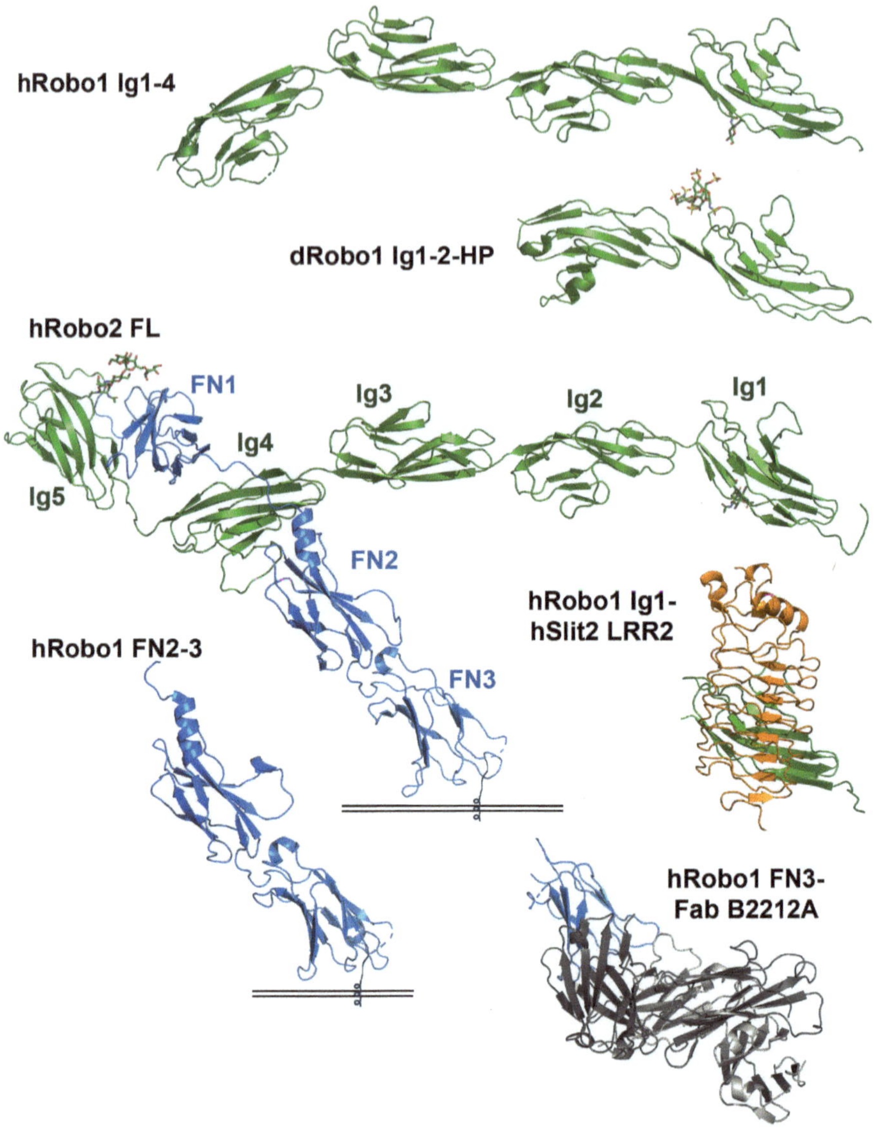

Fig. 9.3 Full length, domain and complex structures for various human (h) and *Drosophila* (d) Robo receptors as determined by X-ray crystallography and currently available in the PDB. The full length Robo2 ecotodomain (PDB 6IAA) is shown in the center with other Robo1 domain and complex structures shown in a similar orientation. A schematic of the membrane is shown to help orientation for Robo1 FN2-3 and Robo2 FL. Domain structures shown include hRobo1 Ig1-4 (PDB 5O5G) and hRobo1 FN2-3-JM (PDB 4HLJ); Complex structures include dRobo Ig1-2-heparin (HP) (PDB 2VRA), hRobo1 Ig1-hSlit2 LRR2 (PDB 2V9T), and hRobo1 FN3-Fab B2212A (PDB 3WIH). Robo Ig and FN domains are coloured green and blue respectively, hSlit2 is coloured orange, and Fab B2212A is coloured gray. Glycosylation and HP moieties are shown in a stick format

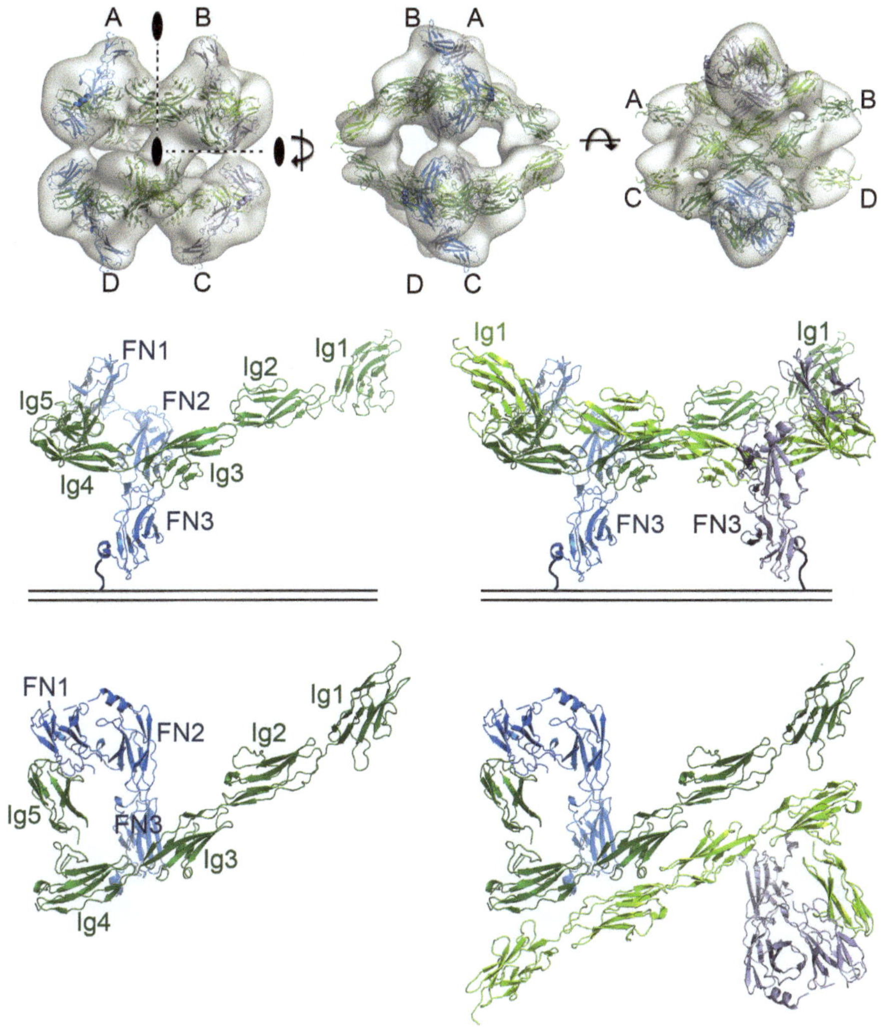

Fig. 9.4 Robo1 forms a 'dimer-of-dimers' tetrameric assembly. Three orientations of the Robo1 domains modelled into the low resolution 3D reconstruction are shown on top to illustrate the 'dimer-of-dimers' (A-B/C-D) configuration. The overall domain organisation is shown below in two orientations (side and top) to illustrate how monomers are arranged as dimers on the cell surface. For illustration purposes the membrane insertion region was included in the middle panel. The Ig and FN domains are labelled and coloured dark or light green and marine or deep blue respectively for each Robo1 monomer

conserved across all Robo receptors except Robo4. The major, and presumably most biologically relevant, dimeric interaction is mediated by the Ig2–4 domains. This is consistent with the role of Robo2 Ig3 in the dimerization and lateral positioning of axons in *Drosophila* (Evans and Bashaw 2010). These dimers at the cell surface can further interact in a 'back-to-back' conformation to complete the tetrameric Robo1 assembly observed (Fig. 9.4).

The hRobo2 ectodomain crystal structure adopts a similar conformation overall to the low resolution hRobo1 EM reconstruction (Figs. 9.3 and 9.4). However, the higher resolution hRobo2 structure provides a much more detailed molecular insight into the unique hairpin domain arrangement observed. The hRobo2 structure determined represents an auto-inhibited Robo receptor conformation, with the Robo Ig4 interface shown to be important for dimerization and signalling being sequestered by the FN2 domain (Barak et al. 2019) (Fig. 9.5). While a similar hRobo1 dimer interaction was not observed in hRobo2, it does form an extended interaction array in the crystallographic lattice. This is mostly mediated by hRobo2 Ig4–Ig5–FN1 domains, with some contributions from Ig3, in a *cis* direction. But a more extensive interaction interface is observed in *trans* from two opposing *cis* layers. This is reminiscent of a cell-cell interaction as the transmembrane proximal domains are oriented in opposite directions. These interactions are primarily mediated by direct contacts between Ig5 and an Ig5 linked glycan branch from one layer with Ig1–2 from an opposing layer. A larger array is assembled by the Ig1–2

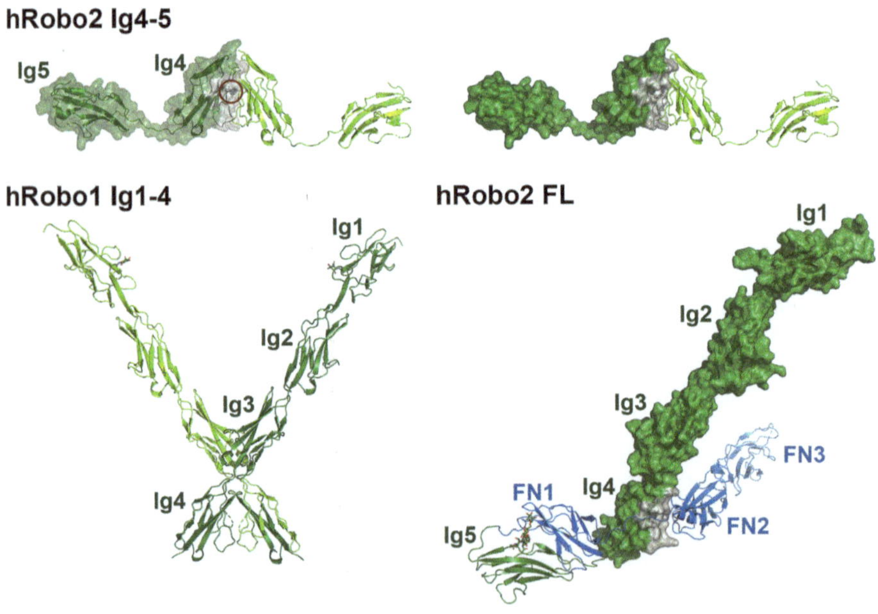

Fig. 9.5 Robo receptors dimerise using a sequence conserved interface sequestered in the full length ectodomain context. hRobo2 Ig4-5 (PDB 5NOI) and hRobo1 Ig1-4 (PDB 5O5G) dimers are shown in a similar orientation to the full Robo2 ectodomain (PDB 6IAA) for easy comparison. The Phe-357 shown to be important for hRobo2 Ig4-5 dimerisation is shown in stick format and circled. The hRobo2 Ig4 dimerisation interface is coloured grey and predominantly precluded from interaction by FN2 in the full hRobo2 ectodomain. The Ig and FN domains are labelled and coloured green and blue, respectively, and glycosylation is shown in stick format

domains from the first molecule interacting with Ig5 and an Ig5 linked glycan of another opposing molecule. The hRobo2 N426 glycosylation site on Ig5, and the residues important for its interaction on Ig1–2, R99 and R132, are sequence conserved across species, suggesting this interaction may be functionally relevant. This was validated by in vitro cellular assays and genetic *C. elegans* experiments showing the importance of Ig5 in fastening an inhibited form of Robo receptors on the cell surface (Barak et al. 2019).

On the cytoplasmic side, the four CC motifs are scaffolding elements responsible for the recruitment of specific proteins, or formation of protein complexes. These adapter proteins are probably shared between various Robo receptors, although not all have been verified. However, the final downstream signalling pathway is largely dependent on the combination of CC motifs present, and the cell type in question. In *Drosophila* CC0 and CC1 are both tyrosine phosphorylation targets of the cytoplasmic Abelson kinase (Abl), which results in Robo inhibition (Bashaw et al. 2000). In the presence of Slit2 and netrin-1, Robo1 CC1 is reported to mediate a cytoplasmic interaction with Deleted in Colorectal Cancer (DCC) that is required to silence netrin-1 attraction (Stein and Tessier-Lavigne 2001). CC2 was shown to interact with Enabled (Ena), an Abl substrate and actin binding protein, which is required for Robo mediated repulsion (Bashaw et al. 2000). In Robo4, CC2 was shown to mediate an interaction with Mena, the mammalian homolog of Ena (Park et al. 2003; Jones et al. 2009). Abl and Ena are long known to interact and regulate reorganization of the actin cytoskeleton (Comer et al. 1998), and in the case of endothelial cells, influence cell migration by promoting filopodia formation. Lastly, CC3 is a polyproline stretch (Kidd et al. 1998) involved in the recruitment of Slit/Robo GTPase activating proteins (srGAP) (Wong et al. 2001; Li et al. 2006) using a two-component molecular mechanism to achieve tighter binding (Guez-Haddad et al. 2015). Another GAP protein, CrossGAP/Vilse, binds the CC2 motif to mediate Robo repulsion (Lundström et al. 2004; Hu et al. 2005). Both GAPs act on small GTPases of the Rho family, RhoA and Cdc42 in the case of SrGAPs, Rac1 and Cdc42 in the case CrossGAP/Vilse, which are known to regulate cytoskeletal dynamics (Chédotal 2007; Ypsilanti et al. 2010).

Slit Proteins Are Robo1/Robo2 Receptor Ligands

Upon midline crossing commissural neurons become sensitive to Slit repulsive cues to ensure they are correctly expelled and prevented from recrossing aberrantly (Long et al. 2004; Zou et al. 2000). Three slit genes (*Slit1*, *Slit2*, *Slit3*) have been identified in mammals (Brose et al. 1999; Holmes et al. 1998; Itoh et al. 1998). Slit1 and Slit2 are known to function as chemorepellents (Nguyen Ba-Charvet et al. 1999; Ringstedt et al. 2000; Hammond et al. 2005). Slit3 is presumed to act similarly but as it exhibits a different neuronal expression pattern (Ringstedt et al. 2000; Hammond et al. 2005) and was shown to be non-repulsive for motor axons (Hammond et al. 2005) this remains to be determined. Slits are large glycosylated proteins secreted by midline glial cells, but often found associated with the extracellular matrix (Brose et al. 1999). Slits are generally composed of four N-terminal leucine rich repeat (LRR) domains (LRR1–4) followed by six epidermal growth factor (EGF)-like domains (EGF1–6), a laminin G–like domain (Lam), another three EGF domains (EGF7–9), and a C-terminal cysteine knot domain (CTCK) (Fig. 9.2).

Slit2 can be cleaved between EGF5 and 6 by an unknown protease to generate N- and C-terminal fragments, Slit2-N and Slit2-C respectively, with distinct properties (Nguyen Ba-Charvet et al. 2001). Most Slits probably undergo proteolysis as the cleavage site is conserved from *Drosophila* to vertebrates (Brose et al. 1999). Slit2-N binds Robo1 and Robo2 to mediate neuronal repulsion (Nguyen Ba-Charvet et al. 2001) while Slit2-C binds PlexinA2 to signal independently (Delloye-Bourgeois et al. 2015). Slit2-C was shown to induce growth cone collapse, and because PlexinA1 expression is upregulated on growth cones at the midline, it might act as a second chemorepellent to reinforce crossing and post-crossing (Delloye-Bourgeois et al. 2015). Slits can homodimerise through their LRR4 domains (Howitt et al. 2004; Seiradake et al. 2009), but LRR1–4 has also been reported to be monomeric (Hohenester 2008).

An analysis of the *Drosophila* Slit LRR domains identified the LRR2 domain as the binding site for Robo receptors (Howitt et al. 2004), while the first two Ig domains of Robo1 were similarly shown to be important (Liu et al. 2004). The structure of the minimal hRobo1-Ig1-hSlit2-LRR2 complex (PDB 2V9T) confirmed that Robo1 Ig1 interacts with the highly conserved concave face of Slit2 LRR2 (Morlot et al. 2007). Their binding buries a 1380 Å^2 solvent accessible area in two distinct regions (Fig. 9.6). The first is electrostatic in nature and mediated by salt bridges, hydrogen bonds, and bridging water molecules with the side-chains undergoing conformational changes upon complex formation. The second is largely hydrophobic in composition and most of this surface is buried by crystallographic contacts in the hRobo1 Ig1–2, hRobo1 Ig1–4 and dRobo Ig1–2–heparin complex crystal structures.

Heparan Binding

Heparan sulphate proteoglycans (HSPGs) are extracellular matrix associated core proteins containing one or more heparan sulphate (HS) glycosaminoglycan chains. HSPGs are known to play a role in neural patterning, and can be divided into three major classes, membrane bound syndecans and glypicans, and secreted molecules such as perlecan, collagen and agrin (Saied-Santiago and Bulow 2018). Full length Slit2 and Slit2-N are long known to tightly associate with cell membranes and can be disassociated by treatment with either high salt (1 M NaCl) or heparin (Brose et al. 1999). A HS dependent binding of Slit to glypican-1 was reported soon after (Liang et al. 1999; Ronca et al. 2001), and enzymatic removal of HS from the cell surface not only decreased the affinity of Slit2 for Robo1, but also abolished its repulsive activity on olfactory neurons (Hu 2001).

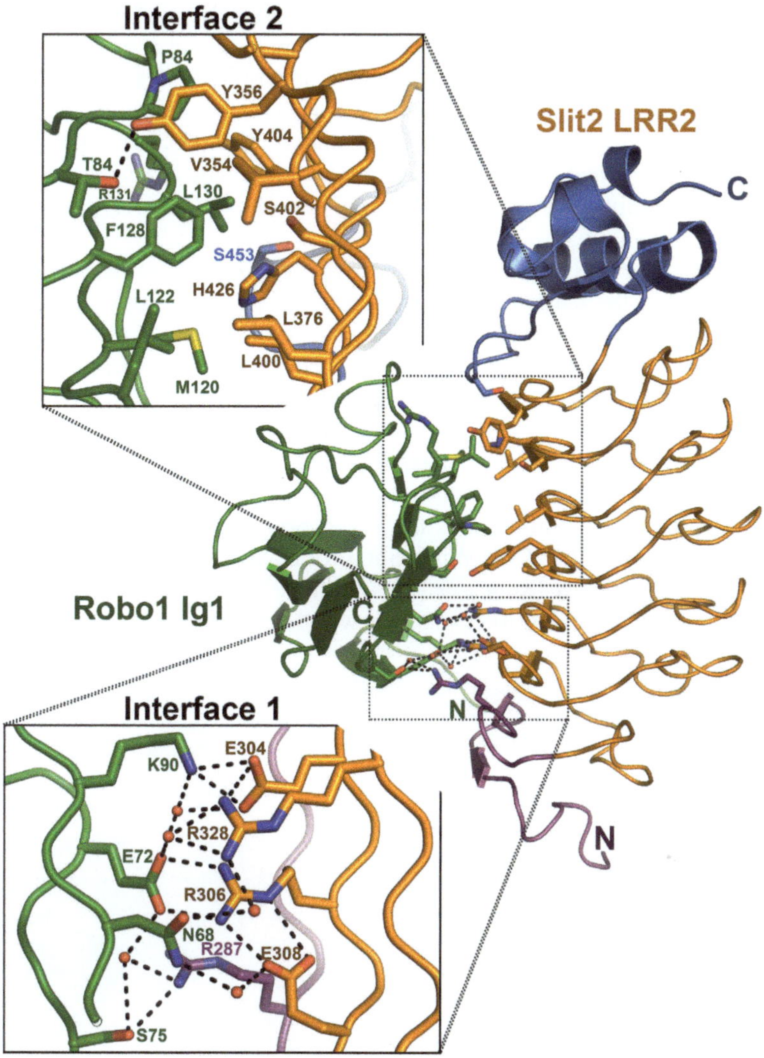

Fig. 9.6 The human Slit2 LRR2-Robo1 Ig1 minimal complex (PDB 2V9T). hRobo1 Ig1 is coloured green, and hSlit2 LRR2 N- and C-terminal caps are coloured magenta and blue, respectively, while the LRRs are coloured orange. Interacting residues are shown in stick representation with water molecules coloured red

This important role in Slit/Robo signalling was further supported when a genetic interaction between Slit and exostosin 1 (*Ext1*), a glycosyltransferase involved in HS biosynthesis, in Slit–mediated retinal axon guidance was observed (Inatani et al. 2003). Syndecan was subsequently shown to be a critical genetic component of Slit-Robo signalling to stabilise Slit-Robo interactions (Steigemann et al. 2004; Johnson et al. 2004). Interestingly, while Slit was still present in midline glia cells it was absent on axon fascicles in *sdc* mutants, indicating syndecan might play an additional regulatory role in the extracellular localisation of Slit on axons (Johnson et al. 2004). The role of HS was further strengthened in genetic *C. elegans* studies, where it was shown to be not only important, but that particular HS sulphation patterns played a key role in certain cellular contexts (Bülow and Hobert 2004). Furthermore, this is likely to be conserved because a genetic link between *Slit1* and/or *Slit* and HS sulphation modifications was also shown in mice (Conway et al. 2011).

Heparin, a highly sulphated form of HS, was observed to bind between Ig1 and 2 of Robo1, and at the C-terminal cap region of Slit2 LRR2 (Hussain et al. 2006; Fukuhara et al. 2008). This interaction can enhance Slit binding to Robo (Hussain et al. 2006), which together form a continuous HS binding interface to increase the stability of the Slit-Robo complex for signalling (Fig. 9.7). The ability of Slit2 conditioned media or Slit2 LRR2 to induce collapse of *Xenopus* retinal growth cones in a heparin/HS dependent manner provides direct evidence for the importance of this interaction in Slit-Robo signalling (Piper et al. 2006; Hussain et al. 2006). Additional HS binding sites have also been identified at the hRobo1 N–terminus (Li et al. 2015); *Drosophila* Slit LRR1 and C–terminus (Hussain et al. 2006); hSlit2 C-terminus (Ronca et al. 2001) and LRR4 (Seiradake et al. 2009). These additional HS binding sites together with LRR2 on Slit have been proposed

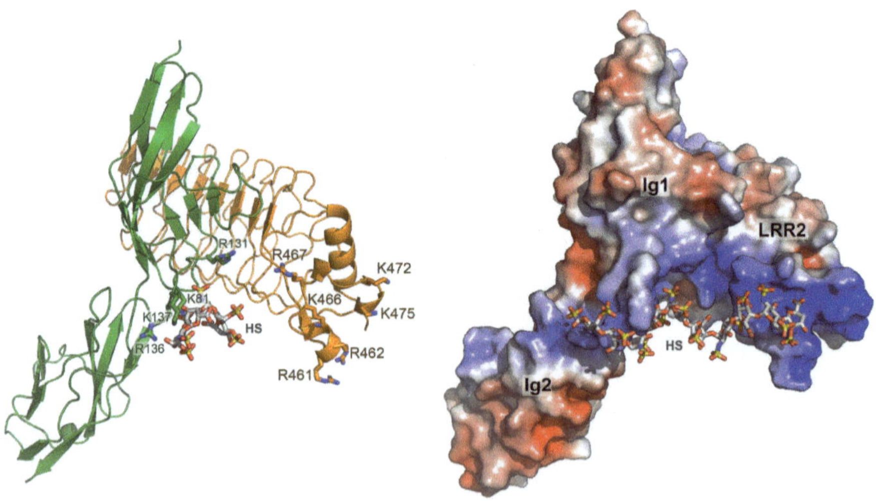

Fig. 9.7 hRobo1 Ig1-2-hSlit2 LRR2 form a composite heparin (HP) binding site. On the left is a model of a tripartite hRobo1 Ig1-2-hSlit2 LRR2-HP (dp10) complex obtained by superimposing the hRobo1 Ig1-hSlit2 LRR2 (PDB 2V9T) and dRobo Ig1-2-HP (PDB 2VRA) complexes onto hRobo1 Ig1-2 (PDB 2V9R) and extending the HP molecule. On the right is an electrostatic surface representation of the same complex (*red*, negative potential; *blue*, positive potential). The positions of HP binding residues in hRobo1 Ig1-2 and hSlit2 LRR2 are shown as sticks and labelled. The hRobo1 Ig and hSlit2 LRR2 domains are coloured green and orange, respectively, and the modelled HP moiety is shown in a grey stick representation

to play a secondary role by concentrating Slit on target cell surfaces, perhaps in a sulphation dependent pattern, or by modulating its diffusion properties (Hussain et al. 2006).

Robo3

Commissureless (Comm) is a transmembrane protein expressed in the commissural neurons of flies that directs Robo receptors to the endocytic pathway to allow midline attraction (Keleman et al. 2002; Georgiou and Tear 2002). After midline crossing Comm is downregulated to ensure newly synthesized Robo receptors can reach the cell surface for interaction with Slit to mediate repulsion. However, no functional Comm homologue was ever identified in vertebrates until the recent discovery of PRRG4 (Justice et al. 2017). But Robo3 was shown to be important for commissural axon midline attraction by blocking Slit mediated Robo1 repulsion (Marillat et al. 2004), suggesting Robo3 might play an attractive rather than repulsive role in midline crossing (Domyan et al. 2013).

To add further complexity four vertebrate Robo3 alternative mRNA splicing isoforms with distinct properties have been identified (Camurri et al. 2005; Chen et al. 2008). Robo3A and Robo3B are evolutionarily conserved and differ by 26 amino acids at the N-terminus (Camurri et al. 2005). Interestingly, in biochemical pull-down assays the shorter Robo3B isoform was shown to bind Slit2 while the longer Robo3A was unable to. In addition, Robo3A was shown to interact homophilically and heterophilically with Robo1. At the C-terminus Robo3 alternative splicing produces Robo3.1 and Robo3.2, which have sequential and opposing roles (Chen et al. 2008). Robo3.1 is observed on pre-crossing axons to suppress Slit-mediated repulsion while Robo3.2 is upregulated upon midline crossing and contributes to repulsion. *Robo3.2* mRNA differs from *Robo3.1* by intron retention that results in a premature stop codon, making it target for nonsense–mediated mRNA decay (NMD) (Colak et al. 2013). Furthermore, while *Robo3.2* mRNA is transported to the axon, possibly due a localisation element in the retained intron, *Robo3.1* mRNA remains in the cell body. This difference in RNA transcript location is believed to facilitate midline crossing by ensuring a sharp and transient spike in Robo3.2 to prevent re-crossing of commissural axons.

More recently it was conclusively shown that mammalian Robo3 receptors do not bind Slit (Zelina et al. 2014). This is primarily due to three mutations (N84P, K86R, and in particular P126L) acquired in the Slit binding Ig1 region of Robo3 during evolution. Interestingly, these mutations were not acquired by non-mammalian vertebrates, which maintain Slit binding. These results support a functional change for Robo3 in mammalian and non-mammalian species. Furthermore, mammalian Robo3 gained the ability to interact with DCC via its cytoplasmic region (Fig. 9.8), possibly via an adaptor protein, and that Netrin-1 binding to DCC can induce Robo3 phosphorylation to function as a chemoattractive receptor complex (Zelina et al. 2014). The subtle changes that occurred in Robo3 to

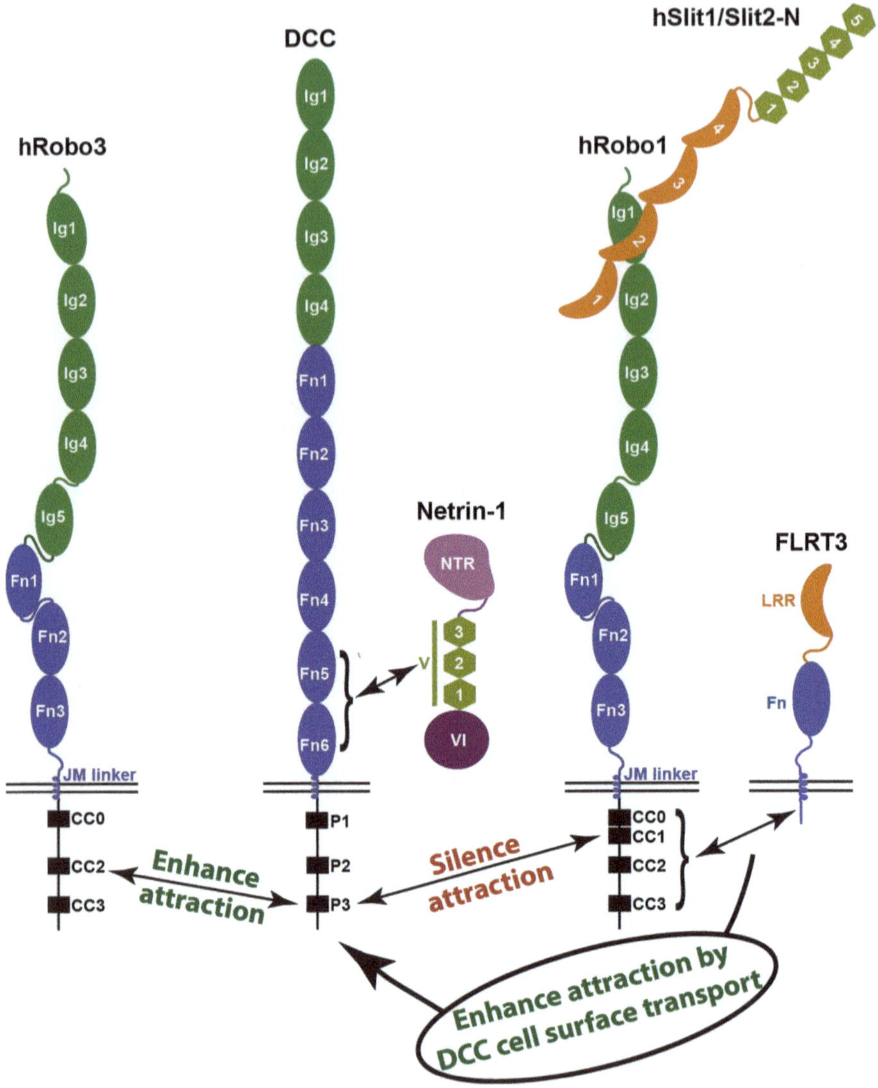

Fig. 9.8 Intracellular Robo co-receptor interactions. Robo3 and Robo1 interact with DCC in a netrin-1 dependent manner using different CC domains to enhance or silence axonal attraction signals. The Robo and DCC receptor Ig and Fn domains are coloured green and blue, respectively. The cytosolic Robo (CC0-3) and DCC (P1-3) motifs are shown as black boxes and labelled. Netrin-1 N-terminal laminin VI, EGF 1-3 (or V domain), and C-terminal netrin-like (NTR) domains are coloured dark purple, green and light purple, respectively. Slit LRR and EGF domains are coloured orange and green, respectively. *Note* Slit is shown as a monomer but likely forms dimers via LRR4. FLRT3 LRR and Fn domains are coloured orange and blue, respectively

switch from a repulsive cue via Slit to an attractive cue via Netrin-1/DCC suggest that this family member may have made a significant contribution to the evolution of mammalian neuronal development.

Intriguingly, an extracellular protein microarray screen for Robo3 binding partners identified an interaction with Neural epidermal growth factor-like-like 2 (NELL2) (Jaworski et al. 2015). NELL2 is the human ortholog of chick Neural EGF-like (Nel), which has been shown to inhibit retinal axon outgrowth and induce growth cone collapse and axon retraction (Jiang et al. 2009). NELL2 is a large secreted glycoprotein containing a laminin G-like domain, six EGF-like domains, and five von Willebrand factor (VF) C domains. The Robo3 FN and NELL2 EGF-like domains were shown to mediate this interaction. Furthermore, NELL2 was shown to repel mouse spinal cord commissural axons in a Robo3 dependent manner while acting as a midline attractive cue in vivo (Jaworski et al. 2015). These findings show how Robo3 can perform a multifunctional receptor axon guidance role to inhibit Slit repulsion, facilitate Netrin attraction, and mediate NELL2 repulsion at the midline. Together all these studies underline the complex roles Robo3 plays in neuronal development. Structural information on the Robo3-NELL2 interaction would likely provide some valuable insights on these mechanisms.

Robo4

Unlike other members of its family, Robo4, is primarily expressed by endothelial cells at sites of active angiogenesis (Park et al. 2003), proliferating cell types such as hematopoietic stem cells and vascular smooth muscle cells (Shibata et al. 2009; Smith-Berdan et al. 2011; Liu et al. 2006), and is involved in cell migration processes (Park et al. 2003; Sheldon et al. 2009; Yadav and Narayan 2014). The knockdown of Robo4 in Zebrafish lethally disrupts vessel sprouting during early embryonic development, underlying its important role in angiogenesis (Bedell et al. 2005). However, Robo4 was also shown to be important for the migration of neuronal cells in mice during development (Zheng et al. 2012). If this is the result of Robo4 alone, or is dependent on interaction with other Robo receptors or partners, remains unclear.

Robo4 expression is regulated by an upstream promoter region that contains binding sites for specificity protein 1 (SP1) (Okada et al. 2007), GA-binding protein (GABP) (Okada et al. 2007, 2008), Sry-related high mobility box (SOX) (Samant et al. 2011), and activator protein-1 (AP-1) transcription factors. This region is demethylated when induced pluripotent stem (iPS) cells undergo differentiation into endothelial cells following a recently described regulatory mechanism, which promotes transcription of downstream genes (Tanaka et al. 2018).

As mentioned previously, the structural organisation of Robo4 is distinctly different from the other Robo receptors. Both its shorter extracellular region (composed of only two Ig and two FN3 domains), and cytoplasmic region (containing only CC0 and CC2 motifs) influence its ability to interact with canonical Slit ligands and their downstream signalling pathways (Fig. 9.2). For instance, Robo1 and Robo2 are unable to rescue Robo4 activity (Bedell et al. 2005), and many of the Slit2 binding residues identified in Robo1 are different (specifically: E72V, N88S, K90Q, F128L and R131Q) (Morlot et al. 2007). Furthermore, Robo4 was shown to be monomeric (Bisiak 2018), lacking the third and fourth Ig domains, which are important for homophilic interactions (Yom-Tov et al. 2017; Aleksandrova et al. 2018).

Despite what is already known of the common extracellular and intracellular Robo receptor partners, the interaction network of Robo4 is quite different, and still poorly mapped. For instance, while Slit2 mediated effects through Robo4 have been observed despite a lack of key interaction residues (Jones et al. 2009; Park et al. 2003), some studies report a complete absence of direct interaction with Slit2 (Suchting et al. 2006; Koch et al. 2011). This is still debated, where, to cite just one example, one study determined that Slit2 acts as a strong angiogenic inducer through Robo4 in primary human endothelial cells (Park et al. 2003). While in a second study, the authors showed Robo4 inhibiting a Slit2-Robo1 mediated primary human umbilical vein endothelial (HUVEC) cell migration through an unknown intracellular mechanism (Enomoto et al. 2016). As the expression of different Robo receptors can overlap in some cell types, most modern models tend to suggest that Slit functions on Robo4 are mediated through receptor heterodimers. To sustain this line of thought, Robo1/Robo4 heterodimers where shown to promote cell migration in vitro by the Robo4 sequestering of Robo1 in intracellular vesicles, although the specific domains involved were not identified (Sheldon et al. 2009).

UNC5B is a member of the UNC5 receptor family that plays a major role in neuronal guidance via Netrin ligand binding (Seiradake et al. 2016). However, UNC5B is especially involved in angiogenetic processes, being highly expressed in developing blood vessels (Lu et al. 2004; Navankasattusas et al. 2008; Tai-Nagara et al. 2017). UNC5B acts as a repulsive receptor upon netrin-1 binding in vascular morphogenesis (Lu et al. 2004) or FLRT2 binding in the placental labyrinth (Tai-Nagara et al. 2017). UNC5B is also activated upon Robo4 binding, which inhibits angiogenesis to maintain vascular integrity (Koch et al. 2011), and does not require the Robo4 cytoplasmic region (Zhang et al. 2016). UNC5 receptors are type I transmembrane proteins, containing two Ig and two thrombospondin (TSP)-like ectodomains (Seiradake et al. 2014; Bisiak 2018), while the intracellular region is composed of a death and regulatory domain (Wang et al. 2009). The UNC5 Ig1 domain and Ig1–2 domains are required for interaction with FLRT (Seiradake et al. 2014) and netrin-1 (Grandin et al. 2016), respectively. Deletion studies showed that the UNC5B TSP domains are sufficient for interaction with the Robo4 Ig1–2 domains (Koch et al. 2011).

At the cytoplasmic level, Robo4 encodes variants of the CC0 and CC2 motifs (Legg et al. 2008). As such Robo4 can recruit other proteins apart from the classical Abl and Mena (in mammals), or Enabled (in *Drosophila*), commonly observed throughout the Robo family. The WASP family proteins, WASP and NWASP, interact with Robo4 through their own polyproline stretches (Sheldon et al. 2009), and are directly responsible for filipodia formation induced by recruitment in a larger complex of the regulatory proteins WIP and Syndapin (Martinez-Quiles et al. 2001; Kaur et al. 2006). Yet another mechanism triggered in the presence of Slit2, includes the recruitment of the cytoplasmic protein Paxillin by Robo4, which inhibits protrusive activity (Jones et al. 2009; Sherchan et al. 2017). Paxillin is a well-known cell matrix adhesion protein found at focal adhesion points, which through regulation of the GTPase Arf6 activity, increases vascular stability (Turner et al. 1990; Deakin and Turner 2008; Jones et al. 2009).

Robo Co-receptors

Slit proteins are long established Robo1 and 2 receptor ligands, while NELL2 was recently shown to act as a Robo3 receptor ligand for midline repulsion (Jaworski et al. 2015). However, other direct and indirect Robo co-receptors have been identified (Fig. 9.8 and 9). Early studies with *Xenopus* spinal axons, complemented by cellular assays, showed how an interaction between the cytoplasmic domains of DCC and Robo upon Slit activation is necessary to silence netrin-1 attractive signalling during midline crossing (Stein and Tessier-Lavigne 2001). Later, this same interaction was shown to attenuate Slit2 mediated repulsion in neocortical axons (Fothergill et al. 2014). Conversely, an intracellular interaction between Robo3 and DCC was shown to enhance netrin-1 attractive signalling (Zelina et al. 2014). Moreover, while the same cytosolic P3 domain of DCC is required for Robo receptor binding, Robo1 and Robo3 utilize different intracellular regions. The Robo1 interaction was mapped to the CC1 region missing in Robo3 (Stein and Tessier-Lavigne 2001), and the Robo3 interaction was mapped to CC2, or between CC2 and CC3 (Zelina et al. 2014), which may account for their different responses to netrin-1 (Fig. 9.8). Both of these interactions are hierarchical signalling events, where one cue can suppress or potentiate the effect of another. Away from the midline, a fine–tuned axonal guidance can occur via a new context–dependent signalling mechanism, as reported for rostral thalamocortical axons (rTCAs) (Leyva-Diaz et al. 2014). Here, a *cis* interaction between the FLRT3 and Robo1 cytoplasmic domains was shown to be necessary in the presence of netrin-1 and Slit2 guidance cues (Fig. 9.8). In this case FLTR3 modifies a Slit1 mediated Robo1 signalling by activating protein kinase A to promote vesicular transport of DCC to the growth cone surface for netrin-1 mediated attraction.

The co-receptor interactions discussed above involve intracellular *cis* interactions, presumably mediated by cytosolic adaptor proteins binding to the CC domains of Robo receptors. But extracellular co-receptor interactions have also been reported (Fig. 9.9). The migration of cortical interneurons through the developing striatum is known to be mediated by chemorepulsive class 3 semaphorin ligands (Sema3A and Sema3F) acting on cell surface neuropilin (Nrp1 and Nrp2) receptors (Marin et al. 2001). It was later shown that cortical interneurons lacking Robo1 were less responsive to semaphorins due to a reduction in Nrp1 and

PlexinA1 receptor levels (Hernandez-Miranda et al. 2011). This is not Slit1/Slit2 dependent, and careful biochemical studies identified a direct interaction between the Robo1 Ig1–2 domains and Nrp1 (Fig. 9.9). This extracellular *cis* interaction shows how Robo1 can modulate the response of interneurons to semaphorin signalling.

Slit was shown to interact with Down syndrome cell adhesion molecule 1 (Dscam1) on mechanosensory neurons for specific axon collateral branch formation in a Robo independent manner (Dascenco et al. 2015). Later it was reported that Dscam1 forms a co-receptor complex with Robo1 that is dependent on Slit-N to promote longitudinal axonal growth in *Drosophila* (Alavi et al. 2016). Immunoprecipitation assays showed Slit cleavage was necessary for Dscam1-Robo1 complex formation as Dscam1 only binds Slit-N. Further biochemical assays identified that the EGF1–3 domains of Slit-N were required to bind two sites

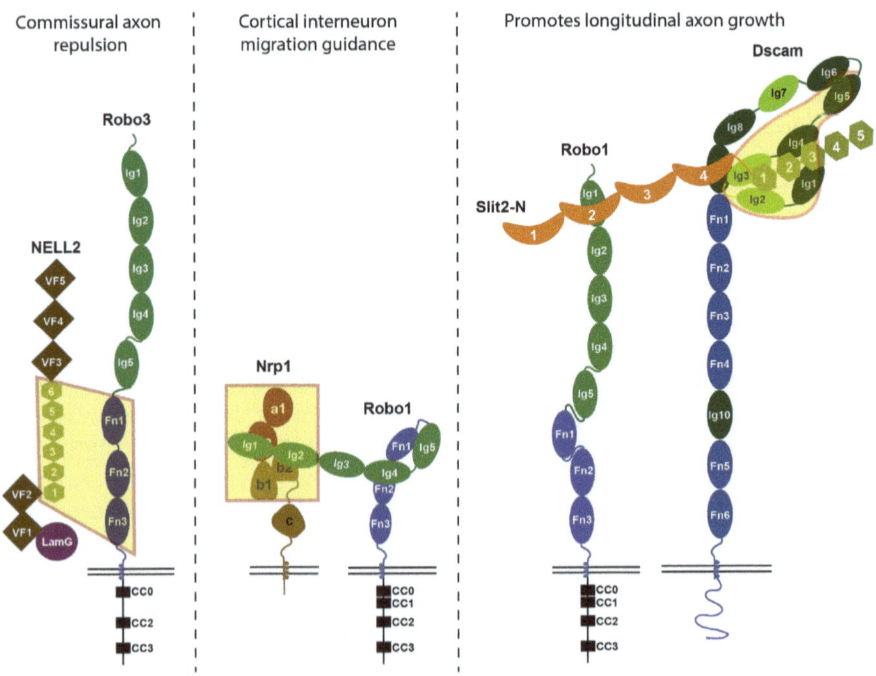

Fig. 9.9 Extracellular Robo co-receptor interactions. NELL2 interacts with Robo3 to mediate axonal repulsion (left). Nrp1 and Robo1 form a co-receptor complex to mediate axon migration (middle). Slit2-N enables a Robo1-Dscam co-receptor complex to promote axon growth (right). The minimal interaction regions determined are highlighted in yellow. NELL2 Laminin G (LamG), von Willebrand factor C (VF), and EGF domains are coloured magenta, brown and green, respectively. Nrp1 a1, a2, b1, b2, and c domains are coloured brown. Robo and Dscam receptor Ig and Fn domains are coloured green and blue, respectively. The cytosolic Robo (CC0-3) motifs are shown as black boxes and labelled. The constant and variable Dscam1 Ig domains are coloured dark and light green, respectively. *Note* Slit is shown as a monomer but likely forms dimers via LRR4

on the N-terminal region of Dscam1 somewhere between Ig1–5 (Fig. 9.9). All these recent examples illustrate how Robo co-receptor interactions can modify cell signals in a hierarchical or context dependent manner. Structural information on these newly identified signalling complexes would likely provide invaluable insights on how multiple guidance effects during neuronal development are integrated.

Robo Signalling Mechanism

Although Slit–Robo signalling has been intensely studied for over three decades there was, until recently, a clear lack of knowledge on how their interaction is relayed across the membrane to mediate intracellular signalling. A major insight into Robo activation was provided by elucidation of the auto-inhibited Robo2 ectodomain structure (Barak et al. 2019). Complementary cellular and in vivo genetic studies further showed the importance of a highly conserved Robo Ig4 domain dimerization interface for Robo signalling. Taken together with previous studies (Zakrys et al. 2014; Aleksandrova et al. 2018) there is now a consensus that Robo receptors on the cell surface probably undergo a conformation change upon Slit binding that is required for dimerisation (or oligomerisation) and subsequent intracellular signalling (Fig. 9.10).

Early studies reported Robo1 shedding as a potential heptocellular carcinoma marker for liver cancer (Ito et al. 2006). Later it was shown that Robo can undergo proteolytic processing by the Kuzbanian (Kuz) Adam family metalloprotease (ADAM10 in mammals) in *Drosophila* (Coleman et al. 2010). Because an uncleavable form of Robo is unable to maintain midline repulsion, this suggested that Kuz mediated ectodomain shedding may play an important role in signalling (Fig. 9.10). For this a cytoskeletal rearrangement upon cleavage by the Slit dependent recruitment of the downstream signalling molecule son of sevenless (Sos) was proposed as a possible mechanism (Coleman et al. 2010). This mechanism is supported by structural studies on the Robo1 juxtamembrane domain region, showing an enhanced ectodomain shedding upon the exposure, or relief, of this structured region (Barak et al. 2014). In this case immobilized Slit on one extracellular cell surface is proposed to create a tension upon Robo1 binding on an approaching axon, leading to cleavage site exposure and intracellular signalling (Barak et al. 2014). The Robo1 cleavage site was located between Q888 and Q889 at juxta-membrane (JM) domain region using mass spectroscopy in human cancer cells (Seki et al. 2010). In addition, following ectodomain shedding Robo1 was observed to be further processed by γ-secretase, with the resulting C-terminal fragment translocated to the nucleus (Seki et al. 2010). However, the function of this C-terminal fragment, and whether ectodomain shedding plays a role in vertebrate neuronal development remain open questions.

Commissural neurons are known to accumulate high levels of Robo only after midline crossing in *Drosophila* (Kidd et al. 1998). Further studies showed how Comm plays an important role by intercepting Robo in the endoplasmic reticulum/

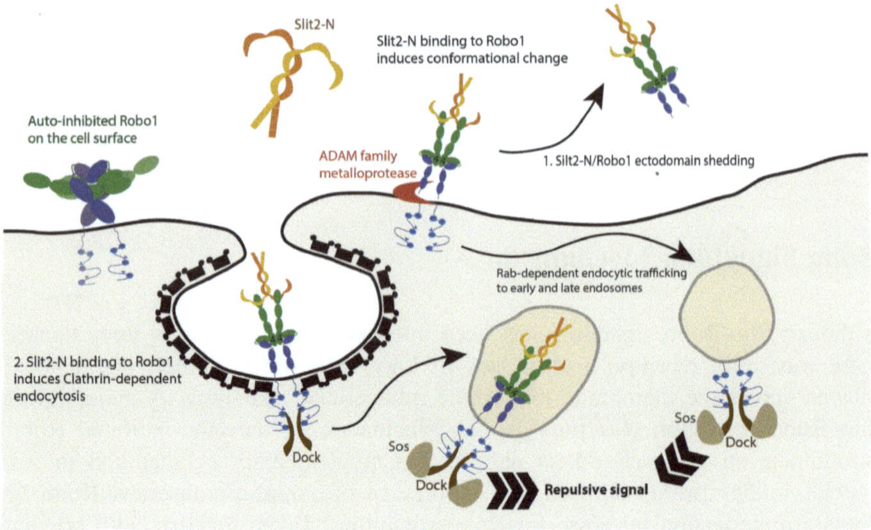

Fig. 9.10 Possible mechanism for Slit2-N mediated Robo1 signalling. Robo1 adopts a compact auto-inhibited dimeric (or oligomeric) assembly on the cell surface. Slit2-N binding induces a conformational rearrangement of Robo1 without dramatically changing its' oligomerization state. This can either induce (1) Slit2-N/Robo1 ectodomain shedding followed by endocytosis of intracellular domains, or (2) Slit2-N mediated endocytosis of the entire Slit2-N/Robo1 complex to the late endosome for Sos recruitment and subsequent cell signalling

Golgi for trafficking away from the growth cone surface to the late endosomes, thus enabling midline crossing (Keleman et al. 2002, 2005). Once this was achieved, the downregulation of Comm results in a surge of cell surface Robo for Slit binding to prevent the axons from re-crossing. While Slit-Robo signalling is conserved in bilateral species, and endocytosis was shown to be important for Slit2 mediated collapse of *Xenopus* retinal growth cones (Piper et al. 2006), no Comm proteins were identified in vertebrates. However, a proline rich and Gla domain protein, PRRG4, containing the L/PPxY motifs found in Comm was recently shown to affect a hRobo1 expression phenotype in flies (Justice et al. 2017). In addition, PRRG4 is able to re-localise hRobo1 from cell surfaces, which taken together strongly suggest it's a functional homologue of Comm (Justice et al. 2017). The role of endocytosis in Slit-Robo signalling is further supported by *Drosophila* genetic experiments and complementary in vitro data (Chance and Bashaw 2015). Two putative YXXϕ sequence motifs, YLQY and YQAGL, known to be important for interaction with the Clathrin adaptor complex, were shown to be important for Robo1 endocytosis (Fig. 9.10). Furthermore, dRobo trafficking to late endosomes is required for recruitment of Sos, mediated by Dock binding to CC2 and CC3, and repulsive signalling. From sequence analyses the endosomal trafficking of Robo1 upon interaction with Slit2 is likely conserved in vertebrates. Only Robo1 homologues have this particular sequence so it will be important to investigate if other

Robo receptors can also undergo endocytosis or require a heterophilic interaction with a Robo1 homologue.

Concluding Remarks

Recent Robo receptor structures have provided valuable insights for the design of complementary biochemical, genetic, and cell based assays to probe Slit mediated Robo signaling (Aleksandrova et al. 2018; Barak et al. 2019). Taken together with previous studies these provide important mechanistic details on Robo mediated signaling pathways. Robo receptors on the cell surface primarily exist in an auto-inhibited conformation as dimers, or higher oligomers, that can be further stabilized in *trans*. Once exposed to Slit emanating from the midline Robo receptors auto-inhibition is relaxed and they must undergo a conformational change to allow Robo Ig4 mediated dimerisation for subsequent intracellular signaling (Fig. 9.10). While the field is still lacking structural details on a Slit-N-Robo ectodomain receptor complex, it is likely that Slit-N mediated endocytosis of Robo1 receptors is required for internal cell signalling (Chance and Bashaw 2015). However, many open questions still remain unresolved. For example, is Robo ectodomain shedding required for signalling? Can hRobo2 (and 3) also undergo endocytosis, or are heterophilic interactions with Robo1 required, because it alone contains endocytic sequence motifs? Lastly, what is the precise role of PRRG4 in hRobo1 cell surface localisation (Justice et al. 2017)?

More recent studies have shown that Slit mediated Robo receptor signalling during neuronal development is not always a simple ligand-receptor mediated event. Here, the ability of Robo receptors to modify axonal growth, as well as attractive or repulsive neuronal signals is often dependent on their interaction with other cell surface receptors and ligands (Stein and Tessier-Lavigne 2001; Hernandez-Miranda et al. 2011) (Figs. 9.8 and 9.9). These, and newer interaction networks (Leyva-Diaz et al. 2014; Jaworski et al. 2015) are continually being discovered, and how they can all coordinate in such a highly cooperative manner is fascinating. In addition, the role of proteoglycans in Slit-Robo and development signaling pathways adds yet another layer of complexity (Townley and Bülow 2018). The interplay and communication between the major neuronal receptor classes is fast becoming an important research focus in the neurobiology field, and many examples now exist, such as the recently reported Robo1-Dscam1 co-receptor complex that is mediated by Slit2-N (Alavi et al. 2016). Structural studies on these complex signalling hubs can provide molecular details that will enable the design of complementary biochemical, genetic, and cell based assays to probe their signaling roles. A collaborative research effort is therefore necessary to provide further insights into the assembly, and signaling, of larger complexes on the cell surface for neuronal guidance. These will also help elucidate the role of Slit-Robo signaling in angiogenesis, organ development and cancer progression. Indeed, once enough experimental results are available on these signalling pathways one could envisage the design of computer algorithms to model such developmental and disease processes.

References

Alavi M, Song M, King GL, Gillis T, Propst R, Lamanuzzi M, Bousum A, Miller A, Allen R, Kidd T (2016) Dscam1 forms a complex with Robo1 and the N-terminal fragment of Slit to promote the growth of longitudinal axons. PLoS Biol 14(9):e1002560. https://doi.org/10.1371/journal.pbio.1002560

Aleksandrova N, Gutsche I, Kandiah E, Avilov SV, Petoukhov MV, Seiradake E, McCarthy AA (2018) Robo1 forms a compact dimer-of-dimers assembly. Structure 26(2):320–328 e 324. https://doi.org/10.1016/j.str.2017.12.003

Ballard MS, Hinck L (2012) A roundabout way to cancer. Adv Cancer Res 114:187–235. https://doi.org/10.1016/B978-0-12-386503-8.00005-3

Barak R, Lahmi R, Gevorkyan-Airapetov L, Levy E, Tzur A, Opatowsky Y (2014) Crystal structure of the extracellular juxtamembrane region of Robo1. J Struct Biol 186(2):283–291. https://doi.org/10.1016/j.jsb.2014.02.019

Barak R, Yom-Tov G, Guez-Haddad J, Gasri-Plotnitsky L, Maimon R, Cohen-Berkman M, McCarthy AA, Perlson E, Henis-Korenblit S, Isupov MN, Opatowsky Y (2019) Structural principles in Robo activation and auto-inhibition. Cell

Bashaw GJ, Kidd T, Murray D, Pawson T, Goodman CS (2000) Repulsive axon guidance: abelson and enabled play opposing roles downstream of the roundabout receptor. Cell 101(7):703–715

Bedell VM, Yeo SY, Park KW, Chung J, Seth P, Shivalingappa V, Zhao J, Obara T, Sukhatme VP, Drummond IA, Li DY, Ramchandran R (2005) Roundabout4 is essential for angiogenesis in vivo. Proc Natl Acad Sci USA 102(18):6373–6378. https://doi.org/10.1073/pnas.0408318102

Bisiak F (2018) Structural studies of the Roundabout protein family Université Grenoble Alpes, Grenoble

Blockus H, Chedotal A (2016) Slit-Robo signaling. Development 143(17):3037–3044. https://doi.org/10.1242/dev.132829

Brose K, Bland KS, Wang KH, Arnott D, Henzel W, Goodman CS, Tessier-Lavigne M, Kidd T (1999) Slit proteins bind Robo receptors and have an evolutionarily conserved role in repulsive axon guidance. Cell 96(6):795–806

Bülow HE, Hobert O (2004) Differential sulfations and epimerization define heparan sulfate specificity in nervous system development. Neuron 41(5):723–736

Camurri L, Mambetisaeva E, Davies D, Parnavelas J, Sundaresan V, Andrews W (2005) Evidence for the existence of two Robo3 isoforms with divergent biochemical properties. Mol Cell Neurosci 30(4):485–493. https://doi.org/10.1016/j.mcn.2005.07.014

Chance RK, Bashaw GJ (2015) Slit-dependent endocytic trafficking of the Robo receptor is required for son of sevenless recruitment and midline axon repulsion. PLoS Genet 11(9):e1005402. https://doi.org/10.1371/journal.pgen.1005402

Chang J, Lan T, Li C, Ji X, Zheng L, Gou H, Ou Y, Wu T, Qi C, Zhang Q, Li J, Gu Q, Wen D, Cao L, Qiao L, Ding Y, Wang L (2015) Activation of Slit2-Robo1 signaling promotes liver fibrosis. J Hepatol 63(6):1413–1420. https://doi.org/10.1016/j.jhep.2015.07.033

Chédotal A (2007) Slits and their receptors. In: Bagnard D (ed) Axon growth and guidance. Advances in experimental medicine and biology, vol 621, 1 edn. Springer-Verlag, New York, pp 65–80. https://doi.org/10.1007/978-0-387-76715-4

Chédotal A, Richards LJ (2010) Wiring the brain: the biology of neuronal guidance. Cold Spring Harb Perspect Biol 2(6):a001917. https://doi.org/10.1101/cshperspect.a001917

Chen Z, Gore BB, Long H, Ma L, Tessier-Lavigne M (2008) Alternative splicing of the Robo3 axon guidance receptor governs the midline switch from attraction to repulsion. Neuron 58 (3):325–332. https://doi.org/10.1016/j.neuron.2008.02.016

Clark K, Hammond E, Rabbitts P (2002) Temporal and spatial expression of two isoforms of the Dutt1/Robo1 gene in mouse development. FEBS Lett 523(1–3):12–16

Colak D, Ji SJ, Porse BT, Jaffrey SR (2013) Regulation of axon guidance by compartmentalized nonsense-mediated mRNA decay. Cell 153(6):1252–1265. https://doi.org/10.1016/j.cell.2013.04.056

Coleman HA, Labrador JP, Chance RK, Bashaw GJ (2010) The Adam family metalloprotease Kuzbanian regulates the cleavage of the roundabout receptor to control axon repulsion at the midline. Development 137(14):2417–2426. https://doi.org/10.1242/dev.047993

Comer AR, Ahern-Djamali SM, Juang JL, Jackson PD, Hoffmann FM (1998) Phosphorylation of Enabled by the *Drosophila* Abelson tyrosine kinase regulates the in vivo function and protein-protein interactions of Enabled. Mol Cell Biol 18(1):152–160

Conway CD, Howe KM, Nettleton NK, Price DJ, Mason JO, Pratt T (2011) Heparan sulfate sugar modifications mediate the functions of slits and other factors needed for mouse forebrain commissure development. J Neurosci 31(6):1955–1970. https://doi.org/10.1523/JNEUROSCI.2579-10.2011

Dalkic E, Kuscu C, Sucularli C, Aydin IT, Akcali KC, Konu O (2006) Alternatively spliced Robo2 isoforms in zebrafish and rat. Dev Genes Evol 216(9):555–563. https://doi.org/10.1007/s00427-006-0070-y

Dascenco D, Erfurth ML, Izadifar A, Song M, Sachse S, Bortnick R, Urwyler O, Petrovic M, Ayaz D, He H, Kise Y, Thomas F, Kidd T, Schmucker D (2015) Slit and receptor tyrosine phosphatase 69D confer spatial specificity to axon branching via Dscam1. Cell 162(5):1140–1154. https://doi.org/10.1016/j.cell.2015.08.003

Deakin NO, Turner CE (2008) Paxillin comes of age. J Cell Sci 121(Pt 15):2435–2444. https://doi.org/10.1242/jcs.018044

Delloye-Bourgeois C, Jacquier A, Charoy C, Reynaud F, Nawabi H, Thoinet K, Kindbeiter K, Yoshida Y, Zagar Y, Kong Y, Jones YE, Falk J, Chedotal A, Castellani V (2015) PlexinA1 is a new Slit receptor and mediates axon guidance function of Slit C-terminal fragments. Nat Neurosci 18(1):36–45. https://doi.org/10.1038/nn.3893

Domyan ET, Branchfield K, Gibson DA, Naiche LA, Lewandoski M, Tessier-Lavigne M, Ma L, Sun X (2013) Roundabout receptors are critical for foregut separation from the body wall. Dev Cell 24(1):52–63. https://doi.org/10.1016/j.devcel.2012.11.018

Enomoto S, Mitsui K, Kawamura T, Iwanari H, Daigo K, Horiuchi K, Minami T, Kodama T, Hamakubo T (2016) Suppression of Slit2/Robo1 mediated HUVEC migration by Robo4. Biochem Biophys Res Commun 469(4):797–802. https://doi.org/10.1016/j.bbrc.2015.12.075

Evans TA, Bashaw GJ (2010) Functional diversity of Robo receptor immunoglobulin domains promotes distinct axon guidance decisions. Curr Biol 20(6):567–572. https://doi.org/10.1016/j.cub.2010.02.021

Fothergill T, Donahoo AL, Douglass A, Zalucki O, Yuan J, Shu T, Goodhill GJ, Richards LJ (2014) Netrin-DCC signaling regulates corpus callosum formation through attraction of pioneering axons and by modulating Slit2-mediated repulsion. Cereb Cortex 24(5):1138–1151. https://doi.org/10.1093/cercor/bhs395

Fujisawa K, Wrana JL, Culotti JG (2007) The slit receptor EVA-1 coactivates a SAX-3/Robo mediated guidance signal in *C. elegans*. Science 317 (5846):1934–1938

Fukuhara N, Howitt JA, Hussain SA, Hohenester E (2008) Structural and functional analysis of Slit and heparin binding to immunoglobulin-like domains 1 and 2 of *Drosophila* Robo. J Biol Chem 283(23):16226–16234

Georgiou M, Tear G (2002) Commissureless is required both in commissural neurones and midline cells for axon guidance across the midline. Development 129(12):2947–2956

Grandin M, Meier M, Delcros JG, Nikodemus D, Reuten R, Patel TR, Goldschneider D, Orriss G, Krahn N, Boussouar A, Abes R, Dean Y, Neves D, Bernet A, Depil S, Schneiders F, Poole K, Dante R, Koch M, Mehlen P, Stetefeld J (2016) Structural decoding of the Netrin-1/UNC5 interaction and its therapeutical implications in cancers. Cancer Cell 29(2):173–185. https://doi.org/10.1016/j.ccell.2016.01.001

Grieshammer U, Le M, Plump AS, Wang F, Tessier-Lavigne M, Martin GR (2004) SLIT2-mediated ROBO2 signaling restricts kidney induction to a single site. Dev Cell 6 (5):709–717

Guez-Haddad J, Sporny M, Sasson Y, Gevorkyan-Airapetov L, Lahav-Mankovski N, Margulies D, Radzimanowski J, Opatowsky Y (2015) The neuronal migration Factor srGAP2 achieves specificity in ligand binding through a two-component molecular mechanism. Structure 23(11):1989–2000. https://doi.org/10.1016/j.str.2015.08.009

Hammond R, Vivancos V, Naeem A, Chilton J, Mambetisaeva E, Andrews W, Sundaresan V, Guthrie S (2005) Slit-mediated repulsion is a key regulator of motor axon pathfinding in the hindbrain. Development 132(20):4483–4495. https://doi.org/10.1242/dev.02038

Hao JC, Yu TW, Fujisawa K, Culotti JG, Gengyo-Ando K, Mitani S, Moulder G, Barstead R, Tessier-Lavigne M, Bargmann CI (2001) *C. elegans* slit acts in midline, dorsal-ventral, and anterior-posterior guidance via the SAX-3/Robo receptor. Neuron 32 (1):25–38

Hernandez-Miranda LR, Cariboni A, Faux C, Ruhrberg C, Cho JH, Cloutier JF, Eickholt BJ, Parnavelas JG, Andrews WD (2011) Robo1 regulates semaphorin signaling to guide the migration of cortical interneurons through the ventral forebrain. J Neurosci 31(16):6174–6187. https://doi.org/10.1523/JNEUROSCI.5464-10.2011

Hivert B, Liu Z, Chuang CY, Doherty P, Sundaresan V (2002) Robo1 and Robo2 are homophilic binding molecules that promote axonal growth. Mol Cell Neurosci 21(4):534–545

Hohenester E (2008) Structural insight into Slit-Robo signalling. Biochem Soc Trans 36(Pt 2):251–256. https://doi.org/10.1042/BST0360251

Holmes GP, Negus K, Burridge L, Raman S, Algar E, Yamada T, Little MH (1998) Distinct but overlapping expression patterns of two vertebrate slit homologs implies functional roles in CNS development and organogenesis. Mech Dev 79(1–2):57–72

Howitt JA, Clout NJ, Hohenester E (2004) Binding site for Robo receptors revealed by dissection of the leucine-rich repeat region of Slit. EMBO J 23(22):4406–4412

Hu H (2001) Cell-surface heparan sulfate is involved in the repulsive guidance activities of Slit2 protein. Nat Neurosci 4(7):695–701

Hu H, Li M, Labrador JP, McEwen J, Lai EC, Goodman CS, Bashaw GJ (2005) Cross GTPase-activating protein (CrossGAP)/Vilse links the Roundabout receptor to Rac to regulate midline repulsion. Proc Natl Acad Sci USA 102(12):4613–4618. https://doi.org/10.1073/pnas.0409325102

Huang T, Kang W, Cheng AS, Yu J, To KF (2015) The emerging role of Slit-Robo pathway in gastric and other gastro intestinal cancers. BMC Cancer 15:950. https://doi.org/10.1186/s12885-015-1984-4

Huminiecki L, Gorn M, Suchting S, Poulsom R, Bicknell R (2002) Magic roundabout is a new member of the roundabout receptor family that is endothelial specific and expressed at sites of active angiogenesis. Genomics 79(4):547–552

Hussain SA, Piper M, Fukuhara N, Strochlic L, Cho G, Howitt JA, Ahmed Y, Powell AK, Turnbull JE, Holt CE, Hohenester E (2006) A molecular mechanism for the heparan sulfate dependence of slit-robo signaling. J Biol Chem 281(51):39693–39698

Hwang DY, Kohl S, Fan X, Vivante A, Chan S, Dworschak GC, Schulz J, van Eerde AM, Hilger AC, Gee HY, Pennimpede T, Herrmann BG, van de Hoek G, Renkema KY, Schell C, Huber TB, Reutter HM, Soliman NA, Stajic N, Bogdanovic R, Kehinde EO, Lifton RP, Tasic V, Lu W, Hildebrandt F (2015) Mutations of the SLIT2-ROBO2 pathway genes SLIT2 and SRGAP1 confer risk for congenital anomalies of the kidney and urinary tract. Hum Genet 134(8):905–916. https://doi.org/10.1007/s00439-015-1570-5

Inatani M, Irie F, Plump AS, Tessier-Lavigne M, Yamaguchi Y (2003) Mammalian brain morphogenesis and midline axon guidance require heparan sulfate. Science 302(5647):1044–1046

Ito H, Funahashi S, Yamauchi N, Shibahara J, Midorikawa Y, Kawai S, Kinoshita Y, Watanabe A, Hippo Y, Ohtomo T, Iwanari H, Nakajima A, Makuuchi M, Fukayama M, Hirata Y,

Hamakubo T, Kodama T, Tsuchiya M, Aburatani H (2006) Identification of ROBO1 as a novel hepatocellular carcinoma antigen and a potential therapeutic and diagnostic target. Clin Cancer Res 12(11 Pt 1):3257–3264

Itoh A, Miyabayashi T, Ohno M, Sakano S (1998) Cloning and expressions of three mammalian homologues of *Drosophila* slit suggest possible roles for Slit in the formation and maintenance of the nervous system. Brain Res Mol Brain Res 62(2):175–186

Jaworski A, Tom I, Tong RK, Gildea HK, Koch AW, Gonzalez LC, Tessier-Lavigne M (2015) Operational redundancy in axon guidance through the multifunctional receptor Robo3 and its ligand NELL2. Science 350(6263):961–965. https://doi.org/10.1126/science.aad2615

Jiang Y, Obama H, Kuan SL, Nakamura R, Nakamoto C, Ouyang Z, Nakamoto M (2009) In vitro guidance of retinal axons by a tectal lamina-specific glycoprotein Nel. Mol Cell Neurosci 41 (2):113–119. https://doi.org/10.1016/j.mcn.2009.02.006

Johnson KG, Ghose A, Epstein E, Lincecum J, O'Connor MB, Van Vactor D (2004) Axonal heparan sulfate proteoglycans regulate the distribution and efficiency of the repellent slit during midline axon guidance. Curr Biol 14(6):499–504. https://doi.org/10.1016/j.cub.2004.02.005

Jones CA, Nishiya N, London NR, Zhu W, Sorensen LK, Chan AC, Lim CJ, Chen H, Zhang Q, Schultz PG, Hayallah AM, Thomas KR, Famulok M, Zhang K, Ginsberg MH, Li DY (2009) Slit2-Robo4 signalling promotes vascular stability by blocking Arf6 activity. Nat Cell Biol 11 (11):1325–1331. https://doi.org/10.1038/ncb1976

Justice ED, Barnum SJ, Kidd T (2017) The WAGR syndrome gene PRRG4 is a functional homologue of the commissureless axon guidance gene. PLoS Genet 13(8):e1006865. https://doi.org/10.1371/journal.pgen.1006865

Kania A, Klein R (2016) Mechanisms of ephrin-Eph signalling in development, physiology and disease. Nat Rev Mol Cell Biol 17(4):240–256. https://doi.org/10.1038/nrm.2015.16

Kaur S, Castellone MD, Bedell VM, Konar M, Gutkind JS, Ramchandran R (2006) Robo4 signaling in endothelial cells implies attraction guidance mechanisms. J Biol Chem 281 (16):11347–11356. https://doi.org/10.1074/jbc.M508853200

Keleman K, Rajagopalan S, Cleppien D, Teis D, Paiha K, Huber LA, Technau GM, Dickson BJ (2002) Comm sorts robo to control axon guidance at the *Drosophila* midline. Cell 110(4):415–427

Keleman K, Ribeiro C, Dickson BJ (2005) Comm function in commissural axon guidance: cell-autonomous sorting of Robo in vivo. Nat Neurosci 8(2):156–163. https://doi.org/10.1038/nn1388

Kidd T, Bland KS, Goodman CS (1999) Slit is the midline repellent for the Robo receptor in *Drosophila*. Cell 96(6):785–794

Kidd T, Brose K, Mitchell KJ, Fetter RD, Tessier-Lavigne M, Goodman CS, Tear G (1998) Roundabout controls axon crossing of the CNS midline and defines a novel subfamily of evolutionarily conserved guidance receptors. Cell 92(2):205–215

Koch AW, Mathivet T, Larrivee B, Tong RK, Kowalski J, Pibouin-Fragner L, Bouvree K, Stawicki S, Nicholes K, Rathore N, Scales SJ, Luis E, del Toro R, Freitas C, Breant C, Michaud A, Corvol P, Thomas JL, Wu Y, Peale F, Watts RJ, Tessier-Lavigne M, Bagri A, Eichmann A (2011) Robo4 maintains vessel integrity and inhibits angiogenesis by interacting with UNC5B. Dev Cell 20(1):33–46. https://doi.org/10.1016/j.devcel.2010.12.001

Kolodkin AL, Tessier-Lavigne M (2011) Mechanisms and molecules of neuronal wiring: a primer. Cold Spring Harbor Perspect Biol 3(6). https://doi.org/10.1101/cshperspect.a001727

Koropouli E, Kolodkin AL (2014) Semaphorins and the dynamic regulation of synapse assembly, refinement, and function. Curr Opin Neurobiol 27:1–7. https://doi.org/10.1016/j.conb.2014.02.005

Lee JS, Ray R, Chien CB (2001) Cloning and expression of three zebrafish roundabout homologs suggest roles in axon guidance and cell migration. Dev Dyn 221(2):216–230. https://doi.org/10.1002/dvdy.1136

Legg JA, Herbert JM, Clissold P, Bicknell R (2008) Slits and Roundabouts in cancer, tumour angiogenesis and endothelial cell migration. Angiogenesis 11(1):13–21

Leyva-Diaz E, del Toro D, Menal MJ, Cambray S, Susin R, Tessier-Lavigne M, Klein R, Egea J, Lopez-Bendito G (2014) FLRT3 is a Robo1-interacting protein that determines Netrin-1 attraction in developing axons. Curr Biol 24(5):494–508. https://doi.org/10.1016/j.cub.2014.01.042

Li X, Chen Y, Liu Y, Gao J, Gao F, Bartlam M, Wu JY, Rao Z (2006) Structural basis of Robo proline-rich motif recognition by the srGAP1 Src homology 3 domain in the Slit-Robo signaling pathway. J Biol Chem 281(38):28430–28437. https://doi.org/10.1074/jbc.M604135200

Li Z, Moniz H, Wang S, Ramiah A, Zhang F, Moremen KW, Linhardt RJ, Sharp JS (2015) High structural resolution hydroxyl radical protein footprinting reveals an extended Robo1-heparin binding interface. J Biol Chem 290(17):10729–10740. https://doi.org/10.1074/jbc.M115.648410

Liang Y, Annan RS, Carr SA, Popp S, Mevissen M, Margolis RK, Margolis RU (1999) Mammalian homologues of the *Drosophila* slit protein are ligands of the heparan sulfate proteoglycan glypican-1 in brain. J Biol Chem 274(25):17885–17892

Liu D, Hou J, Hu X, Wang X, Xiao Y, Mou Y, De Leon H (2006) Neuronal chemorepellent Slit2 inhibits vascular smooth muscle cell migration by suppressing small GTPase Rac1 activation. Circ Res 98(4):480–489. https://doi.org/10.1161/01.RES.0000205764.85931.4b

Liu Z, Patel K, Schmidt H, Andrews W, Pini A, Sundaresan V (2004) Extracellular Ig domains 1 and 2 of Robo are important for ligand (Slit) binding. Mol Cell Neurosci 26(2):232–240

Long H, Sabatier C, Ma L, Plump A, Yuan W, Ornitz DM, Tamada A, Murakami F, Goodman CS, Tessier-Lavigne M (2004) Conserved roles for Slit and Robo proteins in midline commissural axon guidance. Neuron 42(2):213–223

Lu X, Le Noble F, Yuan L, Jiang Q, De Lafarge B, Sugiyama D, Breant C, Claes F, De Smet F, Thomas JL, Autiero M, Carmeliet P, Tessier-Lavigne M, Eichmann A (2004) The netrin receptor UNC5B mediates guidance events controlling morphogenesis of the vascular system. Nature 432(7014):179–186. https://doi.org/10.1038/nature03080

Lundström A, Gallio M, Englund C, Steneberg P, Hemphala J, Aspenstrom P, Keleman K, Falileeva L, Dickson BJ, Samakovlis C (2004) Vilse, a conserved Rac/Cdc42 GAP mediating Robo repulsion in tracheal cells and axons. Genes Dev 18(17):2161–2171. https://doi.org/10.1101/gad.310204

Marillat V, Sabatier C, Failli V, Matsunaga E, Sotelo C, Tessier-Lavigne M, Chedotal A (2004) The slit receptor Rig-1/Robo3 controls midline crossing by hindbrain precerebellar neurons and axons. Neuron 43(1):69–79. https://doi.org/10.1016/j.neuron.2004.06.018

Marin O, Yaron A, Bagri A, Tessier-Lavigne M, Rubenstein JL (2001) Sorting of striatal and cortical interneurons regulated by semaphorin–neuropilin interactions. Science 293(5531):872–875. https://doi.org/10.1126/science.1061891

Martinez-Quiles N, Rohatgi R, Anton IM, Medina M, Saville SP, Miki H, Yamaguchi H, Takenawa T, Hartwig JH, Geha RS, Ramesh N (2001) WIP regulates N-WASP-mediated actin polymerization and filopodium formation. Nat Cell Biol 3(5):484–491. https://doi.org/10.1038/35074551

Mommersteeg MT, Andrews WD, Ypsilanti AR, Zelina P, Yeh ML, Norden J, Kispert A, Chedotal A, Christoffels VM, Parnavelas JG (2013) Slit-roundabout signaling regulates the development of the cardiac systemic venous return and pericardium. Circ Res 112(3):465–475. https://doi.org/10.1161/CIRCRESAHA.112.277426

Morlot C, Thielens NM, Ravelli RB, Hemrika W, Romijn RA, Gros P, Cusack S, McCarthy AA (2007) Structural insights into the Slit-Robo complex. Proc Natl Acad Sci USA 104 (38):14923–14928. https://doi.org/10.1073/pnas.0705310104

Nakayama T, Mizohata E, Yamashita T, Nagatoishi S, Nakakido M, Iwanari H, Mochizuki Y, Kado Y, Yokota Y, Satoh R, Tsumoto K, Fujitani H, Kodama T, Hamakubo T, Inoue T (2015) Structural features of interfacial tyrosine residue in ROBO1 fibronectin domain-antibody complex: Crystallographic, thermodynamic, and molecular dynamic analyses. Protein Sci 24 (3):328–340. https://doi.org/10.1002/pro.2619

Navankasattusas S, Whitehead KJ, Suli A, Sorensen LK, Lim AH, Zhao J, Park KW, Wythe JD, Thomas KR, Chien CB, Li DY (2008) The netrin receptor UNC5B promotes angiogenesis in specific vascular beds. Development 135(4):659–667. https://doi.org/10.1242/dev.013623

Nguyen Ba-Charvet KT, Brose K, Ma L, Wang KH, Marillat V, Sotelo C, Tessier-Lavigne M, Chedotal A (2001) Diversity and specificity of actions of Slit2 proteolytic fragments in axon guidance. J Neurosci 21(12):4281–4289

Nguyen Ba-Charvet KT, Brose K, Marillat V, Kidd T, Goodman CS, Tessier-Lavigne M, Sotelo C, Chedotal A (1999) Slit2-Mediated chemorepulsion and collapse of developing forebrain axons. Neuron 22(3):463–473

Okada Y, Jin E, Nikolova-Krstevski V, Yano K, Liu J, Beeler D, Spokes K, Kitayama M, Funahashi N, Doi T, Janes L, Minami T, Oettgen P, Aird WC (2008) A GABP-binding element in the Robo4 promoter is necessary for endothelial expression in vivo. Blood 112(6):2336–2339. https://doi.org/10.1182/blood-2008-01-135079

Okada Y, Yano K, Jin E, Funahashi N, Kitayama M, Doi T, Spokes K, Beeler DL, Shih SC, Okada H, Danilov TA, Maynard E, Minami T, Oettgen P, Aird WC (2007) A three-kilobase fragment of the human Robo4 promoter directs cell type-specific expression in endothelium. Circ Res 100(12):1712–1722. https://doi.org/10.1161/01.res.0000269779.10644.dc

Park KW, Morrison CM, Sorensen LK, Jones CA, Rao Y, Chien CB, Wu JY, Urness LD, Li DY (2003) Robo4 is a vascular-specific receptor that inhibits endothelial migration. Dev Biol 261 (1):251–267

Piper M, Anderson R, Dwivedy A, Weinl C, van Horck F, Leung KM, Cogill E, Holt C (2006) Signaling mechanisms underlying Slit2-induced collapse of *Xenopus* retinal growth cones. Neuron 49(2):215–228. https://doi.org/10.1016/j.neuron.2005.12.008

Placzek M, Briscoe J (2005) The floor plate: multiple cells, multiple signals. Nat Rev Neurosci 6 (3):230–240. https://doi.org/10.1038/nrn1628

Rajagopalan S, Nicolas E, Vivancos V, Berger J, Dickson BJ (2000) Crossing the midline: roles and regulation of Robo receptors. Neuron 28(3):767–777

Ringstedt T, Braisted JE, Brose K, Kidd T, Goodman C, Tessier-Lavigne M, O'Leary DD (2000) Slit inhibition of retinal axon growth and its role in retinal axon pathfinding and innervation patterns in the diencephalon. J Neurosci 20(13):4983–4991

Ronca F, Andersen JS, Paech V, Margolis RU (2001) Characterization of Slit protein interactions with glypican-1. J Biol Chem 276(31):29141–29147. https://doi.org/10.1074/jbc.M100240200

Saied-Santiago K, Bulow HE (2018) Diverse roles for glycosaminoglycans in neural patterning. Dev Dyn 247(1):54–74. https://doi.org/10.1002/dvdy.24555

Samant GV, Schupp MO, Francois M, Moleri S, Kothinti RK, Chun CZ, Sinha I, Sellars S, Leigh N, Pramanik K, Horswill MA, Remadevi I, Li K, Wilkinson GA, Tabatabai NM, Beltrame M, Koopman P, Ramchandran R (2011) Sox factors transcriptionally regulate ROBO4 gene expression in developing vasculature in zebrafish. J Biol Chem 286(35):30740–30747. https://doi.org/10.1074/jbc.M111.220665

Seeger M, Tear G, Ferres-Marco D, Goodman CS (1993) Mutations affecting growth cone guidance in *Drosophila*: genes necessary for guidance toward or away from the midline. Neuron 10(3):409–426

Seiradake E, del Toro D, Nagel D, Cop F, Hartl R, Ruff T, Seyit-Bremer G, Harlos K, Border EC, Acker-Palmer A, Jones EY, Klein R (2014) FLRT structure: balancing repulsion and cell adhesion in cortical and vascular development. Neuron 84(2):370–385. https://doi.org/10.1016/j.neuron.2014.10.008

Seiradake E, Jones EY, Klein R (2016) Structural perspectives on axon guidance. Annu Rev Cell Dev Biol 32:577–608. https://doi.org/10.1146/annurev-cellbio-111315-125008

Seiradake E, von Philipsborn AC, Henry M, Fritz M, Lortat-Jacob H, Jamin M, Hemrika W, Bastmeyer M, Cusack S, McCarthy AA (2009) Structure and functional relevance of the Slit2 homodimerization domain. EMBO Rep 10(7):736–741

Seki M, Watanabe A, Enomoto S, Kawamura T, Ito H, Kodama T, Hamakubo T, Aburatani H (2010) Human ROBO1 is cleaved by metalloproteinases and gamma-secretase and migrates to the nucleus in cancer cells. FEBS Lett 584(13):2909–2915. https://doi.org/10.1016/j.febslet.2010.05.009

Seth P, Lin Y, Hanai J, Shivalingappa V, Duyao MP, Sukhatme VP (2005) Magic roundabout, a tumor endothelial marker: expression and signaling. Biochem Biophys Res Commun 332 (2):533–541. https://doi.org/10.1016/j.bbrc.2005.03.250

Sheldon H, Andre M, Legg JA, Heal P, Herbert JM, Sainson R, Sharma AS, Kitajewski JK, Heath VL, Bicknell R (2009) Active involvement of Robo1 and Robo4 in filopodia formation and endothelial cell motility mediated via WASP and other actin nucleation-promoting factors. FASEB J 23(2):513–522. https://doi.org/10.1096/fj.07-098269

Sherchan P, Huang L, Akyol O, Reis C, Tang J, Zhang JH (2017) Recombinant Slit2 reduces surgical brain injury induced blood brain barrier disruption via Robo4 dependent Rac1 activation in a rodent model. Sci Rep 7(1):746. https://doi.org/10.1038/s41598-017-00827-z

Shibata F, Goto-Koshino Y, Morikawa Y, Komori T, Ito M, Fukuchi Y, Houchins JP, Tsang M, Li DY, Kitamura T, Nakajima H (2009) Roundabout 4 is expressed on hematopoietic stem cells and potentially involved in the niche-mediated regulation of the side population phenotype. Stem Cells 27(1):183–190. https://doi.org/10.1634/stemcells.2008-0292

Simpson JH, Kidd T, Bland KS, Goodman CS (2000) Short-range and long-range guidance by slit and its Robo receptors. Robo and Robo2 play distinct roles in midline guidance. Neuron 28 (3):753–766

Smith-Berdan S, Nguyen A, Hassanein D, Zimmer M, Ugarte F, Ciriza J, Li D, Garcia-Ojeda ME, Hinck L, Forsberg EC (2011) Robo4 cooperates with CXCR107 to specify hematopoietic stem cell localization to bone marrow niches. Cell Stem Cell 8(1):72–83. https://doi.org/10.1016/j.stem.2010.11.030

Steigemann P, Molitor A, Fellert S, Jackle H, Vorbruggen G (2004) Heparan sulfate proteoglycan syndecan promotes axonal and myotube guidance by slit/robo signaling. Curr Biol 14(3):225–230. https://doi.org/10.1016/j.cub.2004.01.006

Stein E, Tessier-Lavigne M (2001) Hierarchical organization of guidance receptors: silencing of netrin attraction by slit through a Robo/DCC receptor complex. Science 291(5510):1928–1938. https://doi.org/10.1126/science.1058445

Suchting S, Bicknell R, Eichmann A (2006) Neuronal clues to vascular guidance. Exp Cell Res 312(5):668–675. https://doi.org/10.1016/j.yexcr.2005.11.009

Tai-Nagara I, Yoshikawa Y, Numata N, Ando T, Okabe K, Sugiura Y, Ieda M, Takakura N, Nakagawa O, Zhou B, Okabayashi K, Suematsu M, Kitagawa Y, Bastmeyer M, Sato K, Klein R, Navankasattusas S, Li DY, Yamagishi S, Kubota Y (2017) Placental labyrinth formation in mice requires endothelial FLRT2/UNC5B signaling. Development 144(13):2392–2401. https://doi.org/10.1242/dev.149757

Tanaka T, Izawa K, Maniwa Y, Okamura M, Okada A, Yamaguchi T, Shirakura K, Maekawa N, Matsui H, Ishimoto K, Hino N, Nakagawa O, Aird WC, Mizuguchi H, Kawabata K, Doi T, Okada Y (2018) ETV2-TET1/TET2 complexes induce endothelial cell-specific Robo4 expression via promoter demethylation. Sci Rep 8(1):5653. https://doi.org/10.1038/s41598-018-23937-8

Tear G, Seeger M, Goodman CS (1993) To cross or not to cross: a genetic analysis of guidance at the midline. Perspect Dev Neurobiol 1(4):183–194

Townley RA, Bülow HE (2018) Deciphering functional glycosaminoglycan motifs in development. Curr Opin Struct Biol 50:144–154. https://doi.org/10.1016/j.sbi.2018.03.011

Turner CE, Glenney JR Jr, Burridge K (1990) Paxillin: a new vinculin-binding protein present in focal adhesions. J Cell Biol 111(3):1059–1068

Vargesson N, Luria V, Messina I, Erskine L, Laufer E (2001) Expression patterns of Slit and Robo family members during vertebrate limb development. Mech Dev 106(1–2):175–180

Wang R, Wei Z, Jin H, Wu H, Yu C, Wen W, Chan LN, Wen Z, Zhang M (2009) Autoinhibition of UNC5b revealed by the cytoplasmic domain structure of the receptor. Mol Cell 33(6):692–703. https://doi.org/10.1016/j.molcel.2009.02.016

Wong K, Ren XR, Huang YZ, Xie Y, Liu G, Saito H, Tang H, Wen L, Brady-Kalnay SM, Mei L, Wu JY, Xiong WC, Rao Y (2001) Signal transduction in neuronal migration: roles of GTPase activating proteins and the small GTPase Cdc42 in the Slit-Robo pathway. Cell 107(2):209–221

Wright KM, Lyon KA, Leung H, Leahy DJ, Ma L, Ginty DD (2012) Dystroglycan organizes axon guidance cue localization and axonal pathfinding. Neuron 76(5):931–944. https://doi.org/10.1016/j.neuron.2012.10.009

Yadav SS, Narayan G (2014) Role of ROBO4 signalling in developmental and pathological angiogenesis. Biomed Res Int 2014:683025. https://doi.org/10.1155/2014/683025

Yom-Tov G, Barak R, Matalon O, Barda-Saad M, Guez-Haddad J, Opatowsky Y (2017) Robo Ig4 is a dimerisation domain. J Mol Bio 429(23):3606–3616. https://doi.org/10.1016/j.jmb.2017.10.002

Ypsilanti AR, Zagar Y, Chedotal A (2010) Moving away from the midline: new developments for Slit and Robo. Development 137(12):1939–1952. https://doi.org/10.1242/dev.044511

Yuan SS, Cox LA, Dasika GK, Lee EY (1999a) Cloning and functional studies of a novel gene aberrantly expressed in RB-deficient embryos. Dev Biol 207(1):62–75. https://doi.org/10.1006/dbio.1998.9141

Yuan W, Zhou L, Chen JH, Wu JY, Rao Y, Ornitz DM (1999b) The mouse SLIT family: secreted ligands for ROBO expressed in patterns that suggest a role in morphogenesis and axon guidance. Dev Biol 212(2):290–306. https://doi.org/10.1006/dbio.1999.9371

Zakrys L, Ward RJ, Pediani JD, Godin AG, Graham GJ, Milligan G (2014) Roundabout 1 exists predominantly as a basal dimeric complex and this is unaffected by binding of the ligand Slit2. Biochem J 461(1):61–73. https://doi.org/10.1042/BJ20140190

Zelina P, Blockus H, Zagar Y, Peres A, Friocourt F, Wu Z, Rama N, Fouquet C, Hohenester E, Tessier-Lavigne M, Schweitzer J, Roest Crollius H, Chedotal A (2014) Signaling switch of the axon guidance receptor Robo3 during vertebrate evolution. Neuron 84(6):1258–1272. https://doi.org/10.1016/j.neuron.2014.11.004

Zhang F, Prahst C, Mathivet T, Pibouin-Fragner L, Zhang J, Genet G, Tong R, Dubrac A, Eichmann A (2016) The Robo4 cytoplasmic domain is dispensable for vascular permeability and neovascularization. Nat Commun 7:13517. https://doi.org/10.1038/ncomms13517

Zheng W, Geng AQ, Li PF, Wang Y, Yuan XB (2012) Robo4 regulates the radial migration of newborn neurons in developing neocortex. Cereb Cortex 22(11):2587–2601. https://doi.org/10.1093/cercor/bhr330

Zou Y, Stoeckli E, Chen H, Tessier-Lavigne M (2000) Squeezing axons out of the gray matter: a role for slit and semaphorin proteins from midline and ventral spinal cord. Cell 102(3):363–375

Chapter 10
Structure and Function of Molecular Chaperones that Govern Immune Peptide Loading

David H. Margulies, Jiansheng Jiang and Kannan Natarajan

Abstract Major histocompatibility class I (MHC-I) molecules bind peptides derived from cellular synthesis and display them at the cell surface for recognition by receptors on T lymphocytes (TCR) or natural killer (NK) cells. Such recognition provides a crucial step in autoimmunity, identification of bacterial and viral pathogens, and anti-tumor responses. Understanding the mechanism by which such antigenic peptides in the ER are loaded and exchanged for higher affinity peptides onto MHC molecules has recently been clarified by cryo-EM and X-ray studies of the multimolecular peptide loading complex (PLC) and a unimolecular tapasin-like chaperone designated TAPBPR. Insights from these structural studies and complementary solution NMR experiments provide a basis for understanding mechanisms related to immune antigen presentation.

Keywords Immune recognition · Major histocompatibility complex · Chaperone · Tapasin · TAPBPR · Protein loading complex · Structural biology · X-ray crystallography

Abbreviations

MHC	Major histocompatibility complex
MHC-I	MHC class I
MHC-II	MHC class II
TCR	T cell receptor
NK	Natural killer
ER	Endoplasmic reticulum
PLC	Peptide loading complex

D. H. Margulies (✉) · J. Jiang · K. Natarajan
Molecular Biology Section, Laboratory of Immune System Biology, National Institute of Allergy and Infectious Diseases, National Institutes of Health, Bethesda, MD, USA
e-mail: dhm@nih.gov

J. Jiang
e-mail: jiangji@niaid.nih.gov

K. Natarajan
e-mail: knatarajan@niaid.nih.gov

J. R. Harris and J. Marles-Wright (eds.), *Macromolecular Protein Complexes II: Structure and Function*, Subcellular Biochemistry 93,
https://doi.org/10.1007/978-3-030-28151-9_10

TAP	Transporter associated with antigen processing
TAPBP or tapasin	TAP binding protein
TAPBPR	TAP binding protein, related
NMR	Nuclear magnetic resonance

Introduction

The critical function of the adaptive immune system is to recognize and eliminate cells whose physiological function has been dysregulated by infection or oncogenesis. Such recognition also demands discrimination of aberrant from normal cells. The ongoing evolutionary solution to this problem includes generation of a large repertoire of T lymphocytes bearing T cell receptors (TCR) in the thymus that are screened quantitatively for their reactivity to self. This is a selective process dependent on major histocompatibility complex (MHC)-encoded molecules that bind peptides and display them at the cell surface. TCRs that react strongly to self-peptide/MHC complexes are deleted in a process called negative selection. To avoid negative selection, cells bearing TCR must react at a quantitatively lower level with the same MHC molecules, a process of positive selection. The net outcome of such thymic education of T cells is the generation of a TCR repertoire that can discriminate between self and non-self. Once in peripheral lymphoid tissues, such T cells are positioned to recognize dysregulated or foreign peptides bound to self-MHC molecules, a process known as MHC-restricted recognition (Archbold et al. 2008; La Gruta et al. 2018).

Mature T cells fall broadly into those that recognize one of two categories of antigen presenting molecules: (1) MHC class I (MHC-I) that are loaded during folding and assembly in the endoplasmic reticulum (ER) with peptides generated in the cytoplasm by the proteasome or imprecise protein translation; or (2) MHC class II (MHC-II) that bind peptides derived from proteolysis in endosomes or lysosomes (Blum et al. 2013; Germain and Margulies 1993; Rock et al. 2016; York and Rock 1996). Each of these classes of MHC molecules then biosynthetically matures and arrives at the cell surface for recognition by T cells ($CD8^+$ for MHC-I, and $CD4^+$ for MHC-II). Although peptide loading in the MHC-I and MHC-II pathways has been the subject of intensive study for over two decades, the past few years have witnessed new insights derived from structural studies and molecular dynamics simulations of multimolecular complexes involved in peptide loading (Zacharias and Springer 2004). In addition to these studies contributing to our understanding of critical aspects of antigen presentation in cell-mediated immunity, they offer new insight into the broader areas of protein folding, chaperone function, and macromolecular assembly.

This review is primarily focused on the MHC-I loading pathway, with particular attention to the role of chaperone proteins that stabilize the MHC-I molecule in a conformation accessible for peptide loading. This mode of stabilization facilitates peptide exchange favoring high affinity peptides. In particular, the chaperones designated tapasin (also known as TAP binding protein, TAPBP), and a tapasin homolog commonly called TAP binding protein, related (TAPBPR), are of current interest, because they have permitted detailed structural analyses of the mechanisms of peptide loading of the MHC-class I loading pathway. We will briefly compare the MHC-I pathway to some aspects of MHC-II loading pathway as well.

MHC Molecules, Peptide Binding Ligands for TCR

The MHC molecules are highly polymorphic heterodimeric cell surface molecules that sample peptides for display at the cell surface where they function as ligands for TCRs. Their extensive polymorphism (over 15,000 MHC-I and 5000 MHC-II alleles have been identified in humans (Robinson et al. 2015)) provides a reservoir of molecules that are capable of binding and presenting peptides derived from established as well as newly evolved infectious organisms. In addition, particular MHC types are frequently tightly linked to autoimmune diseases, immune responsiveness dictating virus resistance or sensitivity, and the resistance to tumor or tumor encoded neoantigens. Thus, understanding the detailed molecular steps involved in MHC biosynthesis, assembly, peptide loading, and cell surface expression leading to TCR engagement and TCR-mediated signaling is crucial to devising interventions to augment or contain the adaptive immune response. At the center of any molecular understanding of protein biochemical events is knowledge of the three-dimensional structure of the molecules involved, the biophysical basis of their interactions, and how those structures and molecular interactions dictate function. Recent years have seen dramatic advances in X-ray, cryo-EM, and NMR structure determination. In particular, there are now available hundreds of high-resolution models based on X-ray data of MHC-I and MHC-II molecules complexed with different antigenic peptides, and dozens of models of peptide/MHC/TCR complexes (Adams and Luoma 2013; Marrack et al. 2008; Margulies et al. 2008). However, all the structures, with some exceptions, recapitulate the original ones determined years ago (Bjorkman et al. 1987; Brown et al. 1993; Fremont et al. 1992; Garboczi et al. 1996). Summary illustrations of a representative MHC-I molecule, complexed with its light chain, the relatively invariant β_2-microglobulin (β_2m) and a virus-derived antigenic peptide (together denoted pMHC) are shown in Fig. 10.1a–d, f. The larger molecular complex of pMHC bound to a cognate $\alpha\beta$ T cell receptor (TCR) is shown in Fig. 10.1e.

The crucial features of the extracellular domain of the pMHC complex, illustrated here by the molecular complex of the mouse MHC-I molecule, H2-D^d, is that the amino terminal domain unit, designated $\alpha1\alpha2$, about 180 amino acids in length, forms a peptide binding site consisting of two α-helices supported by eight strands of anti-parallel β-sheet (Fig. 10.1a). This binding site is supported by the light chain, β_2-microglobulin (β_2m), that provides stability (Fig. 10.1b). The peptide is

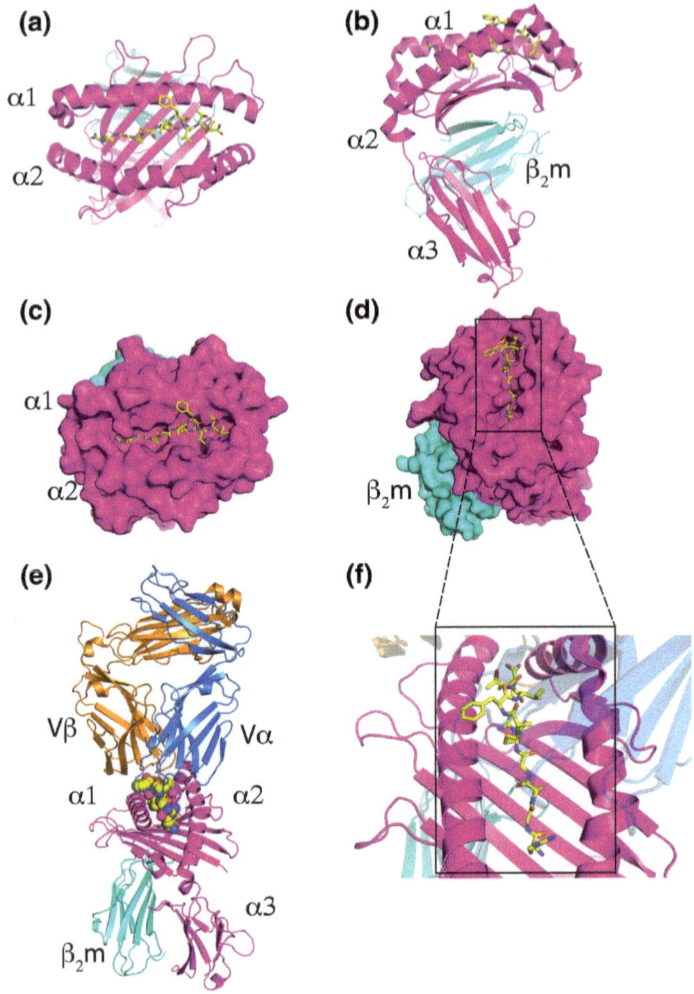

Fig. 10.1 X-ray structure of typical pMHC-I and pMHC/TCR complexes. Ribbon (**a**, **b**, **f**) and space-filling (**c**, **d**) illustrations of the crystal structure of the mouse MHC-I molecule, H2-D[d] complexed with an HIV gp120-derived viral peptide, P18-I10 (RGPGRAFVTI), and β_2m are shown (resolution, 1.7 Å, PDB, 3ECB (Wang et al. 2009)). The complex of the same pMHC with a peptide-specific, MHC-restricted TCR is also shown (2.1 Å, PDB, 5IVX (Natarajan et al. 2017))

bound in a crevasse which contains six pockets (usually for MHC-I molecules designated A through F) that accommodate the amino and carboxyl termini of the peptide as well as several side chains of the peptide (Fig. 10.1c, d, f) (Saper et al. 1991). In general, the most important pockets are at the amino and carboxy termini of the peptide (the A, B and F pockets respectively). The peptide is displayed distal to the cell's plasma membrane, and the complex of the peptide bound to the MHC-I molecule provides a molecular surface available for interaction with the variable (V domains) of the TCR α and β chains (Fig. 10.1e).

Peptide Loading in the MHC-I Pathway

Early studies recognized that cells deficient in peptide transport into the ER (i.e. defective in expression of the heterodimeric ATP-dependent transporter associated with antigen processing, TAP) expressed low levels of surface MHC-I, a phenotype partially reversed by culturing the mutant cells at room temperature rather than at 37 °C (Ljunggren et al. 1990; Spies and DeMars 1991). Defects in the transport of peptides from the cytoplasm to the ER provided a deficiency in available peptides, resulting in fewer available high affinity peptides, and thus a lower steady state concentration of thermally stable cell surface MHC-I molecules. Several decades of biochemical studies lead to the general model illustrated in Fig. 10.2. Peptides generated in the cytoplasm by defective translation or by proteolytic digestion of larger proteins by the proteasome are released to be transported by the TAP1/2 transporter of the PLC into the ER (Fig. 10.2a). There, the peptides may be bound by the peptide-receptive MHC molecule of the PLC, can be trimmed at the amino-terminus by the ERAP endopeptidases (Chen et al. 2016; Serwold et al. 2002), and are submitted to high affinity quality control mediated by the tapasin chaperone. Once a peptide of sufficient affinity is bound, the pMHC-I/β_2m complex proceeds through the Golgi to the cell surface. The glycosylation status of the folded MHC-I serves as an additional quality-control step, mediated by the interaction with UDP-glucose glycoprotein glucosyltransferase (UGT1) (Zhang et al. 2011).

Structure of Tapasin/ERp57

Understanding the role of tapasin as a component of the PLC has been elucidated by several experimental approaches: genetic studies evaluating the function of cells and animals defective in tapasin expression (Copeman et al. 1998; Garbi et al. 2003; Grandea et al. 2000), biochemical studies exploring the role of tapasin in facilitating peptide loading (Sadasivan et al. 1996; Chen and Bouvier 2007; Wearsch and Cresswell 2007), and structural studies of tapasin (Dong et al. 2009). A major step in understanding the structural contribution of tapasin to peptide loading was achieved by the determination of the X-ray crystallographic structure of a molecular complex of tapasin with ERp57, the oxidoreductase component of the PLC (Dong et al. 2009). By coexpressing the lumenal domains of both tapasin and ERp57 as a disulfide-linked unit, the authors successfully produced heterodimers that crystallized, and then determined the X-ray structure at 2.6 Å resolution (Fig. 10.3). Two heterodimers of tapasin/ERp57 are contained in the asymmetric unit, with several

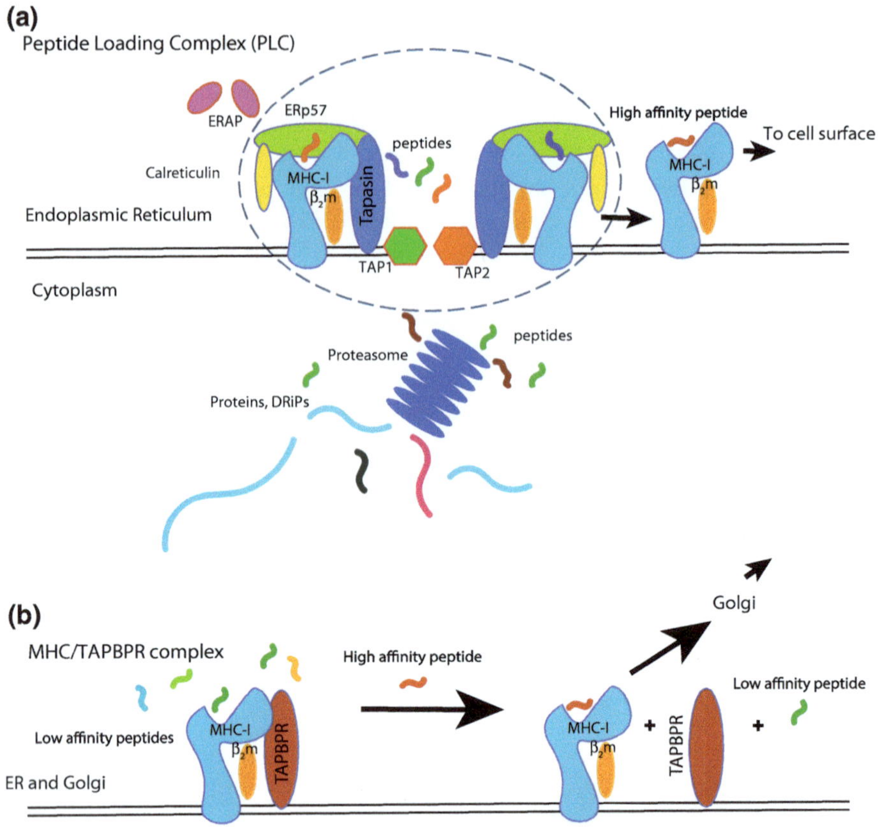

Fig. 10.2 Peptide loading by the PLC or by the PLC-independent TAPBPR. **a** Schematic representation of the key molecules involved in peptide loading. Cytoplasmic proteins or defective ribosomal products (DRiPs (Yewdell and Nicchitta 2006)) are further degraded by the proteasome to a length of about 8–15 amino acids, transported to the ER via the TAP1/2 ATP-dependent transporter, where under the influence of the full PLC they are loaded on MHC-I molecules. Amino terminal trimming by ERAP occurs during this process. MHC-I complexed with high affinity peptides proceed to the Golgi and to the cell surface. **b** the MHC/TAPBPR complex functions independent of the PLC and allows exchange for high affinity peptides. Not shown are additional quality control steps based on interaction with UGT1 (Zhang et al. 2011; Neerincx et al. 2017)

loops of tapasin in regions of poor electron density indicative of disorder in the crystal, and consistent with mobility in solution. Of major interest is the structure of the tapasin component itself, and the structural location of amino acid residues that affect MHC-I interaction. The interacting amino acid residues were established by site-directed mutagenesis and the resulting impairment of peptide loading, the key PLC function. Using a cell-free system (Wearsch and Cresswell 2007), a number of tapasin mutants were identified that defined a conserved surface of the tapasin molecule (Fig. 10.3b). These mutants not only severely impeded PLC function, but

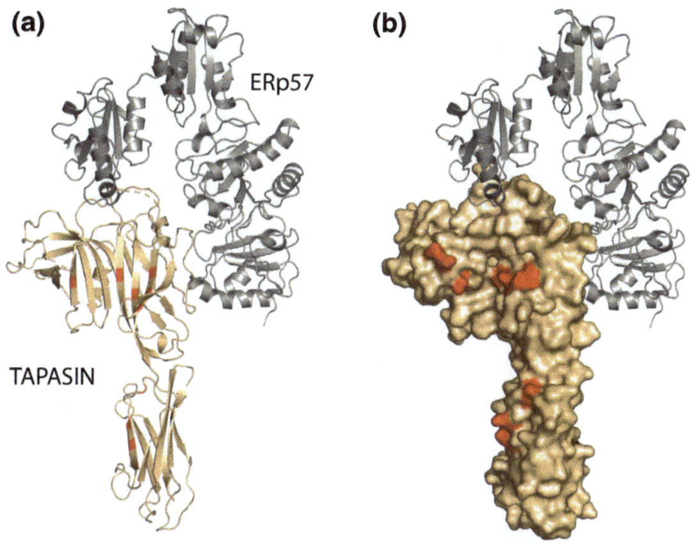

Fig. 10.3 Structure of tapasin/ERp57 reveals location of residues that affect function in MHC-I peptide loading. **a** Ribbon diagram of the tapasin/ERp57 heterodimer (2.6 Å, PDB, 3F8U (Dong et al. 2009)), tapasin, magenta; ERp57, grey. Residues that impaired peptide loading capacity are indicated in red. **b** Surface representation of tapasin, nestled into U-like clasp of ERp57. Same color scheme as (**a**)

also impaired their ability to bind MHC-I molecules. Most strikingly, one mutant also failed to restore cell surface expression of the highly tapasin-dependent HLA-B*44:02 molecule in a tapasin deficient cell line. The structural analysis of the tapasin/ERp57 complex, studies of monoclonal antibody reactivity, and molecular modeling and docking strategies formed the basis of several working models of the structure of the entire PLC (Dong et al. 2009; Koch and Tampe 2006; Panter et al. 2012; van Hateren et al. 2013; Van Hateren et al. 2010). Since the PLC contains not only MHC-I, tapasin and ERp57, but also the lectin-like chaperone calreticulin and the heterodimer TAP, understanding the structural relationship of the MHC-I to tapasin remained a critical point in developing a mechanistic view of the function of the PLC. High resolution appreciation of the interaction of tapasin with MHC-I continues to be an elusive undertaking.

Low Resolution Structures of Tapasin and TAPBPR

When the front door is locked, it is often helpful to see if a side door is open. Difficulties in assessing structural aspects of PLC function and in particular the role played by tapasin in chaperoning peptide-receptive MHC-I have been overcome in part by studies of the tapasin-like molecule, TAP binding protein, related

(TAPBPR). First identified as a gene encoding a molecule with low ($\sim$22%) amino acid sequence identity to tapasin, the TAPBPR protein was initially characterized as a member of the immunoglobulin superfamily. Biochemical studies found TAPBPR predominantly in the endoplasmic reticulum (Teng et al. 2002). Subsequently, Boyle and colleagues explored the molecules associated with TAPBPR, and found that, in contrast to tapasin, it is associated with MHC-I and not with other components of the PLC (Boyle et al. 2013). Also, during biosynthesis and maturation TAPBPR remains in complex with MHC-I past the ER and through the Golgi. Further studies demonstrated the ability of TAPBPR to interact strongly with MHC-I molecules devoid of peptide (Hermann et al. 2015; Morozov et al. 2016), with differential selectivity for varied MHC-I allelic protein products. Mutational analyses of TAPBPR, based on amino acid sequence similarity with tapasin (Hermann et al. 2013; Morozov et al. 2016) supported the view that TAPBPR functioned as a tapasin homolog, and exploited a similar MHC-I binding site. Using human HLA-A*02:01 emptied of bound peptide by photolysis of a UV-sensitive peptide, analysis by size exclusion chromatography, analytical ultracentrifugation, and surface plasmon resonance established the ability of TAPBPR to bind peptide-free MHC-I molecules. TAPBPR interacts with a wide variety of MHC-I allelomorphs, and the human molecule binds some murine MHC-I molecules as well (Morozov et al. 2016).

Efforts to clarify the three-dimensional structure of tapasin bound to MHC-I, however, were initially stymied by the lack of suitable crystals. One alternative approach was to obtain low resolution envelope structures of tapasin and TAPBPR, using small angle X-ray diffraction. The experimentally based models revealed similar solution structure of the two molecules in their lumenal domains (Morozov et al. 2016). The low resolution structure of TAPBPR provided encouragement for the biochemical purification of crystallizable TAPBPR/MHC-I complexes.

X-Ray Structures of TAPBPR

Diffraction quality crystals were obtained in two laboratories, leading to X-ray based models of MHC-I/TAPBPR complexes (Jiang et al. 2017; Thomas and Tampe 2017). Jiang et al. produced the mouse MHC-I molecule, H2-D^d disulfide linked to a carboxyl terminally truncated 5-mer peptide, RGPGRC, that occupied the amino-terminal region of the peptide binding groove, leaving the carboxyl-terminal part stabilized by a glycine-leucine dipeptide (as suggested by refolding studies of Saini et al. 2013). Using surface plasmon resonance, they demonstrated that this molecule, lacking the usual carboxyl terminal anchor in the F pocket, bound rather tightly to human TAPBPR (hTAPBPR). X-ray data sets for crystals of the peptide 5-mer/H2-D^d complex (to 2.7 Å resolution, PDB, 5WES) as well as for the 5-mer/H2-D^d/TAPBPR complex (to 3.4 Å, PDB, 5WER), generated refined models for analyzing the contacts between TAPBPR and the H2-D^d MHC-I molecule as well as the conformational differences distinguishing the bound and unbound MHC-I molecule. Figure 10.4a illustrates the general features of the MHC-I/TAPBPR complex (Fig. 10.4a), the strong structural similarity of TAPBPR to tapasin (Fig. 10.4b), and the conformational changes observed in the MHC-I molecule before (Fig. 10.4d) and after (Fig. 10.4a, c) interaction with the TAPBPR chaperone. Detailed comparisons, based on superposition of the unbound and

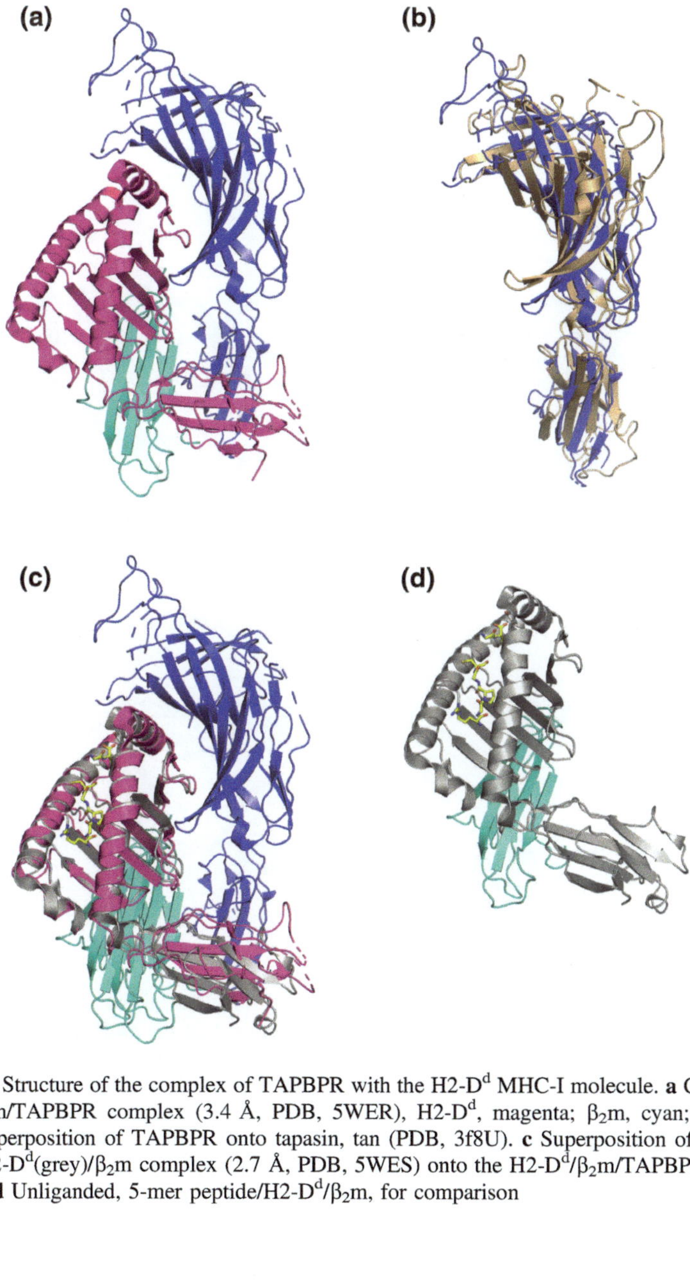

Fig. 10.4 Structure of the complex of TAPBPR with the H2-D^d MHC-I molecule. **a** Complex of H2-D^d/β_2m/TAPBPR complex (3.4 Å, PDB, 5WER), H2-D^d, magenta; β_2m, cyan; TAPBPR, blue. **b** superposition of TAPBPR onto tapasin, tan (PDB, 3f8U). **c** Superposition of the 5-mer peptide/H2-D^d(grey)/β_2m complex (2.7 Å, PDB, 5WES) onto the H2-D^d/β_2m/TAPBPR complex (5WER). **d** Unliganded, 5-mer peptide/H2-D^d/β_2m, for comparison

bound MHC-I molecule (Fig. 10.5) revealed a number of striking details (Fig. 10.5). The α2-1 helix of the MHC molecule is displaced by as much as 3 Å on interaction with TAPBPR. A key amino acid residue of the MHC, Tyr84 is repositioned by interaction with TAPBPR Glu102, drawing it away from its usual position anchoring the carboxylate group of the C-terminal amino acid of the bound peptide (Fig. 10.5a, c). The floor of the peptide binding groove is also displaced by about 1.8 Å (Fig. 10.5b), apparently related to interactions with a Lys211 to Arg213 loop of TAPBPR. The interaction of the membrane proximal Ig-like domain of TAPBPR with both the membrane proximal α3 domain of the MHC-I

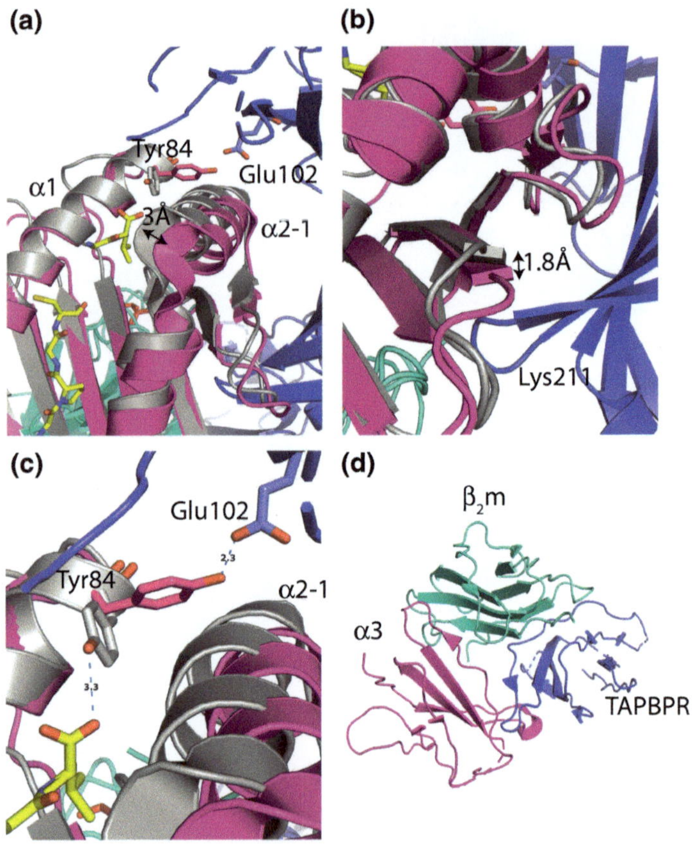

Fig. 10.5 Conformational changes of pMHC accompany TAPBPR interaction. **a** Close-up view of the peptide-binding groove of the pMHC (grey) superposed on the pMHC/TAPBPR complex (α1 and α2, magenta), emphasizing the 3 Å displacement of the α2-1 helix. **b** Close-up of the floor of the binding groove, showing displacement of the β8-strand and the position of the Lys211 to Arg213 (not labelled) loop of TAPBPR. **c** Change in orientation of side chain of Tyr84 of H2-D^d from the unliganded (grey) to the TAPBPR-bound state (magenta) where it forms hydrogen bonds to Glu102 of TAPBPR. **d** Top down view of the membrane proximal Ig-like domains of H2-D^d heavy chain (magenta), β_2m (cyan) and TAPBPR (blue)

heavy chain and with the β_2m subunit, forming a trimeric structure, not predicted by any of the dynamics simulations was another unanticipated finding (Fig. 10.5d). Most remarkably, although the MHC-I molecule used to provide a ligand for TAPBPR in these studies contained a truncated peptide covalently linked via a disulfide bridge to an engineered cysteine residue at position 73, clearly observed in the density map of the unliganded MHC-I molecule, the complex with TAPBPR however showed no consistent electron density for the peptide. Also, the general structure of the peptide binding groove was rearranged to a more open conformation. Importantly, the structure of MHC-I/TAPBPR shows that low affinity peptides are released when chaperone binds to MHC-I, and when a high affinity peptide is bound the chaperone is disassociated from MHC-I. The mechanism of "peptide editing" in the absence of the PLC is shown in Fig. 10.2b.

The studies of Jiang et al. (2017) are remarkably consistent with the structure determination of Thomas and Tampé (Thomas and Tampe 2017) who used a slightly different strategy to produce H2-D^b/TAPBPR complexes and crystals. They expressed and purified the mouse MHC-I molecule H2-D^b with a photolabile viral peptide. Following UV photolysis, the H2-D^b/hTAPBPR complex produced crystals that diffracted to 3.3 Å (PDB, 5OPI). The structural models produced by the two groups are very similar. Superposition of 5OPI (containing one MHC/TAPBPR complex) on 5WER (which has four complexes per asymmetric unit) gives an RMSD of 1.163 Å. The individual complexes of 5WER range in RMSD from 0.934 to 1.467 Å. However, the 5OPI structure was modeled with an α helix for residues 27–34, a region that produced no electron density in the 5WER data sets. Indeed, the lack of density in the 5WER refined data, and large crystallographic *B*-factors for these residues in the 5OPI structure (averaging about 247 Å^2 for backbone atoms, consistent with a thermal motion (u) of such atoms of greater than 1.6 Å) suggest that this is a region of high mobility, and of uncertain structure.

Functional and Structural Examination of the 22 to 35 Loop Region of TAPBPR and Tapasin

The potential role of the loop region encompassing amino acids 22 to 35 of TAPBPR and of the shorter (10 amino acid long) homologous region of tapasin in direct interaction with the peptide-binding groove of MHC-I has been studied in functional studies (for the TAPBPR loop) by Ilca et al. (2018) and for the shorter tapasin loop by X-ray studies (Hafstrand et al. 2019). Deletion and mutagenesis of the TAPBPR loop support its importance in mediating efficient peptide exchange, and focus on the role of a central leucine residue (L30) (Ilca et al. 2018). Structural studies examined a complex between a decamer peptide representative of the tapasin loop bound to the mouse MHC-I molecule H2-D^b containing a truncated viral peptide and compared this to H2-D^b and H2-K^b with the Gly-Leu dipeptide in the C-terminal F pocket (Hafstrand et al. 2019). Although the tapasin derived decamer loop peptide was not fully visualized, all three structures indicated that the leucine side chain was capable of interacting tightly with the F pocket. Taken together, the functional studies and the structural studies are consistent with a role of the 22 to 35 loop of TAPBPR and the shorter tapasin loop in providing some steric effects affecting access of other peptides to the chaperone-stabilized binding groove. Indeed, the lack of definitive electron density for the TAPBPR loop in the MHC-I/TAPBPR structures suggests thermal mobility of the loop when the chaperone is bound to MHC-I.

Cryo-EM Structure of the PLC

Recent improvements in cryo-electron microscopy provided an opportunity to gather a structural view of the entire PLC. Using a TAP-binding herpes virus-encoded protein, ICP47, to pull down the entire PLC, Blees and colleagues isolated detergent–solubilized cross-linked complexes from Burkitt's lymphoma cells (Blees et al. 2017). These complexes consisted of TAP1, TAP2, ERp57, calreticulin, MHC-I heavy chains (likely HLA-A and -B as well as -C) and the MHC-I light chain β_2m (Blees et al. 2017). Though somewhat biochemically heterogenous, the cryo-EM images provided pictures of the macromolecular PLC complex measuring 150 Å × 150 Å across and a full 240 Å through the membrane (see Fig. 10.6). The ER lumenal domains have two editing (i.e. MHC-I containing) modules, consistent with earlier biochemical studies (Panter et al. 2012), and the heterogeneity of the structures suggested considerable dynamic flexibility. The dimeric modules were able to be refined to about 7.2 Å resolution, while careful selection of single modules improved the resolution to 5.8 Å. Looking down on the PLC from the top (Fig. 10.6c, d), one sees the central organization of the complex based on the dimerization of tapasin, centered on the transmembrane TAP1/2 heterodimer, with the other components radiating from this central region. Since high resolution X-ray structures of virtually all the components have been determined (with the exception of the putative long α-helix of calreticulin), a representative high resolution model of the complex was fitted into the lower resolution cryo-EM map (Fig. 10.6b, d, and f). Tapasin, itself bracketed by ERp57, supports the MHC-I, which is engaged by the calreticulin lectin domain, which then extend a long loop towards ERp57. Thus, one can envision TAP delivering peptides from the cytoplasm to the lumenal side of the complex, with rapid access to the tapasin-chaperoned, peptide-receptive, MHC-I binding cleft. Since several different states of assembly were identified in the cryo-EM pictures, some indications of the dynamics of the functional complex are intimated.

Assessing the Dynamics of MHC/TAPBPR Interaction with NMR

With X-ray structures of MHC-I and MHC-I/TAPBPR complexes at hand, it is now possible to couple these data with nuclear magnetic resonance (NMR) analysis of the interactions as they occur in solution at room temperature. To this end, McShan et al. (2018) used selective heavy isotope labelling of subsets of amino acids of the

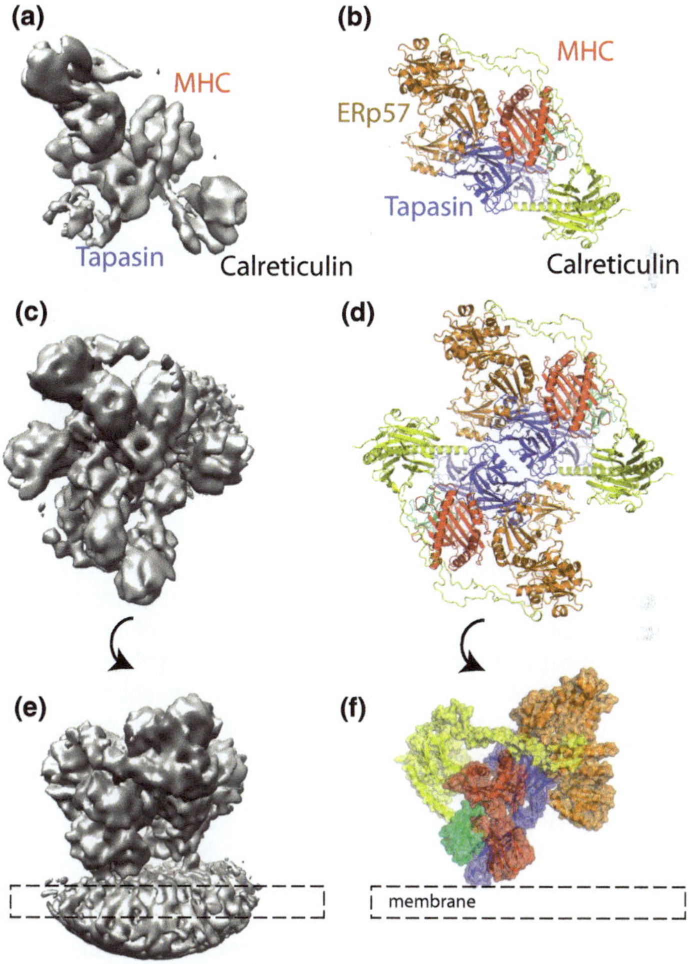

Fig. 10.6 Cryo-EM images and superposition based models reveal organization of the PLC. **a** Cryo-EM based structure taken from the electron microscope database (EMDB) (Lawson et al. 2011) entry 3905 showing the map generated by Blees et al. (Blees et al. 2017). **b** is the high resolution model deposited as pdb, 6ENY, indicating the individual proteins of the PLC. **c**, **e** show the map of the full complex (EMDB 3904), visualized from above (**c**) and from the side (**e**), with corresponding higher resolution images from 6ENY in (**d**) and (**f**)

MHC-I molecule, H2-D^d to experimentally address the dynamics of TAPBPR-dependent chaperoning and peptide exchange. In the peptide/H2-D^d/β_2m complexes, several regions of increased mobility including several loops and the Ig-like α3 domain were observed. Regions of MHC-I flexibility corresponded to regions of TAPBPR interaction. Analysis of Ala, Ile, Leu, and Val (AILV) methyl groups of the MHC-I revealed interaction of TAPBPR with regions completely consistent with the X-ray structures described above. Further analysis indicated that both TAPBPR and incoming peptide suppress the dynamics of the groove of the MHC-I leading to the conformational changes observed in the α2-1 helix. Most remarkably, the NMR analysis permitted detection of conformational intermediates that occurred during peptide-exchange. While peptide is binding, a peptide/MHC/TAPBPR intermediate is detected, and the final peptide/MHC-I conformation releases the TAPBPR. Contributions of interactions of peptide in the A and B pockets as well as evidence for a negative allosteric coupling between the A and F pockets is observed. Thus, NMR spectra provide solution, room temperature characterization of the dynamics of both empty and peptide-loaded MHC-I when bound to TAPBPR, and provide a model that can be extrapolated to the function of tapasin in the PLC.

Conclusions

The biological evolution of the complex cellular phenomena that are required for T cell dependent immune surveillance has been based on a remarkable series of protein/protein and peptide/protein interactions. The MHC-I loading pathway, described above, has exploited general aspects of protein folding and assembly, and the role of chaperones such as tapasin and TAPBPR to stabilize conformations to allow peptide binding. Although the details of the MHC-II loading pathway are quite different, occurring at least in part in acidic endosomal or lysosomal cellular compartments rather than in the ER, and dependent on unique chaperones, the invariant chain (Ii) and the MHC-II-like HLA-DM, the general principle that empty MHC molecules benefit from the stabilization provided by particular chaperones to facilitate peptide loading is preserved (Mellins and Stern 2014; Sadegh-Nasseri 2016; Sadegh-Nasseri et al. 2008). The ongoing improvements in our understanding of peptide loading in both the MHC-I and MHC-II pathways should be expected to not only foster deeper and broader understanding of a variety of processes in protein assembly and folding, but may provide insight into new avenues for effective immunization for infection and treatment for autoimmunity and cancer.

Acknowledgements This research was supported by the Intramural Research Program of the National Institutes of Health, NIAID.

References

Adams EJ, Luoma AM (2013) The adaptable major histocompatibility complex (MHC) fold: structure and function of nonclassical and MHC class I-like molecules. Annu Rev Immunol 31:529–561

Archbold JK, Ely LK, Kjer-Nielsen L, Burrows SR, Rossjohn J, McCluskey J, Macdonald WA (2008) T cell allorecognition and MHC restriction—A case of Jekyll and Hyde? Mol Immunol 45:583–598

Bjorkman PJ, Saper MA, Samraoui B, Bennett WS, Strominger JL, Wiley DC (1987) Structure of the human class I histocompatibility antigen, HLA-A2. Nature 329:506–512

Blees A, Januliene D, Hofmann T, Koller N, Schmidt C, Trowitzsch S, Moeller A, Tampe R (2017) Structure of the human MHC-I peptide-loading complex. Nature 551:525–528

Blum JS, Wearsch PA, Cresswell P (2013) Pathways of antigen processing. Annu Rev Immunol 31:443–473

Boyle LH, Hermann C, Boname JM, Porter KM, Patel PA, Burr ML, Duncan LM, Harbour ME, Rhodes DA, Skjodt K, Lehner PJ, Trowsdale J (2013) Tapasin-related protein TAPBPR is an additional component of the MHC class I presentation pathway. Proc Natl Acad Sci U S A 110:3465–3470

Brown JH, Jardetzky TS, Gorga JC, Stern LJ, Urban RG, Strominger JL, Wiley DC (1993) Three-dimensional structure of the human class II histocompatibility antigen HLA-DR1. Nature 364:33–39

Chen H, Li L, Weimershaus M, Evnouchidou I, van Endert P, Bouvier M (2016) ERAP1-ERAP2 dimers trim MHC I-bound precursor peptides; implications for understanding peptide editing. Sci Rep 6:28902

Chen M, Bouvier M (2007) Analysis of interactions in a tapasin/class I complex provides a mechanism for peptide selection. EMBO J 26:1681–1690

Copeman J, Bangia N, Cross JC, Cresswell P (1998) Elucidation of the genetic basis of the antigen presentation defects in the mutant cell line. 220 reveals polymorphism and alternative splicing of the tapasin gene. Eur J Immunol 28:3783–3791

Dong G, Wearsch PA, Peaper DR, Cresswell P, Reinisch KM (2009) Insights into MHC class I peptide loading from the structure of the tapasin-ERp57 thiol oxidoreductase heterodimer. Immunity 30:21–32

Fremont DH, Matsumura M, Stura EA, Peterson PA, Wilson IA (1992) Crystal structures of two viral peptides in complex with murine MHC class I H-2 Kb. Science 257:919–927

Garbi N, Tiwari N, Momburg F, Hammerling GJ (2003) A major role for tapasin as a stabilizer of the TAP peptide transporter and consequences for MHC class I expression. Eur J Immunol 33:264–73

Garboczi DN, Ghosh P, Utz U, Fan QR, Biddison WE, Wiley DC (1996) Structure of the complex between human T-cell receptor, viral peptide and HLA-A2. Nature 384:134–141

Germain RN, Margulies DH (1993) The biochemistry and cell biology of antigen processing and presentation. Annu Rev Immunol 11:403–450

Grandea AG 3rd, Golovina TN, Hamilton SE, Sriram V, Spies T, Brutkiewicz RR, Harty JT, Eisenlohr LC, Van Kaer L (2000) Impaired assembly yet normal trafficking of MHC class I molecules in Tapasin mutant mice. Immunity 13:213–222

Hafstrand I, Sayitoglu EC, Apavaloaei A, Josey BJ, Sun R, Han X, Pellegrino S, Ozkazanc D, Potens R, Janssen L, Nilvebrant J, Nygren PA, Sandalova T, Springer S, Georgoudaki AM, Duru AD, Achour A (2019) Successive crystal structure snapshots suggest the basis for MHC class I peptide loading and editing by tapasin. Proc Natl Acad Sci U S A 116:5055–5060

Hermann C, Strittmatter LM, Deane, JE, Boyle, LH (2013) The binding of TAPBPR and Tapasin to MHC class I is mutually exclusive. J Immunol 191:5743–5750

Hermann C, van Hateren A, Trautwein N, Neerincx A, Duriez PJ, Stevanovic S, Trowsdale J, Deane JE, Elliott T, Boyle LH (2015) TAPBPR alter MHC class I peptide presentation by functioning as a peptide exchange catalyst. Elife 4

Ilca FT, Neerincx A, Hermann C, Marcu A, Stevanovic S, Deane JE, Boyle LH (2018) TAPBPR mediates peptide dissociation from MHC class I using a leucine lever. Elife 7

Jiang J, Natarajan K, Boyd LF, Morozov GI, Mage MG, Margulies DH (2017) Crystal structure of a TAPBPR-MHC I complex reveals the mechanism of peptide editing in antigen presentation. Science 358:1064–1068

Koch J, Tampe R (2006) The macromolecular peptide-loading complex in MHC class I-dependent antigen presentation. Cell Mol Life Sci 63:653–662

La Gruta NL, Gras S, Daley SR, Thomas PG, Rossjohn J (2018) Understanding the drivers of MHC restriction of T cell receptors. Nat Rev Immunol 18:467–478

Lawson CL, Baker ML, Best C, Bi C, Dougherty M, Feng P, van Ginkel G, Devkota B, Lagerstedt I, Ludtke SJ, Newman RH, Oldfield TJ, Rees I, Sahni G, Sala R, Velankar S, Warren J, Westbrook JD, Henrick K, Kleywegt GJ, Berman HM, Chiu W (2011) EMDataBank.org: unified data resource for CryoEM. Nucleic Acids Res 39:D456–D464

Ljunggren HG, Stam NJ, Ohlen C, Neefjes JJ, Hoglund P, Heemels MT, Bastin J, Schumacher TN, Townsend A, Karre K et al (1990) Empty MHC class I molecules come out in the cold. Nature 346:476–480

Margulies DH, Natarajan K, Rossjohn J, McCluskey J (2008) The major histocompatibility complex. In Paul WE (ed) Fundamental immunology. Wolters Kluwer, Lippincott Williams & Wilkins, Philadelphia, Baltimore

Marrack P, Scott-Browne JP, Dai S, Gapin L, Kappler JW (2008) Evolutionarily conserved amino acids that control TCR-MHC interaction. Annu Rev Immunol 26:171–203

McShan AC, Natarajan K, Kumirov VK, Flores-Solis D, Jiang J, Badstubner M, Toor JS, Bagshaw CR, Kovrigin EL, Margulies DH, Sgourakis NG (2018) Peptide exchange on MHC-I by TAPBPR is driven by a negative allostery release cycle. Nat Chem Biol 14:811–820

Mellins ED, Stern LJ (2014) HLA-DM and HLA-DO, key regulators of MHC-II processing and presentation. Curr Opin Immunol 26:115–122

Morozov GI, Zhao H, Mage MG, Boyd LF, Jiang J, Dolan MA, Venna R, Norcross MA, McMurtrey CP, Hildebrand W, Schuck P, Natarajan K, Margulies DH (2016) Interaction of TAPBPR, a tapasin homolog, with MHC-I molecules promotes peptide editing. Proc Natl Acad Sci U S A 113:E1006–E1015

Natarajan K, McShan AC, Jiang J, Kumirov VK, Wang R, Zhao H, Schuck P, Tilahun ME, Boyd LF, Ying J, Bax A, Margulies DH, Sgourakis NG (2017) An allosteric site in the T-cell receptor Cbeta domain plays a critical signalling role. Nat Commun 8:15260

Neerincx A, Hermann C, Antrobus R, van Hateren A, Cao H, Trautwein N, Stevanovic S, Elliott T, Deane JE, Boyle LH (2017) TAPBPR bridges UDP-glucose: glycoprotein glucosyltransferase 1 onto MHC class I to provide quality control in the antigen presentation pathway. Elife 6

Panter MS, Jain A, Leonhardt RM, Ha T, Cresswell P (2012) Dynamics of major histocompatibility complex class I association with the human peptide-loading complex. J Biol Chem 287:31172–31184

Robinson J, Halliwell JA, Hayhurst JD, Flicek P, Parham P, Marsh SG (2015) The IPD and IMGT/HLA database: allele variant databases. Nucleic Acids Res 43:D423–D431

Rock KL, Reits E, Neefjes J (2016) Present yourself! By MHC Class I and MHC Class II molecules. Trends Immunol 37:724–737

Sadasivan B, Lehner PJ, Ortmann B, Spies T, Cresswell P (1996) Roles for calreticulin and a novel glycoprotein, tapasin, in the interaction of MHC class I molecules with TAP. Immunity 5:103–114

Sadegh-Nasseri S (2016) A step-by-step overview of the dynamic process of epitope selection by major histocompatibility complex class II for presentation to helper T cells. F1000 Res 5

Sadegh-Nasseri S, Chen M, Narayan K, Bouvier M (2008) The convergent roles of tapasin and HLA-DM in antigen presentation. Trends Immunol 29:141–147

Saini SK, Ostermeir K, Ramnarayan VR, Schuster H, Zacharias M, Springer S (2013) Dipeptides promote folding and peptide binding of MHC class I molecules. Proc Natl Acad Sci U S A 110:15383–15388

Saper MA, Bjorkman PJ, Wiley DC (1991) Refined structure of the human histocompatibility antigen HLA-A2 at 2.6 A resolution. J Mol Biol 219:277–319
Serwold T, Gonzalez F, Kim J, Jacob R, Shastri N. (2002) ERAAP customizes peptides for MHC class I molecules in the endoplasmic reticulum. Nature 419:480–483
Spies T, DeMars R (1991) Restored expression of major histocompatibility class I molecules by gene transfer of a putative peptide transporter. Nature 351:323–324
Teng MS, Stephens R, Du Pasquier L, Freeman T, Lindquist JA, Trowsdale J. (2002).A human TAPBP (TAPASIN)-related gene, TAPBP-R. Eur J Immunol 32:1059–1068
Thomas C, Tampe R (2017) Structure of the TAPBPR-MHC I complex defines the mechanism of peptide loading and editing. Science 358:1060–1064
van Hateren A, Carter R, Bailey A, Kontouli N, Williams AP, Kaufman J, Elliott T (2013) A mechanistic basis for the co-evolution of chicken tapasin and major histocompatibility complex class I (MHC I) proteins. J Biol Chem 288:32797–2808
Van Hateren A, James E, Bailey A, Phillips A, Dalchau N, Elliott T (2010) The cell biology of major histocompatibility complex class I assembly: towards a molecular understanding. Tissue Antigens 76:259–275
Wang R, Natarajan K, Margulies DH (2009) Structural basis of the CD8 alpha beta/MHC class I interaction: focused recognition orients CD8 beta to a T cell proximal position. J Immunol 183:2554–2564
Wearsch PA, Cresswell P (2007) Selective loading of high-affinity peptides onto major histocompatibility complex class I molecules by the tapasin-ERp57 heterodimer. Nat Immunol 8:873–881
Yewdell JW, Nicchitta CV (2006) The DRiP hypothesis decennial: support, controversy, refinement and extension. Trends Immunol 27:368–373
York IA, Rock KL (1996) Antigen processing and presentation by the class I major histocompatibility complex. Annu Rev Immunol 14:369–396
Zacharias M, Springer S (2004) Conformational flexibility of the MHC class I alpha1-alpha2 domain in peptide bound and free states: a molecular dynamics simulation study. Biophys J 87:2203–2214
Zhang W, Wearsch PA, Zhu Y, Leonhardt RM, Cresswell P. (2011).A role for UDP-glucose glycoprotein glucosyltransferase in expression and quality control of MHC class I molecules. Proc Natl Acad Sci U S A 108:4956–4961

Chapter 11
Biology and Biochemistry of Bacterial Proteasomes

Samuel H. Becker, Huilin Li and K. Heran Darwin

Abstract Proteasomes are a class of protease that carry out the degradation of a specific set of cellular proteins. While essential for eukaryotic life, proteasomes are found only in a small subset of bacterial species. In this chapter, we present the current knowledge of bacterial proteasomes, detailing the structural features and catalytic activities required to achieve proteasomal proteolysis. We describe the known mechanisms by which substrates are doomed for degradation, and highlight potential non-degradative roles for components of bacterial proteasome systems. Additionally, we highlight several pathways of microbial physiology that rely on proteasome activity. Lastly, we explain the various gaps in our understanding of bacterial proteasome function and emphasize several opportunities for further study.

Keywords Proteasome · Proteolysis · Pupylation · Mycobacterium

Introduction

Proteasomes are a class of protease found in all three domains of life. A minimum proteasome complex is a barrel-shaped, 28-subunit structure known as the 20S core particle (CP, Fig. 1a), named in accordance with its sedimentation coefficient. 20S CPs are gated to prevent uncontrolled proteolysis, and associate with proteasomal activator proteins in order to degrade specific sets of cellular substrates.

S. H. Becker · K. Heran Darwin (✉)
Department of Microbiology, New York University School of Medicine, 430 E. 29th Street, Room 312, New York, NY 10016, USA
e-mail: heran.darwin@med.nyu.edu

S. H. Becker
e-mail: samuel.becker@nyulangone.org

H. Li
Van Andel Research Institute, Cryo-EM Structural Biology Laboratory, 333 Bostwick Ave, NE, Grand Rapids, MI 4950, USA
e-mail: Huilin.Li@vai.org

J. R. Harris and J. Marles-Wright (eds.), *Macromolecular Protein Complexes II: Structure and Function*, Subcellular Biochemistry 93,
https://doi.org/10.1007/978-3-030-28151-9_11

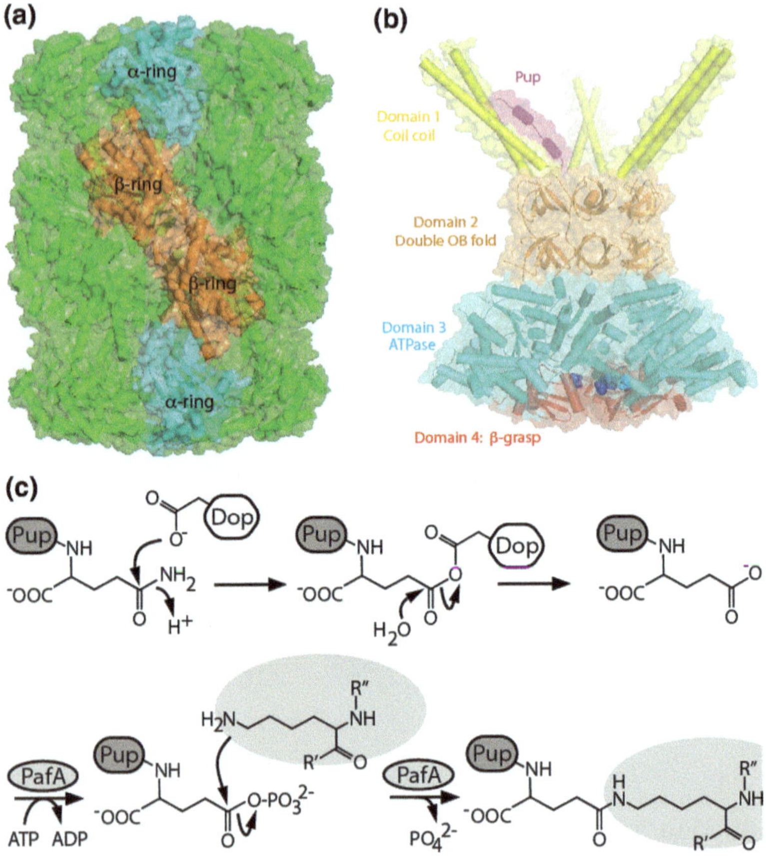

Fig. 1 The ***M. tuberculosis*** **Pup-proteasome system. a** Crystal structure of the *M. tuberculosis* 20S core particle (20S CP) (PBD entry: 2FHG). Individual subunits of PrcA (blue) and PrcB (orange) are highlighted within each α- and β-ring. **b** Model of an Mpa hexamer bound to the N-terminal region of Pup, indicating the four structural domains of Mpa (PDB entries: 3M9D and 5KZF). Note that the last resolved residue Thr-601, shown in spheres, is receded inside the axial chamber. **c** Deamidation of Pup and ligation to a target protein. An aspartate on Dop acts either as a direct nucleophile as shown (Burns et al. 2012) or activates a water molecule (Ozcelik et al. 2012) to attack the C-terminal Gln of Pup, generating a Pup-Dop intermediate; Dop then catalyzes the nucleophilic attack by a water molecule to generate Pup_{Glu}. PafA subsequently uses ATP hydrolysis to phosphorylate Pup_{Glu}, rendering Pup suitable for nucleophilic attack by the side chain of a substrate lysine residue (Guth et al. 2011)

Proteasomes are essential in all eukaryotes and archaea. Among the bacteria, however, 20S CP genes are not ubiquitous, and are found almost exclusively in the *Actinomycetales* and *Nitrospirales* orders (Darwin et al. 2003; De Mot 2007; Lehmann et al. 2017; Lupas et al. 1994; Nagy et al. 1998; Pouch et al. 2000; Tamura et al. 1995). Among the *Actinomycetales* are the mycobacteria, including *Mycobacterium tuberculosis* (*M. tuberculosis*), a major human pathogen. In eukaryotes, proteins are destined for proteasomal degradation primarily through their ligation to a small protein called ubiquitin (Ub) (reviewed in Streich and Lima 2014). Proteasome-bearing bacteria do not have Ub, but instead can modify proteins with Pup (prokaryotic ubiquitin-like protein), after which they are recognized by proteasomal activators for delivery into 20S CPs (Burns et al. 2009, 2010b; Pearce et al. 2008; Striebel et al. 2010). More recent studies have shown that proteins can be sent to bacterial proteasomes independently of pupylation (Delley et al. 2014; Jastrab et al. 2015). Furthermore, there are now data showing that Pup performs degradation-independent functions in several bacterial species (Elharar et al. 2014; Kuberl et al. 2016). Here, we describe the current understanding of bacterial proteasomes and their accessory factors, focusing on both the mechanisms of targeted proteolysis as well as the aspects of bacterial physiology that depend on these macromolecular machines.

20S Proteasome Core Particles

In all domains of life, 20S CPs are composed of four stacked, heptameric rings: two outer rings, composed of α-subunits, act as gated antechambers controlling entry of substrates into the proteasome core, and two inner rings, made up of β-subunits, harbor the active sites that carry out proteolysis (Fig. 1a) (Choi et al. 2016; Groll et al. 1997, 2000; Heinemeyer et al. 1997; Hu et al. 2006; Lin et al. 2006; Lowe et al. 1995; Tamura et al. 1995). All α- and β-subunits belong to the amino (N)-terminal nucleophilic hydrolase protein family (Brannigan et al. 1995). While eukaryotes encode fourteen different α- and β-subunit genes (Groll et al. 1997), most bacteria and archaea have single genes for α- and β-subunits, respectively (Lowe et al. 1995; Tamura et al. 1995).

In contrast to the assembly of eukaryotic and archaeal proteasomes, which requires accessory chaperones (Le Tallec et al. 2007; Ramos et al. 1998), bacterial proteasomes studied to date self-assemble. Bacterial 20S CP assembly initiates with the formation of a half-proteasome consisting of an α- and a β-ring (Groll et al. 2003; Zuhl et al. 1997). Bacterial β-subunits are translated with an N-terminal propeptide that prevents catalytic activity; the association of two half-proteasomes at the β- to β-ring interface initiates autocatalytic cleavage of the propeptide, yielding an N-terminal threonine (Thr1) residue (Li et al. 2010; Pouch et al. 2000; Zuhl et al. 1997; Zwickl et al. 1994). The role of the propeptide appears to differ between bacterial species. In *Rhodococcus erythropolis* (*R. erythropolis*), propeptides promote 20S CP assembly (Zuhl et al. 1997); in contrast, the propeptide of *M. tuberculosis* must be cleaved before half-proteasomes can fully associate (Li et al. 2010; Lin et al. 2006).

The mature β-subunit active site harbors a set of conserved catalytic residues consisting of Thr1 as well as a lysine (Lys) and either an aspartate (Asp) or a glutamate (Glu), a configuration that is present in 20 CPs across all domains of life (Huber et al. 2016). The Lys deprotonates the terminal amine of Thr1, which in turn carries out the nucleophilic attack of an amide bond in a peptide substrate. Meanwhile, an Asp or Glu serves to orient Thr1 within the active site (Hu et al. 2006; Huber et al. 2016; Mc Cormack et al. 1997). Cleavage of the propeptide occurs in an identical manner, where Thr1 attacks the amide bond linking it to the residue in the −1 position (Huber et al. 2016).

Despite a unified catalytic mechanism in 20S CPs, the residues lining β-subunit active sites are variable, and ultimately determine the specificity of proteolytic activity. Archaeal 20S CPs studied to date appear to preferentially cleave following hydrophobic residues within peptides (Hu et al. 2006; Tamura et al. 1995). The three catalytically active β-subunits in eukaryotic 20S CPs have distinct active sites, each with specificity for hydrophobic, basic, or acidic peptide residues (Heinemeyer et al. 1997; Hu et al. 2006). Interestingly, despite only having a single type of β-subunit (PrcB), the *M. tuberculosis* 20S CP can degrade all three peptide classes due to the presence of both hydrophobic and hydrophilic residues lining the active site (Hu et al. 2006; Lin et al. 2006).

To prevent uncontrolled proteolysis, α-subunits have N-termini that gate 20S CPs to hinder proteins or peptides from entering the central channel. In turn, the 20S CP gate must be displaced by a proteasomal activator in order for substrates to be degraded (Groll et al. 2000; Li et al. 2010; Lin et al. 2006). For the *M. tuberculosis* α-subunit PrcA, removal of the N-terminal gating residues ("20S open gate CP") is sufficient to allow for the in vitro degradation of peptide or unfolded protein substrates in the absence of an activator (Lin et al. 2006; Striebel et al. 2010).

ATP-Dependent Proteasome Activation

In order for proteins to be degraded by a proteasome, they must be specifically targeted for unfolding and delivery into the 20S CP. In eukaryotes, the 19S regulatory particle is a heteromeric complex that contains functionally distinct subunits for substrate recognition and unfolding (reviewed in Finley et al. 2016). In contrast, archaeal and bacterial proteasome activators consist of a single gene product that simultaneously recognizes and unfolds substrates. Mpa (mycobacterial proteasome ATPase) in mycobacteria and ARC (AAA ATPase forming ring-shaped complexes) in *R. erythropolis* are currently the most well-characterized bacterial proteasome activators (Darwin et al. 2005; Wolf et al. 1998). Mpa/ARC belongs to the AAA ATPase (ATPases Associated with diverse cellular Activities) family, which contains a variety of proteasomal and non-proteasomal protease activators (Djuranovic et al. 2009; Erzberger and Berger 2006; Striebel et al. 2009b; Wang et al. 2009).

Mpa/ARC monomers consist of four domains, arranged from the N- to carboxyl (C)-terminus: the coiled-coil domain, the interdomain, the ATPase domain, and the β-grasp domain (Djuranovic et al. 2009; Wang et al. 2009; Wu et al. 2017) (Fig. 1b). At the N-termini of Mpa/ARC hexamers, helices from each subunit intercalate to form three coiled-coils, each of which can act as a substrate-binding site (Djuranovic et al. 2009; Wang et al. 2009).

The interdomains of Mpa/ARC proteins contain tandem regions that adopt an oligosaccharide-binding (OB) fold (Murzin 1993). Compared to the highly flexible coiled-coil domain, the series of β-barrels within the OB folds gives the interdomain a rigid structure that is capable of self-assembly in vitro (Wang et al. 2009). While OB folds are a conserved feature of all proteasomal AAA ATPases, the specific function of the interdomain is unknown, though it has been proposed to promote the stability of the mature hexamer while structural movement is generated during substrate binding and ATPase activity (Djuranovic et al. 2009; Wang et al. 2009).

The ATPase domain is required for protein unfolding using a mechanism conserved across protease activators. The defining characteristic of AAA ATPase family proteins is an aromatic-hydrophobic-glycine motif, also known as the pore loop, which makes direct contact with substrates as they translocate through the ATPase channel (Martin et al. 2008; Striebel et al. 2010; Wang et al. 2009). Studies of AAA ATPases associated with non-proteasomal proteases, as well as the archaeal proteasome activator PAN and the eukaryotic 19S subunit Rpt1, have demonstrated a mechanism by which successive rounds of ATP hydrolysis induce movement of pore loops to pull the substrate through the ATPase channel (Kim et al. 2015; Martin et al. 2008).

The C-termini of Mpa/ARC proteins include a disordered region with the sequence glycine-glutamine-tyrosine-leucine, also known as the GQYL motif, which is essential for interaction with 20S CPs (Darwin et al. 2003; Hu et al. 2018; Jastrab et al. 2015; Pearce et al. 2006). A conserved feature of bacterial proteasome activators, the GQYL motif is thought to act similarly to the C-terminal HbYX (hydrophobic-tyrosine-any amino acid) motif on PAN as well as the 19S RP ATPases (Delley et al. 2014; Jastrab et al. 2015; Smith et al. 2007). The HbYX motif inserts into a pocket in between α-subunits, inducing a conformational change that opens the 20S CP gate (Rabl et al. 2008; Smith et al. 2007; Yu et al. 2010). Notably, a peptide of just the GQYL sequence is sufficient to activate *M. tuberculosis* 20S CPs for peptide degradation in vitro (Jastrab et al. 2015).

It has long been observed that Mpa associates weakly with 20S CPs in vitro, preventing the robust degradation of substrates (Striebel et al. 2010; Wang et al. 2009). The observation that Mpa and 20S CPs interact poorly could be explained by the discovery of a new domain in *M. tuberculosis* Mpa, between the ATPase domain and the C-terminus, forming a β-grasp fold (Burroughs et al. 2007; Wu et al. 2017). This feature so far appears to be unique to bacterial proteasome activators and prevents the GQYL motif from interacting with PrcA subunits of a 20S CP. Artificial extension of the Mpa C-terminus ("$Mpa_{C\text{-}ext}$") allows for increased binding between Mpa and 20S CPs. While $Mpa_{C\text{-}ext}$ could not activate proteolysis by wild type 20S CPs, the use of "open gate" 20S CPs, in which the N-termini of the PrcA subunits are truncated, allows for the in vitro degradation of a model proteasome substrate, Pup ~ FabD (Wu et al. 2017). It is therefore likely that other factors regulate the ability of Mpa to activate proteasomal degradation in vivo.

Recently, another bacterial proteasome interactor was identified. Rv0435c, or *cpa* (Cdc48-like protein of actinobacteria), genetically co-occurs with bacterial 20S CP genes, and encodes a protein with structural homologs in eukaryotes and prokaryotes. *M. tuberculosis* Cpa forms a hexamer that has ATPase activity in vitro and that can associate with the *R. erythropolis* 20S CPs. While disruption of *cpa* in *M. smegmatis* leads to broad changes in the proteome, it remains to be determined if Cpa associates with 20S CPs in vivo and what its role is in the physiology of bacteria (Ziemski et al. 2018).

Pup: A Degradation Signal, and More

Pupylation

One mechanism by which bacterial proteins are targeted for proteasomal degradation is through post-translational modification with Pup. Pup is a small protein that is covalently linked by its C-terminus to the side chain of a lysine on target proteins (Fig. 1c) (Burns et al. 2009; Pearce et al. 2008). In the Pup-proteasome system (PPS), pupylated proteins are specifically recognized by Mpa/ARC, which unfolds and delivers them into 20S CPs (Burns et al. 2010b; Striebel et al. 2010; Wang et al. 2010). While functionally similar to Ub, Pup is mostly disordered, unlike the highly structured Ub (Chen et al. 2009; Sutter et al. 2009).

Mycobacterium Pup is translated with a C-terminal glutamine (Gln) residue. However, initial characterizations of pupylated proteins in *M. tuberculosis* and *M. smegmatis* found that lysine-bound Pup contains a C-terminal Glu (Pup_{Glu}) (Burns et al. 2009; Pearce et al. 2008). It was found that, following translation, Pup_{Gln} must be deamidated by Dop (deamidase of Pup) to generate Pup_{Glu} in a reaction that is dependent on the presence, but not hydrolysis of ATP (Fig. 1c) (Imkamp et al. 2010a; Striebel et al. 2009a). Covalent linkage of Pup_{Glu} to a lysine on a target protein is achieved by a second enzyme, PafA (proteasome accessory factor A) (Striebel et al. 2009a). Upon binding to PafA, Pup adopts a helical conformation, positioning the Pup C-terminus within the PafA active site (Barandun et al. 2013). Upon ATP hydrolysis, PafA phosphorylates Pup_{Glu}, preparing it for nucleophilic attack by the ε-amino group of a lysine in a target protein (Guth et al. 2011) (Fig. 1c). Dop, PafA, and Pup_{Gln} together are necessary and sufficient to pupylate proteins in vitro, and the need for Dop can be bypassed if Pup_{Glu} is used (Striebel et al. 2009a).

In eukaryotes, chains of Ub (polyubiquitylation) on protein substrates are generally required for recognition and degradation by proteasomes (Thrower et al. 2000). While polypupylation has been observed in vitro and upon overproduction of Pup in bacteria (Chen et al. 2016; Elharar et al. 2014; Regev et al. 2015), it has never been observed under physiologic conditions. Although a role for polypupylation may exist in certain circumstances, it appears that only a single pupylation event is sufficient to target a substrate for degradation (Burns et al. 2010b). The majority of natively pupylated proteins are modified at a single, specific lysine, although some proteins have up to four different lysines that can be targeted (Fascellaro et al. 2016; Festa et al. 2010; Kuberl et al. 2014; Poulsen et al. 2010; Watrous et al. 2010). There are little reliable data to suggest substrates are pupylated at more than one lysine on the same molecule; i.e., a single polypeptide can only be pupylated once. However, an oligomeric complex can have more than one Pup attached: *Mycobacterium* Ino1 (inositol 1-phosphate synthetase) forms tetramers and can be pupylated on two monomers within the same complex (Burns et al. 2010a). These observations suggest that the Pup ligase PafA is hindered from interacting with a polypeptide or complex once it is pupylated.

In Pup-bearing bacteria, specific proteins are pupylated at steady state, and these proteins are likely to vary under different laboratory conditions (Fascellaro et al. 2016; Festa et al. 2010; Kuberl et al. 2014, 2016; Pearce et al. 2008; Poulsen et al. 2010; Samanovic et al. 2015; Watrous et al. 2010). While numerous targets of pupylation have been identified, there is no unifying model to explain how proteins are selected for pupylation. Furthermore, there is no known consensus sequence that is targeted for pupylation. This observation may not be surprising in light of data showing that free lysine can be pupylated in vitro (Guth et al. 2011) and that the production of the *M. tuberculosis* pupylation system genes in *E. coli*, which does not have a PPS, results in the pupylation of over 50 different proteins (Cerda-Maira et al. 2011). While it may seem that PafA is highly promiscuous, it has a higher affinity for at least one native pupylation substrate than for free lysine (Guth et al. 2011). Furthermore, the presence of surface-exposed lysines is insufficient to target a protein for pupylation, suggesting that there are intrinsic features of specific proteins that make them suitable pupylation targets. Pertinent to this observation, a study has identified a loop region of PafA near its active site that appears to be important for the efficient binding of PafA to pupylation substrates (Regev et al. 2016).

Delivery of Pupylated Proteins to the Proteasome

In mycobacteria, pupylated proteins interact with Mpa in a manner that facilitates their unfolding and entry into a 20S CP. Upon binding to the coiled-coil domain of Mpa, the N-terminal region of Pup adopts a helical structure similar to that observed in Pup-PafA interactions (Barandun et al. 2013; Striebel et al. 2010; Wang et al. 2010) (Fig. 1b). The flexible N-terminal region of Pup is required to achieve unfolding by Mpa, leading to a model by which the N-terminus of Pup is delivered "headfirst" into the Mpa ATPase domain (Maldonado et al. 2013; Striebel et al. 2010; Wang et al. 2009). The precise manner by which the pupylated proteins themselves are translocated through Mpa is unknown. In the case of other protease activator ATPases, this is activity is enhanced by adaptor proteins that interact with both the ATPase and the substrate (reviewed in Sauer and Baker 2011), though no such adaptors have been identified for bacterial proteasomes.

Fate of Pup: Depupylation and Transpupylation

Besides deamidating Pup, Dop also functions to remove Pup from substrates (Burns et al. 2010a; Imkamp et al. 2010b). There is substantial evidence that depupylation is required for the normal function of the PPS. Notably, many bacterial species, including *Streptomyces coelicolor* (*S. coelicolor*), encode Pup_{Glu}, which suggested Dop has functions beyond the deamidation of Pup_{Gln} (Compton et al. 2015; Striebel et al. 2009a). Although the production of Pup_{Glu} can bypass the requirement of Dop in *M. smegmatis*, robust pupylation in *M. tuberculosis* requires Dop activity, even if Pup_{Glu} is ectopically over-produced (Cerda-Maira et al. 2010, 2011; Imkamp et al. 2010a). These observations led to the hypothesis that a depupylase activity by Dop is required to maintain steady-state levels of Pup (Cerda-Maira et al. 2010). Depupylation by Dop has since been directly demonstrated using a mechanism that appears similar to Pup deamidation: a conserved aspartate residue is proposed to act as a nucleophile that attacks either the Gln side-chain amide to release ammonium (for deamidation) or the isopeptide bond between Pup and a substrate lysine (for depupylation) (Burns et al. 2012; Ozcelik et al. 2012).

The observation that some bacterial species encode Pup_{Glu} instead of Pup_{Gln} suggests that depupylation is a major function of Dop. Furthermore, depupylation likely contributes to the cellular levels of proteasome substrates by rescuing them from degradation. Along these lines, it is unclear whether Pup is recycled by Dop following engagement of a pupylated substrate with Mpa, or is instead degraded along with a doomed protein. In in vitro experiments, Mpa and 20S CPs can degrade a model pupylated substrate where Pup is also degraded (Striebel et al. 2010), but it remains to be determined if Pup is degraded with a linked substrate in vivo.

Recently, a second function of PafA was identified in its ability to transfer Pup from one substrate to another. The "transpupylation" activity of PafA proceeds first through the depupylation of a donor substrate, followed by the ligation of the freed Pup to a new, recipient protein (Zhang et al. 2017). Importantly, transpupylation may help to explain why some pupylated proteins are not conspicuously degraded in vivo under several conditions tested (Fascellaro et al. 2016; Festa et al. 2010). It is possible that some proteins function as Pup "donors" from which Pup may be transferred to other proteins that must be degraded, rather than as proteasome substrates themselves. While this is a provocative idea, it will be challenging to determine if transpupylation occurs in vivo and has physiologic consequences.

Degradation-Independent Functions of Pupylation

Some bacterial species encode components of the PPS, but are missing genes for a 20S CP, suggesting roles for pupylation that do not involve proteolysis. In *Corynebacterium glutamicum* (*C. glutamicum*), which encodes Pup_{Glu-}, PafA, Dop, and ARC, disruption of *pup* or *arc* renders this species sensitive to iron starvation. It was shown that ferritin, a protein that oligomerizes to sequester intracellular iron stores, is pupylated. A model was thus proposed whereby dissociation of pupylated ferritin complexes by ARC allows for the release of iron stores necessary during iron-depleted conditions (Kuberl et al. 2016).

There is also evidence that Pup has degradation-independent functions in bacteria that have 20S CPs. In *S. coelicolor*, mutant strains that are unable to pupylate proteins but still have 20S CPs are defective in sporulation and have altered levels of several metabolites (Boubakri et al. 2015). In another example, an *M. smegmatis prcSBA* (*pup* is also known as *prcS* in *M. smegmatis*) mutant is deficient in survival during nitrogen starvation; this phenotype is almost fully complemented by *pup* alone, suggesting that degradation by proteasomes is either not involved or plays a minor role in the ability of this species to withstand nitrogen starvation (Elharar et al. 2014). It is possible that Pup functions in the proteolysis of proteins by another protease in *M. smegmatis*. Alternatively, there may be non-degradative functions of Pup that facilitate survival during nitrogen starvation. For example, several *M. smegmatis* proteins involved in nitrogen metabolism are targets of pupylation, which perhaps may affect their activities or interactions (Fascellaro et al. 2016). Thus it remains to be determined how pupylation affects nitrogen metabolism in *M. smegmatis*.

ATP-Independent Proteasome Activation

An *M. tuberculosis* mutant lacking 20S CP genes has a severe growth defect compared to wild type, *dop*, *pafA* or *mpa* strains, an observation that suggested substrate recognition by Mpa is not the only route for proteasomal degradation in this species (Cerda-Maira et al. 2010; Darwin et al. 2003; Gandotra et al. 2010). Studies seeking novel interaction partners with *M. tuberculosis* 20S CPs identified a second proteasomal activator, PafE (proteasome accessory factor E), also known as Bpa (bacterial proteasome activator) (Delley et al. 2014; Jastrab et al. 2015). PafE operates separately from the PPS, and is required for the proteasomal degradation of an apparently independent set of substrates (Jastrab et al. 2015).

PafE is unique among protease activators as it assembles into dodecameric rings (Bai et al. 2016; Jastrab et al. 2015) (Fig. 2a). Furthermore, like Mpa, PafE has a canonical C-terminal GQYL motif required for its ability to open the 20S CP gate. Unlike the subunits of an Mpa hexamer, the C-termini of PafE subunits extend outwards from the main structure of the dodecamer (Bai et al. 2016; Bolten et al.

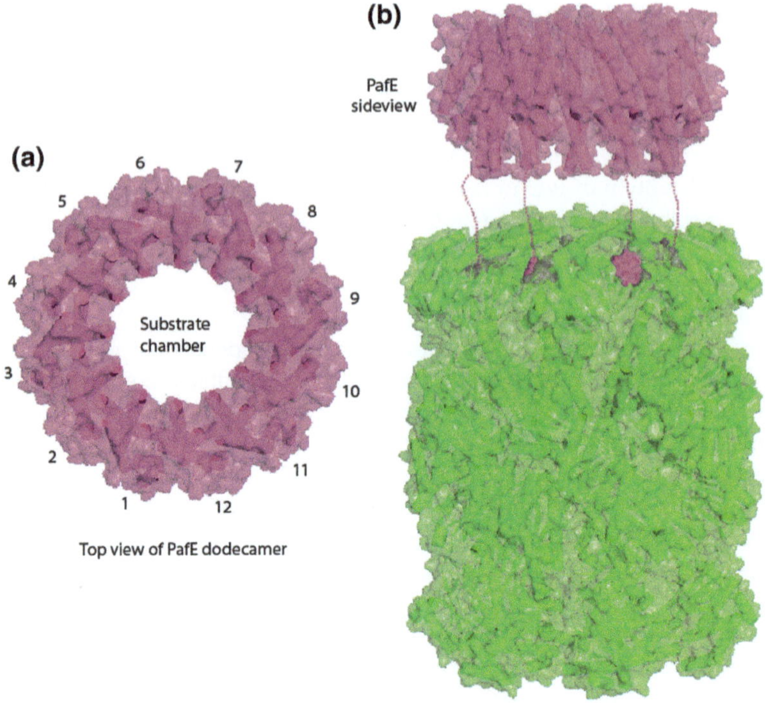

Fig. 2 Structural features of the proteasome activator PafE. a Crystal structure of *M. tuberculosis* PafE (top view) (PDB entry: 5IET). **b** Model of PafE association with a 20S CP (side view). The GQYL motif of PafE inside the top alpha-ring of the 20S core particle is shown as purple spheres. Dashed lines approximate the disordered C-terminal peptide connecting the last α-helix and the terminal GQYL of PafE

2016) (Fig. 2b). This structure is in direct contrast to Mpa, which has buried C-termini (Wu et al. 2017). This major difference likely explains why PafE, but not Mpa, can activate wild type 20S CPs in vitro (Bolten et al. 2016; Delley et al. 2014; Hu et al. 2018; Jastrab et al. 2015).

In contrast to Mpa/ARC, PafE does not hydrolyze ATP. Rather than functioning as an unfoldase, PafE appears to facilitate the translocation of proteins into 20S CPs by acting simply as a gate-opener. Remarkably and consistent with this model, PafE-proteasomes can robustly degrade a model unfolded protein (β-casein) as well as native, folded *M. tuberculosis* proteins without any apparent post-translational modifications in vitro (Delley et al. 2014; Hu et al. 2018; Jastrab et al. 2015). There are many questions about how specific substrates are degraded by PafE-proteasomes, including how native proteins are translocated into the proteasome in the absence of ATP. It also remains to be determined how PafE-dependent proteolysis is regulated in vivo in the absence of post-translational modifiers.

Roles of the Proteasome in Bacterial Physiology

Nitric Oxide Resistance

Significant attention was given to the bacterial proteasome upon the recognition of its essentiality for *M. tuberculosis* to resist nitric oxide (NO) toxicity and cause lethal infections in animals. NO is an antimicrobial molecule produced by macrophages upon infection with *M. tuberculosis*; mice unable to produce NO rapidly succumb to *M. tuberculosis* infections (MacMicking et al. 1997). However, in wild type mice, *M. tuberculosis* is able to avoid complete killing by NO, persisting in a host for years after initial infection. *M. tuberculosis* strains with mutations in *mpa*, *pafA*, *dop*, and *prcBA* are hypersensitive to NO in vitro and are highly attenuated in mice (Cerda-Maira et al. 2010; Darwin et al. 2003; Gandotra et al. 2010). It was since discovered that the failure to degrade a single PPS substrate, Log (lonely guy), renders *M. tuberculosis* more susceptible to NO-mediated killing in vitro (Fig. 3a) (Samanovic et al. 2015).

Log is an orthologue of LONELY GUY, a phosphoribohydrolase that was first identified in plants as an enzyme required for the biosynthesis of a class of adenine-based hormone molecules called cytokinins (Kurakawa et al. 2007). *M. tuberculosis* Log synthesizes cytokinins, and an *M. tuberculosis mpa* mutant makes more cytokinins than a wild type strain due to a failure to degrade Log. In turn, aldehydes derived from cytokinins synergize with NO to kill *M. tuberculosis* through an unknown mechanism (Fig. 3a). Importantly, while disruption of *log* fully restored NO resistance to an *M. tuberculosis mpa* mutant, it did not fully rescue the virulence defect of this strain in mice (Samanovic et al. 2015). This result

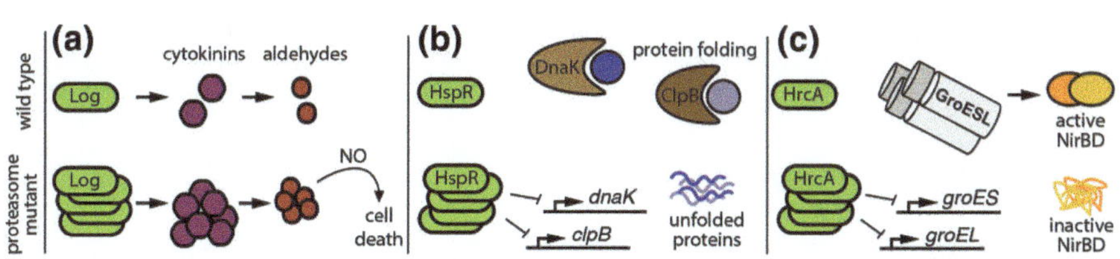

Fig. 3 Regulatory roles of the *M. tuberculosis* proteasome. a Log is a pupylated proteasome substrate that synthesizes cytokinins, which break down into aldehydes. In the absence of proteasomal degradation, accumulation of Log leads to the overproduction of cytokinins, and the ensuing accumulation of aldehydes synergizes with nitric oxide (NO) to kill *M. tuberculosis* (Samanovic et al. 2015). **b** HspR, a repressor of *M. tuberculosis* genes encoding the protein-folding chaperones DnaK and ClpB, is normally degraded by the PafE-proteasome. In the absence of PafE, accumulation of HspR leads to the repression of chaperone genes, which is associated with defective growth and an inability to fully withstand heat shock (Jastrab et al. 2015, 2017). **c** *M. tuberculosis* HrcA represses the genes encoding the GroES/GroEL chaperonins, which are required for the full activity of the nitrite reductase NirBD. It is proposed that failure to pupylate and degrade HrcA prevents full chaperonin expression, leading to a loss of NirBD activity; this prohibits *M. tuberculosis* from using nitrate as a nitrogen source (Becker et al. 2019)

suggests that there are additional proteins whose degradation by the PPS is required for the full pathogenicity of *M. tuberculosis*.

Copper Homeostasis

An early attempt to characterize PPS-dependent gene expression in *M. tuberculosis* identified a previously unrecognized transcriptional repressor, RicR (regulated in copper repressor, Rv0190). RicR represses *M. tuberculosis* genes including the metallothionein-encoding *mymT* as well as *mmcO*, which encodes a copper-detoxifying enzyme (Festa et al. 2011). *mymT* and *mmcO* are each required for the resistance of *M. tuberculosis* to copper in vitro (Shi et al. 2014). While growth of wild type bacteria in copper is associated with de-repression of the RicR regulon, RicR-regulated genes are constitutively repressed in PPS mutants. These observations suggested that RicR itself is a PPS substrate; however, there is currently no evidence that RicR is pupylated or that its levels increase in the absence of proteasomal degradation (Festa et al. 2011). An alternative explanation is that RicR acutely senses changes in the intracellular availability of copper: the accumulation of one or more copper-binding proteins or molecules in a PPS mutant could titrate copper away from RicR, preventing its dissociation from RicR-repressed promoters.

Deletion and disruption of individual RicR-regulated genes does not affect the ability of *M. tuberculosis* to cause lethal infections; however, an *M. tuberculosis* strain with a constitutively active RicR variant is attenuated in mice (Shi et al. 2014). These observations suggest that *M. tuberculosis* copper homeostasis is maintained by a combination of proteins whose levels are controlled by RicR, and that PPS-dependent de-repression of the RicR regulon may be important for the full virulence of this pathogen.

Protein Quality Control

Initial efforts to characterize the function of PafE in *M. tuberculosis* identified HspR (heat shock protein repressor) as a PafE-proteasome substrate (Jastrab et al. 2015). HspR is a transcriptional repressor conserved in many bacteria as a regulator of genes encoding molecular chaperones and proteases (Narberhaus 1999; Roncarati et al. 2007). In *M. tuberculosis*, HspR represses the expression of *dnaKJ, clpB* and *hsp*, which encode chaperones (Stewart et al. 2001, 2002). ClpB, DnaKJ, and Hsp are highly conserved proteins that help disaggregate and refold misfolded proteins that form under conditions such as heat shock (reviewed in Kim et al. 2013). The accumulation of HspR in the absence of PafE-mediated degradation results in reduced expression of *dnaK* and *clpB* in *M. tuberculosis*; accordingly, a *pafE*-mutant has a growth defect and is sensitive to heat shock compared to a parental strain (Fig. 3b) (Jastrab et al. 2015). Perhaps unsurprisingly, point mutations in *hspR* that abolish repressor activity are frequently acquired in a *pafE* strain; analysis of these *hspR* loss-of-function mutations showed that they suppressed the growth defect and heat shock sensitivity caused by disruption of *pafE* (Jastrab et al. 2017).

The observations that the PafE-proteasome can degrade an unfolded protein in vitro (Delley et al. 2014; Jastrab et al. 2015) and that PafE-dependent degradation of HspR is required for the robust production of an essential protein quality control system (Jastrab and Darwin 2015; Jastrab et al. 2017) suggest that the *M. tuberculosis* proteasome has a regulatory function in mitigating proteotoxic stress. This activity may be critical in *M. tuberculosis*, which lacks Lon protease that is important for the degradation of misfolded proteins in *E. coli* (Gottesman and Zipser 1978). It is possible that the PafE-proteasome fulfills some functions of Lon in *M. tuberculosis*. Because PafE-dependent proteolysis does not require ATP, unlike the ClpXP, HslUV, and Lon proteases, the PafE-proteasome may be particularly useful when bacteria are metabolically stressed, a condition that could lead to both ATP depletion and the accumulation of improperly folded proteins.

Nitrogen Metabolism

In eukaryotic cells, amino acid recycling through proteasomal degradation of proteins is essential for life (Suraweera et al. 2012; Vabulas and Hartl 2005). In *Mycobacteria*, amino acids are an optimal source of nitrogen (Gouzy et al. 2013, 2014a, b; Song and Niederweis 2012), and it was thus proposed that bacterial proteasomes are important for nutrient homeostasis by turning over proteins into amino acids. This hypothesis was tested in *M. smegmatis*, where a *prcSBA* strain was found to be deficient in survival during carbon or nitrogen starvation compared to a wild type strain. Importantly, sensitivity of the *prcSBA* mutant to nitrogen limitation is almost fully rescued by the re-introduction of *prcS* into this strain (Elharar et al. 2014), suggesting that while the PPS is important for maintaining nitrogen homeostasis, its role is primarily in the regulation of individual substrates through pupylation, rather than amino acid turnover by proteasomal proteolysis. Alternatively, pupylation may be targeting proteins to another protease as suggested above. In an important distinction, pupylation does not support *M. tuberculosis* survival during nitrogen starvation (Becker et al. 2019), suggesting that *M. smegmatis* and *M. tuberculosis* have different nutrient requirements and may differ in the proteins that are targeted for pupylation or proteasomal degradation.

Supporting the hypothesis that the PPS regulates nitrogen metabolism in *M. smegmatis*, a second study observed that *M. smegmatis* proteins involved in nitrogen transport and assimilation are pupylated during nitrogen starvation. Interestingly, only a subset of these proteins accumulated in a Pup-deficient strain, further supporting a model whereby pupylation provides both degradation-dependent and -independent functions in *M. smegmatis* (Fascellaro et al. 2016).

Finally, recent work has identified a novel role for the PPS in *M. tuberculosis* nitrate metabolism. *M. tuberculosis* mutants lacking an intact PPS cannot use nitrate as a nitrogen source for in vitro growth. A screen to identify proteins whose pupylation and degradation is required for *M. tuberculosis* nitrate metabolism identified the transcriptional repressor HrcA as a putative proteasome substrate (Becker et al. 2019). In *M. tuberculosis*, HrcA represses genes encoding GroES/GroEL chaperonins (Stewart et al. 2002), a class of essential molecular chaperones that facilitate the folding of many bacterial proteins (reviewed in Hayer-Hartl et al. 2016). It was determined that chaperonin production promotes the activity of the nitrite reductase NirBD, an enzyme that is essential for the assimilation of nitrogen from nitrate in *M. tuberculosis*. Chaperonin gene expression is impaired in strains that are unable to degrade pupylated proteins, and a model was proposed that pupylation and degradation of HrcA, leading to de-repression of the chaperonin genes, is required for the full activity of NirBD (Fig. 3c) (Becker et al. 2019). It is unclear whether this novel mechanism of control over *M. tuberculosis* chaperonin production is required during conditions other than growth in nitrate, or if there are additional *M. tuberculosis* proteins whose function strictly depends on robust chaperonin expression.

Remaining Questions

While many aspects of bacterial proteasome systems have been uncovered over the past several decades, important questions remain to be answered. Structural studies of purified proteasome components suggest that in vivo association of activators with 20S CPs may require additional factors. The C-terminus of PafE is twice as long as those found in ATP-independent proteasome activators from eukaryotes, leading to a very weak interaction between PafE and 20S CPs (Bai et al. 2017). Meanwhile, in the case of Mpa/ARC, the C-terminal 20S-interacting peptide appears to be receded inside the Mpa/ARC central chamber, due to the presence of a β-grasp domain that is absent from eukaryotic or archaeal proteasomal ATPase activators (Wu et al. 2017). Accordingly, PafE and Mpa/ARC cannot efficiently activate protein degradation in vitro (Bai et al. 2016; Hu et al. 2018; Wang et al. 2009; Wu et al. 2017). It is possible that post-translational modifications or interactions with additional proteins alter the structure of bacterial proteasome activators or 20S CPs to make them more conducive to binding in vivo.

Another poorly-understood aspect of bacterial proteasomes is the manner in which substrates are selected for pupylation, depupylation, and transpupylation. Because abundant pupylation occurs upon production of PafA and Pup in *E. coli* (Cerda-Maira et al. 2011), it may be that the structures of individual proteins simply makes them inherently more or less prone to pupylation. However, it is possible that bacteria that have pupylation may harbor additional mechanisms of substrate selectivity. Additionally, the breadth of substrate selection by the PafE-proteasome system is currently unknown. Because PafE can facilitate the degradation of unfolded proteins (Delley et al. 2014; Jastrab et al. 2015), the PafE-proteasome may be important for the clearance of un- or mis-folded proteins in bacteria under certain proteotoxic stress conditions.

Lastly, repeated observations that bacteria with an intact PPS only appear to degrade a subset of pupylated proteins under routine culture conditions suggest that there are yet-to-be-discovered mechanisms of control over proteasomal proteolysis. In *M. smegmatis*, along with the observation that the PPS is important for survival during nitrogen starvation (Elharar et al. 2014), it was also shown that the number of proteins with altered abundance in a pupylation mutant compared to a wild type strain was much greater upon total nitrogen starvation than during nitrogen-replete conditions (Fascellaro et al. 2016). Along similar lines in *M. tuberculosis*, the steady-state levels of pupylated proteins is lower when bacteria are grown in nitrate, a relatively poor source of nitrogen for *M. tuberculosis*, than in a more ideal nitrogen source (Becker et al. 2019). These observations together offer an intriguing hypothesis that for *M. smegmatis, M. tuberculosis* and possibly other bacteria, the capacity of the PPS to degrade proteins is determined in some manner by nutrient availability. Therefore, obtaining a better understanding of the precise roles of pupylation and degradation in maintaining nitrogen homeostasis may lead the way towards new regulatory functions mediating proteasome activity.

Acknowledgements Proteasome research in the Darwin lab is supported by NIH grants HL92774 and AI088075 awarded to K. H. D. NIH grant T32AI007180 supported S. H. B. HL is supported by AI070285.

References

Bai L, Hu K, Wang T, Jastrab JM, Darwin KH, Li H (2016) Structural analysis of the dodecameric proteasome activator PafE in Mycobacterium tuberculosis. Proc Natl Acad Sci U S A. https://doi.org/10.1073/pnas.1512094113

Bai L et al (2017) Structural analysis of *Mycobacterium tuberculosis* homologues of the eukaryotic proteasome assembly chaperone 2 (PAC2). J Bacteriol 199. https://doi.org/10.1128/jb.00846-16

Barandun J, Delley CL, Ban N, Weber-Ban E (2013) Crystal structure of the complex between prokaryotic ubiquitin-like protein and its ligase PafA. J Am Chem Soc 135:6794–6797. https://doi.org/10.1021/ja4024012

Becker SH et al (2019) The *Mycobacterium tuberculosis* Pup-proteasome system regulates nitrate metabolism through an essential protein quality control pathway. Proc Natl Acad Sci U S A (In press)

Bolten M, Delley CL, Leibundgut M, Boehringer D, Ban N, Weber-Ban E (2016) Structural analysis of the bacterial proteasome activator Bpa in complex with the 20S proteasome. Structure 24:2138–2151. https://doi.org/10.1016/j.str.2016.10.008

Boubakri H et al. (2015) The absence of pupylation (prokaryotic ubiquitin-like Protein modification) affects morphological and physiological differentiation in *Streptomyces coelicolor*. J Bacteriol 197:3388–3399 https://doi.org/10.1128/jb.00591-15

Brannigan JA, Dodson G, Duggleby HJ, Moody PC, Smith JL, Tomchick DR, Murzin AG (1995) A protein catalytic framework with an N-terminal nucleophile is capable of self-activation. Nature 378:416–419. https://doi.org/10.1038/378416a0

Burns KE et al (2012) Mycobacterium tuberculosis prokaryotic ubiquitin-like protein-deconjugating enzyme is an unusual aspartate amidase. J Biol Chem 287:37522–37529. https://doi.org/10.1074/jbc.M112.384784

Burns KE, Liu WT, Boshoff HI, Dorrestein PC, Barry CE 3rd (2009) Proteasomal protein degradation in Mycobacteria is dependent upon a prokaryotic ubiquitin-like protein. J Biol Chem 284:3069–3075. https://doi.org/10.1074/jbc.M808032200

Burns KE, Cerda-Maira FA, Wang T, Li H, Bishai WR, Darwin KH (2010a) "Depupylation" of prokaryotic ubiquitin-like protein from mycobacterial proteasome substrates. Mol Cell 39:821–827. https://doi.org/10.1016/j.molcel.2010.07.019

Burns KE, Pearce MJ, Darwin KH (2010b) Prokaryotic ubiquitin-like protein provides a two-part degron to Mycobacterium proteasome substrates. J Bacteriol 192:2933–2935. https://doi.org/10.1128/JB.01639-09

Burroughs AM, Balaji S, Iyer LM, Aravind L (2007) Small but versatile: the extraordinary functional and structural diversity of the beta-grasp fold. Biol Direct 2:18. https://doi.org/10.1186/1745-6150-2-18

Cerda-Maira FA, McAllister F, Bode NJ, Burns KE, Gygi SP, Darwin KH (2011) Reconstitution of the Mycobacterium tuberculosis pupylation pathway in *Escherichia coli*. EMBO Rep 12:863–870. https://doi.org/10.1038/embor.2011.109

Cerda-Maira FA, Pearce MJ, Fuortes M, Bishai WR, Hubbard SR, Darwin KH (2010) Molecular analysis of the prokaryotic ubiquitin-like protein (Pup) conjugation pathway in *Mycobacterium tuberculosis*. Mol Microbiol 77:1123–1135. https://doi.org/10.1111/j.1365-2958.2010.07276.x

Chen X, Li C, Wang L, Liu Y, Li C, Zhang J (2016) The mechanism of mycobacterium smegmatis PafA self-pupylation. PLoS ONE 11:e0151021. https://doi.org/10.1371/journal.pone.0151021

Chen X, Solomon WC, Kang Y, Cerda-Maira F, Darwin KH, Walters KJ (2009) Prokaryotic ubiquitin-like protein pup is intrinsically disordered. J Mol Biol 392:208–217. https://doi.org/10.1016/j.jmb.2009.07.018

Choi WH et al (2016) Open-gate mutants of the mammalian proteasome show enhanced ubiquitin-conjugate degradation. Nat Commun 7:10963. https://doi.org/10.1038/ncomms10963

Compton CL, Fernandopulle MS, Nagari RT, Sello JK (2015) Genetic and proteomic analyses of pupylation in *Streptomyces coelicolor*. J Bacteriol 197:2747–2753. https://doi.org/10.1128/JB.00302-15

Darwin KH, Ehrt S, Gutierrez-Ramos JC, Weich N, Nathan CF (2003) The proteasome of *Mycobacterium tuberculosis* is required for resistance to nitric oxide. Science 302:1963–1966. https://doi.org/10.1126/science.1091176

Darwin KH, Lin G, Chen Z, Li H, Nathan CF (2005) Characterization of a *Mycobacterium tuberculosis* proteasomal ATPase homologue. Mol Microbiol 55:561–571. https://doi.org/10.1111/j.1365-2958.2004.04403.x

Delley CL, Laederach J, Ziemski M, Bolten M, Boehringer D, Weber-Ban E (2014) Bacterial proteasome activator bpa (rv3780) is a novel ring-shaped interactor of the mycobacterial proteasome. PLoS One 9:e114348. https://doi.org/10.1371/journal.pone.0114348

Djuranovic S et al (2009) Structure and activity of the N-terminal substrate recognition domains in proteasomal ATPases. Mol Cell 34:580–590. https://doi.org/10.1016/j.molcel.2009.04.030

Elharar Y et al (2014) Survival of mycobacteria depends on proteasome-mediated amino acid recycling under nutrient limitation. EMBO J 33:1802–1814. https://doi.org/10.15252/embj.201387076

Erzberger JP, Berger JM (2006) Evolutionary relationships and structural mechanisms of AAA+ proteins. Annu Rev Biophys Biomol Struct 35:93–114. https://doi.org/10.1146/annurev.biophys.35.040405.101933

Fascellaro G et al (2016) Comprehensive proteomic analysis of nitrogen-starved *Mycobacterium smegmatis* deltapup reveals the impact of pupylation on nitrogen stress response. J Proteome Res 15:2812–2825. https://doi.org/10.1021/acs.jproteome.6b00378

Festa RA et al (2011) A novel copper-responsive regulon in *Mycobacterium tuberculosis*. Mol Microbiol 79:133–148. https://doi.org/10.1111/j.1365-2958.2010.07431.x

Festa RA, McAllister F, Pearce MJ, Mintseris J, Burns KE, Gygi SP, Darwin KH (2010) Prokaryotic ubiquitin-like protein (Pup) proteome of *Mycobacterium tuberculosis* [corrected]. PLoS One 5:e8589 https://doi.org/10.1371/journal.pone.0008589

Finley D, Chen X, Walters KJ (2016) Gates, channels, and switches: elements of the proteasome machine. Trends Biochem Sci 41:77–93. https://doi.org/10.1016/j.tibs.2015.10.009

Gandotra S, Lebron MB, Ehrt S (2010) The *Mycobacterium tuberculosis* proteasome active site threonine is essential for persistence yet dispensable for replication and resistance to nitric oxide. PLoS Pathog 6:e1001040. https://doi.org/10.1371/journal.ppat.1001040

Gottesman S, Zipser D (1978) Deg phenotype of *Escherichia coli* lon mutants. J Bacteriol 133:844–851

Gouzy A et al (2013) *Mycobacterium tuberculosis* nitrogen assimilation and host colonization require aspartate. Nat Chem Biol 9:674–676. https://doi.org/10.1038/nchembio.1355

Gouzy A et al (2014a) *Mycobacterium tuberculosis* exploits asparagine to assimilate nitrogen and resist acid stress during infection. PLoS Pathog 10:e1003928. https://doi.org/10.1371/journal.ppat.1003928

Gouzy A, Poquet Y, Neyrolles O (2014b) Nitrogen metabolism in *Mycobacterium tuberculosis* physiology and virulence, vol 12. https://doi.org/10.1038/nrmicro3349

Groll M et al (2000) A gated channel into the proteasome core particle. Nat Struct Biol 7:1062–1067. https://doi.org/10.1038/80992

Groll M, Brandstetter H, Bartunik H, Bourenkow G, Huber R (2003) Investigations on the maturation and regulation of archaebacterial proteasomes. J Mol Biol 327:75–83

Groll M, Ditzel L, Lowe J, Stock D, Bochtler M, Bartunik HD, Huber R (1997) Structure of 20S proteasome from yeast at 2.4. A resolution. Nature 386:463–471. https://doi.org/10.1038/386463a0

Guth E, Thommen M, Weber-Ban E (2011) Mycobacterial ubiquitin-like protein ligase PafA follows a two-step reaction pathway with a phosphorylated pup intermediate. J Biol Chem 286:4412–4419. https://doi.org/10.1074/jbc.M110.189282

Hayer-Hartl M, Bracher A, Hartl FU (2016) The GroEL-GroES chaperonin machine: a nano-cage for protein folding. Trends Biochem Sci 41:62–76. https://doi.org/10.1016/j.tibs.2015.07.009

Heinemeyer W, Fischer M, Krimmer T, Stachon U, Wolf DH (1997) The active sites of the eukaryotic 20 S proteasome and their involvement in subunit precursor processing. J Biol Chem 272:25200–25209

Hu K, Jastrab JB, Zhang S, Kovach A, Zhao G, Darwin KH, Li H (2018) Proteasome substrate capture and gate opening by the accessory factor PafE from *Mycobacterium tuberculosis*. J Biol Chem 293:4713–4723. https://doi.org/10.1074/jbc.RA117.001471

Hu G, Lin G, Wang M, Dick L, Xu RM, Nathan C, Li H (2006) Structure of the *Mycobacterium tuberculosis* proteasome and mechanism of inhibition by a peptidyl boronate. Mol Microbiol 59:1417–1428. https://doi.org/10.1111/j.1365-2958.2005.05036.x

Huber EM, Heinemeyer W, Li X, Arendt CS, Hochstrasser M, Groll M (2016) A unified mechanism for proteolysis and autocatalytic activation in the 20S proteasome. Nat Commun 7:10900. https://doi.org/10.1038/ncomms10900

Imkamp F, Rosenberger T, Striebel F, Keller PM, Amstutz B, Sander P, Weber-Ban E (2010a) Deletion of dop in *Mycobacterium smegmatis* abolishes pupylation of protein substrates in vivo. Mol Microbiol 75:744–754. https://doi.org/10.1111/j.1365-2958.2009.07013.x

Imkamp F, Striebel F, Sutter M, Ozcelik D, Zimmermann N, Sander P, Weber-Ban E (2010b) Dop functions as a depupylase in the prokaryotic ubiquitin-like modification pathway. EMBO Rep 11:791–797. https://doi.org/10.1038/embor.2010.119

Jastrab JB et al (2015) An adenosine triphosphate-independent proteasome activator contributes to the virulence of *Mycobacterium tuberculosis*. Proc Natl Acad Sci U S A 112:1763–1772. https://doi.org/10.1073/pnas.1423319112

Jastrab JB, Darwin KH (2015) Bacterial proteasomes. Annu Rev Microbiol 69:109–127. https://doi.org/10.1146/annurev-micro-091014-104201

Jastrab JB, Samanovic MI, Copin R, Shopsin B, Darwin KH (2017) Loss-of-function mutations in HspR rescue the growth defect of a *Mycobacterium tuberculosis* proteasome accessory factor E (pafE) Mutant. J Bacteriol 199. https://doi.org/10.1128/jb.00850-16

Kim YE, Hipp MS, Bracher A, Hayer-Hartl M, Hartl FU (2013) Molecular chaperone functions in protein folding and proteostasis. Annu Rev Biochem 82:323–355. https://doi.org/10.1146/annurev-biochem-060208-092442

Kim YC, Snoberger A, Schupp J, Smith DM (2015) ATP binding to neighbouring subunits and intersubunit allosteric coupling underlie proteasomal ATPase function. Nat Commun 6:8520. https://doi.org/10.1038/ncomms9520

Kuberl A, Franzel B, Eggeling L, Polen T, Wolters DA, Bott M (2014) Pupylated proteins in *Corynebacterium glutamicum* revealed by MudPIT analysis. Proteomics 14:1531–1542. https://doi.org/10.1002/pmic.201300531

Kuberl A, Polen T, Bott M (2016) The pupylation machinery is involved in iron homeostasis by targeting the iron storage protein ferritin. Proc Natl Acad Sci U S A 113:4806–4811. https://doi.org/10.1073/pnas.1514529113

Kurakawa T et al (2007) Direct control of shoot meristem activity by a cytokinin-activating enzyme. Nature 445:652–655. https://doi.org/10.1038/nature05504

Le Tallec B, Barrault MB, Courbeyrette R, Guerois R, Marsolier-Kergoat MC, Peyroche A (2007) 20S proteasome assembly is orchestrated by two distinct pairs of chaperones in yeast and in mammals. Mol Cell 27:660–674. https://doi.org/10.1016/j.molcel.2007.06.025

Lehmann G, Udasin RG, Livneh I, Ciechanover A (2017) Identification of UBact, a ubiquitin-like protein, along with other homologous components of a conjugation system and the proteasome in different gram-negative bacteria. Biochem Biophys Res Commun 483:946–950. https://doi.org/10.1016/j.bbrc.2017.01.037

Li D, Li H, Wang T, Pan H, Lin G, Li H (2010) Structural basis for the assembly and gate closure mechanisms of the Mycobacterium tuberculosis 20S proteasome. EMBO J 29:2037–2047. https://doi.org/10.1038/emboj.2010.95

Lin G et al (2006) *Mycobacterium tuberculosis* prcBA genes encode a gated proteasome with broad oligopeptide specificity. Mol Microbiol 59:1405–1416. https://doi.org/10.1111/j.1365-2958.2005.05035.x

Lowe J, Stock D, Jap B, Zwickl P, Baumeister W, Huber R (1995) Crystal structure of the 20S proteasome from the archaeon *T. acidophilum* at 3.4 A resolution. Science 268:533–539

Lupas A, Zwickl P, Baumeister W (1994) Proteasome sequences in eubacteria. Trends Biochem Sci 19:533–534

MacMicking JD, North RJ, LaCourse R, Mudgett JS, Shah SK, Nathan CF (1997) Identification of nitric oxide synthase as a protective locus against tuberculosis. Proc Natl Acad Sci U S A 94:5243–5248

Maldonado AY, Burz DS, Reverdatto S, Shekhtman A (2013) Fate of pup inside the *Mycobacterium* proteasome studied by in-cell NMR. PLoS ONE 8:e74576. https://doi.org/10.1371/journal.pone.0074576

Martin A, Baker TA, Sauer RT (2008) Pore loops of the AAA+ ClpX machine grip substrates to drive translocation and unfolding. Nat Struct Mol Biol 15:1147–1151. https://doi.org/10.1038/nsmb.1503

Mc Cormack T et al (1997) Active site-directed inhibitors of Rhodococcus 20 S proteasome. Kinetics and mechanism. J Biol Chem 272:26103–26109

De Mot R (2007) Actinomycete-like proteasomes in a Gram-negative bacterium. Trends Microbiol 15:335–338. https://doi.org/10.1016/j.tim.2007.06.002

Murzin AG (1993) OB(oligonucleotide/oligosaccharide binding)-fold: common structural and functional solution for non-homologous sequences. EMBO J 12:861–867

Nagy I, Tamura T, Vanderleyden J, Baumeister W, De Mot R (1998) The 20S proteasome of Streptomyces coelicolor. J Bacteriol 180:5448–5453

Narberhaus F (1999) Negative regulation of bacterial heat shock genes. Mol Microbiol 31:1–8

Ozcelik D et al (2012) Structures of Pup ligase PafA and depupylase Dop from the prokaryotic ubiquitin-like modification pathway. Nat Commun 3:1014. https://doi.org/10.1038/ncomms2009

Pearce MJ, Arora P, Festa RA, Butler-Wu SM, Gokhale RS, Darwin KH (2006) Identification of substrates of the *Mycobacterium tuberculosis* proteasome. EMBO J 25:5423–5432. https://doi.org/10.1038/sj.emboj.7601405

Pearce MJ, Mintseris J, Ferreyra J, Gygi SP, Darwin KH (2008) Ubiquitin-like protein involved in the proteasome pathway of *Mycobacterium tuberculosis*. Science 322:1104–1107. https://doi.org/10.1126/science.1163885

Pouch MN, Cournoyer B, Baumeister W (2000) Characterization of the 20S proteasome from the actinomycete. Frankia Mol Microbiol 35:368–377

Poulsen C et al (2010) Proteome-wide identification of mycobacterial pupylation targets. Mol Syst Biol 6:386. https://doi.org/10.1038/msb.2010.39

Rabl J, Smith DM, Yu Y, Chang SC, Goldberg AL, Cheng Y (2008) Mechanism of gate opening in the 20S proteasome by the proteasomal ATPases. Mol Cell 30:360–368. https://doi.org/10.1016/j.molcel.2008.03.004

Ramos PC, Hockendorff J, Johnson ES, Varshavsky A, Dohmen RJ (1998) Ump1p is required for proper maturation of the 20S proteasome and becomes its substrate upon completion of the assembly. Cell 92:489–499

Regev O, Korman M, Hecht N, Roth Z, Forer N, Zarivach R, Gur E (2016) An extended loop of the Pup ligase, PafA, mediates interaction with protein targets. J Mol Biol 428:4143–4153. https://doi.org/10.1016/j.jmb.2016.07.021

Regev O, Roth Z, Korman M, Khalaila I, Gur E (2015) A kinetic model for the prevalence of mono- over poly-pupylation. FEBS J 282:4176–4186. https://doi.org/10.1111/febs.13413

Roncarati D, Danielli A, Spohn G, Delany I, Scarlato V (2007) Transcriptional regulation of stress response and motility functions in Helicobacter pylori is mediated by HspR and HrcA. J Bacteriol 189:7234–7243. https://doi.org/10.1128/JB.00626-07

Samanovic MI et al (2015) Proteasomal control of cytokinin synthesis protects Mycobacterium tuberculosis against nitric oxide. Mol Cell 57:984–994. https://doi.org/10.1016/j.molcel.2015.01.024

Sauer RT, Baker TA (2011) AAA+ proteases: ATP-fueled machines of protein destruction. Annu Rev Biochem 80:587–612. https://doi.org/10.1146/annurev-biochem-060408-172623

Shi X, Festa RA, Ioerger TR, Butler-Wu S, Sacchettini JC, Darwin KH, Samanovic MI (2014) The copper-responsive RicR regulon contributes to *Mycobacterium tuberculosis* virulence. MBio 5 https://doi.org/10.1128/mbio.00876-13

Smith DM, Chang SC, Park S, Finley D, Cheng Y, Goldberg AL (2007) Docking of the proteasomal ATPases' carboxyl termini in the 20S proteasome's alpha ring opens the gate for substrate entry. Mol Cell 27:731–744. https://doi.org/10.1016/j.molcel.2007.06.033

Song H, Niederweis M (2012) Uptake of sulfate but not phosphate by *Mycobacterium tuberculosis* is slower than that for *Mycobacterium smegmatis*. J Bacteriol 194:956–964. https://doi.org/10.1128/JB.06132-11

Stewart GR et al (2001) Overexpression of heat-shock proteins reduces survival of *Mycobacterium tuberculosis* in the chronic phase of infection. Nat Med 7:732–737. https://doi.org/10.1038/89113

Stewart GR et al (2002) Dissection of the heat-shock response in *Mycobacterium tuberculosis* using mutants and microarrays. Microbiology 148:3129–3138. https://doi.org/10.1099/00221287-148-10-3129

Streich FC Jr, Lima CD (2014) Structural and functional insights to ubiquitin-like protein conjugation. Annu Rev Biophys 43:357–379. https://doi.org/10.1146/annurev-biophys-051013-022958

Striebel F, Hunkeler M, Summer H, Weber-Ban E (2010) The mycobacterial Mpa-proteasome unfolds and degrades pupylated substrates by engaging Pup's N-terminus. EMBO J 29:1262–1271. https://doi.org/10.1038/emboj.2010.23

Striebel F, Imkamp F, Sutter M, Steiner M, Mamedov A, Weber-Ban E (2009a) Bacterial ubiquitin-like modifier Pup is deamidated and conjugated to substrates by distinct but homologous enzymes. Nat Struct Mol Biol 16:647–651. https://doi.org/10.1038/nsmb.1597

Striebel F, Kress W, Weber-Ban E (2009b) Controlled destruction: AAA+ ATPases in protein degradation from bacteria to eukaryotes. Curr Opin Struct Biol 19:209–217. https://doi.org/10.1016/j.sbi.2009.02.006

Suraweera A, Munch C, Hanssum A, Bertolotti A (2012) Failure of amino acid homeostasis causes cell death following proteasome inhibition. Mol Cell 48:242–253. https://doi.org/10.1016/j.molcel.2012.08.003

Sutter M, Striebel F, Damberger FF, Allain FH, Weber-Ban E (2009) A distinct structural region of the prokaryotic ubiquitin-like protein (Pup) is recognized by the N-terminal domain of the proteasomal ATPase Mpa. FEBS Lett 583:3151–3157. https://doi.org/10.1016/j.febslet.2009.09.020

Tamura T et al (1995) The first characterization of a eubacterial proteasome: the 20S complex of *Rhodococcus*. Curr Biol 5:766–774

Thrower JS, Hoffman L, Rechsteiner M, Pickart CM (2000) Recognition of the polyubiquitin proteolytic signal. EMBO J 19:94–102. https://doi.org/10.1093/emboj/19.1.94

Vabulas RM, Hartl FU (2005) Protein synthesis upon acute nutrient restriction relies on proteasome function. Science 310:1960–1963. https://doi.org/10.1126/science.1121925

Wang T et al (2009) Structural insights on the *Mycobacterium tuberculosis* proteasomal ATPase Mpa. Structure 17:1377–1385. https://doi.org/10.1016/j.str.2009.08.010

Wang T, Darwin KH, Li H (2010) Binding-induced folding of prokaryotic ubiquitin-like protein on the *Mycobacterium* proteasomal ATPase targets substrates for degradation. Nat Struct Mol Biol 17:1352–1357. https://doi.org/10.1038/nsmb.1918

Watrous J et al (2010) Expansion of the mycobacterial "PUPylome". Mol Biosyst 6:376–385. https://doi.org/10.1039/b916104j

Wolf S et al (1998) Characterization of ARC, a divergent member of the AAA ATPase family from *Rhodococcus erythropolis*. J Mol Biol 277:13–25. https://doi.org/10.1006/jmbi.1997.1589

Wu Y et al (2017) *Mycobacterium tuberculosis* proteasomal ATPase Mpa has a beta-grasp domain that hinders docking with the proteasome core protease. Mol Microbiol 105:227–241. https://doi.org/10.1111/mmi.13695

Yu Y, Smith DM, Kim HM, Rodriguez V, Goldberg AL, Cheng Y (2010) Interactions of PAN's C-termini with archaeal 20S proteasome and implications for the eukaryotic proteasome-ATPase interactions. EMBO J 29:692–702. https://doi.org/10.1038/emboj.2009.382

Zhang S, Burns-Huang KE, Janssen GV, Li H, Ovaa H, Hedstrom L, Darwin KH (2017) *Mycobacterium tuberculosis* proteasome Accessory factor A (PafA) can transfer prokaryotic ubiquitin-like protein (Pup) between substrates. MBio 8. https://doi.org/10.1128/mbio.00122-17

Ziemski M, Jomaa A, Mayer D, Rutz S, Giese C, Veprintsev D, Weber-Ban E (2018) Cdc48-like protein of actinobacteria (Cpa) is a novel proteasome interactor in mycobacteria and related organisms. Elife 7. https://doi.org/10.7554/elife.34055

Zuhl F, Seemuller E, Golbik R, Baumeister W (1997) Dissecting the assembly pathway of the 20S proteasome. FEBS Lett 418:189–194

Zwickl P, Kleinz J, Baumeister W (1994) Critical elements in proteasome assembly. Nat Struct Biol 1:765–770

Chapter 12
The Kai-Protein Clock—Keeping Track of Cyanobacteria's Daily Life

Joost Snijder and Ilka Maria Axmann

Abstract Life has adapted to Earth's day-night cycle with the evolution of endogenous biological clocks. Whereas these circadian rhythms typically involve extensive transcription-translation feedback in higher organisms, cyanobacteria have a circadian clock, which functions primarily as a protein-based post-translational oscillator. Known as the Kai system, it consists of three proteins KaiA, KaiB, and KaiC. In this chapter, we provide a detailed structural overview of the Kai components and how they interact to produce circadian rhythms of global gene expression in cyanobacterial cells. We discuss how the circadian oscillation is coupled to gene expression, intertwined with transcription-translation feedback mechanisms, and entrained by input from the environment. We discuss the use of mathematical models and summarize insights into the cyanobacterial circadian clock from theoretical studies. The molecular details of the Kai system are well documented for the model cyanobacterium *Synechococcus elongatus*, but many less understood varieties of the Kai system exist across the highly diverse phylum of Cyanobacteria. Several species contain multiple *kai*-gene copies, while others like marine *Prochlorococcus* strains have a reduced *kaiBC*-only system, lacking *kaiA*. We highlight recent findings on the genomic distribution of *kai* genes in Bacteria and Archaea and finally discuss hypotheses on the evolution of the Kai system.

Keywords Circadian clock · Biochemical oscillator · KaiC · KaiB · KaiA · PTO · TTFL · Cyanobacteria · *Synechococcus elongatus* · *Prochlorococcus*

J. Snijder
Snijder Bioscience, Zevenwouden 143, 3524CN Utrecht, The Netherlands
e-mail: j.snijder@uu.nl

Biomolecular Mass Spectrometry and Proteomics, Bijvoet Center for Biomolecular Research, Utrecht University, Padualaan 8, 3584 CH Utrecht, The Netherlands

I. M. Axmann (✉)
Synthetic Microbiology, Biology Department, Heinrich Heine University Düsseldorf, Universitätsstraße 1, 40225 Düsseldorf, Germany
e-mail: ilka.axmann@hhu.de

J. R. Harris and J. Marles-Wright (eds.), *Macromolecular Protein Complexes II: Structure and Function*, Subcellular Biochemistry 93,
https://doi.org/10.1007/978-3-030-28151-9_12

Circadian Clocks

We have an intuitive notion of daytime and organize our daily lives accordingly, even without an exact definition or external measure of time. How do we do this? Our body, down to our individual cells, has inner clocks that enable all our rhythmic biological activities like sleeping, eating, body temperature, hormone levels … thinking.

Not only human life is adapted to the rotation of our planet Earth. Almost all living beings including animals, plants, and fungi possess an inner clock—a circadian clock that is defined by a 24-h period. The naming *circadian* originates from the Latin words *circa* meaning "around" and *dies* meaning "day". A fascinating feature of this biological clock is that it ticks very robustly despite daily changes in temperature, light, and nutrient availability. However, the inner clock can be entrained by exactly these cues, temperature, light, and nutrients. For example, when we travel across time zones, our inner clock stays robustly with the 'old daytime' and we experience jetlag. After a few days our body clock adapts to the new environmental light-dark cycle and jetlag is gone. This example also illustrates very clearly that our well being is strongly affected by the synchrony of our body's clock with environmental cues. When there is a mismatch between our inner biological clock and the external environment, like in the case of jetlag, we don't feel well due to daytime fatigue and sleep disturbances. Moreover, chronic misalignment between our lifestyle and the rhythm dictated by our inner clock is associated with increased risk for various diseases ranging from mental disorders, cardiovascular and metabolic diseases, to cancer (Roenneberg and Merrow 2016).

Since the discovery of the genetic code and its information flow from DNA to RNA to proteins, the idea raised that there might exist a closing loop from the protein expressed in a cell back to its gene encoded on the DNA. Indeed, such a transcription-translation feedback loop, or TTFL, was found for the first-ever discovered circadian gene in fruit flies, named *period*, encoding the PERIOD protein. PER proteins prevent their own synthesis resulting in continuous, 24-h oscillations of PER protein levels in each single *Drosophil*a cell (Hardin et al. 1990; Vosshall et al. 1994). For their discoveries of molecular mechanisms controlling the circadian rhythm, Jeffrey C. Hall, Michael Rosbash, and Michael W. Young were jointly awarded with the 2017 Nobel Prize in Physiology or Medicine. This key mechanistic principle for the biological clock applies not only to humans but to all circadian clocks known so far.

How to test for a true circadian clock? In order to demonstrate an endogenous source for 24-h rhythmicity, the organism of interest is placed in a constant environment, for example constant darkness, excluding any influence of external stimuli. Now, behavior, physiology, gene expression, and any other relevant feature, which is expected to be under the influence of a free-running clock, can be monitored over time. Finally, if a 24 h free-running rhythm is found, it has to be robustly detected under altered external parameters. On the one hand, the rhythm should be resettable by exposure to external stimuli such as light and temperature, a process named entrainment. The external stimulus that is able to entrain a rhythm is called the Zeitgeber. On the other hand, the 24-h rhythm has to maintain over a range of physiological temperatures despite the changing kinetics, a feature termed temperature compensation. In the case that all three criteria are fulfilled: endogenous free-running 24-h period, entrainment, and temperature-compensation, a true circadian clock exists.

Until the 1980s, circadian rhythms were believed to be restricted to eukaryotes, even though there was already strong evidence that cyanobacteria also schedule their physiological behavior into daily cycles (Wyatt and Silvey 1969; Gallon et al. 1974; Weare and Benemann 1974; Millineaux et al. 1981). Cyanobacteria are widespread in marine, freshwater and terrestrial environments, and many of them are capable of fixing atmospheric nitrogen making a major contribution to the global nitrogen cycle. Like Plants, Cyanobacteria depend on light to perform photosynthesis during the day, a process that provides the primary source of energy for almost all forms of life on Earth. Due to their capability of using light as the only energy source and CO_2 as feedstock, Cyanobacteria attract growing attention for the production of clean and sustainable energy and other valuable products. As a by-product of the photosynthetic light reaction Cyanobacteria produce oxygen, which interferes with certain biological processes like oxygen-sensitive nitrogen fixation. Cyanobacteria have solved this problem by separating the interfering processes in space and/or in time. For example, several filamentous cyanobacteria are able to differentiate specialized cells, named heterocysts that do not evolve oxygen and, thus, are able to fix molecular nitrogen. Unicellular Cyanobacteria, like those drawn in Fig. 12.1, usually schedule nitrogen fixation at night when oxygen is not being produced by photosynthesis. Here, it is important to note that the circadian model cyanobacterium *Synechococcus elongatus* sp. PCC 7942 (*Synechococcus elongatus*) is not able to fix dinitrogen. However, even under continuous illumination, a temporal separation of disparate processes persists providing the first strong evidence for an internal timing system—a circadian clock.

In the 1980s, several experiments cultivating cyanobacteria under constant light demonstrated 24-h rhythms of not only nitrogen fixation and photosynthesis, but also cell division at different temperatures (Grobbelaar et al. 1986; Stal and Krumbein 1987; Sweeney and Borgese 1989). Thus, cyanobacteria satisfy all criteria of a temperature-compensated circadian clock, although many species are unicellular and can divide faster than once a day (Johnson et al. 1996; Kondo et al. 1997). Today we know that nearly all metabolic activity in a cyanobacterial cell is orchestrated by a circadian clock to overcome simultaneous occurrence of incompatible processes. Obviously, even simple cyanobacteria have evolved an inner timing system to foresee the accompanying daily changes of light and temperature and regulate their physiology in daily cycles (Beck et al. 2014; Cerveny and Nedbal 2009; Diamond et al. 2015; Saha et al. 2016). Thus, the added efficiency provides an important benefit to life, and circadian clocks can be found in all domains of life.

The molecular mechanism underlying a circadian clock in Cyanobacteria remained elusive for another 20 years. First attempts to understand the molecular details of the cyanobacterial clock pointed to a cluster of three genes, *kaiA*, *kaiB* and

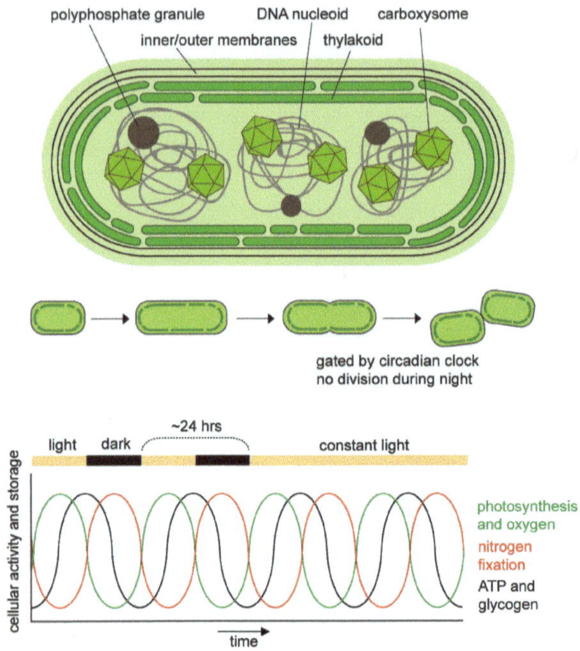

Fig. 12.1 A day in the life of a cyanobacterium. Schematic cellular structure of a single-celled cyanobacterium, like the genus *Synechococcus* (*top*). Schematic of cell division (*middle*). Physiological activities of a cyanobacterial cell under light-dark cycles, followed by constant light (*bottom*). Shown are photosynthesis, oxygen production, cellular ATP levels, glycogen storage, and nitrogen fixation. Please note that the circadian pattern of nitrogen fixation only applies to strains that fix dinitrogen from the environment. Circadian activities persist under constant light: a hallmark of circadian clocks

kaiC (Ishiura et al. 1998). In agreement with the well-established eukaryotic model, a TTFL was suggested for cyanobacteria. Negative feedback control of *kaiC* expression by KaiC itself was suggested to generate the circadian oscillation. Almost 10 years later, in 2005, the TTFL model for cyanobacteria was completely revised: firstly, it was demonstrated that circadian oscillations continue without transcription and translation in darkness (Tomita et al. 2005). Secondly, a very elegant in vitro experiment mixing the three purified recombinant Kai proteins, KaiA, KaiB, and KaiC, together with ATP unambiguously illustrated beautiful 24-h oscillations of KaiC phosphorylation over several days at different temperatures in a test tube (Nakajima et al. 2005). Thus, the underlying oscillatory mechanism in cyanobacteria relies on protein-phosphorylation and is evolutionary not related to transcriptional-translational based circadian systems prevailing in human, animals, plants, and other eukaryotes. Cyanobacteria's biochemical, post-translational oscillator is instead made of only three proteins: KaiA, KaiB, and KaiC, which share no sequence similarity to any known eukaryotic clock component and are sufficient to generate temperature-compensated circadian oscillations of protein phosphorylation in vivo and even in vitro (Nakajima et al. 2005).

In the following chapter, we provide insights into the Kai-protein system of *Synechococcus elongatus*—the best-studied circadian clock of Cyanobacteria, Sect. 2. We illustrate structural details of the three Kai proteins, KaiC, KaiB, and KaiA, Sect. 3. In Sect. 4, we explain how the time signal is transferred by protein factors from the protein clock to the DNA, orchestrating global patterns of gene expression, and how the clock receives input from the environment. The cyanobacterial protein clock is intertwined with several feedback mechanisms that allow for robust 24-h behavior. However, the clock can be entrained by environmental signals. Mathematical models allow us to describe and understand this complex network, and we summarize insights from theoretical studies of the cyanobacterial system, Sect. 5. Although the unique three-protein oscillator is well documented for the model cyanobacterium *Synechococcus elongatus*, information is largely missing for the highly diverse phylum of Cyanobacteria. Several species contain multiple *kai*-gene copies, others like marine *Prochlorococcus* strains harbor only a reduced *kaiBC* system, thus, lacking the *kaiA* gene. We highlight recent findings on the genomic distribution of *kai* genes in Bacteria and Archaea. Finally, we discuss hypotheses on the evolution of the Kai-protein system, Sect. 6.

The Kai-Protein System of *Synechococcus Elongatus*

The circadian clock of cyanobacteria drives expression of most of the cell's gene products, including the clock genes themselves (Guerreiro et al. 2014; Ito et al. 2009; Markson et al. 2013; Xu et al. 2013). Dozens of those gene products are directly involved in sensing light, nutrients, or temperature and relay their input to get the biological clock more in tune with the environment. Transcription and translation feedback on the clock genes is therefore an important part of maintaining a robust circadian rhythm, but at its core the biological clock of cyanobacteria is a purely post-translational oscillator (Axmann et al. 2014; Hosokawa et al. 2013; Kitayama et al. 2008; Qin et al. 2010b; Teng et al. 2013; Tomita et al. 2005). It is known as the Kai system, derived from the Japanese *kai*, meaning 'cycle'.

The oscillator consists of just three gene products: KaiA, KaiB, and KaiC derived from the *kaiA* gene and a *kaiBC* operon (Ishiura et al. 1998). The three components of the Kai system go through a twenty-four-hour cycle of assembly-disassembly and post-translational modifications that determines the phase of the circadian rhythm (Kageyama et al. 2003; Kitayama et al. 2003; Nishiwaki et al. 2004; Xu et al. 2004). With only three components the Kai system is remarkably simple. So simple in fact, that it can be reconstituted in a test tube, just by mixing the purified Kai proteins in the presence of ATP and magnesium ions (Nakajima et al. 2005). Even outside the cell, in the test tube, the purified Kai proteins will run through their twenty-four-hour cycle of assembly and post-translational modifications (see Fig. 12.2). The cyclic reaction will persist for weeks, provided that there is enough ATP. This test-tube oscillator presents an amazing opportunity to study the molecular details of biological timekeeping in cyanobacteria and it has taught us much of what we know about the Kai system today.

KaiC is the central component of the oscillator. It is composed of two internally duplicated domains: KaiC-CI at the N-terminus and KaiC-CII at the C-terminus. Both domains bind ATP (Hayashi et al. 2003, 2004, 2006; Mori et al. 2002; Nishiwaki et al. 2000; Pattanayek et al. 2004). Binding of ATP is required for the domains to form hexamers: one ring of KaiC-CI and another of KaiC-CII. The rings

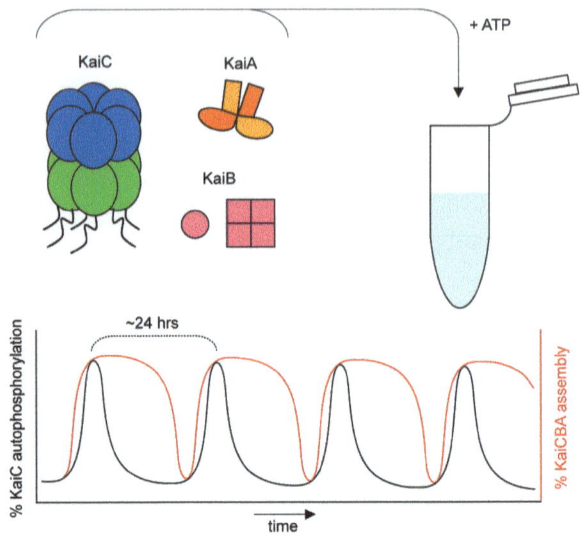

Fig. 12.2 A circadian clock in a test-tube. Kondo and colleagues demonstrated in 2005 that the KaiABC system can be reconstituted in vitro from the purified recombinant protein products. Mixing KaiC, KaiB, and KaiA in the presence of magnesium and ATP produces a temperature-compensated circadian oscillation of KaiC autophosphorylation and KaiCBA complex assembly. This circadian clock in a test-tube has since become an invaluable tool to researchers across the globe studying the molecular details of Kai

are stacked on top of each other and connected by flexible linkers between the two domains of each subunit. The KaiC-CII domain has autokinase and autophosphatase activity (Iwasaki et al. 2002; Nishiwaki et al. 2000, 2004; Pattanayek et al. 2004; Williams et al. 2002; Xu et al. 2004). It sequentially phosphorylates a Threonine (T432), then a Serine residue (S431) across the interface of the subunits, followed by sequential dephosphorylation of the Threonine and Serine residues in the same order (Hayashi et al. 2006; Kitayama et al. 2013; Nishiwaki et al. 2007; Rust et al. 2007). The resulting levels of autophosphorylation set the phase of the circadian oscillation.

The KaiC-CI domain has an ATPase activity that tracks the circadian rhythm of KaiC-CII autophosphorylation (Abe et al. 2015; Murakami et al. 2008; Terauchi et al. 2007). This is mediated by the stacking interactions between the two rings (Chang et al. 2011, 2012; Oyama et al. 2018). Whereas the KaiC-CI domain always forms a tight hexamer in the presence of ATP, the KaiC-CII domains are more loosely associated and interact strongest when S431 is phosphorylated. As a result, the KaiC-CII domains form a tighter hexamer during one-half of the phosphorylation cycle, during which the two rings can also stack together more efficiently to promote the ATPase activity in the KaiC-CI domains. Together, phosphorylation and ATPase activity of KaiC determine how the hexamer interacts with the KaiA and KaiB components. In turn, the interactions with KaiA and KaiB modulate both the autophosphorylation and ATPase activities of the KaiC hexamer. This feedback between the enzymatic activities of KaiC and the interactions with KaiA and KaiB ultimately drive the circadian oscillations of the system (see Fig. 12.3) (Brettschneider et al. 2010; Clodong et al. 2007; Iwasaki et al. 1999; Kageyama et al. 2003; Kitayama et al. 2003; Nishiwaki et al. 2004; Qin et al. 2010a; Rust et al. 2007; Murakami et al. 2016).

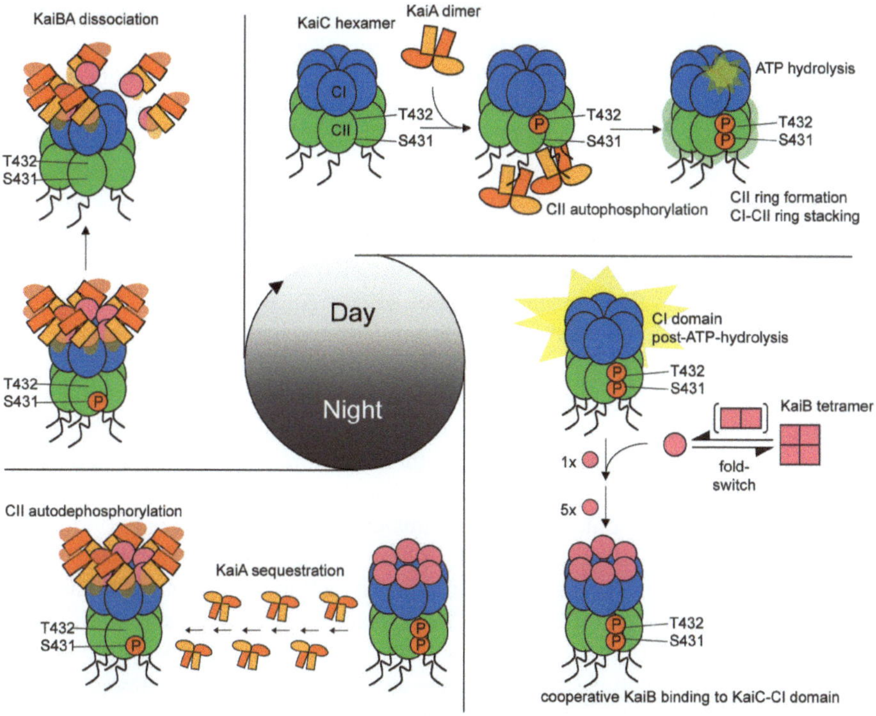

Fig. 12.3 Molecular mechanisms of the Kai system. During the subjective day, KaiA triggers sequential autophosphorylation of KaiC hexamers at T432, followed by S431 in the KaiC-CII domain. KaiC-CII autophosphorylation enhances CII ring formation and stacking interactions with the CI ring. These stacking interactions between the CII and CI rings couple ATPase activity in KaiC-CI to KaiC-CII autophosphorylation. The KaiC-CI domains in their post-hydrolysis conformation cooperatively bind six KaiB monomers in a fold-switched state. Fold-switched KaiB in the KaiCB complexes presents an alternative binding site for KaiA, sequestering the pool of free KaiA. While sequestered KaiA is no longer available to promote KaiC autophosphorylation, the KaiC hexamers revert to sequential autodephosphorylation of T432, followed by S431. Dephosphorylation of S431 releases KaiBA complexes from the KaiC hexamers and brings the circadian oscillation full circle

The autokinase and -phosphatase activities of KaiC are always at play simultaneously. Phosphatase activity of KaiC will dominate by default, resulting in progressively lower levels of autophosphorylation in the KaiC-CII domain. In the unphosphorylated state, KaiC can engage in an interaction with KaiA that promotes the autokinase activity of the KaiC-CII domain (Iwasaki et al. 2002; Williams et al. 2002). This results in sequential autophosphorylation of T432, followed by S431. Phosphorylation of S431 will promote the formation of the KaiC-CII ring to enhance stacking interactions with the KaiC-CI ring and elevate ATPase activity in this domain. In turn, ATP hydrolysis in the KaiC-CI domain triggers an interaction with KaiB (Chang et al. 2012; Mukaiyama et al. 2018; Mutoh et al. 2013;

Phong et al. 2013; Snijder et al. 2017; Tseng et al. 2017). When KaiB binds to KaiC, it exposes a new binding site for KaiA that outcompetes the autokinase-stimulating interaction with the KaiC-CII domain. By depleting the pool of KaiA, KaiB brings KaiC back to its default dephosphorylation activity, resulting in sequential dephosphorylation of T432, followed by S431. The interaction with KaiB will persist until S431 is finally also dephosphorylated. Dephosphorylation of S431 results in dissociation of KaiB and KaiA, bringing KaiC full circle to a free and unphosphorylated state that is capable of interacting with free KaiA again. In short, KaiA stimulates autophosphorylation of KaiC, which triggers a sequestration feedback mechanism that depletes the pool of free KaiA in KaiCB complexes to initiate a phase of dephosphorylation that brings the reaction full circle.

By stimulating the autokinase activity of unphosphorylated KaiC, KaiA drives the free-running circadian rhythm of autophosphorylation. KaiA is a two-domain protein that forms dimers through a domain-swapping interaction (Ye et al. 2004). It has a pseudoreceiver (PsR) domain on the N-terminus that is connected with a linker region to an alpha-helical C-terminal domain (Garces et al. 2004; Uzumaki et al. 2004; Vakonakis et al. 2004b; Williams et al. 2002). KaiA has two modes of interaction with KaiC, depending on the phase of the autophosphorylation cycle. The swapped C-terminal domains of KaiA form a cleft in the dimer that binds to the extended C-terminus of the KaiC-CII domain (Pattanayek and Egli 2015; Vakonakis et al. 2004b). This mode of binding is available to KaiA at all phases of the circadian rhythm and is required for the autokinase-stimulating interaction with KaiC. During most of the cycle, a KaiC hexamer can bind a single KaiA dimer at the C-termini of the KaiC-CII domains, except during the phosphorylation phase, when a second KaiA dimer can bind, promoting the autokinase activity of KaiC-CII (Brettschneider et al. 2010; Hayashi et al. 2004; Snijder et al. 2017). While S431 is phosphorylated and the two KaiC rings stack together, promoting ATP hydrolysis in the KaiC-CI domain and KaiB binding, a second mode of binding becomes available to KaiA. During this dephosphorylation phase of the cycle, KaiB exposes a new binding site that asymmetrically binds one subunit of a KaiA dimer at the linker region between the PsR- and the C-terminal alpha-helical domains (Snijder et al. 2017; Tseng et al. 2017). This mode of binding outcompetes the autokinase-stimulating interaction at the KaiC-CII termini, giving way to the dominant phosphatase activity of KaiC-CII again.

The interaction of KaiB with KaiC marks the dephosphorylation phase of the circadian oscillator. KaiB is a single-domain protein, in equilibrium between a monomeric and tetrameric form (Snijder et al. 2014). It has the remarkable ability to switch between two folded states (Chang et al. 2015). In the tetramer it adopts a unique KaiB fold, whereas it switches to a thioredoxin-like fold when it is bound to KaiC (Garces et al. 2004; Hitomi et al. 2005; Iwase et al. 2005; Pattanayek et al. 2008; Snijder et al. 2014; Villarreal et al. 2013; Tseng et al. 2017). Six KaiB monomers bind cooperatively to the six KaiC-CI domains of the hexamer (Lin et al. 2014; Snijder et al. 2014). The KaiB tetramer is therefore thought to act as a sink that dampens fluctuations in KaiB concentration. KaiB interacts with the KaiC-CI domains in their post-hydrolysis conformation, which is how the interaction is triggered by ATPase activity of KaiC. Fold-switched KaiB bound to KaiC-CI exposes a new binding site for KaiA dimers, away from the KaiC-CII domain where KaiA would be able to stimulate autokinase activity. The KaiB-KaiA complex remains bound for as long as S431 is phosphorylated. When S431 is finally also dephosphorylated, the KaiB-KaiA complex is released from the KaiC hexamer and subsequently dissociates into the individual KaiB and KaiA components.

It is important to note that the Kai system is not a single-molecule timing mechanism. It is rather a biochemical network that produces circadian oscillations based on the ensemble behavior of the tens-of-thousands of copies of the Kai components that are present inside a single cyanobacterial cell. We have described an important succession of events on the molecular level, but it is unlikely that the signal from a single KaiC hexamer going through those steps will rise above the noise to produce robust circadian rhythms of gene expression. A typical cell will carry on the order of 10^4 copies of KaiB and KaiC, together with 10^3 copies of KaiA (Chew et al. 2018; Kitayama et al. 2003). This large excess of KaiBC over KaiA will ensure efficient KaiA sequestration during the dephosphorylation phase of the cycle.

The system has also evolved a mechanism to ensure synchronicity between the vast numbers of copies of the Kai components. During the dephosphorylation phase of the cycle, KaiC hexamers will exchange subunits to result in more homogeneous levels of autophosphorylation in the ensemble (Brettschneider et al. 2010; Clodong et al. 2007; Ito et al. 2007; Lin et al. 2014; Mori et al. 2007). This is especially relevant in the context of translation feedback on the circadian clock. The monomer shuffling will ensure that newly translated KaiC subunits mix with the existing pool of Kai components, thereby factoring into the phase of autophosphorylation in the entire ensemble of KaiC hexamers. Furthermore, all three Kai components also exhibit circadian patterns of subcellular localization (Cohen et al. 2014; Kitayama et al. 2003). Based on the workings of the test-tube oscillator such localization patterns are not essential for circadian oscillations per se but might be required in the context of other cellular processes like photosynthesis, nitrogen fixation, and cell division.

Our description of the Kai system so far corresponds to the biological clock as it is observed in *Synechococcus elongatus*, which has historically been an important model system for cyanobacterial research. We will discuss the evolutionary origins and diversity of Kai systems in more detail later, but would like to highlight one important variation on the Kai system here. Whereas in *Synechococcus elongatus* cells, KaiA is responsible for driving a free-running circadian rhythm of autophosphorylation in KaiC, there are other cyanobacteria that lack the KaiA component altogether (Schmelling et al. 2017). These include the ecologically widespread and important genus of *Prochlorococcus*. In these Kai systems that lack a KaiA component, the biological timer works via a fundamentally different mechanism. Rather than generating free-running rhythms of KaiC autophosphorylation with KaiA, which are rather robust to environmental stimuli and continue oscillating even in constant dark or light conditions, the KaiBC-only systems work more like an hourglass, which has to be reset every day.

In the KaiBC-only systems, environmental stimuli like light or temperature will set off a period of KaiC autophosphorylation, followed by a period of KaiB binding and a drop in the levels of autophosphorylation (Chew et al. 2018). These two periods together last for approximately twenty-four hours. The KaiBC timer will set off a twenty-four-hour global gene expression pattern, enough to anticipate a new dawn, but will require the stimuli of a new day to persist in a circadian rhythm. The KaiBC systems are more in tune with the environment, but therefore simultaneously less robust in this regard. They are also less sensitive to relative noise that arises from the overall expression levels of the clock component, compared to KaiA-based systems, which require higher expression levels to maintain a robust twenty-four-hour cycle (Chew et al. 2018). It is thought that the trade-off between KaiBC and KaiBC-KaiA systems lies in their relative robustness to environmental stimuli and the costs of producing higher levels of KaiBC and KaiA. The KaiBC-only systems are also more common in genera that live in open-ocean surface waters in tropical and subtropical regions, whereas the KaiA-based systems are more common in genera that live in temperate climates that experience larger seasonal variations.

Structures of the Kai Proteins

KaiC

The two-domain structure of KaiC is the result of an internal gene duplication event (Ishiura et al. 1998). Even though the KaiC-CI and -CII domains have diverged substantially on a functional level, they remain remarkably similar on a structural level (Pattanayek et al. 2004). The similarity of the domains relates to their common ATP binding activity, with conserved Walker A motifs (also known as P-loops) and downstream Walker B motifs present in both KaiC-CI and KaiC-CII domains, responsible for the ATPase and autokinase/-phosphatase activities (Nishiwaki et al. 2000). In addition to their shared structural motifs, the domains each contain characteristic regions, and they are connected with a conserved flexible linker (see Fig. 12.4a) (Schmelling et al. 2017). The KaiC-CI domain contains a B-loop that changes conformation upon ATP hydrolysis and is a crucial component of the KaiB binding site (Abe et al. 2015). Compared to KaiC-CI, the KaiC-CII domain has an extended C-terminus containing two additional functional motifs. These mediate the interaction with KaiA and its autokinase-stimulating activity. It consists of a so-called 'A-loop', followed by the KaiA-binding region at the very end of the C-terminus.

The ATP-binding pockets in KaiC are formed at the subunit interface of the assembled hexamer (see Fig. 12.4b–d) (Abe et al. 2015; Egli et al. 2012, 2013;

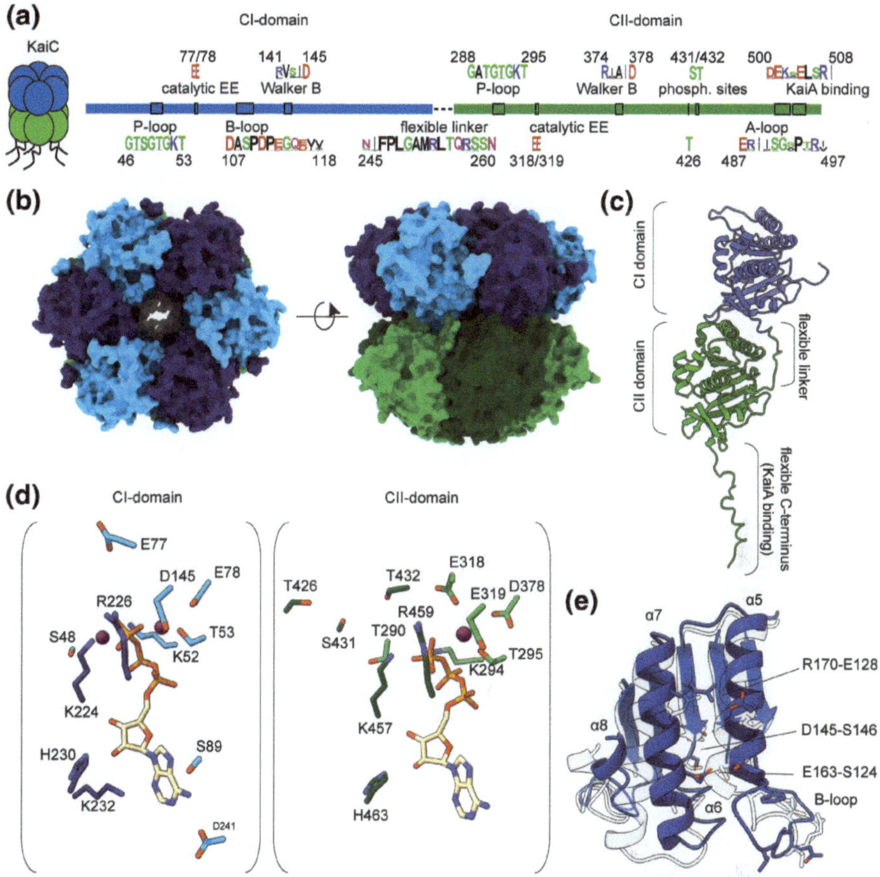

Fig. 12.4 Structure of the central clock component KaiC. **a** Domain structure and conserved sequence motifs in *Synechococcus elongatus* KaiC homologs. Sequence logos with residue numbers are shown for important functional motifs. **b** Surface representation of KaiC hexamers (PDB ID 1TF7). KaiC is shown along the central channel of the CI-domain hexamer (*left*) and in a 90-degree rotation showing stacking of the CI and CII domains (*right*). The flexible C-termini of the CII domain are not displayed. **c** Ribbon diagram of a KaiC subunit, including the flexible C-terminus of the CII domain (PDB ID 3DVL). **d** Detailed view of ATP binding in the CI and CII domains. Light and dark colors represent side chains from two different neighboring subunits. Magnesium ions are shown as purple spheres. Based on PDB ID 1TF7 and 4O0 M. **e** The KaiC-CI domain in the post-hydrolysis conformation (solid blue, PDB ID 4TLA) overlaid with the pre-hydrolysis conformation (transparent, PDB ID 1TF7)

Pattanayek et al. 2004, 2006, 2009, 2011, 2014; Xu et al. 2004). The ATPase and autokinase/-phosphatase activities of KaiC are thereby coupled to oligomerization. The triphosphate groups of the ATP molecules are fixed in position by coordinating interactions with two catalytic magnesium ions on either side of the neighboring subunits. The subunits themselves coordinate the magnesium ions through interactions with P-loop residues (S48/K52/T53 in KaiC-CI, T290/K294/T295 in KaiC-CII) and three additional acidic residues (E77/E78/D145 in KaiC-CI, E318/E319/D378 in KaiC-CII). Residues E77 and E318 in the KaiC-CI and KaiC-CII domains, respectively, are crucial catalytic sites for ATP hydrolysis. Additional hydrogen bonds directly between the gamma-phosphate and two basic residues of the neighboring subunit (K224/R226 in KaiC-CI, K457/R459 in KaiC-CII) further stabilize the triphosphate group of ATP. The 2′-hydroxyl group of the ribose moiety is contacted by H230 in KaiC-CI, H463 in KaiC-CII. In the KaiC-CI domain, additional residues (S89/K232/D241) interact with the nucleobase portion of the ATP's through hydrogen bonds, but similar stabilizing interactions are missing in the KaiC-CII domain.

In the KaiC-CII domain, the gamma-phosphates of the ATP molecules are positioned within hydrogen-bonding distance of the phosphorylation sites that define the phase of the circadian rhythm. Note that most of the stabilizing interactions with ATP and in particular the interactions with the catalytic residues come from a neighboring subunit with respect to the phosphorylation sites. The circadian rhythm is defined by sequential autophosphorylation of T432 first, then S431, and followed by autodephosphorylation in the same order. A third adjacent residue, T426, can be transiently phosphorylated and is crucial to facilitate the circadian oscillation of autophosphorylation at T432/S431 (Pattanayek et al. 2009; Xu et al. 2004, 2009). It forms hydrogen bonds with phosphorylated S431 and shuttles the phosphate group during dephosphorylation, which happens through an ATP synthase and subsequent ATPase mechanism (Egli et al. 2012; Nishiwaki and Kondo 2012). Autophosphorylation of T432 results in additional intersubunit contacts between the transferred phosphate group and R385, as well as the catalytic E318 side chain. Phosphorylation of S431 results in new hydrogen-bonding interactions with H429 of the same subunit, which is in turn hydrogen bonded to D427 of the adjacent subunit. The autophosphorylation reaction thus changes the subunit interface, providing a structural basis for its effect on KaiC-CII hexamerization and the stacking interactions between the KaiC-CI and KaiC-CII rings.

Whereas the autophosphorylation cycle in the KaiC-CII domain results in no noticeable conformational changes beyond side chain rearrangements, ATP hydrolysis in the KaiC-CI domain results in allosteric conformational changes through long stretches of the polypeptide backbone (see Fig. 12.4e) (Abe et al. 2015). One of the three acidic residues that coordinate the triphosphate group of ATP, D145, undergoes a cis-to-trans isomerization with S146 in the ADP bound state. This changes the conformation of alpha helices α6 and α7 that lie directly beyond the D145-S146 peptide. Helix α6 further connects to α5, which follows the conformational changes upon ATP hydrolysis. Helix α5 forms the KaiB-binding site together with the directly preceding B-loop. Helices α6 and α5 run in antiparallel direction on the outer face of the KaiC-CI domain. They connect through Van der Waals interactions as well as hydrogen bonds between E163-S124 and a salt bridge between R170-E128. The movement of α5 changes the conformation of the preceding B-loop, resulting in large side chain rearrangements for Q115, V117, V118, and F121. This is how ATP hydrolysis in the KaiC-CI domain allosterically triggers KaiB-binding. Helix α6 further connects to α8, which also changes backbone conformation after ATP hydrolysis. It is positioned on the opposite subunit interface compared to the D145-S146 peptide and in close proximity to the neighboring ATP triphosphate. The functional role of α8 has not yet been explored experimentally, but it likely also contributes to the allosteric changes upon ATP hydrolysis that trigger the binding of KaiB.

KaiB

KaiB has the astonishing ability to switch between two different folded states: a unique KaiB fold and a thioredoxin-like fold (see Fig. 12.5a, b) (Chang et al. 2015; Garces et al. 2004; Hitomi et al. 2005; Iwase et al. 2005; Pattanayek et al. 2008; Snijder et al. 2017; Tseng et al. 2017; Villarreal et al. 2013). The unique KaiB-fold is found in the tetrameric form of KaiB, whereas the thioredoxin-like fold is found in the KaiC-bound monomeric form. It is not clear whether KaiB switches fold as a free monomer, or whether KaiC-binding induces the transition. The two folds share a common strand-helix-strand motif in the N-terminal half of the polypeptide chain, with both β-strands contributing to the same central β-sheet. In the unique KaiB-fold, the second β-strand leads into a long loop running across the central β-sheet, followed by a very short strand-helix motif, the long α3 helix running antiparallel to the α1 helix, leading into the β4 strand on the C-terminus that completes the central β-sheet. In the thioredoxin-like fold, the common strand-helix-strand motif is followed by the α2 helix running across the central β-sheet, followed by two short antiparallel β-strands that complete the central β-sheet and lead into a long α3 helix on the C-terminus, running parallel to the α1 helix. The role of the unique KaiB-fold and tetramer is currently not well understood, but it has been suggested that the KaiB-tetramer can act as a sink to produce a more stable free KaiB-monomer concentration over a wider range of KaiB expression levels (Nakajima et al. 2010; Snijder et al. 2014).

Six KaiB monomers with the thioredoxin-like fold bind cooperatively to the KaiC-CI domain in the post-hydrolysis conformation (Snijder et al. 2017; Tseng et al. 2017). The newly formed α2 helix of the thioredoxin-like fold is an important part of the KaiC-KaiB interface, sitting inside the B-loop of KaiC-CI and directly contacting the α5 helix that changes conformation in the post-hydrolysis state (see Fig. 12.5c). There are direct contacts between α1 of KaiB and α3 of the neighboring KaiB subunit. These contacts could provide a potential mechanism for the cooperative binding, though these interactions have yet to be explored experimentally on a functional level. KaiC-CI intersubunit interactions may just as well contribute to the cooperative binding observed for KaiB, and the relative contribution of these factors remains to be determined.

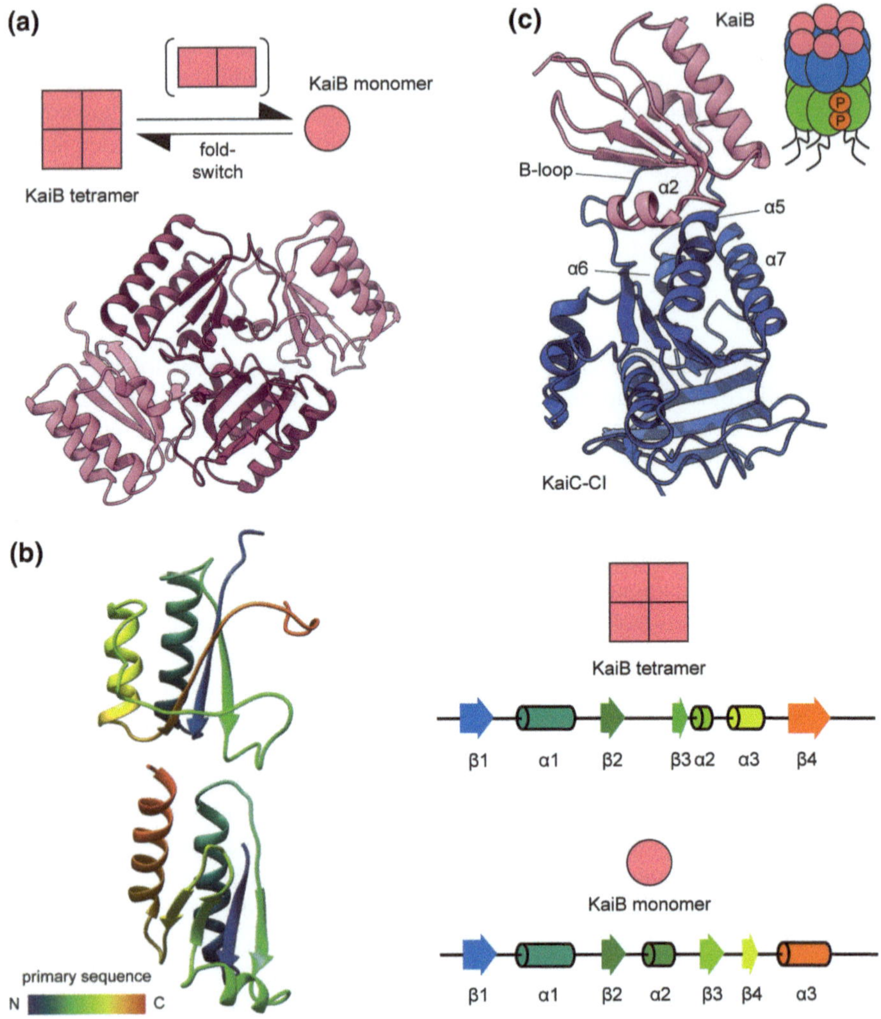

Fig. 12.5 Structure of KaiB and interactions with KaiC. **a** Ribbon diagram of the KaiB tetramer (PDB ID 4KSO). **b** Comparison of KaiB folds of the tetramer (top) and thioredoxin-like monomer (bottom). Ribbon diagrams are colored in rainbow gradients from N- to C-terminus, with corresponding diagrams of their secondary structures on the right. **c** Fold-switched KaiB in complex with the KaiC-CI domain (PDB ID 5JWO)

KaiA

KaiA is a two-domain protein that forms homodimers through a domain-swapping interaction (see Fig. 12.6a) (Ye et al. 2004). At the N-terminus is a Pseudo-Receiver domain, followed by a linker region and an alpha-helical domain on the C-terminus. The C-terminal alpha-helical domains are packed alongside each

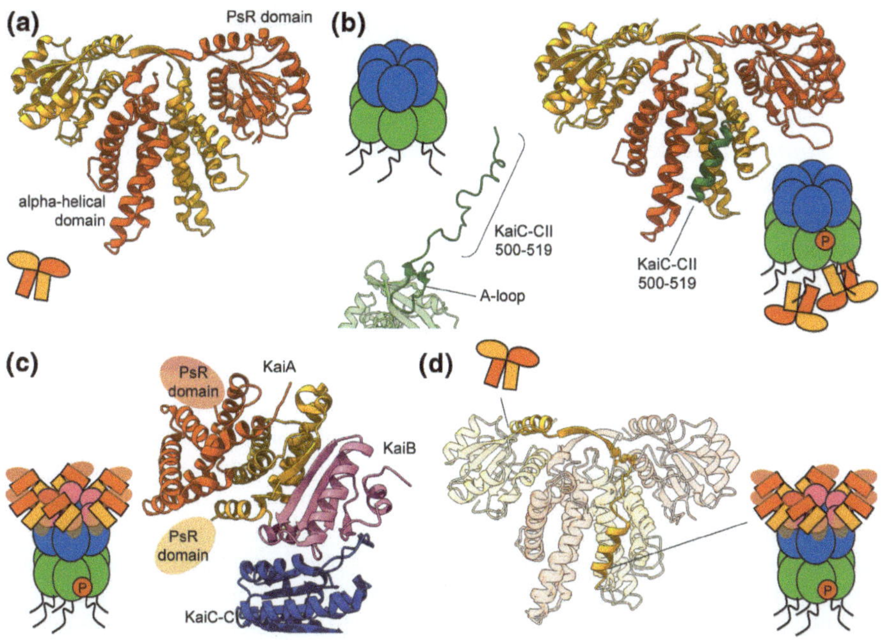

Fig. 12.6 Structure of KaiA and interactions with KaiC and KaiB. **a** Ribbon diagram of the KaiA dimer (PDB ID 1R8J). **b** Detailed view of the KaiC A-loop and KaiA-binding C-terminus (PDB ID 3DVL). Ribbon diagram of KaiA dimer in complex with a peptide derived from the KaiC C-terminus (PDB ID 5C5E). **c** Ribbon diagram of KaiA dimer in complex with KaiCB (PDB ID 5JWR). The KaiA dimer binds asymmetrically with a single subunit to KaiB. KaiCB binding displaces the PsR domains of KaiA, which are no longer visible in reconstructions of the KaiCBA complexes. **d** Detailed view of the conformational change of the PsR domain in KaiA upon KaiCB binding (PDB ID 1R8J and 5JWR). The linker region between the PsR domain and alpha-helical domain swings out in front of the KaiC-binding cleft of the KaiA dimer

other in a V-shape, forming an open cleft at one end of the dimer. On the opposite end, where the two helical bundles meet at the hinge of the V-shape, the subunits crossover with the linkers to the PsR domain running in antiparallel direction. The linkers consist of short β-strands, together forming a small antiparallel β-sheet that leads into the PsR domains on either side.

The cleft formed by the alpha-helical domains of the KaiA-dimer constitutes the KaiC-binding region (see Fig. 12.6b) (Pattanayek and Egli 2015; Vakonakis et al. 2004b). The long flexible C-terminus of the KaiC-CII domain contains the corresponding KaiA-binding region at residues 499–513, which is directly preceded by the so-called A-loop at residues 487–497. In the absence of KaiA-binding, the A-loops dip into the central channel of the KaiC-CII ring. When the A-loop is deleted, it results in constitutive hyperphosphorylation of KaiC (Kim et al. 2008). This effect is mediated through putative interactions of the A-loop with P-loop residues and a loop directly preceding the phosphorylation sites T432/S431/T426 (Egli et al. 2013). It has been suggested that KaiA-binding to the KaiC-CII domain results in extension of the A-loop, such that it is threaded out from the central channel of the KaiC-CII domain. It supposedly activates KaiC autophosphorylation activity in a similar manner to deletion of the A-loop. Consistent with this model, the A-loop sequence has diverged in KaiBC-only systems like those found in *Prochlorococcus* (Axmann et al. 2009; Schmelling et al. 2017). Though these observations clearly point to an important role for the A-loop, the molecular mechanism by which KaiA binding activates KaiC autophosphorylation is not well understood at present.

Autophosphorylation of KaiC triggers a sequestration feedback on KaiA through the phosphorylation-dependent ATPase activity of the KaiC-CI domain, which triggers KaiB binding (Brettschneider et al. 2010; Snijder et al. 2017; Tseng et al. 2017). KaiB in the thioredoxin-like fold bound to KaiC-CI domain presents an alternative binding site to KaiA that outcompetes the interaction on the termini of the KaiC-CII domains. KaiB interacts with KaiA primarily through contacts with the β2 strand of the central β-sheet (see Fig. 12.6c). It binds asymmetrically to a single subunit of the KaiA dimer, contacting the β-strand linker that connects the PsR and alpha-helical domains. The β-strand linker and connected PsR domains of KaiA undergo a large displacement that facilitates binding to the β2 strand of KaiB (see Fig. 12.6d). Rather than crossing over at the hinge of the V-shape formed by the alpha-helical domains, the β-strand linker, and PsR domain now swing over in front of the central cleft of the KaiA dimer that binds the KaiC-CII termini. It has been proposed that this conformation of KaiA represents an autoinhibited state since the β-strand linker and PsR domain block the KaiC-CII binding site in the dimer. It should also be noted that without the PsR domain displacement, binding of the β-strand linker to KaiB would result in steric clashes of the PsR domain with neighboring KaiB subunits. It is currently not clear if the conformational change in KaiA is only required to sterically fit KaiA dimers to a KaiCB complex, or whether it indeed represents a functionally autoinhibited state that further prevents KaiA-stimulated autophosphorylation of KaiC beyond the sequestration mechanism.

Input and Output Protein Factors

We have discussed in detail how interactions between the Kai components produce circadian oscillations of KaiC autophosphorylation, but how does this signal translate to circadian rhythms of global gene expression? And how are environmental cues relayed to the circadian clock to ensure synchronicity between internal and external day-night cycles? The Kai system acts on global gene expression primarily as a suppressor of high amplitude promoters (Nakahira et al. 2004; Qin et al. 2010b). Kai is known to produce 24-h rhythms of chromosome compaction, which may act as a general mechanism to globally suppress transcription (Smith and Williams 2006; Woelfle et al. 2007). However, the output from the Kai system is also specifically targeted through signaling pathways that modulate transcription factors, which in turn regulate the multitude of sigma factors that are characteristic of cyanobacteria. One of the foremost signaling pathways to mediate output from the Kai system consists of SasA, CikA, and RpaA (Iwasaki et al. 2000; Gutu and O'Shea 2013; Hertel et al. 2013; Kageyama et al. 2003; Markson et al. 2013; Nishiwaki et al. 2004; Takai et al. 2006). The circadian oscillation of Kai is relayed to the transcription factor RpaA through direct interactions of SasA and CikA with the Kai components. Following the circadian rhythms of KaiC autophosphorylation, these signaling components regulate global transcription in the cell, including feedback on the Kai components themselves (see Fig. 12.7a).

RpaA is a two-domain transcription factor, with a so-called response regulatory domain on the N-terminal half of the protein and an OmpR/PhoB-type DNA-binding domain on the C-terminal half. The response regulatory domain has a putative phosphorylation site at D53, which becomes phosphorylated by the histidine kinase SasA around subjective dusk (Gutu and O'Shea 2013; Hertel et al. 2013; Markson et al. 2013; Takai et al. 2006; Taniguchi et al. 2010). The SasA-RpaA interaction resembles that of the two-component signaling systems that are widespread in prokaryotes. Phosphorylation of RpaA follows the circadian rhythm of KaiC autophosphorylation with a -4-h phase shift. Phosphorylated RpaA binds directly to DNA approximately 20–50 base pairs upstream of the transcription

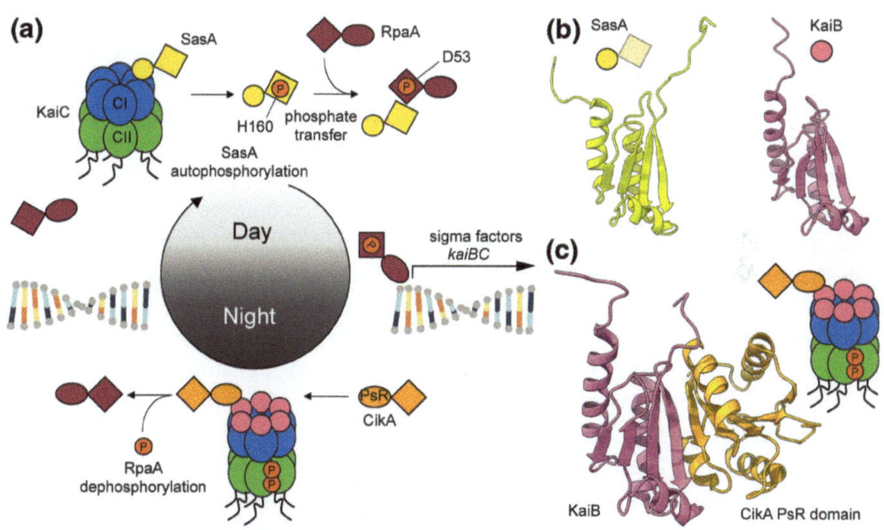

Fig. 12.7 The major output pathway of the Kai system. **a** Schematic of the SasA-RpaA-CikA system and interactions with Kai. During subjective day SasA binds to the same site as KaiB on KaiC-CI, resulting in autophosphorylation of SasA at H160. SasA autophosphorylation is relayed to transcription factor RpaA at D53. RpaA modulates transcription of many sigma factors and the *kaiBC* operon. During the subjective night, phosphorylated KaiCB complexes bind CikA resulting in dephosphorylation of RpaA, relieving its transcriptional regulation. **b** Ribbon diagrams of the thioredoxin-like folds of SasA (PDB ID 5JYU) and fold-switched KaiB (PDB ID 5JYT). Both domains bind to KaiC-CI, competing for the same binding site. **c** Ribbon diagram of KaiB in complex with the PsR domain of CikA (PDB ID 5JYV)

start sites, including the promoters of the *kaiBC* operon and the four sigma factor genes *rpoD2*, *rpoD5*, *rpoD6,* and *sigF2*. Some one hundred RpaA binding sites have currently been identified in the *Synechococcus elongatus* genome, at least half of which contain a characteristic A/T-rich motif. RpaA activates expression of a subset of its targets to peak at subjective dusk, whereas it *suppresses* activation of a different subset of targets to peak at subjective dawn. CikA gradually dephosphorylates RpaA during this phase of the circadian oscillator, thereby relieving RpaA's transcription activation by subjective dawn. RpaA further also regulates the transcription suppression activity of its paralogue RpaB (Hanaoka et al. 2012; Hertel et al. 2013). RpaB also binds to the promoters of the *kaiBC* operon and sigma factor genes *rpoD6* and *sigF2*, but at subjective dawn, suppressing their transcription. RpaB's displacement by RpaA further promotes expression of the circadian clock genes and downstream transcription regulators to produce circadian oscillations of global gene expression.

SasA and CikA relay the circadian oscillation of KaiC autophosphorylation to RpaA at subjective dawn and dusk, respectively. SasA is a two-domain histidine kinase that is responsible for phosphorylating RpaA at the putative D53 phosphorylation site (Markson et al. 2013; Takai et al. 2006). SasA has a thioredoxin-like domain on the N-terminal half of the protein and its kinase domain near the C-terminus (Vakonakis et al. 2004a). The thioredoxin-like domain is homologous to KaiB in the fold-switched, KaiC-CI bound state and it competes with KaiB for the same binding site (see Fig. 12.7b) (Murakami et al. 2012; Pattanayek et al. 2011; Tseng et al. 2014, 2017; Iida et al. 2015). SasA binds unphosphorylated KaiC at subjective dawn, resulting in autophosphorylation of SasA at H160 (Valencia et al. 2012). Autophosphorylated SasA dissociates from KaiC, oligomerizes, and subsequently relays its phosphate group to D53 of RpaA. As KaiC autophosphorylation levels rise and KaiCB complexes are formed, CikA binds to KaiCB complexes with its C-terminal Pseudo Receiver domain, interacting with the same β2 strand of KaiB that binds KaiA (see Fig. 12.7c) (Tseng et al. 2017). The RpaA-dephosphorylation activity of CikA is stimulated by the CikA interaction with KaiCB complexes (Gutu and O'Shea 2013).

Coupling of the Kai system with SasA-RpaA-CikA signaling modulates most physiological activity in cyanobacterial cells. Moreover, through SasA-RpaA-CikA signaling the Kai system also applies a gate on cell division (Dong et al. 2010; Yang et al. 2010). Whereas the inherent pace of the cell cycle is more or less independent from the circadian rhythm, the cell cycle does slow down during subjective night, when Kai suppresses cell division. Cyanobacteria can, therefore, divide multiple times per day, but only during a certain window set by the Kai system.

Feedback between the Kai oscillator and RpaA signaling regulates the circadian rhythm of the cell internally, but external signals also play a pivotal role in maintaining an adaptive biological clock. Both light and temperature are environmental cues that entrain the Kai system, which we will discuss in detail later. We would like to note here that even though light is such an important factor in cyanobacterial daily life, there is currently no known light-sensing protein that directly interacts with the

Kai oscillator. Rather, indirect effects from light via photosynthesis, such as the ATP/ADP ratio and redox state of the cell, are known to modulate the circadian rhythm of the Kai complex. The ATP/ADP ratio works directly on the KaiC ATPase and autophosphorylation activities, but input from the redox state of the cell requires at least two known sensing mechanisms. First, a redox-sensitive protein called LdpA interacts directly with KaiA and CikA (Ivleva et al. 2005). LdpA contains iron-sulfur clusters that are involved in sensing the redox state of the cell and the protein mediates depletion of KaiA and CikA under reducing conditions, effectively extending the period of the oscillation. In addition, KaiA and CikA are sensitive to the redox state of the cell through direct interactions of their PsR domains with oxidized plastoquinones that occur at the onset of darkness (Ivleva et al. 2006; Kim et al. 2012; Pattanayek et al. 2012; Wood et al. 2010). Quinone binding to KaiA or CikA results in aggregation and degradation of the proteins and causes a phase delay to the circadian rhythm of KaiC autophosphorylation.

Other known input and output pathways of the circadian clock operate mainly on the transcriptional level, with no known direct physical interactions with the Kai components. One of the earliest known examples of a transcriptional regulator of Kai is Pex (Arita et al. 2007; Kutsuna et al. 1998, 2007; Kurosawa et al. 2009). It forms a dimer with a winged-helix motif that recognizes a 25 base pair motif in the *kaiA* promoter region. Pex is expressed during prolonged dark periods and when it binds the *kaiA* promoter, it suppresses transcription. The net effect of Pex is to extend the period of the circadian oscillator. Other genes that are known to be involved with transcriptional regulation of the circadian clock include *labA, lalA, cpmA,* and *crm*, which act on either *kaiA* or *kaiBC* promoters. However, molecular details of these interactions still remain elusive (Boyd et al. 2013; Katayama et al. 1999; Taniguchi et al. 2001, 2007, 2010).

Feedback Mechanisms and Entrainment

The negative feedback loop arising from the self-repression of a gene by its gene product is a common motif found in prokaryotic as well as eukaryotic oscillators. Powerful oscillations arise as soon as the feedback loop contains a certain delay. Negative feedback loops lie at the core of many rhythmic processes within cells (DeWoskin et al. 2014), including those involved in circadian timekeeping, embryogenesis, cell cycle, and DNA damage repair providing several major advantages: temporal organization, spatial organization, prediction of repetitive events, efficiency, and precision of control (Rapp 1987).

Circadian clocks responsible for endogenous 24-h oscillations are divided into two groups in regard to their functionality, either based on a transcription-translation feedback loop (TTFL) or a post-translational oscillator (PTO). Eukaryotic circadian clocks are mainly based on TTFL, whereas cyanobacterial circadian clocks, as the only known example for prokaryotic circadian clocks, comprise a PTO (Johnson et al. 2011). Protein sequence analyses of the clock components suggest a convergent evolution with multiple origins of circadian clocks, which is supported by the different functionality of known circadian clocks (Johnson et al. 2011; Rosbash 2009).

A decade of intensive research on the mechanisms and functionality of the KaiABC system—the cyanobacterial PTO-based clock—using computational and mathematical approaches in addition to the detailed biochemical and biophysical understanding make this the best understood circadian clock. Various parts of the KaiABC system were identified and described in detail through computational modeling: The ordered phosphorylation of KaiC and temperature compensation of the clock (Brettschneider et al. 2010; Rust et al. 2007), the stimulating interaction with the other core factors and effects on gene expression (Clodong et al. 2007; Kurosawa et al. 2006; van Zon et al. 2007), as well as the influence of varying ATP/ADP ratios (Rust et al. 2011). Further, mathematical models have identified different strategies for period robustness against internal and external noise, reviewed in Schmelling and Axmann (2018).

The remarkably robust high-amplitude phosphorylation cycles are stabilized by an intracellular synchronization of different KaiC hexamers: This is on the one hand achieved by an exchange of KaiC monomers (Emberly and Wingreen 2006; Mori et al. 2007; Yoda et al. 2007). On the other hand, mathematical modeling demonstrated that sequestration of free KaiA mediates synchronization (Brettschneider et al. 2010; Clodong et al. 2007; Qin et al. 2010a; van Zon et al. 2007). Together, synchronization by KaiA-sequestration and KaiC-monomer exchange leads to robust intracellular oscillations. In addition, the three-protein clock is embedded in a TTFL, similar to eukaryotic clock systems (Hertel et al. 2013; Zwicker et al. 2010). A two-loop transcriptional feedback mechanism could be identified in which only one phosphorylation form of KaiC suppresses *kaiBC* expression while two other forms activate its own expression. In particular: KaiC-S/pT and KaiC-pS/pT activate and KaiC-S/T suppress *kaiBC* transcription (Hertel et al. 2013). Due to this further layer of control, the PTO becomes insensitive to the manifold perturbations in living cells; reviewed in Johnson and Egli (2014).

Besides the circadian clock as seen in *Synechococcus elongatus*, which functions similar to a limit cycle oscillator (Gan and O'Shea 2017; Pittayakanchit et al. 2018), there exists a reduced form of timing mechanism in cyanobacteria, as described before: An hourglass timing system as seen in *Prochlorococcus*, which works more like a point attractor that stops oscillating and relaxes to a stable fixed point in the absence of an exogenous cycle (Pittayakanchit et al. 2018). Fluctuations of external cues have different effects on the circadian clock systems in cyanobacteria. Whereas, the limit cycle oscillator (*Synechococcus*) is almost unaffected (Paijmans et al. 2017; Pittayakanchit et al. 2018), the point attractor (*Prochlorococcus*) is set in free fall towards the night state (Pittayakanchit et al. 2018). On the other side, when considering only internal noise due to finite numbers of proteins the point attractor outperforms the limit cycle in regard to precision due to its ability to change faster between both states (Pittayakanchit et al. 2018).

One defining characteristic of a circadian clock is the ability to be entrained to an exogenous cycle. Thus, a circadian system can be synchronized to an exogenous stimulus in order to be most useful for the organism in anticipating recurring patterns. However, the clock also has to be robust against naturally occurring fluctuation of the input signal. There are two strategies for entraining the clock to an exogenous stimulus: (i) direct sensing of light, a strategy that is normally used by eukaryotic circadian clock systems (Rosbash 2009) and (ii) indirect sensing of light through changes in the metabolic state of the cell, i.e. redox state or ATP/ADP ratio, which is used by cyanobacteria (Rust et al. 2011; Shultzaberger et al. 2015). It has experimentally been shown that the phase of the circadian clock is affected by the ATP/ADP ratio, which is a result of the cellular catabolic metabolism and the photosynthetic apparatus (Shultzaberger et al. 2015). Dark phases cause a drop in the ATP/ADP ratio, which shifts the clock into the dephosphorylation phase. Even the in vitro clock reacts differently to the ATP/ADP ratio (Rust et al. 2011). During the phosphorylation phase (subjective day), the oscillator is most susceptible to changes in the ATP/ADP ratio, whereas in the dephosphorylation phase (subjective night) the oscillator is almost insensitive. The underlying mechanism for this phase shift might be explained by the ADP/ATP exchange in the CII domains that changes during the cycle (Nishiwaki-Ohkawa et al. 2014). Thus, the entrainment mechanism for the KaiABC clock works in vitro and does not rely on an additional signaling pathway, resembling the direct effect of the metabolic state of the cell on the phase of the circadian clock (Rust et al. 2011).

Diversity and Evolution of Kai-Protein System

The unique three-protein oscillator is well documented for the model cyanobacterium *Synechococcus elongatus* PCC 7942. The complex formation between the Kai proteins has been discovered in an exquisitely high resolution emphasizing the importance of the stoichiometry between the three Kai proteins in maintaining robust circadian oscillations. Thus, it is puzzling that several cyanobacteria, e.g. *Synechocystis* sp. PCC 6803, a model organism for photosynthesis and industrial applications, contain multiple *kai*-gene copies in the genome (Schmelling et al. 2017; Wiegard et al. 2013). Other cyanobacteria like marine *Prochlorococcus* strains harbor only a reduced *kaiBC* system, thus, lacking the *kaiA* gene.

Cyanobacteria represent one of the most diverse bacterial phyla regarding their genomes, morphology, and physiology, which is reflected by the diversity of their timing systems. The enormous amount of available genomic data allows a comprehensive overview of the distribution and conservation of the clock factors. Surprisingly, orthologs for multiple clock components, including the core factors KaiB and KaiC, are present in a large number of bacterial and archaeal genera (Schmelling et al. 2017; Dvornyk et al. 2003, 2004). However, the presence of *kaiA* in the genome is absolutely restricted to Cyanobacteria. The high conservation of motifs for the interaction of KaiA and KaiB in Cyanobacteria, where the β2 sheet represents the most conserved part of the KaiB protein (Snijder et al. 2017), underlines the special role and evolution of KaiA being the key component for a true circadian clock. Even in light of the huge diversity of timing systems among Cyanobacteria, two protein sets can be suggested: (i) factors that are found in almost all cyanobacteria, especially KaiB, KaiC, LdpA, SasA, RpaA, and RpaB, indicating that those factors comprise the core set needed for timing; (ii) factors only found in a set of cyanobacteria like KaiA, CikA, and LabA which are important for circadian regulation in *Synechococcus elongatus* PCC 7942 (Schmelling et al. 2017). *Prochlorococcus* strains are lacking genes for KaiA, CikA, and LabA (Axmann et al. 2009). They show stable oscillations of gene expression under light-dark cycles, however, these oscillations disappear under continuous conditions (Holtzendorff et al. 2008). Thus, the reduced KaiBC-timing system in *Prochlorococcus* is not considered as a circadian clock, since oscillations cease without an exogenous stimulus. Instead, it is called an hourglass-like clock (Axmann et al. 2009).

Kai proteins as well as in- and output factors of the clock are well conserved among Cyanobacteria, however, they are also present in other bacteria and Archaea (Schmelling et al. 2017). Interestingly, they share no sequence similarity to any known eukaryotic clock component. In general, the presence of homologous genes decreases from Cyanobacteria to Proteobacteria to Archaea. However, the key component KaiA, as well as the input factors Pex, LdpA, and CdpA, are absent outside the cyanobacterial phylum (Schmelling et al. 2017). Although known input factors are absent in non-cyanobacteria and Archaea, there might exist input pathways enabling entrainment of the endogenous timing system with the environment. A direct input by the ATP/ADP ratio might be a primary mechanism to transmit changes and synchronize the inner timing system with the internal metabolism and the external environment (Schmelling et al. 2017). Further different but species-specific input factors might exist. In the photoheterotrophic proteobacterium *Rhodobacter sphaeroides* for example, which displays 24-h gene expression rhythms, a histidine kinase is encoded in an operon with *kaiBC* and was suggested as a candidate for transducing the redox signal to KaiBC (Min et al. 2005).

Reduced KaiBC-based or even simpler, solely KaiC-based timing systems might exist in other bacteria and Archaea. For example, in the halophilic archaeon *Haloferax volcanii* the transcripts of four *kaiC* homologs display diurnal accumulation profiles, and those profiles are abolished by deletion of one of the *kaiC*-like genes (Maniscalco et al. 2014). Whether the encoded KaiC proteins in Bacteria and Archaea are able to measure time via phosphate uptake can be elucidated by biochemical approaches. Analyzing the kinase activity of KaiC using purified proteins from two hyperthermophilic Archaea, *Thermococcus litoralis* and *Pyrococcus horikoshii*, and several cyanobacterial species demonstrated that kinase activity of distantly related KaiC proteins is well conserved (Schmelling et al. 2017).

Regarding the evolution of KaiC, two valid hypotheses exist, both of which state that KaiC arose from a shorter ancestral *recA* gene followed by gene duplication and fusion (Dvornyk et al. 2003; Leipe et al. 2000). However, on the one hand, it is hypothesized that an ancestral single-domain KaiC originated in Bacteria, was transferred into Archaea, where its two-domain version originated, and a second lateral transfer event introduced the double domain KaiC into cyanobacteria

(Leipe et al. 2000). On the other hand, it is argued that KaiC has to be of cyanobacterial origin (Dvornyk et al. 2003). Given the amount of new genomic data, further studies are promising to unravel the evolutionary history of KaiC in the near future.

Among the three *kai* genes, *kaiC* is evolutionarily the oldest, and *kaiA* is the youngest. A major event was the origin of the *kaiB* gene and the formation of the *kaiBC* cluster between 3,500 and 2,320 million years ago (Dvornyk et al. 2003). This time corresponds to the period when a reducing geochemical environment, which had existed from the beginning on, was replaced by an oxidizing environment produced by cyanobacteria themselves. Given that *kaiA* can be found today in cyanobacteria only, it is very likely that it evolved there as well. Marine *Prochlorococcus* species lost *kaiA* during evolution. The three-gene *kaiABC* cluster itself evolved around 1,000 million years ago—a key event in the evolution of cyanobacteria, which ensured their domination in Earth's ecosystems (Dvornyk et al. 2003).

Circadian clocks appear to be a conserved trait in evolution providing a fitness advantage by two directions: by synchronizing processes to external, environmental factors, and by coordinating internal processes optimally. Predicting repetitive events provides a major advantage for organisms enabling higher efficiency and precision of control. Summarizing our knowledge on circadian clocks suggests a convergent evolution. The invention of time-keeping mechanisms occurred several times independently in evolution and gave rise to diverse clockworks and timers. However, reports on ubiquitous metabolic cycles accumulated in the last years constituting a universal marker for circadian rhythms in all domains of life, Bacteria, Archaea and Eukaryota. An intimate co-evolution of cellular timekeeping has been suggested with redox homeostatic mechanisms after the Great Oxidation Event about 2.5 billion years ago (Edgar et al. 2012), when photosynthetic cyanobacteria started to evolve oxygen from water increasing atmospheric oxygen dramatically. Rhythms of photosynthetically produced oxygen and corresponding reactive oxygen species (ROS) made by sunlight might be a key driving force in the coevolution of the clock and ROS removal systems that would enable anticipation and resonation with externally driven redox cycles (Reddy and Rey 2014). However, many questions remain on how evolution invented time-measuring systems, and what selective pressures were needed to create a clock.

Outlook

The physiological diversity and genetic accessibility of phototrophic cyanobacteria has recently attracted growing attention to using them as promising biological chassis for the synthesis of a variety of natural products. This includes bioactive metabolites like cytotoxins and potential pharmaceutical lead compounds, food supplement, animal feed, pigments, as well as biofuels (Chlipala et al. 2011; Knoop and Steuer 2015; Loeschcke et al. 2017). Due to the ability of cyanobacteria to convert sunlight and atmospheric CO_2 directly into valuable organic compounds like sugars, they could provide sustainable alternatives to fuel biotech processes (Ducat et al. 2011; Hays et al. 2017; Weiss et al. 2017). Moreover, cyanobacteria do not compete for arable land or drinking water—they can be cultivated in open ponds and space-saving bioreactors. This could be of global importance and solve one of the biggest problems humanity is facing in the 21st century—worldwide CO_2 emissions are increasing while natural carbon and drinking water resources are being exhausted.

Understanding how we could control and optimize metabolic processes in light-dark cycles will be important for further establishing phototrophic organisms like cyanobacteria and microalgae in biotechnological applications (Welkie et al. 2019). Indeed, the circadian clock influences metabolism drastically. In particular, RpaA seems to critically influence growth rates by 60% (Ungerer et al. 2018). Thus, manipulating components of the circadian clock leads to higher productivity regarding biotechnologically relevant products (Osanai et al. 2015).

Further, the cyanobacterial KaiABC clock is transferrable to a heterologous organism, which could be demonstrated for *E. coli* (Chen et al. 2015), opening synthetic circadian design in a broad range of organisms. Overall the insights gained by studying the KaiABC clock of cyanobacteria have substantial impact on the field of circadian research, which can be used for the design of synthetic switches, oscillators, and clocks in the future.

References

Abe J, Hiyama TB, Mukaiyama A, Son S, Mori T, Saito S, Osako M, Wolanin J, Yamashita E, Kondo T, Akiyama S (2015) Circadian rhythms. Atomic-scale origins of slowness in the cyanobacterial circadian clock. Science 349(6245):312–316. https://doi.org/10.1126/science.1261040

Arita K, Hashimoto H, Igari K, Akaboshi M, Kutsuna S, Sato M, Shimizu T (2007) Structural and biochemical characterization of a cyanobacterium circadian clock-modifier protein. J Biol Chem 282(2):1128–1135. https://doi.org/10.1074/jbc.M608148200

Axmann IM, Dühring U, Seeliger L, Arnold A, Vanselow JT, Kramer A, Wilde A (2009) Biochemical evidence for a timing mechanism in *Prochlorococcus*. J Bacteriol 191(17):5342–5347. https://doi.org/10.1128/JB.00419-09

Axmann IM, Hertel S, Wiegard A, Dörrich AK, Wilde A (2014) Diversity of KaiC-based timing systems in marine cyanobacteria. Marine genomics 14:3–16. https://doi.org/10.1016/j.margen.2013.12.006

Beck C, Hertel S, Rediger A, Lehmann R, Wiegard A, Kölsch A, Heilmann B, Georg J, Hess WR, Axmann IM (2014) A daily expression pattern of protein-coding genes and small non-coding RNAs in *Synechocystis* sp. PCC 6803. Appl Environ Microbiol. https://doi.org/10.1128/aem.01086-14

Boyd JS, Bordowitz JR, Bree AC, Golden SS (2013) An allele of the crm gene blocks cyanobacterial circadian rhythms. Proc Natl Acad Sci U S A 110(34):13950–13955. https://doi.org/10.1073/pnas.1312793110

Brettschneider C, Rose RJ, Hertel S, Axmann IM, Heck AJR, Kollmann M (2010) A sequestration feedback determines dynamics and temperature entrainment of the KaiABC circadian clock. Mol Syst Biol 6:1–10. https://doi.org/10.1038/msb.2010.44

Červený J, Nedbal L (2009) Metabolic rhythms of the cyanobacterium *Cyanothece* sp. ATCC 51142 correlate with modeled dynamics of circadian clock. J Biol Rhythms 24(4):295–303. https://doi.org/10.1177/0748730409338367

Chang YG, Cohen SE, Phong C, Myers WK, Kim YI, Tseng R, Lin J, Zhang L, Boyd JS, Lee Y, Kang S, Lee D, Li S, Britt RD, Rust MJ, Golden SS, LiWang A (2015) Circadian rhythms. A protein fold switch joins the circadian oscillator to clock output in cyanobacteria. Science 349 (6245):324–328. https://doi.org/10.1126/science.1260031

Chang YG, Kuo NW, Tseng R, LiWang A (2011) Flexibility of the C-terminal, or CII, ring of KaiC governs the rhythm of the circadian clock of cyanobacteria. Proc Natl Acad Sci U S A 108(35):14431–14436. https://doi.org/10.1073/pnas.1104221108

Chang YG, Tseng R, Kuo NW, LiWang A (2012) Rhythmic ring-ring stacking drives the circadian oscillator clockwise. Proc Natl Acad Sci U S A 109(42):16847–16851. https://doi.org/10.1073/pnas.1211508109

Chen AH, Lubkowicz D, Yeong V, Chang RL, Silver PA (2015) Transplantability of a circadian clock to a noncircadian organism. Sci Adv 1(5). https://doi.org/10.1126/sciadv.1500358

Chew J, Leypunskiy E, Lin J, Murugan A, Rust MJ (2018) High protein copy number is required to suppress stochasticity in the cyanobacterial circadian clock. Nature communications 9 (1):3004. https://doi.org/10.1038/s41467-018-05109-4

Chlipala GE, Mo S, Orjala J (2011) Chemodiversity in freshwater and terrestrial cyanobacteria—a source for drug discovery. Curr Drug Targets 12(11):1654–1673

Clodong S, Dühring U, Kronk L, Wilde A, Axmann I, Herzel H, Kollmann M (2007) Functioning and robustness of a bacterial circadian clock. Mol Syst Biol 3:90. https://doi.org/10.1038/msb4100128

Cohen SE, Erb ML, Selimkhanov J, Dong G, Hasty J, Pogliano J, Golden SS (2014) Dynamic localization of the cyanobacterial circadian clock proteins. Curr Biol 24(16):1836–1844. https://doi.org/10.1016/j.cub.2014.07.036

DeWoskin D, Geng W, Stinchcombe AR, Forger DB (2014) It is not the parts, but how they interact that determines the behaviour of circadian rhythms across scales and organisms. Interface focus 4(3):20130076. https://doi.org/10.1098/rsfs.2013.0076

Diamond S, Jun D, Rubin BE, Golden SS (2015) The circadian oscillator in *Synechococcus elongatus* controls metabolite partitioning during diurnal growth. Proc Natl Acad Sci U S A 112(15):E1916–E1925. https://doi.org/10.1073/pnas.1504576112

Dong G, Yang Q, Wang Q, Kim YI, Wood TL, Osteryoung KW, van Oudenaarden A, Golden SS (2010) Elevated ATPase activity of KaiC applies a circadian checkpoint on cell division in *Synechococcus elongatus*. Cell 140(4):529–539. S0092-8674(09)01628-6 [pii]. https://doi.org/10.1016/j.cell.2009.12.042

Ducat DC, Way JC, Silver PA (2011) Engineering cyanobacteria to generate high-value products. Trends Biotechnol 29(2):95–103. https://doi.org/10.1016/j.tibtech.2010.12.003

Dvornyk V, Deng HW, Nevo E (2004) Structure and molecular phylogeny of *sasA* genes in cyanobacteria: insights into evolution of the prokaryotic circadian system. Mol Biol Evol 21 (8):1468–1476

Dvornyk V, Vinogradova O, Nevo E (2003) Origin and evolution of circadian clock genes in prokaryotes. Proc Natl Acad Sci U S A 100(5):2495–2500

Edgar RS, Green EW, Zhao Y, van Ooijen G, Olmedo M, Qin X, Xu Y, Pan M, Valekunja UK, Feeney KA, Maywood ES, Hastings MH, Baliga NS, Merrow M, Millar AJ, Johnson CH, Kyriacou CP, O'Neill JS, Reddy AB (2012) Peroxiredoxins are conserved markers of circadian rhythms. Nature 485(7399):459–464. https://doi.org/10.1038/nature11088

Egli M, Mori T, Pattanayek R, Xu Y, Qin X, Johnson CH (2012) Dephosphorylation of the core clock protein KaiC in the cyanobacterial KaiABC circadian oscillator proceeds via an ATP synthase mechanism. Biochemistry 51(8):1547–1558. https://doi.org/10.1021/bi201525n

Egli M, Pattanayek R, Sheehan JH, Xu Y, Mori T, Smith JA, Johnson CH (2013) Loop-loop interactions regulate KaiA-stimulated KaiC phosphorylation in the cyanobacterial KaiABC circadian clock. Biochemistry 52(7):1208–1220. https://doi.org/10.1021/bi301691a

Emberly E, Wingreen NS (2006) Hourglass model for a protein-based circadian oscillator. Phys Rev Lett 96(3):038303

Gallon JR, LaRue TA, Kurz WG (1974) Photosynthesis and nitrogenase activity in the blue-green alga *Gloeocapsa*. Can J Microbiol 20(12):1633–1637

Gan S, O'Shea EK (2017) An unstable singularity underlies stochastic phasing of the circadian clock in individual cyanobacterial cells. Mol Cell 67(4):659–672, e612. https://doi.org/10.1016/j.molcel.2017.07.015

Garces RG, Wu N, Gillon W, Pai EF (2004) Anabaena circadian clock proteins KaiA and KaiB reveal a potential common binding site to their partner KaiC. EMBO J 23(8):1688–1698. https://doi.org/10.1038/sj.emboj.7600190

Grobbelaar N, Huang TC, Lin HY, Chow TJ (1986) Dinitrogen-fixing endogenous rhythm in synechococcus Rf-1. FEMS Microbiol Lett 37(2):173–177. https://doi.org/10.1111/j.1574-6968.1986.tb01788.x

Guerreiro AC, Benevento M, Lehmann R, van Breukelen B, Post H, Giansanti P, Altelaar AF, Axmann IM, Heck AJ (2014) Daily rhythms in the cyanobacterium Synechococcus elongatus probed by high-resolution mass spectrometry based proteomics reveals a small-defined set of cyclic proteins. Mol Cell Proteomics. https://doi.org/10.1074/mcp.m113.035840

Gutu A, O'Shea EK (2013) Two antagonistic clock-regulated histidine kinases time the activation of circadian gene expression. Mol Cell 50 (2):288–294. S1097-2765(13)00178-0 [pii]. https://doi.org/10.1016/j.molcel.2013.02.022

Hanaoka M, Takai N, Hosokawa N, Fujiwara M, Akimoto Y, Kobori N, Iwasaki H, Kondo T, Tanaka K (2012) RpaB, another response regulator operating circadian clock-dependent transcriptional regulation in *Synechococcus elongatus* PCC 7942. J Biol Chem 287(31):26321–26327. https://doi.org/10.1074/jbc.M111.338251

Hardin PE, Hall JC, Rosbash M (1990) Feedback of the Drosophila period gene product on circadian cycling of its messenger RNA levels. Nature 343(6258):536–540. https://doi.org/10.1038/343536a0

Hayashi F, Itoh N, Uzumaki T, Iwase R, Tsuchiya Y, Yamakawa H, Morishita M, Onai K, Itoh S, Ishiura M (2004) Roles of two ATPase-motif-containing domains in cyanobacterial circadian clock protein KaiC. J Biol Chem 279(50):52331–52337. https://doi.org/10.1074/jbc.M406604200

Hayashi F, Iwase R, Uzumaki T, Ishiura M (2006) Hexamerization by the N-terminal domain and intersubunit phosphorylation by the C-terminal domain of cyanobacterial circadian clock protein KaiC. Biochem Biophys Res Commun 348(3):864–872. https://doi.org/10.1016/j.bbrc.2006.07.143

Hayashi F, Suzuki H, Iwase R, Uzumaki T, Miyake A, Shen JR, Imada K, Furukawa Y, Yonekura K, Namba K, Ishiura M (2003) ATP-induced hexameric ring structure of the cyanobacterial circadian clock protein KaiC. Genes Cells 8(3):287–296

Hays SG, Yan LLW, Silver PA, Ducat DC (2017) Synthetic photosynthetic consortia define interactions leading to robustness and photoproduction. J Biol Eng 11:4. https://doi.org/10.1186/s13036-017-0048-5

Hertel S, Brettschneider C, Axmann IM (2013) Revealing a two-loop transcriptional feedback mechanism in the cyanobacterial circadian clock. PLoS Comput Biol 9(3):e1002966. https://doi.org/10.1371/journal.pcbi.1002966

Hitomi K, Oyama T, Han S, Arvai AS, Getzoff ED (2005) Tetrameric architecture of the circadian clock protein KaiB. A novel interface for intermolecular interactions and its impact on the circadian rhythm. J Biol Chem 280(19):19127–19135

Holtzendorff J, Partensky F, Mella D, Lennon JF, Hess WR, Garczarek L (2008) Genome streamlining results in loss of robustness of the circadian clock in the marine cyanobacterium *Prochlorococcus marinus* PCC 9511. J Biol Rhythms 23(3):187–199

Hosokawa N, Kushige H, Iwasaki H (2013) Attenuation of the posttranslational oscillator via transcription-translation feedback enhances circadian-phase shifts in Synechococcus. Proc Natl Acad Sci U S A 110(35):14486–14491. https://doi.org/10.1073/pnas.1302243110

Iida T, Mutoh R, Onai K, Morishita M, Furukawa Y, Namba K, Ishiura M (2015) Importance of the monomer-dimer-tetramer interconversion of the clock protein KaiB in the generation of circadian oscillations in cyanobacteria. Genes Cells 20(3):173–190. https://doi.org/10.1111/gtc.12211

Ishiura M, Kutsuna S, Aoki S, Iwasaki H, Andersson CR, Tanabe A, Golden SS, Johnson CH, Kondo T (1998) Expression of a gene cluster *kaiABC* as a circadian feedback process in cyanobacteria. Science 281(5382):1519–1523

Ito H, Kageyama H, Mutsuda M, Nakajima M, Oyama T, Kondo T (2007) Autonomous synchronization of the circadian KaiC phosphorylation rhythm. Nat Struct Mol Biol 14 (11):1084–1088

Ito H, Mutsuda M, Murayama Y, Tomita J, Hosokawa N, Terauchi K, Sugita C, Sugita M, Kondo T, Iwasaki H (2009) Cyanobacterial daily life with Kai-based circadian and diurnal genome-wide transcriptional control in *Synechococcus elongatus*. Proc Natl Acad Sci U S A 106(33):14168–14173. https://doi.org/10.1073/pnas.0902587106

Ivleva NB, Bramlett MR, Lindahl PA, Golden SS (2005) LdpA: a component of the circadian clock senses redox state of the cell. EMBO J 24(6):1202–1210

Ivleva NB, Gao T, LiWang AC, Golden SS (2006) Quinone sensing by the circadian input kinase of the cyanobacterial circadian clock. Proc Natl Acad Sci U S A 103(46):17468–17473. https://doi.org/10.1073/pnas.0606639103

Iwasaki H, Nishiwaki T, Kitayama Y, Nakajima M, Kondo T (2002) KaiA-stimulated KaiC phosphorylation in circadian timing loops in cyanobacteria. Proc Natl Acad Sci U S A 99 (24):15788–15793

Iwasaki H, Taniguchi Y, Ishiura M, Kondo T (1999) Physical interactions among circadian clock proteins KaiA, KaiB and KaiC in cyanobacteria. EMBO J 18(5):1137–1145

Iwasaki H, Williams SB, Kitayama Y, Ishiura M, Golden SS, Kondo T (2000) A kaiC-interacting sensory histidine kinase, SasA, necessary to sustain robust circadian oscillation in cyanobacteria. Cell 101(2):223–233

Iwase R, Imada K, Hayashi F, Uzumaki T, Morishita M, Onai K, Furukawa Y, Namba K, Ishiura M (2005) Functionally important substructures of circadian clock protein KaiB in a unique tetramer complex. J Biol Chem 280(52):43141–43149

Johnson CH, Egli M (2014) Metabolic compensation and circadian resilience in prokaryotic cyanobacteria. Annu Rev Biochem 83:221–247. https://doi.org/10.1146/annurev-biochem-060713-035632

Johnson CH, Golden SS, Ishiura M, Kondo T (1996) Circadian clocks in prokaryotes. Mol Microbiol 21(1):5–11

Johnson CH, Stewart PL, Egli M (2011) The cyanobacterial circadian system: from biophysics to bioevolution. Annual review of biophysics 40:143–167. https://doi.org/10.1146/annurev-biophys-042910-155317

Kageyama H, Kondo T, Iwasaki H (2003) Circadian formation of clock protein complexes by KaiA, KaiB, KaiC, and SasA in cyanobacteria. J Biol Chem 278(4):2388–2395

Katayama M, Tsinoremas NF, Kondo T, Golden SS (1999) *cpmA*, a gene involved in an output pathway of the cyanobacterial circadian system. J Bacteriol 181(11):3516–3524

Kim Y-I, Dong G, Carruthers CW Jr, Golden SS, LiWang A (2008) The day/night switch in KaiC, a central oscillator component of the circadian clock of cyanobacteria. Proc Natl Acad Sci U S A 105(35):12825–12830

Kim YI, Vinyard DJ, Ananyev GM, Dismukes GC, Golden SS (2012) Oxidized quinones signal onset of darkness directly to the cyanobacterial circadian oscillator. Proc Natl Acad Sci U S A 109(44):17765–17769. https://doi.org/10.1073/pnas.1216401109

Kitayama Y, Iwasaki H, Nishiwaki T, Kondo T (2003) KaiB functions as an attenuator of KaiC phosphorylation in the cyanobacterial circadian clock system. EMBO J 22(9):2127–2134

Kitayama Y, Nishiwaki-Ohkawa T, Sugisawa Y, Kondo T (2013) KaiC intersubunit communication facilitates robustness of circadian rhythms in cyanobacteria. Nature communications 4:2897. https://doi.org/10.1038/ncomms3897

Kitayama Y, Nishiwaki T, Terauchi K, Kondo T (2008) Dual KaiC-based oscillations constitute the circadian system of cyanobacteria. Genes Dev 22(11):1513–1521

Knoop H, Steuer R (2015) A computational analysis of stoichiometric constraints and trade-offs in cyanobacterial biofuel production. Front Bioeng Biotechnol 3:47. https://doi.org/10.3389/fbioe.2015.00047

Kondo T, Mori T, Lebedeva NV, Aoki S, Ishiura M, Golden SS (1997) Circadian rhythms in rapidly dividing cyanobacteria. Science 275(5297):224–227

Kurosawa G, Aihara K, Iwasa Y (2006) A model for the circadian rhythm of cyanobacteria that maintains oscillation without gene expression. Biophys J 91(6):2015–2023. S0006-3495(06)71917-8 [pii]. https://doi.org/10.1529/biophysj.105.076554

Kurosawa S, Murakami R, Onai K, Morishita M, Hasegawa D, Iwase R, Uzumaki T, Hayashi F, Kitajima-Ihara T, Sakata S, Murakami M, Kouyama T, Ishiura M (2009) Functionally important structural elements of the cyanobacterial clock-related protein Pex. Genes Cells 14 (1):1–16. https://doi.org/10.1111/j.1365-2443.2008.01245.x

Kutsuna S, Kondo T, Aoki S, Ishiura M (1998) A period-extender gene, *pex*, that extends the period of the circadian clock in the cyanobacterium *Synechococcus* sp. strain PCC 7942. J Bacteriol 180(8):2167–2174

Kutsuna S, Kondo T, Ikegami H, Uzumaki T, Katayama M, Ishiura M (2007) The circadian clock-related gene *pex* regulates a negative cis element in the *kaiA* promoter region. J Bacteriol 189(21):7690–7696. https://doi.org/10.1128/JB.00835-07

Leipe DD, Aravind L, Grishin NV, Koonin EV (2000) The bacterial replicative helicase DnaB evolved from a RecA duplication. Genome Res 10(1):5–16

Lin J, Chew J, Chockanathan U, Rust MJ (2014) Mixtures of opposing phosphorylations within hexamers precisely time feedback in the cyanobacterial circadian clock. Proc Natl Acad Sci U S A 111(37):E3937–E3945. https://doi.org/10.1073/pnas.1408692111

Loeschcke A, Dienst D, Wewer V, Hage-Hülsmann J, Dietsch M, Kranz-Finger S, Huren V, Metzger S, Urlacher VB, Gigolashvili T, Kopriva S, Axmann IM, Drepper T, Jaeger KE (2017) The photosynthetic bacteria *Rhodobacter capsulatus* and *Synechocystis* sp. PCC 6803 as new hosts for cyclic plant triterpene biosynthesis. PLoS One 12 (12):e0189816. https://doi.org/10.1371/journal.pone.0189816

Maniscalco M, Nannen J, Sodi V, Silver G, Lowrey PL, Bidle KA (2014) Light-dependent expression of four cryptic archaeal circadian gene homologs. Frontiers in microbiology 5:79. https://doi.org/10.3389/fmicb.2014.00079

Markson JS, Piechura JR, Puszynska AM, O'Shea EK (2013) Circadian control of global gene expression by the cyanobacterial master regulator RpaA. Cell 155(6):1396–1408. https://doi.org/10.1016/j.cell.2013.11.005

Millineaux PM, Gallon JR, Chaplin AE (1981) Acetylene reduction (nitrogen fixation) by cyanobacteria grown under alternating light-dark cycles. FEMS Microbiol Lett 10(3):245–247. https://doi.org/10.1111/j.1574-6968.1981.tb06249.x

Min H, Guo H, Xiong J (2005) Rhythmic gene expression in a purple photosynthetic bacterium. Rhodobacter sphaeroides. FEBS Lett 579(3):808–812. https://doi.org/10.1016/j.febslet.2005.01.003

Mori T, Saveliev SV, Xu Y, Stafford WF, Cox MM, Inman RB, Johnson CH (2002) Circadian clock protein KaiC forms ATP-dependent hexameric rings and binds DNA. Proc Natl Acad Sci U S A 99(26):17203–17208

Mori T, Williams DR, Byrne MO, Qin X, Egli M, McHaourab HS, Stewart PL, Johnson CH (2007) Elucidating the ticking of an in vitro circadian clockwork. PLoS Biol 5(4):e93

Mukaiyama A, Furuike Y, Abe J, Yamashita E, Kondo T, Akiyama S (2018) Conformational rearrangements of the C1 ring in KaiC measure the timing of assembly with KaiB. Scientific reports 8(1):8803. https://doi.org/10.1038/s41598-018-27131-8

Murakami R, Miyake A, Iwase R, Hayashi F, Uzumaki T, Ishiura M (2008) ATPase activity and its temperature compensation of the cyanobacterial clock protein KaiC. Genes Cells 13(4):387–395. https://doi.org/10.1111/j.1365-2443.2008.01174.x

Murakami R, Mutoh R, Ishii K, Ishiura M (2016) Circadian oscillations of KaiA-KaiC and KaiB-KaiC complex formations in an in vitro reconstituted KaiABC clock oscillator. Genes Cells 21(8):890–900. https://doi.org/10.1111/gtc.12392

Murakami R, Mutoh R, Iwase R, Furukawa Y, Imada K, Onai K, Morishita M, Yasui S, Ishii K, Valencia Swain JO, Uzumaki T, Namba K, Ishiura M (2012) The roles of the dimeric and tetrameric structures of the clock protein KaiB in the generation of circadian oscillations in cyanobacteria. J Biol Chem 287(35):29506–29515. https://doi.org/10.1074/jbc.M112.349092

Mutoh R, Nishimura A, Yasui S, Onai K, Ishiura M (2013) The ATP-mediated regulation of KaiB-KaiC interaction in the cyanobacterial circadian clock. PLoS ONE 8(11):e80200. https://doi.org/10.1371/journal.pone.0080200

Nakahira Y, Katayama M, Miyashita H, Kutsuna S, Iwasaki H, Oyama T, Kondo T (2004) Global gene repression by KaiC as a master process of prokaryotic circadian system. Proc Natl Acad Sci U S A 101(3):881–885

Nakajima M, Imai K, Ito H, Nishiwaki T, Murayama Y, Iwasaki H, Oyama T, Kondo T (2005) Reconstitution of circadian oscillation of cyanobacterial KaiC phosphorylation in vitro. Science 308(5720):414–415

Nakajima M, Ito H, Kondo T (2010) In vitro regulation of circadian phosphorylation rhythm of cyanobacterial clock protein KaiC by KaiA and KaiB. FEBS Lett 584(5):898–902. https://doi.org/10.1016/j.febslet.2010.01.016

Nishiwaki-Ohkawa T, Kitayama Y, Ochiai E, Kondo T (2014) Exchange of ADP with ATP in the CII ATPase domain promotes autophosphorylation of cyanobacterial clock protein KaiC. Proc Natl Acad Sci U S A 111(12):4455–4460. https://doi.org/10.1073/pnas.1319353111

Nishiwaki T, Iwasaki H, Ishiura M, Kondo T (2000) Nucleotide binding and autophosphorylation of the clock protein KaiC as a circadian timing process of cyanobacteria. Proc Natl Acad Sci U S A 97(1):495–499

Nishiwaki T, Kondo T (2012) Circadian autodephosphorylation of cyanobacterial clock protein KaiC occurs via formation of ATP as intermediate. J Biol Chem 287(22):18030–18035. https://doi.org/10.1074/jbc.M112.350660

Nishiwaki T, Satomi Y, Kitayama Y, Terauchi K, Kiyohara R, Takao T, Kondo T (2007) A sequential program of dual phosphorylation of KaiC as a basis for circadian rhythm in cyanobacteria. EMBO J 26(17):4029–4037

Nishiwaki T, Satomi Y, Nakajima M, Lee C, Kiyohara R, Kageyama H, Kitayama Y, Temamoto M, Yamaguchi A, Hijikata A, Go M, Iwasaki H, Takao T, Kondo T (2004) Role of KaiC phosphorylation in the circadian clock system of *Synechococcus elongatus* PCC 7942. Proc Natl Acad Sci U S A 101(38):13927–13932

Osanai T, Shirai T, Iijima H, Kuwahara A, Suzuki I, Kondo A, Hirai MY (2015) Alteration of cyanobacterial sugar and amino acid metabolism by overexpression hik8, encoding a KaiC-associated histidine kinase. Environ Microbiol 17(7):2430–2440. https://doi.org/10.1111/1462-2920.12715

Oyama K, Azai C, Matsuyama J, Terauchi K (2018) Phosphorylation at Thr432 induces structural destabilization of the CII ring in the circadian oscillator KaiC. FEBS Lett 592(1):36–45. https://doi.org/10.1002/1873-3468.12945

Paijmans J, Lubensky DK, Ten Wolde PR (2017) Period robustness and entrainability of the Kai system to changing nucleotide concentrations. Biophys J 113(1):157–173. https://doi.org/10.1016/j.bpj.2017.05.048

Pattanayek R, Egli M (2015) Protein-protein interactions in the cyanobacterial circadian clock: structure of KaiA dimer in complex with C-terminal KaiC peptides at 2.8 a resolution. Biochemistry 54(30):4575–4578. https://doi.org/10.1021/acs.biochem.5b00694

Pattanayek R, Mori T, Xu Y, Pattanayek S, Johnson CH, Egli M (2009) Structures of KaiC circadian clock mutant proteins: a new phosphorylation site at T426 and mechanisms of kinase. ATPase and phosphatase. PLoS One 4(11):e7529. https://doi.org/10.1371/journal.pone.0007529

Pattanayek R, Sidiqi SK, Egli M (2012) Crystal structure of the redox-active cofactor dibromothymoquinone bound to circadian clock protein KaiA and structural basis for dibromothymoquinone's ability to prevent stimulation of KaiC phosphorylation by KaiA. Biochemistry 51(41):8050–8052. https://doi.org/10.1021/bi301222t

Pattanayek R, Wang J, Mori T, Xu Y, Johnson CH, Egli M (2004) Visualizing a circadian clock protein: crystal structure of KaiC and functional insights. Mol Cell 15(3):375–388

Pattanayek R, Williams DR, Pattanayek S, Mori T, Johnson CH, Stewart PL, Egli M (2008) Structural model of the circadian clock KaiB-KaiC complex and mechanism for modulation of KaiC phosphorylation. EMBO J 27(12):1767–1778. https://doi.org/10.1038/emboj.2008.104

Pattanayek R, Williams DR, Pattanayek S, Xu Y, Mori T, Johnson CH, Stewart PL, Egli M (2006) Analysis of KaiA-KaiC protein interactions in the cyano-bacterial circadian clock using hybrid structural methods. EMBO J 25(9):2017–2028

Pattanayek R, Williams DR, Rossi G, Weigand S, Mori T, Johnson CH, Stewart PL, Egli M (2011) Combined SAXS/EM based models of the *S. elongatus* post-translational circadian oscillator and its interactions with the output His-kinase SasA. PLoS One 6(8):e23697. https://doi.org/10.1371/journal.pone.0023697

Pattanayek R, Xu Y, Lamichhane A, Johnson CH, Egli M (2014) An arginine tetrad as mediator of input-dependent and input-independent ATPases in the clock protein KaiC. Acta Crystallogr D Biol Crystallogr 70(Pt 5):1375–1390. https://doi.org/10.1107/S1399004714003228

Phong C, Markson JS, Wilhoite CM, Rust MJ (2013) Robust and tunable circadian rhythms from differentially sensitive catalytic domains. Proc Natl Acad Sci U S A 110(3):1124–1129. https://doi.org/10.1073/pnas.1212113110

Pittayakanchit W, Lu Z, Chew J, Rust MJ, Murugan A (2018) Biophysical clocks face a trade-off between internal and external noise resistance. eLife 7. https://doi.org/10.7554/elife.37624

Qin X, Byrne M, Mori T, Zou P, Williams DR, McHaourab H, Johnson CH (2010a) Intermolecular associations determine the dynamics of the circadian KaiABC oscillator. Proc Natl Acad Sci U S A 107(33):14805–14810. https://doi.org/10.1073/pnas.1002119107

Qin X, Byrne M, Xu Y, Mori T, Johnson CH (2010b) Coupling of a core post-translational pacemaker to a slave transcription/translation feedback loop in a circadian system. PLoS Biol 8 (6):e1000394. https://doi.org/10.1371/journal.pbio.1000394

Rapp PE (1987) Why are so many biological systems periodic? Prog Neurobiol 29(3):261–273

Reddy AB, Rey G (2014) Metabolic and nontranscriptional circadian clocks: eukaryotes. Annu Rev Biochem 83:165–189. https://doi.org/10.1146/annurev-biochem-060713-035623

Roenneberg T, Merrow M (2016) The circadian clock and human health. Curr Biol 26(10):R432–R443. https://doi.org/10.1016/j.cub.2016.04.011

Rosbash M (2009) The implications of multiple circadian clock origins. PLoS Biol 7(3):e62. https://doi.org/10.1371/journal.pbio.1000062

Rust MJ, Golden SS, O'Shea EK (2011) Light-driven changes in energy metabolism directly entrain the cyanobacterial circadian oscillator. Science 331(6014):220–223. https://doi.org/10.1126/science.1197243

Rust MJ, Markson JS, Lane WS, Fisher DS, O'Shea EK (2007) Ordered phosphorylation governs oscillation of a three-protein circadian clock. Science 318(5851):809–812

Saha R, Liu D, Hoynes-O'Connor A, Liberton M, Yu J, Bhattacharyya-Pakrasi M, Balassy A, Zhang F, Moon TS, Maranas CD, Pakrasi HB (2016) Diurnal regulation of cellular processes in the cyanobacterium synechocystis sp. Strain PCC 6803: insights from transcriptomic, fluxomic, and physiological analyses. mBio 7(3). https://doi.org/10.1128/mbio.00464-16

Schmelling NM, Axmann IM (2018) Computational modelling unravels the precise clockwork of cyanobacteria. Interface focus 8(6):20180038. https://doi.org/10.1098/rsfs.2018.0038

Schmelling NM, Lehmann R, Chaudhury P, Beck C, Albers SV, Axmann IM, Wiegard A (2017) Minimal tool set for a prokaryotic circadian clock. BMC Evol Biol 17(1):169. https://doi.org/10.1186/s12862-017-0999-7

Shultzaberger RK, Boyd JS, Diamond S, Greenspan RJ, Golden SS (2015) Giving time purpose: the synechococcus elongatus clock in a broader network context. Annu Rev Genet 49:485–505. https://doi.org/10.1146/annurev-genet-111212-133227

Smith RM, Williams SB (2006) Circadian rhythms in gene transcription imparted by chromosome compaction in the cyanobacterium *Synechococcus elongatus*. Proc Natl Acad Sci U S A 103 (22):8564–8569. 0508696103 [pii]. https://doi.org/10.1073/pnas.0508696103

Snijder J, Burnley RJ, Wiegard A, Melquiond AS, Bonvin AM, Axmann IM, Heck AJ (2014) Insight into cyanobacterial circadian timing from structural details of the KaiB-KaiC interaction. Proc Natl Acad Sci U S A 111(4):1379–1384. https://doi.org/10.1073/pnas.1314326111

Snijder J, Schuller JM, Wiegard A, Lössl P, Schmelling N, Axmann IM, Plitzko JM, Förster F, Heck AJ (2017) Structures of the cyanobacterial circadian oscillator frozen in a fully assembled state. Science 355(6330):1181–1184. https://doi.org/10.1126/science.aag3218

Stal LJ, Krumbein WE (1987) Temporal separation of nitrogen fixation and photosynthesis in the filamentous, non-heterocystous cyanobacterium *Oscillatoria* sp. Arch Microbiol 149(1):76–80. https://doi.org/10.1007/bf00423140

Sweeney BM, Borgese MB (1989) A circadian rhythm in cell division in a prokaryote, the cyanobacterium *Synechococcus* WH7803. J Phycol 25(1):183–186

Takai N, Nakajima M, Oyama T, Kito R, Sugita C, Sugita M, Kondo T, Iwasaki H (2006) A KaiC-associating SasA-RpaA two-component regulatory system as a major circadian timing mediator in cyanobacteria. Proc Natl Acad Sci U S A 103(32):12109–12114

Taniguchi Y, Katayama M, Ito R, Takai N, Kondo T, Oyama T (2007) *labA*: a novel gene required for negative feedback regulation of the cyanobacterial circadian clock protein KaiC. Genes Dev 21(1):60–70

Taniguchi Y, Takai N, Katayama M, Kondo T, Oyama T (2010) Three major output pathways from the KaiABC-based oscillator cooperate to generate robust circadian kaiBC expression in cyanobacteria. Proc Natl Acad Sci U S A. 0909924107 [pii]. https://doi.org/10.1073/pnas.0909924107

Taniguchi Y, Yamaguchi A, Hijikata A, Iwasaki H, Kamagata K, Ishiura M, Go M, Kondo T (2001) Two KaiA-binding domains of cyanobacterial circadian clock protein KaiC. FEBS Lett 496(2–3):86–90

Teng SW, Mukherji S, Moffitt JR, de Buyl S, O'Shea EK (2013) Robust circadian oscillations in growing cyanobacteria require transcriptional feedback. Science 340(6133):737–740. 340/6133/737 [pii]. https://doi.org/10.1126/science.1230996

Terauchi K, Kitayama Y, Nishiwaki T, Miwa K, Murayama Y, Oyama T, Kondo T (2007) ATPase activity of KaiC determines the basic timing for circadian clock of cyanobacteria. Proc Natl Acad Sci U S A 104(41):16377–16381

Tomita J, Nakajima M, Kondo T, Iwasaki H (2005) No transcription-translation feedback in circadian rhythm of KaiC phosphorylation. Science 307(5707):251–254. https://doi.org/10.1126/science.1102540

Tseng R, Chang YG, Bravo I, Latham R, Chaudhary A, Kuo NW, Liwang A (2014) Cooperative KaiA-KaiB-KaiC interactions affect KaiB/SasA competition in the circadian clock of cyanobacteria. J Mol Biol 426:389–402. https://doi.org/10.1016/j.jmb.2013.09.040

Tseng R, Goularte NF, Chavan A, Luu J, Cohen SE, Chang YG, Heisler J, Li S, Michael AK, Tripathi S, Golden SS, LiWang A, Partch CL (2017) Structural basis of the day-night transition in a bacterial circadian clock. Science 355(6330):1174–1180. https://doi.org/10.1126/science.aag2516

Ungerer J, Wendt KE, Hendry JI, Maranas CD, Pakrasi HB (2018) Comparative genomics reveals the molecular determinants of rapid growth of the cyanobacterium Synechococcus elongatus UTEX 2973. Proc Natl Acad Sci U S A 115(50):E11761–E11770. https://doi.org/10.1073/pnas.1814912115

Uzumaki T, Fujita M, Nakatsu T, Hayashi F, Shibata H, Itoh N, Kato H, Ishiura M (2004) Crystal structure of the C-terminal clock-oscillator domain of the cyanobacterial KaiA protein. Nat Struct Mol Biol 11(7):623–631. https://doi.org/10.1038/nsmb781

Vakonakis I, Klewer DA, Williams SB, Golden SS, LiWang AC (2004a) Structure of the N-terminal domain of the circadian clock-associated histidine kinase SasA. J Mol Biol 342 (1):9–17. https://doi.org/10.1016/j.jmb.2004.07.010

Vakonakis I, Sun J, Wu T, Holzenburg A, Golden SS, LiWang AC (2004b) NMR structure of the KaiC-interacting C-terminal domain of KaiA, a circadian clock protein: implications for KaiA-KaiC interaction. Proc Natl Acad Sci U S A 101(6):1479–1484. https://doi.org/10.1073/pnas.0305516101

Valencia SJ, Bitou K, Ishii K, Murakami R, Morishita M, Onai K, Furukawa Y, Imada K, Namba K, Ishiura M (2012) Phase-dependent generation and transmission of time information by the KaiABC circadian clock oscillator through SasA-KaiC interaction in cyanobacteria. Genes Cells 17(5):398–419. https://doi.org/10.1111/j.1365-2443.2012.01597.x

van Zon JS, Lubensky DK, Altena PR, ten Wolde PR (2007) An allosteric model of circadian KaiC phosphorylation. Proc Natl Acad Sci U S A 104(18):7420–7425

Villarreal SA, Pattanayek R, Williams DR, Mori T, Qin X, Johnson CH, Egli M, Stewart PL (2013) CryoEM and molecular dynamics of the circadian KaiB-KaiC complex indicates KaiB monomers interact with KaiC and block ATP binding clefts. J Mol Biol 425:3311–3324. https://doi.org/10.1016/j.jmb.2013.06.018

Vosshall LB, Price JL, Sehgal A, Saez L, Young MW (1994) Block in nuclear localization of period protein by a second clock mutation, timeless. Science 263(5153):1606–1609

Weare NM, Benemann JR (1974) Nitrogenase activity and photosynthesis in Plectonema boryanum. J Bacteriol 119(1):258–265

Weiss TL, Young EJ, Ducat DC (2017) A synthetic, light-driven consortium of cyanobacteria and heterotrophic bacteria enables stable polyhydroxybutyrate production. Metab Eng 44:236–245. https://doi.org/10.1016/j.ymben.2017.10.009

Welkie DG, Rubin BE, Diamond S, Hood RD, Savage DF, Golden SS (2019) A hard day's night: cyanobacteria in diel cycles. Trends Microbiol 27(3):231–242. https://doi.org/10.1016/j.tim.2018.11.002

Wiegard A, Dörrich AK, Deinzer HT, Beck C, Wilde A, Holtzendorff J, Axmann IM (2013) Biochemical analysis of three putative KaiC clock proteins from *Synechocystis* sp. PCC 6803 suggests their functional divergence. Microbiology 159:948–958. https://doi.org/10.1099/mic.0.065425-0

Williams SB, Vakonakis I, Golden SS, LiWang AC (2002) Structure and function from the circadian clock protein KaiA of *Synechococcus elongatus*: a potential clock input mechanism. Proc Natl Acad Sci U S A 99(24):15357–15362. https://doi.org/10.1073/pnas.232517099

Woelfle MA, Xu Y, Qin X, Johnson CH (2007) Circadian rhythms of superhelical status of DNA in cyanobacteria. Proc Natl Acad Sci U S A 104:18819–18824

Wood TL, Bridwell-Rabb J, Kim YI, Gao T, Chang YG, LiWang A, Barondeau DP, Golden SS (2010) The KaiA protein of the cyanobacterial circadian oscillator is modulated by a redox-active cofactor. Proc Natl Acad Sci U S A 107(13):5804–5809. https://doi.org/10.1073/pnas.0910141107

Wyatt JT, Silvey JK (1969) Nitrogen fixation by gloeocapsa. Science 165(3896):908–909. https://doi.org/10.1126/science.165.3896.908

Xu Y, Mori T, Pattanayek R, Pattanayek S, Egli M, Johnson CH (2004) Identification of key phosphorylation sites in the circadian clock protein KaiC by crystallographic and mutagenetic analyses. Proc Natl Acad Sci U S A 101(38):13933–13938

Xu Y, Mori T, Qin X, Yan H, Egli M, Johnson CH (2009) Intramolecular regulation of phosphorylation status of the circadian clock protein KaiC. PLoS ONE 4(11):e7509. https://doi.org/10.1371/journal.pone.0007509

Xu Y, Weyman PD, Umetani M, Xiong J, Qin X, Xu Q, Iwasaki H, Johnson CH (2013) Circadian yin-yang regulation and its manipulation to globally reprogram gene expression. Curr Biol 23(23):2365–2374. https://doi.org/10.1016/j.cub.2013.10.011

Yang Q, Pando BF, Dong G, Golden SS, van Oudenaarden A (2010) Circadian gating of the cell cycle revealed in single cyanobacterial cells. Science 327(5972):1522–1526. https://doi.org/10.1126/science.1181759

Ye S, Vakonakis I, Ioerger TR, LiWang AC, Sacchettini JC (2004) Crystal structure of circadian clock protein KaiA from Synechococcus elongatus. J Biol Chem 279(19):20511–20518. https://doi.org/10.1074/jbc.M400077200

Yoda M, Eguchi K, Terada TP, Sasai M (2007) Monomer-shuffling and allosteric transition in KaiC circadian oscillation. PLoS ONE 2(5):e408

Zwicker D, Lubensky DK, Ten Wolde PR (2010) Robust circadian clocks from coupled protein-modification and transcription-translation cycles. Proc Natl Acad Sci U S A 107 (52):22540–22545. 1007613107 [pii]. https://doi.org/10.1073/pnas.1007613107

Chapter 13
Frataxin Structure and Function

Ignacio Hugo Castro, María Florencia Pignataro, Karl Ellioth Sewell, Lucía Daniela Espeche, María Georgina Herrera, Martín Ezequiel Noguera, Liliana Dain, Alejandro Daniel Nadra, Martín Aran, Clara Smal, Mariana Gallo and Javier Santos

Abstract Mammalian frataxin is a small mitochondrial protein involved in iron sulfur cluster assembly. Frataxin deficiency causes the neurodegenerative disease Friedreich's Ataxia. Valuable knowledge has been gained on the structural dynamics of frataxin, metal-ion-protein interactions, as well as on the effect of mutations on protein conformation, stability and internal motions. Additionally, laborious studies concerning the enzymatic reactions involved have allowed for understanding the capability of frataxin to modulate Fe–S cluster assembly function. Remarkably, frataxin biological function depends on its interaction with some proteins to form a supercomplex, among them NFS1 desulfurase and ISCU, the scaffolding protein. By combining multiple experimental tools including high

---

I. H. Castro · M. F. Pignataro · K. E. Sewell · M. G. Herrera · M. E. Noguera · L. Dain · A. D. Nadra · J. Santos (✉)
Departamento de Fisiología y Biología Molecular y Celular, Facultad de Ciencia Exactas y Naturales, Instituto de Biociencias, Biotecnología y Biomedicina (iB3), Universidad de Buenos Aires, Intendente Güiraldes 2160—Ciudad Universitaria, 1428EGA C.A.B.A, Argentina
e-mail: javiersantosw@gmail.com

I. H. Castro
e-mail: ignacas22@gmail.com

M. F. Pignataro
e-mail: mariaflorenciapignataro@gmail.com

K. E. Sewell
e-mail: sewell144@gmail.com

M. G. Herrera
e-mail: geor.herr@gmail.com

M. E. Noguera
e-mail: mnoguera.unq@gmail.com

L. Dain
e-mail: lbdain@gmail.com

A. D. Nadra
e-mail: alenadra@gmail.com

J. R. Harris and J. Marles-Wright (eds.), *Macromolecular Protein Complexes II: Structure and Function*, Subcellular Biochemistry 93,
https://doi.org/10.1007/978-3-030-28151-9_13

resolution techniques like NMR and X-ray, but also SAXS, crosslinking and mass-spectrometry, it was possible to build a reliable model of the structure of the desulfurase supercomplex NFS1/ACP-ISD11/ISCU/frataxin. In this chapter, we explore these issues showing how the scientific view concerning frataxin structure-function relationships has evolved over the last years.

Keywords Frataxin · Structural dynamics · Structure-Function relationships · Iron binding · Iron–Sulfur cluster assembly · Conformational stability

Abbreviations

4'PPT	4'-phosphopantetheine
ACP	Acyl carrier protein
AFM	Atomic Force Microscopy
CD	Circular dichroism
CPMG	The Carr-Purcell-Meiboom-Gill pulse sequence
CTR	C-terminal region
CyaY	Frataxin from *E. coli*
DLS	Dynamic light scattering
DOSY	Diffusion order spectroscopy

I. H. Castro · M. F. Pignataro · K. E. Sewell · M. E. Noguera · J. Santos
Intituto de Química y Fisicoquímica Biológicas, Dr. Alejandro Paladini Universidad de Buenos Aires, CONICET, Junín 956, 1113AAD C.A.B.A, Argentina

L. D. Espeche · L. Dain
Departamento de Diagnóstico Genético, Centro Nacional de Genética Médica "Dr. Eduardo E. Castilla"—A.N.L.I.S, Av. Las Heras 2670, C1425ASQ C.A.B.A, Argentina
e-mail: luciadespeche@gmail.com

M. E. Noguera
Departamento de Ciencia y Tecnología, Universidad Nacional de Quilmes, Roque Sáenz Peña 352, B1876BXD Bernal, Provincia de Buenos Aires, Argentina

A. D. Nadra
Instituto de Química Biológica de la Facultad de Ciencias Exactas y Naturales (IQUIBICEN), Consejo Nacional de Investigaciones Científicas y Técnicas, Buenos Aires, Argentina

M. Aran · C. Smal
Fundación Instituto Leloir E IIBBA-CONICET, Av. Patricias Argentinas 435, C1405BWE Buenos Aires, Argentina
e-mail: MAran@leloir.org.ar

C. Smal
e-mail: clarasmal@gmail.com

M. Gallo
IRBM Science Park S.p.A, Via Pontina km 30,600, 00071 Pomezia, RM, Italy
e-mail: marianagalletita@gmail.com

EM	Electron microscopy
FDX	Ferredoxin
Fe–S	Iron–sulfur
FRDA	Friedreich's Ataxia
FXN	Frataxin
HPLC	High-performance liquid chromatography
HSQC	Heteronuclear single quantum coherence spectroscopy
ISCU	Iron–sulfur cluster assembly enzyme
ISD11	NFS1 interacting protein
ITC	Isothermal titration calorimetry
NFS1	Mitochondrial desulfurase enzyme
NMR	Nuclear magnetic resonance
NOE	Nuclear Overhauser effect
PAGE	Polyacrylamide gel electrophoresis
PDB	Protein Data Bank
RMSD	Root-mean-square deviation
SAXS	Small-angle X-ray scattering
SDS	Sodium dodecyl sulfate
SEC	Size exclusion chromatography
SUF	Sulfur assimilation

Friedreich's Ataxia and the History of Frataxin: Why This Chapter is Focused on Frataxin

Friedreich's Ataxia is a rare autosomal recessive genetic disease (ORPHA95), which is characterized by progressive degeneration and damage of the central and peripheral nervous systems. It has a prevalence of approximately 1:50,000, representing half of all the genetic ataxias and three-quarters of the inherited ataxias in individuals younger than 25 years of age (Pandolfo 2009). The first symptoms occur typically in 8–15-year-olds and some of the most common are ataxia of gait and limbs, absence of tendon reflexes, loss of position sense and dysarthria (alteration in the articulation of words). Remarkably, two-thirds of patients have hypertrophic cardiomyopathy at the time of diagnosis (Durr et al. 1996). The name of this ataxia comes from Dr. Nikolaus Friedreich (1825–1882) who described the disease in five articles published during the late 19th century (Koeppen 2013), presenting the clinical findings of patients with "a severe hereditary disorder of the nervous system". In these works, Dr. Friedreich identified the main features of the disease and highlighted that it is hereditary and that its onset usually occurs before puberty. It is worthy of mention that Dr. Friedreich described the existence of hypertrophic cardiomyopathy, which is now recognized as the main cause of death in Friedreich's ataxia.

A hundred years later, in 1988, the genetic alteration responsible for this disease was localized in the long arm of chromosome 9 by Chamberlain and coworkers (Chamberlain et al. 1988, 1989); whereas in 1996, a gene termed *X25*, was identified in the critical region for the FRDA locus by Campuzano and coworkers (Campuzano et al. 1996). The *X25* gene codes for a small protein of 210 amino acid residues that was named frataxin (FXN).

In this seminal work, Campuzano and collaborators described the structure of the FXN gene and found that most patients with Friedreich's Ataxia (98%) were homozygous for a GAA trinucleotide expansion of 700–800 repeats in most of the cases, localized in the first intron of the gene. Nowadays we also know that the expansion ranges from 70 to more than 1000 GAA triplets in patients, while the first intron of the FXN gene in normal chromosomes contains up to 35 to 40 GAA triplets (Pandolfo 2006; Patel and Isaya 2001). Additionally, close to 2–5% of patients exhibit a compound genotype having an expansion in one allele and a point mutation in the other. The triplet expansion alters FXN gene transcription, decreasing its protein expression, which is essential for viability in mammals. Moreover, FXN has homologs in very distant organisms, like bacteria and yeast. Even if it was soon discovered that FXN is involved in assembly of iron-sulfur clusters, the biological function of FXN have remained elusive for long time. Structural biology and biochemistry concerning this protein have advanced notably in several directions during the last two decades. In particular, during the last 10 years, there have been extremely interesting findings that have changed the paradigm regarding the function of FXN, placing FXN in a new macromolecular context.

The Long Path from Structure to Function: FXN Is Involved in Iron–Sulfur Cluster Assembly

After unambiguously linking the FXN gene to Friedreich's Ataxia, Campuzano and collaborators were able to predict the secondary structure for the *X25*-encoded protein: an alpha helical structure for the regions between residues 90–110 and 185–195, with a possible beta sheet region around residues 125–145 and 175–180 (Campuzano et al. 1996). Furthermore, they suggested that the first 22 amino acids might constitute a cleavable N-terminal signal peptide and, furthermore, no transmembrane domain was identified, indicating that FXN is a soluble protein. During the same year, it was shown that FXN has mitochondrial localization (Koutnikova et al. 1997; Priller et al. 1997).

The first experimental structure for FXN was obtained four years later by Dhe-Paganon and coworkers (PDBID: 1EKG). The crystallographic structure (with a 1.8 Å resolution) provided new insights (Dhe-Paganon et al. 2000). First, it revealed a novel fold and confirmed Campuzanos' secondary structure predictions: two parallel alpha-helical elements supported by a platform, which is provided by a five-stranded, antiparallel beta sheet (Fig. 13.1). By examining the crystallographic structure (PDB ID: 1EKG), one may find that the correct boundaries for the helical

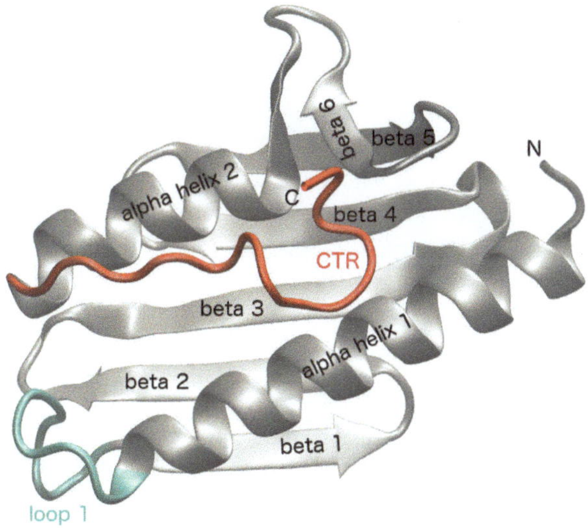

Fig. 13.1 The human FXN globular domain. The ribbon model of PDB ID: 1EKG is shown. Loop-1 (Cyan), as well as α-helices and β-strands are indicated. Additionally, CTR is depicted in red. N and C indicate the N- and the C-terminal residues, respectively

stretches are Glu92 to Ala114 for the first and Leu182 to Leu194 for the second alpha helix. Additionally, a C-terminal segment of 15 residues, with non-periodic structure lays over the alpha helices. Throughout this chapter, this segment will be called the C-terminal region (CTR). Second, the authors identified two different protein surfaces, both very conserved through evolution. One of them, the region known as the "acidic ridge", is formed by several acidic residues of the alpha helix 1, loop-1 and the beta strand 1 (Fig. 13.2) and it comprises ~915 Å^2, accounting for ~20% of the total accessible surface area of the protein. The other conserved surface forms an extended patch of ~1000 Å^2 involving apolar and positively charged residues from the beta-sheet. Interestingly, considering the amino acid conservation profile, the researchers predicted that this surface must be involved in protein-protein interaction processes. Given that some mutations that result in FRDA are in the anionic (D122Y) or in the apolar/positive (G130 V, W155R, and R165C) surfaces, the authors suggested that these surfaces may be critical for FXN function (Fig. 13.3). Moreover, structure-function-phenotype relationships were suggested because some of these mutations result in specific FRDA phenotypes. Without any doubt, the interpretation of these findings was confirmed some years later when FXN was put in context and its essential interaction with other proteins, by means of these surfaces, was established. We will analyze this issue in detail later in this chapter.

Some controversies have emerged regarding the processing of FXN inside the mitochondria and as to which is the functional species. Nevertheless, it was found that FXN has a disordered N-terminal region (Popovic et al. 2015; Prischi et al. 2009; Castro et al. 2018), with a mitochondrial translocation signal that spans the first 41 residues. FXN is processed at least in two steps, the first producing an intermediary form (FXN 42-210) and the second producing the mature (FXN 81-210) and full-functional form (Koutnikova et al. 1998; Schmucker et al. 2008;

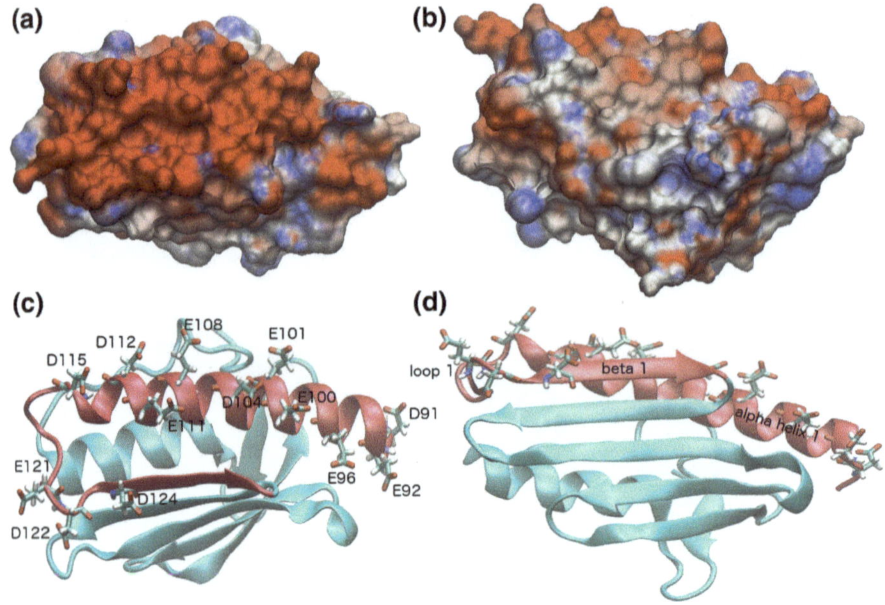

Fig. 13.2 The negative charged surface of the acidic ridge of FXN. **a** and **b** are two different views of the representation of the FXN electrostatic surface. **c** and **d** are in the same orientation as panels (**a**) and (**b**), respectively, but show the topology of the protein. In (**c**), Glu and Asp residues present in the acidic ridge (alpha helix 1, loop-1 and beta strand 1, in red ribbon) are shown. Electrostatics was calculated using APBS and the color scale ranges from +5 (blue) to −5 kT/e (red)

Cavadini et al. 2000a) (Fig. 13.4). Additionally, the first ~ten residues of the mature form (residues 81 to 90) are disordered, though they may be functionally relevant (Gentry et al. 2013).

It is worthy of mention that at least three different isoforms of FXN (type I, II, III) result from alternative splicing (Xia et al. 2012). Type I is the canonical isoform that is observed in mitochondria and presents 130-aminoacid residues with a molecular weight of 14.2 kDa. Type II is 135-residue long, with an expected molecular mass of 14.9 kDa, meanwhile FXN III has 164 residues with a molecular mass of 18.2 kDa. FXN II and III are located mostly in the cytosol and nucleus, respectively. The expression of these isoforms is tissue-specific. Whereas FXN type II is abundant in the nervous system, type III is abundant in the heart (Xia et al. 2012). Type III showed great iron binding capacity (reducing its toxicity) and it promoted Fe-S production and the increase of aconitase activity (Xia et al. 2012).

Recent reports have described that erythrocytes have a FXN isoform called isoform E (Guo et al. 2018). This protein is 135-amino acids long (similar to type II isoform), lacks the mitochondrial targeting sequence and is N-terminally acetylated, in the first methionine residue. This isoform is three times more abundant in

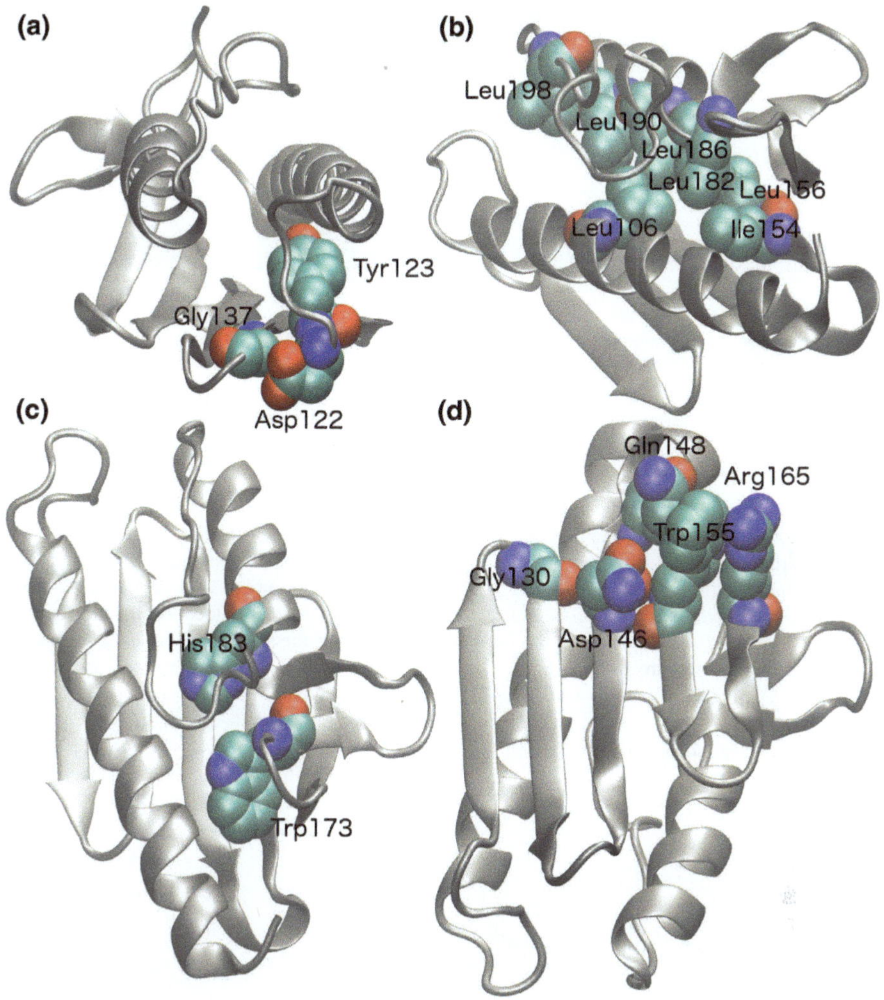

Fig. 13.3 Crucial Residues for FXN Stability and Function. Panels **a**, **b**, **c** and **d** show some of the residues involved in FRDA. Mutations that result in the FRDA phenotype affect stability and/or protein function. They are the following: in the region of loop-1 and β-turn β2-β3: G137V, D122Y, Y122D; residues located in the core region of the protein: L106S, I154F, L156P, W173G, L182F, H183R, L182H, L186R, L190P, L198R; residues located on the accessible face of the β-sheet and β-turn β1-β2: G130V, N146K, Q148R, W155R, R165C, R165D, R165N, R165P

erythrocytes than the mature mitochondrial FXN present in other blood cells (20.9 ± 6.4 ng/mL vs. 7.1 ± 1.0 ng/mL, respectively) (Guo et al. 2018).

It was initially proposed that FXN was involved in iron homeostasis because iron concentrations increase in the mitochondria of FRDA cells. In this context, the presence of the extended acidic ridge suggested that, as for other proteins like Ferritins, iron binding would be mediated by clusters of acidic side-chains of

Fig. 13.4 The Sequence of the Precursor Form of Human FXN (residues 1–210). MISP is the mitochondrial import signal peptide. In addition, boxes corresponding to α-helices, loops and β-strands are represented accordingly to PDB ID: 1EKG in magenta, cyan and blue, respectively. The C-terminal region (CTR) of FXN is highlighted in red. The amino acid sequence 42–56 is highlighted in gray. Points of processing during the import to the mitochondria: the first cut between residues 41 and 42 yielding the intermediate form and the second cut between residue 80 and 81 yielding the mature form of FXN

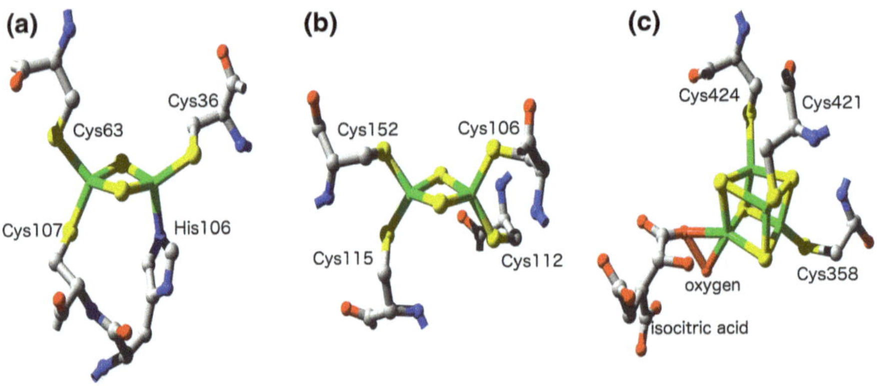

Fig. 13.5 The atomic structure **of clusters**. Rhombic iron–sulfur clusters (**[2Fe–2S]**) from **a** PDB ID: 2Z7E, mounted on three Cys and one His residue (crystal structure of *Aquifex aeolicus* ISCU) or mounted on Cys residues **(b)** from PDB ID: 3P1M (crystal structure of human FDX1). In **c** the structure of a cubane-type Fe–S cluster from PDB ID: 1C97 is shown (the crystal structure of mutant S642A of mitochondrial aconitase from *Bos taurus* in complex with isocitrate and oxygen). Iron atoms are represented in green

glutamic and aspartic residues. In addition, the presence of acidic residues suggested that FXN might play a role in regulating mitochondrial iron transport. Interestingly, knocking-out the FXN homolog gene in *E. coli* did not affect cellular iron content and sensitivity to oxidants (Li et al. 1999). This fact indicates that at least bacterial FXN may have a different function than iron binding and/or protection against the oxidative stress. Because of the deficiency in the activity of the aconitase enzyme and the respiratory complexes I, II and III (all of them containing Fe–S clusters), iron–sulfur (Fe–S) proteins were regarded as the targeted proteins in FRDA (Rotig et al. 1997). However, as the Fe–S cluster is highly sensitive to free radicals, FRDA was linked to iron accumulation.

Fe–S clusters are inorganic cofactors of proteins with unusual electronic configurations, critical for electron transfer reactions involving single electrons (Fig. 13.5). Rhombic clusters ([2Fe–2S]) are found in proteins with reductase activity such as ferredoxins and glutaredoxins. The coalescence of two clusters of [2Fe–2S] type yields a cubane-type cluster [4Fe–4S]. This type of cluster is present, for example, in mitochondrial aconitase. The two iron atoms in the upper part of the cubane-type Fe S cluster share a hybrid orbital in which a single electron is delocalized. In the lower plane a similar delocalization occurs yielding reactivity (Rouault 2015).

Wong and collaborators showed that the absence of FXN confers cellular sensitivity to oxidative stress (Wong et al. 1999). Even though iron chelators rescued FRDA fibroblasts from oxidant-induced death (iron is a key reagent in the Fenton reactions), they found only a slight increase of 40% in the mean mitochondrial iron content of FRDA cells.

Remarkably, using a yeast FXN deletant strain as a model and iron chelator in the culture medium, the mitochondrial iron concentration was found to be similar to the one observed for the wild-type yeast, whereas the activity of the respiratory complexes was restored. However, the activity of the mitochondrial aconitase remained low, suggesting that incorporation of the iron–sulfur cluster into some enzymes might be impaired in a way that is not linked to iron toxicity (Foury 1999). Also, in the absence of yeast FXN, the activity of succinate dehydrogenase (another Fe–S-containing enzyme) proved to be diminished, whereas the activity of a non-Fe–S-containing mitochondrial enzymes, the malate dehydrogenase, was unaffected, suggesting a direct role of FXN in the formation of Fe-S clusters (Chen et al. 2002). The direct link between the lack of FXN function and Fe–S cluster assembly deficiency was proved shortly after that observation.

The Gibson's group studied the phylogenetic distribution of FXN and the co-occurrence of other genes in several genomes and found that FXN has the same distribution as the molecular chaperones hscA and hscB/JAC1, specialized forms of Hsp70 and their J protein co-chaperone, respectively, thus suggesting a direct role in the iron–sulfur cluster protein assembly rather than in iron homeostasis (Huynen et al. 2001). From an experimental viewpoint, but in the same vein, Duby and coworkers showed that even though FXN was not essential for Fe–S protein assembly in yeast, its presence improved the efficiency of the process (Duby et al. 2002). Remarkably, Foury's group demonstrated, by means of an in vivo screening, that yeast FXN and the ISU1 (which is the mitochondrial scaffold protein for the Fe-S cluster assembly), have strongly linked functions, acting in concert to promote Fe-S cluster assembly (Ramazzotti et al. 2004). Cross-linking experiments in isolated mitochondria provided the evidence for a direct interaction. Meanwhile, by using a mitochondrial glutathione-S transferase fusion protein of yeast FXN, Gerber and collaborators showed that FXN formed a complex with ISU1 and the cysteine desulfurase NFS1, which is the key enzyme that provides the persulfur group, necessary for cluster assembly (Muhlenhoff et al. 2003; Gerber et al. 2003). Moreover, interaction between FXN and these proteins was strikingly increased by iron. This dependence on iron suggested a role for FXN in iron loading of the ISU1.

In fact, it was shown that human FXN is able to bind Fe^{2+} and Fe^{3+} metal ions with moderate affinity (10 and 50 μM, respectively) (Yoon and Cowan 2003) and provision of desulfurase as a sulfur donor and holo FXN was sufficient for cluster assembly, obtaining holo ISCU with good yields (Nuth et al. 2002).

More importantly, Foury's group showed that alteration of the electrostatic properties of the acidic ridge of FXN (substituting two or four acidic residues by Lys or Ala) compromised Fe-S cluster assembly, weakening interaction between FXN and ISU1, suggesting that the acidic ridge is essential for FXN function and is likely to be involved in protein-protein interactions (Foury et al. 2007). In addition, Wang and Craig found that yeast FXN and ISU1 interact through the beta-sheet platform. Interestingly, they reported mutations on residues Asn122, Lys123, Gln124 (Asn151, Lys152 and Gln153 in the human variant) that have detrimental effects on Fe-S cluster biogenesis and on the interaction with ISU1, without altering iron binding (Wang and Craig 2008). In agreement with the involvement of the beta-sheet in the ISU1 binding process, Foury's group found a significant impact of residues from beta-strands 3 and 4, whereas the aromatic side-chain of Trp131 (Trp155 in the human variant) is one of the major contributors to binding energy (Leidgens et al. 2010). It is worthy of note that mutation of Trp155 to arginine (W155R) results in FRDA.

One important aspect was whether interaction of FXN, ISU1 and NFS1 was cooperative and if mutation of the acidic ridge region may alter interaction with NFS1, thus altering interaction with ISU1. In this regard, Pastore's group, by means of a combination of biophysical methods, such as ITC, SAXS and NMR, complemented with mutagenesis analysis (Prischi et al. 2010), defined the structural bases of the interaction of the bacterial orthologue of FXN with the IscS/IscU complex (Fig. 13.6, IscS/IscU complex is named NFS1/ISU1 in yeast and NFS1/ISCU in humans). They showed that monomeric FXN binds desulfurase in a region between its active site and the dimer interface. Mutations at the complex interface altered binding and the rates of enzymatic cluster formation. FXN binding increased the affinity of the other members of the protein complex. Furthermore, the triple mutation R220E/R223E/R225E designed to invert the charge of a positively charged patch formed by residues Arg55, Arg67, Arg219, Arg220, Arg223, Arg225 and Arg237 (close to the dimer interface) was enough to abolish interaction with the complementary charged surface of FXN.

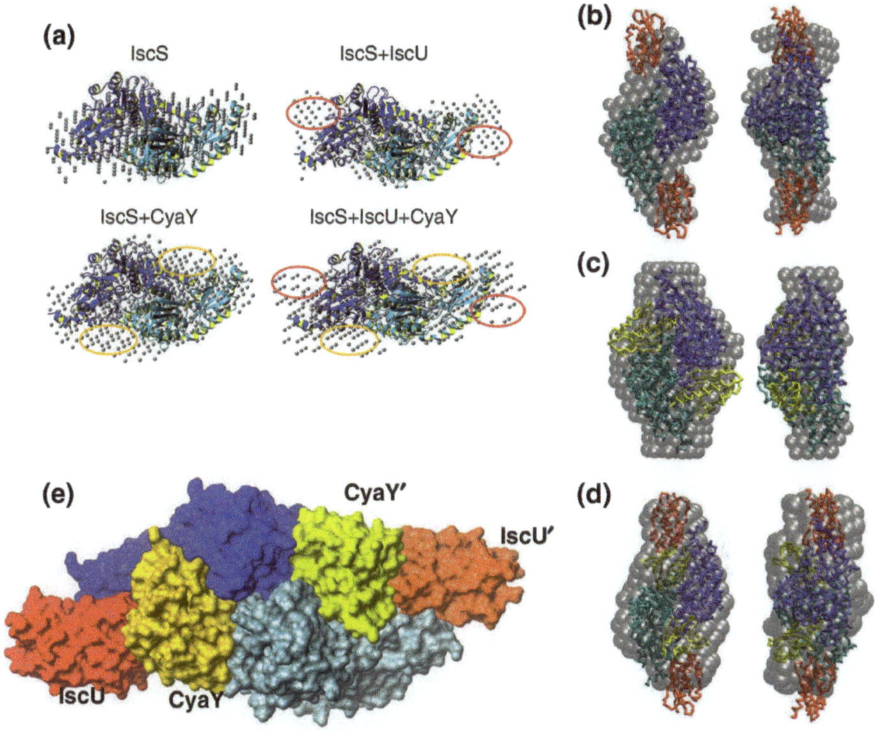

Fig. 13.6 Structural bases of the interaction of FXN from the bacterial orthologue with desulfurase IscS and IscU. **a** Small-angle X-ray scattering (SAXS) shapes corresponding to the different protein complexes: IscS alone, IscS/IscU, FXN/IscS and FXN/IscS/IscU (bacterial FXN is known as CyaY). In all cases the structure of IscS desulfurase (PDB ID: 1P3W is superimposed. These experiments were carried out by Pastore's group. Regions with additional densities, which are not explained by IscS alone, in the binary and ternary complexes, are highlighted in yellow or red ovals. The envelopes of the complexes are markedly different from those of the isolated desulfurase, allowing the authors to identify the excess volumes in which monomeric IscU, FXN or both fitted. Molecular modelling based on SAXS (ab initio bead models) of IscU/IscS (**b**), CyaY/IscS (**c**) and CyaY/IscS/IscU (**d**) complexes superimposed with rigid body models. Each model is left-handed rotated around the vertical y axis. **e** Surface representation of the ternary complex obtained by combining data from SAXS and NMR experiments. IscS subunits are shown in blue and cyan; in red and orange the IscU; subunits and in gold and yellow FXN (CyaY). Reproduced from Figs. 3, 4 and 6 in reference Prischi et al. (2010)

On the other hand, Puccio's group studied interaction of mammalian FXN with the Fe-S assembly protein complex by means of immunoprecipitation experiments. They found that the main endogenous interactors of a recombinant mature human FXN are ISCU, NFS1 and also the small protein ISD11, the components of the core Fe-S assembly complex. Furthermore, using a heterologous expression system (*E. coli*) and a series of mutants, they demonstrated that mammalian FXN interacts through the acidic ridge and the beta sheet with the preformed core complex, rather than with the individual components. Mutations W155R and the N146K (two pathogenic mutations) strongly affected interaction with ISCU and NFS1. Moreover, some mutations directed against the residues from the acidic ridge (E108K, E111K and D124K) drastically disrupted interaction of FXN with the complex (Schmucker et al. 2011). By contrast, mutation of residues E96, D104, D115 and D122, as well as G130V (a pathogenic mutation associated with a milder phenotype), had no significant impact on the interaction (Schmucker et al. 2011).

Conformational Stability, Internal Motions and Folding Dynamics

Understanding the Effect of Mutations on Structural Dynamics

Even though in the enormous majority of cases patients with FRDA are homozygotes for the expansion, about 2–8% of them exhibit a point mutation in the FXN coding sequence that alters one or more of FXN's features needed for its biological function.

Some mutations yield a decrease in the conformational stability of FXN: in equilibrium conditions, the difference in free energy between the folded and unfolded state is lower for these mutants compared to that measured for the wild-type variant. Usually, this information is provided by the analysis of in vitro experiments in which the stability of the isolated FXN is perturbed by denaturants like urea or guanidinium hydrochloride. The unfolded state is stabilized at high denaturant concentrations (Fig. 13.7). Alternatively, the perturbing agent may be by temperature (and much more rarely, pressure or even stretching). The variation in the folded fraction of the protein (and the equilibrium constant K_{NU}) is very commonly monitored by changes in spectroscopic signals like tryptophan fluorescence (330–350 nm) that reports alterations in the environment of the aromatic residues. In particular, tryptophan residues are monitored because they are specifically excited at 295 nm and emit in the range of 310–450 nm and their maximal wavelength and quantum yield are sensitive to the chemical environment. Emission from an apolar environment occurs in the range of 320–340 nm, whereas from a polar and solvated environment it occurs at 350 nm. Additionally, circular dichroism (CD) in the far-UV region (190–240 nm), which reports changes in the secondary structure content (Fig. 13.8), is an important complement.

The biophysical characterization of the protein variants include the change in the hydrodynamic radius, hydrodynamic behavior or compactness, which can be monitored by light scattering (DLS), SEC-FPLC, or NMR (DOSY), and the study of the changes in the tertiary structure that may be followed by monitoring CD in the near-UV region and NMR spectroscopy (e.g., ^{1}H-^{15}N-HSQC, which, by means of the chemical shifts, monitors the environment of backbone amide groups). Figure 13.8 shows some results for wild-type FXN.

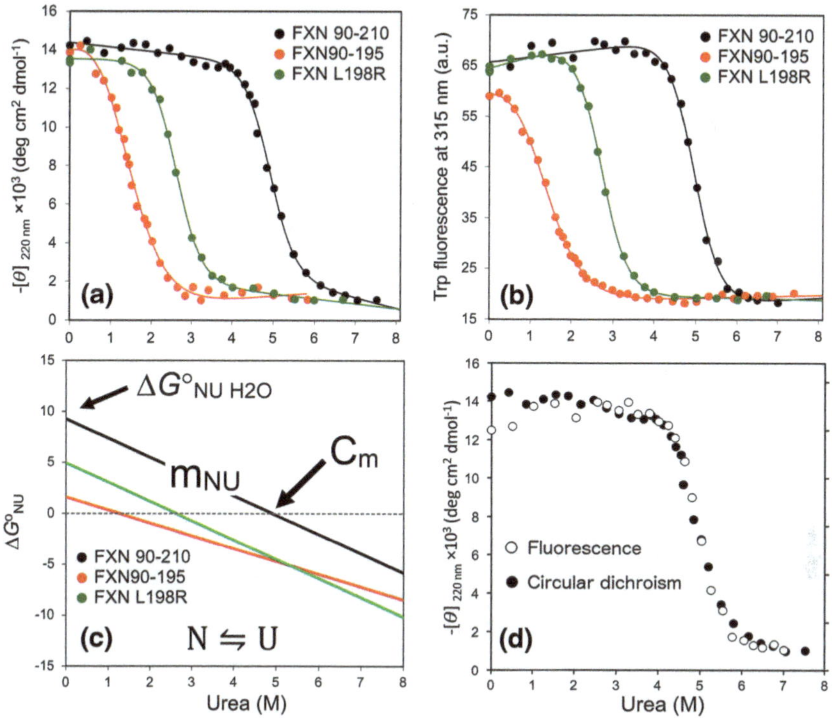

Fig. 13.7 Characterization of the conformational stability of human FXN. The stability of the wild-type (black), the L198R pathogenic mutant (green) and the 90–195 truncated variant (red) were perturbed by the addition of a chaotropic agent (urea). Two different probes were used to characterize folding under equilibrium, **a** the circular dichroism (CD) signal at 220 nm that monitors the secondary structure of the protein, and specifically the alpha helical content, and **b** tryptophan fluorescence intensity that monitors the tertiary structure of the protein, and specifically the environment of tryptophan residues of FXN. For this protein, the fluorescence intensity of the native state is higher than that of the unfolded state, whereas the CD spectrum of native state is characterized by a negative band at 220 nm (it is worthy of note that the sign of the CD signal at 220 nm was changed). **c** A two-state model was fitted to the data obtaining the difference in free energy of unfolding in the absence of denaturant ($\Delta G°_{NU\ H2O,}$ or, for simplicity, $\Delta G°_{NU}$ throughout this chapter) and the m_{NU} value, which is the dependence of $\Delta G°_{NU}$ on the denaturant concentration ($\partial \Delta G°_{NU}/\partial$[urea]). Buffer was 20 mM Tris-HCl, 100 mM, 1 mM EDTA, 7.0. **d** The superimposition of CD and fluorescence data. An adequate superimposition between conformational probes indicates that secondary and tertiary structures are cooperatively stabilized suggesting a two-state mechanism. Experimental details may be found in Faraj et al. (2014)

In particular, residues Leu106, Tyr123, Ile154, Leu156, Trp173, Leu182, His183, Leu190 and Leu198 conform the hydrophobic core and they are tightly packed, suggesting that their mutations cause complementary destabilization of the fold. In Table 13.1, a list of mutants is presented together with data regarding experimental and predicted conformational stabilities.

This explains the effect of FRDA mutations p.L106S, p.Y123D, p.I154F, p. L156P, p.W173G p.L182F, p.L182H, p.L182F, p.H183R, p.L186R, p.L190P and

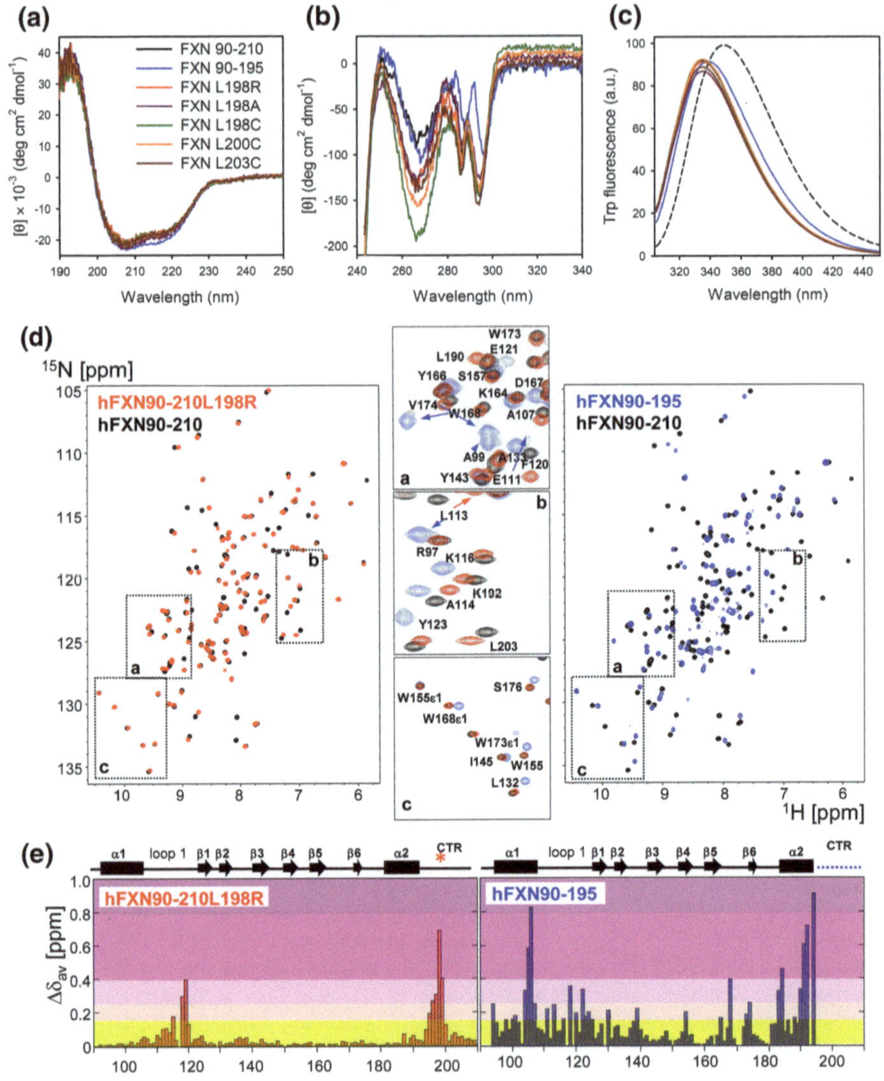

Fig. 13.8 Characterization of FXN conformation by using spectroscopic probes. A series of FXN mutants (FXN 90-210) were prepared and studied. **a** Far-UV CD, **b** Near-UV CD, **c** Tryptophan fluorescence (7.0 M urea unfolded state in black dashed line), **d** ^{1}H–^{15}N HSQC NMR and **e** the chemical shift perturbation per residue respect to the wild-type protein (extracted from the analysis of panel **d**). Even though the truncated variant and the pathogenic L198R mutant are highly unstable, they exhibit some average conformational features (CD and Trp fluorescence spectra) similar to the wild-type FXN. On the other hand, NMR spectra showed significant differences in chemical shifts for these variants. Panel **a**–**c** and **d**, **e** are reproduced from Figs. 2 and 3 in reference Faraj et al. (2014)

Table 13.1 FRDA and Laboratory FXN variants

Variant	ASA[a] (%)	Secondary Structure	Predicted[c] (and experimental[d]) stability (kcal mol^{-1})	ΔTm (°C)
L106S	0	α1	**D**	
D122Y	43	Loop-1	**N** (2.1; (Correia et al. 2008))	8.0 ± 0.3[e] 15.9 ± 0.2 (Correia et al. 2008)
Y123D	10	Loop-1	**M/D**	
G130V	15	Turn β1–β 2	**N** (2.9; (Correia et al. 2008))	14.6 ± 0.3[e] 23.1 ± 0.2 (Correia et al. 2008)
G137V	3	Turn β2–β 3	**N/M**	20.3 (Faggianelli et al. 2015)
N146K	22	β3	**N**	−4.5 ± 0.2[e]
Q148R	22	β3	**N**	
N146A (related)[f]	22	β3	**N**	
Q148G (related)	22	β3	**M**	
Q153A (related)	12	β4	**N/M**	
I154F	0	β4	**M** (1.7; (Correia et al. 2008))	15.6 ± 0.2 (Correia et al. 2008)
W155R	27	β4	**M** (1.4, (Correia et al. 2008))	4.9 ± 0.5 (Correia et al. 2008)
W155A (related)	27	β4	**M**	
W155F (related)	27	β4	**N**	
L156P	3	β4	**M/D**	
R165P	48	β5	**N/M**	
R165C	48	β5	**N**	
R165D	48	β5	**N**	
R165N	48	β5	**N/M**	
R165P	48	β5	**N/M**	
W173G	0	β6	**D**	
L182F	0	β2	**N**	
L182H	0	α2	**M**	
H183R	2	α2	**N**	
L186R	0	α2	**M**	
L190P	0	α2	**M/D**	
L198R	10	CTR[b]	**M** (4.2; (Faraj et al. 2014, 2016)	15.3 ± 1.2 (Faraj et al. 2014, 2016) 10.1 ± 0.2[e]

(continued)

Table 13.1 (continued)

Variant	ASA[a] (%)	Secondary Structure	Predicted[c] (and experimental[d]) stability (kcal mol^{-1})	ΔTm (°C)
L198A (related)	10	CTR	**M/D** (2.6; (Faraj et al. 2014, 2016)	8.3 ± 1.2 (Faraj et al. 2014, 2016)
L198C (related)	10	CTR	**M** (1.0; (Faraj et al. 2014, 2016))	3.3 ± 0.7 (Faraj et al. 2014, 2016)
L200C (related)	1	CTR	**M** (2.0; (Faraj et al. 2014, 2016))	7.1 ± 0.7 (Faraj et al. 2014, 2016)
S202C (related)	78	CTR	**N**	
L203C (related)	12	CTR	**M** (−1; (Faraj et al. 2014, 2016))	−1.6 ± 0.8 (Faraj et al. 2014, 2016)
90–195 (related)	–	CTR	**D** (7.5; (Faraj et al. 2014, 2016))	29.5 ± 0.4 (Faraj et al. 2014, 2016)

[a]The accessible surface area of each residue of wild-type human FXN was determined using PDBID: 1EKG

[b]CTR is the C-terminal region

[c]Predicted stabilities S, N, M, D, indicate a stabilizing, neutral, moderately destabilizing or highly destabilizing, respectively. The stabilities were predicted using iStable, PopMusic, FOLDX and Dynamut programs. The result shown here is the average between these different predictions. Stability takes a D if $\Delta\Delta G^{\circ}_{NU} > 2.5$ kcal mol^{-1}; M if $1 < \Delta\Delta G^{\circ}_{NU} < 2.5$ kcal mol^{-1}; N if $-1 < \Delta\Delta G^{\circ}_{NU} < 1$ kcal mol^{-1} and S if $\Delta\Delta G^{\circ}_{NU} < -1$ kcal mol^{-1}

[d]The experimental differences in stability were determined for a small set of variants and the corresponding values are shown in brackets. $\Delta\Delta G^{\circ}_{NU} = \Delta\Delta G^{\circ}_{NU(wild-type)} - \Delta\Delta G^{\circ}_{NU(mutant)}$

[e]Tm value was determined following unfolding by Sypro-orange probe, which binds to the unfolded state (results from our laboratory). $\Delta Tm = T_{m\ (wild\text{-}type)} - T_{m\ (mutant)}$

[f]Related mutants are variants prepared in the laboratory

p.L198R, which alter FXN packing by means of steric clashes and steric strain, packing defects like internal cavities, or result in the location of a charged residue in an unfavorable environment.

Additionally, the processing of some pathological variants G130V and I154F has shown to be affected (Clark et al. 2017). In both cases, it was detected that the levels of mature FXN were reduced in cells due to an accumulation of the intermediate and an enhancing of MPP binding to these protein variants with a less efficient processing (Clark et al. 2017).

Some pathogenic FRDA mutations have drastic consequences on the internal motions of FXN. These variants may exhibit altered conformational ensembles at physiological conditions of pH and ionic strength. Usually relaxation NMR strategies, which cover the time scale from picoseconds to milliseconds, are applied to investigate the molecular dynamics. Typically, R_1, R_2 and hetero nuclear NOE (hnNOE) experiments are carried out (Fig. 13.9). Relaxation is sensitive to fast-internal motions (picosecond to nanosecond timescale) but are also sensitive to slow motions (microsecond to millisecond timescale) produced by conformational exchanges, protein aggregation and molecular tumbling (this is the case of R_2,

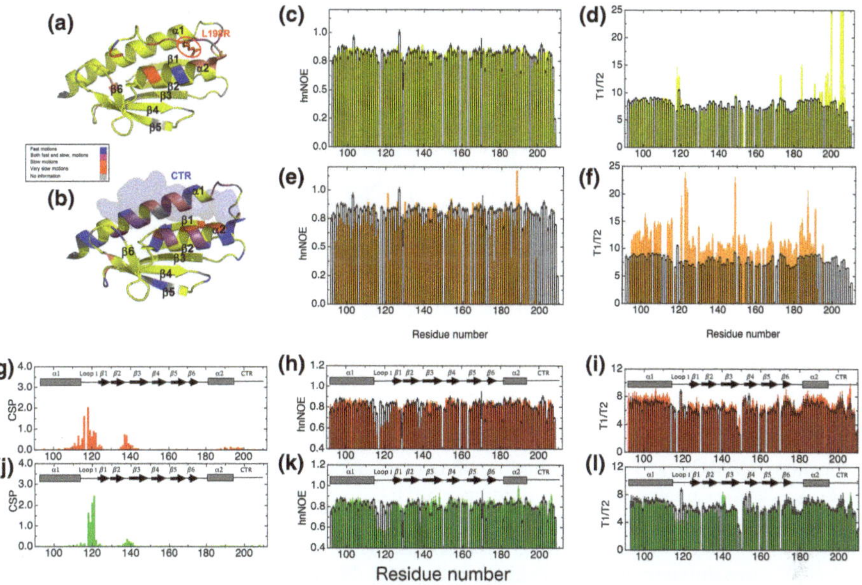

Fig. 13.9 Internal motions studied by NMR relaxation. Internal motions of L198R and the truncated variant were mapped on the PDB structure, **a** and **b**, respectively. Fast motions (ps-ns) hnNOE values <0.7 (blue); slow motions (μs-ms): residues showing T1/T2 higher than the mean plus 1SD (red); residues for which we do not have dynamic information are shown in gray. The L198R mutant (**c** and **d**) and the 90-195 truncated variant (E and F) show local and long-range effects of mutations on the internal motions, respectively. **Loop-1 Mutants:** Two variants in which the loop-1 of the *KPYTFED* sequence was mutated to the *GGGYTFGGG* sequence (**g**, **h** and **i**, red, variant 1L2) or to the *GGGGGGG* sequence (**j**, **k** and **l**, green, 1L3 variant). Panels **g** and **j** show the chemical shift perturbations (for each variant, each cross-peak was compared to that of the wild-type FXN). Panels H and K correspond to the heteronuclear NOE, whereas panels **i** and **l** show T1/T2 ratios. The wild-type is relatively rigid with unvarying hnNOE and relaxation values (black bars). As expected for the 1L2 and 1L3 mutations, which replace a stretch by a poly-Gly, analysis of hnNOE showed that fast motions were locally increased in the mutants' loop-1. The observed effects of loop-1 mutations on T1 and T2 relaxation rates are subtle and very localized. T1/T2 profiles for the 1L2 and 1L3 mutants are mainly similar to that of the wild-type protein, as expected for a highly conserved tertiary structure. Additionally, the T1/T2 values are sensitive to the rotational diffusion anisotropy of FXN. This is due to the deviation of the globularity of the protein: helical elements (and their amide groups) are oriented to the longer axis whereas the beta strands (and their amide groups) are oriented to the short axis. This Figure is a partial reproduction of Figs. 5 and 7, and supplementary Figure S5 is from reference Noguera et al. (2017) and Fig. 8 from reference Faraj et al. (2014)

transverse relaxation). For some pathogenic mutants of FXN, relaxation dispersion (R_{ex}, CPMG) experiments, which focus on the contribution of the conformational exchange to transverse relaxation, were also done, adding the timescale of microseconds to milliseconds (Fig. 13.10). Motions in this range are usually associated with conformational exchanges and function and, therefore, these experiments are very revealing. Wild-type FXN exhibits a pattern of R_{ex}

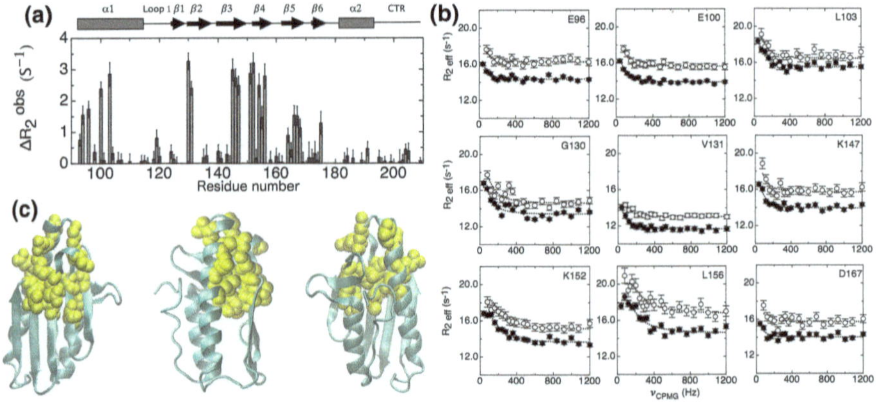

Fig. 13.10 Wild-type human FXN conformational exchange studied by NMR relaxation dispersion (CPMG). **a** The contribution of conformational exchange to transverse relaxation ($\Delta R_{2\ obs}$) was estimated from the difference in $R_{2\ obs}$ at the lowest and the highest CPMG frequencies and plotted along the FXN sequence. **b** CPMG relaxation dispersion profiles for the backbone ^{15}N amide of residues E96, E100, L103, G130, V131, K147, K152, L156 and D167, collected at two different magnetic field strengths, 14.1 (■) and 16.5 T (○) (600 and 700 MHz ^{1}H Larmor frequencies, respectively) and 22 °C. Lines indicate best-fit curves obtained with global values of the exchange rate, k_{ex} and state populations pA and pB, according to Eqs. (8–12) in reference Noguera et al. (2017). **c** Residues mentioned above mapped on the FXN structure. This Figure is a partial reproduction of Figs. 9 and S7 from reference Noguera et al. (2017)

characterized by higher Rex values for residues 94, 96, 100, 103, 130, 131, 145, 146, 147, 151,152, 154, 156 and 166 (Faraj et al. 2014; Noguera et al. 2017). These amino acids are clustered in the three-dimensional structure and most of them establish a network of contacts linking numerous secondary structure elements. As this region of the protein is involved in interactions with ISCU, the scaffold protein in the biosynthesis of iron–sulfur clusters, mutations that alter this network of motions might alter FXN biological function. After the analysis of relaxation dispersion profiles in two different magnetic fields, the populations of each native substate exhibiting conformational exchange may be obtained. In the case of wild-type FXN, a two-state model was fitted to the data, the most abundant native substrate (A) and the excited state (B) have populations of approximately 0.98 and 0.02, respectively. Additionally, the rate coefficients for the exchange processes may be obtained: for wild-type FXN k_{AB} and k_{BA} they are 3.5 and 213 s^{-1}, respectively (Noguera et al. 2017). Moreover, hydrogen-deuterium experiments allowed access to milliseconds or even more (seconds to minutes and hours). Therefore, by complementing these experiments, one may gain access to a broad range of timescales and different processes.

Resistance-to-proteolysis assays are also used to test the flexibility of the protein chain. In this case, the analysis of the peptides and the resulting fragments can be carried out by reverse face HPLC, mass spectrometry, SDS-PAGE, among others. Although wild-type FXN is very resistant to proteolysis, some variants are

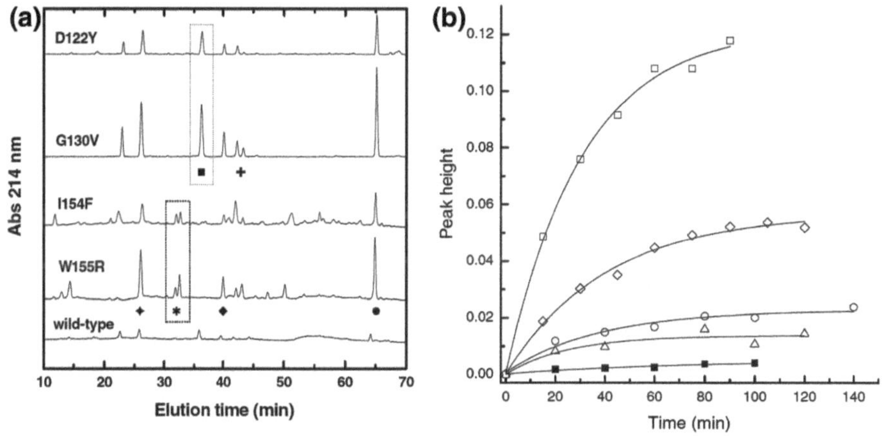

Fig. 13.11 Proteolysis experiments. **a** Peptide maps resulting from controlled proteolysis (90 min at 37 °C) of FXN at pH8.5 using trypsin for wild-type, D122Y, G130V, I154F and W155R variants. Symbols indicate: D91-R97 (*), L136-K147 (◆), Q153-K164 (■), Y166-K171 (✦), N172-K192 (●) and L198-K208 (+) peptides. **b** Time-course of controlled proteolysis monitored by the quantification of N172-K192 peptide for wildtype (filled squares) and D122Y (unfilled diamonds), G130 V (unfilled squares), I154F (unfilled circles) and W155R (unfilled triangles) variants. This Figure is a reproduction of Figs. 5 and 6 from reference Correia et al. (2008)

particularly sensitive to protease treatment. Trypsin that cuts the protein backbone specifically at positively charged residues (Lys or Arg residues) and chymotrypsin, which cuts at aromatic residues (Trp, Phe or Tyr), are commonly used to test flexibility. In fact, when the reaction is carried out using chymotrypsin, only Tyr205, located in the C-terminal region (CTR), is susceptible to proteolysis and this occurs only after long incubation times. However, for some mutants, Tyr205 was a suitable probe for CTR flexibility enhancements (Faraj et al. 2014). Figure 13.11 shows the results of proteolysis experiments carried out by Correia and coworkers for a set of pathogenic variants: D122Y, G130V, I154F and W155R (Correia et al. 2008). The variants exhibit proteolysis profiles reflecting local instabilities (Fig. 13.11a) and, accordingly, the variants show altered proteolysis kinetics (Fig. 13.11b).

Even though FXN is completely soluble with a very low propensity to forming aggregates, some mutations increase the tendency of FXN to aggregate. Aggregation may be monitored by turbidimetry, SDS-PAGE, ultracentrifugation and light scattering (DLS), among other techniques. The effect of some mutations was studied in detail.

Here we have selected some mutations, which exhibit a broad range of alterations. It is worthy of mention that it is not an exhaustive description of all mutants found and studied so far. A very complete review regarding a complete list of compound heterozygous FXN mutations was recently published (Galea et al. 2016).

Two interesting mutants studied in detail are p.G130V and p.G137V. In both cases, the side chain of residue Gly, a hydrogen, is replaced by the isopropyl group of Val, a non-polar branched-side-chain amino acid.

Mutation G130V

Mutation p.G130V results in a large shift in Tm value ($\Delta T_m = 23$ °C) and in a significant difference of free energy compared with the wild-type variant ($\Delta\Delta G°_{NU} \sim 2.9$ kcal/mol). Additionally, as proven by proteolysis, the flexibility of this variant is enhanced. G130V is highly flexible in the loop between strands β3 and β4, as shown by the appearance of a peak matching the Q153-K164 fragment (Correia et al. 2008). Conformation of the turn formed by residues Gly128, Ser129 and Gly130 between strands β1 and β2 is disrupted by mutation G130V and, consequently, an increment in the local strain should take place (Correia et al. 2008). As the backbone carbonyl oxygen atom of Gly130 forms a hydrogen bond with the backbone amide hydrogen of residue Lys147, which is involved in binding to ISCU, mutation of Gly130 is predicted to alter FXN function (Galea et al. 2016). Moreover, given that Lys147 is the in vivo ubiquitination site (Rufini et al. 2011) for the protein degradation pathway via proteasome, mutation of Gly by Val might result in higher degradation inside the cell. Residue Gly130 is involved in slow conformational exchanges as determined by relaxation dispersion (CPMG) experiments (Noguera et al. 2017), and motions are typically related to protein function. Altogether, these data make clear that a number of different aspects should be considered in order to understand in detail the effect of a point mutation.

G137V Mutation

In the case of p.G137V, its structure and activity are preserved; however, the variant exhibits significantly reduced conformational stability (Faggianelli et al. 2015). Probably, this mutation causes misfolding with the consequent reduction of protein concentration in the patient. Indicative of the lower folding efficiency compared with the wild-type is the fact that when expressed in *E. coli* only a low amount of the G137V variant is obtained in the soluble fraction, whereas 100% of the wild-type FXN is obtained in the soluble fraction. Even though all the spectroscopic features are preserved (^{1}H-^{15}N HSQC-NMR chemical shifts, CD, etc.), the melting temperature of the mutant drastically decreases compared to that of the wild-type ($\Delta T_m \sim 20$ °C). On the other hand, this variant is active as an allosteric activator for in vitro cluster biosynthesis, suggesting that this mutation has no direct effect on protein function. Consequently, lower concentrations of FXN found in this case may be the result of low stability of p.G137V and the cause of a particular FRDA form with an onset at the boundary between a classical and a late onset phenotype.

L198R Mutation

Our group studied the L198R mutation in detail because we were interested in understanding the role of CTR (residues 196–210) in FXN's folding and dynamics. In this case, we observed that p.L198R destabilized the native state in a significant fashion ($\Delta\Delta G^{\circ}_{NU} \sim 4.2$ kcal/mol). Other mutations at this position, which only truncate the side chain, were also destabilizing (L198A and L198C), pointing to the relevance of CTR and, in particular, pointing to Leu198 as a key side chain for FXN stability; the stabilities of FXN variants are L198R < L198A < L198C < wild-type. This result is compatible with the fact that the Cys residue can establish interactions via its thiol group, whereas the L198A variant carries a shorter side chain, deleting a number of native contacts. Additionally, kinetic studies of folding dynamics showed that L198A, like other mutations located in this stretch, only alters the stability of the native state but not the stability of the transition, intermediate or unfolded states (Fig. 13.12) (Faraj et al. 2016). As the contribution of the interactions mediated by CTR to the energetics of the transition states and intermediate ensembles are negligible, one can infer that CTR is unstructured in these ensembles but structured in the native state.

It is worthy of note that in p.L198R a charged residue is located in an apolar environment and this feature promotes alterations in the native ensemble, with particular altered motions near the mutation site. Although wild-type FXN shows a high resistance to proteolysis, and after 4 h of incubation a single cutting site at the Y205 position is detected, the L198R mutation enhances sensitivity to the chymotrypsin. Even at a very short incubation time (30 s), 50% of the protein is cut at the Y205 position. In agreement with this result, NMR relaxation dispersion (CPMG) experiments showed that L198R exhibits more dynamic behavior in the microsecond/millisecond timescale in localized regions near the mutation (Faraj et al. 2014). On the other hand, the absence of CTR results in a very destabilizing variant with a global effect on the dynamics and a 10% of protein populating the unfolded ensemble at physiological conditions of ionic strength, pH and temperature (Faraj et al. 2014). Remarkably, the finding that a frame shift that produces the truncation of FXN at residue 193 determines FRDA with a rapid disease progression indicates that CTR is crucial for the biological function of the protein (Sacca et al. 2013).

W155R Mutation

As we mentioned above, some mutations linked to FRDA occur in the beta-sheet, the region involved in protein-protein interactions. This is the case of W155R. Tryptophan 155 is an exposed residue that, most likely, participates in the interaction with ISCU. This variant is slightly destabilizing, showing a decrease of only ~ 5 °C in T_m (Correia et al. 2006). Destabilization is expected from the obliteration

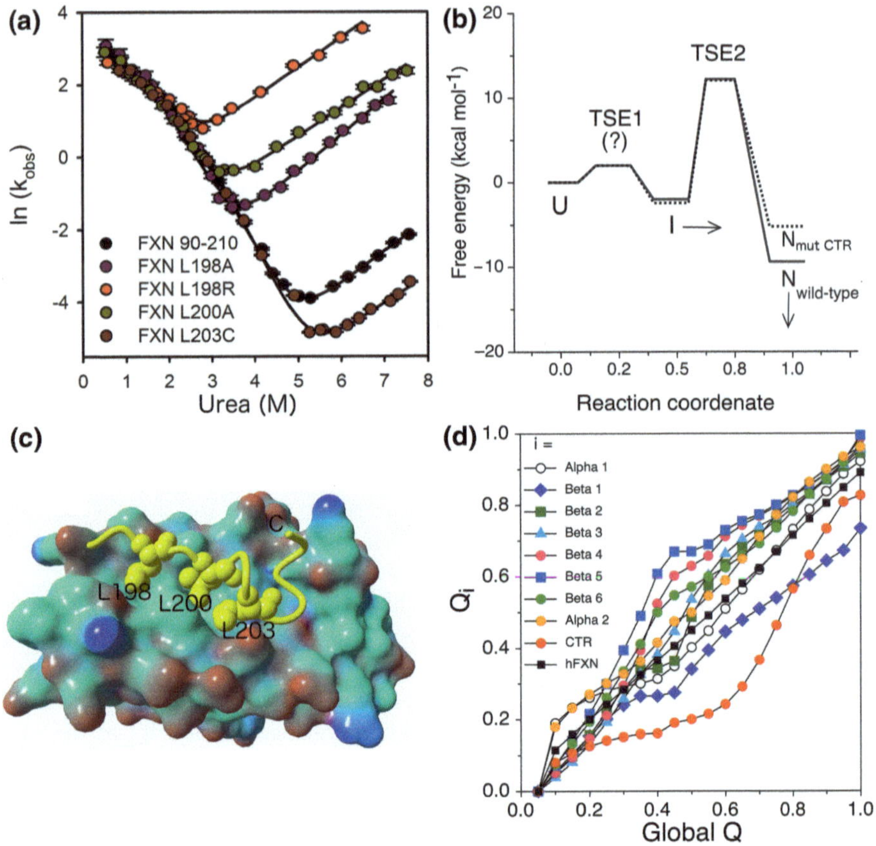

Fig. 13.12 Human FXN folding dynamics. The stability of the native state and the CTR. **a** Chevron plots showing that differences in kinetics among the mutants of CTR of FXN 90-210 come from the unfolding branch (right); on the other hand, the refolding branch (left) among the variants superimposes very well. This effect can be interpreted as a difference in the stability of the native state. **b** Reaction coordinate for folding of human FXN (90-210): the difference in free energy between $N_{\text{wild-type}}$ and $N_{\text{mut CTR}}$ without the effect of mutation on the transition state ensembles (TSE1 and TSE2) or in the intermediate state (I). In panel (**c**) a model of FXN structure is shown in which the CTR is depicted in yellow and key residues L198, L200 and L203 that stabilized the native state are modeled using 50% of the van der Waals volumes. For the rest of the protein, the molecular surface is modeled. **d** Structure-based molecular dynamic simulations using a funneled energy landscape of protein folding (native–centric GO model) allowed the study of the folding state of each one of the secondary structure elements along the reaction coordinate. The fraction of the global number of contacts is plotted. This Figure is a partial reproduction of Figs. 1 and 8 and S1 from reference Faraj et al. (2016)

of a π-cation interaction between a tryptophan residue and the contiguous Arg165 residue, and the electrostatic repulsion that results from the insertion of an additional positively charged residue close to the Arg165. Whereas the ^{1}H–^{15}N HSQC NMR spectrum of W155R is very similar to that of the wild-type FXN, relaxation NMR experiments indicated that this mutant has a significantly larger correlation time (τ_c) compared to that measured for the wild-type variant (7.9 and 9.2 ns, respectively). This may be indicative of the mutant's high tendency towards aggregation. Interestingly, the W155R mutant is proteolyzed at a fast rate at the protein C-terminus and at the loop between strands β2 and β3, but also the first α-helix shows a proteolytic site (Correia et al. 2008), which is indicative of a decreased rigidity and some disorder or local conformational exchange. On the contrary, proteolysis within a rigid canonic secondary structure element would be very energetically unfavorable and unlikely, as in the case of the wild-type variant.

N146K Mutation

By contrast with other mutations mentioned above, p.N146K mutation resulted in a significant stabilization of the native conformation, as suggested by the T_m values (64.9 ± 0.2 and 69.4 ± 0.1 °C for the wild-type and the mutant variants, respectively, non-published results from our group) and computational predictions ($\Delta\Delta G°_{NU}$ are -0.7 and −1.3 kcal/mol, using FOLDX and Dynamut, respectively, and $\Delta\Delta G°_{NU}$ is defined as $\Delta G°_{NU,\ wild\text{-}type} - \Delta G°_{NU,\ mutant}$). On the other hand, this mutation alters interaction with ISCU and, therefore, NFS1 supercomplex activity. Residue N146 is at van der Waals distance (~5 Å) of Trp155 and a charged residue located at this position may imbalance the electrostatics of the FXN surface. In this regard, using a series of mutants, Barondeau's group carried out a key work on structure-function relationships revealing determinants of FXN-mediated activation of the Fe–S assembly complex (Bridwell-Rabb et al. 2011). Interestingly, whereas the wild-type FXN exhibits a $K_D = 0.22 \pm 0.05$ μM for the complex NFS1/ISD11-ACP/ISCU, the K_D for the N146K mutant is significantly higher (6.25 ± 1.40 μM). Moreover, NFS1 activity, cluster formation activity and catalytic efficiency (k_{cat}/K_M) are approximately 8, 4 and 3 times higher for wild-type than for this mutant, respectively. Besides, weaker binding to the complex NFS1/ISD11-ACP/ISCU observed for the N146K variant compared to the one observed for N146A (a variant designed in the laboratory to dissect the effect of the charge) is consistent with results from pull-down experiments performed by Puccio's group (Schmucker et al. 2011). Additionally, Puccio's laboratory generated murine fibroblast cell lines stably expressing full-length N146K and N146A (among other variants) and observed that N146A clones did not present any gross phenotype or sensitivity to oxidative stress. By contrast, the N146K clone showed strong growth defects and displayed the classical features of FRDA (degenerating mitochondria, electron dense deposits, deficit in Fe–S enzyme activities).

D122Y Mutation

D122Y mutation is located in the acidic ridge, in the loop 1 region (residues Asp115 to Tyr123). Although this mutation is predicted to have little effect on the stability of the native conformation of the protein (Galea et al. 2016), the experimental T_m value for this variant decreased in a significant fashion compared to that of the wild-type ($\Delta T_m \sim 15°$ C monitored by circular dichroism (Correia et al. 2008), or $\Delta T_m \sim 8$ °C, unfolding monitored by Sypro-orange probe, unpublished results from our group), in agreement with a significant decrease in conformational stability ($\Delta\Delta G°_{NU} = 2.1$ kcal mol^{-1}, urea-induced unfolding (Correia et al. 2008)). The side-chain of Asp122 establishes two interactions with neighbor residues: one of them with the amine group of the side chain of Lys135 from β2 and the second with the amide group of backbone Gly138, from the β2-β3 turn. These interactions are removed in p.D122Y. In the same way, the p*K*a for Asp122 is predicted by PROPKA (Olsson et al. 2011) as more acidic than expected for a free aspartic acid side-chain (pKa_{D122} and pKa_{model} are 2.90 and 3.80, respectively), suggesting its involvement in intramolecular interactions. Variant D122Y exhibited a lower iron binding stoichiometry (Correia et al. 2008).

FXN Degradation Inside the Cell

Studies on FXN degradation inside the cell have shown that precursor concentration may be regulated by the proteasome system. In this context, the transfection of cell lines with the FXN precursor showed that the precursor is ubiquitinated. When HEK-293 cells were treated with proteasome inhibitors such as MG132, the accumulation of the precursor was detected and, more importantly, an increment of 2.5 times the mature form was observed (Rufini et al. 2011; Benini et al. 2017).

It was observed that Lys147 of the FXN precursor is the key residue that is modified by ubiquitination. The K147R mutation showed an increment of the lifetime with higher levels of the precursor and the mature form. Interestingly, Lys147 is located in a cleft where some target molecules might bind blocking degradation (Rufini et al. 2011). It has been detected that this post-translational modification is performed by the RING E3 ligase RNF 126, which specifically recognize and modifies FXN. Its inhibition may be a key point for incrementing FXN levels in Friedreich ataxia patients. Ubiquitin-competing molecules inhibited FXN degradation, promoting its accumulation and rescuing aconitase activity in model cell lines (Benini et al. 2017; Rufini et al. 2015).

Other reports have shown that modulation of FXN degradation could be triggered by phosphorylation (Cherubini et al. 2015). It has been shown that when the FXN precursor is co-expressed with the Src kinase domains, the precursor is modified in a tyrosine residue. To get information about which is the target residue, the eight Tyr residues of FXN were mutated (Tyr74, Tyr95, Tyr118, Tyr123,

Tyr143, Tyr166, Tyr175 and Tyr205) to Phe and each variant was transfected into the HEK-293 line in the presence of an active Src-Y527F kinase. As a control, the same was done with an inactive Src-Y527F kinase. A total abrogation of protein modification was detected in the case of the Y118F FXN mutant, suggesting that this residue is the one that is mostly modified (Cherubini et al. 2015). When the ubiquitination of the Y118F variant was evaluated, a decrease in the modification and degradation was observed. To confirm the hypothesis that phosphorylation of the Tyr118 residue triggers FXN degradation, the HEK-293 line expressing the wild-type protein was treated with different Src inhibitors. These drugs promoted FXN accumulation in a dose-dependent form. Interestingly, when these inhibitors were used in a FXN deficient B cell line derived from patients of Friedreich's ataxia, an accumulation of FXN and a recovery of aconitase activity were detected, suggesting that FXN phosphorylation is a key pathway for regulating the levels of this protein in the cell (Cherubini et al. 2015).

In summary, FXN presents at least three isoforms, being type I the most important in the mitochondrial and it is a process in the 81–210 mature form. Additionally, FXN can suffer several post-translational modifications that modulate its concentration and function inside the cell.

Iron Binding and Function

FXN may have a role in iron sensing and the control of cluster biosynthesis. It is well known that the presence of iron accelerates Cys desulfurase reaction catalyzed by the NFS1 supercomplex. Whether FXN is the protein that donates iron atoms to the cluster assembly is under debate. As we mentioned before, it was determined that FXN binds Fe^{2+} and Fe^{3+} with moderated affinity; however, some controversy exists regarding the latter (Cai et al. 2018b; Faraj et al. 2014; Yoon and Cowan 2003), most likely due to the extremely low solubility of the metal ion in a buffer with physiological conditions. Nevertheless, experiments using minimalist peptide models showed that motifs present in the FXN surface are able to bind both (Vazquez et al. 2015). Thus, experiments concerning Fe^{3+} can be performed under non-physiological conditions like pH 4–5 or under non-equilibrium conditions. The insolubility of Fe^{3+} was used to separate bound from non-bound metal ions after incubation with the protein, and iron concentration can be accurately measured by the phenanthroline method.

In this context, it is worthy of mention that some pathogenic variants fail to bind these metal ions. Variant D122Y as well as G130V and L198R exhibited a lower iron binding capability (Correia et al. 2008; Faraj et al. 2014).

Chemical shift perturbation and paramagnetic properties of iron allowed monitoring of iron binding to FXN. In particular, ^{1}H-^{15}N cross peaks corresponding to the amide groups of residues that directly interact with metal ions in ^{1}H–^{15}N-HSQC NMR experiments undergo a significant intensity decrease, making the cross peaks undetectable. Therefore, binding sites can be inferred at the residue level

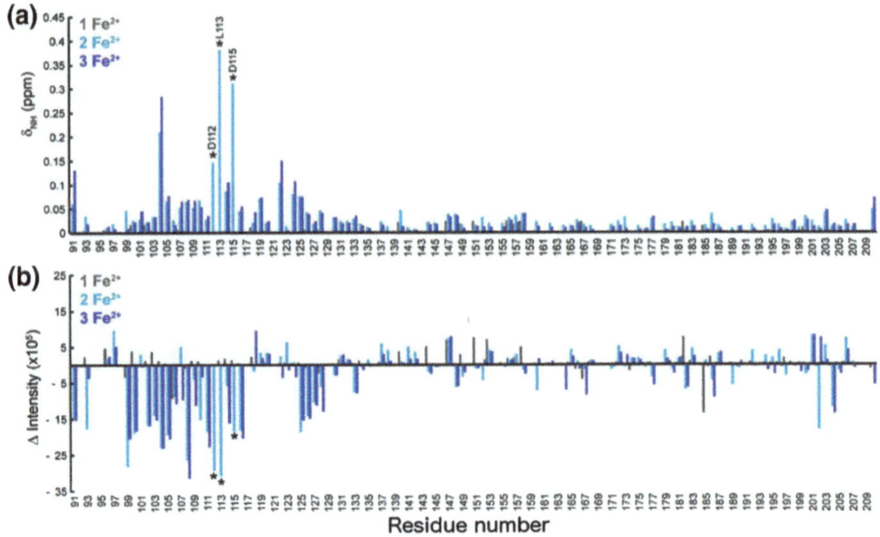

Fig. 13.13 Direct interaction between Fe^{2+} and human FXN monitored by ^{1}H-^{15}N-HSQC NMR experiments. **a** The change in the normalized amide chemical shift (δNH) and **b** the change in intensity of ^{1}H–^{15}N-FXN resonances upon addition of 1, 2 and 3 molar equivalents of Fe^{2+} (gray, cyan, and dark blue, respectively). Asterisks indicate the residues whose resonances broadened beyond detection during the titrations. Protein concentration was 560 μM and samples were prepared anaerobically. This Figure partially reproduces Fig. 5 from reference Gentry et al. (2013)

(Fig. 13.13). This procedure is performed under anaerobic conditions in order to maintain Fe^{2+}concentrations in solution. In the case of human FXN, after the addition of two equivalents of Fe^{2+}, significant changes in chemical shifts were observed for residues of the acidic ridge: Asp104, Ser105, Asp112, Leu113, Ala114, Asp115, Thr119, Asp122, Asp124 and Val125. After the addition of three equivalents of Fe^{2+}, Asp91, Asp104, and Asp122, the three underwent further perturbations of their chemical shifts (Gentry et al. 2013). Signals corresponding to Asp112, Leu113 and Asp 115 broadened, making them undetectable at three molar equivalents of iron (Gentry et al. 2013). Remarkably, His86 located in the N-terminal disorder stretch of the mature form of human FXN was essential for high affinity iron coordination (Gentry et al. 2013).

Whether FXN undergoes conformational changes or changes in the internal motions upon iron binding is a key issue because iron binding may exert an effect on function via the modulation of the interaction between FXN and NFS1 and/or ISCU proteins. Hydrogen-deuterium exchange experiments coupled to mass spectrometry (HDX-MS experiments) were performed by Gentry and coworkers (Gentry et al. 2013) for FXN in the presence of different ratios of protein/metal ion for Co^{2+}, Fe^{2+} or Fe^{3+}. By means of HDX-MS, they monitored the rate of exchange between H and D as the incorporation of deuterium in the backbone of FXN (Fig. 13.14). The differences in mass between deuterated and non-deuterated

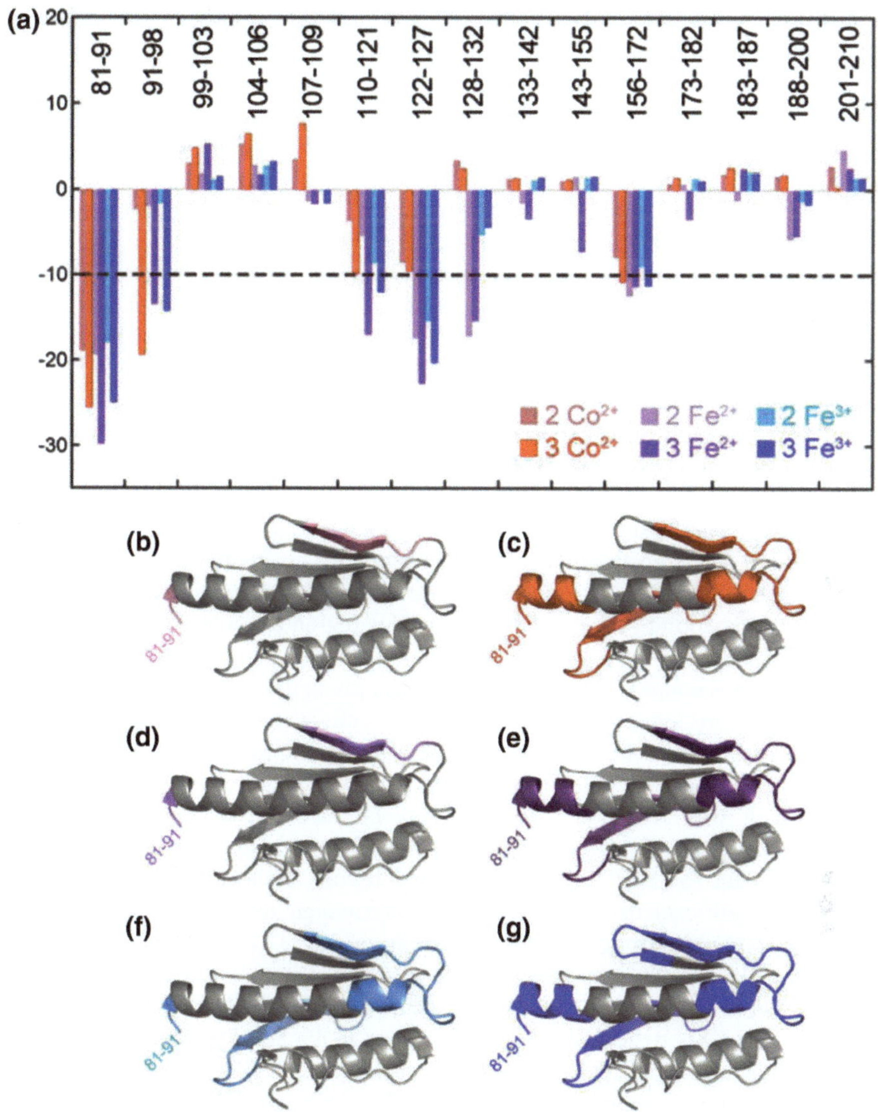

Fig. 13.14 Hydrogen-deuterium exchange of FXN backbone modulated by metal interaction. **a** Protection determined by the percentage of amide groups exchanging for apo-FXN was subtracted from those values obtained with 2 or 3 equivalents of metal: 2 equivalents of Co^{2+} (pink); 3 equivalents of Co^{2+} (red); 2 equivalents of Fe^{2+} (light purple); 3 equivalents of Fe^{2+} (purple); 2 equivalents Fe^{3+} (cyan); 3 equivalents of Fe^{3+} (blue). Panels **b–g** show those FXN peptides that exhibited HDX changes greater than 10% at three equivalents of metal. These peptides are mapped to the structure of FXN 90–210 (PDB:1EKG) using the colors as used in panel A. It is worthy of note that this HDX data included peptide 81–91, which is not observed in the structure. This Figure is a reproduction of Fig. 6 from reference Gentry et al. (2013)

proteolytical peptides can be determined by mass spectrometry. The exchange during the first 15 s reports accessibility of the H-N groups to D_2O and the amide groups compromised in hydrogen bonding, in secondary structure elements, are protected from the exchange. On the other hand, slower exchanges can be the result of local unfolding events or conformational exchanges, even involving the global unfolded state.

Whereas the addition of one equivalent of metal ion to the FXN only produced a slight decrease in deuterium incorporation, the addition of two and three equivalents of metal showed a lower incorporation. On the other hand, no further differences were observed for higher metal-to-protein ratios, suggesting the distribution among three binding sites. Peptides from the N-terminal region (residues 81–91), in strand β1 (residues 122–127) and in the β4–β5 strands (residues 156–172) showed reduced deuterium incorporation for all metal ions in comparison to that observed for the apo form of FXN. Incubation with the third equivalent of metals promotes additional protection of the H-N- amide groups in α1 and β1. These results indicated that interaction with the metal may produce some changes in FXN local stabilities, in some cases in regions near the metal binding sites; however, other amide groups far from these sites are also protected, in particular, amide groups located in strands β4–β5. The authors suggested that this fact may be relevant from a functional viewpoint (Gentry et al. 2013). This region can interact with the ISCU protein (see below).

Taking into account FXN iron capability, and given that yeast cells lacking FXN gene showed very low cytochrome content and cells were defective in iron use by ferrochelatase, a mitochondrial membrane-associated protein that catalyzes the terminal step of heme synthesis, consisting in the insertion of Fe^{2+} into protoporphyrin IX, it was proposed that FXN may be involved in heme biosynthesis (Lesuisse et al. 2003). In the same fashion, combining mouse and human microarray data for FXN-deficient cells and tissues, Schoenfeld and coworkers detected a significant decrease in coproporphyrinogen oxidase (the enzyme that converts coproporphyrinogen III to protoporphyrinogen IX) and down regulation of ferrochelatase, a result consistent with an observed increase in cellular protoporphyrin IX levels and the reduction of mitochondrial heme (A and C) levels, and indicative that FXN deficiency causes alterations in the heme pathway (Schoenfeld et al. 2005). Additional results coming from a quantitative proteomic analysis of Friedreich's ataxia B-lymphocytes showed a 2.5 decrease in the expression of the protoporphyrinogen oxidase, the penultimate enzyme in the heme biosynthetic pathway (Telot et al. 2018).

On the other hand, interesting data were obtained in recent work, employing whole cell systems. Steinkellner and coworkers found no significant differences in heme synthesis levels in normal and FRDA-patient progenitor-erythroid cells (Steinkellner et al. 2017). This result is in line with the fact that most FRDA patients do not suffer anemia.

Direct interaction between human FXN and ferrochelatase was studied by Yoon and coworkers by means of isothermal titration calorimetry ($K_D \sim 17$ nM at 25 °C and pH 8, (Yoon and Cowan 2004)). Interaction between both proteins only occurred in the presence of iron, suggesting that iron may mediate interaction. Heme production activity (monitored by a change in absorbance at 506 nm) increased with FXN concentration being optimal at a ratio of one FXN molecule per ferrochelatase dimer.

More recently, Söderberg and coworkers studied the interaction between yeast FXN and ferrochelatase by means of cross-linking in combination with mass spectroscopic analysis and single-particle reconstruction from electron microscopic images; these data were used together with computational docking to model the structure of the complex (Soderberg et al. 2016). The analysis of the model suggested that both proteins interact creating a path for iron transfer from the FXN metal binding site to the ferrochelatase reaction center. Additionally, kinetics experiments showed that the rate of heme synthesis was slightly accelerated when iron-loaded yeast FXN was included in the assay (Soderberg et al. 2016).

Remarkably, FXN from plants exhibits ferrochelatase activity. Armas et al. studied possible functions of FXN from *Arabidopsis thaliana* and found a potential role as an iron donor for heme biosynthesis, suggesting that plant FXN is able to directly catalyze the incorporation of iron into protoporphyrin IX in vitro (Armas et al. 2019). The efficiency of this process significantly increases in the presence of plant desulfurase NFS1/ISD11. Remarkably, the same group found that *Arabidopsis thaliana* lines deficient in FXN showed a substantial reduction in heme content in flowers and leaves and that it is correlated with a decreased content of several transcripts related to the heme pathway (Maliandi et al. 2011).

It is worthy of mention that at least for FXN from *Vibrio cholerae* it was shown that the protein exhibits the capability of heme binding with an apparent K_D value of 21 ± 6 nM (Uchida et al. 2017). Given that heme binding resulted in a decrease in the affinity of the protein for iron, it may be considered a point of regulation/control of the iron transport dependent on the heme group (Uchida et al. 2017), Possibly, an evidence of a cross-talk between pathways.

Significant evidence is still required to elucidate the details and physiological relevance of the interactions between FXN and ferrochelatase in mammals. In this respect, further kinetic and metabolic analysis regarding FXN, the heme biosynthetic pathway and iron sulfur assembly would shed light over conjectures attempting to eventually find the relationships that might link them.

A Macromolecular Context for FXN in Iron–Sulfur Cluster Biosynthesis

As mentioned before, FXN's activity depends on the interaction with at least two proteins: ISCU and NFS1, of the supercomplex NFS1/ACP-ISD11/FXN/ISCU. How the supercomplex is stabilized by the subunits and how FXN modulates the kinetics of reactions for persulfur synthesis and cluster assembly are key questions. From a functional point of view, it is clear now that FXN controls the reaction by accelerating the rate-limiting sulfur transfer step in the biosynthesis of [2Fe–2S] clusters, on the mammalian iron–sulfur assembly supercomplex (Fox et al. 2015). From an experimental viewpoint, cluster assembly is typically detected, under anaerobic conditions, by an increase of absorbance at 456 nm and the increase of circular dichroism signals at 330 and 430 nm, which are characteristic of the [2Fe–2S]-ISCU complex formation (Fox et al. 2015).

Currently, we know that the supercomplex is formed by two subunits of the NFS1 desulfurase, which are critically stabilized by two subunits of the ISD11 protein (Fig. 13.15). From an evolutionary perspective, ISD11 is only found in eukaryotic organisms. Richards and coworkers concluded that the origin of this protein occurred during or just after the single endosymbiotic event, giving rise to the mitochondria, and ISD11 evolved as an exclusively eukaryotic addition to the pathway of derived iron–sulfur cluster biogenesis of α-proteobacterium (Richards and van der Giezen 2006).

Friemel and coworkers studied some ISD11 mutations; among them, a mutation at position Y26A affects the interaction with the NFS1 protein yielding a twofold increase in K_M and a threefold decrease in k_{cat} (Friemel et al. 2017). They also showed that NFS1 dimer formation is very sensitive to the Y26A mutation, resulting in an impaired complex formation.

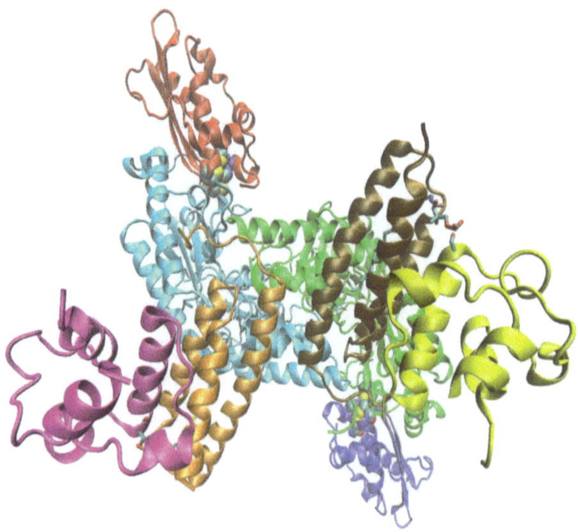

Fig. 13.15 The X-ray structure of supercomplex NFS1/ACP-ISD11/ISCU (PDB ID: 5WLW, (Boniecki et al. 2017)). A dimer of desulfurase NFS1 (cyan and green) is stabilized by the interaction of two subunits of the SD11 (orange and brown), which is a small tutor protein that, in fact, stabilizes the dimer and active conformation of the desulfurase. Interestingly, each one of the ISD11 subunits is solubilized and stabilized by the mitochondrial acyl carrier protein (ACP, magenta and yellow). Additionally, two ISCU subunits (red and blue), the proteins that will function as the scaffold to assemble the clusters [Fe–S], interact with NFS1 subunits. This complex also interacts with FDX and FXN proteins

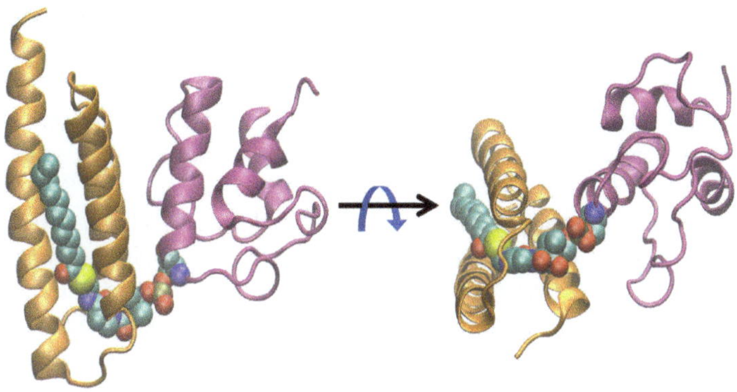

Fig. 13.16 The structure of the complex ACP and ISD11. Two different views of *E. coli* ACP and human ISD11extracted from the supercomplex PDB ID: 5WGB (ribbon representation, ACP in magenta and ISD11 in orange) showing the phosphoSer36 4'PPT and the acyl chain in van der Waals volumes (C, O, N, S and P elements in cyan, red, blue, yellow, and orange respectively)

Moreover, we also know that ISD11 folding likely depends on the interaction with the holo-mitochondrial acyl-carrier protein (holo-ACP) (Herrera et al. 2018), the phosphopantetheine (4'PPT) moiety and the fatty acid chain that the latter carries. In fact, the fatty acid attached to the 4'PPT of ACP penetrates the core of ISD11 and extensively interacts with the protein (Fig. 13.16).

ISCU may populate two interconverting conformational states: one structured and the other disordered (Markley et al. 2013). Although there are some controversies concerning which is the conformation that forms the supercomplex (Cai et al. 2013), by applying NMR strategies to study large complexes Dr. Annalisa Pastore's group showed that ISCU binds in its structured form to the desulfurase homolog from *E. coli* (Yan et al. 2014). Crystallographic data confirmed this result for mammalian ISCU and NFS1 (Fig. 13.15) (Boniecki et al. 2017). These studies carried out by Boniecki and coworkers provided a much more complete picture concerning the structure of the supercomplex and the details at the catalytic center defined by the active-site Cys 381 of NFS1 and conserved Cys, Asp, and His residues of ISCU.

It is worthy of mention that in the supercomplex, electrostatic contribution to protein-protein interactions seems to be critical and coevolving. On the one hand, an extensive positive surface of ISD11 interacts with the large negatively charged surface of ACP. On the other hand, the NFS1 positive surface may interact with the broad negative acidic ridge of FXN.

The acidic ridge of FXN interacts not only with NFS1 but also with iron. The fact that FXN binds iron is very well known; however, much less known is what exactly the role of this interaction is.

As we mentioned above, NMR studies showed, with an amino acid level of resolution, which residues of the acid ridge form the iron-binding site in the FXN surface. At least two different possibilities for the iron binding function have been suggested: (i) that FXN may be the first iron donor for cluster assembly; or, (ii) rather than being the iron donor, that FXN may control the entry of iron to ISCU.

Remarkably, Markley's group (Cai et al. 2018b) carried out a series of NMR experiments that allowed the study of FXN in the context of the supercomplex, which exhibits a very high correlation time. FXN is labeled with ^{15}N (whereas the rest of the proteins of the supercomplex are not) and the 2D $^{1}H,^{15}N$ TROSY-HSQC spectra of ^{15}N-labeled FXN are acquired in the presence or in the absence of the supercomplex. It was observed that the complex formed between FXN and Fe^{2+} binds to the supercomplex without iron release (Cai et al. 2018b). Given the paramagnetic properties that this metal ion displays, iron interaction yields the disappearance of specific cross peaks corresponding to the protein-iron interaction sites, and especially, the absence of these cross peaks persists in the context of the supercomplex. Conversely, the initiation of cluster assembly by addition of L-cysteine and reductant (DTT or Ferredoxin protein) leads to the dissociation of Fe^{2+} from FXN, the absent peaks reappear, and other chemical shifts perturbations that result from iron binding return to their positions in the spectrum of iron-free FXN (Cai et al. 2018b).

Significant efforts were made to establish, with high resolution, how mammalian FXN interacts in the supercomplex with the rest of the proteins. How FXN is docked is important to understand how the activity of the supercomplex is allosterically controlled by FXN.

Very recently, Markley's group, by means of chemical crosslinking experiments combined with mass spectrometry, small-angle X-ray scattering and distance constraints from nuclear magnetic resonance, concluded that the supercomplex involves closed NFS1 dimer conformations for both the NFS1/ACP-ISD11/ISCU and the NFS1/ACP-ISD11/ISCU/FXN complexes.

The resulting model showed interactions between the highly conserved Arg-rich loop of NFS1 (the stretch of RRRPRVR sequence) and the conserved acidic ridge of FXN. Additionally, ISCU and FXN interact by means of the β3–β5 strands of the latter with the conserved CLPPVKLH sequence of ISCU, located between helix 4 and helix 5, the C-terminal helix. In this model, FXN fits into a cleft between NFS1 desulfurase and the scaffolding protein ISCU (Fig. 13.17), and Trp155 of FXN is at van der Waals distance (3.6 Å) of His112 of ISCU, two key residues for super-complex activity. The catalytic center is in fact defined by the Cys381 of NFS1 and the conserved Cys44, Cys70, Cys113, Asp46, and His112 residues of ISCU. Interestingly, this experimental model of the supercomplex suggests that FXN's Asn146 and Gln148 may modulate the activity of the Fe-S cluster assembly. This model elucidates why W155R, N146K or Q148R (Gomes and Santos 2013; Bridwell-Rabb et al. 2011) have a negative effect on function by placing positive charges at these sites, resulting in pathogenic mutations.

There are still other proteins that interact with the supercomplex. Among them, ferredoxins (FDX) play a key role as electron donors for cluster biosynthesis. Mitochondrial FDX1 and FDX2 interact with the supercomplex as judged by the

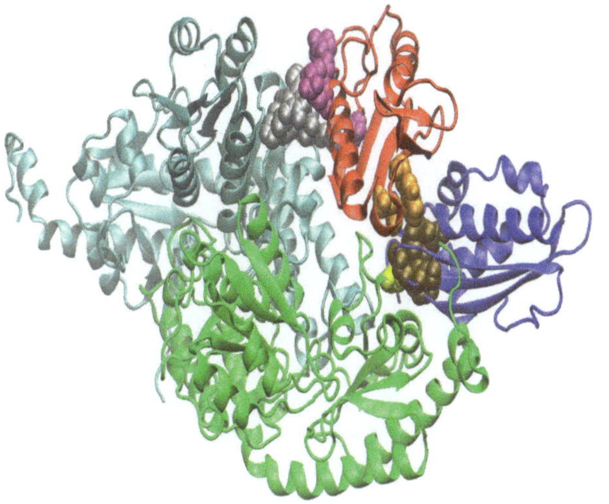

Fig. 13.17 An experimental model of the supercomplex including human FXN. A structural model constructed by Dr. Markley's group (Cai et al. 2018a). Chemical crosslinking experiments were combined with mass spectrometry, small-angle X-ray scattering and distance constraints from nuclear magnetic resonance of the supercomplex. For clarity, only both NFS1 subunits (green and cyan), one ISCU subunit (blue) and one FXN subunit (red) in ribbon are shown. The interaction between the NFS1 stretch of sequence RRRPRVR (gray) and the acidic ridge of FXN (magenta) is shown in van der Waals. FXN fits into a cleft between desulfurase NFS1 and ISCU. Q148 and Trp155 and (orange) of FXN are very close to the active site of ISCU (Cys44, Cys70, Cys113, Asp46, and His112, brown). Cys 381 (yellow) from Cys-loop in the active site of NFS1 is at van der Waals distance of the active center of ISCU

analysis of signal broadening and chemical shift perturbations (experiments ^{1}H-^{15}N TROSY-HSQC) for residues Phe11, Asn13, Arg14, Leu57-Ile63, Ile71-Leu90, Leu96, Thr97 and Asn102 and residues Leu38-Ile40, Leu50, Tyr64, Glu81-Leu97 for FDX1 and FDX2, respectively (Cai et al. 2017); most of the mentioned residues are close to the FDX cluster [2Fe–2S]. However, apo-FDXs (FDXs without cluster) do not interact with the supercomplex (Cai et al. 2017).

Additionally, the release of the cluster [2Fe–2S] from ISCU is facilitated by a specific chaperone system formed by a dedicated HSP70 chaperon (mtHSP70) and a J-type co-chaperone (HscB/Jac1) (Wachnowsky et al. 2018; Maio and Rouault 2015). When the chaperon-co-chaperon complex binds to the holo-ISCU, the latter may change its conformation, releasing the cluster to specific [2Fe–2S] cluster carrier proteins (Wachnowsky et al. 2018).

Activator in Eukaryotes, Inhibitor in Prokaryotes: Differences in the Role of FXN in the Fe-S Clusters Biosynthesis

It has been demonstrated that in eukaryotes the fundamental role of FXN is in iron homeostasis through the pathway of biosynthesis of the Fe–S clusters. In contrast, the deletion of the FXN gene in *Bacteria* essentially has no consequences, even when FXN is a highly conserved protein among these organisms. Additionally, a number of experiments have shown that, while in eukaryotes FXN increases the synthesis speed of Fe–S clusters by NFS1 desulfurase, the FXN homologue in bacteria called CyaY has the opposite effect, that is, in eukaryotes FXN is an activator and in *Bacteria* a desulfurase inhibitor. In *E. coli* and *S. enterica* it was found that CyaY deletion does not lead to a marked phenotype, although this may happen under certain growth conditions or in conjunction with mutations of other genes associated with Fe–S cluster metabolism (Yoon et al. 2012; Lesuisse et al. 2003; Duby et al. 2002; Vivas et al. 2006). It is important to consider that in most bacteria there is an alternative pathway for Fe–S cluster synthesis, known as SUF (sulfur assimilation), which is activated under conditions of oxidative stress or lack of iron.

Thus, several questions arise: Is the role of FXN conserved among these organisms? Why is FXN vital in eukaryotes and unessential in prokaryotes? Why is this protein an activator of the synthesis of Fe-S clusters in eukaryotes and an inhibitor in prokaryotes?

As we will review in this section, the differences in the activity of FXN on the synthesis of Fe-S clusters between eukaryotes and prokaryotes does not seem to reside in FXN itself, but mainly in the desulfurase and in the complexes that the latter forms with other key proteins of the biosynthetic pathway.

Although the three-dimensional structures of the eukaryotic and prokaryotic desulfurase enzymes are very similar (RMSD = 1.9 Å between the human variant and *E. coli*), as well as their sequences ($\sim$60% identity), there are very important differences from the functional viewpoint (Table 13.2): *E. coli* desulfurase is very active itself (k_{cat} = 7.5–8.5 min^{-1} (Urbina et al. 2001; Bridwell-Rabb et al. 2012), while eukaryotic desulfurase requires accessory proteins ACP-ISD11 to be stable and exhibits significant lower enzymatic activity (k_{cat} = 0.9 min^{-1} for the human complex NFS1/ISD11-ACP, (Bridwell-Rabb et al. 2014)), requiring FXN and ISCU to have a comparable enzymatic activity (k_{cat} = 6.7 min^{-1} for the NFS1/ACP-ISD11/ISCU/FXN complex, (Bridwell-Rabb et al. 2014)) to the prokaryotic variant.

On the other hand, some aspects of the interaction mode between FXNs and desulfurase from eukaryotic or prokaryotic organisms might be similar because both CyaY and human FXN complement yeast strains, in which the FXN gene was deleted (Δyfh1), in terms of the restitution of enzyme activities for Fe-S cluster biosynthesis and iron homeostasis (Bedekovics et al. 2007; Cavadini et al. 2000b). To investigate this fact in detail, in vitro experiments were carried out to measure

Table 13.2 Catalytic activity of desulfurase enzymes and desulfurase complexes from different organisms

Organism	Proteins	k_{cat} (min^{-1})	Effect
Bacteria	IscS	7.5–8.5 (Urbina et al. 2001; Bridwell-Rabb et al. 2012)	–
	IscS/IscU	4.7 (Bridwell-Rabb et al. 2012)	↓
	IscS/CyaY	–	↓ (Iannuzzi et al. 2011; Adinolfi et al. 2009)
	IscS/IscU/CyaY	5.8 (Bridwell-Rabb et al. 2012)	↓
Yeast	Nfs1	–	–
	Nfs1/Acp-Isd11	–	↑ (Pandey et al. 2012, 2013)
	Nfs1/Acp-Isd11/Yfh1	–	↑↑ (Pandey et al. 2013)
	Nfs1/Acp-Isd11/Isu1	–	↓ (Pandey et al. 2013)
	Nfs1/Acp-Isd11/Isu1/FXN	–	↑↑(Pandey et al. 2013)
Human	NFS1	–	–
	NFS1/ACP-ISD11	0.9–1.5 (Bridwell-Rabb et al. 2012, 2014)	↑ (Pandey et al. 2012)
	NFS1/ACP-ISD11/FXN	1.7 (Tsai and Barondeau 2010)	↑↑
	NFS1/ACP-ISD11/ISCU	0.8–1.0 (Bridwell-Rabb et al. 2012, 2014)	↓
	NFS1/ACP-ISD11/ISCU/FXN	6.7–7.8 (Bridwell-Rabb et al. 2012, 2014)	↑↑

–, not determined or not effect; ↓, negative effect; ↑, positive effect

desulfurase activity and assembly of the Fe-S clusters, where it was found that the effects of these proteins on desulfurase are interchangeable (Bridwell-Rabb et al. 2012). When ISCU (either human or from *E. coli*) was added to the purified complex ACP-ISD11-NFS1, and also human FXN or CyaY from *E. coli* was included, the result in all cases was an increase in desulfurase activity, which is revealed by an increase in k_{cat}. On the other hand, when the desulfurase was from *E. coli*, the activity markedly decreases after adding bacterial ISCU and had a small additional decrease when human FXN or CyaY were added. Regarding the cluster assembly activity, in the absence of FXN the activity of the human NFS1/ACP-ISD11/ISCU complex is essentially null. On the other hand, after the addition of FXN, the activity strongly increases, while the increase is moderate when adding

CyaY. The equivalent experiment with bacterial desulfurase and bacterial ISCU showed total inhibition with CyaY and partial inhibition with human FXN.

The three-dimensional structure of the dimeric IscS desulfurase from *E. coli* was determined by crystallography (PDB ID: 1P3W, (Cupp-Vickery et al. 2003) and 3LVM, (Shi et al. 2010)) and the monomers are arranged so that the active site lies between the two subunits of IscS (closed conformation). On the other hand, a very different dimeric arrangement was found for the human NFS1/ACP-ISD11, where the active sites are exposed (open conformation, PDB ID: 5USR, (Cory et al. 2017)). As we mentioned above in the previous section, ACP was from *E. coli.*

Almost simultaneously, the structure of the same complex of human NFS1 (without or with ISCU) was independently determined by crystallography (PDB ID: 5WGB, 5WLW and 5WKP), finding a closed conformation, similar to that observed for IscS. Subsequently, the structure of the complex was studied in solution by SAXS and crosslinking coupled to MS for the ACP-ISD11-NFS1-ISCU complex alone and with FXN (Cai et al. 2018a). From XL-MS the authors obtained crosslinks compatible with the closed conformation but incompatible with the open form; also, the profiles and envelopes obtained by SAXS were better adjusted by the structure of NFS/ACP-ISD11/ISCU in the closed conformation than by a model derived from the open conformation docked with ISCU.

Even so, complex 2[NFS1/ACP-ISD11] may exist as two alternative conformations that might interconvert; indeed, this balance might be relevant to explain the functional difference between NFS1 and IscS.

Finally, the differences observed for FXN activity in the eukaryotic and prokaryotic systems may be originated from the transient complexes that the desulfurase enzymes form with other proteins. In *E. coli*, IscS desulfurase interacts with a number of proteins, including those involved in the Fe-S cluster synthesis pathway: IscU, CyaY, FDX and IscX; and thiolation and the thiamin biosynthesis or molybdenum cofactor, such as TusA (PDB ID: 3LVJ) and ThiI (Hidese et al. 2011; Leimkuhler et al. 2017). Several of the binding sites have been mapped in the structure of IscS, some of these sites are independent and others are overlapping (Kim et al. 2013; Yan et al. 2013). On the other hand, in humans, NFS1 interacts with different proteins in the mitochondrial matrix and in the cytosol. In the mitochondria, its role is the synthesis of the Fe-S clusters, for which it requires interaction with the ISCU, FXN, FDX, and ACP-ISD11 proteins, although it also transfers the persulfur group to TUM1 for tRNA modification (Noma et al. 2009).

In particular, interaction with ferredoxin is different between eukaryotes and prokaryotes. In *E. coli*, CyaY and ferredoxin (FDX) compete for the same site (Yan et al. 2013; Kim et al. 2013). By contrast, in the human biosynthetic complex, it was found that kinetic experiments carried at different concentrations of the components are better explained when considering that both ferredoxin and FXN are simultaneously bound in a complex with NFS1/ACP-ISD11/ISCU (Webert et al. 2014).

On the other hand, IscX, a small protein of 66 amino acid residues which is absent in almost all eukaryotes (Pastore et al. 2006), is able to bind iron by means of a negatively charged surface and, also, it binds to IscS. Notably, IscX, CyaY and

FDX compete for the same site in IscS, as was suggested by pull-down experiments using mutants affecting the interaction surfaces (Shi et al. 2010); although recent studies showed that IscX binds to two different sites in IscS (Adinolfi et al. 2017), binding of IscX to one of them can displace CyaY from the complex.

It has been observed that cluster synthesis inhibition exerted by CyaY increases with iron concentration (Adinolfi et al. 2009). The addition of IscX attenuates this effect at low iron concentrations; however, at higher iron concentrations, IscX has no effect on the inhibition of CyaY. While the interaction of CyaY with IscS is dependent on iron concentration, the interaction of desulfurase with IscX is not. Based on these results, it was suggested that IscX may modulate CyaY such that at low concentrations of iron, the IscX junction displaces CyaY competing for the same site where IscX has higher affinity, but at a higher iron concentration, the affinity between CyaY and IscS increases, contributing to the modulation of cluster biosynthesis.

We conclude that as more insight into the complexes formed by desulfurase is gained, our understanding about the differential effect of FXN between eukaryotes and prokaryotes will be enhanced.

Conflicting results have been obtained about the effect of FXN in the NFS1 persulfide formation step of the cluster assembly. Studying the yeast cluster biosynthetic system using ^{35}S labelling of L-cysteine, it was found that the persulfide is not formed in NFS1 in the absence of ACP-ISD11(Pandey et al. 2013). This fact was used to develop an assay to separately monitor persulfide formation in NFS1 from its transfer to ISCU. They found that FXN is not necessary for persulfide formation in NFS1, but it is strongly enhanced by FXN binding, probably as a result of a conformational change induced by FXN binding, which exposes additional binding sites for cysteine. However, this latter result was challenged by another study with the human proteins, where no effect was observed for FXN in persulfide formation in NFS1 (Parent et al. 2015). These experiments were based in a novel assay of maleimide labeling of both persulfides and thiols. They noted that FXN enhances persulfur transfer from NFS1 to ISCU. These emerging mechanistic insights may probably reveal further differences in FXN regulation within eukaryotes.

FXN as a Scaffold to Form Oligomers and Nanoparticles

It was found that some FXN variants are able to form high molecular weight arrangements. In particular, this behavior was first observed for yeast FXN. In this case, the multimeric form is stabilized in the presence of iron. Even though yeast FXN is a soluble monomeric protein of ~ 14 kDa, the addition of Fe^{2+} results in the assembly of a spherical and homogeneous multimer with a molecular mass of $\sim 1.1 \times 10^6$ Da and a diameter of 13 ± 2 nm, as judged by Atomic Force Microscopy (AFM) and Electron microscopy (EM) (Adamec et al. 2000). Each multimer is formed by several subunits of FXN and it may bind a large quantity of iron (Gakh et al. 2002). Additionally, size exclusion chromatography results

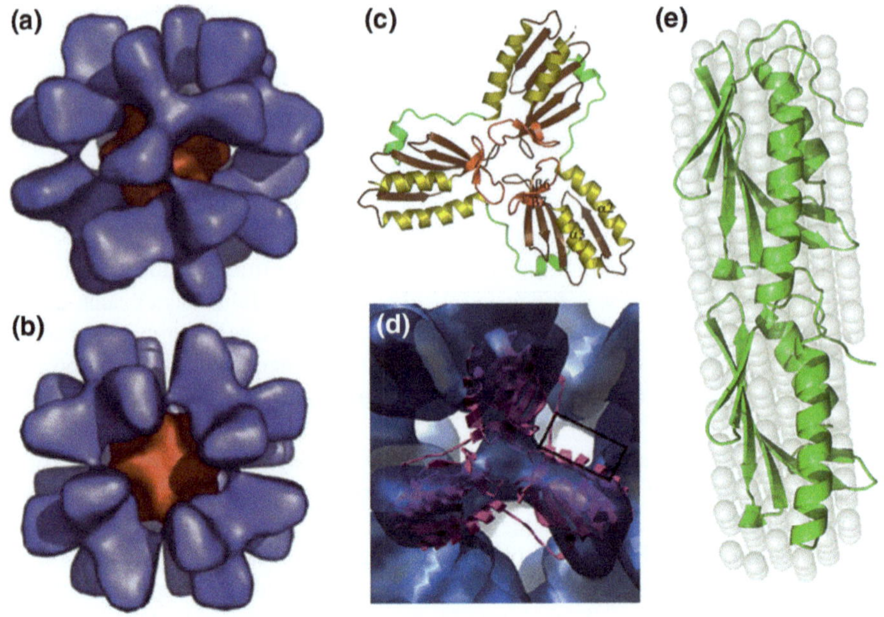

Fig. 13.18 FXN as a scaffold for generating small oligomers and nanoparticles. **a** and **b** Yeast FXN can form large oligomers (24 subunits) in the presence of iron, cryoEM structures, and single-particle reconstruction. **c** On the other hand, mutant Y73A forms a trimer, crystallographic structure PDB ID: 2FQL. In (**d**), the crystallographic structure of the trimeric form is docked into the iron-loaded FXN reconstructions. **e** *E. coli* FXN may be stabilized as a dimer (rigid-body modeling for iron-induced CyaY dimer, SAXS results). This Figure is a partial reproduction of Fig. 5 from reference Schagerlof et al. (2008), Fig. 1 from reference Karlberg et al. (2006) and Fig. 4 from reference Fekry et al. (2017)

suggest that there is a starting point in which the monomer is assembled first into a trimer and afterward two trimers may form a hexamer to subsequently form a dodecamer and, finally, dodecamers may form the multimer. In addition, the mature form of yeast Y73A FXN mutant assembles into trimers and 24-subunit oligomers, in an iron-independent fashion (Fig. 13.18). Remarkably, the multimer may acquire and store iron. Schagerlöf and coworkers by means of Cryo-EM and single-particle reconstruction techniques obtained important information regarding multimer structure (Schagerlof et al. 2008), showing that packing within the particles is stabilized by weak interactions between trimers. This model added to the information provided by the X-ray structure of the yeast FXN trimer variant (Y73A) (Karlberg et al. 2006) supports a key role for the N-terminal region in the oligomerization process.

Oligomerization was also demonstrated for *E. coli* FXN. In the presence of iron, the protein mostly forms a dimer; nevertheless, it was shown that trimers, tetramers, pentamers and possibly hexamers may exist in solution (Fekry et al. 2017). Modeling of the dimer using data provided by small-angle X-ray scattering (SAXS) experiments confirms a head-to-tail packing arrangement of the monomers (Fekry et al. 2017), as previously assessed by cross-linking, mass spectrometry and docking studies (Ahlgren et al. 2017) (Fig. 13.18d).

It is worthy of mention that some human FXN also forms oligomers (Faraj et al. 2013). This is the case of the variant that includes an N-terminal stretch from residue 56. The overexpression of this form in *E. coli* yields both the oligomer and the monomeric forms. After size exclusion chromatography, most of the protein was found in the high molecular weight fractions. The N-terminal moiety is an essential element for the assembly of this high molecular weight multimer. Whereas biophysical characterization by fluorescence and circular dichroism showed that subunits are well folded, sharing similar features to the mature form of FXN, controlled proteolysis of the oligomer indicated that the N-terminal segment is labile in the context of the oligomer and the yielded FXN 81-210 that persisted is strongly resistant (Faraj et al. 2013).

In vitro studies have shown that low concentration of guanidinium hydrochloride disrupts intermolecular interactions, shifting the ensemble toward the monomeric form and showing that the protein-protein interactions that stabilize the multimer are easily weakened.

Whereas functionality of the oligomers inside the cell and inside the mitochondria, in the case of eukaryotic FXN variants, remains under debate, all these results suggest FXN is a malleable scaffold and building block that may be used in nanoparticle design to obtain different species from monomer, dimers, trimers to oligomers.

Conclusions

To understand what the relationships between protein structure and function of mammalian FXN are, and with the aim of finding a cure for Friedreich's Ataxia, it was essential to gain knowledge concerning the tridimensional structure of FXN, its internal motions, interaction with iron, and the effect of mutations of its conformation and stability.

For understanding FXN function, however, the study of its interaction with the mitochondrial enzyme NFS1 and its allosteric modulation, interaction with the scaffolding protein ISCU and involvement of FXN in mitochondrial iron–sulfur cluster assembly were all of great importance.

The combination of multiple experimental tools including high resolution techniques like NMR and X-ray, but also SAXS, crosslinking and mass-spectrometry was necessary to build reliable models of the structure of the desulfurase complex NFS1/ACP-ISD11/ISCU/FXN. Finally, the study of the enzymatic reactions involved will help to understand the capability of FXN to modulate function. Furthermore, knowledge regarding the specific control inside the cell, of conformational stability, protein degradation and protein complex assembly/disassembly dynamics in both normal and pathological contexts will provide new strategies to act on cell function.

Acknowledgements This work was supported by the Agencia Nacional de Promoción Científica y Tecnológica (ANPCyT PICT2016-2280), the Consejo Nacional de Investigaciones Científicas y Técnicas (CONICET), the Universidad de Buenos Aires and FARA, Friedreich's Ataxia Research Alliance.

Notes

The authors declare no competing financial interest. While this chapter was being edited, Fox and et al. presented a cryo-electron microscopy structure (3.2 Å resolution, PDB ID: 6NZU) of the human frataxin-bound iron-sulfur cluster assembly complex, containing two copies of the NFS1/ISD11-ACP/ISCU/FXN hetero-pentamer. A key feature of FXN binding is its simultaneous interactions with both NFS1 protomers of the complex and with ISCU (Fox et al. 2019).

References

Adamec J, Rusnak F, Owen WG, Naylor S, Benson LM, Gacy AM, Isaya G (2000) Iron-dependent self-assembly of recombinant yeast frataxin: implications for Friedreich ataxia. Am J Hum Genet 67(3):549–562

Adinolfi S, Iannuzzi C, Prischi F, Pastore C, Iametti S, Martin SR, Bonomi F, Pastore A (2009) Bacterial frataxin CyaY is the gatekeeper of iron–sulfur cluster formation catalyzed by IscS. Nat Struct Mol Biol 16(4):390–396

Adinolfi S, Puglisi R, Crack JC, Iannuzzi C, Dal Piaz F, Konarev PV, Svergun DI, Martin S, Le Brun NE, Pastore A (2017) The MOLECULAR bases of the dual regulation of bacterial iron sulfur cluster biogenesis by CyaY and IscX. Front Mol Biosci 4:97

Ahlgren EC, Fekry M, Wiemann M, Soderberg CA, Bernfur K, Gakh O, Rasmussen M, Hojrup P, Emanuelsson C, Isaya G, Al-Karadaghi S (2017) Iron-induced oligomerization of human FXN81-210 and bacterial CyaY frataxin and the effect of iron chelators. PLoS ONE 12(12): e0188937

Armas AM, Balparda M, Terenzi A, Busi MV, Pagani MA, Gomez-Casati DF (2019) Ferrochelatase activity of plant frataxin. Biochimie 156:118–122

Bedekovics T, Gajdos GB, Kispal G, Isaya G (2007) Partial conservation of functions between eukaryotic frataxin and the *Escherichia coli* frataxin homolog CyaY. FEMS Yeast Res 7 (8):1276–1284

Benini M, Fortuni S, Condo I, Alfedi G, Malisan F, Toschi N, Serio D, Massaro DS, Arcuri G, Testi R, Rufini A (2017) E3 Ligase RNF126 Directly ubiquitinates frataxin, promoting its degradation: identification of a potential therapeutic target for Friedreich ataxia. Cell Rep 18 (8):2007–2017

Boniecki MT, Freibert SA, Muhlenhoff U, Lill R, Cygler M (2017) Structure and functional dynamics of the mitochondrial Fe/S cluster synthesis complex. Nat Commun 8(1):1287

Bridwell-Rabb J, Fox NG, Tsai CL, Winn AM, Barondeau DP (2014) Human frataxin activates Fe–S cluster biosynthesis by facilitating sulfur transfer chemistry. Biochemistry 53(30):4904–4913

Bridwell-Rabb J, Iannuzzi C, Pastore A, Barondeau DP (2012) Effector role reversal during evolution: the case of frataxin in Fe–S cluster biosynthesis. Biochemistry 51(12):2506–2514

Bridwell-Rabb J, Winn AM, Barondeau DP (2011) Structure-function analysis of Friedreich's ataxia mutants reveals determinants of frataxin binding and activation of the Fe–S assembly complex. Biochemistry 50(33):7265–7274

Cai K, Frederick RO, Dashti H, Markley JL (2018a) Architectural features of human mitochondrial cysteine desulfurase complexes from crosslinking mass spectrometry and small-angle X-ray scattering. Structure 26(8):1127–1136 e 1124

Cai K, Frederick RO, Kim JH, Reinen NM, Tonelli M, Markley JL (2013) Human mitochondrial chaperone (mtHSP70) and cysteine desulfurase (NFS1) bind preferentially to the disordered conformation, whereas co-chaperone (HSC20) binds to the structured conformation of the iron–sulfur cluster scaffold protein (ISCU). J Biol Chem 288(40):28755–28770

Cai K, Frederick RO, Tonelli M, Markley JL (2018b) Interactions of iron-bound frataxin with ISCU and ferredoxin on the cysteine desulfurase complex leading to Fe–S cluster assembly. J Inorg Biochem 183:107–116

Cai K, Tonelli M, Frederick RO, Markley JL (2017) Human mitochondrial ferredoxin 1 (FDX1) and ferredoxin 2 (FDX2) both bind cysteine desulfurase and donate electrons for iron–sulfur cluster biosynthesis. Biochemistry 56(3):487–499

Campuzano V, Montermini L, Molto MD, Pianese L, Cossee M, Cavalcanti F, Monros E, Rodius F, Duclos F, Monticelli A, Zara F, Canizares J, Koutnikova H, Bidichandani SI, Gellera C, Brice A, Trouillas P, De Michele G, Filla A, De Frutos R, Palau F, Patel PI, Di Donato S, Mandel JL, Cocozza S, Koenig M, Pandolfo M (1996) Friedreich's ataxia: autosomal recessive disease caused by an intronic GAA triplet repeat expansion. Science 271 (5254):1423–1427

Castro IH, Ferrari A, Herrera MG, Noguera ME, Maso L, Benini M, Rufini A, Testi R, Costantini P, Santos J (2018) Biophysical characterisation of the recombinant human frataxin precursor. FEBS Open Bio 8(3):390–405

Cavadini P, Adamec J, Taroni F, Gakh O, Isaya G (2000a) Two-step processing of human frataxin by mitochondrial processing peptidase. Precursor and intermediate forms are cleaved at different rates. J Biol Chem 275(52):41469–41475

Cavadini P, Gellera C, Patel PI, Isaya G (2000b) Human frataxin maintains mitochondrial iron homeostasis in *Saccharomyces cerevisiae*. Hum Mol Genet 9(17):2523–2530

Chamberlain S, Shaw J, Rowland A, Wallis J, South S, Nakamura Y, von Gabain A, Farrall M, Williamson R (1988) Mapping of mutation causing Friedreich's ataxia to human chromosome 9. Nature 334(6179):248–250

Chamberlain S, Shaw J, Wallis J, Rowland A, Chow L, Farrall M, Keats B, Richter A, Roy M, Melancon S et al (1989) Genetic homogeneity at the Friedreich ataxia locus on chromosome 9. Am J Hum Genet 44(4):518–521

Chen OS, Hemenway S, Kaplan J (2002) Inhibition of Fe–S cluster biosynthesis decreases mitochondrial iron export: evidence that Yfh1p affects Fe–S cluster synthesis. Proc Natl Acad Sci U S A 99(19):12321–12326

Cherubini F, Serio D, Guccini I, Fortuni S, Arcuri G, Condo I, Rufini A, Moiz S, Camerini S, Crescenzi M, Testi R, Malisan F (2015) Src inhibitors modulate frataxin protein levels. Hum Mol Genet 24(15):4296–4305

Clark E, Butler JS, Isaacs CJ, Napierala M, Lynch DR (2017) Selected missense mutations impair frataxin processing in Friedreich ataxia. Ann Clin Transl Neurol 4(8):575–584

Correia AR, Adinolfi S, Pastore A, Gomes CM (2006) Conformational stability of human frataxin and effect of Friedreich's ataxia-related mutations on protein folding. Biochem J 398(3):605–611

Correia AR, Pastore C, Adinolfi S, Pastore A, Gomes CM (2008) Dynamics, stability and iron-binding activity of frataxin clinical mutants. FEBS J 275(14):3680–3690

Cory SA, Van Vranken JG, Brignole EJ, Patra S, Winge DR, Drennan CL, Rutter J, Barondeau DP (2017) Structure of human Fe–S assembly subcomplex reveals unexpected cysteine desulfurase architecture and acyl-ACP-ISD11 interactions. Proc Natl Acad Sci U S A 114(27):E5325–E5334

Cupp-Vickery JR, Urbina H, Vickery LE (2003) Crystal structure of IscS, a cysteine desulfurase from *Escherichia coli*. J Mol Biol 330(5):1049–1059

Dhe-Paganon S, Shigeta R, Chi YI, Ristow M, Shoelson SE (2000) Crystal structure of human frataxin. J Biol Chem 275(40):30753–30756

Duby G, Foury F, Ramazzotti A, Herrmann J, Lutz T (2002) A non-essential function for yeast frataxin in iron–sulfur cluster assembly. Hum Mol Genet 11(21):2635–2643

Durr A, Cossee M, Agid Y, Campuzano V, Mignard C, Penet C, Mandel JL, Brice A, Koenig M (1996) Clinical and genetic abnormalities in patients with Friedreich's ataxia. N Engl J Med 335(16):1169–1175

Faggianelli N, Puglisi R, Veneziano L, Romano S, Frontali M, Vannocci T, Fortuni S, Testi R, Pastore A (2015) Analyzing the Effects of a G137V Mutation in the FXN Gene. Front Mol Neurosci 8:66

Faraj SE, Gonzalez-Lebrero RM, Roman EA, Santos J (2016) Human frataxin folds via an intermediate state. Role of the C-terminal region. Sci Rep 6:20782

Faraj SE, Roman EA, Aran M, Gallo M, Santos J (2014) The alteration of the C-terminal region of human frataxin distorts its structural dynamics and function. FEBS J 281(15):3397–3419

Faraj SE, Venturutti L, Roman EA, Marino-Buslje CB, Mignone A, Tosatto SC, Delfino JM, Santos J (2013) The role of the N-terminal tail for the oligomerization, folding and stability of human frataxin. FEBS Open Bio 3:310–320

Fekry M, Alshokry W, Grela P, Tchorzewski M, Ahlgren EC, Soderberg CA, Gakh O, Isaya G, Al-Karadaghi S (2017) SAXS and stability studies of iron-induced oligomers of bacterial frataxin CyaY. PLoS ONE 12(9):e0184961

Foury F (1999) Low iron concentration and aconitase deficiency in a yeast frataxin homologue deficient strain. FEBS Lett 456(2):281–284

Foury F, Pastore A, Trincal M (2007) Acidic residues of yeast frataxin have an essential role in Fe–S cluster assembly. EMBO Rep 8(2):194–199

Fox NG, Das D, Chakrabarti M, Lindahl PA, Barondeau DP (2015) Frataxin accelerates [2Fe–2S] cluster formation on the human Fe–S assembly complex. Biochemistry 54(25):3880–3889

Fox NG, Yu X, Feng X, Bailey HJ, Martelli A, Nabhan JF, Strain-Damerell C, Bulawa C, Yue WW, Han S (2019) Structure of the human frataxin-bound iron-sulfur cluster assembly complex provides insight into its activation mechanism. Nat Commun 10(1)

Friemel M, Marelja Z, Li K, Leimkuhler S (2017) The N-terminus of iron–sulfur cluster assembly factor ISD11 Is crucial for subcellular targeting and interaction with l-cysteine desulfurase NFS1. Biochemistry 56(12):1797–1808

Gakh O, Adamec J, Gacy AM, Twesten RD, Owen WG, Isaya G (2002) Physical evidence that yeast frataxin is an iron storage protein. Biochemistry 41(21):6798–6804

Galea CA, Huq A, Lockhart PJ, Tai G, Corben LA, Yiu EM, Gurrin LC, Lynch DR, Gelbard S, Durr A, Pousset F, Parkinson M, Labrum R, Giunti P, Perlman SL, Delatycki MB, Evans-Galea MV (2016) Compound heterozygous FXN mutations and clinical outcome in friedreich ataxia. Ann Neurol 79(3):485–495

Gentry LE, Thacker MA, Doughty R, Timkovich R, Busenlehner LS (2013) His86 from the N-terminus of frataxin coordinates iron and is required for Fe–S cluster synthesis. Biochemistry 52(35):6085–6096

Gerber J, Muhlenhoff U, Lill R (2003) An interaction between frataxin and Isu1/Nfs1 that is crucial for Fe/S cluster synthesis on Isu1. EMBO Rep 4(9):906–911

Gomes CM, Santos R (2013) Neurodegeneration in Friedreich's ataxia: from defective frataxin to oxidative stress. Oxid Med Cell Longev 2013:487534

Guo L, Wang Q, Weng L, Hauser LA, Strawser CJ, Mesaros C, Lynch DR, Blair IA (2018) Characterization of a new N-terminally acetylated extra-mitochondrial isoform of frataxin in human erythrocytes. Sci Rep 8(1):17043

Herrera MG, Pignataro MF, Noguera ME, Cruz KM, Santos J (2018) Rescuing the rescuer: on the protein complex between the human mitochondrial acyl carrier protein and ISD11. ACS Chem Biol 13(6):1455–1462

Hidese R, Mihara H, Esaki N (2011) Bacterial cysteine desulfurases: versatile key players in biosynthetic pathways of sulfur-containing biofactors. Appl Microbiol Biotechnol 91(1):47–61

Huynen MA, Snel B, Bork P, Gibson TJ (2001) The phylogenetic distribution of frataxin indicates a role in iron–sulfur cluster protein assembly. Hum Mol Genet 10(21):2463–2468

Iannuzzi C, Adinolfi S, Howes BD, Garcia-Serres R, Clemancey M, Latour JM, Smulevich G, Pastore A (2011) The role of CyaY in iron sulfur cluster assembly on the *E. coli* IscU scaffold protein. PLoS One 6 (7):e21992

Karlberg T, Schagerlof U, Gakh O, Park S, Ryde U, Lindahl M, Leath K, Garman E, Isaya G, Al-Karadaghi S (2006) The structures of frataxin oligomers reveal the mechanism for the delivery and detoxification of iron. Structure 14(10):1535–1546

Kim JH, Frederick RO, Reinen NM, Troupis AT, Markley JL (2013) [2Fe–2S]-ferredoxin binds directly to cysteine desulfurase and supplies an electron for iron–sulfur cluster assembly but is displaced by the scaffold protein or bacterial frataxin. J Am Chem Soc 135(22):8117–8120

Koeppen AH (2013) Nikolaus Friedreich and degenerative atrophy of the dorsal columns of the spinal cord. J Neurochem 126(Suppl 1):4–10

Koutnikova H, Campuzano V, Foury F, Dolle P, Cazzalini O, Koenig M (1997) Studies of human, mouse and yeast homologues indicate a mitochondrial function for frataxin. Nat Genet 16(4): 345–351

Koutnikova H, Campuzano V, Koenig M (1998) Maturation of wild-type and mutated frataxin by the mitochondrial processing peptidase. Hum Mol Genet 7(9):1485–1489

Leidgens S, De Smet S, Foury F (2010) Frataxin interacts with Isu1 through a conserved tryptophan in its beta-sheet. Hum Mol Genet 19(2):276–286

Leimkuhler S, Buhning M, Beilschmidt L (2017) Shared sulfur mobilization routes for tRNA Thiolation and molybdenum cofactor biosynthesis in prokaryotes and eukaryotes. Biomolecules 7(1)

Lesuisse E, Santos R, Matzanke BF, Knight SA, Camadro JM, Dancis A (2003) Iron use for haeme synthesis is under control of the yeast frataxin homologue (Yfh1). Hum Mol Genet 12(8):879–889

Li DS, Ohshima K, Jiralerspong S, Bojanowski MW, Pandolfo M (1999) Knock-out of the cyaY gene in *Escherichia coli* does not affect cellular iron content and sensitivity to oxidants. FEBS Lett 456(1):13–16

Maio N, Rouault TA (1853) Iron–sulfur cluster biogenesis in mammalian cells: New insights into the molecular mechanisms of cluster delivery. Biochim Biophys Acta 1853(6):1493–1512

Maliandi MV, Busi MV, Turowski VR, Leaden L, Araya A, Gomez-Casati DF (2011) The mitochondrial protein frataxin is essential for heme biosynthesis in plants. FEBS J 278(3):470–481

Markley JL, Kim JH, Dai Z, Bothe JR, Cai K, Frederick RO, Tonelli M (2013) Metamorphic protein IscU alternates conformations in the course of its role as the scaffold protein for iron–sulfur cluster biosynthesis and delivery. FEBS Lett 587(8):1172–1179

Muhlenhoff U, Gerber J, Richhardt N, Lill R (2003) Components involved in assembly and dislocation of iron–sulfur clusters on the scaffold protein Isu1p. EMBO J 22(18):4815–4825

Noguera ME, Aran M, Smal C, Vazquez DS, Herrera MG, Roman EA, Alaimo N, Gallo M, Santos J (2017) Insights on the conformational dynamics of human frataxin through modifications of loop-1. Arch Biochem Biophys 636:123–137

Noma A, Sakaguchi Y, Suzuki T (2009) Mechanistic characterization of the sulfur-relay system for eukaryotic 2-thiouridine biogenesis at tRNA wobble positions. Nucleic Acids Res 37(4): 1335–1352

Nuth M, Yoon T, Cowan JA (2002) Iron–sulfur cluster biosynthesis: characterization of iron nucleation sites for assembly of the [2Fe–2S]2 + cluster core in IscU proteins. J Am Chem Soc 124(30):8774–8775

Olsson MH, Sondergaard CR, Rostkowski M, Jensen JH (2011) PROPKA3: consistent treatment of internal and surface residues in empirical pKa predictions. J Chem Theory Comput 7(2): 525–537

Pandey A, Golla R, Yoon H, Dancis A, Pain D (2012) Persulfide formation on mitochondrial cysteine desulfurase: enzyme activation by a eukaryote-specific interacting protein and Fe–S cluster synthesis. Biochem J 448(2):171–187

Pandey A, Gordon DM, Pain J, Stemmler TL, Dancis A, Pain D (2013) Frataxin directly stimulates mitochondrial cysteine desulfurase by exposing substrate-binding sites, and a mutant Fe–S cluster scaffold protein with frataxin-bypassing ability acts similarly. J Biol Chem 288(52): 36773–36786

Pandolfo M (2006) Friedreich ataxia: detection of GAA repeat expansions and frataxin point mutations. Methods Mol Med 126:197–216

Pandolfo M (2009) Friedreich ataxia: the clinical picture. J Neurol 256(Suppl 1):3–8

Parent A, Elduque X, Cornu D, Belot L, Le Caer JP, Grandas A, Toledano MB, D'Autreaux B (2015) Mammalian frataxin directly enhances sulfur transfer of NFS1 persulfide to both ISCU and free thiols. Nat Commun 6:5686

Pastore C, Adinolfi S, Huynen MA, Rybin V, Martin S, Mayer M, Bukau B, Pastore A (2006) YfhJ, a molecular adaptor in iron–sulfur cluster formation or a frataxin-like protein? Structure 14(5):857–867

Patel PI, Isaya G (2001) Friedreich ataxia: from GAA triplet-repeat expansion to frataxin deficiency. Am J Hum Genet 69(1):15–24

Popovic M, Sanfelice D, Pastore C, Prischi F, Temussi PA, Pastore A (2015) Selective observation of the disordered import signal of a globular protein by in-cell NMR: the example of frataxins. Protein Sci 24(6):996–1003

Priller J, Scherzer CR, Faber PW, MacDonald ME, Young AB (1997) Frataxin gene of Friedreich's ataxia is targeted to mitochondria. Ann Neurol 42(2):265–269

Prischi F, Giannini C, Adinolfi S, Pastore A (2009) The N-terminus of mature human frataxin is intrinsically unfolded. FEBS J 276(22):6669–6676

Prischi F, Konarev PV, Iannuzzi C, Pastore C, Adinolfi S, Martin SR, Svergun DI, Pastore A (2010) Structural bases for the interaction of frataxin with the central components of iron–sulphur cluster assembly. Nat Commun 1:95

Ramazzotti A, Vanmansart V, Foury F (2004) Mitochondrial functional interactions between frataxin and Isu1p, the iron–sulfur cluster scaffold protein, in *Saccharomyces cerevisiae*. FEBS Lett 557(1–3):215–220

Richards TA, van der Giezen M (2006) Evolution of the Isd11-IscS complex reveals a single alpha-proteobacterial endosymbiosis for all eukaryotes. Mol Biol Evol 23(7):1341–1344

Rotig A, de Lonlay P, Chretien D, Foury F, Koenig M, Sidi D, Munnich A, Rustin P (1997) Aconitase and mitochondrial iron–sulphur protein deficiency in Friedreich ataxia. Nat Genet 17 (2):215–217

Rouault TA (2015) Mammalian iron–sulphur proteins: novel insights into biogenesis and function. Nat Rev Mol Cell Biol 16(1):45–55

Rufini A, Cavallo F, Condo I, Fortuni S, De Martino G, Incani O, Di Venere A, Benini M, Massaro DS, Arcuri G, Serio D, Malisan F, Testi R (2015) Highly specific ubiquitin-competing molecules effectively promote frataxin accumulation and partially rescue the aconitase defect in Friedreich ataxia cells. Neurobiol Dis 75:91–99

Rufini A, Fortuni S, Arcuri G, Condo I, Serio D, Incani O, Malisan F, Ventura N, Testi R (2011) Preventing the ubiquitin-proteasome-dependent degradation of frataxin, the protein defective in Friedreich's ataxia. Hum Mol Genet 20(7):1253–1261

Sacca F, Marsili A, Puorro G, Antenora A, Pane C, Tessa A, Scoppettuolo P, Nesti C, Brescia Morra V, De Michele G, Santorelli FM, Filla A (2013) Clinical use of frataxin measurement in a patient with a novel deletion in the FXN gene. J Neurol 260(4):1116–1121

Schagerlof U, Elmlund H, Gakh O, Nordlund G, Hebert H, Lindahl M, Isaya G, Al-Karadaghi S (2008) Structural basis of the iron storage function of frataxin from single-particle reconstruction of the iron-loaded oligomer. Biochemistry 47(17):4948–4954

Schmucker S, Argentini M, Carelle-Calmels N, Martelli A, Puccio H (2008) The in vivo mitochondrial two-step maturation of human frataxin. Hum Mol Genet 17(22):3521–3531

Schmucker S, Martelli A, Colin F, Page A, Wattenhofer-Donze M, Reutenauer L, Puccio H (2011) Mammalian frataxin: an essential function for cellular viability through an interaction with a preformed ISCU/NFS1/ISD11 iron–sulfur assembly complex. PLoS ONE 6(1):e16199

Schoenfeld RA, Napoli E, Wong A, Zhan S, Reutenauer L, Morin D, Buckpitt AR, Taroni F, Lonnerdal B, Ristow M, Puccio H, Cortopassi GA (2005) Frataxin deficiency alters heme pathway transcripts and decreases mitochondrial heme metabolites in mammalian cells. Hum Mol Genet 14(24):3787–3799

Shi R, Proteau A, Villarroya M, Moukadiri I, Zhang L, Trempe JF, Matte A, Armengod ME, Cygler M (2010) Structural basis for Fe–S cluster assembly and tRNA thiolation mediated by IscS protein-protein interactions. PLoS Biol 8(4):e1000354

Soderberg C, Gillam ME, Ahlgren EC, Hunter GA, Gakh O, Isaya G, Ferreira GC, Al-Karadaghi S (2016) The Structure of the Complex between Yeast Frataxin and Ferrochelatase: characterization and pre-steady state reaction of ferrous iron delivery and heme synthesis. J Biol Chem 291(22):11887–11898

Steinkellner H, Singh HN, Muckenthaler MU, Goldenberg H, Moganty RR, Scheiber-Mojdehkar B, Sturm B (2017) No changes in heme synthesis in human Friedreich s ataxia erythroid progenitor cells. Gene 621:5–11

Telot L, Rousseau E, Lesuisse E, Garcia C, Morlet B, Leger T, Camadro JM, Serre V (2018) Quantitative proteomics in Friedreich's ataxia B-lymphocytes: a valuable approach to decipher the biochemical events responsible for pathogenesis. Biochim Biophys Acta Mol Basis Dis 1864(4 Pt A):997–1009

Tsai CL, Barondeau DP (2010) Human frataxin is an allosteric switch that activates the Fe–S cluster biosynthetic complex. Biochemistry 49(43):9132–9139

Uchida T, Kobayashi N, Muneta S, Ishimori K (2017) The iron chaperone protein CyaY from Vibrio cholerae Is a heme-binding protein. Biochemistry 56(18):2425–2434

Urbina HD, Silberg JJ, Hoff KG, Vickery LE (2001) Transfer of sulfur from IscS to IscU during Fe/S cluster assembly. J Biol Chem 276(48):44521–44526

Vazquez DS, Agudelo WA, Yone A, Vizioli N, Aran M, Gonzalez Flecha FL, Gonzalez Lebrero MC, Santos J (2015) A helix-coil transition induced by the metal ion interaction with a grafted iron-binding site of the CyaY protein family. Dalton Trans 44(5):2370–2379

Vivas E, Skovran E, Downs DM (2006) *Salmonella enterica* strains lacking the frataxin homolog CyaY show defects in Fe–S cluster metabolism in vivo. J Bacteriol 188(3):1175–1179

Wachnowsky C, Fidai I, Cowan JA (2018) Iron–sulfur cluster biosynthesis and trafficking—impact on human disease conditions. Metallomics 10(1):9–29

Wang T, Craig EA (2008) Binding of yeast frataxin to the scaffold for Fe–S cluster biogenesis, Isu. J Biol Chem 283(18):12674–12679

Webert H, Freibert SA, Gallo A, Heidenreich T, Linne U, Amlacher S, Hurt E, Muhlenhoff U, Banci L, Lill R (2014) Functional reconstitution of mitochondrial Fe/S cluster synthesis on Isu1 reveals the involvement of ferredoxin. Nat Commun 5:5013

Wong A, Yang J, Cavadini P, Gellera C, Lonnerdal B, Taroni F, Cortopassi G (1999) The Friedreich's ataxia mutation confers cellular sensitivity to oxidant stress which is rescued by chelators of iron and calcium and inhibitors of apoptosis. Hum Mol Genet 8(3):425–430

Xia H, Cao Y, Dai X, Marelja Z, Zhou D, Mo R, Al-Mahdawi S, Pook MA, Leimkuhler S, Rouault TA, Li K (2012) Novel frataxin isoforms may contribute to the pathological mechanism of Friedreich ataxia. PLoS ONE 7(10):e47847

Yan R, Kelly G, Pastore A (2014) The scaffold protein IscU retains a structured conformation in the Fe–S cluster assembly complex. ChemBioChem 15(11):1682–1686

Yan R, Konarev PV, Iannuzzi C, Adinolfi S, Roche B, Kelly G, Simon L, Martin SR, Py B, Barras F, Svergun DI, Pastore A (2013) Ferredoxin competes with bacterial frataxin in binding to the desulfurase IscS. J Biol Chem 288(34):24777–24787

Yoon H, Golla R, Lesuisse E, Pain J, Donald JE, Lyver ER, Pain D, Dancis A (2012) Mutation in the Fe–S scaffold protein Isu bypasses frataxin deletion. Biochem J 441(1):473–480

Yoon T, Cowan JA (2003) Iron–sulfur cluster biosynthesis. Characterization of frataxin as an iron donor for assembly of [2Fe–2S] clusters in ISU-type proteins. J Am Chem Soc 125(20): 6078–6084

Yoon T, Cowan JA (2004) Frataxin-mediated iron delivery to ferrochelatase in the final step of heme biosynthesis. J Biol Chem 279(25):25943–25946

Chapter 14
Crystallins and Their Complexes

Kalyan Sundar Ghosh and Priyanka Chauhan

Abstract The crystallins (α, β and γ), major constituent proteins of eye lens fiber cells play their critical role in maintaining the transparency and refractive index of the lens. Under different stress factors and with aging, β- and γ-crystallins start to unfold partially leading to their aggregation. Protein aggregation in lens basically enhances light scattering and causes the vision problem, commonly known as cataract. α-crystallin as a molecular chaperone forms complexes with its substrates (β- and γ-crystallins) to prevent such aggregation. In this chapter, the structural features of β- and γ-crystallins have been discussed. Detailed structural information linked with the high stability of γC-, γD- and γS-crystallins have been incorporated. The nature of homologous and heterologous interactions among crystallins has been deciphered, which are involved in their molecular association and complex formation.

Keywords α-crystallin · β-crystallin · γ-crystallin · Crystallin complexes · Aggregation of crystallins

Crystallin Proteins of Eye Lens

Vertebrate lens is composed of ocular proteins commonly known as crystallins. Crystallins were discovered about 100 years ago by Mörner and these proteins are claimed as the oldest proteins in human body (Ray 2015). Three types (α, β and γ) of crystallins constitute about 90% of the lens protein content (Bloemendal 1981; Bloemendal et al. 2004). Due to high solubility and prolonged stability, crystallins can sustain for the whole life-span of an individual (Bloemendal et al. 2004; Sharma and Santhoshkumar 2009). The content of α, β and γ crystallin in human lens is approximately 28, 43 and 28% respectively (Brakenhoff et al. 1990; Lampi

K. S. Ghosh (✉) · P. Chauhan
Department of Chemistry, National Institute of Technology Hamirpur, Hamirpur 177005, Himachal Pradesh, India
e-mail: kalyan@nith.ac.in

J. R. Harris and J. Marles-Wright (eds.), *Macromolecular Protein Complexes II: Structure and Function*, Subcellular Biochemistry 93,
https://doi.org/10.1007/978-3-030-28151-9_14

et al. 1997; Robinson et al. 2006). The core part of the lens is enriched with γ-crystallins and γC-, γD- and γS-crystallin are found predominantly in human lens (Brakenhoff et al. 1990; Robinson et al. 2006). Being a molecular chaperone, α-crystallin belongs to the sHsp family (de Jong et al. 1993). Whereas β- and γ-crystallins belong to a structural protein family commonly called the βγ-crystallin superfamily (Breitman et al. 1984; Wistow and Piatigorsky 1988). Stability of these proteins is very much essential for their proper physiological function. Their stability is maintained by specific inter and intramolecular atomic contacts in their three dimensional conformation. Different crystallin proteins from eye lens can be separated using size exclusion chromatographic techniques based on the wide difference in their molecular weights (Spector 1964). Formation of polydisperse oligomers is noted in case of α- and β-crystallins, whereas γ-crystallins mostly exist in monomeric form in their native state.

Due to the absence of any repair and recycling system for the proteins in lens, damaged proteins start to accumulate in lens with increase in age. Therefore, environmental stresses and aging cause various structural changes and modifications. This will disrupt homologous and heterologous interactions among crystallins and cause partial unfolding followed by aggregation. Photo-oxidation due to UV radiation, deamidation, disulfide bond formation, cleavage of peptide bonds etc. can occur with aging of the lens and wcan impact deleteriously on the native conformations of crystallins.

βγ-Crystallins

βγ-crystallins act as structural proteins in the lens. Specific packing of βγ-crystallins in the fiber cells results in the short range order interactions required for maintaining lens transparency in the nuclear region (Benedek 1971; Delaye and Tardieu 1983; Stradner et al. 2007). The monomeric structure of βγ-crystallins consists of four Greek key motifs, which are present in two domains. This structural arrangement in βγ-crystallins is due to gene duplication and fusion of a single Greek key motif which occurs prior to vertebrate divergence (Lubsen et al. 1988). A notable similarity in sequence of motif 1 and 3 as well as motif 2 and 4 was also observed (D'Alessio 2002).

At single domain level, the β-sheets interact with each other to form a β-sandwich, which contains two intercalated Greek key motifs (Slingsby and Clout 1999; Bloemendal et al. 2004). The Greek key motif has four β-strands (strands a–d), where the third strand from each motif is swapped with the paired partner giving a total of eight β-strands per domain (Fig. 14.1). The presence of β-hairpins is the characteristic of the βγ-crystallins, which are packed with β-sheets and hence contribute to the stability of the domain (Slingsby and Clout 1999). In some of the βγ-crystallins, a Tyr corner is conserved in each domain along with LXPGXY sequence. The Tyr residue is conserved between strand c and d of Greek key motifs

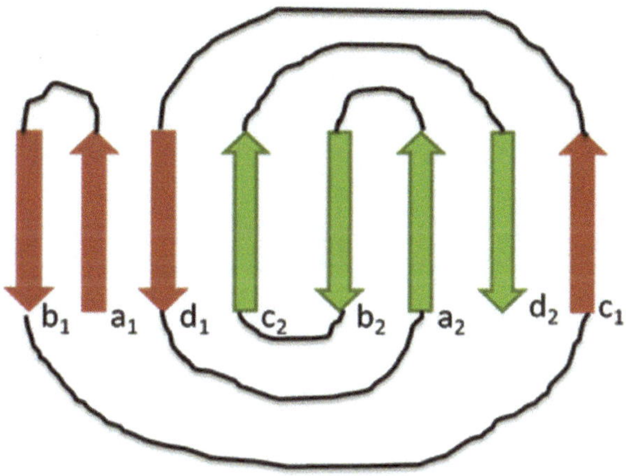

Fig. 14.1 Greek key motif structure consists of four pairs of antiparallel β-strands (strands a–d) and their linking loops, where the third strand of each motif is swapped with the fourth strand of other motif rendering the domain more stable

2 and 4. This Tyr corner act as the folding nucleus of the Greek key β-sandwich domain which is found in the βγ-crystallins family (Bagby et al. 1998).

In βγ-crystallins, the complex topology of their tertiary structure and the Greek key motif feature contribute to the stability of the native state of these proteins (Jaenicke and Slingsby 2001; MacDonald et al. 2005). In addition to this, a hydrophobic cluster is present at the domain interface and it consists of several hydrophobic residues from N-terminal (N-td) and C-terminal (C-td) domains (Jaenicke and Slingsby 2001). The interdomain interface interactions in the β- and γ-crystallins also provide additional stability.

β-Crystallins

β-crystallins of the vertebrate eye lens consist of a group of several proteins. These are categorized into two sub-groups based on their pIs: acidic (βA1, βA3, and βA4) and basic (βB1, βB2, and βB3) in human (Lampi et al. 1997). The acidic β-crystallins contain extensions of N and C-terminal whereas the basic β-crystallins contain N-terminal extension only. The monomeric structure of β-crystallins contains two domains having four Greek key motifs in each. The structures of human βA4- and βB2-crystallins are shown in Fig. 14.2. All β-crystallins bear a high degree of sequence homology. The molecular mass of the β-crystallin monomer ranges from 22 to 28 kDa. The βB1 and βB3-crystallin are present in the nuclear region and the dimeric βB2-crystallin is found in the cortex and nuclear regions (Lampi et al. 1998; Cvekl and Duncan 2007). βB2-crystallin is most stable among all the β-crystallins. It participates in extensive protein-protein interactions. It is the predominant subunit and can interact with the acidic and basic subunits of β-crystallin and form the dimers, which leads to the formation of larger aggregates (Slingsby and Bateman 1990). β-crystallins are usually less soluble, but

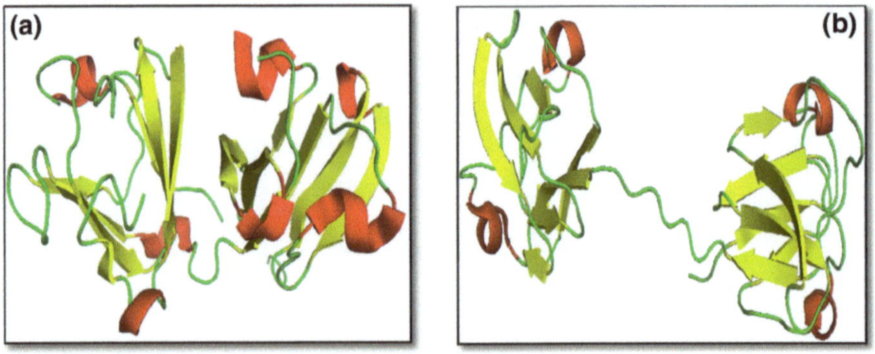

Fig. 14.2 Crystal structure of human β-crystallins: **a** βA4 (PDB ID: 3LWK), acidic β-crystallin containing extensions of N and C-terminal, and **b** βB2 (PDB ID: 1YTQ), a basic β-crystallin containing only N-terminal extension reported by Smith et al. (2007)

βB2-crystallin has higher solubility. It is involved in non-covalent association with the less soluble β-crystallins and thereby increases their solubility (Bateman and Slingsby 1992). Under physiological conditions, βA and βB-crystallins also demonstrate the phenomenon of subunit exchange (Hejtmancik et al. 1997, 2004). The β-crystallins are found to form oligomers. Dimer formation occurs in the βB2 by domain swapping. The domain swapped conformation contains the N-terminal domain of first molecule, which interacts with the C-terminal domain of the second molecule. The intermolecular interaction is governed by the domain linker (Bax et al. 1990). In βB1-crystallin, the linker is bent and the interaction between the monomers occurs intramolecularly. Hetero-oligomeric and homo-oligomeric complexes have been formed by different β-crystallins. Bateman et al. (2003) had reported that the oligomerization occurs in βA3 and βA4 crystallins through subunit exchange. The polydispersity of these crystallins is probably due to the formation of these homo- and hetero-oligomeric complexes.

γ-Crystallins

γ-crystallins, another major constituent of βγ-family is claimed to be evolved from primordial sources, as supported by the existence of microbial βγ-crystallins reported by Mishra et al. (2014) and Barnwal et al. (2009). With respect to evolution, in amphibians γ-crystallins appeared before α-crystallins. γ-crystallins are the smallest and simplest monomeric proteins having a molecular weight of 20 kDa. They are the basic crystallin proteins present in mammals, having isoelectric points ranging from 7.1 to 8.6. They often contain several Cys residues (4–7 per molecule) which are present free and make mixed disulfide bonds under oxidative stress. Their various structural features and associated properties had been reviewed by Serebryany and King (2014) and Vendra et al. (2016). γ-crystallins contain ~175 residues. The members of γ-crystallins subfamily exhibit different behavior in phase separation. The phase separation temperature is low for γA/B/C-crystallins whereas it is high for the γD/E/F-crystallins. This might be due to different states of hydration found in different γ-crystallins. Distinct hydration states arise due to the presence of Leu 51, Ile 103 and His 115 residues in γD-crystallin, which makes substantial changes in hydrophobicity/hydrophilicity of γD-crystallin in comparison to γB-crystallin (Slingsby et al. 1997). Abundantly expressed γ-crystallins are γD, γS and γC-crystallin, which constitute 7, 9, and 11% respectively to the total crystallin content in lens (Robinson et al. 2006; Lampi et al. 1997). γA and γB crystallin are present in low level in human lens. The nuclear region of the lens contains γD and γC-crystallin and they are expressed in early stage of lens development. The γE and γF-crystallin are pseudogenes of human eye lens (Siezen et al. 1987; Robinson et al. 2006). The cortical region of human lens contains γS-crystallin (Siezen et al. 1987). A special type of secondary structural signature called the Greek key motif as defined by Richardson (1977) is generally found in γ-crystallins. This structural pattern is preserved through phylogeny. Greek key motifs consist of four β-sheets, which are folded in an antiparallel arrangement and make a compact folded domain, as shown in Fig. 14.1.

In natively-folded state, the presence of the Greek key motifs contributes towards various features and properties of γ-crystallins, determining their specific roles in the lens. A monomeric globule of ~5 nm is formed by intramolecular packing of N and C-terminal domains. For γ-crystallins, the Stokes radius is 2.13 nm (Zhao et al. 2014) and the frictional ratio is ~1.21 (slightly higher than the value of 1.12 found for a perfect compact smooth protein sphere). This suggests low hydration of γ-crystallins and their low propensity for sticky interactions with solvent and other proteins (Chen et al. 2014). These proteins are present at a high concentration in the lens (~400 mg/ml). Short range order homologous and heterologous interactions among them maintain the transparency of the eye lens (Delaye and Tardieu 1983). Moreover, it has been found that weak attractive interactions also exist between γ-crystallins. The crystal structural analysis of γ-crystallins has shown the presence of nonpolar residues in the interior core and the polar ones on the surface. Solution state studies corroborate the results of structural analysis. This can also explain the kinetic and thermodynamic stability of these proteins. Various methods, like NMR, thermodynamic and optical spectroscopic techniques have been used to study the roles played by specific residues of γ-crystallins and the intra- and inter-molecular packing features of the molecule (particularly human γC, γD and γS). This has been well reported in various publications from King's group (http://web.mit.edu/king-ab/www/publications/pubs_cat.htm#Lens), Pande's group (http://www.albany.edu/chemistry/jpande.shtml#recentpub) and others. These studies have presented various interesting structural and functional aspects of γ-crystallins. Strong denaturants like guanidinium hydrochloride (GdnHCl) can denature γ-crystallins only at a concentration above 2.5 M. The conventional denaturing agent urea is not capable of denaturing gamma crystallins (Kosinski-Collins and King 2003). In contrast, βA1, βB1, βB2 and βA3 crystallins can easily be denatured by urea (Bateman et al. 2003, Lampi et al. 2014). For human γD-crystallin, the free energy of denaturation is about 8. 9 kcal/mol (Flaugh et al. 2005a) and that of γS is found to be 10.5 ± 0.9 kcal/mol (Mills et al. 2007). Whereas for αA and αB-crystallin the values are 6.38 kcal/mol and 5.04 kcal/mol respectively (Sun et al. 1999). The denaturation temperature of γS-crystallin is found to be 74.1 °C and that of γD is 83.8 °C (Mills et al. 2007) and this is pH dependent (Wu et al. 2014). For βB2 and βB1-crystallins, the denaturation temperature is 58 °C and ~67 °C respectively, which are much lower than that of gamma crystallins (Lampi et al. 2002).

γC-Crystallin

γC-crystallin has 173 amino acid residues and is the most abundant gamma crystallin expressed during embryonic development in young human lens. It is found in the nuclear region of the lens (Robinson et al. 2006). Its primary sequence shares 71% identity with human γD-crystallin. Dixit et al. (2016) reported the solution structure of human γC-crystallin (Fig. 14.3). Earlier, Purkiss et al. (2007) reported the crystal structure of mouse γC-crystallin. Four highly-conserved Trp residues were found to be present in both human and mouse γC-crystallin (Wang et al. 2010). The solubility of γC-crystallin is found to be low, unlike other γ-crystallins (Purkiss et al. 2007). In γC-crystallin, at the 79th residue Cys is present as a conserved surface residue, whereas, Arg is present at the same position in γD-crystallin. The inter-motif ion pair interaction between Asp 21—Arg 79 is likely to be disrupted in γC-crystallin.

Therefore, the presence of Cys 79 causes a decrease in the dipole moment and affects the aqueous interactions of the protein. This would decrease the aqueous solubility of the protein at higher concentration (Purkiss et al. 2007). Human γC-crystallin shows high conformational stability, regardless of its lower solubility (Purkiss et al. 2007). The thermodynamic stability of human γC-crystallin was

Fig. 14.3 Crystal structure of human γC-crystallin (PDB ID: 2NBR) reported by Dixit et al. (2016) showing the compact structures of its N-terminal and C-terminal domains

studied by Fu and Liang (2002). γC-crystallin exhibited its unfolding through a three-state mechanism, as demonstrated by equilibrium unfolding experiments using Trp fluorescence (Fu and Liang 2002).

γD-Crystallin

Human γD-crystallin (HGD) is another abundant γ-crystallin present in the nucleus of embryonic lens (Robinson et al. 2006). It has 173 amino acid residues having molecular mass ~20 kDa (Bloemendal et al. 2004). The X-ray diffraction crystal structure of HGD (Basak et al. 2003) demonstrated various structural features of this protein (Fig. 14.4). Partially folded intermediates of HGD were found to be formed during unfolding/refolding experiments carried out in presence of GdnHCl (Kosinski-Collins and King 2003). A similar equilibrium unfolding/refolding experiments was also carried out with a HGD mutant, which contains a single Trp residue and the three other Trp residues were mutated with the Phe. This experiment concluded that the folding intermediate state contains unfolded N-td and folded C-td (Kosinski-Collins et al. 2004). Experiments using individual N-td and C-td of HGD substantiated the above conformational model during unfolding. Individually, C-td of HGD was found to be more stable than N-td as observed in the equilibrium folding/refolding experiments (Mills et al. 2007). Interface interactions thus contribute to the conformational stability of HGD (Flaugh et al. 2005a, b). A comparison of unfolding free energies for HGD and its two isolated domains showed that the interface interactions contributed ~4.2 kcal/mol to the stability of HGD (Mills et al.

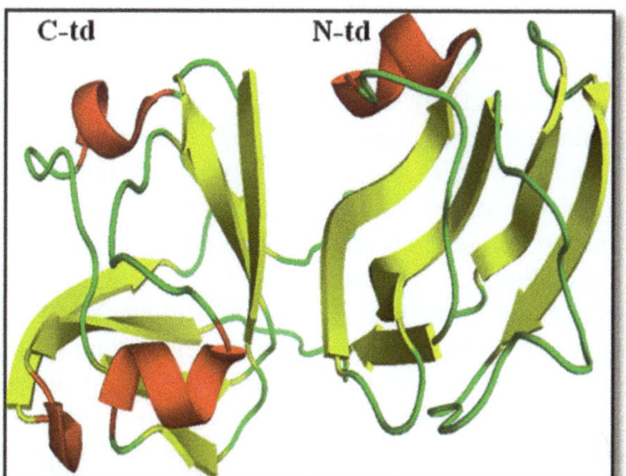

Fig. 14.4 Crystal structure of human γD-crystallin (PDB ID: 1HK0) reported by Basak et al. (2003). The N-terminal and C-terminal domains have been marked. Various structural features along with the compactness of the domains are responsible for the stability of this long-lived protein. The hydrophobic interface shown in the structure between these two domains acts as the nucleation center for the folding of the N-terminal domain during the overall folding process of this protein

2007). A hydrophobic cluster (Met 43, Phe 56, Ile 81, Val 132, Leu 145 and Val 170) is formed at the interface of the two domains of HGD, along with two pairs of flanking polar residues (Gln 54/Gln 143 and Arg 79/Met 147) (Basak et al. 2003). The thermodynamic and kinetic destabilization of the N-td occurs upon substitution of the hydrophobic cluster and flanking polar pair residues (Flaugh et al. 2005a, b). The above results suggested that the C-td interface basically acts as a nucleation center for the folding of the N-td in HGD. Earlier studies also pointed towards a high kinetic barrier for the unfolding of N-td of HGD (Flaugh et al. 2005a; Mills-Henry 2007). These observations suggest that the kinetic and thermodynamic stability of HGD is contributed by the interface interactions in a major way (Mills-Henry 2007). The kinetic unfolding experiments on HGD carried out by Mills-Henry (2007) proved that the higher kinetic stability against unfolding allows HGD to retain its folded form throughout adult lifespan, even if there is a destabilizing environment and accumulation of covalent damage. Das et al. (2010) used all-atom explicit solvent models to explain that the stability of N-td increases with interface interactions. Simulation studies (Das et al. 2010) showed that N-td unfolds prior to C-td. This will lead to the formation of partially folded intermediates having folded C-td and unfolded N-td (Serebryany and King 2014).

To assess the role of some specific residues to stabilize the native state conformation of HGD, site directed mutagenesis had been done. These studies pointed out the role of hydrophobic residues to stabilize the domain interface (Flaugh et al. 2005b, 2006; Mills et al. 2007). Mutation of tyrosines in the N-td affects the stability of N-td only, whereas Tyr mutations in C-td result a global destabilization of HGD with cooperative unfolding of both C-td and N-td (Kong and King 2011; Yang et al. 2014). Development of dominant cataract was reportedly due to mutations of specific hydrophobic core residues of murine γD-crystallin (Sinha et al. 2001; Graw et al. 2002, 2004). Similar mutations in HGD were also found to destabilize the protein through faster unfolding of N-td as compared to the native protein (Moreau and King 2009). Congenital cataract was also found to be caused due to V76D mutation in mice (Graw et al. 2002). This mutation in HGD (V75D) destabilizes the N-td leading to the aggregation of the mutated protein even at physiological conditions (Moreau and King 2009). This mutated protein is also able to escape from chaperone recognition and therefore suppression of its aggregation by chaperone is also ineffective (Moreau and King 2012). W42R mutation also causes congenital cataract in humans by destabilizing the N-td (Ji et al. 2013).

A major deleterious factor for cataract development is the exposure of the eye to UV-B light (West et al. 1998). Due to the exposure to UV radiations, the indole ring of tryptophan side chains starts to break and form an oxidation product called kynurenine (Hains and Truscott 2007). The contribution of tryptophan residues in the protection of HGD from UV-induced photo-aggregation had been reported (Schafheimer and King 2013). A Tyr-Trp-Tyr sandwich-like hydrophobic cluster also contributes to the stability of HGD and the photooxidation of Trp to kynurenine greatly reduces the stability of the protein (Xia et al. 2013). To determine the effect of such photodamage, the molecular dynamics simulation technique was adopted on HGD, where four tryptophan residues were replaced with kynurenine

(Xia et al. 2013). This caused the melting of the double Greek key domains by the leakage of water molecules into the hydrophobic domain cores. Replacement of the tyrosine residue with alanine also resulted a similar perturbation (Kong and King 2011; Yang et al. 2014).

As aggregation of crystallins is associated with cataractogenesis, hence aggregation inhibition can also be targeted to prevent the formation of cataract and it could be an alternative to cataract surgery. Our group is engaged in the design and development of small molecules and nanomaterails to achieve inhibition of unordered and ordered aggregation of human γD-crystallin (Chauhan et al. 2017a, b; Sharma and Ghosh 2017, 2019).

γS-Crystallin

γS-crystallin is also a predominant γ-crystallin protein expressed in human lens (Siezen et al. 1987). γS-crystallin has a 5 residue N-terminal extension. The solution state NMR structure for human γS-crystallin is shown in Fig. 14.5 (Kingsley et al. 2013). Wenk et al. (2000) studied the structural properties of isolated C-td and N-td of bovine and human γS-crystallin using Trp fluorescence. Two-state transition for γS-crystallin was observed in GdnHCl unfolding-refolding studies on γS-crystallin at pH 7.0 (Mills et al. 2007). The two-state transition was supposed to be due to comparable stabilities of the C-td and N-td of the γS-crystallins. Equilibrium unfolding-refolding experiments by Mills et al. (2007) showed that the stability of individual N-td is less than that of individual C-td of γS-crystallin. These results indicated that the overall stability of the protein is also contributed by the interface interactions, but these are not as significant as in case of bovine γB-crystallin and HGD. Human γS-crystallin is found to be less stable than HGD as evident from their melting temperature values under similar experimental conditions in vitro (Mills et al. 2007).

Fig. 14.5 Crystal structure of human γS-crystallin (PDB ID: 2M3T) with 5 residues as N-terminal extension reported by Kingsley et al. (2013)

Detailed studies on primary and tertiary structures of HGD, human and murine γS-crystallin did not show any major difference (Mills et al. 2007). The irregular conformation of Tyr residue in Tyr corner was seen in the crystal structure of the C-td of human γS-crystallin. The lower stability of γS-crystallin is attributed to the changes in Tyr corners in each domain (Purkiss et al. 2002; Wu et al. 2005; Mills et al. 2007). The kinetic stability of human γS-crystallin was also found to be lower than that of human γD-cystallin (Mills et al. 2007).

α-Crystallin

α-crystallin belongs to the family of small heat shock proteins commonly abbreviated as sHsp (de Jong et al. 1993). Alpha-crystallin contributes ~40% of total protein content of lens (Bloemendal et al. 2004). α-crystallin importantly maintains the lens transparency. Considering very negligible turnover of crystallins in lens, maintenance of the lens transparency is a very challenging task. Many covalent and non-covalent modifications can occur in eye lens proteins throughout the life-span, which serve as a major reason for protein aggregation. α-crystallin provides protection to the lens by prevention of aggregation of other lens proteins (Klemenz et al. 1991; Horwitz 1992).

Structure of α-Crystallins

α-crystallin exists as hetero-oligomer consisting of two subunits: αA and αB (Bloemendal et al. 2004). Around 60% of identity in the sequences of the two subunits was observed (Bloemendal and de Jong 1991). αA to αB ratio in human lenses is found to 3:1 (Horwitz et al. 1999). αA-crystallin is present only in eye lens whereas αB-crystallin is found in various mammalian tissues including brain, heart, liver and skeletal muscles as well as in eye lens (Graw 2009). α-crystallin generally forms polydisperse ensembles. These ensembles consist of several oligomers and exchange of subunits among them is also apparent (de Jong et al. 1998; Haslbeck et al. 2005; Kriehuber et al. 2010; Basha et al. 2012; Garrido et al. 2012; Delbecq and Klevit 2013; Bakthisaran et al. 2015; Treweek et al. 2015). α-crystallin as a whole exists as a polydisperse assembly of homo- and hetero-oligomers. The αA and αB-subunits contains 173 and 175 amino acids respectively and the average molecular mass of oligomers is ~800 kDa (Reddy et al. 2006). The αA-crystallin contains immunoglobulin-like folding pattern (Bloemendal et al. 2004). The far-UV CD spectra of bovine α-crystallin have shown that secondary structure is comprised of 14% α-helix, 35% β-sheet and the rest in the form of random-coil (Farnsworth et al. 1997). The tertiary and quaternary structures of α-crystallin are still unclear. The C-terminal of αA and αB–crystallin is found to be more flexible than the N-terminal as revealed by nuclear magnetic resonance studies (Carver and Lindner 1998).

Domain Organization

The presence of the tripartite domain in αA and αB leads toward formation of large oligomers. Similar domain organization of a tripartite domain occurs in all sHsps (de Jong et al. 1998; Basha et al. 2012; Delbecq and Klevit 2013; Treweek et al. 2015). The α-crystallin domain (ACD) is the central element and is known as the hallmark of the sHsp family (Kappé et al. 2010). It spans the N-terminal region (NTR) and a short C-terminal region (CTR) (Caspers et al. 1995; Kriehuber et al. 2010; Basha et al. 2012). The α-crystallin domain consists of residues 63–144 and 68–148 residues in αA and αB respectively (Reddy et al. 2006). α-crystallin has also the property to exchange subunits and accordingly is considered as a dynamic oligomer. The dissociation and reassociation of subunits occurs continously. The chaperone activity of the α-crystallin arises due to the exchange of these subunits. Truncation (Aziz et al. 2007; Chaves et al. 2008) and age-related modifications on α-crystallin can alter the rate of subunit exchange (Gupta and Srivastava 2004a, b; Chaves et al. 2008).

The sHsps shows three major regions: the N and C-terminal extension and a P-sandwich α-crystallin domain. The P-sandwich α-crystallin domain conformation with N-terminal and C-terminal regions has been shown by circular dichroism (CD) and infrared spectroscopy (IR) analyses (Bloemendal et al. 2004). Cryo-electron microscopic images of αB-crystallin indicates the formation a micelle-like structure having a diameter of ~19 nm with a central cavity of ~8 nm in diameter (Haley et al. 1998). Structure elucidation and chaperone assays on α-crystallins had helped to understand the plausible mechanism of action of α-crystallins.

α-Crystallin Domain (ACD)

ACD consists of ~80 residues in human αA and αB-crystallin. This is comparably shorter than the ACD length found in other proteins of sHsp family (~90–100 residues) (Kriehuber et al. 2010). The ACD of human αA and αB crystallin share a high degree of sequence identity. Techniques like circular dichroism (CD), X-ray crystallography and solution as well as solid state NMR have shown the prevalence of β-strands in the ACD of both αA (Laganowsky and Eisenberg 2010; Laganowsky et al. 2010) and αB (Bagnéris et al. 2009; Jehle et al. 2009, 2010; Laganowsky et al. 2010; Clark et al. 2011). The β-strands in the ACD of αA and αB are present as a immunoglobulin-like β-sandwich fold, which contains seven such strands (β3–β9).

X-ray crystallographic analysis of the ACD of αA and αB suggested that the dimer interface is flexible in terms of the protomer positions (Bagnéris et al. 2009; Laganowsky and Eisenberg 2010; Laganowsky et al. 2010; Clark et al. 2011; Hochberg et al. 2014). Several histidines are present at the dimer interface and electrostatic interactions play significant role in its stabilization (Clark et al. 2011). The ACD can bind divalent metal cations (Ganadu et al. 2004; Ahmad et al. 2008;

Mainz et al. 2012). For example copper (II) can coordinates with αB-crystallin at the edges of the dimer interface. This coordination enhances the dynamics of the dimeric substructure, heterogeneity and oligomer plasticity (Mainz et al. 2012). The glutamates and histidines from three protomers have also been shown to form an intermolecular zinc-binding motif (Laganowsky et al. 2010) in truncated αA protein (residues from 59 to 163).

N-Terminal Region (NTR)

The NTR in αA and αB is ~65 residues long. The hydrophobicity and conformational mobility of NTR make it interesting for biochemical and structural studies. Insoluble fractions of the NTR were recovered when it was expressed individually. This shows its high propensity for aggregation (Merck et al. 1992; Asomugha et al. 2011). The weak resonances in the ssNMR spectra for the terminal five residues of the NTR of αB-crystallin were accounted for their high flexibility (Jehle et al. 2010). No resonances were found in the of αA-crystallin NTR. The model based on ssNMR revealed that the NTR of αB-crystallin consists of two helices (spanning residue number 14–17 and 27–31) and antiparallel beta sheets (48–50 and 61–63) linked through a long loop (Jehle et al. 2011). The higher-order oligomer assembly formation, ensemble composition and the dynamics of α-crystallins are all mostly contributed by their NTR. Human αA (62–173) (Kundu et al. 2007), truncated rat αA (64–173) (Merck et al. 1992), human αB (67–175) (Peschek et al. 2013) and bovine αB (70–175) (Merck et al. 1993) can form heterogeneous ensemblies of oligomers. These oligomers are considerably smaller than that of the full proteins. The residues susceptible for phosphorylation are also located within the NTR of α-crystallin. Three residues in αB (Ser 19, Ser 45, and Ser 59) (Chiou et al. 2010) and two residues in αA (Ser 45 and Ser 122) (Voorter et al. 1986; Chiou et al. 2010) were found to be responsible for such post-translational modification in α-crystallin.

C-Terminal Region (CTR)

The CTR contains ~20 amino acids. The residues Ile 159, Pro 160 and Ile 161 in αB and Ile 159, Pro 160 and Val 161 in αA-crystallin play the role of an anchor for the formation of oligomers. These residues contribute to the α-crystallin polydispersity (Delbecq and Klevit 2013).

The CTR of αA and αB-crystallin is flexible, solvent accessible and polar (Fig. 14.6). The flexibility is found to be maximum at the C-terminus which covers the residues from Glu 164 to Ser 173 in αA and from Glu 164 to Lys 175 in αB-crystallin. The last nine amino acids in αA (Glu 165 to Ser 173) and the last eleven amino acids in αB (Glu 165 to Lys 175) show their flexibility and solvent accessibility (Carver et al. 1992; Carver and Lindner 1998). Their chemical shifts indicate that these residues possess small well-defined secondary structure as compared to isolated peptides of the same length (Carver et al. 1992; Carver and

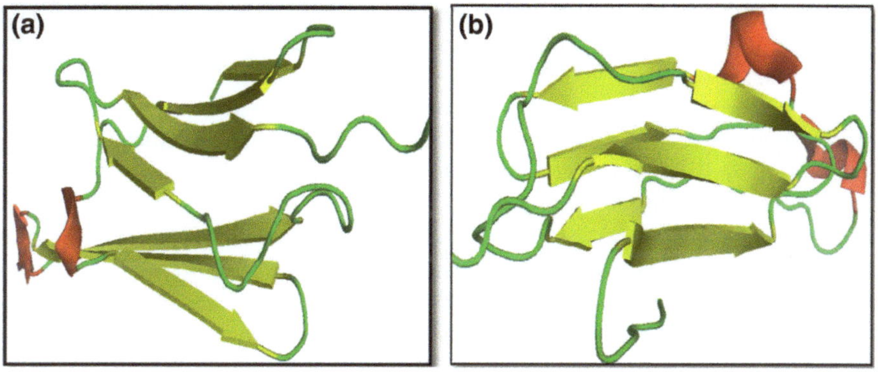

Fig. 14.6 Crystal structure of α-crystallins: **a** bovine αA-crystallin (PDB ID: 3L1F), and **b** human αB-crystallin (PDB ID: 3L1G) as reported by Laganowsky et al. (2010). For both the proteins, the crystal structure of only the truncated proteins (residue number 62–163 for αA-crystallin and residue number 68–162 for αB-crystallin) are reported. The above structures include mostly the α-crystallin domain (commonly found in small heat shock proteins) and part of C-terminal region. The structural features can explain the mechanism of polydispersity and chaperone activity of these proteins

Lindner 1998; Carver 1999). NMR experiments were also performed to monitor the backbone dynamics. This suggested a greater flexibility of the CTR of αA-crystallin as compared to αB-crystallin (Treweek et al. 2010). The CTR of α-crystallins are hydrophilic and contribute to the solubility of the oligomers, as the solubility of human αB-crystallin decreases with the deletion of the CTR residues from 155 to 165 (Ghosh et al. 2006). The susceptibility of α-crystallins towards proteolytic cleavage is also likely to be due to the solvent accessibility of the CTR of α-crystallins (Siezen and Hoenders 1979; Yoshida et al. 1986).

Crystallin Complexes

The eye lens can be considered as a model system for studying protein-protein interactions and formation of homologous and heterologous protein complexes. Various protein–protein interactions in lens maintain its transparency and such interactions were reviewed earlier in detail (Takemoto and Sorensen 2008). Existence of short-range order interactions among crystallins was reported earlier (Benedek 1971; Delaye and Tardieu 1983; Bettelheim 1985). Several investigations were carried out earlier to investigate the interactions involved in crystallin complexes with different compositions and stability. The composition of protein-protein complexes constituted by crystallins also varies with the age of an individual. The process of complexation between α-crystallin and its substrates (β and γ-crystallins) initiates in the lens and propagates with age. Size-exclusion chromatography was used early on to fractionate high molecular weight (molecular weight >300,000) lens proteins into free α-crystallin and α-crystallin complexed with β and γ-crystallins. By using an anti-α-crystallin immune-adsorbent, these complexes were dissociated, which yielded free β- and γ-crystallins. These results indeed suggest the formation of complexes among crystallins (Manski et al. 1979). Solvent ionic composition was found to influence protein-protein complex formation. Different compositions of α-β-γ-crystallins were determined in such type of different complexes (Manski et al. 1979).

To understand the process of homologous and heterologous complexation among crystallins, it is essential to study the short-range order interactions among them. Based on X-ray scattering and osmotic pressure measurement, the interactions between γ-crystallins were predicted by Tardieu et al. (1992), which was affirmed later using ^{13}C-NMR by Stevens et al. (1995). Light scattering experiments (Bettelheim and Chen 1998) also provided convincing evidences for the existence of heterologous interactions among crystallin molecules and α-γ interactions were found to be the strongest. Another study (Stradner et al. 2007) based on neutron scattering and molecular dynamics simulations suggested that though the interactions are weak in nature, but they are very important for maintaining lens transparency. Specifically, the interactions between α- and γ-crystallins are more important for that purpose. This was also proved later in case of a genetic cataract (Banerjee et al. 2011). For a singly mutated human γD-crystallin (E107A), the structure, stability and solubility of the native and mutated protein were identical but the interactions between α- and γ-crystallin were found to be altered in case of the mutated γ-crystallin, leading to cataractogenesis (Banerjee et al. 2011). Microequilibrium dialysis based experiments also suggested interactions between alpha and gamma crystallins (Ponce and Takemoto 2005), which decreases due to aging (Takemoto and Ponce 2006; Takemoto et al. 2008). Another study demonstrated that mutation in either alpha or gamma crystallin may cause a change in their interactions (Fu and Liang 2003). Any change in interactions between the same crystallins or between different crystallins may lead to non-uniform changes in protein density. This will in turn result in non-uniform refractive index and subsequent increase in light scattering. Additional studies based on light scattering (Thurston et al. 1999), NMR (Carver et al. 1994) and filtration (Biswas and Das 2004) also confirmed the existence of interactions and complexation between homo and hetero crystallins. Attempts were also made to study protein-protein interactions in crystallins by using various spectroscopic and biochemical techniques at high protein concentration (Siezen and Owen 1983; Liang and Li 1991; Koenig et al. 1992; Liang and Chakrabarti 1998; Beaulieu et al. 1988; Morgan et al. 1989; Cooper et al. 1994; Veretout et al. 1989). Mach et al. (1990) reported the interactions between α-crystallin molecules based on non-conformational techniques like dynamic light scattering and size exclusion chromatography and it was also found that the presence of low molecular weight β- or monomeric γ-crystallin can diminish the association of α-crystallins. Enhanced thermodynamic stability of α-crystallin was also noted in presence of a mixture of β- and γ-crystallin or γ-crystallin alone (Bettelheim and Chen 1998).

Srivastava et al. (2008) studied the crystallin complexes (held noncovalently) found in water soluble fractions of normal and cataractous human eye lenses of different ages. The protein complexes were isolated and subjected to electrospray tandem mass spectrometry after trypsin digestion. The normal lenses aged 20 years and 60–70 years contain the protein complexes comprised of α-, β- and γ-crystallin species and beaded filament protein (e.g. filensin), though the number of species varies among them. But in the age-matched cataractous lens, the complexes consist of the above crystallins without any beaded filament protein.

It was also found that the aggregation of partially unfolded human γ-crystallins (γC, γD and γS) during their refolding pathway was suppressed by the molecular chaperone human αB-crystallin through the formation of stable α-γ crystallin complexes (Acosta-Sampson and King 2010). By measuring tryptophan fluorescences in that work, it was concluded that the bound γ-crystallin substrate in the complex exists in a partially folded conformation. The complexation between α-crystallin and its substrate, such as β- and γ-crystallins, also alters the metal ion binding capability of α-crystallin (Ghosh et al. 2011). In general, α-crystallin binds Cu^{2+} and Zn^{2+} ions and as a chaperone protects other crystallins from metal mediated oxidative stress (Ahmad et al. 2008). Not only that, the binding of these metal ions with α-crystallin enhances the chaperone function of α-crystallin (Ganadu et al. 2004; Biswas and Das 2008). But, due to the binding of γ-crystallin substrate with the chaperone α-crystallin, the metal ion binding by the α-crystallin was found to be diminished (Ghosh et al. 2011). As a result, in an aged lens, the concentration of free metal ions will be increased and that will cause oxidative damage of crystallins and proceed towards aggregation.

References

Acosta-Sampson L, King J (2010) Partially folded aggregation intermediates of human γD-, γC-, and γS-crystallin are recognized and bound by human αB-crystallin chaperone. J Mol Biol 401:134–152

Ahmad MF, Singh D, Taiyab A, Ramakrishna T, Raman B, Rao CM (2008) Selective Cu^{2+} binding, redox silencing, and cytoprotective effects of the small heat shock proteins αA and αB-crystallin. J Mol Biol 382:812–824

Asomugha CO, Gupta R, Srivastava OP (2011) Structural and functional properties of NH2-terminal domain, core domain, and COOH-terminal extension of αA- and αB-crystallins. Mol Vis 17:2356–2367

Aziz A, Santhoshkumar P, Sharma KK, Abraham EC (2007) Cleavage of the C-terminal serine of human alphaA-crystallin produces alphaA1-172 with increased chaperone activity and oligomeric size. Biochemistry 46:2510–2519

Bagby S, Go S, Inouye S, Ikura M, Chakrabartty A (1998) Equilibrium folding intermediates of a Greek key beta-barrel protein. J Mol Biol 276:669–681

Bagnéris C, Bateman OA, Naylor CE, Cronin N, Boelens WC, Keep NH, Slingsby C (2009) Crystal structures of α-crystallin domain dimers of αB-crystallin and Hsp20. J Mol Bio 392:1242–1252

Bakthisaran R, Tangirala R, Rao CM (2015) Small heat shock proteins: role in cellular functions and pathology. Biochim Biophys Acta 1854:291–319

Banerjee PR, Pande A, Patrosz J, Thurston GM, Pande J (2011) Cataract-associated mutant E107A of human γD-crystallin shows increased attraction to α-crystallin and enhanced light scattering. Proc Natl Acad Sci U S A 108:574–579

Barnwal RP, Jobby MK, Devi KM, Sharma Y, Chary KVR (2009) Solution structure and calcium binding properties of M-crystallin, a primordial βγ-crystallin from archaea. J Mol Biol 386:675–689

Basak AK, Bateman O, Slingsby C, Pande A, Asherie N, Ogun O, Benedek G, Pande J (2003) High-resolution X-ray crystal structures of human γD-crystallin (1.25A) and the R58H mutant (1.15A) associated with aculeiform cataract. J Mol Biol 328:1137–1147

Basha E, O'Neill H, Vierling E (2012) Small heat shock proteins and α-crystallins: dynamic proteins with flexible functions. Trends Biochem Sci 37:106–117

Bateman OA, Slingsby C (1992) Structural studies on βH-crystallin from bovine eye lens. Exp Eye Res 55:127–133

Bateman OA, Sarra A, Van Genesan ST, Kappe G, Lubsen NH, Slingsby C (2003) The stability of human acidic beta-crystallin oligomers and hetero-oligomers. Exp Eye Res 77:409–422

Bax B, Lapatto R, Nalini V, Driessen H, Lindley PF, Mahadevan D, Blundell TL, Slingsby C (1990) X-ray analysis of beta-B2-crystallin and evolution of oligomeric lens proteins. Nature 347:776–780

Beaulieu CF, Clark JI, Brown RD III et al (1988) Relaxometry of calf lens homogenates, including cross-relaxation by crystallin NH groups. Magn Reson Med 8:45–57

Benedek GB (1971) Theory of transparency of the eye. Appl Opt 10:459–473

Bettelheim FA (1985) The ocular lens: structure, function, and pathology. In: Maisel H (ed) Marcel Dekker, Inc., New York, pp 265–300

Bettelheim FA, Chen A (1998) Thermodynamic stability of bovine alpha-crystallin in its interactions with other bovine crystallins. Int J Biol Macromol 22:247–252

Biswas A, Das KP (2004) Role of ATP on the interaction of alpha-crystallin with its substrates and its implications for the molecular chaperone function. J Biol Chem 279:42648–42657

Biswas A, Das KP (2008) Zn^{2+} enhances the molecular chaperone function and stability of alpha-crystallin. Biochemistry 47:804–816

Bloemendal H (1981) The lens proteins. In: Bloemendal H (ed) Molecular and cellular biology of the eye lens. Willey, New York, NY, pp 1–49

Bloemendal H, de Jong WW (1991) Lens proteins and their genes. Prog Nucleic Acid Res Mol Biol 41:259–281

Bloemendal H, de Jong W, Jaenicke R, Lubsen NH, Slingsby C, Tardieu A (2004) Ageing and vision: structure, stability and function of lens crystallins. Prog Biophy Mol Biol 86:407–485

Brakenhoff RH, Aarts HJ, Reek FH, Lubsen NH, Schoenmakers JG (1990) Human γ-crystallin genes: a gene family on its way to extinction. J Mol Biol 216:519–532

Breitman ML, Lok S, Wistow G, Piatigorsky J, Treton JA, Gold RJ, Tsui LC (1984) Gamma-crystallin family of the mouse lens: structural and evolutionary relationships. Proc Natl Acad Sci U S A 81:7762–7766

Carver JA (1999) Probing the structure and interactions of crystallin proteins by NMR spectroscopy. Prog Retin Eye Res 18:431–462

Carver JA, Lindner RA (1998) NMR spectroscopy of α-crystallin, Insights into the structure, interactions and chaperone action of small heat-shock proteins. Int J Biol Macromol 22:197–209

Carver JA, Aquilina JA, Truscott RJW, Ralston GB (1992) Identification by 1H NMR spectroscopy of flexible C-terminal extensions in bovine lens α-crystallin. FEBS Lett 311:143–149

Carver JA, Aquilina JA, Cooper PG, Williams GA, Truscott RJ (1994) Alpha crystallin: molecular chaperone and protein surfactant. Biochim Biophys Acta 1205:195–206

Caspers GJ, Leunissen JA, de Jong WW (1995) The expanding small heat-shock protein family, and structure predictions of the conserved α-crystallin domain. J Mol Evol 40:238–248

Chauhan P, Muralidharan SB, Velappan AB, Datta D, Pratihar S, Debnath J, Ghosh KS (2017a) Inhibition of copper-mediated aggregation of human γD-crsytallin by Schiff bases. J Biol Inorg Chem 22:505–517

Chauhan P, Velappan AB, Sahoo BK, Debnath J, Ghosh KS (2017b) Studies on molecular interactions between Schiff bases and eye lens chaperone human αA-crystallin. J Lumin 192:148–155

Chaves JM, Srivastava K, Gupta R, Srivastava OP (2008) Structural and functional roles of deamidation and/or truncation of N- or C-termini in human alpha A-crystallin. Biochemistry 47:10069–10083

Chen Y, Zhao H, Schuck P, Wistow G (2014) Solution properties of γ-crystallins: compact structure and low frictional ratio are conserved properties of diverse γ-crystallins. Protein Sci 23:76–87

Chiou SH, Huang CH, Lee IL, Wang YT, Liu NY, Tsay YG, Chen YJ (2010) Identification of in vivo phosphorylation sites of lens proteins from porcine eye lenses by a gel-free phosphoproteomics approach. Mol Vis 16:294–302

Clark AR, Naylor CE, Bagnéris C, Keep NH, Slingsby C (2011) Crystal structure of R120G disease mutant of human αB-crystallin domain dimer shows closure of a groove. J Mol Biol 408:118–134

Cooper PG, Aquilina JA, Truscott RJW, Carver JA (1994) Supramolecular order within the lenses: ^{1}HNMR spectroscopic evidence for specific crystallin-crystallin interactions. Exp Eye Res 59:607–616

Cvekl A, Duncan MK (2007) Genetic and epigenetic mechanisms of gene regulation during lens development. Prog Retin Eye Res 26:555–597

D'Alessio G (2002) The evolution of monomeric and oligomeric betagamma-type crystallins. Facts and hypotheses. Eur J Biochem 269:3122–3130

Das P, King JA, Zhou R (2010) Beta-strand interactions at the domain interface critical for the stability of human lens gammaD-crystallin. Protein Sci 19:131–140

de Jong WW, Leunissen JA, Voorter CE (1993) Evolution of the alpha-crystallin/small heat-shock protein family. Mol Biol Evol 10:103–126

de Jong WW, Caspers GJ, Leunissen JA (1998) Genealogy of the α-crystallin-small heat shock protein superfamily. Int J Biol Macromol 22:151–162

Delaye M, Tardieu A (1983) Short-range order of crystallin proteins accounts for eye lens transparency. Nature 302:415–417

Delbecq SP, Klevit RE (2013) One size does not fit all: the oligomeric states of alpha-B crystallin. FEBS Lett 587:1073–1080

Dixit K, Pande A, Pande J, Sarma SP (2016) Nuclear magnetic resonance structure of a major lens protein, human gamma γC-crystallin: role of the dipole moment in protein solubility. Biochemistry 55:3136–3149

Farnsworth PN, Groth-Vasselli B, Greenfield NJ, Singh K (1997) Effects of temperature and concentration on bovine lens alpha-crystallin secondary structure: a circular dichroism spectroscopic study. Int J of Biol Macromol 20:283–291

Flaugh SL, Kosinski-Collins MS, King J (2005a) Interdomain side-chain interactions in human gammaD-crystallin influencing folding and stability. Protein Sci 14:2030–2043

Flaugh SL, Kosinski-Collins MS, King JA (2005b) Contributions of hydrophobic domain interface interactions to the folding and stability of human gammaD-crystallin. Protein Sci 14:569–581

Fu L, Liang JJ (2002) Unfolding of human lens recombinant betaB2 and gammaC-crystallins. J Struct Biol 139:191–198

Fu L, Liang JJ (2003) Alteration of protein-protein interactions of congenital cataract crystallin mutants. Invest Ophthalmol Vis Sci 44:1155–1159

Ganadu ML, Aru M, Mura GM, Coi A, Mlynarz P, Kozlowski H (2004) Effects of divalent metal ions on the alphaB-crystallin chaperone-like activity: spectroscopic evidence for a complex between copper (II) and protein. J Inorg Biochem 98:1103–1109

Garrido C, Paul C, Seigneuric R, Kampinga HH (2012) The small heat shock proteins family: the long forgotten. Int J Biochem Cell Biol 44:1588–1592

Ghosh JG, Shenoy AK, Clark JI (2006) N- and C-terminal motifs in human αB crystalline play an important role in the recognition, selection, and solubilization of subtrates. Biochemistry 45:13847–13854

Ghosh KS, Pande A, Pande J (2011) Binding of γ-crystallin substrate prevents the binding of copper and zinc ions to the molecular chaperone α-crystallin. Biochemistry 50:3279–3281

Graw J (2009) Genetics of crystallins: cataract and beyond. Exp Eye Res 88:173–178

Graw J, Löster J, Soewarto D, Fuchs H, Reis A, Wolf E, Balling R, Angelis MH (2002) V76D mutation in a conserved γD-crystallin region leads to dominant cataracts in mice. Mamm Genome 13:452–455

Graw J, Neuhäuser-Klaus A, Klopp N, Selby PB, Löster J, Favor J (2004) Genetic and allelic heterogeneity of *Cryg* mutations in eight distinct forms of dominant cataract in the mouse. Invest Ophthalmol Vis Sci 45:1202–1213

Gupta R, Srivastava OP (2004a) Deamidation affects structural and functional properties of human alphaA-crystallin and its oligomerization with alphaB-crystallin. J Biol Chem 279:44258–44269

Gupta R, Srivastava OP (2004b) Effect of deamidation of asparagine 146 on functional and structural properties of human lens alphaB-crystallin. Invest Ophthalmol Vis Sci 45:206–214

Hains PG, Truscott RJW (2007) Post-translational modifications in the nuclear region of young, aged, and cataract human lenses. J Proteome Res 6:3935–3943

Haley DA, Horwitz J, Stewart PL (1998) The small heat-shock protein, alpha B-crystallin, has a variable quaternary structure, J Mol. Biol. 277:27–35

Haslbeck M, Franzmann T, Weinfurtner D, Buchner J (2005) Some like it hot: the structure and function of small heat-shock proteins. Nat Struct Mol Biol 12:842–846

Hejtmancik JF, Wingfield PT, Chambers C, Russell P, Chen HC, Sergeev YV, Hope JN (1997) Association properties of beta B2- and beta A3-crystallin: ability to form dimmers. Protein Eng 10:1347–1352

Hejtmancik JF, Wingfield PT, Sergeev YV (2004) Beta-crystallin association. Exp Eye Res 79:377–383

Hochberg GK, Ecroyd H, Liu C, Cox D, Cascio D, Sawaya MR, Collier MP, Stroud J, Carver JA, Baldwin AJ, Robinson CV, Eisenberg DS, Benesch JL, Laganowsky A (2014) The structured core domain of αB-crystallin can prevent amyloid fibrillation and associated toxicity. Proc Natl Acad Sci U S A 111:1562–1570

Horwitz J (1992) α-Crystallin can function as a molecular chaperone. Proc Natl Acad Sci U S A 89:10449–10453

Horwitz J, Bova MP, Ding L, Haley DA, Stewart PL (1999) Lens α-crystallin: function and structure. Eye 13:403–408

Jaenicke R, Slingsby C (2001) Lens crystallins and their microbial homologs: structure, stability, and function. Crit Rev Biochem Mol Biol 36:435–499

Jehle S, van Rossum B, Stout JR, Noguchi SM, Falber K, Rehbein K, Oschkinat H, Klevit RE, Rajagopal P (2009) αB-crystallin: a hybrid solid-state/solution-state NMR investigation reveals structural aspects of the heterogeneous oligomer. J Mol Biol 385:1481–1497

Jehle S, Rajagopal P, Bardiaux B, Markovic S, Kühne R, Stout JR, Higman VA, Klevit RE, van Rossum BJ, Oschkinat H (2010) Solid-state NMR and SAXS studies provide a structural basis for the activation of αB-crystallin oligomers. Nat Struct Mol Bio 1371:1037–1042

Jehle S, Vollmar BS, Bardiaux B, Dove KK, Rajagopal P, Gonen T, Oschkinat H, Klevit RE (2011) N-terminal domain of αB-crystallin provides a conformational switch for multimerization and structural heterogeneity. Proc Natl Acad Sci U S A 108:6409–6414

Ji F, Jung J, Koharudin LMI, Gronenborn AM (2013) The human W42R gamma D-crystallin mutant structure provides a link between congenital and age related cataracts. J Biolog Chem 288:99–109

Kappé G, Boelens WC, de Jong WW (2010) Why proteins without an α-crystallin domain should not be included in the human small heat shock protein family HSPB. Cell Stress Chaperones 15:457–461

Kingsley CN, Brubaker WD, Markovic S, Diehl A, Brindley AJ, Oschkinat H, Martin RW (2013) Preferential and specific binding of human alpha B-crystallin to a cataract-related variant of gamma S-crystallin. Structure 21:2221–2227

Klemenz R, Fröhli E, Steiger RH, Schäferand R, Aoyama A (1991) αB-crystallin is a small heat shock protein. Proc Natl Acad Sci U S A 88:3652–3656

Koenig SH, Brown RD III, Spiller M et al (1992) Intermolecular protein interactions in solutions of calf lens-crystallin: results from 1/T1 nuclear magnetic relaxation dispersion profiles. Biophys J 61:776–785

Kong F, King J (2011) Contributions of aromatic pairs to the folding and stability of long-lived human γD-crystallin. Protein Sci 20:513–528

Kosinski-Collins MS, King J (2003) In vitro unfolding, refolding, and polymerization of human gammaD crystallin, a protein involved in cataract formation. Protein Sci 12:480–490

Kosinski-Collins MS, Flaugh SL, King JA (2004) Probing folding and fluorescence quenching in human gammaD crystallin Greek key domains using triple tryptophan mutant proteins. Protein Sci 13:2223–2235

Kriehuber T, Rattei T, Weinmaier T, Bepperling A, Haslbeck M, Buchner J (2010) Independent evolution of the core domain and its flanking sequences in small heatshock proteins. FASEB J. 24:3633–3642

Kundu M, Sen PC, Das KP (2007) Structure, stability, and chaperone function of αAcrystallin: role of N-terminal region. Biopolymers 86:177–192

Laganowsky A, Eisenberg D (2010) Non-3D domain swapped crystal structure of truncated zebrafish alpha-crystallin. Protein Sci 19:1978–1984

Laganowsky A, Benesch JLP, Landau M, Ding L, Sawaya MR, Cascio D, Huang Q, Robinson C, Horwitz J, Eisenberg D (2010) Crystal structures of truncated alphaA and alphaB-crystallins reveal structural mechanisms of polydispersity important for eye lens function. Protein Sci 19:1031–1043

Lampi KJ, Ma Z, Shih M, Shearer TR, Smith JB, Smith DL, David LL (1997) Sequence analysis of βA3, βB3, and βA4 crystallins completes the identification of the major proteins in young human lens. J Biol Chem 272:2268–2275

Lampi KJ, Ma Z, Hanson SR, Azuma M, Shih M, Shearer TR, Smith DL, Smith JB, David LL (1998) Age-related changes in human lens crystallins identified by two dimensional electrophoresis and mass spectrometry. Exp Eye Res 67:31–43

Lampi KJ, Kim YH, Bachinger HP, Boswell BA, Linder RA, Carver JA, Shearer TR, David LL, Kapfer DM (2002) Decreased heat stability and increased chaperone requirement of modified human betaB1-crystallins. Mol. Vis. 8:359–366

Lampi KJ, Wilmarth PA, Murray MR, David LL (2014) Lens β-crystallins: the role of deamidation and related modifications in aging and cataract. Prog Biophys Mol Biol 115:21–31

Liang JJN, Chakrabarti B (1998) Intermolecular interaction of lens crystallins: from rotationally mobile to immobile states at high protein concentrations. Biochem Biophys Res Commun 246:441–445

Liang JN, Li XY (1991) Interaction and aggregation of lens crystallins. Exp Eye Res 53:61–66

Lubsen NH, Aarts HJ, Schoenmakers JG (1988) The evolution of lenticular proteins: the beta- and gamma-crystallin super gene family. Prog Biophys Mol Biol 51:47–76

MacDonald JT, Purkiss AG, Smith MA, Evans P, Goodfellow JM, Slingsby C (2005) Unfolding crystallins: the destabilizing role of a beta-hairpin cysteine in betaB2-crystallin by simulation and experiment. Protein Sci 14:1282–1292

Mach H, Trautman PA, Thomson JA, Lewis RV, Middaugh CR (1990) Inhibition of alpha-crystallin aggregation by gamma-crystallin. J Biol Chem 265:4844–4848

Mainz A, Bardiaux B, Kuppler F, Multhaup G, Felli IC, Pierattelli R, Reif B (2012) Structural and mechanistic implications of metal binding in the small heat-shock protein αB-crystallin. J Biol Chem 287:1128–1138

Manski W, Naliitowski K, Boxitsis G (1979) Immunochemical studies on lens protein-protein complexes I. the heterogeneity and structure of complexed α-crystallin. Exp Eye Res 29:625–635

Merck KB, de Haard-Hoekman WA, Oude Essink BB, Bloemendal H, de Jong WW (1992) Expression and aggregation of recombinant αA-crystallin and its two domains. Biochim Biophys Acta 1130:267–276

Merck KB, Horwitz J, Kersten M, Overkamp P, Gaestel M, Bloemendal H, de Jong WW (1993) Comparison of the homologous carboxy-terminal domain and tail of α-crystallin and small heat shock protein. Mol Biol Rep 18:209–215

Mills IA, Flaugh SL, Kosinski-Collins MS, King JA (2007) Folding and stability of the isolated Greek key domains of the long-lived human lens proteins γD-crystallin and γS-crystallin. Protein Sci 16:2427–2444

Mills-Henry IAR (2007) Stability, unfolding, and aggregation of the gamma D and gamma S human eye lens crystallins. Ph.D. thesis from Department of Biology, MIT, Cambridge, USA

Mishra A, Krishnan B, Swaroop SS, Sharma Y (2014) Microbial βγ-crystallins. Prog Biophys Mol Biol 115:42–51

Moreau KL, King J (2009) Hydrophobic core mutations associated with cataract development in mice destabilize human gamma D-crystallin. J Biol Chem 284:33285–33295

Moreau KL, King JA (2012) Cataract-causing defect of a mutant gamma crystallin proceeds through an aggregation pathway which bypasses recognition by the alpha-crystallin chaperone. PLoS ONE 7:e37256

Morgan CF, Schleich T, Caines GH, Farnsworth PN (1989) Elucidation of intermediate (mobile) and slow (solid like) protein motions in bovine lens homogenates by carbon-13 NMR spectroscopy. Biochemistry 28:5065–5074

Peschek J, Braun N, Rohrberg J, Back KC, Kriehuber T, Kastenmüller A, Weinkauf S, Buchner J (2013) Regulated structural transitions unleash the chaperone activity of αB-crystallin. Proc Natl Acad Sci U S A 110:3780–3789

Ponce A, Takemoto L (2005) Screening of crystallin-crystallin interactions using microequilibrium dialysis. Mol Vis 11:752–757

Purkiss AG, Bateman OA, Goodfellow JM, Lubsen NH, Slingsby C (2002) The X-ray crystal structure of human gamma S-crystallin C-terminal domain. J Biol Chem 277:4199–4205

Purkiss AG, Bateman OA, Wyatt K, Wilmarth PA, David LL, Wistow GJ, Slingsby C (2007) Biophysical properties of γC-crystallin in human and mouse eye lens: the role of molecular dipoles. J Mol Biol 372:205–222

Ray NJ (2015) Biophysical chemistry of the ageing eye lens. Biophys Rev 7:353–368

Reddy GB, Kumar PA, Kumar MS (2006) Chaperone-like activity and hydrophobicity of alpha-crystallin. IUBMB Life 58:632–641

Richardson JS (1977) Beta-sheet topology and the relatedness of proteins. Nature 268:495–500

Robinson NE, Lampi KJ, Speir JP, Kruppa G, Easterling M, Robinson AB (2006) Quantitative measurement of young human eye lens crystallins by direct injection Fourier transform ion cyclotron resonance mass spectrometry. Mol Vis 12:704–711

Schafheimer N, King J (2013) Tryptophan cluster protects human γD-crystallin from ultraviolet radiation-induced photoaggregation in vitro. Photochem Photobiol 89:1106–1115

Serebryany E, King JA (2014) The βγ-crystallins: native state stability and pathways to aggregation. Prog Biophys Mol Biol 115:32–41

Sharma V, Ghosh KS (2017) Inhibition of amyloid fibrillation and destabilization of fibrils of human γD-crystallin by direct red 80 and orange G. Intl J Biol Macromol 105:956–964

Sharma V, Ghosh KS (2019) Inhibition of amyloid fibrillation by small molecules and nanomaterials: strategic development of pharmaceuticals against amyloidosis. Prot Pept Lett (accepted)

Sharma KK, Santhoshkumar P (2009) Lens aging: effects of crystallins. Biochim Biophys Acta 1790:1095–1108

Siezen RI, Hoenders HJ (1979) The quaternary structure of bovine α-crystallin. Surface probing by limited proteolysis in vitro. Eur J Biochem 96:431–440

Siezen RJ, Owen EA (1983) Interactions of lens proteins: self-association and mixed-association studies of bovine α -crystallin and γ-crystallin. Biophys Chem 18:181–194

Siezen RJ, Thomson JA, Kaplan ED, Benedek GB (1987) Human lens gamma crystallins: isolation, identification, and characterization of the expressed gene products. Proc Natl Acad Sci U S A 84:6088–6092

Sinha D, Wyatt MK, Sarra R, Jaworski C, Slingsby C, Thaung C, Pannell L, Robison WG, Favor J, Lyon M, Wistow G (2001) A temperature-sensitive mutation of *Crygs* in the murine *Opj* cataract. J Biol Chem 276:9308–9315

Slingsby C, Bateman OA (1990) Quaternary interactions in eye lens beta-crystallins: basic and acidic subunits of beta-crystallins favor heterologous association. Biochemistry 29:6592–6599

Slingsby C, Clout NJ (1999) Structure of the crystallins. Eye (Lond) 13:395–402

Slingsby C, Norledge B, Simpson A, Bateman OA, Wright G, Driessen HPC, Lindley PF, Moss DS, Bax B (1997) X-ray diffraction and structure of crystallins. Prog Ret Eye Res 16:3–29

Smith MA, Bateman OA, Jaenicke R, Slingsby C (2007) Mutation of interfaces in domain-swapped human βB2-crystallin. Protein Sci 16:615–625

Spector A (1964) Methods of isolation of alpha, beta, and gamma crystallins and their subgroups. Invest Ophthalmol 3:182–193

Srivastava K, Chaves JM, Srivastava OP, Kirk M (2008) Multi-crystallin complexes exist in the water-soluble high molecular weight protein fractions of aging normal and cataractous human lenses. Exp Eye Res 87:356–366

Stevens A, Wang SX, Caines GH, Schleich T (1995) ^{13}C-NMR off-resonance rotating frame spin-lattice relaxation studies of bovine lens gamma-crystallin self association: effect of macromolecular crowding. Biochim Biophys Acta 1246:82–90

Stradner A, Foffi G, Dorsaz N, Thurston G, Schurtenberger R (2007) New insight into cataract formation: enhanced stability through mutual attraction. Phys Rev Lett 99:198103-1–198103-4

Sun TX, Akhtar NJ, Liang JJ (1999) Thermodynamic stability of human lens recombinant alphaA- and alphaB-crystallins. J Biol Chem 274:34067–34071

Takemoto LJ, Ponce A (2006) Decreased association of aged alpha-crystallins with gamma crystallins. Exp Eye Res 83:793–797

Takemoto L, Sorensen CM (2008) Protein-protein interactions and lens transparency. Exp Eye Res 87:496–501

Takemoto L, Ponce A, Sorensen CM (2008) Age-dependent association of gamma crystallins with aged alpha crystallins from old bovine lens. Mol. Vis. 14:970–974

Tardieu A, Veretout F, Krop B, Slingsby C (1992) Protein interactions in the calf eye lens: interactions between alpha-crystallins are repulsive whereas in gamma-crystallins they are attractive. Eur Biophys J 21:1–12

Thurston GM, Pande J, Ogun O, Benedek GB (1999) Static and quasielastic light scattering and phase separation of concentrated ternary mixtures of bovine alpha and gammaB crystallins. Invest Ophthalmol Vis Sci 40:S299

Treweek TM, Rekas A, Walker MJ, Carver JA (2010) A quantitative NMR spectroscopic examination of the flexibility of the C-terminal extensions of the molecular chaperones, αA- and αB-crystallin. Exp Eye Res 91:691–699

Treweek TM, Meehan S, Ecroyd H, Carver JA (2015) Small heat-shock proteins: important players in regulating cellular proteostasis. Cell Mol Life Sci 72:429–451

Vendra VPR, Khan I, Chandani S, Muniyandi A, Balasubramanian D (2016) Gamma crystallins of the human eye lens. Biochim Biophys Acta 1860:333–343

Veretout F, Delaye M, Tardieu A (1989) Molecular basis of eye lens transparency: osmotic pressure and X-ray analysis of α-crystallin solutions. J Mol Biol 205:713–728

Voorter CEM, Mulders JW, Bloemendal H, de Jong WW (1986) Some aspects of the phosphorylation of α-crystallin A. Eur J Biochem 160:203–210

Wang Y, Petty SA, Trojanowski AT, Knee KM, Goulet DR, Mukerji I, King JA (2010) Formation of amyloid fibrils in vitro from partially unfolded intermediates of human gamma C-crystallin. Invest Ophthalmol Vis Sci 51:672–678

Wenk M, Herbst R, Hoeger D, Kretschmar M, Lubsen NH, Jaenicke R (2000) Gamma S crystallin of bovine and human eye lens: solution structure, stability and folding of the intact two-domain protein and its separate domains. Biophys Chem 86:95–108

West SK, Duncan DD, Munoz B, Rubin GS, Fried LP, Bandeen-Roche K, Schein OD (1998) Sunlight exposure and risk of lens opacities in a population-based study: the Salisbury Eye Evaluation project. J Am Med Assoc 280:714–718

Wistow GJ, Piatigorsky J (1988) Lens crystallins: the evolution and expression of proteins for a highly specialized tissue. Annu Rev Biochem 57:479–504

Wu Z, Delaglio F, Wyatt K, Wistow G, Bax A (2005) Solution structure of gamma-S-crystallin by molecular fragment replacement NMR. Protein Sci 14:3101–3114

Wu JW, Chen ME, Wen WS, Chen WA, Li CT, Chang CK, Lo CH, Liu HS, Wang SS (2014) Comparative analysis of human γD-crystallin aggregation under physiological and low pH conditions. PLoS ONE 9(11):e112309. https://doi.org/10.1371/journal.pone.0112309

Xia Z, Yang ZX, Huynh T, King JA, Zhou RH (2013) UV-radiation induced disruption of dry-cavities in human γD-crystallin results in decreased stability and faster unfolding. Sci Rep 3:1560

Yang Z, Xia Z, Huynh T, King JA, Zhou R (2014) Dissecting the contributions of β-hairpin tyrosine pairs to the folding and stability of long-lived human γD-crystallins. Nanoscale 6:1797–1807

Yoshida H, Yumoto N, Tsukahara I, Murachi T (1986) The degradation of α-crystallin at its carboxyl-terminal portion by calpain in bovine lens. Invest Ophthalmol Vis Sci 27:1269–1273

Zhao H, Chen Y, Rezabkova L, Wu Z, Wistow G, Schuck P (2014) Solution properties of γ-crystallins: hydration of fish and mammal γ-crystallins. Protein Sci 23:88–99

Chapter 15
Structure and Function of the TREX-2 Complex

Murray Stewart

Abstract The Three prime repair exonuclease 2 (TREX-2) complex functions as a platform to which many of the components of the nuclear mRNA processing machinery bind, facilitating integration of this phase of the gene expression pathway, as well as mediating the re-positioning of highly regulated actively transcribing genes (such as *GAL1*) to nuclear pores (NPCs) to accelerate their activation. In *Saccharomyces cerevisiae* the TREX-2 complex is based on a Sac3 scaffold to which Thp1, Sem1, Cdc31 and two Sus1 chains are bound. A combination of X-ray crystallography and electron microscopy studies have established the structure of two major regions of this complex: the M-region that functions to bind nucleic acids and the CID region that functions to link the complex to nuclear pores. These structures have facilitated the engineering of mutants that have been used to define the contributions made by the TREX-2 complex to locating high-expressed genes to nuclear pores and the contributions made to mRNA nuclear export.

Keywords TREX-2 complex · mRNA processing · Nuclear pore · NPC · Karyopherin · Chaperonin · Nucleoporin

Biological Function of the TREX-2 Complex

The nuclear phase of the gene expression pathway culminates with export of mature mRNAs through nuclear pores (NPCs) for translation in the cytoplasm. The structure of NPCs has been established at atomic resolution by a combination of X-ray crystallography and cryoEM combined with crosslinking and mass spectrometry (Beck and Hurt 2017; Kim et al. 2018; Lin and Hoelz 2019) and the

---

M. Stewart (✉)
MRC Laboratory of Molecular Biology, Francis Crick Avenue, Cambridge Biomedical Campus, Cambridge CB2 0QH, UK
e-mail: ms@mrc-lmb.cam.ac.uk

J. R. Harris and J. Marles-Wright (eds.), *Macromolecular Protein Complexes II: Structure and Function*, Subcellular Biochemistry 93,
https://doi.org/10.1007/978-3-030-28151-9_15

nuclear export and import of proteins and small RNAs by karyopherins is understood in considerable detail (Stewart 2007; Kelly and Corbett 2009; Katahira 2015). However, the export of mRNAs and how this function is integrated into the gene expression pathway is more complicated because it is crucial to complete nuclear processing steps (such as 5′ capping, splicing, and 3′ polyadenylation) before transcripts are exported (Kohler and Hurt 2007; Stewart 2010; Nino et al. 2013; Katahira 2015). The nuclear export of mRNA differs from other nuclear trafficking pathways by using the Mex67:Mtr2 export factor rather than a karyopherin and in employing a series of changes in RNA structure to mediate its attachment in the nucleus and release in the cytoplasm after passing through the NPCs (Kohler and Hurt 2007; Stewart 2010; Nino et al. 2013; Katahira 2015). The generation of export-competent mRNPs is a key step in the gene expression pathway and employs a series of checkpoints to ensure processing has been completed. Moreover, nuclear pores also function to coordinate many of the nuclear steps of the gene expression pathway, often by using the TREX-2 complex (also called THSC complex) that acts as a scaffold to which components of the processing machinery bind (reviewed by Garcia-Oliver et al. 2012) and which may also function to alleviate collisions between the transcription and replication machineries (Santos-Pereira et al. 2014). The TREX-2 complex is conserved across Eukaryotes and has been implicated in a spectrum of biological roles including preventing genome instability, transcription initiation, gene-gating, and mRNA export (reviewed by Garcia-Oliver et al. 2012). Furthermore, in budding yeast, in addition to integrating mRNA export into the gene expression pathway, the TREX-2 complex also mediates the location of actively-expressed genes such as *GAL1* to the nuclear envelope (reviewed by Garcia-Oliver et al. 2012). In Metazoans, the corresponding complex has been proposed to also function as a chaperonin, contributing to providing a fast track route to nuclear pores and nuclear export for stress-related transcripts that have a small number of introns (Jani et al. 2012).

Composition and Structure of the TREX-2 Complex

The TREX-2 complex in *S. cerevisiae* is based on a Sac3 scaffold to which Thp1, Sem1, Cdc31, and two copies of Sus1 bind (Fig. 15.1). The corresponding human complex is based on the Sac3 homologue GANP to which PCID2 (Thp1 homologue), DSS1 (Sem1 homologue) and ENY2 (Sus1 homologue) are bound in a similar way (Jani et al. 2012), whereas the *Drosophila* complex is based on the Sac3 homologue, XMAS-2, and *Arabadopsis* on SAC3B. Structurally, the TREX-2 complex can be subdivided into modules (Fig. 15.1): an N-module (Sac3 residues ~ 1–100) that harbors degenerate FG-like repeats similar to those seen in many nuclear pore proteins (FG-nucleoporins) and which can bind to the Mex67: Mtr2 export factor (Fischer et al. 2002; Dimitrova et al. 2015); a central M- or

PCI-module, containing Sac3 residues ~ 100–551 bound to Thp1 and Sem1, that binds nucleic acids (Ellisdon et al. 2012) and which has also been proposed to function as docking site for components of the Mediator complex (Schneider et al. 2015); and a C-terminal CID-region in which Cdc31 and two Sus1 chains bind to Sac3 residues ~ 720–805 (Jani et al. 2009) and which, in *S. cerevisiae*, binds to the nuclear pore (NPC) basket protein, Nup1 (Jani et al. 2014), to tether the complex close to the nuclear basket to facilitate localization of genes such as *GAL1* (reviewed by Garcia-Oliver et al. 2012). In the M-module the Sac3 chain and Thp1 both have similar structures that are based on a PCI fold (Ellisdon et al. 2012; Ellisdon and Stewart 2012; Gordon et al. 2017), which is based on a series of TPR-like α-helical repeats capped by a winged helix motif. The juxtaposition of the winged helix motifs of Sac3 and Thp1, together with a surface patch of positive charge, contributes to the formation of a RNA-binding region in the TREX-2 complex (Ellisdon et al. 2012; Schneider et al. 2015; Valkov and Stewart 2015; Gordon et al. 2017).

Structure of the TREX-2 C-Terminal CID Domain

Crystal structures have been determined for the *S. cerevisiae* (Jani et al. 2009), *C. thermophilium* (Dimitrova et al. 2015), and human CID complexes (Jani et al. 2012). These structures show that in this region the Sac3 scaffold forms an extraordinarily long α-helix around which Cdc31 and the Sus1 chains are wrapped. Two Sus1 chains (Sus1A and Sus1B) are bound to Sac3 in the *S. cerevisiae* (Fig. 15.1b) and human complexes, whereas only one Sus1 chain is bound in *C. thermophilium.* Sus1 is constructed from five α-helices that are arranged in a hairpin-like conformation that allows it to wrap around the Sac3 helix by binding to a predominately hydrophobic stripe (Jani et al. 2009). Sus1 is also a component of the SAGA complex (Rodriguez-Navarro et al. 2004) in which it binds to an analogous helix in the Sgf11 component (Ellisdon et al. 2010). Mutagenesis experiments indicated that the presence of the Sus1B chain was more important for TREX-2 function than the Sus1A chain (Jani et al. 2009).

The *S. cerevisiae* TREX-2 complex is localized at nuclear pores and binding to nucleoporin Nup1, a component of the pore nuclear basket, contributes to this process (Jani et al. 2014). The human homologue similarly is located at the pore basket though an interaction with nucleoporin Nup153 (Umlauf et al. 2013) although it also appears to be located transiently at the nuclear speckles at which many mRNA processing steps are performed and may shuttle between these sites to chaperone mature mRNPs to provide a fast-track for nuclear export of transcripts that contain a small number of introns (Wickramasinghe et al. 2010). The crystal structure of a Nup1 fragment bound to part of the CID domain (Jani et al. 2014) has provided details of the interaction interface that is formed primarily between Sac3 and Sus1B and relies on burying two key Nup1 Phe side chains (Fig. 15.2). This structural information was used to engineer a Sac3 variant (*sac3 L768A-H774D*) that impaired the interaction with of TREX-2 Nup1 while retaining all the other interactions present in the CID region. In vivo, this variant inhibited TREX-2 binding to nuclear pores when *GAL1* expression was either activated (by using galactose as the carbon source) or de-repressed (by using raffinose as the carbon source) and showed reduced growth rates as well as leading to an increase in nuclear poly(A)+ RNA, consistent with bulk mRNA export being impaired (Jani et al. 2014). The re-positioning of *GAL1* to NPCs that is required for transcriptional control requires TREX-2 being bound to Nup1 at the pore (Texari et al. 2013). Repression of the *GAL1-GAL10* cluster by glucose is mediated by Mig1 that functions together with Ssn6 and Tup1, two co-repressors that bind to the gene when they are modified by SUMOylation. The SUMO protease Ulp1, that is bound to nuclear pores, mediates de-repression of the *GAL* cluster by removing the SUMO modification from Ssn6 and Tup1, releasing them from the promoter. Removal of Ssn6 and Tup1 in turn leads to chromatin remodeling and binding of SAGA complex (Texari et al. 2013). Thus, when glucose is replaced by another carbon source under de-repressive (raffinose) or inducing (galactose) conditions, localizing the *GAL* cluster to nuclear pores facilitates de-repression and subsequent transcription by removing SUMO from the chromatin-bound gene regulators (Texari et al. 2013). This contrasts to the glucose-induced repressive state in which inducible genes such as the *GAL* cluster are thought to be located away from NPCs and so inaccessible to Ulp1. Consequently, in the presence of glucose, Ssn6 and Tup1 retain their SUMO modification and inhibit transcription (Texari et al. 2013). Although movement of the *GAL* locus to nuclear pores requires the TREX-2 complex to be attached to the nuclear basket, it is not clear whether TREX-2 mediates the actual movement of the *GAL* cluster to the pores or whether, alternatively, the TREX-2 complex could function by retaining de-repressed genes at nuclear pores to prevent their becoming repressed by binding Ssn6 and Tup1 that had been SUMOylated elsewhere in the nucleus (Jani et al. 2014).

Fig. 15.1 **a** Schematic illustration of the structural modules of the *S. cerevisiae* TREX-2 complex. ► The N-terminal Sac3 region contains FG repeats and interacts with the mRNA nuclear export factor Mex67:Mtr2, whereas the central M-region contains Thp1, Sem1 and the region of Sac3 that contains a series of TPR repeats and a winged helix (WH) domain. Thp1 has an analogous structure based on TPR repeats and a winged helix domain. The M-module binds to RNA and may also bind to the Mediator complex. In the C-terminal CID domain, two Sus1 chains and a Cdc31 chain wrap around an extraordinarily long Sac3 α-helix and function to bind the TREX-2 complex to the nuclear basket via Nup1. TREX-2 analogues in Metazoans have analogous arrangements, although some only have a single Sus1 chain in the CID domain. **b** Crystal structure of the CID domain of the *S. cerevisiae* TREX-2 complex (based on PDB 3FWC). Two Sus1 chains (Sus1A—light blue—and Sus1B—yellow) together with one Cdc31 chain (gold) wrap around an extensive Sac3 α-helix (red) that spans residues 725–805 (Jani et al. 2009). **c** Structure of the M- domain of the *S. cerevisiae* TREX-2 complex (based on PDB 5UBP). Both Thp1 (green) and residues 92–500 of Sac3 (red) have folds based on a series of α-helical TPR repeats capped with a winged-helix fold (WH). Sem1 (blue) binds to one side of Thp1 (Gordon et al. 2017)

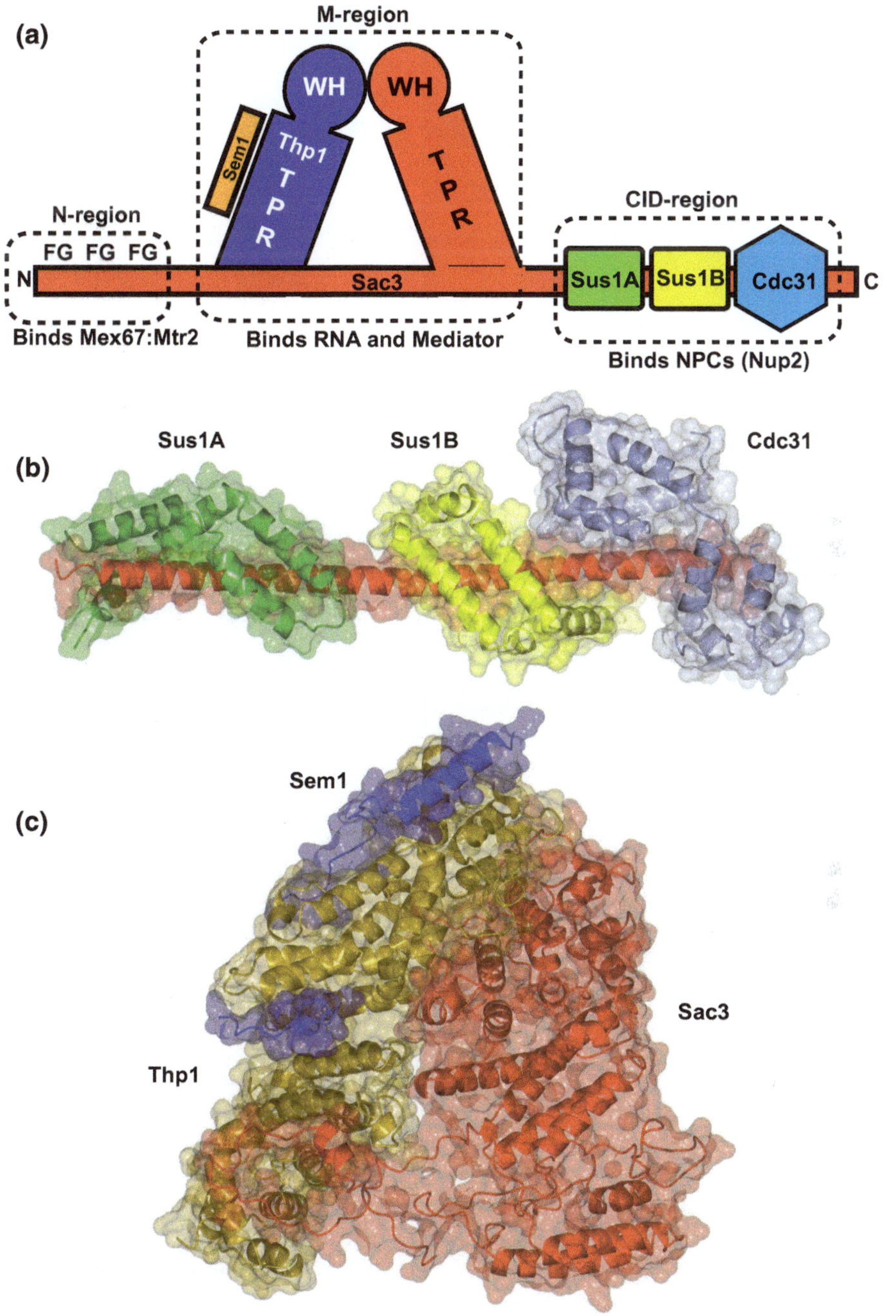
(a)
M-region
WH
WH
Sem1
Thp1
TPR
TPR
N-region
FG FG FG
N
Sac3
CID-region
Sus1A
Sus1B
Cdc31
C
Binds Mex67:Mtr2
Binds RNA and Mediator
Binds NPCs (Nup2)
(b)
Sus1A
Sus1B
Cdc31
(c)
Sem1
Thp1
Sac3

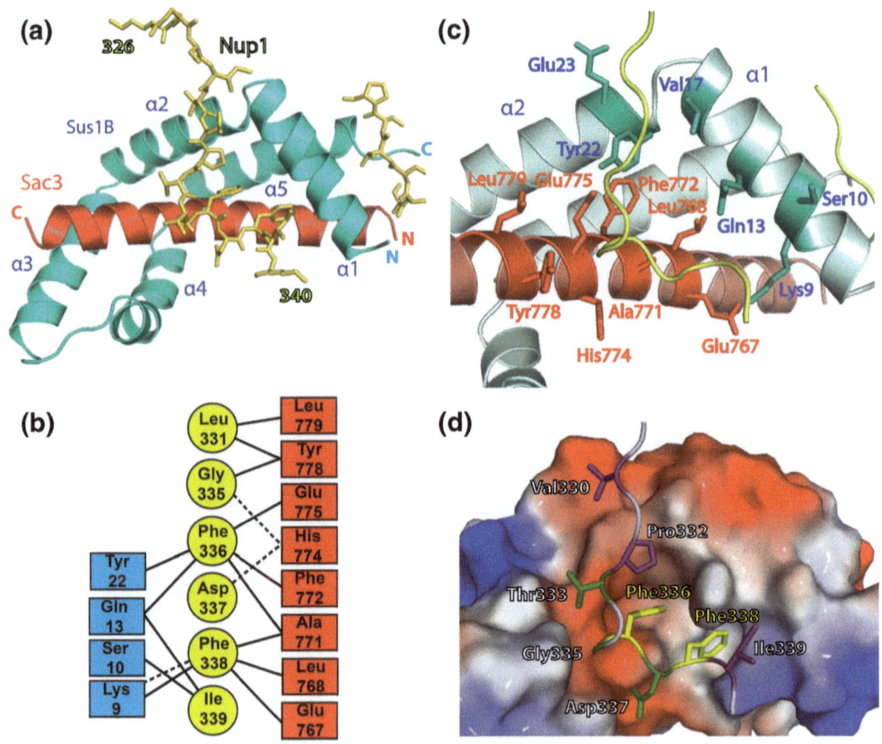

Fig. 15.2 Structure of the Nup1:Sac3757–787:Sus1B complex. **a** Overview of the arrangement of the chains in the complex. **b** Schematic representation of the interactions formed between Nup1, Sac3 and Sus1B upon complex formation. Dashed lines represent putative hydrogen bonds or salt bridges. **c** Principal interacting residues in Sac3 (highlighted in red) and Sus1B (highlighted in dark cyan) that are buried in the interface with Nup1. (D) Electrostatic surface representation of Sac3 and Sus1B in the same view as in (C) showing the cavity to which Nup1 binds. Nup1 residues interacting with only Sus1 are shown in purple; those interacting with only Sac3 in green and those interacting with both are shown in yellow. Reproduced from Jani et al. (2014). Reproduced by permission of Oxford University Press

The TREX-2 Central M-Region

The M-region of the *S. cerevisiae* TREX-2 complex comprises residues 90–550 of Sac3 to which Thp1 and Sem1 are bound (Fig. 15.1c). The crystal structure of Sac3 residues 253–551 complexed with full-length Thp1 and Sem1 (Ellisdon et al. 2012) showed that both the Sac3 fragment and Thp1 have PCI folds that are based on as stack of α-helical TPR repeats followed by a winged helix domain (Ellisdon and Stewart 2012). The winged helix domains, together with a prominent positively-charged patch formed of the surface of the structure, provided a platform to which nucleic acids could bind. The structure of a smaller fragment, consisting of Sac3 residues 222–572, Thp1 residues 170–455 and Sem1 had a similar structural

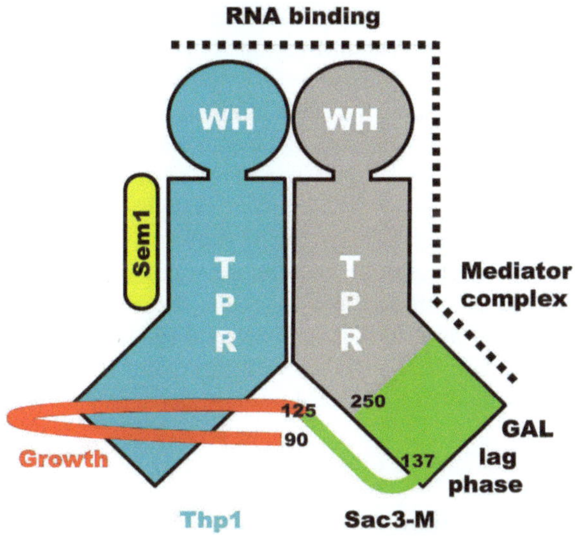

Fig. 15.3 Schematic illustration of the structure of the TREX-2 M-region. Both Sac3 and Thp1 have PCI folds based on a series of α-helical TPR repeats together with a winged helix (WH) domain, whereas Sem1 lies along one side of Thp1. The crystal structure of the Sac360–556: Thp1:Sem1 complex identified an extension of the Sac3 TPR region from residue 250 to residue 127 (green) together with an extended loop comprising residues 90–125 that links the distal portion of the Sac3 TPR region with Thp1. A cluster of positively-charged residues that extends from the Thp1 winged helix domain and down one side of Sac3 appears to be important for binding RNA. The extended loop (residues 90–125) is important for growth, whereas residues 126–250 are important for the induction of *GAL* genes when cells are moved to galactose as a carbon source. Other studies (Schneider et al. 2015) have indicated that positive residues in the Sac3 TPR region also function in conjunction with the mediator complex. Reproduced from Gordon et al. (2017). Reproduced by permission of Oxford University Press

organization (Schneider et al. 2015). An electron microscopic study of the full-length *C. thermophilum* TREX-2 complex (Dimitrova et al. 2015) together with a cryoEM study of the *S. cerevisiae* complex (Aibara et al. 2016) indicated that the M-domain might be more extensive. Although three tubes of electron density consistent with helices could be observed, the limited resolution of these studies frustrated identification of the precise residues involved and was insufficient to enable the chain to be traced reliably outside the region already established by X-ray crystallography (Fig. 15.3).

A 2.3 Å resolution crystal structure of the *S. cerevisiae* M-region complex formed by Sac3 residues 60–550 bound to Thp1 and Sem1 identified two additional structural features in Sac3 that are important for the function of the TREX-2 complex in yeast (Gordon et al. 2017). This structure indicated that the Sac3 TPR-like repeats extended to residue 137 (Fig. 15.1c), confirming the indications from cryoEM (Aibara et al. 2016) that indicated that this region could be more extensive. The crystal structure also showed the presence of an unanticipated extensive unstructured loop formed by Sac3 residues 90–137 and which extended from the N-terminus of the Sac3 M-region TPR repeats to connect it to the N-terminal region of Thp1. Mutagenesis studies demonstrated the functional importance of both the additional Sac TPR repeats and the loop that linked them to Thp1 (Gordon et al. 2017). When Sac3 mutants in which the loop had been deleted were exchanged for wild-type Sac3 in *S. cerevisiae,* these cells showed impaired mRNA nuclear export together with slower growth at 37 °C, supporting the functional role of the connection between the distal regions of the Sac3 and Thp1 distal regions. When either glucose or raffinose was replaced by galactose as the carbon source in the growth medium to assess the expression of *GAL1*, Sac3 mutants in which the additional TPR region (residues 125–250) had been deleted showed an increased lag phase together with a pronounced mRNA export defect consistent with the induction of the GAL system being impaired.

The TREX-2 N-Terminal Region

The N-terminal region is thought to be primarily unstructured but it contains a number of putative FG repeats that have been proposed to provide a binding site for the principal mRNA nuclear export factor Mex67:Mtr2 (Fischer et al. 2002). Previous work using homologues from the thermophile *C. thermophilum* indicated that the multiple FG-like repeats present on the Sac3 N-terminal region might, in addition to providing a docking site for the principal mRNA export factor Mex67: Mtr2, also function to mediate the generation an annular conformation in the TREX-2 complex by making intra-molecular interactions with the CID-region (Dimitrova et al. 2015). However, a subsequent cryoEM study of the intact *S. cerevisiae* TREX-2 complex expressed in insect cells indicated that here the CID region was orientated randomly relative to the M-region and so did not support an interaction between these two regions mediated by the N-region (Aibara et al. 2016). However, although the FG-repeats in the *S. cerevisiae* Sac3 N-terminal region are fewer and do not have all of the sequence characteristics seen in the FG repeats found in nucleoporins, deletion of residues 1–140 generates growth and mRNA export defects (Dimitrova et al. 2015).

Summary and Questions Outstanding

Complementary structural, molecular and cellular studies have identified the contributions made by the different modules of the TREX-2 complex to the coordination of the nuclear steps in the gene expression pathway and their integration with nuclear mRNA export. However, several details of the molecular mechanisms involved remain to be clarified. An interaction with the Mediator complex has been proposed, but it is not clear whether this is direct or could be a consequence of their being components of a more extended interaction network. Although mutation of Sac3 Arg256 to Asp impaired this interaction (Schneider et al. 2015), this residue is buried in the intact complex and so it remains possible that this mutation could generate widespread conformational changes in TREX-2 that could influence its binding to partners such as Thp1 or Sem1 (Gordon et al. 2017).

Additional work will be required to resolve this aspect. The interaction between the M-module and nucleic acids also remains somewhat unclear. Although this module does bind to a range of RNAs in vitro (Ellisdon et al. 2012) and some hints have been obtained regarding sequence specificity (Valkov and Stewart 2015), the precise way in which TREX-2 functions to relocate active genes such as those of the *GAL* cluster to the nuclear pores in yeast remains unclear.

Acknowledgements Supported in part by MRC grants MC_U105178939, MC-A025-5PL41, MC_UP_1201/6 and a Leverhulme Emeritus Fellowship (EM-2016-062) to MS.

References

Aibara S, Bai XC, Stewart M (2016) The Sac3 TPR-like region in the *Saccharomyces cerevisiae* TREX-2 complex is more extensive but independent of the CID region. J Struct Biol 195:316–324

Beck M, Hurt E (2017) The nuclear pore complex: understanding its function through structural insight. Nature Rev Mol Cell Biol 18:73–89

Dimitrova L, Valkov E, Aibara S, Flemming D, McLaughlin SH, Hurt E, Stewart M (2015) Structural characterization of the *Chaetomium thermophilum* TREX-2 complex and its interaction with the mRNA nuclear export factor Mex67:Mtr2. Structure 23:1246–1257

Ellisdon AM, Dimitrova L, Hurt E, Stewart M (2012) Structural basis for the assembly and nucleic acid binding of the TREX-2 transcription-export complex. Nature Struct Mol Biol 19:328–336

Ellisdon AM, Jani D, Kohler A, Hurt E, Stewart M (2010) Structural basis for the interaction between yeast Spt-Ada-Gcn5 acetyltransferase (SAGA) complex components Sgf11 and Sus1. J Biol Chem 285:3850–3856

Ellisdon AM, Stewart M (2012) Structural biology of the PCI-protein fold. Bioarchitecture 2:118–123

Fischer T, Strasser K, Racz A, Rodriguez-Navarro S, Oppizzi M, Ihrig P, Lechner J, Hurt E (2002) The mRNA export machinery requires the novel Sac3p-Thp1p complex to dock at the nucleoplasmic entrance of the nuclear pores. EMBO J 21:5843–5852

Garcia-Oliver E, Garcia-Molinero V, Rodriguez-Navarro S (2012) mRNA export and gene expression: the SAGA-TREX-2 connection. Biochim Biophys Acta 1819:555–565

Gordon JMB, Aibara S, Stewart M (2017) Structure of the Sac3 RNA-binding M-region in the *Saccharomyces cerevisiae* TREX-2 complex. Nucleic Acids Res 45:5577–5585

Jani D, Lutz S, Hurt E, Laskey RA, Stewart M, Wickramasinghe VO (2012) Functional and structural characterization of the mammalian TREX-2 complex that links transcription with nuclear messenger RNA export. Nucleic Acids Res 40:4562–4573

Jani D, Lutz S, Marshall NJ, Fischer T, Kohler A, Ellisdon AM, Hurt E, Stewart M (2009) Sus1, Cdc31, and the Sac3 CID region form a conserved interaction platform that promotes nuclear pore association and mRNA export. Mol Cell 33:727–737

Jani D, Valkov E, Stewart M (2014) Structural basis for binding the TREX2 complex to nuclear pores, GAL1 localisation and mRNA export. Nucleic Acids Res 42:6686–6697

Katahira J (2015) Nuclear export of messenger RNA. Genes (Basel) 6:163–184

Kelly SM, Corbett AH (2009) Messenger RNA export from the nucleus: a series of molecular wardrobe changes. Traffic 10:1199–1208

Kim SJ, Fernandez-Martinez J, Nudelman I, Shi Y, Zhang W, Raveh B, Herricks T, Slaughter BD, Hogan JA, Upla P, Chemmama IE, Pellarin R, Echeverria I, Shivaraju M, Chaudhury AS, Wang J, Williams R, Unruh JR, Greenberg CH, Jacobs EY, Yu Z, de la Cruz MJ, Mironska R, Stokes DL, Aitchison JD, Jarrold MF, Gerton JL, Ludtke SJ, Akey CW, Chait BT, Sali A, Rout MP (2018) Integrative structure and functional anatomy of a nuclear pore complex. Nature 555:475–482

Kohler A, Hurt E (2007) Exporting RNA from the nucleus to the cytoplasm. Nature Rev Mol Cell Biol 8:761–773

Lin DH, Hoelz A (2019) The structure of the nuclear pore complex (an update). Ann Rev Biochem (in press)

Nino CA, Herissant L, Babour A, Dargemont C (2013) mRNA nuclear export in yeast. Chem Rev 113:8523–8545

Rodriguez-Navarro S, Fischer T, Luo MJ, Antunez O, Brettschneider S, Lechner J, Perez-Ortin JE, Reed R, Hurt E (2004) Sus1, a functional component of the SAGA histone acetylase complex and the nuclear pore-associated mRNA export machinery. Cell 116:75–86

Santos-Pereira JM, Garcia-Rubio ML, Gonzalez-Aguilera C, Luna R, Aguilera A (2014) A genome-wide function of THSC/TREX-2 at active genes prevents transcription-replication collisions. Nucleic Acids Res 42:12000–12014

Schneider M, Hellerschmied D, Schubert T, Amlacher S, Vinayachandran V, Reja R, Pugh BF, Clausen T, Kohler A (2015) The nuclear pore-associated TREX-2 complex employs mediator to regulate gene expression. Cell 162:1016–1028

Stewart M (2007) Molecular mechanism of the nuclear protein import cycle. Nature Rev Mol Cell Biol 8:195–208

Stewart M (2010) Nuclear export of mRNA. Trends Biochem Sci 35:609–617

Texari L, Dieppois G, Vinciguerra P, Contreras MP, Groner A, Letourneau A, Stutz F (2013) The nuclear pore regulates GAL1 gene transcription by controlling the localization of the SUMO protease Ulp1. Mol Cell 51:807–818

Umlauf D, Bonnet J, Waharte F, Fournier M, Stierle M, Fischer B, Brino L, Devys D, Tora L (2013) The human TREX-2 complex is stably associated with the nuclear pore basket. J Cell Sci 126:2656–2667

Valkov E, Stewart M (2015) 1.25 A resolution structure of an RNA 20-mer that binds to the TREX2 complex. Acta Crystallogr F Struct Biol Commun 71:1318–1321

Wickramasinghe VO, Stewart M, Laskey RA (2010) GANP enhances the efficiency of mRNA nuclear export in mammalian cells. Nucleus 1:393–396

Chapter 16
Amyloid Oligomers, Protofibrils and Fibrils

Mohammad Khursheed Siddiqi, Nabeela Majid, Sadia Malik, Parvez Alam and Rizwan Hasan Khan

Abstract Amyloid diseases are of major concern all over the world due to a number of factors including: (i) aging population, (ii) increasing life span and (iii) lack of effective pharmacotherapy options. The past decade has seen intense research in discovering disease-modifying multi-targeting small molecules as therapeutic options. In recent years, targeting the amyloid cascade has emerged as an attractive strategy to discover novel neurotherapeutics. Formation of amyloid species, with different degrees of solubility and neurotoxicity is associated with the gradual decline in cognition leading to dementia/cell dysfunction. Here, in this chapter, we have described the recent scenario of amyloid diseases with a great deal of information about the structural features of oligomers, protofibrils and fibrils. Also, comprehensive details have been provided to differentiate the degree of toxicity associated with prefibrillar aggregates. Moreover, a review of the technologies that aid characterisation of oligomer, protofibrils and fibrils as well as various inhibition strategies to overcome protein fibrillation are also discussed.

Keywords Alzheimer' disease · Amyloid · Prefibrillar aggregates · Cell toxicity

M. K. Siddiqi · N. Majid · S. Malik · P. Alam · R. H. Khan (✉)
Interdisciplinary Biotechnology Unit, Aligarh Muslim University, Aligarh 202002, U.P., India
e-mail: rizwanhkhan1@gmail.com

M. K. Siddiqi
e-mail: khursheedibu@gmail.com

N. Majid
e-mail: nabilamajid1994@gmail.com

S. Malik
e-mail: sadia.alig1@gmail.com

P. Alam
e-mail: parvezalam777@gmail.com

J. R. Harris and J. Marles-Wright (eds.), *Macromolecular Protein Complexes II: Structure and Function*, Subcellular Biochemistry 93,
https://doi.org/10.1007/978-3-030-28151-9_16

Introduction

Amyloid formation refers to the process by which soluble, typically monomeric peptides and proteins misfold and self-assemble involving several intermediate conformations (such as oligomer, protofibrils) into amyloid fibrils. They are thread-like structures typically composed of several protofilaments with a common architecture, termed the cross-β structure. The formation of cross beta sheet rich structure is largely independent of the amino acid sequence and native fold of the protein of which the fibrils are composed (de la Paz and Serrano 2004; Chiti and Dobson 2006; Siddiqi et al. 2017a, b, c, 2018a, b).

Amyloid fibrils have been recognized as a hallmark of a number of medical disorders, importantly neurodegenerative disorders (Table 16.1). Approximately 50 distinct human diseases (such as type 2 diabetes, Alzheimers, Parkinson's, Huntington's disease) with remarkable clinical relevance, are linked with the accumulation of amyloid deposits and have shown dramatic prevalence in the modern world (Selkoe 2002; Seeley et al. 2009; Chaturvedi et al. 2016; Parvez Alam 2017). Neurodegenerative diseases are ubiquitous in nature and the number of cases is increasing. Alzheimer's disease (AD) , for example, affects 1% of people at the age of 60 and the frequency doubles every fifth year of age after 60 indicating that 30–50% of people >85 years are affected with this disease (Braak and Braak 1997). The causes and effects linking the protein aggregation and the degenerative central nervous system are not known, but studies have shown that many neurodegenerative diseases are the consequences of the occurrence of disease-specific

Table 16.1 Summary of amyloid diseases and protein

Disease	Amyloid protein	Location	Affected region
Alzheimer's disease (AD)	β-Amyloid (amyloid plaques), Tau (neurofibrillary tangles)	Extracellular, Intracytoplasmic, Intraneuronal	Brain (cerebral cortex)
Parkinson's disease (PD)	α-synuclein	Intracytoplasmic	Brain (subtantia nigra, Brain stem)
Huntington's disease (HD)	Huntingtin	Intranuclear neuronal	Whole Brain most vulnerable part is neostriatum
Amylotrophic lateral Sclerosis (ALS)	SOD1, TDP-43, FUS, Ubiquitin positive protein	Intraneuronal	Motor neurons
Prion diseases	Prions	Extracellular	Brain (grey matter) and Peripheral nervous system
Transthyretin Amyloidosis	Transthyretin	Extracellular, Intracellular	Peripheral nervous system, Heart, Kidney, Eye
Spinocerebral Ataxia (SCA)	Ataxia	Intranuclear neuronal	Brain (cerebellum, Spinocerebellar)

misfolded and aggregated proteins/peptides and their accumulation in tissues (Mahdavimehr et al. 2018). These deposited amyloids may be found in the extracellular matrix (amyloid) or inside the cells (inclusion bodies, aggresomes) where they impair tissue physiology and the cell viability. Depending on the type of disorders, these amyloid deposits are composed of one specific type of peptide or protein. In spite of such association of amyloids with diseases, much remains to be explored about the molecular mechanism of amyloid fibrillation and the mechanism of cytotoxicity associated with amyloid fibrils.

The toxicity of amyloids is usually categorized as either loss of function or gain of function (Winklhofer et al. 2008). Proteins which are important for the normal function of the cell/tissue/organ, if aggregated, the critical functions associated with these proteins are then hampered (e.g., p53); this is termed as loss of function toxicity, whereas in case of gain of function toxicity, the aggregate itself shows toxicity and leads to cell or tissue degeneration (e.g., beta amyloid, Aβ). Initially, it was hypothesized that fibrillar species, reported to be present in the medial temporal lobe region of brain of the patient suffering from amyloid diseases, were the key toxic agents responsible for the pathogenicity associated with the amyloid diseases. But this is true only for some systemic amyloidosis conditions, in which abundant fibrils are deposited in vital organs. However, in other amyloid disorders including most neurodegenerative diseases, there is a lack of direct correlation between the extent of protein fibrillation and severity of the disease. This suggested that some other amyloid form is responsible for the pathogenesis of neurodegenerative diseases (disease conditions associated with degeneration of neurons arises due to the deposition of fibrils). It becomes increasingly evident from the analysis of human patients that amyloid intermediate species, particularly oligomeric species, play an important role in the progression and severity of these disease (Kirkitadze et al. 2002; Haass and Selkoe 2007). This conclusion is also consistent with the hypothesis of the likely effect of pre-fibrillar aggregates, even when the aggregates are not related with any known disorders. This chapter will focus on the importance of the biochemical and biophysical features of prefibrillar species and mature fibrils.

Protein Folding and Aggregation Are Competing Processes

In cellular environment, newly synthesized polypeptides undergo the process of folding with the help of molecular chaperones to attain a properly folded and functional state of the protein (Hartl et al. 2011; Frydman 2001). But under some conditions, these polypeptides form a structure, which completely differs from the compactly folded protein, termed as amyloids (Damaschun et al. 1999). Interestingly, proteins which are not associated with any amyloid disorders may also form amyloid fibrils. This suggests that, in addition to folding of protein, information to form amyloids is also present in its sequence. In other words, we can state that the ability to form amyloid fibrils is an inherent property of polypeptides and depends upon the intermolecular hydrogen bonds established between the peptide bonds in parallel or anti-parallel polypeptide stretches in the beta-strand conformation (Trovato et al. 2006). It suggests, and the environmental conditions decide which will be favored for a given polypeptide.

There are several physicochemical factors, including macromolecular crowding, which affects both protein folding and protein aggregation (van den Berg et al. 1999). For instance, if macromolecule are present in high concentration (ranging from 300–400 mg/mL), they will create a condition which will affect the choice between folding and aggregation. This condition is termed as macromolecular crowding, that affects thermodynamically the conformation states of proteins. Other factors may include gain in secondary structure, low net charge and high hydrophobicity. It has been confirmed that peptides or proteins having these physicochemical properties can favor its aggregation kinetically from the unfolded structure, in dynamic equilibrium with folded structure. This information suggests that both process, either folding or aggregation, is governed by the same physicochemical feature/medium in which the polypeptide is present and hence confirms that both pathways are in close competition.

A recent study showed that natural protein where they are synthesized and perform function, by evolution, prefer to fold over aggregation, although latter is thermodynamically more favored. Since the protein folding and aggregation are in competition with same processes, they rely on a general energy landscape encompassing different intermediate states not involved in protein folding (Fig. 16.1). The intermolecular (aggregation) and intramolecular (folding)

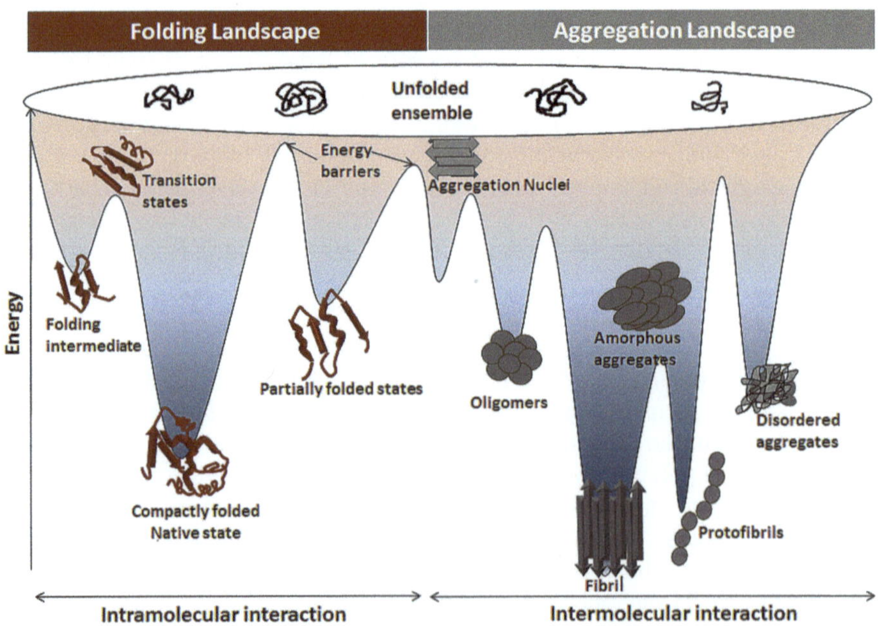

Fig. 16.1 The energy landscape of protein folding and protein aggregation phenomena

interaction highlights the protein folding and aggregation energy landscape. Figure 16.1 shows that a polypeptide may attain different conformational state during protein folding and aggregation. Although the folded state of protein is functional according to the energy landscape, but the mature fibrils are relatively thermodynamically more favored.

General Structural Features of Amyloid Oligomers, Protofibrils and Fibrils

Pre-fibrillar entities of amyloid type have been reported to be present in in vitro and/ or in vivo in the aggregation path of many disease associated proteins such as amyloid beta, alpha synuclein (αS), amylin, prion, transthyretin, huntingtin protein, serum amyloid and others (Stefani 2004). Similar pre-fibrillar assemblies have also been found in aggregation path of proteins which are not associated with amyloid disease. However these entities finally result in the formation of mature amyloid fibrils (Stefani 2012).

Considerable information has recently been gained on the hierarchical formation of amyloid fibrils from structurally more simple precursors through a number of steps, as well as on the structural features of the ordered beta-sheet-rich core of amyloid fibrils and the supramolecular organization of the latter (Fig. 16.2). Amyloid fibrils are the consequences of self-assembly of protein molecules, characterized by the presence of cross beta sheet rich structure in which beta sheet runs perpendicular to the fibril axis, generating intermolecular beta sheet that runs along the length of aggregate (Eisenberg and Jucker 2012; Tycko 2004). Morphologically, amyloids are unbranched and thread-like structures, just a few nanometers in diameter, and are composed of several protofilaments that wrap around each other (Otzen 2013). The structural difference between amyloid fibrils arises due to different polypeptide chains, consequence of the manner in which different side chains of amino acid residues are incorporated into the cross beta sheet structure.

Characterization of structural features of amyloid fibrils also helps in exploring the details of the seeding mechanism which occurs at the end of fibrils. While the molecular structure of amyloid fibrils is not strongly dependent on polypeptide sequence, since the main interaction that stabilize the core of fibril made up of an array of inter-backbone hydrogen bonds. Self-complementary interdigitation of the side chains between polypeptides within the same beta sheet and between beta sheet have been proposed to be important to create the hydrophobic core (Lewandowski et al. 2011; Siddiqi et al. 2017a, b, c). Additionally, cross beta sheet structure also show the multitude of different morphologies of amyloid fibrils, attained by a particular protein in response to different aggregating conditions and mutations in its sequence.

In addition to mature fibrils, understanding the structural information of oligomers and protofibrils is also significant (Fig. 16.2). These species are frequently

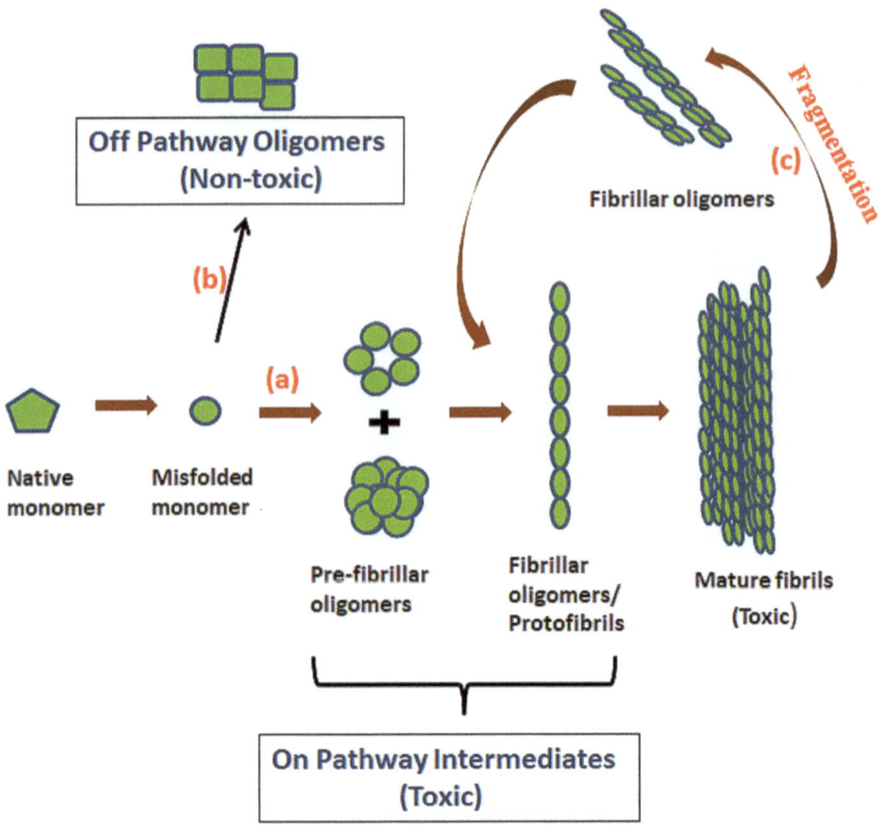

Fig. 16.2 Schematic representation of the protein fibrillation process, involving different intermediate species

observed to accumulate during the process of fibrils formation. Although trapping these intermediate species during the path of amyloid fibrillation is challenging, study of oligomers and protofibrils forms has opened up the possibility of gaining insights into the nature and structure of these species. Various experimental procedures have been used to isolate these species (Bitan et al. 2001).

The very early species in path of amyloid formation are found to be dimers, trimers, tetramers [collectively termed as oligomers (Podlisny et al. 1995)]. Oligomers may have variable sizes, with an average of approximately 25–30 protein molecules. These oligomers are able to interact with the cell membrane, causing cell death by permeabilizing the membrane. The oligomeric form of αS possesses beta sheet structure formed in between the monomeric and mature fibrils of αS. Oligomers range in size from 20 to >50 kDa. Further addition of monomeric unit to oligomers results in the formation of bead like structures up to 200 nm in length (4–11 nm in diameter) called protofibrils (Wong et al. 1997). Study on insulin and α-synuclein fibril suggests that three intertwining protofibrils constitute mature insulin fibril whereas one for α-synuclein fibril respectively. Each protofibril is assumed to consist of two intertwining proto-filaments (Vestergaard et al. 2007).

Another striking difference between the αS oligomer and mature fibrils is that mature fibrils consist of parallel beta sheet arrangement whereas antiparallel beta sheet dominates in oligomeric form, however, the core residues of both forms are almost similar (Kim et al. 2009). Importantly, a further study also suggested that information obtained for αS oligomer such as pathogenicity, physico-chemical properties possess high degree of similarity with oligomeric species obtained from different protein and peptides associated with amyloid diseases. Additionally, variable parallel and anti parallel beta sheet are also observed for other amyloidogenic proteins such as Aβ-peptide, lysozyme, β2-microglobulin and a prion-related peptide (Burdicket al. 1992; Forloni et al. 1993; Wadai et al. 2005; Westermark 2005). Further data showed that oligomers which differ in size also differ in their beta sheet content, larger size possess more beta sheet structure (Glabe 2008). The similarities in the architecture of the fibrillar and the oligomeric forms, despite the differences in the organization of their β-sheet structure, suggest that similar types of interactions to those that stabilize amyloid fibrils are likely to be responsible for the initial acquisition of cross-β structure in the oligomeric species.

Characterization of Amyloid Oligomers, Protofibrils and Fibrils

Protein aggregation actually is the result of protein-protein interaction and the direct analysis of protein-protein interaction requires them to be soluble in nature (Fink 1998). However protein aggregates are insoluble in nature especially those adopting amyloid conformation thus posing a challenge in their characterization (Knowles et al. 2014). But in many cases the conformational states and the interactions of native or near-native still soluble proteins are known that finally fall into the end stage forming mature amyloid fibrils rich in cross-β sheet structure via the formation of intermediate structures. Thus, the characterization of folding intermediate and oligomerization dynamics that connects the above mentioned extremes is the persistent analytical challenge.

Protein aggregation is a highly complex process in itself so it should be kept in mind that a single technique cannot be enough for the complete characterization of the chain of events because any step in the process of aggregation may give rise to a number of oligomeric species with a distribution of size and conformational properties and thus the analytical techniques should be applied accordingly (Ross and Poirier 2004). Here we describe the various analytical techniques that can be used for the characterization of amyloid oligomers, protofibrils and mature amyloid fibrils (Fig. 16.3), which are intermediates and end products of the protein aggregation phenomenon.

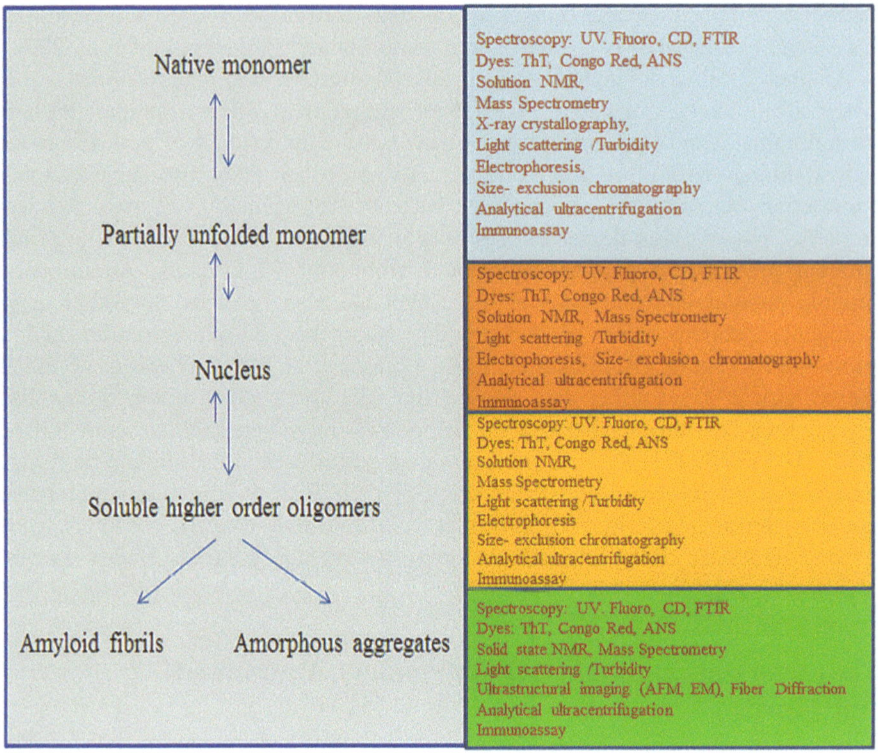

Fig. 16.3 Illustration of techniques used to characterize prefibrillar structures as well as mature fibrils

Electrophoresis

In electrophoresis, the separation of molecules take place on the basis of size, shape, and charge; hence electrophoresis can be used for characterizing protein oligomers, provided that they are soluble and stable to the separation conditions and are analyzed in pure form or detectable when present in complex mixtures such as biofluids. For example, Aβ monomers and oligomers are readily separated after covalent cross-linking via polyacrylamide gel electrophoresis (PAGE) (Levine Iii 2004). SDS-PAGE (Sodium Dodecyl Sulfate-PAGE) cannot be imply used as in some studies, because it has been shown to be responsible for the induction of amyloid oligomerization (Bitan et al. 2005). Capillary electrophoresis (CE) can also be used to monitor the changes in the concentration of both monomers and initial soluble oligomers over time during kinetic analysis, as demonstrated for both Aβ (Pedersen et al. 2011) and insulin oligomers (Pryor et al. 2011). Alternatively, if CE is coupled with specialized detectors, e.g., laser-induced fluorescence anisotropy (LIFA) detectors, an online assessment of the size/shape of the separated Aβ

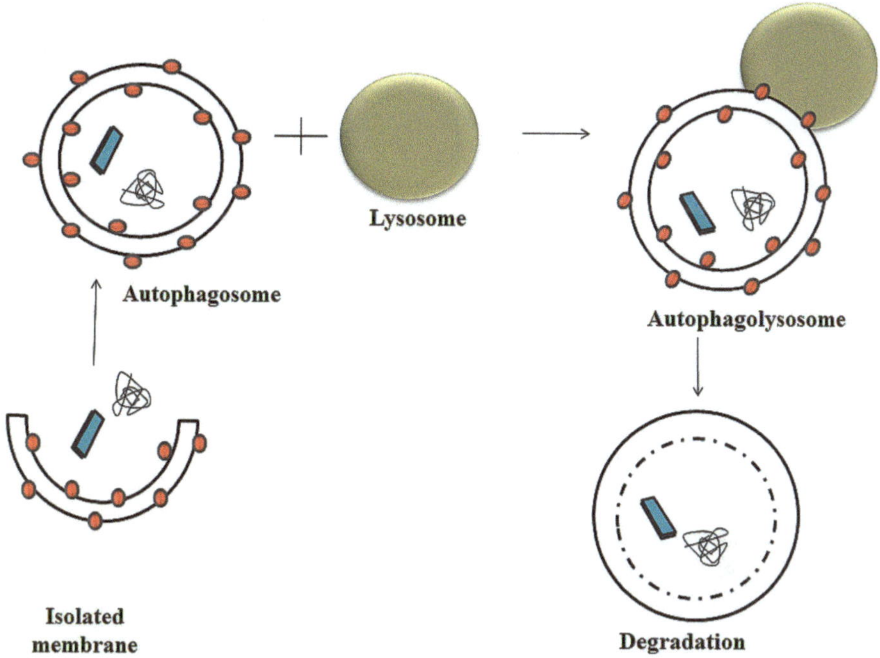

Fig. 16.4 Illustration of steps involved in autophagy to degrade misfolded proteins

species can be simultaneously obtained (Picou et al. 2012). Although CE is convenient for monitoring the contents of small biological samples, including micro dialysis fluids, over time, due to its very low sample volume consumption. Generally electrophoretic methods are not able to provide much structural information about separated analytes. In native gel electrophoresis (without anionic detergent), aggregates are more likely to be stable than in denaturing electrophoresis, and this concept has been exploited to analyze Aβ, yeast prion aggregates (Bagriantsev et al. 2006) and oligomer formation of neuroserpin variants (Miranda et al. 2008). However, native gel electrophoresis is not reliable because in the absence of anionic detergent the separation is dependent on a combination of analyte charge, shape and mass, and it cannot separate oligomers with an identical size-to charge ratio. Thus for strongly complexed aggregates denaturing gel electrophoresis may be useful (Lawrence and Payne 1983; Bullock 1993). For example, soluble 2010Aβ oligomers and prion protein aggregates (SDS-resistant) are characterized using sodium dodecyl sulfate SDS-PAGE but not the non-SDS resistant oligomers (Mc Donald et al. 2010; Coalier et al. 2013).

Analytical Ultracentrifugation

Analytical ultracentrifugation (AUC) is a versatile and powerful hydrodynamic technique used for the analysis of macromolecular properties (Cantor and Schimmel 1981). Different sedimentation coefficients are responsible here for the separation of macromolecules, which directly correlates with the molecular mass and inversely with the frictional coefficient (Schuck 2000). Since such an experiment covers a size range in molar mass of 3 orders of magnitude, thus making sedimentation velocity AUC suitable for characterizing the molecular weight distribution of proteins in relevant solutions where the quantity of biological material is not the limiting factor (Gabrielson et al. 2011; Brown et al. 2008). Moreover, AUC can also be used for evaluating noncovalent interactions in a 2 (maximum 3) component protein solution under equilibrium conditions, if provided with enough material and less complex mixtures. Sedimentation equilibrium AUC has been reported to provide data suggesting tetramer assembly of native α-synuclein, which contrasts with the prevailing view of this Parkinson's disease related aggregating protein, normally found as an unstructured monomer. But this technique has a drawback associated with it, that it may take several days of work.

Size Exclusion Chromatography

Size exclusion chromatography (SEC) is a form of partition chromatography where the hydrodynamic volume (i.e., analyte size and shape) determines individual eluate path length and thus the time to emerge at the detector gives the information about the molecular mass of the analyte. Smaller the analyte, the greater the time it will take to emerge at the detector and vice versa as the small analyte will pass through the pores in the column beads, while the larger analytes will pass through in the void volume. SEC has been reported for the analysis of Aβ oligomers in several studies. Analytical SEC can also be used for the separation of Aβ monomers from dimers and higher order oligomers, and hence may prove useful for the preparative isolation of well-defined monomers or oligomers for further studies (Welzel et al. 2012). Conflicting results were always obtained while characterizing α- synuclein from various sources using SEC (Fauvet et al. 2012; Podlisny et al. 1998). SEC columns are still being designed for different molecular size ranges although, the approach will normally not be useful in the case of higher order oligomers, polymers, and fibrillar species which will either not enter the bead of the column at all or elute with the void volume without size separation.

Mass Spectrometry and Ion Mobility Mass Spectrometry

Mass spectrometry (MS) separates the molecules on the basis of their mass to charge ratio and the samples are either ionized and volatilized by matrix assisted laser desorption ionization of sample crystallized with matrix molecules or by electrospray ionization of liquid samples from direct injection. MS provides data on aggregate stoichiometry and mass distribution hence it is widely used to characterize and monitor the formation of ionizable and non-covalent oligomers and aggregates in vitro. MS offers certain advantages such as very high selectivity, requires little material, can be used for mapping post-translational modifications and may also be performed directly on tissues including neuropathological samples containing Aβ deposits (Caprioli et al. 1997; Luxembourg et al. 2004; Stoeckli et al. 2006). Electrospray ionization (ESI) and desorption-ESI techniques require molecules to be in solution, which may pose a problem since amyloidogenic proteins aggregate and precipitate out of solution over time; however by using MALDI-MS (Matrix-assisted laser desorption/ionization-MS) insoluble species can also be studied. But sample preparation in MALDI-MS causes disruption of non-covalent complexes and hence yields imprecise results as compared to ESI.

MS in combination with gas-phase electrophoretic molecular analysis allows two dimensional separation approaches where ESI-generated ions are charge-reduced and separated according to their size dependent electrophoretic mobility in air or a gas such as helium in the first dimension. The second dimension is the time-of-flight MS analysis (Woods et al. 2013; Karasek 1974). Thus, conformers with the same masses but different size (e.g., different conformations of the same species) can be separated. Different oligomeric species with identical m/z (mass/charge) value but different absolute size and charge can also be separated, hence this technique is very useful for characterizing very early unfolding events and the prefibrillar aggregates formed during aggregation. It is also useful for studying the interaction of small molecules with Aβ isoforms.

Hydrogen-deuterium (H/D) exchange monitored by MS is a very sensitive approach, used to measure the dynamics of conformational fluctuations in proteins. It exploits the fact that the hydrogens that are shielded by structure or are present in the core, e.g., involved in hydrogen bonding, and exchange much more slowly than the accessible free hydrogens of backbone amides. When any conformational change takes place the hydrogens participating in hydrogen bonding are set free, having been previously inaccessible, and are exposed to solvent and then readily exchange with deuterium in the solvent. This approach helps to differentiate between states with or without protecting structures and has been reported for investigating the unstructured/structured core state of soluble and fibrillar α-synuclein (Vilar et al. 2008; Paslawski et al. 2014) and it was observed that fibrils are more dynamic structures than previously assumed.

Turbidity and Light Scattering Analysis

Turbidity is an optical kinetic method that indicates the presence of suspended particles and can be used to investigate the formation of protein aggregates (with variable size) with a hydrodynamic radius (R_H) greater than the wavelength of the incident light (λ_i) (Mahler et al. 2009). A decrease in the transmitted light/an increase in absorbance or turbidity is observed at wavelength of around 350 nm (since soluble monomeric proteins do not absorb at this wavelength) when pre-fibrillar aggregates and aggregates that are large enough to scatter light are present in the solution. Turbidity measurement offers a single advantage that the protein need not to be labeled, but it is not reliable as it cannot differentiate between different types of aggregates, or even detect the oligomeric intermediates, and hence it must be supported by some other biophysical techniques.

Dynamic light scattering (DLS) is a sensitive and label-free method where the sample is irradiated by a light source and the fluctuations of the scattered light intensity is recorded, to determine the size distribution of formed oligomers, protofibrils and aggregates, as well as aggregation kinetics (Siddiqi et al. 2017a, b, c). The size of aggregates is measured in terms of hydrodynamic radius (R_H) ranging from 1 nm to 4 μm. The fluctuation in the scattering pattern is caused by the Brownian motions of the molecules which are a measure of the molecular diffusion coefficient as is dependent upon the size and shape of the molecule and thus the small molecules diffuse faster than larger molecules so the small molecules will cause less fluctuation in the intensity of the scattered light (Hassan et al. 2014). This important information can be exploited to study the production of oligomeric species with time; to determine the elongation rate and the fibril length as has been reported for Aβ and the huntingtin protein.

Dyes and Extrinsic Fluorescent Probes

Congo Red (CR) dye has been used for almost 100 years for histological detection of amyloid fibrils which exhibits apple-green birefringence under polarized light when bound to amyloid fibrils (Groenning 2010). Although the exact binding mechanism between CR and amyloid fibrils is not well understood and may in fact differ for different proteins, CR can interfere with the aggregation of different amyloid proteins, hence it cannot be used in straightforward manner for in situ monitoring of protein aggregation. This information has been exploited to develop the CR mimics as amyloidosis inhibitors for animal models, to investigate the refining effects of CR in different neurodegenerative diseases (Lorenzo and Yankner 1994; Frid et al. 2007).

The fluorescent dye thioflavin T (ThT) has also been widely used for detecting the presence of amyloid fibrils and exhibits a marked increase in fluorescence upon binding to amyloid fibrils with an excitation maximum at around 440 nm and an emission maximum at around 480 nm, and generally does not affect the aggregation kinetics (Levine Iii 1993; Groenning 2010; Alam et al. 2017a, b). The formation of amyloid fibrils can be probed by monitoring the increase in fluorescence as a function of time. This approach has certain disadvantages as ThT fluorescence intensity is not an absolute measure of the amount of formed amyloid (and indeed is not completely specific for amyloid folds). At high concentration of protein, a saturation effect may be observed and importantly the overall aggregate structure may also be modified due to the interaction between ligand and the protein, such as metal ions and β-sheet antagonists which can alter the overall fluorescence intensity (Fodera et al. 2008). Hence, linearity between fluorescence intensity and protein concentration must be verified by a standard curve to quantify the total amount of aggregated material and the data must be supported by some other techniques.

ANS (1-anilino-naphthalene 8-sulfonate) is a fluorescent dye which is frequently used to characterize the presence of hydrophobic patches on protein molecules as well as for the characterization of protein folding and aggregation intermediates (Nilsson 2004). ANS exhibits increase in the fluorescence intensity accompanied by a blue-shift of the spectral maximum upon binding to the exposed hydrophobic, which are formed due to the conformational changes during protein aggregation. Native proteins exhibit insignificant ANS fluorescence intensity due to the fact that compactly folded native protein typically has hydrophobic regions hidden in the protein core which are inaccessible to the ANS, as reported for human insulin and human serum albumin (Siddiqi et al. 2018a, b).

Imaging Method

Light microscopy readily reveals CR-stained amyloid deposits but the visualization of morphology of insoluble protein deposits needs higher resolution, that can be obtained using ultrastructural imaging techniques, including electron microscopy (EM) and Atomic Force Microscopy (AFM) (Shirahama and Cohen 1967; Stine et al. 1996). The electron microscope was used for the first imaging of amyloid deposits indicating the presence of long unbranched fibrils. Two types of EM have been used, the transmission electron microscope (TEM) and the scanning electron microscope (SEM), which are useful for examining the morphology of amyloid aggregates, especially ordered aggregates such as oligomers, protofibrils and mature fibrillar species (Shirahama and Cohen 1967; Goldsbury et al. 2011).

In classical TEM, a focused electron beam, accelerated at high-voltage, is passed through the specimen in vacuum, resulting in the formation of an image that depends upon the scattering of the electrons by the specimen. Negative staining is sometimes done to enhance the imaging but it gives limited resolution. Cryo-EM can also be utilized which allows the examination of unstained and unfixed specimens, rapidly frozen-hydrated so preserved in vitrified (non-crystalline) water at −180 °C. Using low radiation dose the CryoEM and it has the potential to provide near to atomic resolution in 3D image reconstructions. CryoEM has been utilized for characterization of amyloid fibrils of Aβ, α-synuclein, yeast prion, and the protein polymers of the serpinopathies.

In SEM, the sample is irradiated to a subnanometer 100 kV electron beam that scans the specimen in a systematic fashion which provides scanning EM with analytical capabilities. Thus, the scattered electrons are recorded to get intensities that are directly proportional to the mass of the irradiated region. In such a manner protein masses and mass-per-length data can be determined.

Direct visualization of the 3D-structure (topography) of a solid sample is achieved by AFM. A cantilever with a tip of radius in nanometer under piezo-electrical control scans the surface of a sample either continuously in contact (static mode) or as an oscillating probe (non-contact or tapping mode). The deflections of cantilever are recorded by optical interferometry, laser reflection, or other means and combined into a 3D-image of the specimen surface (Gosal et al. 2006; Goldsbury et al. 2011). AFM provides resolution at subnanometer level and time-lapse investigation of morphology and growth can be performed on amyloid (Aβ, α-synuclein, etc.) aggregates, nanostructures, oligomers, protofibrils, and fibrils. The main disadvantage of AFM is the limited scan area compared to TEM and SEM and slow scanning speed (several minutes vs. seconds).

Circular Dichroism Measurements

Circular dichrosim (CD) is a reliable and sensitive technique, enabling determinations of the conformational changes in the structure of protein and it can give information about the intermediates that are formed during the aggregation process, in terms of increase and decrease in the ellipticity. To monitor the secondary and tertiary structural change of protein during aggregation, CD spectra are recorded in far- (200–250 nm) and near-UV (250–300 nm) range, respectively (Kelly and Price 2000; Siddiqi et al. 2017a, b, c). Usually amyloids are characterized by the presence of a beta sheet rich structure (Sabate and Ventura 2013).

In the far-UV region, the secondary structural changes are indicated in the form of peaks e.g., alpha helical proteins give rise to two negative minima around 208 and 222 nm while beta sheet rich proteins give a single minima around 218 nm (Siddiqi et al. 2018c).

In the near-UV region the two minima are found around 262 and 268 nm in case of alpha helical protein whereas the ellipticity of these minima got decreased in case of beta sheet rich proteins indicating the tertiary structural changes (Siddiqi et al. 2018a, b).

Nuclear Magnetic Resonance Spectroscopy

Nuclear magnetic resonance (NMR) spectroscopy is a versatile and powerful spectroscopic technique which exploits the fact that many atomic nuclei behave like small electromagnets when placed in a strong external magnetic field for probing structures of isolated homogeneous species in protein aggregation pathways. NMR may provide structural information about monomeric, secondary (e.g., identification of β-strand segments and quantification of backbone torsion angles), tertiary (e.g., alignment of β-strands in parallel or antiparallel β-sheets), and quaternary (e.g., relative orientation of β-sheets) structure, dynamics, folding kinetics of purified proteins and the residue-specific structural changes during self-assembly can also be obtained by using different and often complex radiofrequency pulse sequences (Chen 2015).

NMR relaxation is an important solution NMR approach which may provide information on the dynamics and kinetics of the aggregation events, the exchange kinetics between monomeric Aβ and oligomeric Aβ (protofibrils), but it is associated with certain limitations including the size limit which is around 40 kD, and is time consuming for large proteins (Fawzi et al. 2010).

Solid state NMR (ssNMR) is primarily used for the samples that cannot be analyzed by X-ray crystallography because they cannot be crystallized, e.g., for detailed structural determination of insoluble homogenous amyloid fibrils including Aβ, α-Synuclein, and prion protein and ssNMR provides good quality data (Tycko 2006).

Immunochemical Methods

Antibodies have been widely used to quantify specific proteins in complex biological samples by ELISA (enzyme-linked immunosorbent assay), RIA (radioimmunoassay), western blot etc., and the tunable properties of antibodies have also been useful for specific detection of protein oligomers, protofibrils, fibrils and conformers associated with neurodegenerative diseases (MacBeath 2002). As we know that monoclonal antibodies bind to a single epitope on the target protein, thus they can be used for the detection of specific type of conformation (e.g., oligomers). In this way only dimeric or higher order oligomers will bind to the antibody recognition region unless the self-assembly leads to occlusion of the antibody epitope. Monoclonal antibodies can also be designed for protofibrils and amyloid fibrils; e.g., mAb158 for protofibrils and MABN687 for amyloid-β fibrils has already been reported (Haass and Selkoe 2007; Tucker et al. 2015). This method provides us with the advantage that it requires low sample volume and has high specificity.

Toxicity of Oligomer, Protofibrils and Mature Fibrils

Protein aggregation is a dynamic process which involves the formation of transient intermediate structures, as well as dissociation of mature fibrils, making it challenging to define the real culprit leading to toxicity (Verma et al. 2015). The close link of protein aggregation with neurodegeneration was supported by the presence of amyloid plaques rich in Aβ fibrils in the brains of AD patients (Irvine et al. 2008; Guivernau et al. 2016). According to initial studies amyloid fibrils were supposed to be the primary cause of cell death and disease pathogenesis (Pimplikar 2009). It was demonstrated by some experimental approaches, involving the application of extracellular Aβ fibrils which induced AD-like changes in cultured neurons, including the increased frequency of action potential, membrane depolarization and reduced cell viability (Kress and Mennerick 2009). Cognitive decline and the neuronal cell death was also observed as a result of injecting Aβ fibrils into primary hippocampal neurons in rat-impaired synaptic transmission (Stephan et al. 2001). However, there has always been a controversy regarding the correlation between amyloid plaque burden and severity of neuronal loss and other AD-symptoms (Morris et al. 2014). Increasing number of recent studies suggests that prefibrillar species (especially oligomers) formed during early fibrillization are more toxic species than the mature fibrils, as reported by Hey et al. who demonstrated the significant impairment of learning and memory functions on infusing oligomeric Aβ into left ventricles of brain, while no significant effect was observed due to Aβ fibrils (Dasari et al. 2011; Kitazawa et al. 2012). Moreover, Aβ oligomers were shown to suppress synaptic plasticity by specifically inhibiting presynaptic P/Q calcium currents, unlike protein monomers and fibrils (Mezler et al. 2012). Instead in some instances, formation of mature fibrils is believed to be protective by acting as a harmless reservoir of toxic oligomers, but it can also result in the release of off pathway oligomers which are themselves toxic (Hubin et al. 2014). Hence, the original amyloid cascade hypothesis has been modified to include the intermediary species formed during the aggregation process and is named as oligomer hypothesis (Musiek and Holtzman 2015). Thus we can say that prefibrillar species (such oligomers and protofibrils) are relatively more toxic than mature fibrils.

The differential toxicity of amyloid oligomers and fibrils can be the result of their different structural arrangement:

1. Hydrophobic surfaces are exposed in oligomers comprising of β-sheets while they are hidden inside the interacting stacks in fibrils (Stefani 2010a, b).
2. Because of their small size oligomers can easily diffuse in tissues as compared to larger fibrils (Sengupta et al. 2016).
3. Improved interaction of oligomers with cellular target is attributed to the presence of a greater number of open active ends compared to the fibrils (Stefani 2012).

4. Oligomers are rich in disordered structures hence unstable whereas fibrils are stable organized molecules (Chen et al. 2015).

From the research that has been reviewed here, it can be concluded that there is a link between protein aggregation and disease pathogenesis (Valastyan and Lindquist 2014; Spires-Jones et al. 2017). However, to get a clear and detailed insight into the protein aggregation process and to provide the definite clinical correlation of disease stage or severity associated with oligomeric intermediates and mature fibrils there is an urgent need to developing of in vivo imaging techniques.

Cellular Mechanism of Toxicity

Membrane Interaction

The research done todate indicates that oligomers, soluble in intercellular fluid and forming at an earlier stage, are the most toxic form of the peptide (Li et al. 2008; Valincius et al. 2008). These toxic oligomers may cross the blood brain barrier and hence invade the brain long before large-scale brain damage becomes apparent (Zheng 2001; Brkic et al. 2015). But the question arises concerning the mechanism involved behind the toxicity of amyloid oligomers (Kayed and Lasagna-Reeves 2013). It is actually the cell membrane which is the main target of these oligomers. These oligomers have the ability to penetrate and disrupt the cell membrane and thus induce toxicity by directly interacting with the lipid bilayers and this depends upon the physicochemical properties of both the oligomers and the membranes (Table 16.2) i.e., lipid composition of the latter (Arispe and Doh 2002). For example, in case of Alzheimer's disease GM1 (ganglioside) has been observed to be an important factor (Yanagisawa 2007). A variety of functions including, cell type specific markers, as differentiation and developmental markers, and as receptors and mediators of cell adhesion have been reported to be served by gangliosides which comprise about 5–10% of the outer membrane leaflet

Table 16.2 Oligomer with different property shows variable cytotoxicity depending upon the abundance of membrane cholesterol and GM1

		Oligomer type	
		Loose oligomers, high hydrophobic exposure	Compact oligomers, low hydrophobic exposure
Membrane cholesterol	Low	Highly toxic	Toxic
	Basal	Toxic	Non toxic
	High	Non toxic	Non toxic
Membrane GM1	Low	Non toxic	Non toxic
	Basal	Toxic	Non toxic
	High	Highly toxic	Toxic

(Robert et al. 2012). According to previous studies, it has been found that there is an abundance of GM1, together with cholesterol and sphingomyelin in lipid rafts, domains within the cell membrane that contain a vast array of membrane proteins including channels and receptors (Crane and Tamm 2004; de Chaves and Sipione 2010). Neuronal dysfunction and neurodegenerative condition has actually been associated with the perturbation of membrane regions containing lipid rafts, as the neuronal membranes are enriched with the latter (Kakio et al. 2002). Alteration in the membrane distribution of lipids such as GM1 and GM2 has recently been reported in case of Alzheimer's condition indicating that lipid raft distribution plays an important role in order to provide protection to the mature neurons against oligomer cytotoxicity, as they can exhibit their detergent like behavior by directly removing lipid molecules from the membrane hence destabilizing it (Pernber et al. 2012). As a result of such strong interaction of oligomers with the cell membrane the leakage of nerve cells take place making them unable to maintain the ion potential between the inside and outside of cells, thus hampering some essential processes including information processing, and consequently the central nervous system function (Lodish et al. 2000). Amyloid fibrils are now considered less toxic to cells than oligomers and other prefibrillar aggregates, but the toxicity can emerge at any point of time as a result of the fibril breakage with the generation of shorter fibrils or leakage of some toxic species depending upon their structural and stability properties and the environmental conditions where they grow and become deposited, and the same has been observed under in vivo conditions (Xue et al. 2009; Stefani 2010a, b).

Fibril formation occurs as a result of the rearrangement of the hydrogen bonds along with the backbone of the protein which may lead to the exposure of the hydrophobic interior and other reactive groups including disulfides to the environment (Wang et al. 2014). These exposed domains enable fibrils to interact with the lipid bilayer and to induce oxidative stress leading to cell membrane disruption (Huang et al. 2009). Fibrils and protofibrils also cause the aggregation of the membrane protein via intermolecular disulfide cross linking which is a very strong interaction that cannot be reversed even by treating with any of the anti-aggregating agents, including metal ion chelators, radical scavengers and antioxidants (Schild and Reiser 2005; Jomova and Valko 2011).

Perturbation of Calcium Homeostasis

Although much of the work has been done on the role of oligomers in inducing neurotoxicity, the underlying mechanism is still not fully understood. However a glimpse of the whole mechanism has been revealed including mitochondrial dysfunction and lysosomal failure that may or may not directly convoy a general neuronal cell derangement (a common effect of Firstly, (Marambaud et al. 2009; Kawamata and Manfredi 2010).

Two main mechanisms have been proposed in order to explain the inception of calcium homeostasis dysregulation associated with various protein misfolding diseases.

Firstly, *Pre-existing ion channel activation*—Amyloid proteins have the ability to interact with the calcium permeable channels causing a rise in the intracellular calcium level (Lal et al. 2007; Hermes et al. 2010). Glutamate receptors have been shown to be involved in the excitotoxic neuronal cell damage as their overstimulation is the reason for intracellular calcium rise and has been reported for amyloid beta protein oligomer. The same is true for the other amyloidogenic proteins too (Dong et al. 2009; Kondratskyi et al. 2013). Amyloids also causes the degradation of dynamin 1 that hampers the synaptic integrity as a result of NMDA receptor activation (Malchiodi-Albedi et al. 2011; Sebollela et al. 2012) and secondly, the *Calcium permeable pore hypothesis*—Amyloids such as protofibrils and oligomers are surface active moieties, and they are more susceptible towards forming pore-like assemblies in the plasma membrane of neuronal cells (Di Scala et al. 2016). These pores act as calcium selective channels responsible for dysregulated entry of calcium into the cytoplasm, causing rise in the intracellular calcium level. The mechanism behind pore formation is still unknown but cholesterol has been observed to play an important role in their formation (Magi et al. 2016).

Excessive calcium in the cell causes the activation of cascades leading to programmed cell death by forming permeable transition pores in the mitochondria. This triggers the collapse of transmembrane potential and thus the release of cytochrome c takes place ultimately resulting in apoptosis (Lemasters et al. 1998). Excessive Ca^{2+} signals can also cause the degradation of structural and enzymatic proteins by directly activating Ca^{2+}-binding calpain proteases (Ono and Sorimachi 2012). Cell death can be initiated after a prolonged period of subtle changes in Ca^{2+} homeostasis as well as rapidly. Ca^{2+} dysregulation is a complex, dynamic process that moderately impairs neuronal mitochondria, the ER (endoplasmic reticulum), the plasma membrane and signal transduction processes specifying the interplay between Ca^{2+} homeostasis, signaling and other signal transduction networks (Gleichmann and Mattson 2011; Missiroli et al. 2018).

Intermediary Oligomers as Potential Biomarkers

The debate on the correlation between oligomers or mature fibrils with disease severity in amyloidopathies has led to the conclusion that amyloid oligomers are more toxic than mature fibrils and play an important role in the propagation of disease. They are the intermediates formed during the process of protein folding and their presence is witnessed in case of Parkinson's disease and Alzheimer's disease (Duan and Kollman 1998; Haass and Selkoe 2007). Hence oligomer count in the cerebrospinal fluid (CSF) or plasma can be exploited as a potential diagnostic marker for amyloidopathies, because the population of oligomeric species is elevated in the body fluid samples of patients, as compared to controls

(Blennow et al. 2010; Duan and Kollman 1998). Currently, the development of delicate and specific assays to detect the presence of oligomers has become the center of attraction for biophysicists, in order to design strategies to combat the amyloidopathies, either by halting the propagation of disease (growth of oligomers) or by clearing oligomers from the body. Some approaches were mentioned in the above sections, including immunoassay using monoclonal antibodies, etc., but still there is an immense need to readily develop less time consuming, sensitive and specific techniques to benefit mankind.

Amyloid Inhibition Strategies

In general all the steps falling in the pathway of protein aggregation can be targeted for blocking disease onset and or progression, but still no precise target has been identified that once inhibited will quantitatively block the progression of protein aggregation. This is due to the certain factors making it more complex, including (1) lack of knowledge of the molecular mechanisms for protein aggregation, (2) few promoting agents (either exogenous or endogenous) or events have been identified, (3) enzymes and chaperones involved in disease onset also have physiological roles, precluding their inhibition, (4) more than one single specific event could be involved in the initial and subsequent steps, and (5) the lack of specific in vitro and in vivo assays suitable for mimicking the complex human pathological scenario and for monitoring the formation of different prefibrillar species (Bartolini and Andrisano 2010; Guise et al. 1996; Thomas et al. 1997; Morais-de-Sa et al. 2006). However, as a consequence of growing efforts and interest towards understanding the aggregation process, some potential targets have been identified and characterized and so several new approaches are now available.

Inhibition of the Accumulation of Peptide/Protein Monomers

In the development of aggregation process, the first and foremost risk factor is an increase in the monomer concentration which includes: (1) naturally occurring proteins that can accumulate in an unusual manner under pathological conditions (e.g., β-amyloid peptides in sporadic forms of AD), and also the abnormal forms of native proteins pathologically generated either: (2) by specific enzymes (for example, hyperphosphorylated tau protein in tauopathies) or: (3) as a result of point mutations in the encoding genes. If the mature protein is released as a result of the cleavage of large precursor protein, inhibiting the cleaving enzyme may prove to be a potential therapeutic strategy for preventing pathological protein accumulation and aggregation. One example is amyloid β-peptide, generated as a result of cleavage of amyloid precursor protein (APP) by enzymes including alpha secretase, beta secretase and gamma secretase and thus the toxicity is determined depending on the enzymes involved in APP processing (Vassar et al. 1999).

Targeting Protein Misfolding

As reported earlier the aggregation diseases are due to the conformational changes (a key feature) from native state to the cross beta sheet rich structure (Chaudhuri et al. 1996). It is the earliest step in protein aggregation and hence inhibition of such changes that can provide us with a promising strategy for blocking the same. A number of potential strategies have been applied.

Stabilization of Native Conformation (Induction of Physiological Chaperones)

The normal folding and transfer of misfolded proteins to the proteasome for degradation is the physiological function of chaperones in cells and thus their activation may enhance the cell's natural ability to clear damaged or misfolded proteins. For example, in vitro studies have demonstrated that heat shock protein 70 (HSP70) is able to interact with soluble α-synuclein monomers thus preventing their Oligomerization (Huang et al. 2006).

Interference with the Transition Process (Peptide Based Inhibitors and Small Synthetic Compounds)

Beta sheet breakers are actually a modified core sequences of the target protein, having an affinity for the native protein and block conformational changes. The interaction of beta sheet breaker with the target sequence should prevent the formation of oligomeric species and the progression of the diseases (Viet et al. 2011). These are obtained as a result of some modifications of core domain such as by proline insertion, conjugation with a polycationic sequence, insertion of non-natural amino acids and backbone modification through alternation of amide hydrogen atoms with methyl groups. But their use is associated with a limitation that they have a very short half-life in the bloodstream, because of enzymatic degradation.

The use of small organic compounds that are capable of stabilizing and destabilizing unfolded monomer conformation and misfolded small oligomers, respectively, overcomes the problem of stability and their modes of action is similar to those of natural chaperones, and they are able to cross the blood brain barrier (for example, Epigallocatechin-3-gallate) (Brown et al. 1996).

Inhibition of Pathological Chaperones

Pathological chaperones are a group of proteins that promote the conformational changes of disease specific proteins and stabilize the aggregation prone conformations (Hartl et al. 2011). Inhibition of their action is also a potential strategy for blocking aggregation at an early stage. Chaperone mimetics and peptide mimetics have similarly been explored to interfere competitively with the interactions between the soluble monomers and the pathological chaperones.

Reduction of Soluble Oligomers

Inhibition of Oligomerisation Pathway

Exploration of the key role played by prefibrillar soluble aggregates has led to the development of therapeutic approaches focusing on the direct interference with the protein oligomerization and aggregation. However, blocking the oligomerization path will stop the initial protein misfolding and may in turn intensify the accumulation of toxic oligomers and thus the "correct" inhibition of protein oligomerization still poses a challenging task for scientists. But, the development of immunotherapy against the fibrillar forms and large oligomers blockage has been a revolution, as it represents a general approach to boost the clearance of protein aggregates. Antibodies against the pathological conformers may clear the soluble toxic species, preventing their accumulation and prevent cell death (Lindstrom et al. 2014).

Over-Acceleration of Oligomer Aggregation

A new strategy has been explored based on the idea that accelerating, rather than inhibiting the process of oligomerization, and thus inducing inclusion formation could be beneficial, as it will reduce the level of soluble toxic intermediate available for propagation of the disease process, so reducing the extent of toxicity. The potential positive effects of such a strategy were evaluated for the cytoprotective role of inclusions in HD and PD (Bodner et al. 2006; Lindstrom et al. 2014; Sengupta et al. 2016). This led to the selection of the compound 5-[4-(4-chlorobenzoyl)piperazin-1-yl]-8-nitroquinoline, which promoted the formation of larger inclusions, while reducing toxicity in cellular disease models of HD and PD simultaneously.

Promotion of Fibril Clearance

Regulation of Ubiquitin Proteasome System and Autophagy

In general misfolded cytosolic proteins are removed from the cells via two main pathways: the ubiquitin-proteasome system (UPS) (Fig. 16.5) and the autophagy-lysosomal pathways (Ding and Yin 2008). Degradation of short-lived misfolded or damaged soluble proteins is mainly done by the UPS and it may play a major role in removing aberrant monomers, whereas the degradation of long-lived, stable proteins is primarily handled by autophagy (Fig. 16.4). Therefore when refolding is not possible, the UPS is the ideal target for the removal of soluble short misfolded proteins/peptides, whereas enhancers of autophagy favors the clearance of larger toxic oligomeric complexes that accumulate with the progression of the fibrillation pathway. Indeed, both the UPS and autophagy can be targeted directly or indirectly to overcome misfolded protein accumulation.

Activation of Specific Proteases or Clearing Enzymes

Clearance of specific deposits can be achieved by certain specific proteolytic enzymes. For example, in case of polyQ diseases accumulation of polyQ peptides in the cell can be prevented by up-regulation of puromycin-sensitive aminopeptidase (PSA), the major peptidase responsible for the digestion of polyQ sequences released by proteasomes during protein degradation (Bhutani et al. 2007). The same has been observed in case of Huntington disease where PSA causes the degradation

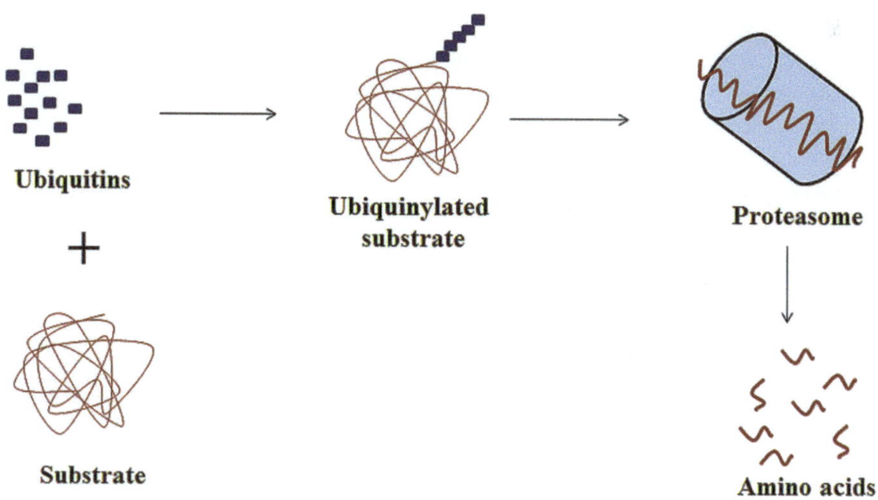

Fig. 16.5 Illustration of proteasome mediated degradation of misfolded proteins

of tau proteins in human brains whereas in Alzheimer's disease several proteases have been identified in the clearance of Aβ aggregates, including neprilysin (NEP), the insulin degrading enzyme (IDE), endothelin-converting enzyme and plasmin (McDermott and Gibson 1997; Hersh and Rodgers 2008). Their important role has been highlighted through the experimental evidence revealing the Aβ accumulation and aggregate formation in vivo under decreased protease expression and their up-regulation may provide us with a novel and viable therapeutic strategy for AD.

Therapeutics Under Clinical Trials

At present, the drugs that are available to treat the amyloid diseases are symptomatic in nature, i.e. they improve the symptoms but do not cure the disease (Salomone et al. 2012; Schelterns and Feldman 2003; Porat et al. 2006). Therefore, disease-modifying therapies (DMTs) that will prevent or delay the onset or slow the progression of amyloid diseases, particularly AD, are urgently needed. Medications to effectively improve cognition or ameliorate neuropsychiatric symptoms of patients in the symptomatic phases of AD are needed to improve memory and behavior (Cummings et al. 1998). However, understanding the amyloid cascade with respect to different intermediate species such as monomers, oligomers, protofibrils and mature fibrils, led to the development of new strategies for therapeutics and diagnostics. Here we have provided a summary of the some of the test agent (based on their mechanism of action, Tables 16.3, 16.4 and 16.5) which are under clinical trial (I, II and III) for the treatment of AD (Cummings et al. 2018).

Table 16.3 Summary of test agent in phase III of Alzheimer's disease drug development (Cummings et al. 2018)

	Mechanism of action	Therapeutic purpose
Aducanumab	Monoclonal antibody	Remove amyloid
Albumin + immunoglobulin	Polyclonal antibody	Remove amyloid
Crenezumab	Monoclonal antibody	Remove amyloid
Escitalopram	Serotonin reuptake inhibition	Improve neuropsychiatric symptoms (agitation)
Gantenerumab and solanezumab and JNJ-54861911	Monoclonal antibody, BACE inhibitor	Remove amyloid/reduce amyloid production
Icosapent ethyl (IPE)	Purified form of the omega-3 fatty acid EPA	Protect neurons from disease pathology
Insulin intranasal (Humulin)	Replace insulin in the brain	Enhance cell signaling and neurogenesis (cognitive enhancer)
Nabilone	Cannabinoid (receptor agent)	Improve neuropsychiatric symptoms (agitation)

(continued)

Table 16.3 (continued)

	Mechanism of action	Therapeutic purpose
Octohydroaminoacridine succinate	Acetylcholinesterase inhibitor	Improve acetylcholine signaling (cognitive enhancer)
GV-971 (Sodium Oligo-mannurarate)	Inhibit amyloid aggregation	Remove amyloid plaque load
TRx0237 (LMTX)	Tau protein aggregation inhibitor	Reduce tau-mediated neuronal damager
TTP488 (azeliragon)	RAGE antagonist	Reduce amyloid uptake in brain and lower inflammation in glial cells (DMT/cognitive enhancer)
Zolpidem	Positive allosteric modulator of GABA-A receptors	Improve neuropsychiatric symptoms (sleep disorders)

Table 16.4 Summary of the test agent in phase II of Alzheimer's disease drug development (Cummings et al. 2018)

	Mechanism of action	Therapeutic purpose
Methylene Blue	Tau protein aggregation inhibitor	Reduce neurofibrillary tangle formation
AstroStem	Autologous adipose tissue derived mesenchymal stem cells	Regenerate neurons
Atomoxetine	Norepinephrine reuptake inhibitor	Improve neurotransmission (cognitive enhancer) and improve behavioral symptoms
Benfotiamine	Synthetic thiamine (B1)	Improve multiple cellular processes (cognitive enhancer)
Bryostatin 1	Protein kinase C modulator	Improve cellular processes (cognitive enhancer) and reduce amyloid pathology (DMT)
Candesartan	Angiotensin receptor blocker	Improve vascular functioning and effects on amyloid pathology (DMT)
Cilostazol	Phosphodiesterase 3 antagonist	Regulate cAMP and improve synaptic function (cognitive enhancer)
Crenezumab	Monoclonal antibody	Remove amyloid
Deferiprone	Iron chelating agent	Reduce reactive oxygen species that damage neurons; effect on amyloid and BACE pathology
Dronabinol	CB1 and CB2 endocannabinoid receptor partial agonist	Improve neuropsychiatric symptoms (agitation)
Formoterol	b2 adrenergic receptor agonist	Effects on multiple cellular pathways

(continued)

Table 16.4 (continued)

	Mechanism of action	Therapeutic purpose
Insulin detemir (intranasal)	Increases insulin signaling in the brain	Enhance cell signaling and growth (DMT)
Lithium	Ion channel modulator	Improve neuropsychiatric symptoms (agitation, mania, psychosis)
L-Serine Amino acid	Amino acid	Stabilize protein misfolding

Table 16.5 Summary of test agent in phase I of Alzheimer's disease drug development (Cummings et al. 2018)

	Mechanism of action	Therapeutic purpose
Aducanumab	Monoclonal antibody	Remove amyloid
Bisnorcymserine	Butyrylcholinesterase inhibitor	Acetylcholine neurotransmision
Crenezumab	Monoclonal antibody	Remove amyloid
Insulin aspart (Intranasal)	Increases insulin signaling in the brain	Enhance cell signaling and growth (cognitive enhancer)
Oxaloetate (OAA)	Mitochondrial enhancer	Enhance multiple cellular processes
Salsalate	Non-steroidal anti-inflammatory	Reduce neuronal injury
Telmisartan	Angiotensin II receptor blocker, PPAR-gamma Agonist	Improve vascular functioning and effects on amyloid pathology (DMT)
Vorinostat	Histone deacetylase Inhibitor	Enhance multiple cellular processes including tau aggregation and amyloid deposition
NDX-1017	Hepatocyte growth factor	Regenerate neurons
NPT088	IgG1 Fc-GAIM fusion protein	Clear amyloid and tau

Conclusion

In summary, the information reviewed in this article indicates that several intermediate species appear along the path of amyloid fibrillation and specific conformational structures are necessary to form entities rich in cross-beta-sheets. Now it is well established that oligomers are the main toxic species. Therefore, detailed oligomer characterization is the most requirement to develop novel therapeutics. Recent advancement in microscopic techniques including cryo-EM has led to the definition of oligomeric/protofibrills structures in atomic details, some proteins. Designing of inhibitory molecules that target the oligomeric/protofibrill species and reduction of their toxicity or increase the process of fibrils formation remain to be investigated in greater detail. There is a long way to go before we can find a permanent treatment for amyloid diseases.

Acknowledgements Facilities provided by Interdisciplinary Biotechnology Unit, Aligarh Muslim University, Aligarh are gratefully acknowledged. For providing financial assistance, M.K. S. is thankful to Department of Biotechnology (DBT), New Delhi, India, P.A. and N.M. are thankful to Council of Scientific and Industrial Research (CSIR), New Delhi, India, S.M. is thankful to Indian Council of Medical Research (ICMR), New Delhi, India. R.H.K. is thankful to CSIR and UGC for project referenced as 37(1676)/17/EMR—II and F. 19-219/2018, respectively.

Conflict of interest
The authors declare that they have no conflicts of interest with the contents of this article.

References

Alam P, Beg AZ et al (2017a) Ascorbic acid inhibits human insulin aggregation and protects against amyloid induced cytotoxicity. Arch Biochem Biophys 621:54–62

Alam P, Siddiqi MK et al (2017b) Vitamin B12 offers neuronal cell protection by inhibiting Aβ-42 amyloid fibrillation. Int J Biol Macromol 99:477–482

Arispe N, Doh M (2002) Plasma membrane cholesterol controls the cytotoxicity of Alzheimer's disease AβP (1–40) and (1–42) peptides. FASEB J 16(12):1526–1536

Bagriantsev SN, Kushnirov VV et al (2006) Analysis of amyloid aggregates using agarose gel electrophoresis. Methods Enzymol 412:33–48

Bartolini M, Andrisano V (2010) Strategies for the inhibition of protein aggregation in human diseases. ChemBioChem 11(8):1018–1035

Bhutani N, Venkatraman P et al (2007) Puromycin-sensitive aminopeptidase is the major peptidase responsible for digesting polyglutamine sequences released by proteasomes during protein degradation. EMBO J 26(5):1385–1396

Bitan G, Lomakin A et al (2001) Amyloid β-protein oligomerization. I. Prenucleation interactions revealed by photo-induced cross-linking of unmodified proteins. J Biol Chem

Bitan G, Fradinger EA et al (2005) Neurotoxic protein oligomers—what you see is not always what you get. Amyloid 12(2):88–95

Blennow K, Hampel H et al (2010) Cerebrospinal fluid and plasma biomarkers in Alzheimer disease. Nat Rev Neurol 6(3):131

Bodner RA, Outeiro TF et al (2006) Pharmacological promotion of inclusion formation: a therapeutic approach for Huntington's and Parkinson's diseases. Proc Natl Acad Sci 103 (11):4246–4251

Braak H, Braak E (1997) Frequency of stages of Alzheimer-related lesions in different age categories. Neurobiol Aging 18(4):351–357

Brkic M, Balusu S et al (2015) Amyloid β oligomers disrupt blood's CSF barrier integrity by activating and detection. Process Biochem 51(9):1183–1192

Brown CR, Hong-Brown LQ et al (1996) Chemical chaperones correct the mutant phenotype of the Î"F508 cystic fibrosis transmembrane conductance regulator protein. Cell Stress Chaperon 1(2):117

Brown PH, Balbo A et al (2008) Characterizing protein - protein interactions by sedimentation velocity analytical ultracentrifugation. Curr Protoc Immunol 81(1):18.15.1–18.15.39

Bullock J (1993) Application of capillary electrophoresis to the analysis of the oligomeric distribution of polydisperse polymers. J Chromatogr A 645(1):169–177
Burdick D, Soreghan B et al (1992) Assembly and aggregation properties of synthetic Alzheimer's A4/beta amyloid peptide analogs. J Biol Chem 267(1):546–554
Cantor CR, Schimmel PR (1981). Biophysical chemistry: part II - Techniques for the study of biological structure and function. Biochem Educ 1:157–157
Caprioli RM, Farmer TB, Gile J (1997) Molecular imaging of biological samples: localization of peptides and proteins using MALDI-TOF MS. Anal Chem 69(23):4751–4760
Chaudhuri JB, Batas B et al (1996) Improving protein refolding yields by minimizing aggregation. Ann N Y Acad Sci 782(1):495–505
Chaturvedi SK, Siddiqi MK, Alam P, Khan RH (2016) Protein misfolding and aggregation: mechanism, factors and detection. Process Biochemi 51(9):1183–1192
Chen L (2015) De novo protein structure modeling and energy function design. Old Dominion University
Chen SW, Drakulic S et al (2015) Structural characterization of toxic oligomers that are kinetically trapped during α-synuclein fibril formation. Proc Nat Acad Sci 201421204
Chiti F, Dobson CM (2006) Protein misfolding, functional amyloid, and human disease. Annu Rev Biochem 75:333–366
Coalier KA, Paranjape GS et al (2013) Stability of early-stage amyloid-β (1–42) aggregation species. Biochim et Biophys Acta (BBA)-Proteins Proteomics 1834(1):65–70
Crane JM, Tamm LK (2004) Role of cholesterol in the formation and nature of lipid rafts in planar and spherical model membranes. Biophys J 86(5):2965–2979
Cummings J, Lee G et al (2018) Alzheimer's disease drug development pipeline: 2018. Alzheimer's & Dementia: Translational Research & Clinical Interventions
Cummings JL, Vinters HV et al (1998) Alzheimer's disease etiologies, pathophysiology, cognitive reserve, and treatment opportunities. Neurology 51(1 Suppl 1):S2–S17
Damaschun G, Damaschun H et al (1999) Proteins can adopt totally different folded conformations. J Mol Biol 291(3):715–725
Dasari M, Espargaro A et al (2011) Bacterial inclusion bodies of Alzheimer's Disease β-amyloid peptides can be employed to study native-like aggregation intermediate states. ChemBioChem 12(3):407–423
de Chaves EP, Sipione S (2010) Sphingolipids and gangliosides of the nervous system in membrane function and dysfunction. FEBS Lett 584(9):1748–1759
de la Paz ML, Serrano L (2004) Sequence determinants of amyloid fibril formation. Proc Natl Acad Sci 101(1):87–92
Di Scala C, Yahi N et al (2016) Common molecular mechanism of amyloid pore formation by Alzheimer's β-amyloid peptide andα-synuclein. Sci Rep 6:28781
Ding W-X, Yin X-M (2008) Sorting, recognition and activation of the misfolded protein degradation pathways through macroautophagy and the proteasome. Autophagy 4(2):141–150
Dong X-X, Wang Y et al (2009) Molecular mechanisms of excitotoxicity and their relevance to pathogenesis of neurodegenerative diseases. Acta Pharmacol Sin 30(4):379
Duan Y, Kollman PA (1998) Pathways to a protein folding intermediate observed in a 1-microsecond simulation in aqueous solution. Science 282(5389):740–744
Eisenberg D, Jucker M (2012) The amyloid state of proteins in human diseases. Cell 148(6):1188–1203
Fauvet B, Kamdem MM et al (2012) Alpha-synuclein in the central nervous system and from erythrocytes, mammalian cells and *E. coli* exists predominantly as a disordered monomer. J Biol Chem: jbc. M111 318949
Fawzi NL, Ying J et al (2010) Kinetics of amyloid β monomer-to-oligomer exchange by NMR relaxation. J Am Chem Soc 132(29):9948–9951
Fink AL (1998) Protein aggregation: folding aggregates, inclusion bodies and amyloid. Fold Des 3 (1):R9–R23
Fodera V, Librizzi F et al (2008) Secondary nucleation and accessible surface in insulin amyloid fibril formation. J Phys Chem B 112(12):3853–3858

Forloni G, Angeretti N et al (1993) Neurotoxicity of a prion protein fragment. Nature 362 (6420):543
Frid P, Anisimov SV et al (2007) Congo red and protein aggregation in neurodegenerative diseases. Brain Res Rev 53(1):135–160
Frydman J (2001) Folding of newly translated proteins in vivo: the role of molecular chaperones. Annu Rev Biochem 70(1):603–647
Gabrielson JP, Arthur KK et al (2011) Precision of protein aggregation measurements by sedimentation velocity analytical ultracentrifugation in biopharmaceutical applications. Anal Biochem 396(2):231–241
Glabe CG (2008) Structural classification of toxic amyloid oligomers. J Biol Chem 283 (44):29639–29643
Gleichmann M, Mattson MP (2011) Neuronal calcium homeostasis and dysregulation. Antioxid Redox Signal 14(7):1261–1273
Goldsbury C, Baxa U et al (2011) Amyloid structure and assembly: insights from scanning transmission electron microscopy. J Struct Biol 173(1):1–13
Gosal WS, Myers SL et al (2006) Amyloid under the atomic force microscope. Protein Pept Lett 13(3):261–270
Groenning M (2010) Binding mode of Thioflavin T and other molecular probes in the context of amyloid fibrils—current status. J Chem Biol 3(1):1–18
Guise AD, West SM et al (1996) Protein folding in vivo and renaturation of recombinant proteins from inclusion bodies. Mol Biotechnol 6(1):53–64
Guivernau B, Bonet J et al (2016) Amyloid-β peptide nitrotyrosination stabilizes oligomers and enhances nmdar-mediated toxicity. J Neurosci 36(46):11693–11703
Haass C, Selkoe DJ (2007) Soluble protein oligomers in neurodegeneration: lessons from the Alzheimer's amyloid β-peptide. Nat Rev Mol Cell Biol 8(2):101
Hartl FU, Bracher A et al (2011) Molecular chaperones in protein folding and proteostasis. Nature 475(7356):324
Hassan PA, Rana S et al (2014) Making sense of Brownian motion: colloid characterization by dynamic light scattering. Langmuir 31(1):3–12
Hermes M, Eichhoff G et al (2010) Intracellular calcium signalling in Alzheimer's disease. J Cell Mol Med 14(12):30–41
Hersh LB, Rodgers DW (2008) Neprilysin and amyloid beta peptide degradation. Curr Alzheimer Res 5(2):225–231
Huang C, Cheng H et al (2006) Heat shock protein 70 inhibits α-synuclein fibril formation via interactions with diverse intermediates. J Mol Biol 364(3):323–336
Huang B, He J et al (2009) Cellular membrane disruption by amyloid fibrils involved intermolecular disulfide cross-linking. Biochemistry 48(25):5794–5800
Hubin E, Van Nuland NAJ et al (2014) Transient dynamics of Aβ contribute to toxicity in Alzheimer's disease. Cell Mol Life Sci 71(18):3507–3521
Irvine GB, El-Agnaf OM et al (2008) Protein aggregation in the brain: the molecular basis for Alzheimer's and Parkinson's diseases. Mol Med 14(7–8):451–464
Jomova K, Valko M (2011) Importance of iron chelation in free radical-induced oxidative stress and human disease. Curr Pharm Des 17(31):3460–3473
Kakio A, Nishimoto S-I et al (2002) Interactions of amyloid β-protein with various gangliosides in raft-like membranes: importance of GM1 ganglioside-bound form as an endogenous seed for Alzheimer amyloid. Biochemistry 41(23):7385–7390
Karasek FW (1974) Plasma chromatography. Anal Chem 46(8):710A–720a
Kawamata H, Manfredi G (2010) Mitochondrial dysfunction and intracellular calcium dysregulation in ALS. Mech Ageing Dev 131(7–8):517–526
Kayed R, Lasagna-Reeves CA (2013) Molecular mechanisms of amyloid oligomers toxicity. J Alzheimers Dis 33(s1):S67–S78
Kelly SM, Price NC (2000) The use of circular dichroism in the investigation of protein structure and function. Curr Protein Pept Sci 1(4):349–384

Kim H-Y, Cho M-K et al (2009) Structural properties of pore-forming oligomers of α-synuclein. J Am Chem Soc 131(47):17482–17489
Kirkitadze MD, Bitan G et al (2002) Paradigm shifts in Alzheimer's disease and other neurodegenerative disorders: the emerging role of oligomeric assemblies. J Neurosci Res 69 (5):567–577
Kitazawa M, Medeiros R et al (2012) Transgenic mouse models of Alzheimer disease: developing a better model as a tool for therapeutic interventions. Curr Pharm Des 18(8):1131–1147
Knowles TPJ, Vendruscolo M et al (2014) The amyloid state and its association with protein misfolding diseases. Nat Rev Mol Cell Biol 15(6):384
Kondratskyi A, Yassine M et al (2013) Calcium-permeable ion channels in control of autophagy and cancer. Front Physiol 4:272
Kress GJ, Mennerick S (2009) Action potential initiation and propagation: upstream influences on neurotransmission. Neuroscience 158(1):211–222
Lal R, Lin H et al (2007) Amyloid beta ion channel: 3D structure and relevance to amyloid channel paradigm. Biochim et Biophys Acta (BBA)-Biomembr 1768(8):1966–1975
Lawrence GJ, Payne PI (1983) Detection by gel electrophoresis of oligomers formed by the association of high-molecular-weight glutenin protein subunits of wheat endosperm. J Exp Bot 34(3):254–267
Lemasters JJ, Nieminen AL et al (1998) The mitochondrial permeability transition in cell death: a common mechanism in necrosis, apoptosis and autophagy. Biochim et Biophys Acta (BBA)-Bioenerg 1366(1–2):177–196
Levine Iii H (1993) Thioflavin T interaction with synthetic Alzheimer's disease β-amyloid peptides: detection of amyloid aggregation in solution. Protein Sci 2(3):404–410
Levine Iii H (2004) Alzheimer's β-peptide oligomer formation at physiologic concentrations. Anal Biochem 335(1):81–90
Lewandowski JZR, van der Wel PCA et al (2011) Structural complexity of a composite amyloid fibril. J Am Chem Soc 133(37):14686–14698
Li D-W, Mohanty S et al (2008) Formation and growth of oligomers: a Monte Carlo study of an amyloid tau fragment. PLoS Comput Biol 4(12):e1000238
Lindstrom V, Fagerqvist T et al (2014) Immunotherapy targeting α-synuclein protofibrils reduced pathology in (Thy-1)-h [A30P] α-synuclein mice. Neurobiol Dis 69:134–143
Lodish H, Berk A et al (2000) The action potential and conduction of electric impulses
Lorenzo A, Yankner BA (1994) Beta-amyloid neurotoxicity requires fibril formation and is inhibited by congo red. Proc Natl Acad Sci 91(25):12243–12247
Luxembourg SL, Mize TH et al (2004) High-spatial resolution mass spectrometric imaging of peptide and protein distributions on a surface. Anal Chem 76(18):5339–5344
MacBeath G (2002) Protein microarrays and proteomics. Nat Genet 32(4s):526
Magi S, Castaldo P et al (2016) Intracellular calcium dysregulation: implications for Alzheimer's disease. BioMed Res Int
Mahdavimehr M, Katebi B et al (2018) Effect of fibrillation conditions on the anti-amyloidogenic properties of polyphenols and their involved mechanisms. Int J Biol Macromol 118:552–560
Mahler H-C, Friess W et al (2009) Protein aggregation: pathways, induction factors and analysis. J Pharm Sci 98(9):2909–2934
Malchiodi-Albedi F, Paradisi S et al (2011) Amyloid oligomer neurotoxicity, calcium dysregulation, and lipid rafts. Int J Alzheimer's Dis
Marambaud P, Dreses-Werringloer U et al (2009) Calcium signaling in neurodegeneration. Mol Neurodegeneration 4(1):20
McDermott JR, Gibson AM (1997) Degradation of Alzheimer's β-Amyloid Protein by Human and Rat Brain Peptidases: involvement of Insulin-Degrading Enzyme. Neurochem Res 22(1):49–56
McDonald JM, Savva GM et al (2010) The presence of sodium dodecyl sulphate-stable Aβ dimers is strongly associated with Alzheimer-type dementia. Brain 133(5):1328–1341
Mezler M, Barghorn S et al (2012) A β-amyloid oligomer directly modulates P/Q-type calcium currents in Xenopus oocytes. Br J Pharmacol 165(5):1572–1583

Miranda E, MacLeod I et al (2008) The intracellular accumulation of polymeric neuroserpin explains the severity of the dementia FENIB. Hum Mol Genet 17(11):1527–1539
Missiroli S, Patergnani S et al (2018) Mitochondria-associated membranes (MAMs) and inflammation. Cell Death Dis 9(3):329
Morais-de-Sa E, Neto-Silva RM et al (2006) The binding of 2, 4-dinitrophenol to wild-type and amyloidogenic transthyretin. Acta Crystallogr D Biol Crystallogr 62(5):512–519
Morris GP, Clark IA et al (2014) Inconsistencies and controversies surrounding the amyloid hypothesis of Alzheimer's disease. Acta Neuropathol Commun 2(1):135
Musiek ES, Holtzman DM (2015) Three dimensions of the amyloid hypothesis: time, space and'wingmen'. Nat Neurosci 18(6):800
Nilsson MR (2004) Techniques to study amyloid fibril formation in vitro. Methods 34(1):151–160
Ono Y, Sorimachi H (2012) Calpains—an elaborate proteolytic system. Biochim et Biophys Acta (BBA)-Proteins Proteomics 1824(1):224–236
Otzen DE (2013) Amyloid fibrils and prefibrillar aggregates: molecular and biological properties. Wiley
Parvez Alam KSSKCRHK (2017) Protein aggregation: from background to inhibition strategies. Int J Biol Macromol 109:208–219
Paslawski W, Mysling S et al (2014) Co-existence of two different α-synuclein oligomers with different core structures determined by hydrogen/deuterium exchange mass spectrometry. Angew Chem Int Ed 53(29):7560–7563
Pedersen JT, Ostergaard J et al (2011) Cu (II) mediates kinetically distinct, non-amyloidogenic aggregation of amyloid-β peptides. J Biol Chem: jbc. M111:220863
Pernber Z, Blennow K et al (2012) Altered distribution of the gangliosides GM1 and GM2 in Alzheimer's disease. Dement Geriatr Cogn Disord 33(2–3):174–188
Picou RA, Schrum DP et al (2012) Separation and detection of individual Aβ aggregates by capillary electrophoresis with laser-induced fluorescence detection. Anal Biochem 425(2):104–112
Pimplikar SW (2009) Reassessing the amyloid cascade hypothesis of Alzheimer's disease. Int J Biochem Cell Biol 41(6):1261–1268
Podlisny MB, Ostaszewski BL et al (1995) Aggregation of secreted amyloid-protein into sodium dodecyl sulfate-stable oligomers in cell culture. J Biol Chem 270(16):9564–9570
Podlisny MB, Walsh DM et al (1998) Oligomerization of endogenous and synthetic amyloid β-protein at nanomolar levels in cell culture and stabilization of monomer by Congo red. Biochemistry 37(11):3602–3611
Porat Y, Abramowitz A et al (2006) Inhibition of amyloid fibril formation by polyphenols: structural similarity and aromatic interactions as a common inhibition mechanism. Chem Biol Drug Des 67(1):27–37
Pryor E, Kotarek JA et al (2011) Monitoring insulin aggregation via capillary electrophoresis. Int J Mol Sci 12(12):9369–9388
Robert KY, Tsai Y-T et al (2012) Functional roles of gangliosides in neurodevelopment: an overview of recent advances. Neurochem Res 37(6):1230–1244
Ross CA, Poirier MA (2004) Protein aggregation and neurodegenerative disease. Nat Med 10(7): S10
Sabate R, Ventura S (2013) Cross-β-sheet supersecondary structure in amyloid folds: techniques for detection and characterization. Protein Supersecondary Structures, Springer, pp 237–257
Salomone S, Caraci F et al (2012) New pharmacological strategies for treatment of Alzheimer's disease: focus on disease modifying drugs. Brit J Clin Pharmacol 73(4):504–517
Scheltens P, Feldman H (2003) Treatment of Alzheimer's disease; current status and new perspectives. Lancet Neurol 2(9):539–547
Schild L, Reiser G (2005) Oxidative stress is involved in the permeabilization of the inner membrane of brain mitochondria exposed to hypoxia/reoxygenation and low micromolar Ca2 +. FEBS J 272(14):3593–3601
Schuck P (2000) Size-distribution analysis of macromolecules by sedimentation velocity ultracentrifugation and lamm equation modeling. Biophys J 78(3):1606–1619

Sebollela A, Freitas-Correa L et al (2012) Amyloid-β oligomers induce differential gene expression in adult human brain slices. J Biol Chem jbc. M111:298471

Seeley WW, Crawford RK et al (2009) Neurodegenerative diseases target large-scale human brain networks. Neuron 62(1):42–52

Selkoe DJ (2002) Alzheimer's disease is a synaptic failure. Science 298(5594):789–791

Sengupta U, Nilson AN et al (2016) The role of amyloid-β oligomers in toxicity, propagation, and immunotherapy. EBioMedicine 6:42–49

Shirahama T, Cohen AS (1967) High-resolution electron microscopic analysis of the amyloid fibril. J Cell Biol 33(3):679–708

Siddiqi MK, Alam P et al (2017a) Probing the interaction of cephalosporin antibiotic—ceftazidime with human serum albumin: a biophysical investigation. Int J Biol Macromol 105:292–299

Siddiqi MK, Alam P et al (2017b) Attenuation of amyloid fibrillation in presence of Warfarin: a biophysical investigation. Int J Biol Macromol 95:713–718

Siddiqi MK, Alam P et al (2017c) Mechanisms of protein aggregation and inhibition. Front Biosci (Elite Ed) 9:1–20

Siddiqi MK, Alam P et al (2018) Stabilizing proteins to prevent conformational changes required for amyloid fibril formation. J Cell Biochem 120(2):2642–2656

Siddiqi MK, Alam P et al (2018a) Capreomycin inhibits the initiation of amyloid fibrillation and suppresses amyloid induced cell toxicity. Biochim et Biophys Acta (BBA)-Proteins Proteomics 1866(4):549–557

Siddiqi MK, Alam P et al (2018b) Elucidating the inhibitory potential of designed peptides against amyloid fibrillation and amyloid associated cytotoxicity. Front Chem 6:311

Siddiqi M, Nusrat S et al (2018c) Investigating the site selective binding of busulfan to human serum albumin: biophysical and molecular docking approaches. Int J Biol Macromol 107:1414–1421

Spires-Jones TL, Attems J et al (2017) Interactions of pathological proteins in neurodegenerative diseases. Acta Neuropathol 134(2):187–205

Stefani M (2004) Protein misfolding and aggregation: new examples in medicine and biology of the dark side of the protein world. Biochim et Biophys Acta (BBA)-Mol Basis Dis 1739(1):5–25

Stefani M (2010a) Structural polymorphism of amyloid oligomers and fibrils underlies different fibrillization pathways: immunogenicity and cytotoxicity. Curr Protein Pept Sci 11(5):343–354

Stefani M (2010b) Biochemical and biophysical features of both oligomer/fibril and cell membrane in amyloid cytotoxicity. FEBS J 277(22):4602–4613

Stefani M (2012) Structural features and cytotoxicity of amyloid oligomers: implications in Alzheimer's disease and other diseases with amyloid deposits. Prog Neurobiol 99(3):226–245

Stephan A, Laroche S et al (2001) Generation of aggregated β-amyloid in the rat hippocampus impairs synaptic transmission and plasticity and causes memory deficits. J Neurosci 21 (15):5703–5714

Stine WB, Snyder SW et al (1996) The nanometer-scale structure of amyloid-β visualized by atomic force microscopy. J Protein Chem 15(2):193–203

Stoeckli M, Knochenmuss R et al (2006) MALDI MS imaging of amyloid. Methods Enzymol 412:94–106

Thomas JG, Ayling A et al (1997) Molecular chaperones, folding catalysts, and the recovery of active recombinant proteins from *E. coli*. Appl Biochem Biotechnol 66(3):197–238

Trovato A, Chiti F et al (2006) Insight into the structure of amyloid fibrils from the analysis of globular proteins. PLoS Comput Biol 2(12):e170

Tucker S, Muller C et al (2015) The murine version of BAN2401 (mAb158) selectively reduces amyloid-β protofibrils in brain and cerebrospinal fluid of tg-ArcSwe mice. J Alzheimers Dis 43 (2):575–588

Tycko R (2004) Progress towards a molecular-level structural understanding of amyloid fibrils. Curr Opin Struct Biol 14(1):96–103

Tycko R (2006) Molecular structure of amyloid fibrils: insights from solid-state NMR. Q Rev Biophys 39(1):1–55

Valastyan JS, Lindquist S (2014) Mechanisms of protein-folding diseases at a glance. Dis Models Mech 7(1):9–14
Valincius G, Heinrich F et al (2008) Soluble amyloid β-oligomers affect dielectric membrane properties by bilayer insertion and domain formation: implications for cell toxicity. Biophys J 95(10):4845–4861
van den Berg B, Ellis RJ et al (1999) Effects of macromolecular crowding on protein folding and aggregation. EMBO J 18(24):6927–6933
Vassar R, Bennett BD et al (1999) β-Secretase cleavage of Alzheimer's amyloid precursor protein by the transmembrane aspartic protease BACE. Science 286(5440):735–741
Verma M, Vats A et al (2015) Toxic species in amyloid disorders: Oligomers or mature fibrils. Ann Indian Acad Neurol 18(2):138
Vestergaard B, Groenning M et al (2007) A helical structural nucleus is the primary elongating unit of insulin amyloid fibrils. PLoS Biol 5(5):e134
Viet MH, Ngo ST et al (2011) Inhibition of aggregation of amyloid peptides by beta-sheet breaker peptides and their binding affinity. J Phys Chem B 115(22):7433–7446
Vilar M, Chou H-T et al (2008) The fold of α-synuclein fibrils. Proc Natl Acad Sci 105(25):8637–8642
Wadai H, Yamaguchi K-I et al (2005) Stereospecific amyloid-like fibril formation by a peptide fragment of β2-microglobulin. Biochemistry 44(1):157–164
Wang X, Li Y et al (2014) Effect of strong electric field on the conformational integrity of insulin. J Phys Chem A 118(39):8942–8952
Welzel AT, Williams AD et al (2012) Human anti-Aβ IgGs target conformational epitopes on synthetic dimer assemblies and the AD brain-derived peptide. PLoS ONE 7(11):e50317
Westermark P (2005) Aspects on human amyloid forms and their fibril polypeptides. FEBS J 272 (23):5942–5949
Winklhofer KF, Tatzelt J et al (2008) The two faces of protein misfolding: gain—and loss—of function in neurodegenerative diseases. EMBO J 27(2):336–349
Wong EW, Sheehan PE et al (1997) Observation of metastable abeta amyloid protofibrils by atomic force microscopy. Science 277:1971–1975
Woods LA, Radford SE et al (2013) Advances in ion mobility spectrometry: mass spectrometry reveal key insights into amyloid assembly. Biochim et Biophys Acta (BBA)-Proteins Proteomics 1834(6):1257–1268
Xue W-F, Hellewell AL et al (2009) Fibril fragmentation enhances amyloid cytotoxicity. J Biol Chem 284(49):34272–34282
Yanagisawa K (2007) Role of gangliosides in Alzheimer's disease. Biochim et Biophys Acta (BBA)-Biomembr 1768(8): 1943–1951
Zheng W (2001) Neurotoxicology of the brain barrier system: new implications. J Toxicol Clin Toxicol 39(7):711–719

Chapter 17
CAD, A Multienzymatic Protein at the Head of de Novo Pyrimidine Biosynthesis

Francisco del Caño-Ochoa, María Moreno-Morcillo and Santiago Ramón-Maiques

Abstract CAD is a 1.5 MDa particle formed by hexameric association of a 250 kDa protein that carries the enzymatic activities for the first three steps in the de novo biosynthesis of pyrimidine nucleotides: glutamine-dependent Carbamoyl phosphate synthetase, Aspartate transcarbamoylase and Dihydroorotase. This metabolic pathway is essential for cell growth and proliferation and is conserved in all living organisms. However, the fusion of the first three enzymatic activities of the pathway into a single multienzymatic protein only occurs in animals. In prokaryotes, by contrast, these activities are encoded as distinct monofunctional enzymes that function independently or by forming more or less transient complexes. Whereas the structural information about these enzymes in bacteria is abundant, the large size and instability of CAD has only allowed a fragmented characterization of its structure. Here we retrace some of the most significant efforts to decipher the architecture of CAD and to understand its catalytic and regulatory mechanisms.

Keywords Glutaminase · Carbamoyl phosphate synthetase · Aspartate transcarbamoylase · Dihydroorotase · Nucleotide metabolism · PALA · Metabolic disease

F. del Caño-Ochoa · M. Moreno-Morcillo · S. Ramón-Maiques (✉)
Department of Genome Dynamics and Function, Centro de Biología Molecular Severo Ochoa (CSIC-UAM), Nicolas Cabrera 1, 28049 Madrid, Spain
e-mail: santiago.ramon@cbm.csic.es

F. del Caño-Ochoa
e-mail: fdelcano@cbm.csic.es

M. Moreno-Morcillo
e-mail: mmoreno@cbm.csic.es

J. R. Harris and J. Marles-Wright (eds.), *Macromolecular Protein Complexes II: Structure and Function*, Subcellular Biochemistry 93,
https://doi.org/10.1007/978-3-030-28151-9_17

The de Novo Biosynthetic Pathway for Pyrimidines

Nucleotides are essential compounds for all forms of life as building blocks for RNA and DNA synthesis. Uracil, thymine and cytosine are the three major pyrimidine bases for the formation of nucleic acids. These pyrimidines also exist as free nucleotides (e.g. uridine 5-phosphate, UMP) or nucleosides (e.g. uridine) in living tissues where they act as allosteric regulators or as coenzymes (Brown 1998). In addition, uridine diphosphate (UDP) and cytidine diphosphate (CDP) play prominent roles as activators and carriers of intermediate metabolites: UDP-glucose is required for protein glycosylation and for the biosynthesis of polysaccharides (e.g. glycogen, cellulose), and CDP-activated compounds (e.g. CDP-choline, CDP-ethanolamine) are important intermediates for the biosynthesis of phospholipids.

The cells require a constant supply of pyrimidines that can be obtained by two different pathways (Fig. 17.1). Pyrimidines can be synthetized de novo from small precursors such as ammonia, bicarbonate, ATP, aspartate (Asp) and 5-phosphoribosyl-1-pyrophosphate (PRPP) (Jones 1980). Alternatively, organisms can employ different recycling or salvage pathways to uptake preformed pyrimidine bases (uracil) or nucleosides (uridine) released during the turnover of nucleic acids or acquired from the diet (Nyhan 2005). The relative contribution of de novo and salvage pathways is dependent upon cell type and developmental stage (Evans and Guy 2004). In general, fully differentiated and quiescent cells obtain pyrimidines through salvage pathways and the de novo synthesis is low. Contrary, the de novo biosynthesis is needed to fuel the high demand of pyrimidines during cell growth and proliferation.

The metabolic pathway leading to de novo synthesis of UMP consists of six sequential enzymatic reactions that are preserved in all living organisms, from microorganisms to humans (Jones 1980). Figure 17.1 depicts this process as an assembly line, where the product of one enzyme acts as the substrate of the next, and the semi-finished intermediates move somehow between active centers. The metabolic pipeline starts with the manufacturing of carbamoyl phosphate (CP), a labile and high-energy phosphate metabolite that is also a common precursor for the de novo biosynthesis of arginine and urea (Jones 1980; Shi et al. 2018). CP is made from bicarbonate, ammonia (obtained from glutamine hydrolysis) and two molecules of ATP, in a four-step reaction catalyzed by carbamoyl phosphate synthetase (CPS; EC 6.3.5.5), an enzyme actually composed of two activities, a glutamine-dependent amidotransferase (GLN) and a synthetase (SYN) (Meister 1989; Simmer et al. 1990). In a second step, CP and Asp are used by the enzyme aspartate transcarbamoylase (ATC; EC 2.1.3.2) to form carbamoyl aspartate. Next, the enzyme dihydroorotase (DHO; EC 3.5.2.3) catalyzes the condensation of carbamoyl aspartate to dihydroorotate, the first cyclic compound of the pathway. Dihydroorotate is further oxidized to orotate by dihydroorotate dehydrogenase (DHODH; EC 1.3.5.2), an enzyme that in eukaryotes is tethered to the outer face of the inner mitochondrial membrane and couples pyrimidine synthesis to the respiratory chain (Fig. 17.1). Orotate is a completed pyrimidine base and is incorporated

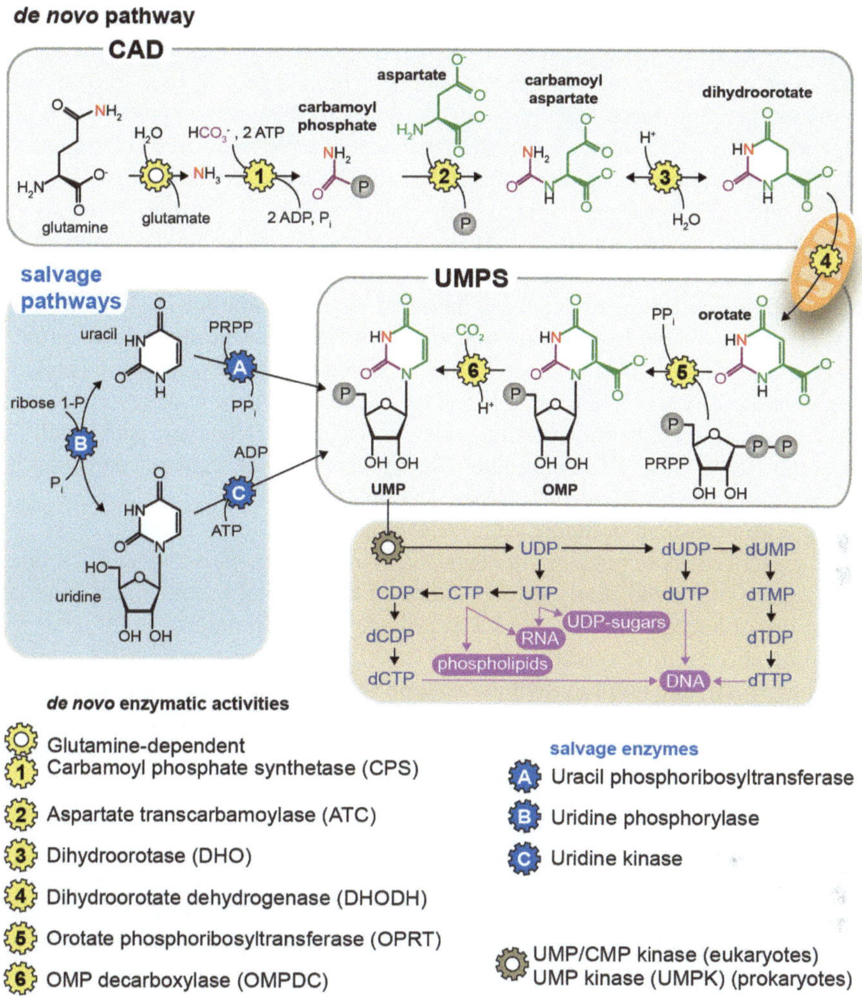

Fig. 17.1 Schematic outline of de novo and salvage pathways for pyrimidine biosynthesis in animals. The reactions catalyzed by the cytosolic multienzymatic proteins CAD and UMPS are framed in grey background, while the intermediate step occurs in the mitochondria. The salvage pathways for recycling of uracil or uridine are indicated in blue background. The numerous reactions that transform UMP into the pool of different pyrimidine nucleotides are represented in brownish background, with final fates for the different nucleotides shown in pink

to the activated sugar PRPP by the enzyme orotate phosphoribosyltransferase (OPRT; EC 2.4.2.10), producing the nucleotide orotidine 5-phosphate (OMP). Lastly, the OMP decarboxylase (OMPDC; EC 4.1.1.23) converts OMP to UMP, which is subsequently phosphorylated to UDP and UTP by different kinase activities. It is from these nucleotides that the other common pyrimidine derivatives arise (Fig. 17.1).

Although the six enzymatic steps for de novo manufacturing of UMP are evolutionary conserved, the organization of the enzymatic activities varies greatly in prokaryotes, protists, plants, fungi, and animals (Fig. 17.2a). In most prokaryotes and plants, the first three reactions are catalyzed by distinct enzymes that work independently or by forming more or less transient complexes (Jones 1980; Zrenner et al. 2006). In these organisms a single CPS, formed by association of two distinct proteins with GLN and SYN activities, makes CP both for the synthesis of pyrimidine and arginine, and thus, ATC becomes the first committed enzyme for de novo pyrimidine synthesis. In some bacteria, like *Escherichia coli*, ATC is under a strict allosteric control, being feedback inhibited by the pyrimidine nucleotides UTP and CTP and activated by a purine nucleotide, ATP [reviewed in (Allewell 1989; Jacobson and Stark 1973)]. The enzyme is formed by two trimers of catalytic subunits related by three dimers of regulatory subunits where nucleotide effectors bind, inducing large conformational changes in the holoenzyme [reviewed in (Lipscomb 1994)] (Fig. 17.2b). In other bacteria, ATC is not regulated and consists of a catalytic trimer without regulatory subunits that functions independently or forming a non-covalent complex with DHO (Fig. 17.2c).

The scenario is more puzzling in other organisms, where the enzymatic activities appear highly organized. Early studies indicated that in simple eukaryotes, such as *Neurospora* or *Saccharomyces*, two CPSs provide distinct intracellular CP pools for pyrimidine and arginine synthesis (Fig. 17.2a). The arginine-specific CPS, formed like in bacteria by the non-covalent association of two proteins with GLN and SYN activities, locates in the mitochondria (Lacroute et al. 1965), whereas the pyrimidine-specific CPS and ATC activities are present in a single bi-functional protein encoded by the genes *pyr-3* in *Neurospora* (Williams et al. 1970; Williams and Davis 1970) or *ura2* in *S. cerevisiae* (Lue and Kaplan 1969). This bi-functional protein also contains an inactive DHO-like domain, and these organisms have a monofunctional DHO enzyme encoded by an independent gene (*ura4* in *S. cerevisiae*) (Denis-Duphil 1989; Souciet et al. 1989) (Fig. 17.2a). Thus, in fungi only five genes code for the six enzymatic activities for UMP synthesis.

Similarly, in animals, distinct mitochondrial and cytosolic CPSs make CP pools for arginine and pyrimidine synthesis, respectively (Jones 1980) (Fig. 17.2a). In most terrestrial vertebrates, there is an arginine-specific CPS (CPS-1; EC. 6.3.4.16) formed by the fusion of an inactive GLN domain and a SYN domain, which is only expressed in the mitochondria of hepatocytes and requires acetylglutamate as a co-factor (Jones 1980; Rubio et al. 1981). CPS-1 cannot hydrolyze glutamine and is used to detoxify free-ammonia directly through the urea cycle. On the other hand, early studies proved the existence of a pyrimidine-specific CPS (CPS-2) that co-purified with the ATC activity in some sort of cytosolic complex (Hoogenraad et al. 1971).[1] Unlike in fungi, this complex also contained DHO activity (Shoaf and

[1]There is yet a CPS-3 in some invertebrates and fish that combines properties of CPS-1 and CPS-2 [Anderson PM (1989) Biochem J 261(2): 523–529]. It consists of a single polypeptide with GLN and SYN domains, requires acetylglutamate as co-factor, is not regulated by nucleotides and hydrolyzes glutamine.

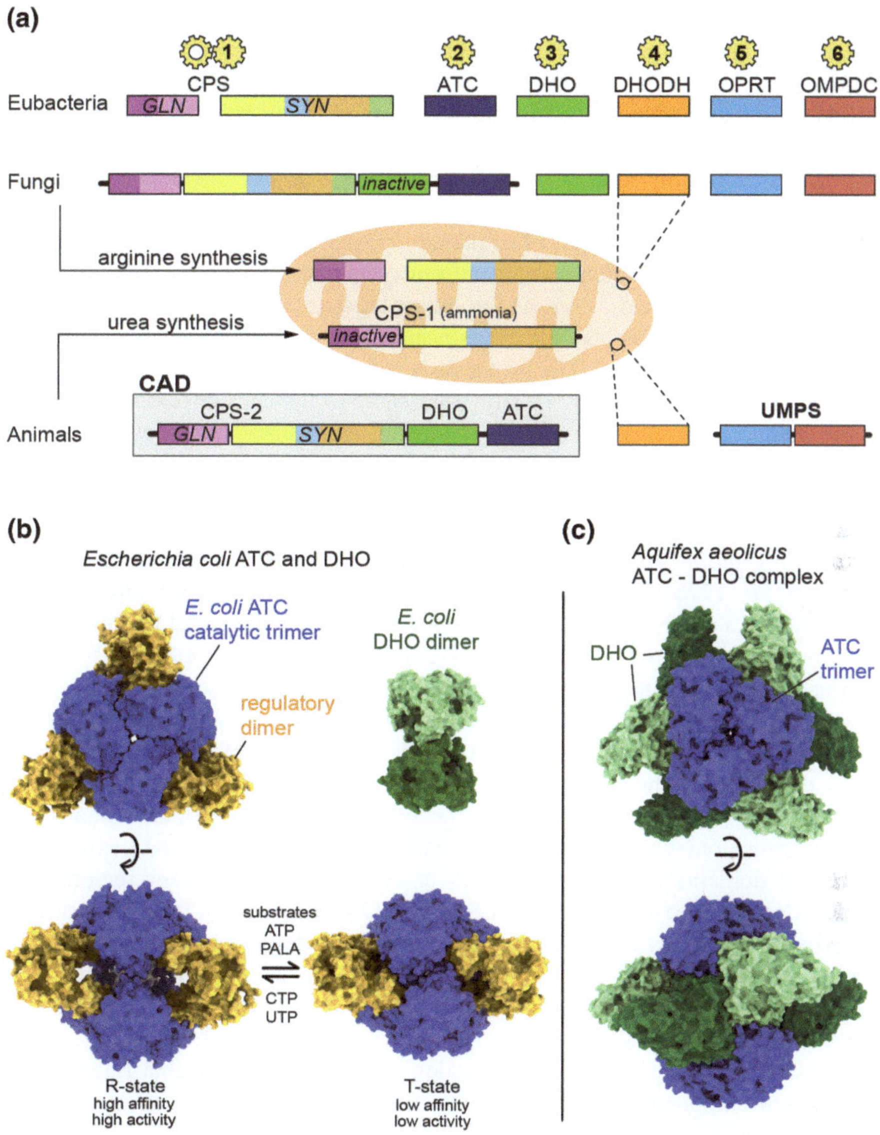

Fig. 17.2 Organization of the enzymes involved in de novo pyrimidine biosynthesis. **a** Schematic representation of the six enzymatic activities needed for de novo pyrimidine biosynthesis of UMP in eubacteria, fungi and animals. The rectangles represent individual proteins or domains within multienzymatic proteins. A second arginine-specific CPS present in the mitochondria of fungi and some animals is also shown. **b**, **c** Space filling representation of the ATC and DHO enzymes in *Escherichia coli* (**b**) and *Aquifex aeolicus* (**c**). *E. coli* ATC is an holoenzyme formed by two trimers of catalytic subunits related by three dimers of regulatory subunits. The enzyme undergoes conformational transitions between active (R) and inactive (T) states regulated by the binding of substrates, inhibitors and allosteric effectors. In *A. aeolicus*, ATC and DHO enzymes are intimately associated forming a heterododecamer

Jones 1971), leading M. E. Jones to postulate that the initial three enzymes for the de novo biosynthesis in mammals could form a stable complex or be part of a single multienzymatic protein. This association was named CAD for the first letters of the distinct enzymatic activities, and was subsequently found in every animal species investigated, from *Dictyostelium* to human [reviewed in (Evans 1986; Jones 1980)].

The nature of this multienzymatic association was not unveiled until it was purified to homogeneity by the group of G. Stark from hamster cells treated with the drug PALA (N-phosphonacetyl-L-aspartate). PALA was synthetized as a potent inhibitor of *E. coli* ATC that combines features of both enzyme substrates (CP and Asp) and was postulated to mimic the transition state of the reaction (Collins and Stark 1971). PALA was also effective in inhibiting the growth of mouse tumors, proving that de novo pyrimidine synthesis is required for cell proliferation (Swyryd et al. 1974; Yoshida et al. 1974). However, some of the cells developed resistance to the inhibitory effect of PALA by producing more than 100 times the original activity of the three-enzyme complex (Kempe et al. 1976). When the complex was isolated from these overproducing cells, the three catalytic activities were found integrated within a single polypeptide of ~250 kDa that self-associated in a mixture of oligomeric forms, mostly trimers and hexamers (Coleman et al. 1977; Kempe et al. 1976). This physical association between the three enzymes was confirmed by a combined genetic, biochemical and immunological approach using CAD defective mutants in CHO cells (Davidson and Patterson 1979).

The attempts to purify CAD revealed its extraordinary susceptibility to proteolytic cleavage, which allowed the isolation of different protein fragments retaining different enzymatic activities (Davidson et al. 1981; Evans 1986; Kim et al. 1992; Mally et al. 1981). The analysis of the proteolytic fragments and of mutants in the CAD locus in *Drosophila* (*rudimentary* gene) provided strong evidence of a simple domain structure, where enzymatic activities were present in CAD as different functional domains connected by more or less unstructured linker regions. The correct order of the domains, GLN-SYN-DHO-ATC (Fig. 17.2a), was not confirmed until the genes from different organisms were fully sequenced [*Drosophila* (Freund and Jarry 1987); hamster (Shigesada et al. 1985); human (Davidson et al. 1990); *Dictiostelyum* (Faure et al. 1989); and *S. cerevisiae ura2* (Denis-Duphil 1989; Souciet et al. 1989)].

In addition, a number of gene dissection experiments proved that fragments of CAD could complement *E. coli* deficient in CPS, DHO or ATC activities, indicating that the isolated enzymatic CAD domains were functional and topologically independent [reviewed in (Davidson et al. 1993)].

CAD is not the only multienzymatic protein in the de novo pyrimidine pathway. In animals (and also in plants), the OPRT and OMPDC activities, are also fused into a single bi-functional protein named UMP synthetase (UMPS) (Fig. 17.2a) (Jones 1980; Traut and Jones 1979). Thus, two cytosolic multienzymatic proteins are responsible for catalyzing five out of six steps in UMP synthesis in animals (Jones 1980; Shoaf and Jones 1971), while the reaction catalyzed by the dehydrogenase occurs inside the mitochondria (Fig. 17.1).

The evolutionary diversity shown in Fig. 17.2a is difficult to reconcile with the view of a metabolic assembly line formed by enzymes functioning as autonomous catalytic units. Enzymes work more efficiently when assembled into complexes that favor the communication, coordination and interdependence of the different activities and have the potential to express unique catalytic and regulatory properties (Gaertner 1978). Likely, the most effective of these interactions is the covalent linkage into a multienzymatic conjugate such as CAD. We are familiar with the dynamics that improve the relation with co-workers in our everyday life at the laboratory or in the office. Now, we want to understand whether and how enzymes apply similar teamwork strategies to make pyrimidines in the most efficient manner.

Although bacterial CPS, ATC and DHO enzymes are well characterized both structurally and biochemically, the large size of CAD and the high sensitivity to proteases hampered many structural characterization attempts so far. In this chapter, we retrace some of the most significant efforts to decipher the structure of CAD and to understand its functioning.

Starting from the Beginning: The GLN and SYN Domains

CP is a small and highly unstable molecule that requires a truly remarkable protein machinery of more than one thousand aminoacids for its synthesis. The reaction involves three discrete chemical steps with canalization of highly reactive and unstable intermediates: carboxyphosphate, ammonia and carbamate (Anderson and Meister 1965; Meister 1989) (Fig. 17.3a). In a first step, a molecule of bicarbonate is phosphorylated at the expense of one ATP molecule to form carboxyphosphate and ADP. Then, a molecule of ammonia obtained directly (urea cycle CPS-1) or hydrolyzed from glutamine reacts with the carboxyphosphate to form carbamate and inorganic phosphate. In a final step, carbamate is phosphorylated by consuming a second ATP molecule to form CP.

All known CPSs share a common reaction mechanism, and present substantial sequence homology and similar multidomain architecture. As previously mentioned, CPS is composed of two parts, a ~40 kDa GLN that delivers ammonia and a ~120 kDa SYN that catalyzes the three-step reaction mentioned above. The GLN and SYN parts exist either as different subunits, such as in bacterial CPSs or in the arginine-specific CPS in fungi, or they are connected by a short linker within a single polypeptide, such as in the urea-cycle CPS-1 (Fig. 17.2a). Alternatively, GLN and SYN can be fused together with ATC and DHO (or an inactive DHO-like domain) forming the CAD (or CAD-like) multienzymatic protein. Currently, only the structures of the CPS from *E. coli* and the ammonia-dependent CPS-1 from human hepatocytes have been characterized.

E. coli CPS has been extensively studied both biochemically and structurally (Holden et al. 1998; Meister 1989; Raushel et al. 1998; Thoden et al. 1998, 1999a, b, c, d, 2002, 2004). The enzyme is a stable heterodimer formed by the small GLN

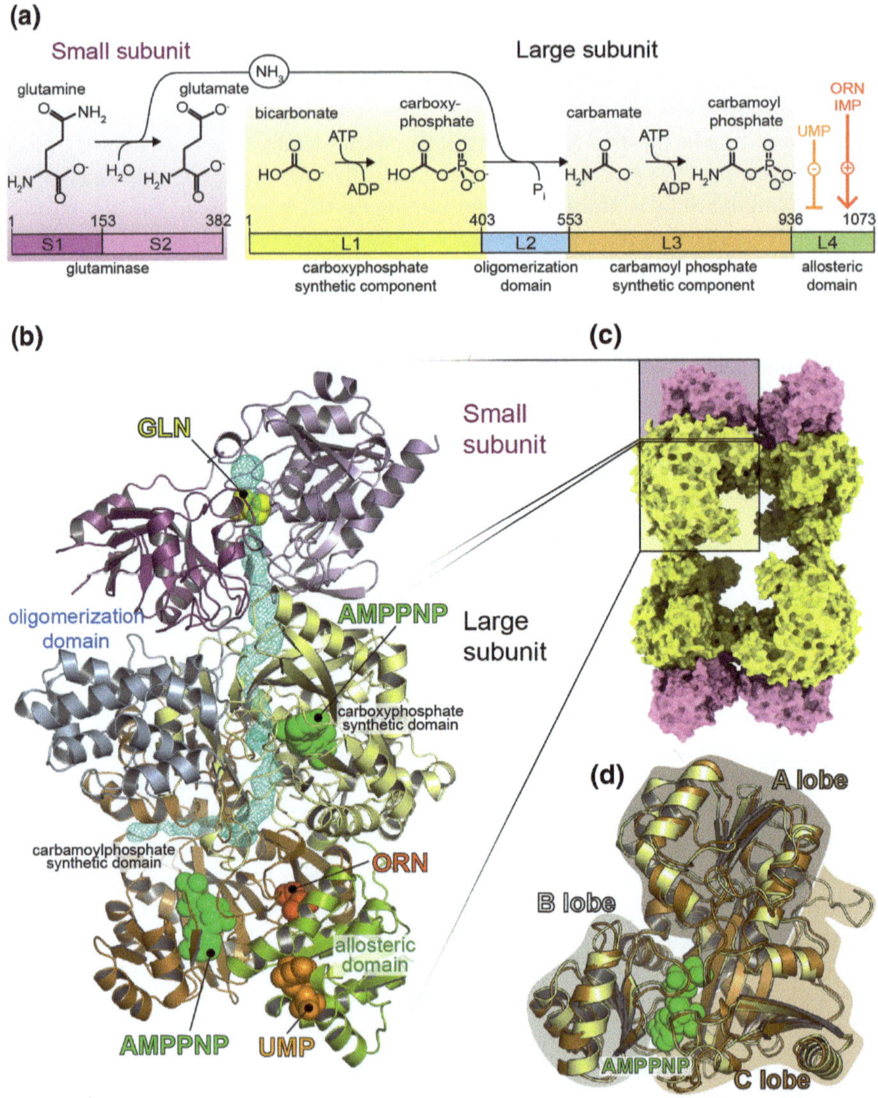

Fig. 17.3 *E. coli* CPS structure. **a** Schematic representation of the small and large subunits of *E. coli* CPS and partial reactions catalyzed by each domain within the protein. The binding of allosteric effectors to the C-terminal region of the protein is indicated. **b** Cartoon representation of the enzyme heterodimer, with domains colored as in (**a**). Glutamine (GLN), AMPPNP (a non-hydrolyzable analog of ATP), UMP and ornithine (ORN) are represented as spheres. The internal tunnel connecting the three active sites is represented as cyan mesh. **c** Space filling representation of *E. coli* CPS tetramer as found in the crystal structure. **d** Superposition of the carboxyphosphate synthetic component (L1; shown in yellow) and the carbamoyl phosphate synthetic component (L3; orange) with AMPPNP represented as spheres. The A, B and C lobes forming the ATP-grasp fold are indicated

subunit and the large SYN subunit (Fig. 17.3a), encoded respectively by the genes *carA* and *carB*. The large subunit, also holds the binding sites for the allosteric regulators. Although the GLN/SYN heterodimer is the functional unit of *E. coli* CPS, the protein readily converts to a dimer or a tetramer with unaffected catalytic properties (Anderson 1986) (Fig. 17.3c).

CPS-1, on the other hand, is a protein of 1500 amino acids divided into an inactive N-terminal GLN domain and a C-terminal SYN domain, which correspond respectively, to the small and large subunits of bacterial CPS (Fig. 17.4a). The structure of human CPS-1 probed to be highly similar to that of *E. coli* CPS (de Cima et al. 2015), as correctly inferred from biochemical studies (Nyunoya et al. 1985; Powers-Lee and Corina 1986). Despite being inactive, the GLN domain is an integral part of the enzyme and is tightly associated with the SYN domain.

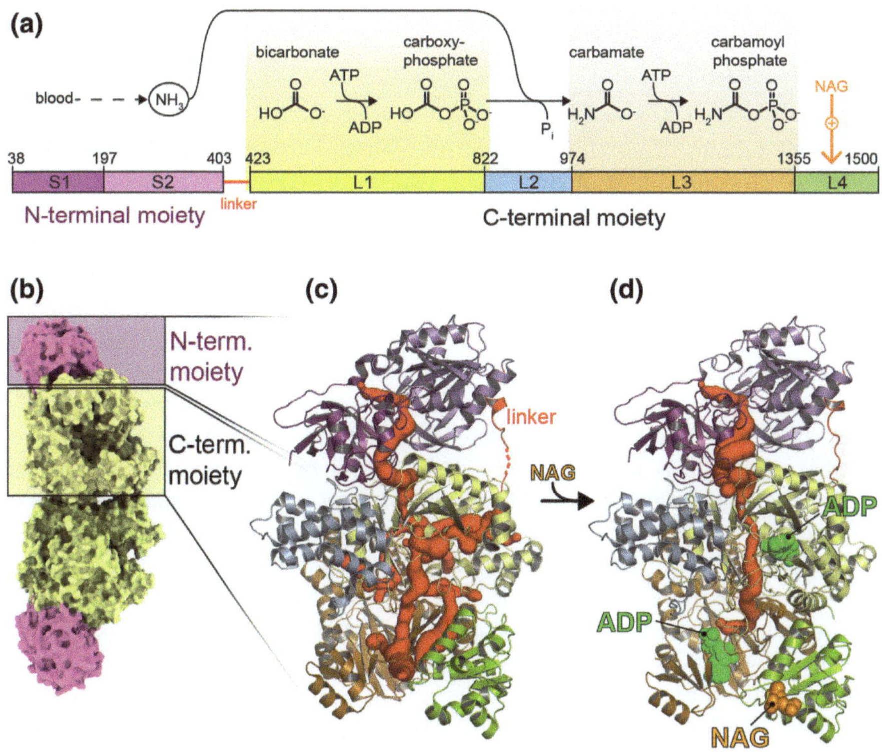

Fig. 17.4 Human CPS-1 structure. **a** Schematic representation of the enzyme with the partial reactions catalyzed by each protein region. The allosteric activation by binding of acetylglutamate (NAG) to the C-terminal region is indicated. **b** Space filling representation of CPS-1 dimer as seen in the asymmetric unit of the crystal. **c**, **d** Cartoon representation of human CPS-1 free (**c**) or bound (**d**) to NAG. Binding of NAG induces conformational changes that define the path of the tunnel - shown in red surface- through the active sites. The N- and C-terminal moieties of the protein are connected by a 20 aa linker that spans 50 Å and appears partially disordered in the structure free of acetylglutamate (indicated by a dashed line)

In solution, CPS-1 exists in a monomer-dimer rapid equilibrium where the monomer predominates (Diez-Fernandez et al. 2013; Pierson and Brien 1980; Rubio et al. 1981) (Fig. 17.4b). Both the monomers and dimers are active upon acetylglutamate binding (Lusty 1981), and there is no evidence of formation of tetramers.

So far, there is no direct structural information about the GLN or SYN domains of CAD. The CPS-2 activity of CAD is of outmost importance since it is the rate-limiting reaction in the biosynthesis of pyrimidines and the point for allosteric control (Coleman et al. 1977; Mori and Tatibana 1978). The difficulty to produce a stable construct for these regions of the protein has precluded many crystallization attempts so far. Nevertheless, the CPS-2 activity of CAD is expected to share similar structural features with *E. coli* CPS and with human CPS-1. The comparison between the structures of these evolutionary distant enzymes should give us some hints on the structure and functioning of CAD.

GLN Domain

The small subunit of *E. coli* CPS and the N-terminal moiety of human CPS-1 exhibit a globular shape divided into two lobes (Figs. 17.3 and 17.4). The N-terminal lobe of ~16 kDa (named S1) is formed by four major α-helices and two perpendicular four stranded β-sheets. The C-terminal lobe of ~25 kDa (S2) presents a central β-sheet of ten elements flanked on both sides by six α-helices, which is characteristic of the class I family of amidotransferases (Nyunoya and Lusty 1984). These glutaminases share a common reaction mechanism that involves the formation of a glutamyl-thioester intermediate between glutamine and a catalytic Cys. A conserved His is also important to activate the Cys for nucleophilic attack. In *E. coli* CPS, the catalytic residues, C269 and H353, locate at the S2 lobe, and the active site is formed at the interface with the S1 lobe (Figs. 17.3b and 17.5a). The reaction process has been delineated by different crystal structures of *E. coli* CPS bound to various ligands and bearing different mutations (Thoden et al. 1997, 1998, 1999a, c, d, 2002, 2004), and the formation of the glutamyl-thioester intermediate has been demonstrated biochemically both for *E. coli* CPS and for CAD (Chaparian and Evans 1991; Wellner et al. 1973). In contrast, human CPS-1 presents a Ser residue (S294) replacing the catalytic Cys (Fig. 17.5a), explaining why this domain is inactive.

The GLN domain of CAD is expected to be structurally similar to the GLN moieties of *E. coli* CPS and CPS-1. Sequence alignment shows the preservation of the secondary structural elements throughout the predicted S1 and S2 lobes and the conservation of active site residues, including the catalytic Cys and His residues (C252 and H336 in human CAD) (Figs. 17.5a and 17.6). Interestingly, the isolated GLN domain of hamster CAD has been produced recombinantly in bacteria but is only active when mixed in stoichiometric amounts with the large subunit of *E. coli* CPS, supporting a high degree of conservation between both enzymes (Guy and Evans 1994). It is expected that, similarly to *E. coli* CPS and human CPS-1, both

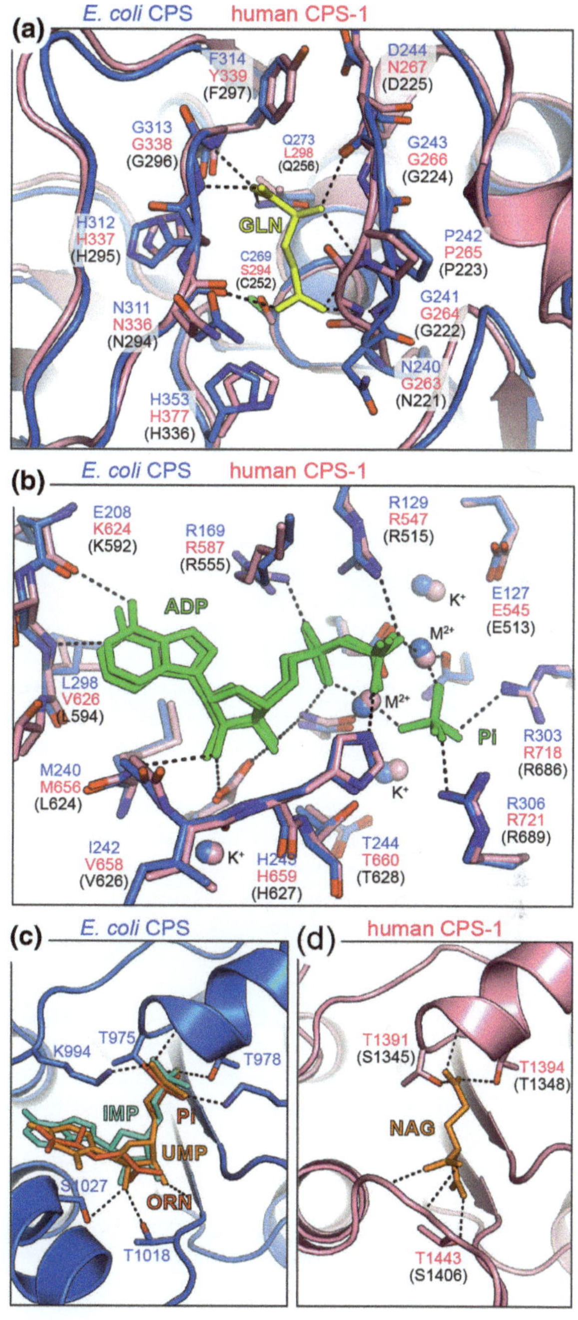

Fig. 17.5 Structural conservation between CPSs. **a** Cartoon representation of the superposition of the active site of *E. coli* CPS small subunit (blue) with the equivalent region in human CPS-1 (pink). A molecule of glutamine (GLN) bound to the *E. coli* enzyme is shown in yellow sticks and the interactions with conserved residues are indicated by dashed lines. **b** Similar representation for the superposition of the active sites of the carboxyphosphate synthetic component of *E. coli* and human CPSs. A molecule of ADP and an inorganic phosphate present in both structures are shown as green sticks. Potassium and divalent metal ions (Mg^{2+} in human CPS-1 or Mn^{2+} in *E. coli* CPS) are shown as spheres. **c** Allosteric domain of *E. coli* CPS showing the overlapping binding sites for UMP and IMP. At high concentrations, ornithine (ORN) and a molecule of inorganic phosphate (Pi) can also bind at this position. **d** Allosteric domain in human CPS-1 with acetylglutamate. In **a**, **b** and **d**, the numbers of the conserved residues in human CAD are indicated in parentheses

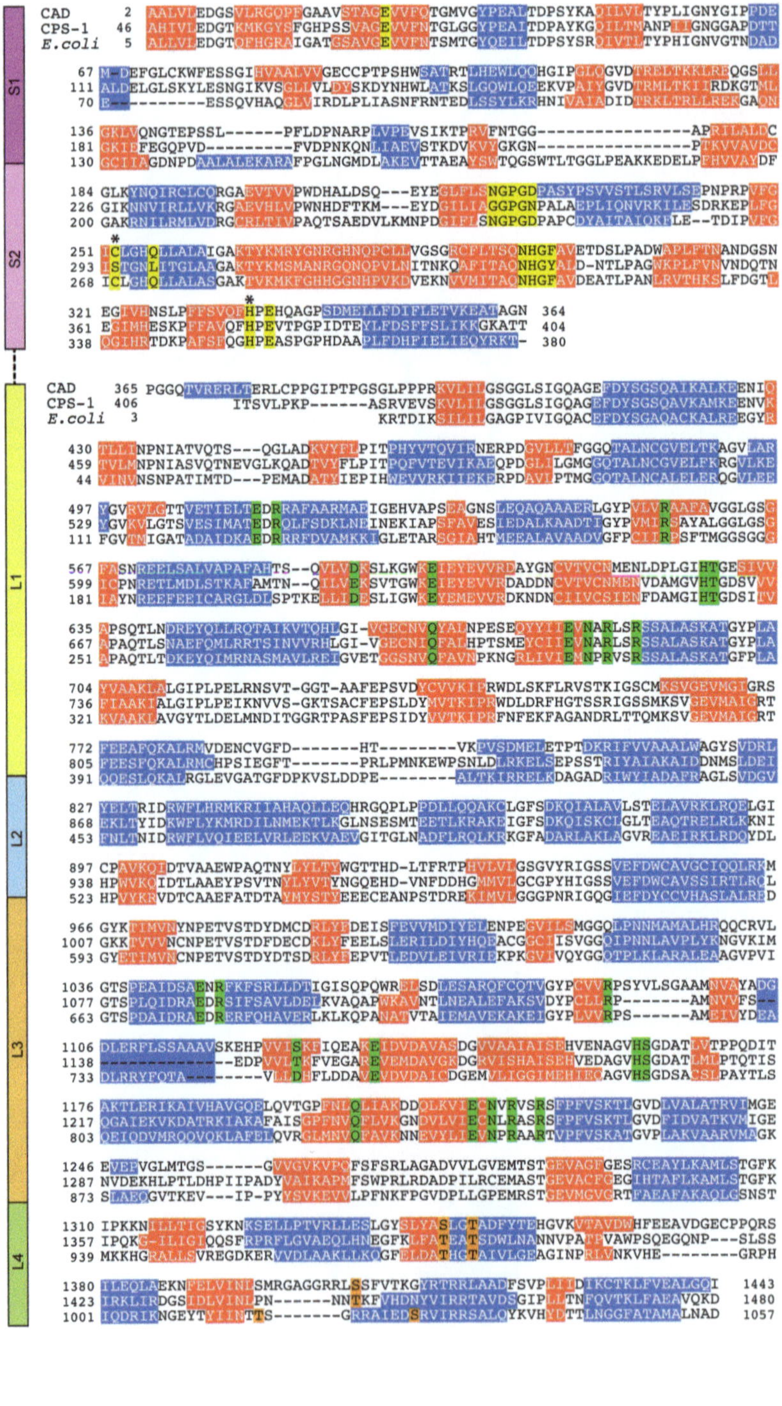
CAD
CPS-1
E.coli
S1
S2
L1
L2
L3
L4

◂**Fig. 17.6** Sequence alignments. Multiple alignment of the CPS-2 in human CAD, human CPS-1 and *E. coli* CPS sequences. A schematic diagram of the protein domains is shown on the left side. The red or blue backgrounds indicate respectively the α-helices and β-strands observed in the crystal structures or predicted from the CAD sequence. Residues involved in binding of glutamine, in the L1 or L3 phosphorylation sites and in the binding of allosteric effectors are highlighted in yellow, green and orange background, respectively. The N-terminal end of human CPS-1 corresponds to a mitochondrial targeting peptide that is cleaved off from the mature protein and is not included in the alignment

the S1 and S2 participate in an extensive interaction with the SYN domain in CAD (Figs. 17.3b and 17.4c). This interaction, which stabilizes both subunits in *E. coli* CPS (Cervera et al. 1993), somehow ensures the synchronization and enhancement of the GLN and SYN activities both in *E. coli* CPS and in CAD, likely to avoid wasteful hydrolysis of glutamine or ATP if the other substrates of the reaction are not available (Hewagama et al. 1999; Miles et al. 1998; Miles and Raushel 2000; Miran et al. 1991).

SYN Domain

The large subunit of *E. coli* CPS and the C-terminal moiety of CPS-1 are divided into four structural units (L1–4; Figs. 17.3a and 17.4a). L1 and L3 correspond to two homologous phosphorylation domains (Britton et al. 1979) that share 40% sequence identity and probably arose by a gene duplication and fusion event (Nyunoya and Lusty 1983). These two synthetic components form a pseudo-homodimer with nearly exact twofold rotational symmetry and are topologically equivalent but not identical, as expected based on their different substrates and interactions with the rest of the protein (Figs. 17.3b, d and 17.4c) (Thoden et al. 1997). L1 is the carboxyphosphate synthetic component responsible for the phosphorylation of bicarbonate and is in contact with the GLN domain. L3 is the CP component that phosphorylates carbamate and interacts with the allosteric region (L4) described below. Both synthetic components are structured in an “ATP-grasp” fold, with three lobes surrounding (“grasping”) the nucleotide (Fawaz et al. 2011) (Fig. 17.3d). The A lobe (*E. coli* CPS L1 residues 1–140 and L3 residues 554–686) exhibits a modified Rossmann-fold with a five stranded β-sheet flanked on both sides by α-helices. The B lobe (*E. coli* CPS L1 residues 141–210 and L3 residues 687–756) folds in a four-stranded antiparallel β-sheet flanked on the solvent side by two α-helices. The C lobe (*E. coli* L1 residues 211–403 and L3 residues 757–936) is nucleated by an antiparallel β-sheet flanked by four helical regions. The active site, located between the B- and C-lobes, is highly similar in *E. coli* CPS and human CPS-1 (Fig. 17.5b). The nucleotide, interacting residues and divalent cations occupy virtually identical positions in both enzymes. Even the potassium ions, which are known to be required for the activity (Anderson and Meister 1966) occupy similar positions at the active sites.

The synthetic L1 and L3 domains are linked by the L2 domain, composed of seven α-helices and a two stranded β-sheet. In *E. coli* CPS, L2 participates in the formation of the tetramers, and thus it was named as the oligomerization domain (Thoden et al. 1997) (Fig. 17.3b, c). The interaction between two oligomerization domains is relatively small, as expected from the readily conversion between monomer and tetramer (Anderson 1986). In CPS-1, L2 is not involved in oligomerization and received the name of the integrating domain, for its role in embracing L1 and connecting the L1 and L3 with the GLN moiety (Fig. 17.4) (de Cima et al. 2015).

Based on the alignment of Fig. 17.6, the predicted synthetic L1 and L3 domains and the connecting L2 domain of CAD are expected to be highly similar to *E. coli* CPS and CPS-1, including the conserved catalytic residues at the phosphorylation sites (Figs. 17.5b and 17.6).

Allosteric Regulation

Most CPSs are under a strict allosteric control. Pyrimidine-specific CPSs are generally activated by the purine nucleotide inosine 5′-monophosphate (IMP) or by PRPP, a substrate for both purine and pyrimidine synthesis, and are feedback inhibited by UMP or UTP, the products of de novo pyrimidine synthesis (Jones 1980), whereas arginine-specific CPSs are generally activated by ornithine, a substrate with CP for the synthesis of arginine. CPS-1, on the other hand, is essentially an inactive enzyme that requires the activation by the co-factor acetyl-glutamate (Rubio et al. 1983).

E. coli CPS makes CP both for the synthesis of arginine and pyrimidines, and thus, it is inhibited by UMP and is activated by IMP and ornithine (Fig. 17.3a) (Meister 1989). These allosteric effectors regulate the activity primarily by raising or lowering the K_M for ATP nearly 10-fold (Braxton et al. 1992, 1996). Biochemical and structural studies have proven that UMP, IMP and ornithine bind at the ~20 kDa C-terminal region of the large subunit, the L4 domain (Fig. 17.3a) (Cervera et al. 1996; Czerwinski et al. 1995; Rubio et al. 1991; Thoden et al. 1999c). This allosteric domain is structured in a Rossmann-fold (nearly identical to the A lobe in the L1 and L3 synthetic components) with a five-stranded parallel β-sheet flanked on either side by α-helices (Fig. 17.3b). UMP and IMP compete for binding in a nearly identical position at the carboxy-edge of the central β-sheet (Fig. 17.5c) (Bueso et al. 1999; Cervera et al. 1996; Mora et al. 1999; Thoden et al. 2004; Thoden et al. 1999c). Ornithine on the other hand, binds in a different pocket, 12 Å apart from the nucleotide binding site, at the interface between the allosteric domain and the CP synthetic domain (Fig. 17.3b) (Thoden et al. 1997). Some crystal structures show a second molecule of ornithine (and a phosphate ion) at the UMP/IMP binding site, likely due to the high concentrations of these ligands used in the crystallization solution (Fig. 17.5c).

Acetylglutamate also binds to the L4 domain of CPS-1 (Figs. 17.4d and 17.5d) (de Cima et al. 2015; Rodriguez-Aparicio et al. 1989). Despite having very different effector molecules, the CPS-1 and *E. coli* CPS allosteric domains are structurally very similar. Indeed, acetylglutamate occupies the equivalent position of UMP/IMP in *E. coli* CPS, and interacts with similar conserved elements (Fig. 17.5c, d).

The CPS-2 activity of CAD is allosterically inhibited by UTP and activated by PRPP, which alter the apparent K_M for ATP (Mori and Tatibana 1978). The binding of the allosteric effectors is also expected to occur at a regulatory domain at the C-terminal region of the SYN domain. This is supported by analogy to other CPSs (Rubio 1994) and because the UTP inhibition is lost when this region is deleted (Liu et al. 1994). Furthermore, E. Carrey discovered that the UTP inhibition is abolished upon phosphorylation by cAMP-dependent protein kinase A (PKA) of residue S1406 at the predicted allosteric domain (Carrey et al. 1985; Carrey and Hardie 1988). Based on sequence conservation and preservation of secondary structural elements, it is likely that the L4 domain of CPS-2 will exhibit the fold observed in *E. coli* CPS and human CPS-1. We propose that residue S1406 will occupy the equivalent position of T1443 in CPS-1, which directly binds to acetylglutamate (Figs. 17.5d and 17.6). Thus, it is feasible that a phosphorylation at residue S1406 in CAD would pose a steric impediment for the binding of UTP, as originally postulated by Carrey (Carrey 1995a).

CAD is also phosphorylated at residue T456 within the L1 domain via the MAP kinase (Erk1/2) signaling cascade (Graves et al. 2000). This modification converts UTP from an inhibitor to a modest activator and stimulates the activation by PRPP. The expected similarity with other CPSs suggests that this residue is far from the allosteric domain and thus, it is difficult to adventure a possible mechanism that explains the effect of the modification on the binding of the allosteric effectors or in the modulation of the activity.

So far, we focused the attention on CAD and ignore the comparison with the CAD-like protein from fungi. Although not shown in the alignment of Fig. 17.6, the GLN and SYN domains of CAD-like proteins are predicted to share the overall structure with other CPSs. In particular, *S. cerevisiae* URA2 has a unique binding site for UTP, likely localized at the putative L4 allosteric domain, which regulates both CPS and ATC activities (Antonelli et al. 1998), and at least in vitro, does not appear to be phosphorylated by PKA (Denis-Duphil et al. 1990). Thus, the regulatory domain appears to be conserved in all CPSs, including some that do not respond to allosteric effectors. Indeed, the unregulated arginine-specific CPS in *S. cerevisiae* (Fig. 17.2a) holds a latent allosteric domain that binds UMP and IMP, although the nucleotides do not seem to have any effect on the activity of the protein (Eroglu and Powers-Lee 2002).

A Reaction Tunnel

The *E. coli* CPS structure revealed an exceptional feature: the GLN active site is 45 Å away from the bicarbonate phosphorylation site, and this one is 35 Å apart from the phosphorylation site for carbamate (Thoden et al. 1997). Thus, the unstable reaction intermediates must be shuttled between the active centers without exposure to the bulk solvent. In *E. coli* CPS, a narrow tunnel, more than 90 Å long, runs through the interior of the protein, connecting all three active sites (Fig. 17.3b). The extensive interaction between the small subunit and the first half of the large subunit provide a safe pathway for the delivery of ammonia to the L1 active site where it will react with carboxyphosphate to form carbamate. This "ammonia" tunnel is lined by non-reactive residues that can form H-bonds with ammonia. The second portion of the tunnel transports carbamate from the L1 to the L3 active sites where it is phosphorylated to CP. This "carbamate" tunnel passes across the twofold rotational axis between the two synthetic domains and is delimited by few charged side chains to prevent the reaction with carbamate.

In human CPS-1, many of the residues that in *E. coli* CPS line the ammonia channel are conserved, but the Gly at the exit of the channel at the interface between the GLN and SYN domains is replaced by a Gln (Q318) that blocks the path (de Cima et al. 2015). This blockage explains why some CPS-1 enzymes have a complete GLN active site (e.g. CPS-1 from *Rana catesbiana*), including the catalytic Cys and His residues, but are still unable to deliver the ammonia hydrolyzed from glutamine to the SYN domain (Saeed-Kothe and Powers-Lee 2003) (de Cima et al. 2015). Likely in CPS-1, ammonia intake follows a different path to that delineated in *E. coli* CPS. It has been proposed that this alternative entry could also exist in other CPSs, including CAD, since they are active with external ammonia although at high concentrations (de Cima et al. 2015).

In *E. coli* CPS, the intramolecular tunnel is formed even in the absence of a full set of substrates. This could be due to the fact that the protein is always crystallized in presence of the allosteric activator, ornithine. In turn, the structures of human CPS-1 free or bound to acetylglutamate show important differences in the tunnel that explain in part the mechanism of activation by the co-factor (Fig. 17.4c, d). Binding of acetylglutamate causes small changes in the allosteric domain that are propagated to the L3 domain through the movement of a potassium binding loop also present in *E. coli* CPS (de Cima et al. 2015). This movement induces long-range conformational changes that stabilize the two phosphorylation sites and promote the correct formation of the reaction tunnel (de Cima et al. 2015). Similar changes triggered by the allosteric effectors are expected to occur both in *E. coli* CPS and in the SYN domain of CAD.

A Cooperative ATC Domain

The arrangement of the domains along the CAD polypeptide does not follow the order of the reactions in de novo pathway (Fig. 17.2a). The ATC domain, located at the C-end of the protein, catalyzes the second step after CPS, the transfer of the carbamoyl group of CP to the α-amino group of aspartate (Asp) to form carbamoyl aspartate (Figs. 17.1 and 17.2a) (Jones 1980).

ATC is likely one of best characterized enzymes ever. In particular, the amount of biochemical and structural information obtained on *E. coli* ATC is impressive, becoming a text-book enzyme and, together with hemoglobin, a paradigm of protein cooperativity and allosteric regulation [reviewed in (Allewell 1989; Jacobson and Stark 1973; Lipscomb 1994)]. *E. coli* ATC consists of twelve polypeptides chains: six catalytic chains of 33 kDa organized as two catalytic trimers, and six regulatory chains of 17 kDa forming three regulatory dimers (Fig. 17.2b). The affinity for the substrates is regulated both by the binding of substrates to other active sites (homotropic effects) and by the binding of nucleotide effectors to the regulatory subunits (heterotropic effects). This regulation requires the communication of conformational changes between chains and subunits (Schachman 1988). Indeed, the dissociation of the holoenzyme results in isolated catalytic trimers that are more active but lack cooperativity and allosteric regulation (Gerhart and Holoubek 1967). Numerous crystal structures of *E. coli* ATC, including the active (R) and inactive (T) states of the holoenzyme (Fig. 17.2b), and the unregulated catalytic trimer, helped to unveil the complexity of the catalytic and regulatory mechanisms (Lipscomb and Kantrowitz 2011). There are also crystal structures for other forms of ATC found in bacteria, such as the non-regulated catalytic trimers of *Bacillus subtilis* or the DHO-associated ATC from *Aquifex aeolicus* (Fig. 17.2c) (Stevens et al. 1991; Zhang et al. 2009). However, until recently, there were no structures of any eukaryotic ATC.

The ATC domain of human CAD was the first eukaryotic ATC to be structurally characterized (Ruiz-Ramos et al. 2013, 2016). As predicted (Scully and Evans 1991), the domain shows high similarity with the catalytic subunits of bacterial ATCs. The enzyme is a homotrimer with equilateral triangular appearance, and three active sites located in between subunits (Fig. 17.7a). Each subunit is divided in an N-terminal domain and a C-terminal domain of similar size, both structured by a central β-sheet of five parallel strands flanked by α-helices (Fig. 17.7b). The active site locates at the cleft between the N- and C-domains with participation of a loop (CP-loop) from the adjacent subunit. The N-terminal domain provides most of the contacts with the other subunits and holds the binding site for CP, whereas the C-domain occupies an external position at the trimer and provides the binding site for Asp. There are two mobile loops, the CP-loop at the N-domain (named loop 80s in *E. coli* ATC) and the Asp-loop at the C-domain (loop 200s in *E. coli* ATC) that appear flexibly disordered in the absence of substrates (Fig. 17.7b).

ATC catalyzes an ordered reaction, with CP binding before Asp, and carbamoyl aspartate leaving before phosphate (Collins and Stark 1969; Porter et al. 1969). CP

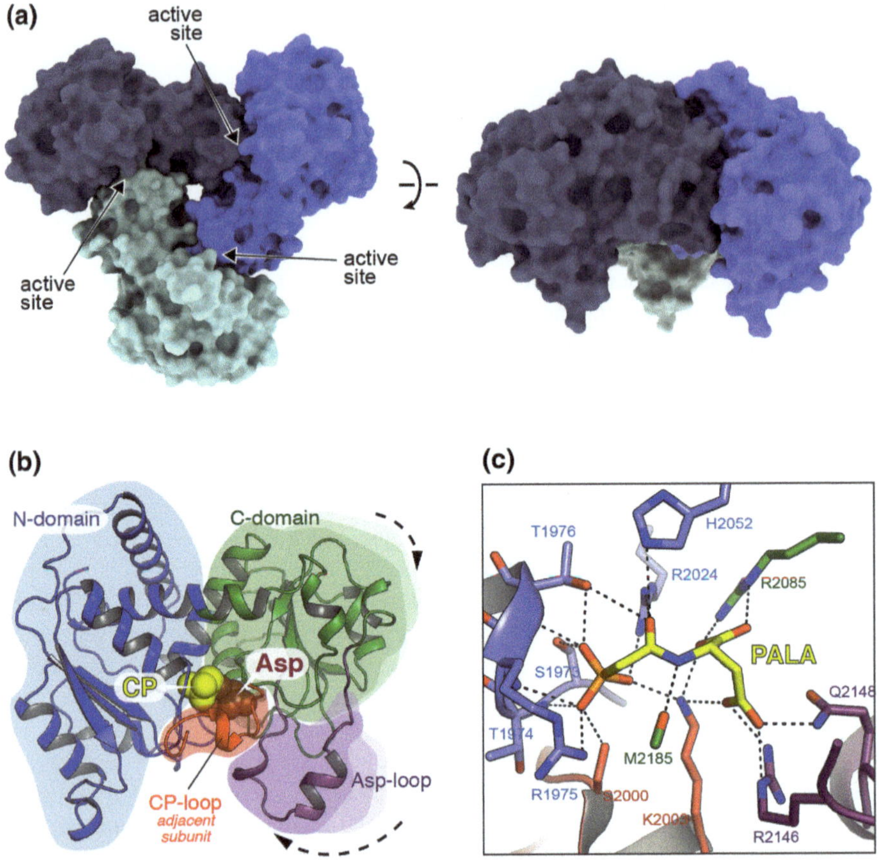

Fig. 17.7 ATC domain of human CAD. **a** Space filling representation of the trimer formed by the isolated ATC domain of human CAD in two perpendicular orientations. Each subunit is represented in a different color. **b** Cartoon representation of human ATC subunit, with the N- and C-domains represented in blue and green, respectively, and the Asp-loop depicted in purple. The arrows indicate the closure movement of the subunit upon PALA binding. PALA is shown in spheres, with the CP and the Asp moieties colored in yellow and red, respectively. The CP-loop from the adjacent subunit is depicted in red. **c** Detail of the interactions between PALA and residues at the active site

binding at the N-domain induces the positioning of the CP-loop from the adjacent subunit at the active site, and promotes a partial hinge-closure of the C-domain. These conformational changes favor the binding of Asp to a contiguous pocket in the active site. Binding of both substrates induces a further approximation between the N- and C-domains and a rigid body rotation of the Asp-loop that closes the active site (Fig. 17.7b). Overall, these movements are proposed to compress the substrates and favor the reaction (Collins and Stark 1969; Ruiz-Ramos et al. 2016).

The active sites of human and *E. coli* ATCs are indistinguishable, confirming a common reaction mechanism for both enzymes (Collins and Stark 1969; Gouaux et al. 1987; Ruiz-Ramos et al. 2016). Catalysis occurs by nucleophilic attack of the amino nitrogen of Asp on the amide carbon of CP. The phosphate group of CP or the side chain of residue H2052 (H134 in *E. coli* ATC) are likely candidates to act as general base, deprotonating and increasing the nucleophilicity of the α-amino group of aspartate, whereas the positive dipole of the nearby α-helix and the side chains of residues R2024 (R105) and H2052 (H134) act in concert to polarize the carbonyl group of CP and to stabilize the transition state of the reaction (Fig. 17.7c).

Figure 17.7c shows how every polar atom of PALA interacts with the protein, explaining the nanomolar affinity for the inhibitor (Newell et al. 1989; Ruiz-Ramos et al. 2016). The dissociation of PALA from this highly stable complex is impossible without reversing all the conformational changes, and this is a slow process that explains why the molecule acts as a nearly irreversible inhibitor. An unexpected fact is that PALA binds to human ATC with negative cooperativity: the binding of PALA to one subunit decreases the affinity for the inhibitor in the other active sites (Ruiz-Ramos et al. 2016). Indeed, only two subunits of human ATC show high affinity for PALA, while affinity for the third site is 100-fold lower. This difference with the *E. coli* ATC catalytic trimer, is likely due to a communication of conformational changes between the subunits in the human enzyme. Since PALA is proposed to resemble the transition state of the reaction, the negative cooperativity effect also suggests that in CAD, the ATC trimer might work more efficiently with only two active sites catalyzing the reaction at a time. Indeed, at high substrate concentrations, the activity of ATC is partially inhibited, suggesting that a trimer with the three subunits forced to work simultaneously might present additional intersubunit interactions that slow down the conformational movements required for catalysis (LiCata and Allewell 1997; Ruiz-Ramos et al. 2016).

In CAD, ATC catalyzes the reaction ~50-fold faster than CPS, and it has a low K_M for CP (Coleman et al. 1977; Qiu and Davidson 2000). This should facilitate that as soon as the unstable CP molecule is delivered by CPS, it will be trapped and converted by ATC. A possible asymmetry in the functioning of ATC is interesting to envision the performance of the overall activity of CAD. Perhaps the ATC subunits are fired one at a time, in some sequential process coordinated with CPS.

A DHO Domain in the Midst of CAD

DHO is a ubiquitous Zn metalloenzyme catalyzing the reversible interconversion of carbamoyl aspartate to dihydroorotate –the precursor of the pyrimidine ring– in the third step of the de novo pyrimidine biosynthesis (Fig. 17.1). Despite full conservation of the reaction, this enzyme adopts an intriguing number of different forms (Fields et al. 1999; Grande-Garcia et al. 2014). DHOs can be monomeric or dimeric proteins that function independently (e.g. *E. coli* DHO) or in association with ATC

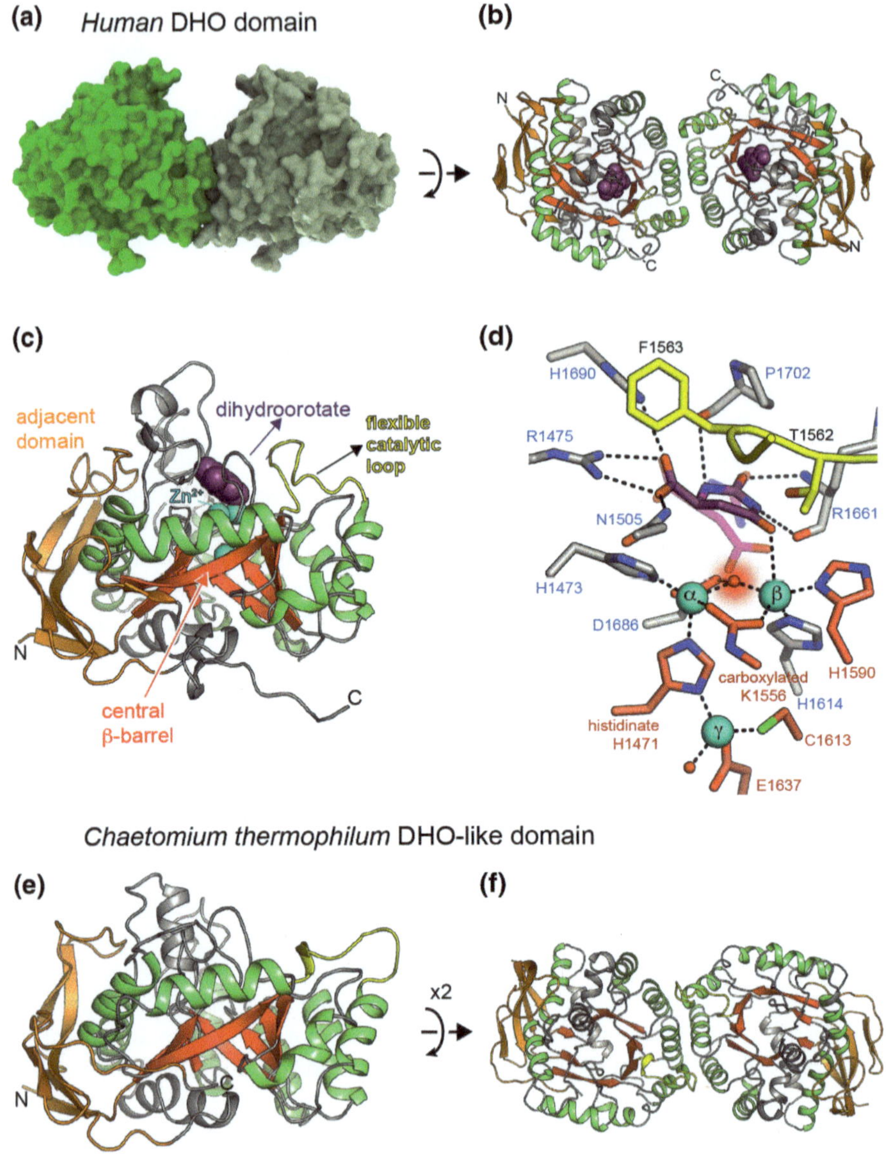

(e.g. *A. aeolicus* DHO) (Fig. 17.2b, c), or alternatively, DHO can be fused as a functional domain within CAD (Fig. 17.2a). Moreover, some DHOs seem to play a non-enzymatic role, such as the inactive DHO-like domain linking the CPS and ATC activities in the fungal CAD-like proteins (Fig. 17.2a) (Souciet et al. 1989), or the inactive DHOs that in some bacteria (e.g. *Pseudomonas aeruginosa*) form non-covalent complexes with ATC (Schurr et al. 1995). Thus, besides the well-characterized enzymatic function, DHOs seem to perform other less understood functions in the midst of the pyrimidine enzymatic machinery.

◀**Fig. 17.8** DHO domain of human CAD. **a** Space filling representation of the dimer formed by the isolated DHO domain of human CAD. **b** Cartoon view of the dimer with two molecules of dihydroorotate bound at the active sites. **c** Representation of the DHO subunit with Zn^{2+} ions and a dihydroorotate molecule represented as cyan and purples spheres, respectively. **d** Detail of the active site with dihydroorotate bound. A molecule of carbamoyl aspartate is shown in semi-transparent to show that the β-carboxylate occupies the position of the bridging water. Residues from the central β-barrel and from the loops above the barrel are colored with carbons in red and grey, respectively. The flexible loop is depicted in yellow. **e** Cartoon representation of *C. thermophilum* inactive DHO-like subunit (**e**) and dimer (**f**). The orientations and colors are as in (**b**) and (**c**) to highlight similarities with human DHO

Although the crystal structures of different bacterial DHOs were known (Fig. 17.2b, c) (Thoden et al. 2001; Zhang et al. 2009), there were no structures of any eukaryotic counterpart. The DHO domain of human CAD was the first eukaryotic DHO for which the crystal structure was solved (Grande-Garcia et al. 2014; Lallous et al. 2012). In agreement with studies reporting that the proteolytic fragment of CAD retaining DHO activity forms dimers (Davidson et al. 1981), the isolated human DHO domain was shown to be a homodimer in solution (Fig. 17.8a, b) (Lallous et al. 2012). The overall subunit fold is similar to bacterial counterparts, and is also shared with a large number of enzymes –most of which catalyze the hydrolysis of substrates at amide or ester groups– forming the amidohydrolase superfamily of proteins (Holm and Sander 1997). Each subunit is structured in a "TIM" barrel with eight strands of parallel β-sheet flanked on the outer surface by α-helices, and a smaller adjacent β-stranded subdomain composed of the N- and C-terminal regions of the protein (Fig. 17.8c). This adjacent subdomain is continued by a C-terminal extension that stretches through the bottom of the barrel, and places the N- and C-ends on opposite sides of the globular domain, a convenient arrangement for the intercalation of DHO in the middle of CAD.

As in other members of the amidohydrolase superfamily, the active center of the DHO domain of CAD locates on the C-terminal edge of the β-barrel, a cavity shaped by the loops connecting the β-strands with the outer α-helices (Fig. 17.8c). There are two Zn^{2+} ions (Zn-α and Zn-β) coordinated by four His and one Asp at conserved positions in the active site (Fig. 17.8d). The metals are bridged by the side chain of a carboxylated Lys and by a water molecule that is activated by the metal ions for nucleophilic attack (Porter et al. 2004). Human DHO has a third Zn^{2+} ion (Zn-γ) approximately at the center of the β-barrel, which is not found in bacterial DHOs or in other members of the amidohydrolase superfamily (Fig. 17.8c, d). This Zn-γ interacts with Zn-α through the side chain of a rare histidine (H1471) with negative charge (called histidinate anion). Although Zn-γ appears to be too far to participate directly in catalysis, mutations that impede the binding of this metal are shown to reduce the activity of the protein to half (Grande-Garcia et al. 2014). Interestingly, the introduction by site-directed mutagenesis of the third Zn^{2+} into a bacterial DHO, increased the activity and stability of the protein (Huang and Huang 2015). These results suggest that Zn-γ could play a role in the stabilization of the DHO domain of CAD, and perhaps influence the electrostatic environment at the active site (Grande-Garcia et al. 2014).

Despite having a low sequence identity (15%), the human and *E. coli* DHOs active sites are virtually identical and both enzymes are proposed to share a common catalytic mechanism (Grande-Garcia et al. 2014; Porter et al. 2004). The conversion of carbamoyl aspartate to dihydroorotate is reversible and pH-dependent, with the forward and reverse reactions reaching equilibrium at approximately neutral pH (Christopherson and Jones 1980). In the synthesis of dihydroorotate, favored at low pH, the water molecule bridging Zn-α and Zn-β is displaced by the binding of the side chain of carbamoyl aspartate (Fig. 17.8d). The metals neutralize the negative charge of the carboxylate group, increasing its susceptibility to a nucleophilic attack by the amino group of carbamoyl aspartate. To favor the reaction, the amino group is deprotonated by an Asp (D1686 in human DHO) acting as the general base. The reaction proceeds by formation of a tetrahedral intermediate that becomes stabilized by the metal ions. Then, the OH leaving group is protonated, collapsing the transition state and releasing dihydroorotate and a water molecule that is retained between the two metal ions. In turn, the hydrolysis of dihydroorotate is favored above pH 8 and involves the nucleophilic attack of the bridging water (or hydroxide ion) to the amide bond of dihydroorotate (Fig. 17.8d). This reaction, rather than the forward synthesis of dihydroorotate, is favored under physiological conditions (Christopherson and Jones 1980). It is proposed that the equilibrium of the reactions could be displaced toward the synthesis of dihydroorotate by the location of CAD near the mitochondria, which may facilitate the efficient capture of dihydroorotate by DHODH, the enzyme catalyzing the next step in de novo pathway (Evans and Guy 2004) (Fig. 17.1).

The DHO subunit is rigidly built by tight hydrophobic packing of the central β-barrel and the α-helical palisade. Indeed, different crystal structures of human and *E. coli* DHOs show no significant conformational changes upon binding of substrates or different inhibitors at the active site. There is only one exception, a flexible loop that adopts an open solvent-exposed position or a closed conformation whether dihydroorotate or carbamoyl aspartate are bound, respectively, to the active site (del Cano-Ochoa et al. 2018; Grande-Garcia et al. 2014; Lee et al. 2005) (Fig. 17.8c, d). This loop plays an important catalytic role, placing the substrate in correct orientation, stabilizing the transition state of the reaction and excluding the solvent molecules from the active site.

The position of the flexible loops near the dimerization interface of human DHO suggests a possible coordination between both active sites (Fig. 17.8b). However, enzymatic assays have failed to detect signs of cooperativity in human DHO (del Cano-Ochoa et al. 2018; Grande-Garcia et al. 2014). In contrast, in *E. coli* DHO, the subunits dimerize through the loops above the β-barrel, rather than by lateral contacts as in human DHO, and the movements of the flexible loop appear related to the cooperativity effect observed between the two confronted active sites (Lee et al. 2005) (Fig. 17.2b). In the *A. aeolicus* DHO/ATC complex, on the other hand, the DHO subunits make lateral contacts but the subunits are rotated 180° compared to human DHO (Fig. 17.2c). As a result, adjacent DHO subunits direct their active sites towards different ATC trimers. Interestingly, *A. aeolicus* DHO lacks a catalytic flexible loop and requires the interaction with a loop from ATC to complete the active site and achieve maximal activity (Prange et al. 2019; Zhang et al. 2009). It is intriguing how some DHOs equipped with a flexible loop are catalytically independent, whereas others have an unfinished active site that needs to be complemented by a stable association with ATC. Then, we wonder why the DHO domain of CAD, which has a catalytic flexible loop, is covalently engaged to an ATC domain that is not necessary for its activity. The answer to this question is that probably the linkage of both domains is not intended to increase the efficacy of DHO and must fulfill other purposes. Indeed, the recent crystal structure of the inactive DHO-like domain from the CAD-like protein of the fungus *Chaetomium thermophilum* (Fig. 17.8e) proved that as predicted (Souciet et al. 1989), the catalytic residues and Zn^{2+} ions are missing from the active site (Moreno-Morcillo et al. 2017). However, this inactive domain forms homodimers in an identical quaternary arrangement as human DHO (Fig. 17.8f). The preservation of the dimer architecture strongly suggests that the DHO domain plays a conserved structural role within the context of the full-length protein oligomerization. The linkage with ATC must harness both domains for the assembly of a multienzymatic protein machinery that appears conserved both in animals and fungi (Moreno-Morcillo et al. 2017).

Putting the Pieces Together for the Pyrimidine Factory

Based on the expected similarity of the GLN and SYN domains of CAD with *E. coli* CPS and human CPS-1 and on the crystal structures of the human DHO and ATC structures, we can build a hypothetical model of the full-length protein with the four domains arranged as beads on a string (Fig. 17.9a). However, this model is utterly useless unless we define how this protein self-assembles into larger particles that explain the communication and coordination between the different activities.

Early studies reported that CAD self-assembles into a mixture of oligomers (Coleman et al. 1977), mostly hexamers of ~1.5 MDa in size (Lee et al. 1985). Since the proteolytic fragments retaining DHO and ATC activities were found to form dimers and trimers, respectively (Davidson et al. 1981; Hemmens and Carrey 1994; Kelly et al. 1986), an idea prevailed, that the CAD hexamers could result from the association of three proteins through their respective ATC domains, and that two of these trimers could further dimerize through their DHO domains (Carrey 1995b; Evans 1986). A similar organization was proposed for the architecture of the CAD-like protein in *Neurospora* (Makoff et al. 1978), although in this model the dimerization of the trimers was proposed to be mediated by the CPS region.

The key role of ATC in the molecular organization of CAD was demonstrated by Qiu and Davidson, who proved that mutations in the predicted ATC trimer interface

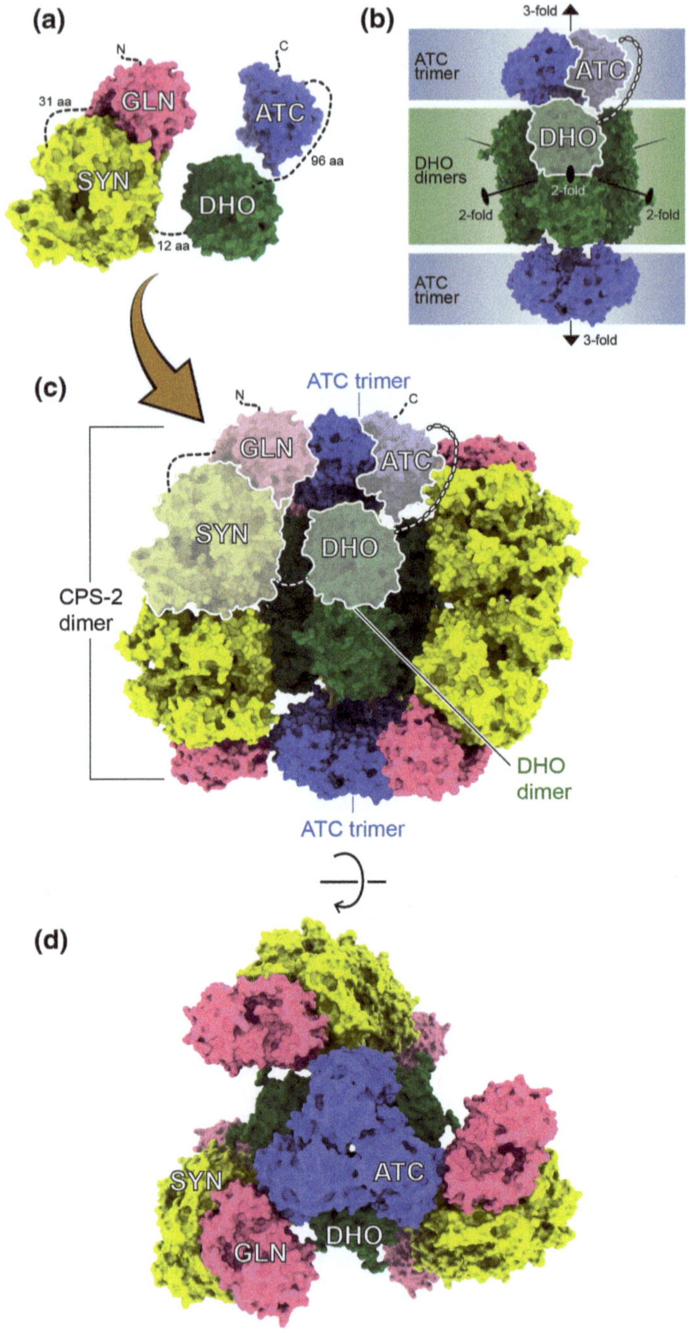

(a)
N
31 aa
GLN
SYN
C
ATC
96 aa
DHO
12 aa
(b)
3-fold
ATC trimer
ATC
DHO dimers
DHO
2-fold
2-fold
2-fold
ATC trimer
3-fold
(c)
ATC trimer
N
C
GLN
ATC
SYN
DHO
CPS-2 dimer
DHO dimer
ATC trimer
(d)
SYN
ATC
GLN
DHO

◂**Fig. 17.9** Model of the architecture of CAD. **a** Model of CAD full-length protein with the GLN, SYN, DHO and ATC domains as beads on a string. Dashed lines indicate linker regions of different lengths. **b** Model of the bi-functional DHO-ATC construct forming an hexamer (or "dimer of trimers"). **c**, **d** Hypothetical model of CAD hexameric particle in two perpendicular orientations

caused the dissociation of CAD hexamers (Qiu and Davidson 1998, 2000). However, this study ruled out the possibility that the DHO or CPS domains participated directly in the oligomerization of the particle.

To challenge the idea proposed by Carrey that CAD could assemble as a "dimer of trimers" (Carrey 1995b), our group made a construct spanning the DHO and ATC domains of human CAD, including the long linker in between. This bi-functional construct was shown to form stable homo-hexamers in solution, and point mutations in the ATC or DHO oligomerization interfaces resulted in the formation of dimers and trimers, respectively (Moreno-Morcillo et al. 2017). These results indicate that CAD indeed assembles as a "dimer of trimers" and suggest that the DHO and ATC domains provide a central scaffold for the architecture of the particle (Fig. 17.9b). Moreover, we proved that despite having an inactive DHO-like domain, a similar construct from the CAD-like protein from *C. thermophilum* behaves in the same manner, supporting that CAD and CAD-like particles share a common architecture (Moreno-Morcillo et al. 2017).

Taking into consideration previous models (Carrey 1995b; Evans 1986; Makoff et al. 1978) and gathering all the current structural information, we propose a plausible blueprint for the architecture of CAD (Fig. 17.9c, d). The DHO and ATC domains form the central axis of the particle. The ATC monomers are arranged into two trimers occupying apical positions and with the active sites face-to-face. Three DHO dimers interpose in between the ATC trimers, with their long axes in parallel to the threefold axis of the ATC domains and with the active sites facing inwards. This results in a closed hexameric structure of 190 × 100 Å, with D3 symmetry (a threefold axis with three perpendicular twofold axes) and a delimited inner space of ~120 × 50 Å (Moreno-Morcillo et al. 2017; Moreno-Morcillo and Ramon-Maiques 2017). This disposition of the ATC and DHO domains is somehow reminiscent of the noncovalent association between the regulatory and catalytic subunits of *E. coli* ATC or between the ATC and DHO enzymes of *A. aeolicus* (Fig. 17.2b, c).

Negative staining electron microscopy data indicated that the hexamer formed by the DHO-ATC construct is quite flexible (Moreno-Morcillo et al. 2017). The long sequence connecting the DHO and ATC domains likely restrains the relative position of both domains, but allows certain flexibility between them. Perhaps the DHO dimers could tilt around their twofold axes, allowing the separation or approximation of the ATC trimers, into tighter or more relaxed states that limit the conformational changes required for catalysis, in a similar manner to the allosteric transitions described in the *E. coli* ATC holoenzyme (Fig. 17.2b) (Lipscomb 1994; Lipscomb and Kantrowitz 2011; Ruiz-Ramos et al. 2016).

Some regions of the long linker connecting DHO and ATC could make interactions with both domains, and also perhaps with the GLN/SYN domains, adding stability to the assembly. However, a part of the linker must be exposed to the solvent to explain the high sensitivity to proteases and also, for being phosphorylated. Two different studies reported that residue S1859 at the linker sequence in human CAD is phosphorylated by S6 kinase in the downstream of mTORC1 pathway, and that this modification increases the rate of de novo pyrimidine synthesis (Ben-Sahra et al. 2013; Robitaille et al. 2013). Interestingly, the phosphorylation also favors the oligomerization of CAD (Robitaille et al. 2013), in agreement with a possible role of the linker in the stabilization of the particle.

To complete the model, six CPS-2, formed by the GLN and SYN domains, need to be added to the central DHO-ATC framework. It is clear by similarity to other CPSs and by biochemical data that the GLN and SYN domains will form a tight heterodimer. Also, the observation that proteolytic fragments of *Neurospora* CAD-like protein with CPS activity form dimers (Makoff et al. 1978), suggests that CPS-2 could further dimerize through the allosteric regions, similarly to the dimers observed in the crystal structures of *E. coli* CPS and human CPS-1 (Figs. 17.3c and 17.4b). We propose that the six CPS-2 could be organized as three dimers surrounding the central DHO-ATC assembly, with the twofold axes in the equatorial plane of this globular complex of $\sim 200 \times 200$ Å (Fig. 17.9c, d).

Despite the symmetrical beauty, this model could be wrong and needs more structural data to be validated. A detailed characterization of the particle, probably by using high resolution cryo-microscopy, will shed light on the evolutionary advantages of a single multienzymatic protein versus the mono-functional homologs in prokaryotes. Till then, it can only be hypothesized that the engineering of several enzymatic activities into a single polypeptide likely improves the metabolic efficiency in a number of ways (Davidson et al. 1993; Stark 1977). Firstly, similar to a bacterial operon, the fusion of the enzymes into a single polypeptide ensures the coordinated and stoichiometric production of the enzymatic activities. In addition, the covalent linkage favors the co-localization of the different activities within the cell. The proximity between domains might also enhance the specificity for their association, and this could also result in increased stability compared to the isolated proteins. Another advantage for the formation of a complex such as the one proposed in Fig. 17.9, is the close proximity between active sites, which should favor the transfer or "channeling" of intermediate metabolites. The concept of channeling was introduced by Davis to explain the existence of two different pools of CP in the cell for the de novo biosynthesis of pyrimidines and arginine (Davis 1972). The fact that in de novo pyrimidine synthesis there is no significant accumulation of any of the intermediates between bicarbonate and UMP, led M. E. Jones to consider that the enzymes could have a special structural conformation that ensures products and substrates to reach their destination (Jones 1971, 1980; Shoaf and Jones 1971). The proposed model accounts for the partial channeling of CP reported by different studies in mammalian CAD and in the CAD-like from yeast (Christopherson and Jones 1980; Mally et al. 1980; Otsuki et al. 1982; Penverne et al. 1994). Channeling starts at the GLN domain following the ~ 90 Å tunnel running through the interior of the SYN domain. The orientation of the CPS-2 dimers in the particle would be such that the exit of this intramolecular tunnel could face the gaps between the DHO dimers. In this manner, the CP would be delivered from the SYN domains directly into the central cavity of the particle where it would diffuse a short distance to reach the ATC active site without exposure to the bulk solvent.

This large multi-subunit and multi-domain assembly might also allow for a more complex mechanism of control. It is possible that binding of the allosteric effectors, UTP and PRPP, at the regulatory domains could induce conformational changes as those observed in CPS-1 upon acetylglutamate binding (de Cima et al. 2015). Changes in the relative orientation of the CPS-2 dimer could be transmitted to the rest of the protein, inducing the rotation and translation of the DHO dimers and ATC trimers along their respective symmetry axes, and reversely, conformational changes in the ATC trimer due to substrate or PALA binding, could be communicated to the outer CPS-2 dimers. This conformational cross-talk would explain the mutual effects between both enzymatic activities that E. Carrey described as "reciprocal allostery" (Irvine et al. 1997). The concerted movements of the domains could modulate the affinity for the substrates, the rate and the coupling of the reactions, the channeling of intermediates as well as the flux of substrates and products in and out the particle. In a stretch of imagination, one can envision that the hexameric particle could transit through a series of conformational states with different overall activity, and that a number of external inputs such as allosteric effectors, phosphorylations and other posttranslational modifications or interactions with other cellular components could displace the conformational equilibrium, thus contributing to the regulation of the de novo pyrimidine pathway.

CAD in Human Diseases

Whereas a complete lack of activity in any of the enzymatic activities for de novo biosynthesis of pyrimidines appears to be incompatible with life, mutations causing impairment of the activities associate with different human disorders. Defects in DHODH activity are associated with Miller syndrome, a malformation disorder with no intellectual damage (Ng et al. 2010). Also, UMPS malfunctions cause orotic aciduria, a condition in which the accumulation of orotate produces intellectual and motor impairment, seizures and immunodeficiency (Imaeda et al. 1998; Loffler et al. 2015). However, until recently, no human disease had been related to a partial loss of CAD activity. Mutations compromising CAD function were known to cause severe malformations and death in model organisms such as *Drosophila* (rudimentary phenotype) (Norby 1970), zebrafish (perplexed phenotype) (Willer et al. 2005) and *C. elegans* (Franks et al. 2006). These results led to the assumption that a reduced function of CAD would be lethal and explained that no human diseases were associated with failures in this protein.

However, in 2015, H. Freeze and colleagues described the first case of a 4-year old boy with a severe impairment of protein glycosylation caused by a partial deficit in CAD (Ng et al. 2015). The identified mutation, R2024Q, maps in the active site of ATC, and it was shown to severely reduce the activity of the isolated ATC domain and its affinity for the substrates (Ruiz-Ramos et al. 2016). One year later, a second study reported three families with young children carrying mutations in CAD that caused developmental delay, epileptic encephalopathy, anemia and seizures (Koch et al. 2017). While the untreated disease course is lethal in early childhood, both studies reported that patients survived upon an oral treatment with uridine, which compensates by salvage the defects of the de novo pyrimidine synthesis. Since then, new cases of children with potential CAD-deficits have continued appearing (H. Freeze personal communication).

The identification of CAD-deficiency as a treatable metabolic disorder stresses the importance of an early detection in newborns and young children (Koch et al. 2017). However, the diagnosis is difficult (e.g. numerous symptoms, pyrimidine levels in urine might seem normal) and so far, patients were only diagnosed after complete exome sequencing (Koch et al. 2017; Ng et al. 2015). As these sequencing techniques become more affordable, a major challenge that remains, giving our fragmented knowledge about CAD, is the distinction between disease-causing mutations and simple protein polymorphisms. Thus, deciphering the architecture of CAD is not only key to unveil the catalytic and regulatory mechanisms of a central metabolic machinery and to guide in the design of specific inhibitors with a potential chemotherapeutic use, but also is of paramount importance for helping in the correct diagnosis and treatment of patients.

References

Allewell NM (1989) *Escherichia coli* aspartate transcarbamoylase: structure, energetics, and catalytic and regulatory mechanisms. Annu Rev Biophys Biophys Chem 18:71–92

Anderson PM (1986) Carbamoyl-phosphate synthetase: an example of effects on enzyme properties of shifting an equilibrium between active monomer and active oligomer. Biochemistry 25(19):5576–5582

Anderson PM, Meister A (1965) Evidence for an activated form of carbon dioxide in the reaction catalyzed by *Escherichia coli* carbamyl phosphate synthetase. Biochemistry 4(12):2803–2809

Anderson PM, Meister A (1966) Bicarbonate-dependent cleavage of adenosine triphosphate and other reactions catalyzed by *Escherichia coli* carbamyl phosphate synthetase. Biochemistry 5 (10):3157–3163

Antonelli R, Estevez L, Denis-Duphil M (1998) Carbamyl-phosphate synthetase domain of the yeast multifunctional protein Ura2 is necessary for aspartate transcarbamylase inhibition by UTP. FEBS Lett 422(2):170–174

Ben-Sahra I, Howell JJ, Asara JM et al (2013) Stimulation of de novo pyrimidine synthesis by growth signaling through mTOR and S6K1. Science 339(6125):1323–1328

Braxton BL, Mullins LS, Raushel FM et al (1992) Quantifying the allosteric properties of *Escherichia coli* carbamyl phosphate synthetase: determination of thermodynamic linked-function parameters in an ordered kinetic mechanism. Biochemistry 31(8):2309–2316

Braxton BL, Mullins LS, Raushel FM et al (1996) Allosteric effects of carbamoyl phosphate synthetase from *Escherichia coli* are entropy-driven. Biochemistry 35(36):11918–11924

Britton HG, Rubio V, Grisolia S (1979) Mechanism of carbamoyl-phosphate synthetase. Properties of the two binding sites for ATP. Eur J Biochem 102(2):521–530

Brown EG (1998) Pyrimdines, ring nitrogen and key biomolecules: the biochemistry of N-heterocycles. Springer

Bueso J, Cervera J, Fresquet V et al (1999) Photoaffinity labeling with the activator IMP and site-directed mutagenesis of histidine 995 of carbamoyl phosphate synthetase from *Escherichia coli* demonstrate that the binding site for IMP overlaps with that for the inhibitor UMP. Biochemistry 38(13):3910–3917

Carrey EA (1995a) Key enzymes in the biosynthesis of purines and pyrimidines: their regulation by allosteric effectors and by phosphorylation. Biochem Soc Trans 23(4):899–902

Carrey EA (1995b) The shape of CAD J. N. Davidson Paths to pyrimidines - an international newsletter

Carrey EA, Hardie DG (1988) Mapping of catalytic domains and phosphorylation sites in the multifunctional pyrimidine-biosynthetic protein CAD. Eur J Biochem 171(3):583–588

Carrey EA, Campbell DG, Hardie DG (1985) Phosphorylation and activation of hamster carbamyl phosphate synthetase II by cAMP-dependent protein kinase. A novel mechanism for regulation of pyrimidine nucleotide biosynthesis. EMBO J 4(13B):3735–3742

Cervera J, Conejero-Lara F, Ruiz-Sanz J et al (1993) The influence of effectors and subunit interactions on *Escherichia coli* carbamoyl-phosphate synthetase studied by differential scanning calorimetry. J Biol Chem 268(17):12504–12511

Cervera J, Bendala E, Britton HG et al (1996) Photoaffinity labeling with UMP of lysine 992 of carbamyl phosphate synthetase from *Escherichia coli* allows identification of the binding site for the pyrimidine inhibitor. Biochemistry 35(22):7247–7255

Chaparian MG, Evans DR (1991) The catalytic mechanism of the amidotransferase domain of the Syrian hamster multifunctional protein CAD. Evidence for a CAD-glutamyl covalent intermediate in the formation of carbamyl phosphate. J Biol Chem 266(6):3387–3395

Christopherson RI, Jones ME (1980) The overall synthesis of L-5,6-dihydroorotate by multienzymatic protein pyr1-3 from hamster cells. Kinetic studies, substrate channeling, and the effects of inhibitors. J Biol Chem 255(23):11381–11395

Coleman PF, Suttle DP, Stark GR (1977) Purification from hamster cells of the multifunctional protein that initiates de novo synthesis of pyrimidine nucleotides. J Biol Chem 252(18):6379–6385

Collins KD, Stark GR (1969) Aspartate transcarbamylase. Studies of the catalytic subunit by ultraviolet difference spectroscopy. J Biol Chem 244(7):1869–1877

Collins KD, Stark GR (1971) Aspartate transcarbamylase interaction with the transition state analogue N-(phosphonacetyl)-L-aspartate. J Biol Chem 246(21):6599–6605

Czerwinski RM, Mareya SM, Raushel FM (1995) Regulatory changes in the control of carbamoyl phosphate synthetase induced by truncation and mutagenesis of the allosteric binding domain. Biochemistry 34(42):13920–13927

Davidson JN, Patterson D (1979) Alteration in structure of multifunctional protein from Chinese hamster ovary cells defective in pyrimidine biosynthesis. Proc Natl Acad Sci U S A 76(4):1731–1735

Davidson JN, Rumsby PC, Tamaren J (1981) Organization of a multifunctional protein in pyrimidine biosynthesis. Analyses of active, tryptic fragments. J Biol Chem 256(10):5220–5225

Davidson JN, Rao GN, Niswander L et al (1990) Organization and nucleotide sequence of the 3′ end of the human CAD gene. DNA Cell Biol 9(9):667–676

Davidson JN, Chen KC, Jamison RS et al (1993) The evolutionary history of the first three enzymes in pyrimidine biosynthesis. BioEssays 15(3):157–164

Davis RH (1972) Metabolite distribution in cells. Science 178(4063):835–840

de Cima S, Polo LM, Diez-Fernandez C et al (2015) Structure of human carbamoyl phosphate synthetase: deciphering the on/off switch of human ureagenesis. Sci Rep 5:16950 del Cano-Ochoa F, Grande-Garcia A, Reverte-Lopez M et al (2018) Characterization of the catalytic flexible loop in the dihydroorotase domain of the human multi-enzymatic protein CAD. J Biol Chem 293(49):18903–18913

Denis-Duphil M (1989) Pyrimidine biosynthesis in *Saccharomyces cerevisiae*: the ura2 cluster gene, its multifunctional enzyme product, and other structural or regulatory genes involved in de novo UMP synthesis. Biochem Cell Biol 67(9):612–631

Denis-Duphil M, Lecaer JP, Hardie DG et al (1990) Yeast carbamoyl-phosphate-synthetase–aspartate-transcarbamylase multidomain protein is phosphorylated in vitro by cAMP-dependent protein kinase. Eur J Biochem 193(2):581–587

Diez-Fernandez C, Martinez AI, Pekkala S et al (2013) Molecular characterization of carbamoyl-phosphate synthetase (CPS1) deficiency using human recombinant CPS1 as a key tool. Hum Mutat 34(8):1149–1159

Eroglu B, Powers-Lee SG (2002) Unmasking a functional allosteric domain in an allosterically nonresponsive carbamoyl-phosphate synthetase. J Biol Chem 277(47):45466–45472

Evans DR (1986) CAD, a chimeric protein that initiates de novo pyrimidine biosynthesis in higher eukaryotes. In: Coggings JR, Hardie DG (eds) Multidomain proteins—structure and evolution. Elsevier

Evans DR, Guy HI (2004) Mammalian pyrimidine biosynthesis: fresh insights into an ancient pathway. J Biol Chem 279(32):33035–33038

Faure M, Camonis JH, Jacquet M (1989) Molecular characterization of a *Dictyostelium discoideum* gene encoding a multifunctional enzyme of the pyrimidine pathway. Eur J Biochem 179(2):345–358

Fawaz MV, Topper ME, Firestine SM (2011) The ATP-grasp enzymes. Bioorg Chem 39(5–6):185–191

Fields C, Brichta D, Shepherdson M et al (1999) Phylogenetic analysis and classification of dihydroorotases: a complex history for a complex enzyme. Paths Pyrimidines 7:49–63

Franks DM, Izumikawa T, Kitagawa H et al (2006) *C. elegans* pharyngeal morphogenesis requires both de novo synthesis of pyrimidines and synthesis of heparan sulfate proteoglycans. Dev Biol 296(2):409–420

Freund JN, Jarry BP (1987) The rudimentary gene of *Drosophila melanogaster* encodes four enzymic functions. J Mol Biol 193(1):1–13

Gaertner FH (1978) Unique catalytic properties of enzyme clusters. Trends Biochem Sci 3:63–65

Gerhart JC, Holoubek H (1967) The purification of aspartate transcarbamylase of *Escherichia coli* and separation of its protein subunits. J Biol Chem 242(12):2886–2892

Gouaux JE, Krause KL, Lipscomb WN (1987) The catalytic mechanism of *Escherichia coli* aspartate carbamoyltransferase: a molecular modelling study. Biochem Biophys Res Commun 142(3):893–897

Grande-Garcia A, Lallous N, Diaz-Tejada C et al (2014) Structure, functional characterization, and evolution of the dihydroorotase domain of human CAD. Structure 22(2):185–198

Graves LM, Guy HI, Kozlowski P et al (2000) Regulation of carbamoyl phosphate synthetase by MAP kinase. Nature 403(6767):328–332

Guy HI, Evans DR (1994) Cloning, expression, and functional interactions of the amidotransferase domain of mammalian CAD carbamyl phosphate synthetase. J Biol Chem 269(10):7702–7708

Hemmens B, Carrey EA (1994) Proteolytic cleavage of the multienzyme polypeptide CAD to release the mammalian aspartate transcarbamoylase. Biochemical comparison with the homologous *Escherichia coli* catalytic subunit. Eur J Biochem 225(3):845–853

Hewagama A, Guy HI, Vickrey JF et al (1999) Functional linkage between the glutaminase and synthetase domains of carbamoyl-phosphate synthetase. Role of serine 44 in carbamoyl-phosphate synthetase-aspartate carbamoyltransferase-dihydroorotase (cad). J Biol Chem 274(40):28240–28245

Holden HM, Thoden JB, Raushel FM (1998) Carbamoyl phosphate synthetase: a tunnel runs through it. Curr Opin Struct Biol 8(6):679–685

Holm L, Sander C (1997) An evolutionary treasure: unification of a broad set of amidohydrolases related to urease. Proteins 28(1):72–82

Hoogenraad NJ, Levine RL, Kretchmer N (1971) Copurification of carbamoyl phosphate synthetase and aspartate transcarbamoylase from mouse spleen. Biochem Biophys Res Commun 44(4):981–988
Huang YH, Huang CY (2015) Creation of a putative third metal binding site in type II dihydroorotases significantly enhances enzyme activity. Protein Pept Lett 22(12):1117–1122
Imaeda M, Sumi S, Imaeda H et al (1998) Hereditary orotic aciduria heterozygotes accompanied with neurological symptoms. Tohoku J Exp Med 185(1):67–70
Irvine HS, Shaw SM, Paton A et al (1997) A reciprocal allosteric mechanism for efficient transfer of labile intermediates between active sites in CAD, the mammalian pyrimidine-biosynthetic multienzyme polypeptide. Eur J Biochem 247(3):1063–1073
Jacobson GR, Stark GR (1973) Aspartate transcarbamylases. In: Boyer PD (ed) The enzymes. Academic Press (Elsevier)
Jones ME (1971) Regulation of pyrimidine and arginine biosynthesis in mammals. Adv Enzyme Regul 9:19–49
Jones ME (1980) Pyrimidine nucleotide biosynthesis in animals: genes, enzymes, and regulation of UMP biosynthesis. Annu Rev Biochem 49(1):253–279
Kelly RE, Mally MI, Evans DR (1986) The dihydroorotase domain of the multifunctional protein CAD. Subunit structure, zinc content, and kinetics. J Biol Chem 261(13):6073–6083
Kempe TD, Swyryd EA, Bruist M et al (1976) Stable mutants of mammalian cells that overproduce the first three enzymes of pyrimidine nucleotide biosynthesis. Cell 9(4 Pt 1):541–550
Kim H, Kelly RE, Evans DR (1992) The structural organization of the hamster multifunctional protein CAD. Controlled proteolysis, domains, and linkers. J Biol Chem 267(10):7177–7184
Koch J, Mayr JA, Alhaddad B et al (2017) CAD mutations and uridine-responsive epileptic encephalopathy. Brain 140(Pt 2):279–286
Lacroute F, Pierard A, Grenson M et al (1965) The biosynthesis of carbamoyl phosphate in *Saccharomyces cerevisiae*. J Gen Microbiol 40(1):127–142
Lallous N, Grande-Garcia A, Molina R et al (2012) Expression, purification, crystallization and preliminary X-ray diffraction analysis of the dihydroorotase domain of human CAD. Acta Crystallogr, Sect F: Struct Biol Cryst Commun 68(Pt 11):1341–1345
Lee L, Kelly RE, Pastra-Landis SC et al (1985) Oligomeric structure of the multifunctional protein CAD that initiates pyrimidine biosynthesis in mammalian cells. Proc Natl Acad Sci U S A 82 (20):6802–6806
Lee M, Chan CW, Mitchell Guss J et al (2005) Dihydroorotase from *Escherichia coli*: loop movement and cooperativity between subunits. J Mol Biol 348(3):523–533
LiCata VJ, Allewell NM (1997) Is substrate inhibition a consequence of allostery in aspartate transcarbamylase? Biophys Chem 64(1–3):225–234
Lipscomb WN (1994) Aspartate transcarbamylase from *Escherichia coli*: activity and regulation. Adv Enzymol Relat Areas Mol Biol 68:67–151
Lipscomb WN, Kantrowitz ER (2011) Structure and mechanisms of *Escherichia coli* aspartate transcarbamoylase. Acc Chem Res 45(3):444–453
Liu X, Guy HI, Evans DR (1994) Identification of the regulatory domain of the mammalian multifunctional protein CAD by the construction of an *Escherichia coli* hamster hybrid carbamyl-phosphate synthetase. J Biol Chem 269(44):27747–27755
Loffler M, Carrey EA, Zameitat E (2015) Orotic acid, more than just an intermediate of pyrimidine de novo synthesis. J Genet Genomics 42(5):207–219
Lue PF, Kaplan JG (1969) The aspartate transcarbamylase and carbamoyl phosphate synthetase of yeast: a multi-functional enzyme complex. Biochem Biophys Res Commun 34(4):426–433
Lusty CJ (1981) Catalytically active monomer and dimer forms of rat liver carbamoyl-phosphate synthetase. Biochemistry 20(13):3665–3674
Makoff AJ, Buxton FP, Radford A (1978) A possible model for the structure of the Neurospora carbamoyl phosphate synthase-aspartate carbamoyl transferase complex enzyme. Mol Gen Genet 161(3):297–304

Mally MI, Grayson DR, Evans DR (1980) Catalytic synergy in the multifunctional protein that initiates pyrimidine biosynthesis in Syrian hamster cells. J Biol Chem 255(23):11372–11380

Mally MI, Grayson DR, Evans DR (1981) Controlled proteolysis of the multifunctional protein that initiates pyrimidine biosynthesis in mammalian cells: evidence for discrete structural domains. Proc Natl Acad Sci U S A 78(11):6647–6651

Meister A (1989) Mechanism and regulation of the glutamine-dependent carbamyl phosphate synthetase of *Escherichia coli*. Adv Enzymol Relat Areas Mol Biol 62:315–374

Miles BW, Raushel FM (2000) Synchronization of the three reaction centers within carbamoyl phosphate synthetase. Biochemistry 39(17):5051–5056

Miles BW, Banzon JA, Raushel FM (1998) Regulatory control of the amidotransferase domain of carbamoyl phosphate synthetase. Biochemistry 37(47):16773–16779

Miran SG, Chang SH, Raushel FM (1991) Role of the four conserved histidine residues in the amidotransferase domain of carbamoyl phosphate synthetase. Biochemistry 30(32):7901–7907

Mora P, Rubio V, Fresquet V et al (1999) Localization of the site for the nucleotide effectors of *Escherichia coli* carbamoyl phosphate synthetase using site-directed mutagenesis. FEBS Lett 446(1):133–136

Moreno-Morcillo M, Ramon-Maiques S (2017) CAD: a multifunctionl protein leading de novo pyrimidine biosynthesis. In: Encyclopedia of life sciences. John Wiley and Sons

Moreno-Morcillo M, Grande-Garcia A, Ruiz-Ramos A et al (2017) Structural Insight into the Core of CAD, the multifunctional protein leading de novo pyrimidine biosynthesis. Structure 25 (6):912–923 e915

Mori M, Tatibana M (1978) A multienzyme complex of carbamoyl-phosphate synthase (glutamine): aspartate carbamoyltransferase: dihydoorotase (rat ascites hepatoma cells and rat liver). Methods Enzymol 51:111–120

Newell JO, Markby DW, Schachman HK (1989) Cooperative binding of the bisubstrate analog N-(phosphonacetyl)-L-aspartate to aspartate transcarbamoylase and the heterotropic effects of ATP and CTP. J Biol Chem 264(5):2476–2481

Ng SB, Buckingham KJ, Lee C et al (2010) Exome sequencing identifies the cause of a mendelian disorder. Nat Genet 42(1):30–35

Ng BG, Wolfe LA, Ichikawa M et al (2015) Biallelic mutations in CAD, impair de novo pyrimidine biosynthesis and decrease glycosylation precursors. Hum Mol Genet 24(11):3050–3057

Norby S (1970) A specific nutritional requirement for pyrimidines in rudimentary mutants of Drosophila melanogaster. Hereditas 66(2):205–214

Nyhan WL (2005) Nucleotide synthesis via salvage pathway. In: Encyclopedia of life sciences. John Wiley and Sons

Nyunoya H, Lusty CJ (1983) The carB gene of *Escherichia coli*: a duplicated gene coding for the large subunit of carbamoyl-phosphate synthetase. Proc Natl Acad Sci U S A 80(15):4629–4633

Nyunoya H, Lusty CJ (1984) Sequence of the small subunit of yeast carbamyl phosphate synthetase and identification of its catalytic domain. J Biol Chem 259(15):9790–9798

Nyunoya H, Broglie KE, Widgren EE et al (1985) Characterization and derivation of the gene coding for mitochondrial carbamyl phosphate synthetase I of rat. J Biol Chem 260(16):9346–9356

Otsuki T, Mori M, Tatibana M (1982) Studies on channeling of carbamoyl-phosphate in the multienzyme complex that initiates pyrimidine biosynthesis in rat ascites hepatoma cells. J Biochem 92(5):1431–1437

Penverne B, Belkaid M, Herve G (1994) In situ behavior of the pyrimidine pathway enzymes in *Saccharomyces cerevisiae*. 4. The channeling of carbamylphosphate to aspartate transcarbamylase and its partition in the pyrimidine and arginine pathways. Arch Biochem Biophys 309(1):85–93

Pierson DL, Brien JM (1980) Human carbamylphosphate synthetase I. Stabilization, purification, and partial characterization of the enzyme from human liver. J Biol Chem 255(16):7891–7895

Porter RW, Modebe MO, Stark GR (1969) Aspartate transcarbamylase. Kinetic studies of the catalytic subunit. J Biol Chem 244(7):1846–1859

Porter TN, Li Y, Raushel FM (2004) Mechanism of the dihydroorotase reaction. Biochemistry 43 (51):16285–16292

Powers-Lee SG, Corina K (1986) Domain structure of rat liver carbamoyl phosphate synthetase I. J Biol Chem 261(33):15349–15352

Prange T, Girard E, Fourme R et al (2019) Pressure-induced activation of latent Dihydroorotase from *Aquifex aeolicus* as revealed by high pressure protein crystallography. FEBS J

Qiu Y, Davidson JN (1998) Aspartate-90 and arginine-269 of hamster aspartate transcarbamylase affect the oligomeric state of a chimaeric protein with an *Escherichia coli* maltose-binding domain. Biochem J 329(Pt 2):243–247

Qiu Y, Davidson JN (2000) Substitutions in the aspartate transcarbamoylase domain of hamster CAD disrupt oligomeric structure. Proc Natl Acad Sci 97(1):97–102

Raushel FM, Thoden JB, Reinhart GD et al (1998) Carbamoyl phosphate synthetase: a crooked path from substrates to products. Curr Opin Chem Biol 2(5):624–632

Robitaille AM, Christen S, Shimobayashi M et al (2013) Quantitative phosphoproteomics reveal mTORC1 activates de novo pyrimidine synthesis. Science 339(6125):1320–1323

Rodriguez-Aparicio LB, Guadalajara AM, Rubio V (1989) Physical location of the site for N-acetyl-L-glutamate, the allosteric activator of carbamoyl phosphate synthetase, in the 20-kilodalton COOH-terminal domain. Biochemistry 28(7):3070–3074

Rubio V (1994) Structure-activity correlations in carbamoyl phosphate synthetases. In: Brändén CI, Schneider G (eds) Carbon dioxide fixation and reduction in biological and model systems. Proceedings of the royal swedish academy of sciences nobel symposium 1991. Oxford University Press

Rubio V, Ramponi G, Grisolia S (1981) Carbamoyl phosphate synthetase I of human liver. Purification, some properties and immunological cross-reactivity with the rat liver enzyme. Biochim Biophys Acta 659(1):150–160

Rubio V, Britton HG, Grisolia S (1983) Mitochondrial carbamoyl phosphate synthetase activity in the absence of N-acetyl-L-glutamate. Mechanism of activation by this cofactor. Eur J Biochem 134(2):337–343

Rubio V, Cervera J, Lusty CJ et al (1991) Domain structure of the large subunit of *Escherichia coli* carbamoyl phosphate synthetase. Location of the binding site for the allosteric inhibitor UMP in the COOH-terminal domain. Biochemistry 30(4):1068–1075

Ruiz-Ramos A, Lallous N, Grande-Garcia A et al (2013) Expression, purification, crystallization and preliminary X-ray diffraction analysis of the aspartate transcarbamoylase domain of human CAD. Acta Crystallogr, Sect F: Struct Biol Cryst Commun 69(Pt 12):1425–1430

Ruiz-Ramos A, Velazquez-Campoy A, Grande-Garcia A et al (2016) Structure and functional characterization of human aspartate transcarbamoylase, the target of the anti-tumoral drug PALA. Structure 24(7):1081–1094

Saeed-Kothe A, Powers-Lee SG (2003) Gain of glutaminase function in mutants of the ammonia-specific frog carbamoyl phosphate synthetase. J Biol Chem 278(29):26722–26726

Schachman HK (1988) Can a simple model account for the allosteric transition of aspartate transcarbamoylase? J Biol Chem 263(35):18583–18586

Schurr MJ, Vickrey JF, Kumar AP et al (1995) Aspartate transcarbamoylase genes of *Pseudomonas putida*: requirement for an inactive dihydroorotase for assembly into the dodecameric holoenzyme. J Bacteriol 177(7):1751–1759

Scully JL, Evans DR (1991) Comparative modeling of mammalian aspartate transcarbamylase. Proteins 9(3):191–206

Shi D, Caldovic L, Tuchman M (2018) Sources and fates of carbamyl phosphate: a labile energy-rich molecule with multiple facets. Biology (Basel) 7(2)

Shigesada K, Stark GR, Maley JA et al (1985) Construction of a cDNA to the hamster CAD gene and its application toward defining the domain for aspartate transcarbamylase. Mol Cell Biol 5 (7):1735–1742

Shoaf WT, Jones ME (1971) Initial steps in pyrimidine synthesis in *Ehrlich ascites* carcinoma. Biochem Biophys Res Commun 45(3):796–802

Simmer JP, Kelly RE, Rinker AG Jr et al (1990) Mammalian carbamyl phosphate synthetase (CPS). DNA sequence and evolution of the CPS domain of the Syrian hamster multifunctional protein CAD. J Biol Chem 265(18):10395–10402

Souciet JL, Nagy M, Le Gouar M et al (1989) Organization of the yeast URA2 gene: identification of a defective dihydroorotase-like domain in the multifunctional carbamoylphosphate synthetase-aspartate transcarbamylase complex. Gene 79(1):59–70

Stark GR (1977) Multifunctional proteins: one gene—more than one enzyme. Trends Biochem Sci 2:64–66

Stevens RC, Reinisch KM, Lipscomb WN (1991) Molecular structure of *Bacillus subtilis* aspartate transcarbamoylase at 3.0 A resolution. Proc Natl Acad Sci U S A 88(14):6087–6091

Swyryd EA, Seaver SS, Stark GR (1974) N-(phosphonacetyl)-L-aspartate, a potent transition state analog inhibitor of aspartate transcarbamylase, blocks proliferation of mammalian cells in culture. J Biol Chem 249(21):6945–6950

Thoden JB, Holden HM, Wesenberg G et al (1997) Structure of carbamoyl phosphate synthetase: a journey of 96 A from substrate to product. Biochemistry 36(21):6305–6316

Thoden JB, Miran SG, Phillips JC et al (1998) Carbamoyl phosphate synthetase: caught in the act of glutamine hydrolysis. Biochemistry 37(25):8825–8831

Thoden JB, Huang X, Raushel FM et al (1999a) The small subunit of carbamoyl phosphate synthetase: snapshots along the reaction pathway. Biochemistry 38(49):16158–16166

Thoden JB, Raushel FM, Benning MM et al (1999b) The structure of carbamoyl phosphate synthetase determined to 2.1 A resolution. Acta Crystallogr D Biol Crystallogr 55(Pt 1):8–24

Thoden JB, Raushel FM, Wesenberg G et al (1999c) The binding of inosine monophosphate to *Escherichia coli* carbamoyl phosphate synthetase. J Biol Chem 274(32):22502–22507

Thoden JB, Wesenberg G, Raushel FM et al (1999d) Carbamoyl phosphate synthetase: closure of the B-domain as a result of nucleotide binding. Biochemistry 38(8):2347–2357

Thoden JB, Phillips GN Jr, Neal TM et al (2001) Molecular structure of dihydroorotase: a paradigm for catalysis through the use of a binuclear metal center. Biochemistry 40(24):6989–6997

Thoden JB, Huang X, Raushel FM et al (2002) Carbamoyl-phosphate synthetase. Creation of an escape route for ammonia. J Biol Chem 277(42):39722–39727

Thoden JB, Huang X, Kim J et al (2004) Long-range allosteric transitions in carbamoyl phosphate synthetase. Protein Sci 13(9):2398–2405

Traut TW, Jones ME (1979) Interconversion of different molecular weight forms of the orotate phosphoribosyltransferase.orotidine-5′-phosphate decarboxylase enzyme complex from mouse Ehrlich ascites cells. J Biol Chem 254(4):1143–1150

Wellner VP, Anderson PM, Meister A (1973) Interaction of *Escherichia coli* carbamyl phosphate synthetase with glutamine. Biochemistry 12(11):2061–2066

Willer GB, Lee VM, Gregg RG et al (2005) Analysis of the Zebrafish perplexed mutation reveals tissue-specific roles for de novo pyrimidine synthesis during development. Genetics 170 (4):1827–1837

Williams LG, Davis RH (1970) Pyrimidine-specific carbamyl phosphate synthetase in *Neurospora crassa*. J Bacteriol 103(2):335–341

Williams LG, Bernhardt S, Davis RH (1970) Copurification of pyrimidine-specific carbamyl phosphate synthetase and aspartate transcarbamylase of *Neurospora crassa*. Biochemistry 9 (22):4329–4335

Yoshida T, Stark GR, Hoogenraad J (1974) Inhibition by N-(phosphonacetyl)-L-aspartate of aspartate transcarbamylase activity and drug-induced cell proliferation in mice. J Biol Chem 249(21):6951–6955

Zhang P, Martin PD, Purcarea C et al (2009) Dihydroorotase from the hyperthermophile *Aquifex aeolicus* is activated by stoichiometric association with aspartate transcarbamoylase and forms a one-pot reactor for pyrimidine biosynthesis. Biochemistry 48(4):766–778

Zrenner R, Stitt M, Sonnewald U et al (2006) Pyrimidine and purine biosynthesis and degradation in plants. Annu Rev Plant Biol 57:805–836

Chapter 18
The Anaphase Promoting Complex/ Cyclosome (APC/C): A Versatile E3 Ubiquitin Ligase

Natalie L. Curtis and Victor M. Bolanos-Garcia

Abstract In the present chapter we discuss the essential roles of the human E3 ubiquitin ligase Anaphase Promoting Complex/Cyclosome (APC/C) in mitosis as well as the emerging evidence of important APC/C roles in cellular processes beyond cell division control such as regulation of genomic integrity and cell differentiation of the nervous system. We consider the potential incipient role of APC/C dysregulation in the pathophysiology of the neurological disorder Alzheimer's disease (AD). We also discuss how certain Deoxyribonucleic Acid (DNA) and Ribonucleic Acid (RNA) viruses take control of the host's cell division regulatory system through harnessing APC/C ubiquitin ligase activity and hypothesise the plausible molecular mechanisms underpinning virus manipulation of the APC/C. We also examine how defects in the function of this multisubunit protein assembly drive abnormal cell proliferation and lastly argue the potential of APC/C as a promising therapeutic target for the development of innovative therapies for the treatment of chronic malignancies such as cancer.

Keywords Anaphase Promoting Complex/Cyclosome (APC/C) · E3 ubiquitin ligase · Spindle Assembly Checkpoint (SAC) · Proteasome · Genome instability · Mitosis regulation · Chromosome segregation · Kinetochore

Abbreviations

Aβ	Amyloid beta
Ad	Adenovirus
AD	Alzheimer's disease
ADM	Adriamycin
AML	Acute Myeloid Leukemia
APC/C	Anaphase Promoting Complex/Cyclosome

N. L. Curtis · V. M. Bolanos-Garcia (✉)
Faculty of Health and Life Sciences, Department of Biological and Medical Sciences, Oxford Brookes University, Oxford OX3 0BP, England, UK
e-mail: vbolanos-garcia@brookes.ac.uk

N. L. Curtis
e-mail: 17083926@brookes.ac.uk

J. R. Harris and J. Marles-Wright (eds.), *Macromolecular Protein Complexes II: Structure and Function*, Subcellular Biochemistry 93,
https://doi.org/10.1007/978-3-030-28151-9_18

ATL	Adult T-cell Leukaemia/Lymphoma
ATP	Adenosine Triphosphate
ATR	Ataxia Telangiectasia and Rad3-related protein
Bard1	BRCA1-associated RING domain protein 1
BRSK2	Brain-Specific Kinase 2
Bub1	Budding uninhibited by benzimidazoles 1
Bub3	Budding uninhibited by benzimidazoles 3
BubR1	Budding uninhibited by benzimidazoles Related 1
CAV	Chicken Anemia Virus
CAV-Apoptin	Chicken Anemia Virus-Apoptin
CBP	Cyclic AMP response element Binding Protein
Cdc5	Cell division cycle 5
Cdc6	Cell division cycle 6
Cdc7	Cell division cycle 7
Cdc20	Cell division cycle 20
Cdc23	Cell division cycle 23
Cdc26	Cell division cycle 26
Cdc27	Cell division cycle 27
Cdc42	Cell division cycle 42
Cdc55	Cell division cycle 55
CDDP	Cisplatin
Cdh1	Cadherin 1
Cdk	Cyclin-dependent kinase
Cdk1	Cyclin-dependent kinase 1
Cdk4	Cyclin-dependent kinase 4
Cdk5	Cyclin-dependent kinase 5
Cdt1	Chromatin licensing and DNA replication factor 1
CENP-E	Centromere-associated Protein E
CENP-F	Centromere-associated Protein F
Chk1	Checkpoint kinase 1
Cik1	Chromosome instability and karyogamy protein 1
CK1	Casein Kinase 1
CK1δ	Casein Kinase 1 delta
CKAP2	Cytoskeleton-Associated Protein 2
cKO	conditional Knockout
Cks1	Cyclin-dependent kinases regulatory subunit 1
CNS	Central Nervous System
co-IP	co-Immunoprecipitation
CPC	Chromosomal Passenger Complex
cryo-EM	cryogenic Electron Microscopy
D-box	Destruction box
DDB1	Damaged DNA Binding protein 1
DNA	Deoxyribonucleic Acid
dNTPS	deoxyribonucleotide Triphosphate

dTTP	deoxythymidine Triphosphate
E	Early
E1A	Adenovirus Early region 1A
E2F1	E2F Transcription Factor 1
E2F3	E2F Transcription Factor 3
E4orf4	Early region 4 open reading frame 4
EAA	Excitatory Amino Acid
Ect2	Epithelial cell transforming 2
EHMT2	Euchromatic Histone Lysine Methyltransferase 2
EM	Electron Microscopy
Emi1	Early mitotic inhibitor
ER	Endoplasmic Reticulum
EYA1	Eyes Absent 1
FAN1	FANCD2-Associated Nuclease 1
Fin1	Filaments in between nuclei protein 1
FoxM1	Forkhead box M1
Fzr	Fizzy-related protein homolog
G1	Gap 1
G2	Gap 2
G6PD	Glucose-6-Phosphate Dehydrogenase
GCPs	Granule Cell Progenitors
GLP	Glucagon-Like Peptide
GLP1	Glucagon-Like Peptide-1
GLS1	Glutaminase 1
GluR1	Glutamate Receptor 1
GSCs	Glioblastoma Stem-like Cells
GSH	Glutathione
GSSG	Glutathione disulfide
HAd	Human Adenovirus
HBV	Hepatitis B Virus
HCC	Hepatocellular Carcinoma
HCMV	Human Cytomegalovirus
HECT	Homologous to the E6-AP Carboxyl Terminus
HGyv-Apoptin	Human Gyrovirus-Apoptin
Hmmr	Hyaluronan-mediated motility receptor
HPV	Human Papillomaviruses
Hsl1 from yeast	Histone synthetic lethal 1
HTLV-1	Human T-cell Lymphotropic Virus type 1
HURP	Hepatoma Up-Regulated Protein
Id1	Inhibitor of differentiation 1
Id2	Inhibitor of differentiation 2
IE	Intermediate Early
IE72	Intermediate protein 72
IE86	Intermediate protein 86

IQGAP	IQ motif containing GTPase-Activating Protein 1
IR	Isoleucine Arginine
JAK	Janus Kinase
JNK	C-Jun N-terminal Kinase
kDa	kiloDalton
Kif18A	Kinesin family member 18A
Knl1	Kinetochore scaffold 1
L	Late
LT	Large T antigen
LTP	Long-Term Potentiation
M	Mitosis
Mad1	Mitotic arrest deficient 1
Mad2	Mitotic arrest deficient 2
MAPK	Mitogen-Activated Protein Kinase
MCC	Mitotic Checkpoint Complex
MCPH1	Microcephalin-1
MDC1	Mediator DNA Damage Checkpoint 1
MEFs	Mouse Embryonic Fibroblasts
MgcRacGAP	Male germ cell RacGTPase Activating Protein
MOAP-1	Modulator of Apoptosis Protein 1
Mps1	Monopolar spindle 1
mRNA	messenger RNA
NADPH	Nicotinamide Adenine Dinucleotide Phosphate
Nek2A	NIMA-related kinase 2A
NeuroD2	Neurogenic Differentiation factor 2
NFT	Neurofibrillary Tangles
NIPA	Nuclear Interaction Partner of Alk kinase
Nlp	Ninein-like protein
NMDARs	*N*-methyl-D-aspartate Receptors
NPCs	Neural Progenitor Cells
NS5A	Nonstructural protein 5A
NuSAP	Nucleolar Spindle–Associated Protein
ORFV	Orf Virus
$p21^{Cip1}$	Cyclin dependent kinase inhibitor 1
$p27^{Kip1}$	Cyclin dependent kinase inhibitor 1B
PDAC	Pancreatic Ductal Adenocarcinoma
Pds1	Premature dissociation of sisters
PFKFB3	6-phosphofructo-2-kinase/fructose-2,6-biphosphatase isoform 3
Plk1	Polo-like kinase 1
PML bodies	Promyelocytic Leukaemia bodies
PP2a	Protein Phosphatase 2
PPP	Pentose-Phosphate Pathway
pRb	Retinoblastoma protein
proTAME	pro-tosyl-L-arginine methyl ester

PSD	Postsynaptic Density
Rac1	Ras-related C3 botulinum toxin substrate 1
RAP80	Receptor-Associated Protein 80
Rassf1a	Ras association domain containing family 1 isoform a
RbX1	RING-box protein 1
RbX2	RING-box protein 2
RhoA	Ras homolog gene family member A
RING	Really Interesting New Gene
RNA	Ribonucleic Acid
RNAi	RNA interference
Rock2	Rho protein kinase 2
ROS	Reactive Oxygen Species
S	Synthesis
SAC	Spindle Assembly Checkpoint
S. cerevisiae	*Saccharomyces cerevisiae*
SCF	Skp1-Cullin-1-F-box protein
Sgo1	Shugoshin 1
Shank	SH3 and multiple ankyrin repeat domains protein
siRNA	small interfering RNA
Six1	Sineoculis homeobox homolog 1
Skp1	S-phase-kinase-associated protein 1
Skp2	S-phase kinase-associated protein 2
SnoN	Ski-related novel protein N
SPOP	Speckle-type POZ Protein
STAT	Signal Transducer and Activator of Transcription
SV40	Simian Virus 40
TACC3	Transforming Acidic Coiled-Coil protein 3
TAME	Tosyl-L-arginine methyl ester
TGCTs	Testicular Germ Cell Tumours
TGF-β	Transforming Growth Factor beta
TK1	Thymidine Kinase
TMPK	Thymidylate Protein Kinase
TNM	TNM classification of malignant tumors
TPR	Tetratricopeptide Repeat
Tpx2	Targeting protein for Xklp2
TRB3	Tribbles homolog 3
TRRAP	Transformation/Transcription domain-Associated protein
Tyc1	Tiny yeast comet 1
Ube2C	Ubiquitin-conjugating enzyme E2 C
Ube2D	Ubiquitin-conjugating enzyme E2 D2
UPS1	Ubiquitin carboxyl-terminal hydrolase 1
USP37	Ubiquitin Specific-processing Protease 37
WES	Whole-Exome Sequencing

Introduction

Protein degradation through ubiquitination is mediated by the sequential action of activating (E1), conjugating (E2), and ligase (E3) enzymes to render a poly-ubiquitinated protein substrate that is subsequently recognised and degraded by the proteasome (Nakayama and Nakayama 2006; Nalepa et al. 2006; Pickart 2001). In the human genome, a small number of ubiquitin E1 and E2 enzymes and hundreds of different E3 ubiquitin ligases have been identified. The much larger number of the latter class of enzymes reflects their role of recruiting specific substrate proteins for targeted ubiquitination, thus conferring substrate selectivity to the Ubiquitin/Proteasome System (UPS) (Nagy and Dikic 2010; Deshaies and Joazeiro 2009; Petroski and Deshaies 2005).

Two major classes of E3 ligases occur in eukaryotes: the Homologous to the E6-AP Carboxyl Terminus (HECT) domain and the Really Interesting New Gene (RING) domain (Deshaies and Joazeiro 2009; Bedford et al. 2011). HECT E3s establish a transient, covalent linkage with ubiquitin at a conserved cysteine residue before the ubiquitin molecule is transferred to its substrate. In contrast, RING E3s mediate the transfer of the ubiquitin molecule from the E2 enzyme to a substrate in a process that does not require its physical interaction with the substrate (Metzger et al. 2012; Bedford et al. 2011; Skaar and Pagano 2009; Rotin and Kumar 2009).

The RING ubiquitin E3 ligases are the largest family of E3 ubiquitin ligases and associated to the regulation of cell division (Nakayama and Nakayama 2006; Nagy and Dikic 2010; Deshaies and Joazeiro 2009). APC/C is one of two multi-subunit RING finger ubiquitin ligase families. The other is the Skp1-Cullin-1-F-box protein (SCF) family, which consists of the scaffold protein Cullin-1, a RING-finger protein RING-box protein 1 (RbX1) or RING-box protein 2 (RbX2) (also known as ROC1/2), the adaptor S-phase-kinase-associated protein 1 (Skp1), and a F-box protein that recruits specific proteins substrates for SCF-mediated degradation (Cardozo and Pagano 2004; Jin et al. 2004; Ang and Wade Harper 2005). The two families of RING finger ubiquitin ligases play pivotal roles in the regulation of the cell cycle as they mediate the proteasome-dependent degradation of cell cycle regulators such as Cyclins and Cyclin-dependent kinase (Cdk) inhibitors (Skaar and Pagano 2009; Harper et al. 2002).

Similar to the architecture of the SCF complex, the APC/C assembly contains one Cullin-like subunit (APC2), a RING finger protein (APC11), and an interchangeable substrate recognition subunit (Lipkowitz and Weissman 2011). However, APC/C exhibits a unique molecular architecture as it is composed of 14–16 different subunits, depending on the species (Peters 2006; Pines 2011). In the human, the known 14 protein subunits define a holoenzyme complex with a combined molecular mass of approximately 1.5 mega Daltons, consisting of APC1/Tsg24, APC2, APC3/Cell division cycle 27 (Cdc27), APC4, APC5, APC6/Cell division cycle 6 (Cdc6), APC7, APC8/Cell division cycle 23 (Cdc23), APC10/Doc1, APC11, APC13/Swm1, APC15/Mnd2, APC16 and Cell division cycle 26 (Cdc26) as well as the co-activator subunit Cell division cycle 20 (Cdc20) or Cadherin 1 (Cdh1) (Barford 2011; Foe and Toczyski 2011; McLean et al. 2011).

Structural Architecture

Recent advances in cryogenic Electron Microscopy (cryo-EM) methods have revealed important mechanistic details of APC/C formation and regulation and showed that APC/C consists of three sub-complexes: a scaffolding subcomplex, a catalytic and substrate recognition sub-complex, and an arm defined by multiple Tetratricopeptide Repeat (TPR) motifs (Barford 2015; Chang et al. 2014, 2015; Alfieri et al. 2016; Yamaguchi et al. 2016; Zhang et al. 2016; Foe and Toczyski 2011; Schreiber et al. 2011; Thornton et al. 2006; Vodermaier et al. 2003). The structural architecture of the APC/C is summarised in Figs. 1 and 2. The scaffolding sub-complex is composed of APC1, APC4 and APC5; the catalytic sub-complex of the APC2, APC11 (the RING finger protein) and APC10/Doc1 subunits; and the TPR arm, which is defined by the APC3/Cdc27, APC6/Cdc16 and APC8/Cdc23 subunits and delineates a binding surface for the coactivators Cdc20 or Cdh1 and for the subunits Cdc26, APC13/Swm1, and APC16. The APC/C core subunit

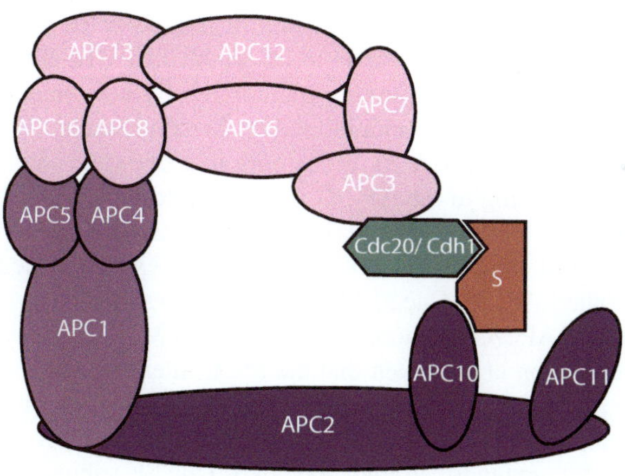

Fig. 1 Schematic structure of the APC/C. The APC/C consists of three distinct subcomplexes; a catalytic subcomplex, scaffolding complex and a specificity arm defined by multiple tetratricopeptide repeat motifs (TPR). For clarity, the subunits that assemble each subcomplex have been colour-coded; the subunits presented are not drawn to scale; and not all the subunits are depicted. The catalytic sub complex is composed of APC2 and APC11 which contain the Cullin and RING domain respectively and APC10/Doc1 which is the subunit that identifies and recruits specific substrates. The scaffolding subcomplex contains APC1, APC4 and APC5 and holds the catalytic subcomplex and the specificity arm together. The specificity TPR arm is defined by APC3/Cdc27, APC6/Cdc16 and APC8/Cdc23 and delineates a binding surface for APC/C coactivators Cdc20 and Cdh1. These coactivators direct the recruitment of substrates to the complex

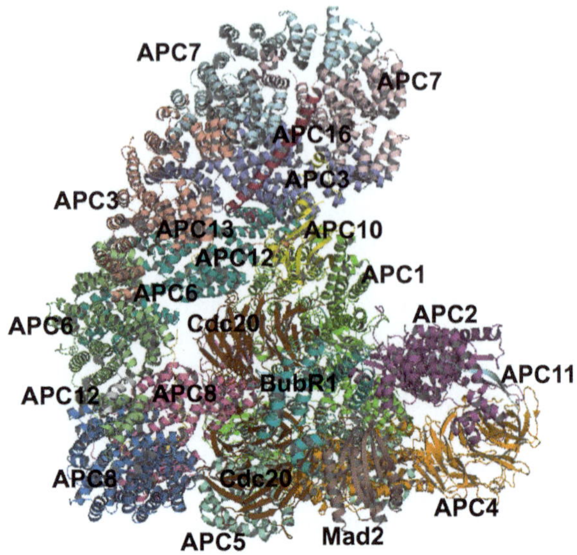

Fig. 2 Structural architecture of the APC/C assembly bound to the Mitotic Checkpoint Complex (MCC) as defined by cryo-EM methods (pdb ID 5KHU). The labels indicate the position and copy number of APC/C subunits, coactivators and MCC proteins

APC10/Doc1 contributes to substrate recruitment via binding to the Destruction-boxes (D-boxes) of protein substrates (da Fonseca et al. 2011; Buschhorn et al. 2011; Frye et al. 2013). Mechanistic details of APC/C assembly and whether there are additional components of this complex in the cell are key aspects of cell cycle control that remain to be established. The Mitotic Checkpoint Complex (MCC) composed of Budding uninhibited by benzimidazoles Related 1 (BubR1), together with Budding uninhibited by benzimidazoles 3 (Bub3), Mitotic arrest deficient 2 (Mad2) and Cdc20, exists both in a free state and stably bound to the APC/C, and its binding to the APC/C is essential to mount an effective Spindle Assembly Checkpoint (SAC) response (Hein and Nilsson 2014; Alfieri et al. 2016). The MCC binds and inhibits Cdc20-bound APC/C (APC/C-Cdc20), forming a large complex called APC/C-Cdc20-MCC that contains two copies of Cdc20.

Cryo-EM data of APC/C structure in complex with MCC has revealed new clues about APC/C regulation and showed that the MCC interacts with an APC/C region located in close proximity to the site where a Cdc20 subunit was previously docked (Yamaguchi et al. 2016; Alfieri et al. 2016). The APC/C-Cdc20-MCC structure shows that BubR1 extends around the two Cdc20 subunits to occupy all degron-binding sites. A consequence of this mode of interaction is that binding of the MCC to APC/C-Cdc20 enables it to act as an allosteric regulator that blocks the access of APC/C substrates. Although it has been established that the WD40 protein Bub3 contributes to the termination of the SAC signal (Alfieri et al. 2016; Vanoosthuyse et al. 2009), the precise role of this MCC component in the regulation of APC/C function remains obscure. In the cryo-EM structures, the lack of well-defined electron density around the anticipated Bub3 location of the APC/C-MCC assembly suggests that Bub3 remains largely flexible upon MCC binding to the APC/C. Whether that is the case and how such mode of interaction contributes to fine tune APC/C functions warrants further investigations. Furthermore, the overall similarity of Electron Microscopy (EM) maps for APC/C-Cdc20-MCC complexes with or without Bub3 raises the questions if MCC subcomplexes lacking Bub3 may be sufficient for the Ube2C-mediated ubiquitination of Cdc20 (Alfieri et al. 2016; Yamaguchi et al. 2016; Chang et al. 2015).

Unlike the major role of APC/C-Cdc20 in promoting the metaphase to anaphase transition, APC/C-Cdh1 plays a key role in governing cell cycle progression through the Gap 1 (G1) phase by sustaining low Cdk activity, which is achieved by mediating the degradation of mitotic Cyclins (Irniger and Nasmyth 1997), Cdc25A (Donzelli et al. 2002), Cyclin-dependent kinases regulatory subunit 1 (Cks1) (Bashir et al. 2004) and S-phase kinase-associated protein 2 (Skp2) (Bashir et al. 2004; Wei et al. 2004). Activation of APC/C-Cdh1 promotes the degradation of Aurora A (Littlepage and Ruderman 2002), Aurora B (Stewart and Fang 2005; Nguyen et al. 2005), Cdc20 (Huang et al. 2001; Hyun et al. 2013), Polo-like kinase 1 (Plk1) (Lindon and Pines 2004), and Targeting protein for Xklp2 (Tpx2) (Stewart and Fang 2005), leading to mitotic exit.

The APC/C is the ultimate effector of the SAC. In simple terms, the SAC can be defined as an evolutionary conserved mechanism of regulation of higher organisms that prolongs mitosis until all chromosomes have established proper attachment to spindle microtubules. SAC signalling involves the physical interaction of certain central protein components with the kinetochore and microtubules. The kinetochore is a proteinaceous framework that assembles onto centromeric DNA to organise SAC activity, chromosome attachment, and cell cycle progression from metaphase to anaphase (*reviewed in* Derive et al. 2015; Kapanidou et al. 2015; Sacristan and Kops 2015).

In addition to Cdc20, the central SAC components Mitotic arrest deficient 1 (Mad1) and the proteins kinases Budding uninhibited by benzimidazoles 1 (Bub1) and Monopolar spindle 1 (Mps1) enhance the rate of MCC formation and amplify the SAC signals (Hardwick et al. 1996; Abrieu et al. 2001; Chung and Chen 2002; De Antoni et al. 2005; Morrow et al. 2005). These core components of the SAC are highly conserved across species and include the Serine/Threonine kinases Bub1 and BubR1, and the dual specificity Serine/Threonine/Tyrosine kinase Mps1. Bub1 kinase is required for the kinetochore localisation of BubR1, Mad2, and the centromere-associated proteins CENP-E and CENP-F (Ciossani et al. 2018; Sivakumar and Gorbsky 2015) as well as for the establishment and maintenance of productive attachment to spindle microtubules (Meghini et al. 2016; Braunstein et al. 2007; Burton and Solomon 2007). Bub1 kinase activity is also necessary for the centromere recruitment of the Chromosomal Passenger Complex (CPC) and Shugoshin 1 (Sgo1) (Williams et al. 2017; Baron et al. 2016). At least in mitosis, BubR1 functions as a pseudokinase that plays important roles in chromosome segregation but also has reported to play important functions in DNA repair, ciliogenesis and neuron differentiation (*reviewed in* Bolanos-Garcia and Blundell 2011). BubR1 together with Bub3, Mad2 and Cdc20, forms part of the MCC, which is assembled in response to improper chromosome attachment to the mitotic spindle to inhibit the APC/C. Mps1 kinase activity is crucial for the conformational activation of Mad2 that is required for Mad2 binding to Cdc20 (Zich and Hardwick 2010), and for the kinetochore recruitment of Mad1 and Mad2 in mitotic cells (Tipton et al. 2013; Hewitt et al. 2010). In addition to the MCC, the protein Early mitotic inhibitor (Emi1) is an important inhibitor of the APC/C. Both, the MCC and Emi1 inhibitors regulate coactivator binding, substrate recognition, and activity/binding of the E2 enzymes.

APC/C Regulation Is Intricate

In vertebrates, full APC/C substrate ubiquitination activity is modulated by two E2 class of ubiquitin ligases: UbcH10 (also named Ubiquitin-conjugating enzyme E2 C (Ube2C)) and UbcH5 (also known as Ubiquitin-conjugating enzyme E2 D (Ube2D)) (Xie et al. 2014; Aristarkhov et al. 1996) as well as the binding of the APC/C coactivators Cdh1 and Cdc20. By using UbcH10 and the Lys11-specific 'elongating' E2 enzyme Ube2S, APC/C builds Lys11-linked ubiquitin chains as a degradation signal (Komander and Rape 2012; Rape 2010; Wickliffe et al. 2011; Ye and Rape 2009). The APC/C binds to substrates that contain D-box [RXXL] and KEN box motifs (Glotzer et al. 1991; Pfleger and Kirschner 2000). These motifs define a class of degrons that play an important role in the regulation of protein degradation. The APC/C coactivators Cdc20 and Cdh1 are characterised by having a structure largely organised as a WD40 domain and the presence of C-box and Isoleucine Arginine (IR)-motifs, all of which mediate direct binding to the APC/C (Chang et al. 2015; Izawa and Pines 2012; Schwab et al. 2001; Passmore et al. 2005; Vodermaier et al. 2003; Visintin et al. 1997). Besides the D-box and KEN box motifs, Cdc20 and Cdh1 bind to proteins that harbour the ABBA motif (for Acm1, Bub1, BubR1, and Cyclin A) which is also known as the Phe box (Di Fiore et al. 2015, 2016; Diaz-Martinez et al. 2015). The region that defines the longitudinal axis of the WD40 repeat mediates binding of KEN box, D-box and ABBA motif-containing substrates (Chao et al. 2012; He et al. 2013; Passmore et al. 2005; Carroll and Morgan 2005; Matyskiela and Morgan 2009; Buschhorn et al. 2011; da Fonseca et al. 2011). The *bona fide* APC/C substrates reported to date are summarised in Table 1.

APC/C E3 ubiquitin ligase activity is tightly regulated by multiple mechanisms including acetylation, phosphorylation, the binding of inhibitors, subcellular localisation, and destabilisation of its activators and/or component subunits acetylation (Glotzer et al. 1991; Touati et al. 2018; Hein et al. 2017; Craney et al. 2016; Höckner et al. 2016). Tight regulation of cell cycle progression by the APC/C E3 ligase requires its physical and sequential interaction with its two co-activators Cdh1 and Cdc20, which form an APC/C-Cdh1 or an APC/C-Cdc20 complex, respectively. Essentially, the APC/C-Cdh1 complex is primarily active during the end of mitotic exit and early G1 phase of the cell cycle whereas the APC/C-Cdc20 complex regulates the transition from metaphase to anaphase and mitotic exit.

Table 1 Bona fide APC/C substrates reported to date

Protein substrate	Function(s)	APC/C co-activator	References
Acm1 from budding yeast	Regulation of APC/C through Cdh1 inhibition	Cdh1/ Cdc20	Enquist-Newman et al. (2008)
Anillin	Completion of cytokinesis	Cdh1	Zhao and Fang (2005)
Aurora A kinase	Regulation of cell cycle progression	Cdh1	Littlepage and Ruderman (2002)
Aurora B kinase	Regulation of spindle assembly, chromosome alignment and segregation	Cdh1	Stewart and Fang (2005), Nguyen et al. (2005)
BRCA1-associated RING domain protein 1 (Bard1)	Regulation of spindle pole formation	Cdh1/ Cdc20	Song and Rape (2010)
Brain-Specific Kinase 2 (BRSK2)	Regulation of BRSK2 centrosome localisation during mitosis	Cdh1	Li et al. (2012)
Budding uninhibited by benzimidazoles 1 (Bub1) kinase	Regulation of chromosome segregation and spindle checkpoint signalling	Cdh1	Qi and Yu (2007)
Cell division cycle 5 (Cdc5) Polo-like kinase	Regulation of Cdc14 localisation	Cdh1	Visintin et al. (2008)
Cell division cycle 6 (Cdc6)	Control of DNA replication initiation	Cdh1	Petersen et al. (2000)
Cell division cycle 7 (Cdc7) kinase	Activator of S-phase Kinase (ASK; also known as Dbf4)	Cdh1	Yamada et al. (2013)
Cell division cycle 20 (Cdc20)	APC/C coactivator; recruitment of APC/C substrates for proteasome-dependent degradation	Cdh1	Huang et al. (2001), Hyun et al. (2013)
Cdc25A phosphatase	Regulation of Cyclin-dependent kinase 1 (Cdk1) activity in mitosis	Cdh1	Donzelli et al. (2002)
Chromatin licensing and DNA replication factor 1 (Cdt1)	Initiation of DNA replication in association with Cdc6	Cdh1	Sugimoto et al. (2008)
Centromere Protein F (CENP-F)	Control of chromosome segregation and kinetochore formation in mitosis	Cdc20	Gurden et al. (2010)
Chromosome instability and karyogamy protein 1 (Cik1) from budding yeast	Regulation of mitotic spindle and microtubule functions	Cdh1	Benanti et al. (2009)
Cytoskeleton-Associated Protein 2 (CKAP2), also known as TMAP	Roles in mitotic spindle assembly and maintenance	Cdh1	Seki and Fang (2007)
Cyclin-Dependent kinases regulatory subunit 1 (Cks1)	Regulation of G1/S transition through interaction with S-phase kinase-associated protein 2 (Skp2)	Cdh1	Bashir et al. (2004)

(continued)

Table 1 (continued)

Protein substrate	Function(s)	APC/C co-activator	References
Claspin	Activation of Checkpoint kinase 1 (Chk1) and regulation of DNA damage repair	Cdh1	Gao et al. (2009), Bassermann et al. (2008)
Clb2	Control of degradation of the mitotic entry regulator Swe1	Cdh1	Simpson-Lavy et al. (2009)
Conductin	Inhibitor of the Wnt signalling pathway	Cdc20	Bassermann et al. (2008)
Cyclin A	Control of S phase and G2/M transition	Cdh1/ Cdc20	den Elzen and Pines (2001), Geley et al. (2001)
Cyclin B	Activation of Cdk1 and regulation of the G2/M transition	Cdh1/ Cdc20	Clute and Pines (1999)
E2F Transcription Factor 1 (E2F1)	Regulation of G1/S transition and apoptosis	Cdh1/ Cdc20	Budhavarapu et al. (2012)
E2F Transcription Factor 3 (E2F3)	Roles in cell cycle exit and neuronal differentiation	Cdh1	Ping et al. (2012)
Epithelial cell transforming 2 (Ect2)	Regulation of Ras homolog gene family member A (RhoA) in mitosis	Cdh1	Liot et al. (2011)
Euchromatic Histone Lysine Methyltransferase 2 (EHMT2)	Regulation of histone H3K9 methylation and senescence	Cdh1	Takahashi et al. (2012)
Eyes Absent 1 (EYA1)	Control of cell proliferation and M-to-G1 transition	Cdh1	Sun et al. (2013)
FANCD2-Associated Nuclease 1 (FAN1)	Confers cellular resistance against DNA inter-strand crosslinking agents	Cdh1	Lai et al. (2012)
Filaments in between nuclei protein 1 (Fin1) from budding yeast	Stability of the mitotic spindle	Cdh1	Woodbury and Morgan (2007)
Forkhead box M1 (FoxM1) transcription factor	Regulation of S and M phase progression	Cdh1	Park et al. (2008)
Geminin	Inhibition of DNA replication	Cdh1	McGarry and Kirschner (1998)
Glucagon-Like Peptide-1 (GLP1)	Regulation of histone H3K9 methylation and senescence	Cdh1	Takahashi et al. (2012)
Glutaminase 1 (GLS1)	Glutaminolysis	Cdh1	Colombo et al. (2011)
Glutamate receptor 1 (GluR1)	Regulation of homeostatic plasticity	Cdh1	Fu et al. (2011)
HEC1	Regulation of kinetochore microtubule dynamics and mitotic exit	Cdh1	Li et al. (2011)

(continued)

Table 1 (continued)

Protein substrate	Function(s)	APC/C co-activator	References
Hyaluronan-mediated motility receptor (Hmmr)	Required for spindle pole localisation of Targeting protein for Xklp2 (Tpx2)	Cdh1/ Cdc20	Song and Rape (2010)
Histone synthetic lethal 1 (Hsl1) from yeast	Regulation of Mitogen-Activated Protein Kinase (MAPK) pathway signalling	Cdh1	Simpson-Lavy et al. (2009)
Hepatoma Up-Regulated Protein (HURP)	Microtubules crosslinking	Cdh1/ Cdc20	Song and Rape (2010)
Inhibitor of differentiation 1 (Id1)	Inhibition of dendrite growth	Cdc20	Kim et al. (2009)
Inhibitor of differentiation 2 (Id2)	Stimulation of axon growth	Cdh1	Lasorella et al. (2006)
IQ motif containing GTPase-Activating Protein 1 (IQGAP) from yeast	Regulation of cytokinesis	Cdh1	Ko et al. (2007)
c-Jun N-terminal Kinase (JNK)	Regulation of cell survival, cell differentiation, and mitotic exit	Cdh1	Gutierrez et al. (2010)
Kid from frog	Regulation of spindle formation and of chromosome movements in mitosis and meiosis	Cdh1	Feine et al. (2007)
Kinesin family member 18A (Kif18A)	Control of chromosome congression	Cdc20	Sedgwick et al. (2013)
Liprin-α	Regulation of synaptic size	Cdh1	Teng and Tang (2005), van Roessel et al. (2004)
Mcl-1	Negative regulation of apoptosis	Cdc20	Harley et al. (2010)
Microcephalin-1 (MCPH1)	Regulation of chromosome condensation and Cdk1 activation	Cdh1	Meyer et al. (2019)
Mes1 from fission yeast	Dual effector of the APC/C (e.g., substrate and inhibitor)	Cdc20	Kimata et al. (2008)
Male germ cell RacGTPase Activating Protein (MgcRacGAP)	Control of cytokinesis and cell cycle progression; regulation of RhoA, Ras-related C3 botulinum toxin substrate 1 (Rac1) and Cell division control protein 42 (Cdc42)	Cdh1	Nishimura et al. (2013)
Modulator of Apoptosis Protein 1 (MOAP-1)	Modulator of apoptosis and DNA damage repair	Cdh1	Huang et al. (2012)

(continued)

Table 1 (continued)

Protein substrate	Function(s)	APC/C co-activator	References
Monopolar spindle 1 (Mps1) kinase	Regulation of the spindle assembly checkpoint and chromosome-microtubule attachments	Cdh1/ Cdc20	Cui et al. (2010)
NEDL2 ubiquitin ligase	Control of p73 and Ataxia Telangiectasia and Rad3-related protein (ATR) kinase levels	Cdh1	Lu et al. (2013)
NIMA-related kinase 2A (Nek2A)	Regulation of centrosome separation and spindle formation	Cdc20	Hayes et al. (2006), Hames et al. (2001)
Neurogenic Differentiation factor 2 (NeuroD2)	Negative regulation of presynaptic differentiation	Cdc20	Yang et al. (2009)
Nuclear Interaction Partner of Alk kinase (NIPA)	Regulation of nuclear Cyclin B1 degradation	Cdh1	Klitzing et al. (2011)
Ninein-like protein (Nlp)	Regulation of centrosome maturation	Cdh1/ Cdc20	Wang and Zhan (2007)
Nrm1 from yeast	Transcriptional activation of MBF and Mcm1 target genes	Cdh1	Ostapenko and Solomon (2011)
Nucleolar Spindle–Associated Protein (NuSAP)	Control of microtubules dynamics	Cdh1/ Cdc20	Song and Rape (2010)
p190	Regulation of cell mobility	Cdh1	Naoe et al. (2010)
Cyclin-dependent kinase inhibitor 1 ($p21^{Cip1}$)	Inhibition of cyclin-dependent kinase activity	Cdc20	Amador et al. (2007)
6-phosphofructo-2-kinase/ fructose-2,6-biphosphatase isoform 3 (PFKFB3)	Regulation of glycolysis	Cdh1	Herrero-Mendez et al. (2009)
PHF8 demethylase	Regulation of histone H3 and H4 expression	Cdc20	Lim et al. (2013)
Polo-like kinase 1 (Plk1)	Control of mitotic spindle formation	Cdh1	Lindon and Pines (2004)
PR-Set7 methyltransferase	Control of cell cycle progression	Cdh1	Herrero-Mendez et al. (2009), Wu et al. (2010)
Rad17	Activation of DNA damage checkpoint mechanisms	Cdh1	Zhang et al. (2010)
Receptor-Associated Protein 80 (RAP80)	Recruitment of BRCA1 to DNA damage sites	Cdh1/ Cdc20	Cho et al. (2012)
RCS1	Regulation of the metaphase-to-anaphase transition	Cdh1	Zhao et al. (2008)
REV1 polymerase	Control of stalled DNA replication	Cdc20	Chun et al. (2013)

(continued)

Table 1 (continued)

Protein substrate	Function(s)	APC/C co-activator	References
Securin	Catalytic activity inhibition of the protease Separase	Cdh1/ Cdc20	Michaelis et al. (1997), Uhlmann et al. (1999)
Shugoshin 1 (Sgo1)	Prevention of premature sister-chromatid separation	Cdh1	Karamysheva et al. (2009), Christensen et al. (2007)
Sineoculis homeobox homolog 1 (Six1)	Regulation of cell development	Cdh1	Christensen et al. (2007)
S-phase kinase-associated protein 2 (Skp2)	Promotes degradation of Cdk inhibitors Cyclin dependent kinase inhibitor 1B ($p27^{Kip1}$) and $p21^{Cip1}$	Cdh1	Wei et al. (2004), Bashir et al. (2004)
Ski-related novel protein N (SnoN)	Inhibition of Transforming Growth Factor beta (TGF-β) signalling and stimulation of axon growth	Cdh1	Stegmuller et al. (2006)
Sp100 scaffold protein	Control of transcription regulation and apoptosis	Cdc20	Wang et al. (2011)
Transforming Acidic Coiled-Coil protein 3 (TACC3)	Regulation of mitotic spindle assembly and chromosome segregation	Cdh1	Jeng et al. (2009)
Thymidine Kinase 1 (TK1)	Control of genome stability	Cdh1	Ke et al. (2005, 2007)
Thymidylate Kinase (TMPK)	Regulation of deoxythymidine Triphosphate (dTTP) production and genome stability	Cdh1	Ke et al. (2005)
Tribbles homolog 3 (TRB3)	Regulation of Endoplasmic Reticulum (ER) stress-induced cell death	Cdh1	Ohoka et al. (2010)
Transformation/ Transcription domain-Associated Protein (TRRAP)	Condensation of chromatin and control of chromosome segregation	Cdh1/ Cdc20	Ichim et al. (2014)
UbcH10 E2 ubiquitin ligase	Essential factor of the APC/C	Cdh1	Rape and Kirschner (2004)
Ubiquitin carboxyl-terminal hydrolase 1 (UPS1)	Regulation of DNA repair and genomic stability	Cdh1	Cotto-Rios et al. (2011)
Yhp1 from yeast	Transcriptional activation of MBF and Mcm1 target genes	Cdh1	Ostapenko and Solomon (2011)

Towards the Gap 2 (G2) phase, Cdc20 is phosphorylated by Cyclin-dependent kinase 1 (Cdk1) and other mitotic kinases, resulting in the activation of the APC/C-Cdc20 complex (Hein et al. 2017; Höckner et al. 2016; Kramer et al. 2000; Kraft et al. 2003). However, spindle misalignment or improper kinetochore attachments to sister chromatids trigger SAC signalling, resulting in the inhibition of APC/C E3 ubiquitin ligase activity through sequestering Cdc20 from the APC/C core complex by the MCC (Yu 2002, 2007; Bharadwaj and Yu 2004; Musacchio and Salmon 2007; Musacchio 2011; Lara-Gonzalez et al. 2012).

Extensive phosphorylation of the APC/C occurs during mitosis, where more than 50 phosphorylation sites have been identified (Touati et al. 2018; Craney et al. 2016). Cdk1, Bub1, and Mitogen-Activated Protein Kinase (MAPK) inhibit APC/C-Cdc20 E3 ligase activity during the spindle checkpoint (Chung and Chen 2003; Tang et al. 2004; Yudkovsky et al. 2000) whereas APC/C activation of the ubiquitin chain-forming E2 subunit of Ube2S is impaired when Cdc20 is phosphorylated by protein phosphatase PP2AB56 (Lee et al. 2017). Furthermore, depletion of Kinetochore scaffold 1 (Knl1), a large kinetochore protein (2342 amino acid residues in the human) that acts as an organising centre and recruits BubR1 to the kinetochore, prevents PP2AB56 and Ube2S binding to Cdc20 (Craney et al. 2016). These features suggest that the regulated phosphorylation/dephosphorylation of Cdc20 acts as a molecular rheostat that contributes to the modulation of APC/C ubiquitin E3-ligase activity through a finely controlled assembly and disassembly of APC/C complexes.

The D-box motifs of the Ras association domain containing family 1 isoform a (Rassf1a) mediate the binding of this protein to the APC/C-Cdc20 complex, resulting in the inhibition of APC/C ubiquitin ligase activity during mitotic entry (Chow et al. 2012; Song and Rape 2010). Similar to the case of MCC-mediated inhibition, Rassf1a only suppresses APC/C-Cdc20 activity. In contrast, the F-box protein Emi1 (also called FBXO5) inhibits the APC/C-Cdc20 complex in Synthesis (S) and G2 phases (Reimann et al. 2001) and the APC/C-Cdh1 complex during the transition from G1 to S phase (Miller et al. 2006; Hsu et al. 2002). Unlike Emi1, the Emi1 homolog Emi2 inhibits APC/C to regulate meiotic cell division in a process that probably involves the Emi2 RL motif and additional residues (Liu et al. 2007; Tung et al. 2007).

An additional layer of regulation of APC/C-Cdc20 activity in response to the status of the mitotic spindle involves fine-tune cycles of Cdc20 synthesis and degradation and the formation of the APC/C-Cdc20-MCC complex (Izawa and Pines 2015; London and Biggins 2014). Indeed, the ubiquitination of the Cdc20 residues Lys485 and Lys490 trigger extensive conformational transitions that enable the adoption of an active, open state that docks the E2 subunit of Ube2C thus fine-tuning APC/C ubiquitination activity that ultimately ensures a tight control of mitosis progression. The fact that Cyclin B and Securin are quickly degraded once the last sister chromatid is properly attached to the kinetochore supports the view that SAC signalling and mitotic control are connected (Hagting et al. 2002; Clute and Pines 1999).

The picture emerging from these studies is that of one sophisticated regulatory mechanism involving extensive cooperative interactions that act in a concerted fashion to ensure proper chromosome segregation and cell cycle progression. This manner, depending on the precise requirements by the cell at a given point in time, involves interactions that control APC/C ubiquitin E3-ligase activity that can act either as an inhibitor or as an activator of the complex in order to maintain a stable genome. The fact that defects in APC/C regulation results in genome instability and aneuploidy provides evidence of the evolutionary advantage of this intricate mode of cell cycle regulation. The fact that most of the germline knockout mice targeting the APC/C pathway are embryonic lethal, confirms further the essential role of APC/C in the control of the eukaryotic cell cycle.

Cell-Cycle-Independent Functions of APC/C E3 Ligase

APC/C has cell-cycle-independent functions in the cell, including regulation of genomic integrity and cell differentiation of the nervous system (Puram and Bonni 2011; Hu et al. 2011; Wasch et al. 2010; Manchado et al. 2010). Protein regulators of DNA damage repair and genomic stability such as Claspin (Gao et al. 2009; Bassermann et al. 2008), Ubiquitin carboxyl-terminal hydrolase 1 (UPS1) (Cotto-Rios et al. 2011), and Rad17 (Zhang et al. 2010) as well as the proteins G9a and Glucagon-Like Peptide (GLP) (Takahashi et al. 2012) have been reported to be *bona fide* Cdh1 substrates. However, a deeper understanding of the physiological role of the APC/C-Cdh1 complex in these cellular processes warrants further investigations. If confirmed, the interactions will expand further the repertoire of APC/C functions to include important roles in apoptosis and senescence.

In the last two decades, the association of the APC/C with numerous crucial functions in the Central Nervous System (CNS), including axon guidance, synaptic development and plasticity, neurogenesis, neuronal morphogenesis and neuron differentiation, survival and metabolism has been ascertained. The APC/C-Cdh1 complex plays a role in the differentiation and function of the nervous system, particularly in the control of axon growth and patterning during normal brain development (Konishi et al. 2004). Such regulatory role in neuronal development is mediated by the targeting of the axon growth-promoting factors Inhibitor of differentiation 2 (Id2) and Ski-related novel protein N (SnoN) by the APC/C-Cdh1 complex, leading to their subsequent degradation by the proteasome (Stegmuller et al. 2006; Lasorella et al. 2006). The APC/C-Cdc20 complex seems to also play important roles in neuronal development as it regulates dendrite morphogenesis and presynaptic differentiation, in a process that involves degradation of the transcription factors NeuroD2 (Yang et al. 2009) and Inhibitor of differentiation 1 (Id1).

Further studies showed that synaptic plasticity, synaptic size and the bioenergetic and antioxidant status of neurons are controlled by APC/C-Cdh1-dependant degradation of Glutamate Receptor 1 (GluR1) (Fu et al. 2011) and Liprin-α

(Teng and Tang 2005; van Roessel et al. 2004). Despite these advances, whether APC/C deficiency is associated with neurological and psychiatric disorders such as learning and memory impairment remains largely obscure.

An Incipient Role of the APC/C in Alzheimer's Disease?

There is accumulating evidence of a direct involvement of APC/C-Cdh1 in numerous Alzheimer's disease (AD) pathophysiological hallmarks with substantial indication that deregulation of APC/C substrates has significant implication in AD. For example, brains of individuals suffering from AD exemplify an accumulation of APC/C substrates, implying APC/C deregulation. Furthermore, it has recently been depicted that the APC/C is inactivated in AD (*reviewed in* Fuchsberger et al. 2017). Key AD pathological hallmarks include senile plaque development comprising the toxic peptide Amyloid beta (Aβ) and production of Neurofibrillary Tangles (NFT) by hyperphosphorylation of the microtubule associated protein tau, with subsequent aggregate formation. These pathophysiological hallmarks have been associated with numerous processes exemplified in AD including ectopic cell cycle re-entry, oxidative stress, excitotoxicity, impaired Long-Term Potentiation (LTP), and deregulation of numerous signalling pathways, collectively resulting in the subsequent development of neurodegeneration in AD. It is postulated that these disease-related processes possibly correlate with deregulation of the APC/C and its substrates (summarised in Fig. 3) (Fuchsberger et al. 2017).

APC/C Association with the Ectopic Cycle in AD

Differentiated cells including neurons which are post-mitotic cells, remain in the G0 phase in a quiescence state, due to an active downregulation of cell cycle-related proteins. In neurons, the cell cycle protein Cyclin B1 and the glycolytic enzyme 6-phosphofructo-2-kinase/fructose-2,6-biphosphatase isoform 3 are actively downregulated by the APC/C-Cdh1 (Almeida 2012). This downregulation is fundamental in averting erroneous re-entry of post-mitotic neurons into the G1 and S phase of the cell cycle and for preserving neurons reduced antioxidant status. However, it has been exemplified that in response to CNS insults, neurons can retain the ability to re-enter the cell cycle under pathological conditions (Almeida 2012). Moreover, it has been depicted that APC/C activity is not only crucial for regulating cell cycle progression in proliferating cells but also required for neuronal survival. APC/C-Cdh1 is active in postmitotic neurons, regulating axonal growth and patterning in the developing brain (Konishi et al. 2004).

The mounting evidence illustrating the accumulation of Cyclin B1 in neurons of brains from individuals suffering AD, demonstrates this APC/C substrate plays an important role in AD pathological hallmarks of neuronal loss and apoptosis. Indeed,

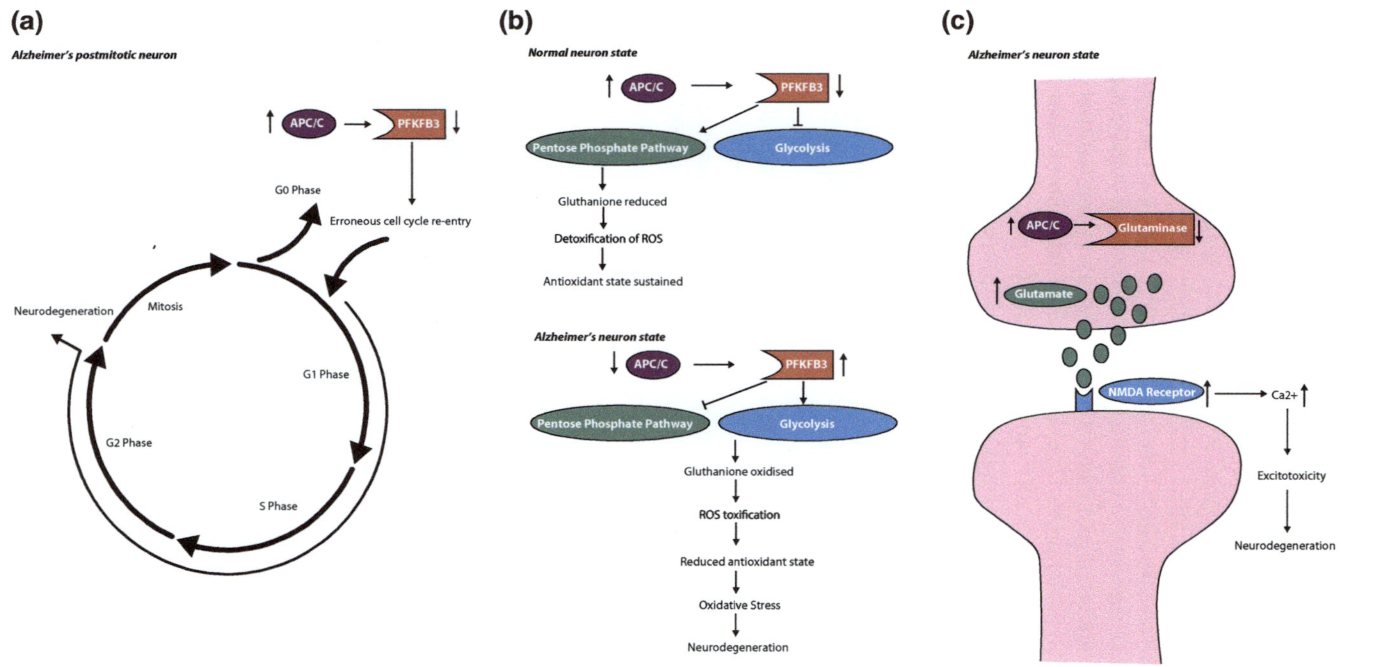

Fig. 3 Association of APC/C dysregulation in Alzheimer's disease. **a** Inactivation of APC/C-Cdh1 causes an upregulation in APC/C substrate Cyclin B1 that allows an erroneous re-entry in the cell cycle which leads to subsequent neurodegeneration. **b** In normal state neuron, APC/C-Cdh1 activity is high and APC/C substrate 6-phosphofructo-2-kinase/fructose-2,6-bisphosphatase isoform 3 (PFKFB3) is low. Glycolysis is switched off and the pentose phosphate pathway is used which in turn reduces glutathione that subsequently detoxifies Reactive Oxygen Species (ROS) and maintains the antioxidant state. In an Alzheimer's neuron, APC/C-Cdh1 activity is low and APC/C substrate PFKFB3 is high. The pentose-phosphate pathway (PPP) is switched off and glycolysis is used which in turn oxidises glutathione and leads to a reduced antioxidant state and subsequent neuronal death. **c** APC/C-Cdh1 downregulation causes an increase in APC/C substrate glutaminase, a critical precursor for synthesis of the neurotransmitter glutamate. The resulting high levels of the latter amino acid overstimulates the *N*-methyl-D-aspartate (NMDA) receptor, which subsequently leads to an increase in Ca2+ thus causing excitotoxicity and neuronal death

Mosch and collaborators have reported an elevated content of DNA and Cyclin B1 in AD neurons, indicating reactivation of their cycle and progression towards the S phase (Mosch et al. 2007). It has been shown that the APC/C activator Cdh1 was predominantly expressed in vivo throughout terminal differentiation of cortical neurons. In contrast, expression of the APC/C activator Cdc20 was severely decreased during neuron differentiation, strongly indicating that Cdh1 is the sole APC/C activator active in postmitotic, terminally differentiated neurons (Almeida et al. 2005). In cortical neurons where Cdh1 was depleted, Cyclin B1 accumulation and subsequent neuronal degeneration ensued, strongly indicating that the increased Cyclin B1 levels are implicated in apoptotic neuronal death (Almeida et al. 2005). Furthermore, Cyclin B1 depletion in Cdh1-depleted cortical neurons significantly increased cell survival, signifying Cdh1 promotes active Cyclin B1 degradation to facilitate postmitotic neuron survival, impeding neurons from entering an erroneous S phase with subsequent apoptotic neuron degeneration.

Collectively, these findings depict APC/C activator Cdh1 is an essential survival protein for postmitotic neurons and its inactivation and deregulation stipulate a potential mechanism for the erroneous reactivation of Cyclin B1 that has been reported to precede ectopic cell cycle and subsequent neuronal death in AD (Almeida et al. 2005).

Under excitotoxic conditions *N*-methyl-D-aspartate Receptor (NMDAR) overactivation occurs in AD and stimulates Cyclin-dependent kinase 5 (Cdk5) activation, which in turn induces triple phosphorylation of Cdh1 residues Ser-40, Thr-121 and Ser-163 with subsequent nuclear export of Cdh1 and APC/C-Cdh1 inactivation. Consequently, nuclear localisation and accumulation of Cyclin B1 in cortical neurons ensues, with successive induction of neuronal apoptotic death (Maestre et al. 2008). Intriguingly, NMDAR residue Ser-1232 is phosphorylated by Cdk5 mediating NMDAR synaptic transmission (Li et al. 2001). It is possible that overactivation of NMDAR, which arises in prolonged release of neurotransmitter glutamate that occurs in AD, potentially instigates a Cdk5-NMDAR activation feedback loop that contributes to neurodegeneration response propagation. If so, inactivation of APC/C-Cdh1 stimulated by Cdk5 with subsequent elevated Cyclin B1 levels potentially provides a novel pathway in neurodegeneration ensuing NMDAR stimulation (Maestre et al. 2008). Furthermore, endogenous glutamatergic stimulation of rat cortical primary neurons induces Cdk5-dependent-mediated phosphorylation of Cdh1 with subsequent Cdh1 accumulation in the cytoplasm and disassociation from the APC3 core protein with consequential APC/C-Cdh1 inactivation (Veas-Pérez de Tudela et al. 2015). This inactivation of APC/C-Cdh1 causes depletion of $p27^{Kip1}$ and switches on a Cyclin D1-Cyclin-dependent kinase 4 (Cdk4)-retinoblastoma protein (pRb) pathway inducing an erroneous S-phase entry of post-mitotic neurons and subsequent neuronal death. These findings indicate APC/C-Cdh1 actively suppresses ectopic cell cycle and neuronal death, emphasising its fundamental neuroprotection function (Veas-Pérez de Tudela et al. 2015).

In addition, it has been exemplified that localisation of Cdk5 not its total amount is a key determinant in neuron postmitotic state. This is because when primary neurons were stimulated to re-enter the cell cycle a loss of nuclear Cdk5 was observed. This feature is found in neurons from individuals suffering AD and AD mouse models with Cdk5 redistributed to the cytoplasm (Zhang et al. 2008a). Consequently, cell cycle suppression is alleviated with subsequent erroneous and lethal cell cycle proceeding, potentially resulting in further neuronal stress (Zhang et al. 2008a). A similar principle applies during normal cell cycle. When a neuron enters S-phase Cdk5 is re-localised to the cytoplasm, where it is unstable and ubiquitinated by APC/C-Cdh1 with subsequent degradation by the proteasome. The Cdk5 ubiquitination site seems to overlap with the binding region implicated in Cdk5 physical interaction with the protein p35. When p35 levels are elevated Cdk5 ubiquitination is impaired, resulting in its attenuated degradation by the proteasome (Zhang et al. 2012). It is postulated that an unsuccessful normal clearance pathway would exacerbate elevated Cdk5 kinase activity outcome, and in conjunction with alleviated cell cycle suppression, indicates that Cdk5 participates in ectopic cell cycle of AD through numerous mechanisms. The timing of Cdk5 ubiquitination in early S-phase is out of phase with APC/C-Cdh1 activity, which is ordinarily inactivated prior to entry into S-phase. This signifies a previously unidentified APC/C-Cdh1 activity in G1/S phase. It is possible that APC/C-Cdh1 subunits change and consequently alter substrate specificity favouring Cdk5 over other APC/C substrates. Alternatively, APC/C is primarily located in the cytoplasm in neurons, however its localisation may be altered (Zhang et al. 2012).

Collectively, these findings provide strong indications that inactivation and downregulation of APC/C-Cdh1 is implicated in the AD ectopic cell cycle. Further findings have corroborated a role of APC/C-Cdh1 in S-phase entry. Ubiquitin ligase complex SCF regulates S-phase entry by inducing Cdk inhibitors $p21^{Cip1}$ and $p27^{Kip1}$ degradation. In G1 phase, SCF complex subunits Skp2 and Cks1 are unstable and their degradation is facilitated by APC/C-Cdh1. It has been exemplified that APC/C-Cdh1 silencing in HeLa cells at G1 phase stabilises the SCF complex with subsequent increased $p21^{Cip1}$ and $p27^{Kip1}$ degradation and an abundance of cells at S-phase. Collectively, it is exemplified in more ways than one, that APC/C-Cdh1 inactivation potentially causes erroneous re-entry at S-phase (Aulia and Tang 2006).

APC/C Involvement in Oxidative Stress in AD

There is mounting evidence illustrating that during AD the brain tissue is subjected to oxidative stress (protein, lipid and DNA oxidation and glycoxidation), which is denoted by an imbalance in homeostasis due to the escalated radical production of Reactive Oxygen Species (ROS) or reduced antioxidative defence. (Gella and Durany 2009; Huang et al. 2016). A significant attribute of the antioxidant defence mechanism is the low molecular weight reducing equivalent Glutathione (GSH), a small molecule that is accountable for the maintenance of an appropriate redox potential in the cell. Glutathione most significant function is the donation of electrons to ROS with the purpose to scavenge the ROS.

The concentration of intracellular GSH declines in aged mammalian brain regions including the hippocampus (Sasaki et al. 2001; Sandhu and Kaur 2002; Calabrese et al. 2004; Zhu et al. 2006), consequently creating a state in which ROS production rate surpasses the capability of the antioxidant defence mechanism, inducing a condition that favours oxidative stress defence (Gella and Durany 2009). Furthermore, ROS can react with proteins, lipids, DNA and other molecules potentially modifying their structure and function, and subsequently affecting organs and tissues. Importantly, the brain is particularly susceptible to ROS due to its constitution of easily oxidisable lipids and activity, which results in a high oxygen consumption rate (Huang et al. 2016).

Oxidative stress partakes in the induction of Aβ deposition, hyperphosphorylation of the protein tau and consequently synaptic and neuronal loss, indicating that oxidative stress is a critical contributor to AD pathophysiology (Gella and Durany 2009). The damage caused by ROS mediates alterations in numerous signalling pathways in neurons that are associated with oxidative stress and neurodegeneration. Among these, APC/C-Cdh1 downregulation stimulates an alteration in the oxidant and antioxidant homeostasis of neurons (Fuchsberger et al. 2017). In comparison to other brain cells including astrocytes, neurons have the highest oxygen consumption rate. However, the glycolysis rate, the metabolic process accountable for producing the majority of energy needs in human cells, is relatively low (Almeida et al. 2004). Glucose-derived glucose-6-phosphate is the crucial metabolite linking glycolysis to the Pentose-Phosphate Pathway (PPP), an alternative pathway to glycolysis for glucose oxidation. Utilising the PPP, neurons metabolise glucose. PPP is the key metabolic pathway accountable for regenerating Nicotinamide Adenine Dinucleotide Phosphate (NADPH), a vital reducing cofactor for many oxidoreductases (Almeida 2012). A central PPP regulatory enzyme is Glucose-6-Phosphate Dehydrogenase (G6PD) that promptly reacts to the surge in demand for NADPH essential for sustaining the neurons antioxidant state. This antioxidant state is crucial in preserving glutathione in its reduced form (GSH), which consequentially is critical for detoxification of ROS. The PPP supplies the majority of reducing equivalents in the form of NADPH, which are required for reduction of Glutathione disulfide (GSSG) to GSH, a reaction catalysed by glutathione reductase (Wamelink et al. 2008). 6-phosphofructo-2-kinase/fructose-2,6-bisphosphatase isoform 3 (PFKFB3) is a central enzyme in regulating and activating glycolysis. Its activation is fundamental for the prompt activation of glycolysis and subsequent energy production from glucose as well as protection from apoptosis. Its virtual depletion in neurons provides clues about neurons severe sensitivity to energy diminution and neurodegeneration (Almeida et al. 2004).

It has been depicted that PFKFB3 contains a ^{142}Lys-Glu-Asn (KEN) box motif that is continually targeted for ubiquitination by APC/C-Cdh1 and subsequent proteasomal degradation (Herrero-Mendez et al. 2009). In neurons APC/C-Cdh1 activity is high and subsequently PFKFB3 abundance is low. In contrast, astrocytes exhibit high PFKFB3 concentration and very low Cdh1 levels, with a concomitant low APC/C-Cdh1 activity (Herrero-Mendez et al. 2009). Moreover, APC/C-Cdh1 activity dictates the regulation of glucose consumption rate through both glycolysis and the PPP. PFKFB3 upregulation through either Cdh1 inhibition or PFKFB3 overexpression instigated glycolysis activation concurrently with a decrease in the PPP. As PPP is critical for glutathione reduction at the expense of NADPH for neuroprotection, this shift in glucose consumption towards glycolysis through the expense of a decrease in PPP, increased the oxidation of glutathione and production of ROS, promoting a state favouring oxidative stress and subsequent neuronal death. These observations depict APC/C-Cdh1 as fundamental for regulating the glycolytic pathway and PPP to generate an antioxidant state for neuroprotection at the expense of glucose utilisation for bioenergetics purposes (Herrero-Mendez et al. 2009), which are met by other sources (Herrero-Mendez et al. 2009).

Moreover, it has been demonstrated in cortical neurons that NMDAR activation through APC/C-Cdh1 inhibition induced stabilisation of PFKFB3, instigating a metabolic alteration of increased glycolysis activity and reduced PPP activity with subsequent oxidative stress and neuronal apoptosis (Rodriguez-Rodriguez et al. 2013). Furthermore, overactivation of NMDAR represses APC/C-Cdh1 activity through a Ca^{2+}-Cdk5-mediated signalling pathway, which causes phosphorylation of Cdh1 and the subsequent inactivation of the APC/C-Cdh1. This inactivation of the APC/C-Cdh1 complex not only stabilises PFKFB3, causing increased glycolysis and reduced PPP action, but also stimulates the accumulation of Cyclin B1, illustrating that the NMDAR-APC/C-Cdh1-PFKFB3 axis triggers oxidative stress and subsequent neuronal death by excitotoxicity as well as neuronal death by ectopic cell cycle (Zhou et al. 2016).

Stabilisation of PFKFB3 and accumulation of Cyclin B1 link APC/C-Cdh1 inactivation/downregulation with neuronal death through oxidative stress and ectopic cell cycle. Future studies are imperative to further comprehend and ascertain whether Cdh1 deficiency or APC/C-Cdh1 inactivation exert a pivotal regulatory role in the mechanisms that trigger neurodegeneration.

APC/C Connection with Excitotoxicity in AD

Excitotoxicity is characterised as consequential cell death from Excitatory Amino Acid (EAA) toxic actions. In the mammalian CNS, the key EAA neurotransmitter is glutamate, thus ordinarily neuronal excitotoxicity alludes to neuronal injury and death ensuing prolonged exposure to synaptic release of glutamate and the successive deregulated Ca^{2+} homeostasis (Dong et al. 2009; Arundine and Tymianski 2003). Glutamate activates postsynaptic ionotropic EAA receptor NMDAR, which opens its associated ion channel enabling Ca^{2+} and Na^{+} ion influx. While normal functioning of a cell requires physiological elevations of intracellular Ca^{2+}, the surplus of Ca^{2+} ions can overwhelm Ca^{2+} regulatory mechanisms and is exceptionally neurotoxic, inducing protein, membrane lipids and nuclei acid degradation, oxidative stress, altered synaptic plasticity and eventual subsequent neuronal necrosis and apoptosis (Arundine and Tymianski 2003; Dong et al. 2009).

Compelling evidence indicates that in AD soluble Aβ peptide oligomers provoke synaptic dysfunction and loss. This Aβ-mediated synaptic toxicity is dependent on NMDAR overstimulation which instigates increase cytoplasmic Ca^{2+} (Tu et al. 2014). Although the molecular mechanisms remain not fully understood, it has been alluded that Aβ induces these events by causing an aberrant increase in extrasynaptic glutamate abundance through impeding uptake of glutamate or stimulating release of glutamate from glial cells (Tu et al. 2014).

In a primary culture of neurons, Aβ caused Cdh1 proteasomal-dependent reduction which subsequently stabilised APC/C-Cdh1 substrate glutaminase, a critical neuronal glutamate production source. Consequently, glutamate levels increased and further intraneuronal Ca^{2+} dysregulation ensued (Fuchsberger et al. 2016). Furthermore, glutamate also reduced Cdh1 levels and caused glutaminase accumulation, indicating a potential positive feedback loop of Cdh1 inactivation. Inhibition of glutaminase averted glutamate excitotoxicity and subsequent apoptosis in Aβ treated neurons. In addition, in vivo findings utilising microinjection of either Aβ or glutamate in the CA1 region of the rat hippocampus, exemplified re-localisation of Cdh1 from the nucleus to the cytoplasm and increased glutaminase levels. Furthermore, elevated levels of glutaminase were depicted in APP/PS1 mice compared to wild-type control mice. These findings collectively indicate that elevated glutaminase and subsequent glutamate generation contribute to AD excitotoxicity (Fuchsberger et al. 2016).

It has been exemplified that both glutamate- and glutaminase-immunoreactive pyramidal neurons in the hippocampus and cerebral cortex illustrate morphological abnormalities with irregular shortened and disorganised dendritic fields in AD individuals compared to age-matched controls. In conjunction with these observed degenerative changes both glutamate- and glutaminase-immunoreactive pyramidal neurons depicted NFT (Kowall and Beal 1991). Collectively, the mounting evidence of glutaminase accumulation and the finding of Aβ-induced Cdh1 reduction stabilises glutaminase, provides a probable vindication for the observations in AD.

Furthermore, under excitotoxic conditions NMDAR hyperactivation inhibits APC/C-Cdh1 function through a Ca^{2+}-Cdk5-mediated signalling pathway. This inactivation results in PFKFB3 stabilisation, increased glycolysis and reduced PPP action, and Cyclin B1 accumulation. These findings of APC/C-Cdh1 inactivation/downregulation during excitotoxicity proposes a link between Ca^{2+} dynamics and glucose metabolism regulation (Zhou et al. 2016). The intracellular Ca^{2+} elevation results in mitochondrial Ca^{2+} overload accountable for production of ROS and release of cytochrome c, both critically contributing to glutamate-induced excitotoxicity (Luetjens et al. 2000). Oxidative stress related to excitotoxicity also causes mitochondrial fragmentation and this imbalance in mitochondrial dynamics can induce overactivation of NMDAR, further promoting excitotoxicity (Nguyen et al. 2011). Thus, it can be depicted that a metabolic imbalance ensuing activation of NMDAR is significantly implicated in excitotoxicity (Rodriguez-Rodriguez et al. 2013).

It is evident that all these intracellular events converge at the ubiquitous pathways of necrosis and apoptosis. While oxidative stress has been associated with glutamate and Aβ-mediated neurotoxicity, ectopic cell cycle and perturbations in systems involving glutamate are potentially implicated in chronic AD neurodegeneration pathogenesis. Aβ-mediated neurotoxicity and neurodegeneration have been related with NMDAR overactivation, with subsequent excitotoxicity ensuing that potentially augments the localised neuronal susceptibility in a manner consistent with the pathogenesis of AD (Dong et al. 2009). Furthermore, Aβ induces surplus ROS production, dependent on NMDAR activation (De Felice et al. 2007). This indicates that NMDAR activity dysregulation and oxidative stress potentially devise dual deleterious roles in AD (Dong et al. 2009). Furthermore, the evidence of Aβ-induced APC/C-Cdh1 degradation and subsequent glutaminase accumulation is postulated as a further molecular link between the AD pathogenic factor Aβ and slow excitotoxicity in which APC/C-Cdh1 inactivation/downregulation appears to be the crucial intermediate mechanism.

APC/C Contribution to Long-Term Potentiation Impairment in AD

Long-Term Potentiation (LTP) is a form of activity-dependent plasticity that triggers continual synaptic transmission augmentation (Bliss and Cooke 2011). APC/C-Cdh1 has been implicated in cortical plasticity, learning and memory (Li et al. 2008). A study analysed *cdh1*-knockout mice, in which a gene-trap construct was inserted into intron 5 of *cdh1*. Cdh1 protein was absent in homozygous *cdh1*-knockout mice and an approximately 50% decrease was observed in heterozygous *cdh1*-knockout mice (Li et al. 2008). In homozygous *cdh1*-knockout mice, the absence of Cdh1 caused early lethality. In heterozygous *cdh1*-knockout mice compared to wild-type, no gross alteration in morphology of hippocampus or other brain regions was observed, however impairment in the late-phase LTP induced by multiple spaced trains of high-frequency stimulation was demonstrated. Furthermore, in hippocampal slices from heterozygous *cdh1*-knockout mice late-phase LTP declined back to baseline levels within 3 h after the final train of high-frequency stimulation compared to the robust late-phase LTP observed in wild-type mice. These results propose that APC/C-Cdh1 potentially plays a vital role in late-phase LTP induction likely through mediating the ubiquitination and subsequent degradation of unidentified proteins that negatively regulate late-phase LTP, corroborating the concept that both protein synthesis and degradation are crucial for the long-lasting form of synaptic plasticity expression and maintenance (Li et al. 2008; Fonseca et al. 2006).

In addition, it has been shown that Cdh1 is essential for amygdala-dependent memory. In *cdh1* conditional Knockout (cKO) mice contextual and cued fear memory, which both require a functioning amygdala, were defective. Furthermore, late-phase LTP impairment was observed in amygdala slices (Pick et al. 2012). In conjunction with these observations, in amygdala *cdh1* cKO mice a Cdh1 expression reduction was noted. In addition, an elevated expression of the NMDAR subunit NR2A and the Synaptic scaffolding protein SH3 and multiple ankyrin repeat domains protein (Shank), which are two postsynaptic density (PSD) proteins, ensued LTP-inducing stimulation in *cdh1* cKO mice amygdala slices. These findings indicate and are consistent with the concept that ordinarily APC/C-Cdh1 is implicated in amygdala synaptic plasticity and memory as well as regulating in the PSD the appropriate protein complexes following neuronal activity (Pick et al. 2012).

Collectively, these findings propose that either at various development stages or in different cell types, Cdh1 has distinctive responsibilities in different regions of the brain. It is probable that Cdh1 activity divergent effects in the hippocampus and amygdala are attributable to the various Cdh1 substrates diverse expression in these brain regions (Pick et al. 2012).

Furthermore, heterozygous *cdh1*-knockout mice compared to wild-type mice performed poorly in the hippocampus-dependent process of contextual fear conditioning, suggesting *cdh1* haploinsufficiency instigates hippocampus-dependent memory impairments and that the ubiquitination and subsequent degradation of APC/C protein substrates mediated by APC/C-Cdh1 is vital in learning and memory (Li et al. 2008). In addition, it has been demonstrated in a Morris water maze test that cKO of APC/C subunit APC2 in excitatory adult forebrain neurons in mice decreases capability in spatial memory formation and abrogates the capability of extinguishing contextual fear conditioning, indicating a hippocampal functional defect (Kuczera et al. 2010).

The stability of the dendrite destabiliser Rho protein kinase 2 (Rock2) is regulated by APC/C-Cdh1. In *cdh1* cKO mice, loss of APC/C-Cdh1 function in adult neurons causes Rock2 accumulation and increased activity with subsequent destabilisation of cortex and hippocampus dendrites, in conjunction with cognitive impairment and neurodegeneration in mice. Elimination of these pathological events ensued Rock2 inhibition. These findings indicate APC/C-Cdh1-mediated degradation of Rock2 regulates the integrity of the dendritic structural and functional network depicting Cdh1 as a crucial molecular factor in AD pathogenesis (Bobo-Jiménez et al. 2017).

Although these findings indicate that the APC/C potentially participates in hippocampus-dependent memory and learning, its implication in synapse function and plasticity remains to be fully investigated. In particular, the underlying molecular mechanisms of APC/C contributions in learning and memory remain elusive (Puram and Bonni 2011).

APC/C Implications in Neurogenesis Impairment in AD

The two hallmark features of neurogenesis are proliferation and differentiation. Although causal links have not been ascertained, a variety of critical molecules implicated in the pathogenesis of AD seem to exert positive or negative regulation of new neurons generation (Mu and Gage 2011).

Studies conducted in Cdh1 knockout mouse model, have depicted that Cdh1 elimination shortens G1 phase duration and extends S-phase duration in Neural Progenitor Cells (NPCs), significantly delaying NPCs cell cycle exit, which promotes replicative stress that consequently initiates NPCs p53-mediated apoptotic death with an overall reduction in newly differentiated cells (Delgado-Esteban et al. 2013). In addition, in vitro investigations into neuronal differentiation suggested the rapid degradation of APC/C-Cdh1 substrate Skp2 with subsequent stabilisation of Cdk inhibitor $p27^{Kip1}$ that ultimately lead to cell cycle exit. Both differentiation of a rat pheochromocytoma cell line PC12 and human neuroblastoma cell line SH-SY5Y to neuronal phenotypes required elimination of Skp2 and stabilisation of $p27^{Kip1}$. This proposes that at least for cell differentiation in vitro, the APC/C-Cdh1-Skp2-$p27^{Kip1}$ axis defines a conserved signalling pathway (Delgado-Esteban et al. 2013). Furthermore, Cdh1 phosphorylation status is crucial for neuronal differentiation and survival in vitro. In culture NPCs, phosphorylated and non-phosphorylated Cdh1 delayed or accelerated cell cycle exit and neurite extension, respectively. As non-phosphorylated Cdh1 does not impede APC/C-Cdh1 activity, APC/C substrate Cyclin B1 degradation is promoted, with subsequent Cdk1 activity inhibition supporting the neuronal phenotype. Besides the negative effect on the progress of neuronal differentiation, phosphorylated Cdh1 also causes NPCs apoptotic death (Delgado-Esteban et al. 2013).

Collectively, these findings indicate that functional APC/C-Cdh1 activity is a necessity for both neurogenesis in vivo and cortical neuronal differentiation in vitro, coupling the NPCs cell exit with neuronal differentiation (Delgado-Esteban et al. 2013).

Furthermore, utilising mouse cerebellar Granule Cell Progenitors (GCPs) as a model of brain neurogenesis, the role of Casein Kinase 1 delta (CK1δ) in the CNS progenitor cell expansion was exemplified. Conditional loss of CK1δ in GCPS or small interfering RNA (siRNA) knockdown of CK1δ in wild-type GCPs demonstrated diminished GCPs proliferation. In addition, administration of a CK1δ specific inhibitor also illustrated GCPs proliferation reduction in vitro and ex vivo. In contrast overexpression of CK1δ resulted in increased proliferation of GCPs (Penas et al. 2015).

Moreover, CK1δ is an APC/C-Cdh1 substrate and conditional Cdh1 deletion in GCPS caused elevated CK1δ levels. CK1δ is the sole isoform of Casein Kinase 1 (CK1) that is targeted by APC/C-Cdh1 for degradation in the developing cerebellum. The mechanism in which APC/C-Cdh1 acquires specificity for CK1δ with regards to GCPs proliferation remains elusive. CK1δ is the only CK1 isoform that comprises a N-terminal D-box motif, all other CK1 isoforms contain D-boxes that could potentially facilitate turnover by the APC/C-Cdh1, however would necessitate upstream signalling pathway activation for recognition by APC/C-Cdh1. Further investigations ascertaining GCP-specific interactors or substrates should help to elucidate the mechanism in which APC/C-Cdh1 regulates CK1δ levels during cell proliferation (Penas et al. 2015).

Collectively, these findings indicate that CK1δ controls neurogenesis of GCPs and that APC/C-Cdh1 mediated degradation of CK1δ regulates GCPs proliferation

(Penas et al. 2015). Cdh1 is essential for appropriate neurogenesis, which consequently has crucial implications in the regulation of mammalian brain size, thus establishing Cdh1 as an important player in the molecular pathogenesis of neurological disorders (Delgado-Esteban et al. 2013).

Manipulation of the APC/C by Viruses

APC/C manipulation by viruses can severely affect cell cycle control and other physiological attributes of animal cells including the human. These qualities in conjunction with many viruses during infection controlling the host cell cycle including oncoviruses, denote the APC/C as an attractive viral target for establishing an intracellular environment favouring viral replication (Mo et al. 2012). There is mounting evidence that numerous viruses target the APC/C. Two excellent reviews examining how viruses control the host cell cycle through manipulating the APC/C listed 7 viruses and identified several viral proteins that target the APC/C (summarised in Table 2 and Fig. 4) (Mo et al. 2012; Fehr and Yu 2013). As discussed below, viruses utilise a diverse collection of molecular mechanisms, accentuating the capability of modulating APC/C function through convergent evolution. Importantly, such manipulation retains the mutual trait of enabling viruses to maintain their replication (Fehr and Yu 2013).

Human T-Cell Lymphotropic Virus Type 1 Tax

The retrovirus Human T-cell Lymphotropic Virus type 1 (HTLV-1), is the etiological agent of the aggressive Adult T-cell Leukaemia/Lymphoma (ATL) and neurological disorder HTLV-1-associated myelopathy/tropical spastic paraparesis. The viral transactivator/oncoprotein HTLV-1 Tax facilitates potent viral transcription activation and targets mechanisms of cell control including cell cycle progression to sustain viral replication and stimulate infected T-cell proliferation (Boxus and Willems 2009).

Tax-expressing cells manipulate the APC/C through provoking a delay in cell cycle phases S/G2/M by directly interacting and prematurely activating APC/C-Cdc20 before mitosis onset. This correlates with decreased levels of APC/C substrates Cyclin A, Cyclin B1, Securin and Skp2 (Liu et al. 2003, 2004). Tax directly interacts with APC/C-Cdh1 and APC/C-Cdc20 complexes, denoting Tax likely binds a core APC/C subunit. The mechanism of action remains elusive, although altering APC/C phosphorylation state is a probable cause (Liu et al. 2004). More recently, an inverse correlation between expression of Tax and numerous Cdc20 substrates abundance including Cyclin B and Bim has been exemplified (Wan et al. 2014). Defective mitosis induced by Tax resulting in chromosome instability and cytokinesis failure that leads to severe aneuploidy as reflected in

Table 2 Proposed mechanisms of APC/C manipulation by viruses

Virus	Viral protein expressed	Mode of APC/C manipulation	Postulated mechanism of APC/C manipulation	Oncovirus	References
HTLV-1	Tax	Direct	Directly interacts with APC/C-Cdh1 and APC/C-Cdc20. Mechanism of action remains elusive although altering APC/C phosphorylation state is a probable cause	Yes	Liu et al. (2003, 2004)
HBV	X	Direct	Binds BubR1 potentially disrupting MCC inhibition towards APC/C-Cdc20 by promoting Cdc20 association and premature activation of APC/C. Binds DDB1, however mechanism of action remains elusive. Impairs APC/C-Cdh1proteolytic machinery	Yes	Kim et al. (2008), Martin-Lluesma et al. (2008), Pandey and Kumar (2012)
ORFV	PACR	Direct	Imitates APC11 interacting with APC2 and appears to be integrated into APC/C in place of APC11. Lacks ubiquitin ligase activity potentially inhibiting E2-E3 interaction crucial for APC/C function. May cause redirection of APC/C activity rather than simple inhibition	No	Mo et al. (2009, 2010a, b)
HPV	E2	Direct	Directly interacts with Cdh1 and Cdc20. Instigates accumulation of Cyclin B and	Yes	Bellanger et al. (2005), Tan et al. (2015)

(continued)

Table 2 (continued)

Virus	Viral protein expressed	Mode of APC/C manipulation	Postulated mechanism of APC/C manipulation	Oncovirus	References
			induces redistribution of Cdh1 to insoluble cytoplasmic aggregates, inhibiting APC/C by refusing it access to Cdh1. Binds BubR1 and Mad2. As stabilises Cyclin B implies potential Mad2/BubR1/Cdc20/E2 complex formation, preventing SAC inactivation and MCC disassembly subsequently inhibiting APC/C		
HPV	E6/E7	Direct	Elevated Cdc20 and UbcH10 levels cause abnormal APC/C activation and mitotic exit. E7 causes transcriptional upregulation and stabilisation of APC/C inhibitor Emi1. Emi1 upregulation may inhibit APC/C substrates degradation and cause prometaphase delay	Yes	Patel and McCance (2010), Yu and Munger (2013)
CAV	Apoptin	Direct	Interacts with APC1 initiating APC/C disruption with APC/C subunits degradation and APC/C substrate stabilisation. Interaction with APC1 combined	No	Teodoro et al. (2004), Coster et al. (2007), Kucharski et al. (2011), Mo et al. (2012), Lee et al. (2009b), Wiebusch and Hagemeier (2010)

(continued)

Table 2 (continued)

Virus	Viral protein expressed	Mode of APC/C manipulation	Postulated mechanism of APC/C manipulation	Oncovirus	References
			with increased Cyclin B1 causes APC/C inactivation sufficient for apoptosis induction. Relocalises MDC1 and APC3 to PML bodies with subsequent degradation and consequential APC/C disassociation and inhibition. APC/C and MDC1 disruption during infection possibly impedes APC/C-Cdh1 which if active would perturb viral replication		
Ad	E1A	Direct	APC5/APC7 and E1A compete with one another in binding CBP/p300. E1A regulates APC/C-CBP/p300 complexes ability by disrupting the interaction consequently impeding their functions. Inhibition of CBP/p300–APC/C-dependent transcriptional activation of genes involved in G1 arrest may ensue and utilisation of CBP/p300 complexes in absence of APC/C promotes S-phase	No	Turnell et al. (2005), Mo et al. (2012)

(continued)

Table 2 (continued)

Virus	Viral protein expressed	Mode of APC/C manipulation	Postulated mechanism of APC/C manipulation	Oncovirus	References
			entry and cellular DNA synthesis. How E1A affects APC/C activity remains elusive		
Ad	E4orf4	Direct	E4orf4 binds PP2A. E4orf4-PP2A complex directly interacts with APC/C altering APC/C activity. Both APC/C-Cdc20 inhibition and activation could involve different potential PP2A targets within APC/C and within its modulators. Binding site and mechanism of action remains elusive. E4orf4 and its effectors Ynd1p and Cdc55 potentially functionally interact with Cdc20 and Cdh1	No	Mui et al. (2010), Kornitzer et al. (2001), Kleinberger (2015), Maoz et al. (2005)
HCMV	pUL97	Direct	Induces phosphorylation of Cdh1 impeding Cdh1 interacting with APC/C and APC/C-Cdh1 activity	Potential association	Tran et al. (2010), Wiebusch et al. (2005)
HCMV	pUL21a	Direct	Interacting partners identified as APC3, APC7 and APC8. Essential and sufficient for inducing APC4 and APC5 degradation with subsequent APC/C	Potential association	Fehr et al. (2012)

(continued)

Table 2 (continued)

Virus	Viral protein expressed	Mode of APC/C manipulation	Postulated mechanism of APC/C manipulation	Oncovirus	References
			disassembly. Which APC/C subunit pUL21a directly binds and precise mechanism inducing APC4 and APC5 degradation remains elusive. Possibly binds neighbouring subunit of APC4 and APC5 or recruits protein degradation enzyme targeting subunits for degradation		
SV40	LT	Indirect	Interacts with Bub1 causing a compromised SAC potentially by disrupting MCC integrity which instigates APC/C deregulation	Yes	Cotsiki et al. (2004), Hein et al. (2008)
HCV	NS5A	Indirect	Cellular factors targeted and mechanism of action remains elusive. Decreased degradation of APC/C substrates Cyclin B and Securin denote elevated Cyclin B and Securin levels perturb the normal timing of the mitotic cell cycle	Yes	Baek et al. (2006)

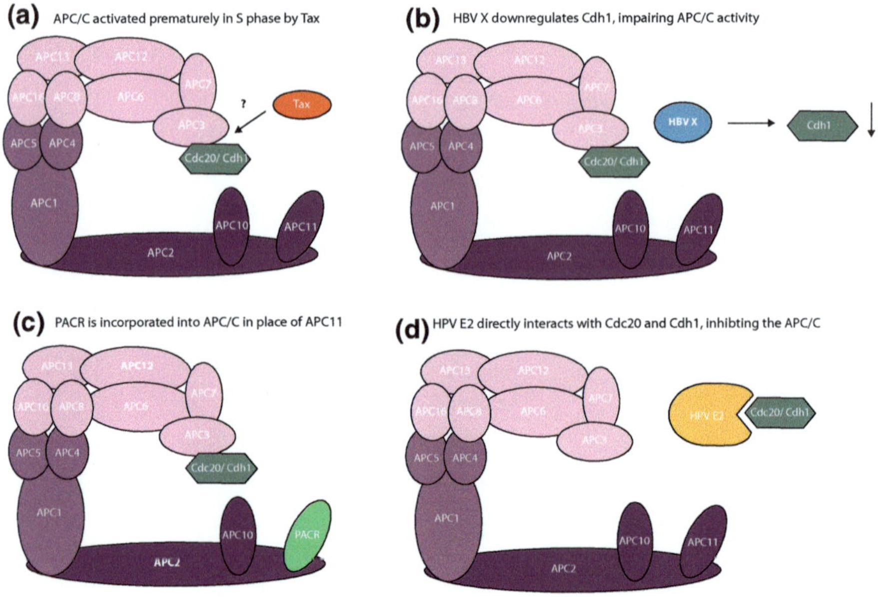

Fig. 4 Proposed mechanisms of human APC/C manipulation by viruses. **a** Human T cell Lymphotropic Virus type 1 (HTLV-1) Tax interacts with APC/C-Cdh1 and APC/C-Cdc20, prematurely activating the APC/C during S phase. Despite these advances the mechanism of action remains elusive. **b** Hepatitis B Virus (HBV) X downregulates Cdh1, impairing APC/C-Cdh1 proteolytic machinery and correlates with an unscheduled entry into S phase. **c** Orf Virus (ORFV) Poxviral Anaphase Promoting Complex/Cyclosome Regulator (PACR) inhibits the APC/C by acting as a non-functional mimic of APC11 and is incorporated into the APC/C in place of APC11. **d** Human Papillomavirus (HPV) E2 directly interacts with Cdh1 and Cdc20 and redistributes Cdh1 to insoluble cytoplasmic aggregates, inhibiting the APC/C through denying the complex access to its coactivator. **e** Human Papillomavirus (HPV) E6 and E7 causes elevated Cdc20 levels resulting in abnormal APC/C activation. **f** Human Papillomavirus (HPV) E7 upregulates APC/C inhibitor Emi1, impeding APC/C function and subsequently APC/C substrate degradation, resulting in a prometaphase delay. **g** Chicken Anemia Virus (CAV) Apoptin interacts with APC1 and relocalises the APC/C to Promyelocytic Leukaemia (PML) bodies causing degradation of APC/C subunits and subsequent APC/C inactivation. **h** Adenovirus (Ad) E1A competes with APC5/APC7 in binding CBP/p300, therefore disrupts the formation of the APC/C-CBP/p300 complex and consequently impedes the complex's functions promoting S-phase entry and aberrant mitoses. **i** Adenovirus (Ad) E4orf4 can have both stimulatory and inhibitory effects on the APC/C in a PP2A-dependent manner, however the binding site and mechanism of action remains elusive. **j** Human Cytomegalovirus (HCMV) pUL97 induces phosphorylation of Cdh1, impeding Cdh1 interacting with the APC/C resulting in APC/C inactivation. **k** Human Cytomegalovirus (HCMV) pUL21a induces APC4 and APC5 proteasomal degradation, resulting in APC/C disassembly and subsequent APC/C inactivation

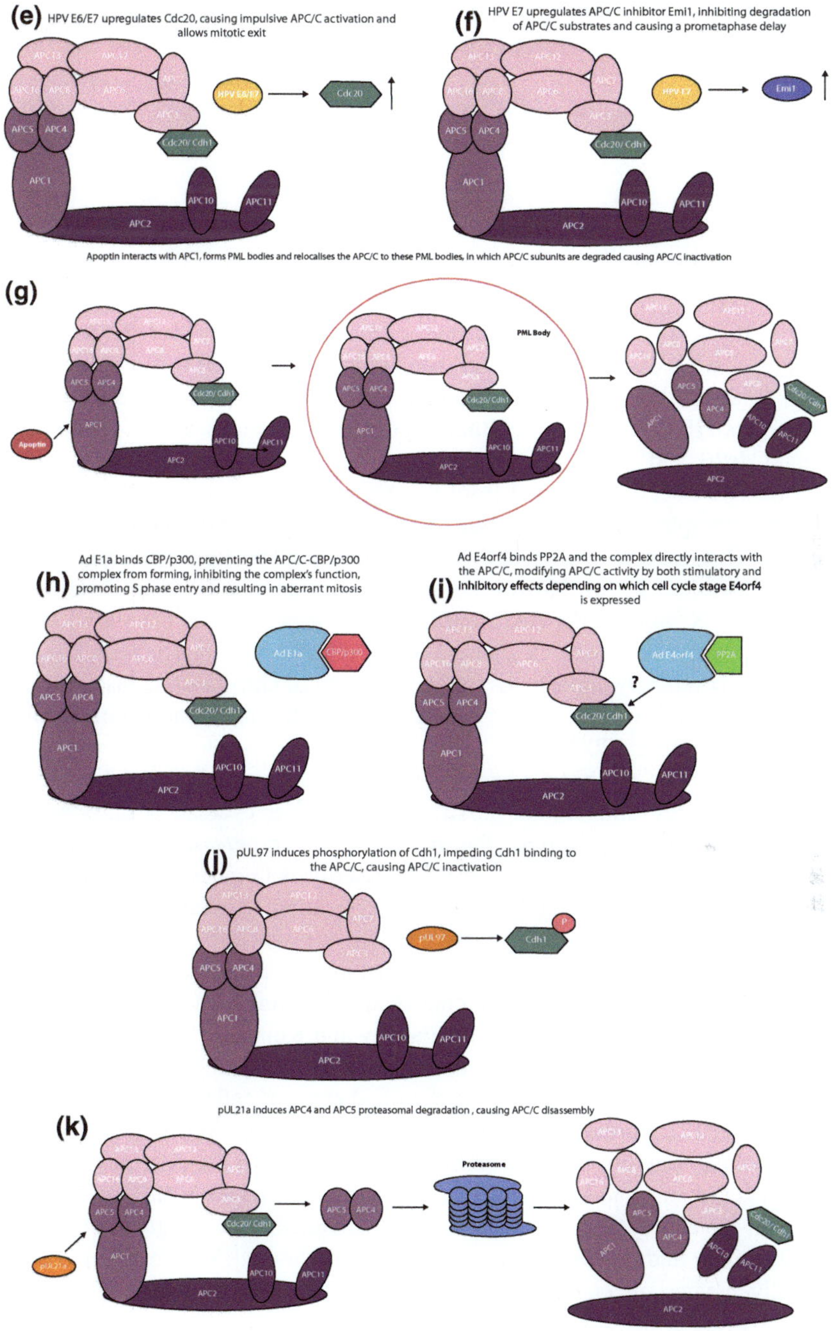

Fig. 4 (continued)

pathological findings of micro- and multinucleated cells, postulates a molecular explanation for recurrent defective chromosomes in infected T-cells and highly aneuploid attribute of ATL cells (Liu et al. 2003; Liang et al. 2002). Collectively, it is depicted that premature APC/C activation by Tax has a probable significant contribution in ATL development.

Hepatitis B Virus X

Hepatitis B Virus (HBV) is a primary etiological factor for Hepatocellular Carcinoma (HCC) development. It has been conveyed that HBV X protein binds to BubR1. Both the HBV X binding domain and Cdc20 binding domain of BubR1 overlap, thus HBV X binding to BubR1 potentially disrupts BubR1 capability of binding Cdc20 and its inhibitory activity towards the APC/C-Cdc20 complex. Consequently, Cdc20 is released from the MCC and can prematurely activate the APC/C before spindle pole alignment, thus potential chromosome missegregation can ensue. This provides a prospective insight into viral pathogen-induced defective mitosis (Kim et al. 2008). However, further investigations have not been able to confirm HBV X disrupts the SAC and MCC but instead it has been depicted that HBV X binds the E3 ubiquitin ligase Damaged DNA Binding protein 1 (DDB1) and that this potentially contributes to the induction of chromosome instability. The molecular mechanisms and effects mediated from this interaction remain elusive (Martin-Lluesma et al. 2008).

More recently it has been depicted HBV X impairs APC/C-Cdh1 proteolytic machinery (Pandey and Kumar 2012). Cell division cycle 6 (Cdc6) is a key component in DNA replication. During cell cycle progression Cdc6 is tightly controlled and a target of APC/C-Cdh1 for proteolysis in quiescence and G1 phase restricting cell proliferation and preventing replication reinitiation (Petersen et al. 2000). It has been demonstrated that in the presence of HBV X, Cdh1 is down-regulated with elevated Cdc6 levels ensuing, accentuating the importance of APC/C-Cdh1 regulation in cell cycle control and the establishment of this system in HBV X-induced oncogenic phenotype (Pandey and Kumar 2012). This impairment of APC/C-Cdh1-dependent protein degradation pathway instigating intracellular stability of Cdc6 correlates with an unscheduled entry of cells into S-phase (Pandey and Kumar 2012). Furthermore, HBV X down-regulating APC/C-Cdh1 results in the intracellular stability of deubiquitinase Ubiquitin Specific processing Protease 37 (USP37) (Saxena and Kumar 2014). In the presence of HBV X, Cyclin A levels throughout the cell cycle are stabilised and maintained, this upregulation in Cyclin A/Cdk2 activity and downregulation of Cdh1 under the HBV X microenvironment potentially creates an environment favouring USP37 upregulation (Saxena and Kumar 2014). Further investigation is imperative for ascertaining to what extent HBV X regulation of the APC/C plays a role in HBV-mediated tumorigenesis.

Orf Virus PACR

Orf Virus (ORFV) is a large double-stranded DNA parapoxvirus that causes an exanthemous disease primarily in sheep and goats and infects differentiated epidermal cells in humans. It has been ascertained that the ORFV gene ORFV014 encodes for a RING-H2 protein called Poxviral Anaphase Promoting Complex/Cyclosome Regulator (PACR) with substantial sequence homology to APC/C core subunit APC11 (Fleming et al. 2015; Mo et al. 2009).

PACR only exists in the parapoxvirus family. PACR expression causes dysregulated cell cycle with fewer cells in G1 phase, more in S, and an accumulation of cells in G2/Mitosis (M) as well as APC/C substrate accumulation collectively in a manner consistent with defective APC/C function (Mo et al. 2009). PACR imitates APC11 in its capability via its N-terminal to interact directly with APC/C core subunit APC2 and appears to be integrated into the APC/C in place of APC11 (Mo et al. 2009, 2010a, b). However, the catalytic residues present in E3 ligases are absent in PACR, resulting in a non-functional enzymatic domain in PACR that lacks ubiquitin ligase activity. This feature correlates with sequence differences in the region between the 6th and 8th Cys/His as evidenced from an APC11 and PACR enzymatic domain swap that converses their ubiquitin ligase activity (Mo et al. 2009). The lacking ubiquitin ligase activity correlates with the absence of residues fundamental for E2-E3 interaction, consequently PACR may inhibit the E2–E3 interaction. This in turn suggests that the integration of PACR into the APC/C in place of APC11 impedes ubiquitin-charged conjugating enzyme E2 recruitment, which is crucial for APC/C function (Mo et al. 2009). Overexpression of PACR demonstrates G2/M phase arrest with an increase in APC/C substrates, thus confirming that PACR impedes the APC/C by imitating a non-functional APC11 (Mo et al. 2009).

It has been demonstrated that ORFV infection may stimulate differentiated epidermal skin cells to enter a S-phase-like state providing cellular factors to sustain viral replication, a consequence of APC/C function disrupted by PACR (Mo et al. 2009). ORFV does not encode its own Thymidine Kinase (TK) and Ribonucleotide Reductase (RR) (Mo et al. 2010a), two enzymes vital for nucleotide metabolism providing nucleotides for DNA replication and the cellular versions are APC/C substrates (Ke and Chang, 2004; Chabes et al. 2003). This depicts the ORFV exploits APC/C inhibition as an alternative method for acquiring specific enzymatic activity to sustain viral replication (Mo et al. 2009, 2010a).

Interestingly, PACR appears to integrate into an otherwise intact APC/C with APC/C subunits APC2, APC3 and APC4 coprecipitating with PACR, indicating that a full complex is assembled with PACR in place of APC11 (Mo et al. 2010a, b). This may result in the redirection of APC/C activity as opposed to simple inhibition. Consequently, PACR is referred to as an APC regulator rather than presuming it is an APC/C inhibitor. Whether PACR is capable of such subtle APC/C activity manipulation is the substantial focus of on-going investigations (Mo et al. 2010a, b).

Human Papillomavirus E2

Human Papillomaviruses (HPVs) are small DNA viruses infecting human epithelial skin cells or mucosal sites specifically the cervix, stimulating hyperproliferative lesions resulting in cervical cancer. Different strains are categorised into two types; high risk comprise viruses with malignant properties and related to cervical cancer and low risk viruses are accountable for benign growths. Viral manipulation of the host cell to endow an environment favouring replication is essential for viral genome replication and high-level amplification in differentiated, non-dividing cells that depends almost entirely on host factors.

The HPV E2 protein as a viral transcription and replication factor has well ascertained roles. It has been depicted that E2 from high risk not low risk HPV strains directly interacts with Cdh1 and Cdc20 disrupting APC/C function, stimulating a mitotic block (a G2/M arrest) and often subsequently metaphase-specific apoptosis (Bellanger et al. 2005). HPV E2-expressing cells that abscond the mitotic block exemplify aberrant long metaphases and subsequently polyploidy, missegregated chromosomes and amplification of centrosomes, concurrently instigating severe genomic instability (Bellanger et al. 2005).

High risk HPV E2 binds Cdh1 and Cdc20 as proficiently as Emi1, a cellular APC/C inhibitor. It was depicted that HPV E2 co-localises with Cdh1 in insoluble intracellular structures, alluding that HPV E2 inhibits the APC/C by refusing it access to Cdh1 (Bellanger et al. 2005). Interestingly, in contrast to Emi1, in an in vitro degradation assay HPV E2 failed to inhibit APC/C activity, signifying that HPV E2 specific mode of action may differ to Emi1.

Furthermore, it has been ascertained that HPV E2 instigates accumulation of APC/C substrate Cyclin B and induces redistribution of Cdh1 to insoluble cytoplasmic aggregates (Bellanger et al. 2005). More recently, it has been depicted that in addition to Cdc20, HPV E2 binds the MCC proteins BubR1 and Mad2 and that these interactions correlate with mitotic block, DNA breaks and aneuploidy. As HPV E2 is known to stabilise APC/C substrates including Cyclin B, this implies that interaction of HPV E2 with MCC proteins potentially enables the formation of Mad2-BubR1-Cdc20-HPV E2 complex, which prevents SAC inactivation and the disassembly of the MCC at the end of metaphase, subsequently obstructing SAC inactivation, with APC/C inhibition ensuing and cells arrested in mitosis (Tan et al. 2015). This advocates that APC/C inhibition is one potential mechanism for HPV to stimulate tumours. However, it is yet to be ascertained whether HPV E2-induced genomic instability directly ensues APC/C inhibition or can be attributed to other functions of this multifunctional protein.

As HPV E2 is essential for viral DNA it is postulated that regulation of the APC/C may also be imperative for viral genome replication, exemplified by increased HPV E2 levels resulting in increased viral episomal copy number (Penrose and McBride 2000). The generation of a mutant E2 protein with its capability of APC/C inhibition specifically abrogated would be useful to dissect more precisely the mechanism of APC/C manipulation by HPV and how harnessing the control of the APC/C is exploited to promote replication and tumorigenesis.

In addition, it has been demonstrated that HPV E6 and HPV E7 gene expression facilitates circumventing the G2 DNA damage checkpoint and the SAC despite DNA damage presence. Elevated Cyclin B, Cdc20 and UbcH10 levels were observed in HPV E6 and HPV E7 expressing cells (Patel and McCance 2010). It is plausible that elevated Cyclin B permits evading G2 DNA damage checkpoint and enables M phase entry. It is also probable that high Cdc20 and UbcH10 levels results in aberrant APC/C activation, ensuing SAC inactivation with Cyclin B degradation allowing mitotic exit.

Furthermore, it has been demonstrated that HPV16 E7 oncoprotein manipulates the APC/C. HPV16 E7 interferes with degradation of APC/C substrates with transcriptional upregulation and stabilisation of APC/C inhibitor Emi1 in HPV16 E7-expressing mitotic cells (Yu and Munger 2013). The subsequent abnormally high levels of Emi1 may consequently inhibit degradation of APC/C substrates and cause the prometaphase delay displayed in HPV16 E7-expressing cells with Emi1 depletion partially reversing inhibition of APC/C substrate degradation. The detection of HPV16 E7 targeting Emi1 suggests APC/C inhibition potentially provides a critical role in HPV E7-induced mitotic abnormalities (Yu and Munger 2013). The correlation between genomic instability and HPV E6 and HPV E7 expression in malignant lesions, in which the HPV E2 open reading frame is disrupted, is a probable consequence of uncontrolled APC/C activation (Patel and McCance 2010; Fehr and Yu 2013). If so, HPV is the first virus revealed to supress and activate the APC/C at different stages of disease progression.

Chicken Anemia Virus Apoptin

Chicken Anemia Virus (CAV) is a small single-stranded virus targeting bone marrow hematopoietic cells and thymus T cell precursors, instigating severe thymocytes and erythroblastoid cells depletion in the bone marrow through apoptosis, inducing severe anemia and immunodeficiency in young chickens (Noteborn 2004; Los et al. 2009).

CAV expresses three proteins, one being a 14 kiloDalton (kDa) non-structural protein referred to as VP3 or Apoptin owing to its capability of selectively stimulating a severe G2/M arrest and ensuing apoptosis in tumour cells rendering this protein a desirable aspirant for novel cancer therapies (Noteborn 2004; Teodoro et al. 2004). Apoptin selective killing correlates with its subcellular localisation; cytoplasmic and nuclear in non-transformed and transformed cells, respectively (Danen-van Oorschot et al. 2003; Teodoro et al. 2004).

The APC/C has been identified as a nuclear target of Apoptin, with Apoptin interacting with APC/C subunit APC1 in transformed cells exemplified in co-Immunoprecipitation (co-IP) mass spectrometry analysis and immunoblot analysis (Teodoro et al. 2004; Kucharski et al. 2016). The APC/C binding domain has been mapped to Apoptin C-terminal domain and is both required and adequate for apoptosis induction. The association of Apoptin with APC1 in transformed cells initiates APC/C disruption with subsequent degradation of some APC/C subunits and APC/C substrate stabilisation, consequently instigating a G2/M phase arrest and inducing apoptosis in a p53-independent fashion (Teodoro et al. 2004). Apoptin induces the formation of Promyelocytic Leukaemia (PML) nuclear bodies and relocalises the APC/C to these PML bodies in which APC/C subunits including APC3 are degraded with consequential APC/C inhibition. This APC/C inhibition is probably due to Apoptin cell-type specific localisation and activity and its capability to shuttle between the nucleus and cytoplasm (Heilman et al. 2006). This eludes that apoptosis is likely influenced by Apoptin capability in controlling the APC/C. This premise is supported by studies where siRNA knockdown of APC1 was carried out to stimulate G2/M arrest and apoptosis, which closely resembled the effect of Apoptin-mediated APC/C inhibition (Teodoro et al. 2004). Further investigations are required to determine how Apoptin provokes APC/C disassociation and if this necessitates APC/C redistribution to PML bodies.

More recently, it has been depicted Human Gyrovirus-Apoptin (HGyv-Apoptin) a homolog of Chicken Anemia Virus-Apoptin (CAV-Apoptin), has a similar mechanism of action in manipulating the cell cycle as CAV-Apoptin through inhibiting APC/C function. Like CAV-Apoptin, HGyv-Apoptin induces G2/M arrest and abnormal spindle in cancer cells. HGyv-Apoptin appears to disrupt the cell cycle most likely by inhibiting APC/C function, as HGyv-Apoptin expressing cells maintain high Cyclin B1 expression, blocking mitotic exit and cell cycle completion selectively in cancer cells, while having no effect on cell cycle progression in normal cells (Chaabane et al. 2017).

Furthermore, the Mediator DNA Damage Checkpoint 1 protein (MDC1), a fundamental protein in the DNA damage response and a protein that associates with the APC/C subunit Cdc27, colocalises in PML bodies with Cdc27 when Apoptin is expressed resulting in the degradation of MDC1 (Coster et al. 2007; Kucharski et al. 2011). As APC/C-Cdh1 activity is upregulated in response to DNA damage and is essential for sustaining the subsequent G2 arrest, APC/C and MDC1 disruption during CAV infection possibly impedes APC/C-Cdh1, which if were active would perturb viral replication (Mo et al. 2012; Kucharski et al. 2011; Lee et al. 2009b; Wiebusch and Hagemeier 2010).

Intriguingly, it has been shown that while Apoptin-null CAV is defective in DNA synthesis, an Apoptin point mutant had the capacity of viral DNA synthesis but failed to support virus particle production (Prasetyo et al. 2009). Thus, further investigations are required to define the potential role of APC/C regulation in CAV replication.

Adenovirus E1A

Adenovirus (Ad) is a class of double-stranded DNA viruses, which cause diverse clinical illnesses such as upper respiratory infections, conjunctivitis, tonsillitis, and gastroenteritis in mammalian species. Interestingly, although it is not known to have oncogenic potential, human Ad (HAd) can stimulate transformation of cells in tissue culture (Fehr and Yu 2013).

The HAd Early region 1A (E1A) gene is accountable for viral gene transcription activation with extensive influences on host cell gene expression and renowned for negatively regulating many cellular transformation pathways including pRb, Cyclic AMP response element Binding Protein (CBP) and p300 (Mo et al. 2012; DeCaprio 2009). The discovery of E1 protein capability in binding and inhibiting pRb was the first demonstration of a direct interaction between an oncogene and a tumour suppressor (Whyte et al. 1988).

CBP/p300 are large lysine acetyltransferases acting as transcriptional coactivators through chromatin modification, controlling numerous transcriptional pathways including p53 and E2F (Goodman and Smolik 2000). It has been established that APC/C subunits APC5 and APC7 interact with CBP/p300 in vivo, correlating the APC/C with transcriptional activity of CBP/p300 (Turnell and Mymryk 2006; Turnell et al. 2005). The interaction between APC5/APC7 and CBP/p300 contributes significantly to their function with APC5/APC7 expression stimulating CBP/p300 activity and CBP knockdown impeding APC/C ubiquitin ligase activity (Turnell et al. 2005). In addition, APC5/APC7-CBP/p300 complex exemplifies ubiquitin ligase activity towards APC/C substrates and it has been demonstrated that both APC5/APC7 enhance p53-dependent transactivation of a $p21^{Cip1/Waf1}$ promoter-tethered reporter and E2F-1–DP-1-dependent transactivation through directly interacting with the respective promoters in a CBP/p300-dependent manner (Turnell et al. 2005). Intriguingly, APC5/APC7 contain within their primary sequences *bona fide* CBP/p300 protein-protein interaction domains illustrating homology to E1A, with APC5/APC7 and E1A competing in vitro with one another in binding CBP/p300. Importantly, wild-type APC5/APC7 overexpression competitively inhibited the binding of E1A to CBP/p300 and consequently impeded E1A transforming ability, while APC5/APC7 mutants did not supress E1A-induced transformation. Furthermore, APC5/APC7 knockdown enhanced the transformation ability of a transformation-defective E1A mutant unable to bind CBP/p300 (Turnell et al. 2005).

Collectively, these findings propose E1A regulates APC/C-CBP/p300 complexes ability by disrupting the interaction, consequently impeding their functions. Subsequently, inhibition of APC/C-CBP/p300-dependent transcriptional activation of genes involved in G1 arrest (e.g. $p21^{Cip1/Waf1}$) may ensue, and the utilisation of CBP/p300 complexes in the absence of APC/C to induce genes supporting progression into S-phase, together promotes S-phase entry and cellular DNA synthesis, with APC/C deregulation ultimately resulting in aberrant mitoses (Turnell et al. 2005; Mo et al. 2012). However, how E1A affects APC/C activity is yet to be directly examined, thus the effects of APC5/APC7 overexpression and knockdown may be indirect. Biochemical data is fundamental in ascertaining if E1A specifically disrupts APC/C-CBP/p300 complex activity.

Adenovirus E4orf4

Another protein HAd encodes is Early region 4 open reading frame 4 (E4orf4), which can stimulate a G2/M phase cell cycle block and acts as effective as Apoptin in its ability to induce apoptosis of transformed cells (Noteborn 2009; Mui et al. 2010). These effects are dependent on E4orf4 binding to Protein Phosphatase 2A (PP2A). The E4orf4-PP2A complex is also associated with manipulating APC/C activation in which the complex directly interacts with APC/C to alter its activity (Mui et al. 2010; Kornitzer et al. 2001).

It has been demonstrated in *Saccharomyces cerevisiae* (*S. cerevisiae*) and in mammalian cells that E4orf4 exerts its effects principally through directly binding to the Cell division cycle 55 (Cdc55) regulatory subunit of PP2A. Expression of E4orf4 induces G2/M cell accumulation with elevated Clb2-Cdc28/Cdk1 activity, alluding E4orf4 disrupts cell cycle timing and regulates mitotic kinase activity (Mui et al. 2010). In yeast, it has been indicated the cell cycle arrest is irreversible (Kornitzer et al. 2001). Overexpression of Clb2 stimulates high Cdk1 kinase activity and severely reduces cell viability, denoting unregulated Clb2-Cdc28/Cdk1 activity is lethal for cells and is a potential mechanism for the Cdc55-dependent toxicity of E4orf4 (Mui et al. 2010). Cdk1 is known to phosphorylate APC/C subunits and participate in APC/C-Cdc20 activation (Ciosk et al. 1998; Cohen-Fix et al. 1996). Premature dissociation of sisters (Pds1)/Securin is a substrate of APC/C-Cdc20 that is ubiquitinated and degraded resulting in sister chromatid segregation necessary at anaphase. During S phase, Pds1 is usually stable, impeding Esp1 protease activity against Scc1. Interestingly, it has been depicted that E4orf4 provokes premature degradation of Pds1 in S-phase arrested cells and consequently seems to induce the premature release of Esp1 with subsequent inappropriate loss of Scc1. It seems plausible that reduced Pds1 and Scc1 levels are caused by the premature APC/C-Cdc20 activation that is observed in E4orf4-expressing cells, which is a probable consequence of Cdc55-dependent increase in Clb2-Cdc28/Cdk1 activity. This supports the hypothesis that increased mitotic kinase activity stimulated by E4orf4 expression is a potential crucial mechanism for inducing apoptosis. APC/C-Cdh1 activity was not altered under these conditions (Mui et al. 2010). Ase1 and Pds1 are substrates of APC/C-Cdh1 and APC/C-Cdc20, respectively. In contrast with studies undertaken by others indicating premature APC/C-Cdc20 activation, stabilisation of Ase1 and increased Pds1 levels were demonstrated in the presence of E4orf4, indicating the latter protein inhibits both APC/C complexes (Kornitzer et al. 2001). A possible explanation for both APC/C-Cdc20 inhibition and activation could involve the existence of different potential PP2A targets within the APC/C and within its modulators that are potentially targeted differentially by E4orf4-P22A complex in S-phase and mitosis. For example, one such target could be APC/C inhibitor Emi1, which is itself inhibited by Cdk1 phosphorylation. Although in S-phase Emi1 is not usually phosphorylated, a Cdc55-dependent E4orf4-induced Cdk1 activation potentially could cause phosphorylation of Emi1 with subsequent premature APC/C activation.

Alternatively, during mitosis Cdc55-dependent APC/C inhibition may be more prominent, while Emi1 is degraded at mitotic entry (Kleinberger 2015).

While APC/C inhibition may account at least in part for the E4orf4-induced G2/M arrest, how premature APC/C-Cdc20 activation in S-phase assists in this process remains elusive as are the binding site and mechanism that E4orf4 utilises for regulating the APC/C (Kleinberger 2015; Kornitzer et al. 2001). Thus, future studies ascertaining a better understanding of the interaction between E4orf4 and the cell cycle are imperative.

A classical genetics approach identifying yeast genes that are required for E4orf4 mediated toxicity in *S. cerevisiae* revealed Ynd1p, a Golgi NTPDase, functionally interacted with Cdc55 and contributed in an additive manner to E4orf4-induced toxicity (Maoz et al. 2005). As full resistance to E4orf4 was illustrated in a double mutant lacking both Ynd1p and Cdc55, it appears Ynd1p and Cdc55 potentially are the only direct effectors of E4orf4. Biochemical studies revealed that Ynd1p associated with Cdc55 and this interaction was disrupted when E4orf4 was expressed. However, it is improbable that removal of Ynd1p or Cdc55 from a complex containing their common target is the only effect that instigates E4orf4 toxicity since E4orf4 expression is not equivalent to Ynd1p or Cdc55 deletion. Instead, overexpression of one protein and deletion of the other is necessary to cause toxicity in yeast. A pathway in which a joint target for PP2A and Ynd1p may be functional has been suggested (Maoz et al. 2005). As previously illustrated, mutants defective in APC/C activating subunits Cdc20 and Cdh1/Hct1 are supersensitive to E4orf4-induced toxicity, with Cdc20 mutant cells supersensitive to Cdc55 overexpression (Kleinberger 2015). It has also been depicted Cdh1/Hct1 mutant cells are more sensitive to Ynd1p and Cdc55 overexpression than wild-type cells and Cdc20 mutant cells are supersensitive to Ynd1p overexpression. Thus, it is postulated that E4orf4, as well as E4orf4 effectors Ynd1p and Cdc55, functionally interact with Cdc20 and Cdh1/Hct1 (Maoz et al. 2005).

Expression of E4orf4 in both yeast and mammalian cells modifies the activity of the APC/C, with effects both stimulatory and inhibitory, depending on which cell cycle stage the protein is expressed (Mui et al. 2010; Kornitzer et al. 2001). Furthermore, it is yet to be ascertained that regulation of the APC/C by E4orf4 occurs during Ad infection. In cultured human cancer or primary cells, E4orf4 is non-essential for viral replication; the influence E4orf4 has on Ad biology remains elusive (Miron et al. 2008). Thus, future studies ascertaining whether E4orf4 regulates the APC/C during infection, elucidating its significance on virus replication, and identifying E4orf4 mechanism of action are imperative.

Human Cytomegalovirus PUL97

Human Cytomegalovirus (HCMV) is a large double-stranded DNA that infects a broad range of cell types and establishes a persistent and latent infection in human hosts (Bhattacharjee et al. 2012). HCMV is a ubiquitous virus asymptomatically infecting the majority of the human population (Ranganathan et al. 2011). While immunocompetent individuals maintain a reservoir of HCMV asymptomatically, it is a frequent source of severe infectious complications in immunocompromised individuals and a leading viral cause of birth defects (Bhattacharjee et al. 2012; Ranganathan et al. 2011). Furthermore, mounting evidence proposes a possible link of HCMV with certain human cancers (Ranganathan et al. 2011; (Soroceanu and Cobbs 2011). Viral infection with HCMV-encoded proteins can produce the hallmarks of cancer including genomic instability (Fortunato et al. 2000), inflammation (Britt 2008), angiogenesis (Caposio et al. 2011), cell migration (Vomaske et al. 2010), proliferation (Hume and Kalejta 2009), inactivation of the immune response (Powers et al. 2008) and apoptotic pathways (Brune 2011) and in conjunction with HCMV specifically detected at low levels in a variety of human cancers, denotes HCMV an intriguing human cancer virus candidate (Ranganathan et al. 2011).

Irrespective of the cell type or strain, there are three classes of gene expression, Immediate-Early (IE), Early (E) and Late (L), which are expressed in a well-co-ordinated manner and fundamental for a productive viral infection. E gene expression is dependent on IE gene expression while L gene expression ensues IE and E gene expression (Bhattacharjee et al. 2012). HCMV stimulates an intracellular environment favouring viral replication through expressing IE proteins including IE72 and IE86 that disable pRb-family proteins with subsequent E2F-responsive S-phase gene expression (Bain and Sinclair 2007). HCMV also expresses factors and utilises various mechanisms to regulate APC/C function and even disassemble the complex (Tran et al. 2007, 2010; Wiebusch et al. 2005). One mechanism is the activity of a viral protein kinase pUL97 during HCMV infection which induces phosphorylation of Cdh1, a modification known in uninfected cells to impede Cdh1 interacting with the APC/C (Tran et al. 2010; Wiebusch et al. 2005). It has been established that the APC/C plays a limiting role in the regulation of G0 maintenance and that HCMV rapidly inactivates the G0/G1 APC/C in early infection of quiescent cells with the untimely stabilisation of APC/C substrates consequently ensuing. This APC/C inactivation is instigated by the dissociation of its positive regulator Cdh1. This dissociation is independent of known Cdh1 inhibitors (Wiebusch et al. 2005). pUL97 is a viral Cdk functional orthologue, thus has the expected capacity to phosphorylate Cdh1, a natural Cdk target (Hume et al. 2008). Compared to wild-type virus infected cells, in pUL97 deletion virus infected cells Cdh1 was hypophosphorylated and accumulations of APC/C substrates were delayed at early time points. However, by 24–36 h post infection (time at which viral DNA synthesis occurs) even in pUL97 deletion virus infected cells, APC/C substrates accumulated efficiently alluding that APC/C activation was still occurring (Tran et al. 2010). Furthermore, in pUL97 deletion virus infected cells and wild-type virus infected cells, Cdh1 and APC1 association with APC3 had severely reduced 16 h post infection and was undetectable 24 h post infection (Tran et al. 2010). This depicts that the disassembly of the APC/C occurs independently from Cdh1 phosphorylation and that HCMV must regulate the APC/C through additional mechanisms. APC/C disassembly during infection with localisation of APC/C subunits APC3, APC7, APC8 and APC10 to the cytosol, could account for the subsequent Cdh1 disassociation in pUL97 deletion virus infected cells (Tran et al. 2007). Retention of APC1 in the nucleus possibly denotes the necessity for this APC/C subunit in viral replication. Indeed, during mitosis different localisation patterns of APC/C subunits have been detected, implying different APC/C subunits possibly function independently of the APC/C (Acquaviva et al. 2004). APC/C disassembly is attributed to the proteasome-dependent degradation of APC4 and APC5 early in infection (6 to 8 h post infection). The fact that E1 (ubiquitin-activating enzyme) inhibition prevented APC4 and APC5 loss adds support to the notion the degradation of these APC/C subunits is ubiquitin-dependent. This alludes that potentially a cellular ubiquitin ligase is implicated in ubiquitination of the subunits as targets for proteasome degradation, as there are no known HCMV-encoded ubiquitin ligases. Perhaps a viral early protein is obligatory for the degradation of these subunits (Tran et al. 2010). Although the primary mechanism of APC/C inactivation appears as the disassembly of the APC/C and targeted APC4 and APC5 degradation, pUL97 inducing Cdh1 phosphorylation possibly endows a small kinetic advantage in inhibiting the APC/C earlier during infection, with complete disassociation of Cdh1 from APC3 occurring before APC1 disassociates from APC3. It is possible that Cdh1 phosphorylation provides another purpose in facilitating viral replication other than mediating APC/C inactivation (Tran et al. 2010). However, whether APC/C inhibition by pUL97 solely contributes to HCMV capability to replicate or causes disease remains elusive.

Furthermore, HCMV does not express APC/C substrates Thymidine Kinase 1 (TK1), and Thymidylate Kinase (TMPK) which are essential for DNA metabolism (Wiebusch et al. 2005). Through disrupting the APC/C, HCMV may enhance accessibility to these cellular enzymes to acquire vital deoxyribonucleotide Triphosphate (dNTPS) for facilitating its own DNA synthesis. Additionally, through targeting the APC/C, HCMV manipulates quiescent cells into a pseudo-S phase supportive of viral replication, while simultaneously impeding cellular DNA synthesis through expressing other viral factors (Qian et al. 2010). Future studies ascertaining these prospects and unravelling the APC/C substrates potentially vital for viral replication are imperative.

Human Cytomegalovirus Virus PUL21a

Another factor HCMV encodes that can also regulate the APC/C through inducing APC/C disassembly is pUL21a. pUL21a is a 14.3 kDa protein sharing no apparent homology with any known cellular or viral protein, alluding its function is likely to be unique to HCMV biology (Fehr and Yu 2009). pUL21a facilitates efficient viral DNA synthesis that is required for efficient late accumulation of IE transcripts and initiating a productive infection, illustrated by a pUL21a-null virus having a marked defect in viral DNA synthesis (Fehr and Yu 2010).

Interacting partners of pUL21a were identified in a proteomics screen as APC/C subunits APC3, APC7 and APC8, in which pUL21a interacts with the APC/C via pUL21a carboxyl terminus which contains the APC/C binding domain and residues $PR_{109-110}$ which are fundamental for this interaction (Fehr et al. 2012). Functional analysis signified pUL21a was essential and sufficient for inducing proteasomal degradation of APC4 and APC5, with subsequent APC/C disassembly. A pUL21a point mutant PR109-110AA (pUL21a$_{PR\text{-}AA}$) virus in which proline-arginine residues were mutated to alanine and incapable of binding to the APC/C was unable to induce degradation of APC4 and APC5, signifying APC4 and APC5 degradation is dependent on pUL21a binding to the APC/C (Fehr et al. 2012). pUL21a does not contain a sequence domain that would imply it is an E3 ligase and it is currently unknown which APC/C subunit pUL21a directly binds to. Consequently, the precise mechanism in which pUL21a induces APC4 and APC5 degradation remains elusive. It is plausible that pUL21a may bind a neighbouring subunit of APC4 and APC5, resulting in APC/C disassembly and subsequent degradation of APC4 and APC5, or possibly pUL21a recruits a protein degradation enzyme such as an E3 ubiquitin ligase to target the subunits for proteasomal degradation (Fehr et al. 2012).

More recently, a Cyclin-binding domain (RxL) within pUL21a that confers its ability to bind and target Cyclin A for proteasomal degradation has been discovered (Caffarelli et al. 2013). pUL21a inhibition of APC/C would result in increased Cyclin A levels, which in turn would be detrimental to HCMV replication. Cyclin A promotes cellular replication which exhausts vital enzymes and metabolites necessary for viral replication, subsequently HCMV blocking Cyclin A is imperative.

Interestingly, pUL21a is highly unstable undergoing proteasome dependent but ubiquitin independent degradation (Fehr and Yu 2009). It is intriguing to deliberate whether pUL21a directly binds APC4 and APC5 targeting them for ubiquitin independent degradation. Thus, future studies ascertaining which APC/C subunit pUL21a directly binds to, the mechanism by which pUL21a induces degradation of APC4 and APC5 and whether the intrinsic instability of pUL21a contributes to its capability in regulating APC/C are imperative. It is intriguing that both CAV Apoptin and HCMV pUL21a target the APC/C bridge subcomplex, inducing its disassembly to regulate the APC/C. This is consistent with the bridge subcomplex essential function in sustaining stability of the APC/C, alluding different viruses have evolved a conserved strategy converging on this subcomplex as an efficient means in APC/C activity regulation (Fehr and Yu 2013).

It does not appear HCMV directly abolishes the APC/C enzymatic unit, thus ascertaining if the APC/C retains some activity or is modulated to target different substrates during infection is of great significance. A double mutant virus that carried both alanine substitution of proline-arginine residues in pUL21a and a pUL97 deletion exemplified significantly more attenuated viral replication than a pUL97 deletion virus alone, signifying disruption of both pUL21a and pUL97 are synthetically lethal to HCMV replication (Fehr et al. 2012). This corroborates the working model that HCMV has evolved a sophisticated strategy by exploiting two viral factors that it encodes to support viral replication through ensuring successful disruption of the APC/C to incapacitate the APC/C restriction on virus infection. It is plausible that pUL97 and pUL21a have differential roles during infection under different conditions or cell types, serving as a substitute for one another or acting synergistically to fully exploit the virus capability to acquire entire control over the APC/C during infection (Fehr et al. 2012). The verity that HCMV utilises various mechanisms to disrupt the APC/C accentuates its critical role in HCMV infection. However, analysing these intertwined viral mechanisms during infection proves overall challenging, due to the existence of other factors targeting the same process (Fehr et al. 2012). Future studies unequivocally ascertaining the critical role of APC/C modulation in HCMV replication and mechanistic insight into how APC/C modulation influences HCMV biology are important and a pending assignment in the field.

Furthermore, the mechanism of APC/C inactivation and the relative contributions of pUL21a and pUL97 were described recently (Clark and Spector 2015). In brief, in uninfected cells only pUL21a alone was able to disrupt APC/C function with subsequent APC/C substrate accumulation ensuing. In contrast, pUL97 alone was insufficient to cause APC/C substrate accumulation despite its ability to phosphorylate Cdh1. Despite Cdh1 phosphorylation the APC/C itself remains intact, thus a relatively small amount of unphosphorylated Cdh1 potentially would be adequate for APC/C coactivation. It was also demonstrated that pUL21a was both sufficient and necessary for instigating degradation of APC1 in addition to APC4 and APC5 (Clark and Spector 2015). Knockdown of any of the bridge subcomplex subunits (APC1, APC4, APC5) by siRNA mediated the destruction of the other bridge subcomplex subunits in uninfected cells. In addition, knockdown of the TPR subunit APC8 caused the destruction of all the APC/C bridge subcomplex subunits. This is not completely surprising as the apparent binding of APC1, APC4, APC5 and APC8 is interdependent in vitro (Thornton et al. 2006) and the solved structure of human APC/C places APC1 beside APC8 (Chang et al. 2014). It remains elusive whether the APC/C subunits are degraded by an active process, or APC/C subunit degradation is a part of normal APC/C turnover, with synthesis of new subunits recognised as improperly folded proteins if they do not associate correctly. While this evidence furthers our understanding of HCMV manipulation of cell cycle machinery, ascertaining the functional relevance of APC/C inhibition in HCMV infection in vivo and more precisely the mechanisms underpinning APC/C and specific APC/C subunit degradation in the presence of pUL21a warrant future investigations.

Other Demonstrations of Viruses Manipulating the APC/C

Other instances of potential viral manipulation of the APC/C have been reported. Simian virus 40 (SV40) is a small DNA tumour virus, belonging to the polyomavirus family, that produces a productive infection in its natural host, the rhesus macaque, but stimulates oncogenic transformation in nonpermissive hosts such as rodent cells. SV40 Large T antigen (LT) manipulates cell cycle regulation through indirectly activating APC/C. LT interacts with Bub1, a central SAC regulator and component of the potent APC/C inhibitory complex MCC. LT expression results in a compromised SAC potentially by disrupting MCC integrity (Cotsiki et al. 2004). SAC disruption instigates APC/C deregulation, allowing genomic instability and transformation to ensue, a phenomenon associated with SV40 infection (Hein et al. 2008). However how precisely unscheduled APC/C activation by these viruses potentially supports viral replication remains elusive.

Another virus reported to manipulate the APC/C indirectly is Hepatitis C, a virus demonstrated to be a major cause of HCC. It has been depicted that cells expressing Hepatitis C protein Nonstructural protein 5A (NS5A) exhibit increased fractions of mitotic cells, reduced G1 cell populations and decreased degradation of APC/C substrates Cyclin B1 and Securin, denoting elevated levels of Cyclin B1 and Securin perturb the normal timing of the mitotic cell cycle. During prolonged mitotic arrest, mitotic checkpoint components have demonstrated inhibiting APC/C averting it from targeting Cyclin B1 and Securin for subsequent degradation prior to the metaphase to anaphase transition. Delay of anaphase onset occurs when mitotic abnormalities arise, triggering apoptosis of the cell to avoid aneuploidy ensuing. However, it has been indicated that NS5A expressing cells by unknown mechanisms adapt to aneuploidy and return to cell cycle following a prolonged mitotic arrest (Baek et al. 2006). Future studies ascertaining the cellular factors targeted by NS5A are imperative.

APC/C Deregulation Drives Carcinogenesis

Precise cell cycle control is fundamental in averting oncogenic transformation and is safeguarded by oscillating activities of Cyclins and Cdks. It is well established that the APC/C is instrumental for regulation of these mitotic Cyclins. An intricate network of signals regulates both the APC/C timely activation and inactivation, in which defective regulation often instigates faulty mitotic checkpoint signalling, uncontrolled genome replication and erroneous mitotic exit, mutually causing genomic instability, a prevalent cancer hallmark (Zhou et al. 2016). Not only have mutations in APC/C subunits been related to oncogenic transformation but also defective activating and inhibitory mechanisms for APC/C. Thus, attributable to its essential responsibility in mitotic regulation, it is not unexpected that both a direct and indirect association of the APC/C with cancer has been elucidated.

Abnormal expression levels of APC/C subunits are an important contributor to genome instability and cancer (Rahimi et al. 2015). A total of 132 missense mutations in APC/C subunits have been reported in cancer, with the majority negatively affecting the structure stability of the subunits (Sansregret et al. 2017). These partial APC/C dysfunctions lengthened mitosis, supressed pharmacologically induced chromosome segregation errors and reduced naturally occurring lagging chromosomes in cancer cell lines (Sansregret et al. 2017). APC/C components are heterogeneously regulated in breast cancer (Park et al. 2005). However, mechanistic details of the process that enables breast cancer cells exhibiting a defective functional APC/C to progress through mitosis remains elusive. In contrast, it has been established that in colon cancer nonsense mutations affect the genes encoding for the APC/C subunits APC4, APC6/Cdc16 and APC8/Cdc23, causing premature stop codons which in turn result in the expression of truncated proteins of impaired stability (Wang et al. 2003). Ectopic expression of the truncation APC8/Cdc23 mutant in colon epithelial cells has confirmed these findings as the APC8/Cdc23 deletion mutant results in a phenotype characterised by an elevated number of cells in G2/M phase after release of G2/M phase arrest with abnormal levels of APC/C substrate Cyclin B1 observed. This indicates that APC8/Cdc23 mutant potentially deregulates cell cycle progression through mitosis, with subsequent delay in mitotic exit through an alteration in the degradation schedule of crucial mitotic regulators (Wang et al. 2003).

Significant overexpression of APC/C subunits APC2 and APC7 occur in numerous cancer cell lines and Acute Myeloid Leukaemia (AML) patients, with APC2 upregulation significantly correlating with splenomegaly (Rahimi et al. 2015). However, various APC/C expression levels have been observed in different cancer cell lines. The dissimilar APC/C subunit expression level found in different tissues, is a contributing factor to the heterogeneity of APC/C subunit expression in different cancers (Rahimi et al. 2015). In lung carcinoma characterised by rapid cell growth, APC7 was produced at high expression levels, while 90% of renal carcinomas, a form of cancer that grow slowly, showed loss of APC7 expression (Park et al. 2005). These findings signify that an APC7 expression loss is unique to some carcinomas and APC/C components are heterogeneously regulated across cancer types.

A significant elevation in APC3/Cdc27 expression has been found in gastric cancer (Xin et al. 2018). High APC3/Cdc27 expression negatively correlated with 5-year overall survival while significantly positively correlated with lymph node metastasis. In addition, Cdc27 silencing effectively inhibited gastric cancer cell proliferation, invasion and metastasis both in vivo and in vitro (Xin et al. 2018). Collectively, these findings illustrate that Cdc27 upregulation mediates gastric cancer progression and that high Cdc27 expression significantly correlates with tumour size, TNM classification of malignant tumours (TNM) stage and distant metastasis (Xin et al. 2018). Cdc27 overexpression has also been noted in breast cancer tissue, where it significantly correlates with disease recurrence, whereas Testicular Germ Cell Tumours (TGCTs) are known to harbour recurrent mutations in Cdc27 (Litchfield et al. 2015). In the latter case, all Cdc27 mutations identified to date were missense variants characterised by a constantly low frequency of mutant allelic reads, consistent with mutations of Cdc27 only present in a subclone of each tumour sample (Litchfield et al. 2015). In addition, Whole-Exome Sequencing (WES) has revealed recurrent mutations of Cdc27 in adrenocortical (Juhlin et al. 2015), osteosarcoma (Reimann et al. 2014), prostate (Lindberg et al. 2013) and lung carcinomas (Ahn et al. 2014). Elevated APC3/Cdc27 protein expression level has been illustrated in colorectal cancer and significantly correlated with tumour size, TNM stage, distant metastasis and a poorer survival rate (Qiu et al. 2016). Cdc27 overexpression promoted proliferation in vitro and in vivo with greater tumour growth capacity exhibited while siRNA knockdown of Cdc27 inhibited proliferation with a significant tumour weight and volume reduction (Qiu et al. 2016). Upregulated Cdc27 expression has been further correlated with colorectal cancer metastasis (Qiu et al. 2017). siRNA knockdown of Cdc27 significantly repressed both migratory and invasive capabilities and sphere-formation capacity of cells while Cdc27 overexpression promoted metastasis and sphere-formation capacity. Furthermore, knockdown of Cdc27 in vivo supressed metastasis with a significant reduction in metastatic nodule numbers (Qiu et al. 2017).

Abnormally high expression levels of APC11 have been reported in lung cancer (Zhou et al. 2018), where high APC11 messenger RNA (mRNA) levels significantly correlated with poorer overall survival for lung adenocarcinoma. siRNA knockdown of APC11 arrested cells in mitosis and significantly reduced cell proliferation, colony formation and migration and invasive capabilities. Conversely, APC11 overexpression promoted cell proliferation and was associated with poorer prognosis. Collectively, these findings show that APC11 crucially contributes to the development of lung adenocarcinoma (Zhou et al. 2018). APC11 overexpression has also been reported in colorectal cancer cell lines, which significantly correlated with chromosomal instability, lymphovascular invasion and residual tumours (Drouet et al. 2018).

APC/C Coactivators Association with Tumorigenesis

The two APC/C co-activators Cdc20 and Cdh1 have opposing functions in tumorigenesis, with Cdc20 and Cdh1 identified as an oncoprotein and a tumour suppressor, respectively. Abnormal Cdc20 expression has been exemplified in diverse cancer types.

Overexpression of Cdc20 in cervical cancer has been reported in high grade squamous intraepithelial lesions and invasive squamous cell carcinoma (Kim et al. 2014). In addition, upregulated and downregulated Cdc20 expression was observed in glioblastoma and low-grade gliomas respectively, illustrating Cdc20 as a potential biomarker for glioblastoma (Marucci et al. 2008). Overexpression of Cdc20 has been reported in high-risk multiple myeloma patients, which seems to correlate with cellular proliferation and overall poor prognosis (Lub et al. 2016).

Cdc20 overexpression also occurs in non-small cell lung and oral squamous cell carcinoma, which results in a significant reduction in overall survival (Kato et al. 2012; Moura et al. 2014). Moreover, Cdc20 elevated expression significantly correlates with tumour size in non-small-cell lung cancer patients and larger primary tumour size, higher DNA ploidy level and poor prognosis in lung adenocarcinoma patients (Shi et al. 2017). Furthermore, gene knockout or knockdown in lung cancer cells causes a decrease in cell proliferation (Yi et al. 2018), whereas Cdc20 upregulation in oral squamous cell carcinomas and primary head and neck tumours seems to deregulate the timing of the APC/C in promoting premature anaphase with consequential genome instability and aneuploidy (Mondal et al. 2007).

Additionally, Cdc20 is overexpressed in colon cancer cell lines and primary tissues. The elevated Cdc20 expression levels significantly correlated with decreased overall survival and additional parameters of clinical stage, N and M classification and pathological differentiation (Wu et al. 2013). However, further investigation comprising more cases are required to ascertain the clinicopathological significance of elevated Cdc20 expression in colon cancer, particularly in early stages of tumour development.

Significantly elevated Cdc20 expression has been reported in pancreatic tumour tissues and associated with poor differentiation and a reduced 5-year recurrence-free survival rate (Chang et al. 2012). Furthermore, knockdown of Cdc20 in pancreatic carcinoma cells enhanced cell cytotoxicity following paclitaxel treatment and increased cell sensitisation to gamma radiation, indicating Cdc20 could provide a promising therapeutic target for treating this devastating disease (Taniguchi et al. 2008). More recently, elevated Cdc20 expression exhibited in Pancreatic Ductal Adenocarcinoma (PDAC) tumours, significantly correlated with additional parameters of advanced tumour stage (Dong et al. 2019).

Furthermore, screening of Cdc20 expression in 445 breast cancer patients with up to 20 years of follow-up, confirmed Cdc20 overexpression. The elevated Cdc20 levels was associated with aneuploid DNA content and an aggressive course of breast cancer with Cdc20 overexpression in prognostic analyses indicating 2-fold risk of breast cancer death (Karra et al. 2014). More recently, siRNA delivery against Cdc20 in triple-negative breast cancer cells inhibited tumour growth, emphasising Cdc20 potential as a valuable therapeutic target for inhibiting tumour growth of highly aggressive metastatic breast cancer cells (Parmar et al. 2018).

Upregulated Cdc20 expression has been reported in 68.18% of HCC tissues in contrast to adjacent non-malignant tissues. This Cdc20 overexpression positively correlated with tumour differentiation and TNM stage (Li et al. 2014). In addition, siRNA knockdown of Cdc20 in HepG2 cells illustrated supressed cell proliferation and induced G2/M phase arrest, (Li et al. 2014). More recently, Cdc20 upregulation has been correlated with the additional parameters of poor histological grade, vascular invasion and poorer overall survival (Menyhárt et al. 2018).Furthermore, elevated Cdc20 expression exhibited in HCC tumours, significantly correlated with additional parameters of advanced tumour stage (Zhuang et al. 2018).

Requirement for APC/C-Cdc20 in sustaining Glioblastoma Stem-like Cells (GSCs) function has been exemplified. Knockdown of Cdc20 through RNA interference (RNAi) considerably diminished GSCs tumourigenicity, while overexpression of Cdc20 augmented GSCs self-renewal capacity and invasiveness, denoting Cdc20 is both necessary and sufficient for GSCs self-renewal, invasion and tumour initiation. Consistent with these findings, treatment with the APC/C-Cdc20 and APC/C-Cdh1 inhibitor pro-Tosyl-L-arginine methyl ester (ProTAME) can abrogate the GSCs phenotypes (Mao et al. 2015). Upregulation of Cdc20 has been identified in glioblastoma-associated stromal cells and proposed to promote glioblastoma tumorigenesis (Gujar et al. 2015).

High Cdc20 expression levels occur in urothelial bladder cancer, with overexpression of Cdc20 associated with high grade, advanced stage, non-papillary growth pattern and distant metastasis. In addition, elevated Cdc20 expression correlates with shorter recurrence-free survival as well as overall survival (Choi et al. 2013).

Cdc20 expression is significantly elevated in gastric cancer tumour tissues which positively correlates with tumour size, histological grade, lymph node involvement, TNM stage and poorer overall survival (Ding et al. 2014).

In prostate cancer, a correlation between Cdc20 overexpression with high Gleason scores and recurrence after prostatectomy has been revealed (Mao et al. 2016). More recently, prostate cancer-derived mutations in Speckle-type POZ Protein (SPOP) a protein binding and promoting Cdc20, impedes SPOP ability to bind Cdc20. Consequently, these prostate cancer cells exhibited elevated Cdc20 expression, rendering them more resistant to Cdc20 inhibitor Apcin (Wu et al. 2017). Furthermore, both in vitro and in vivo Cdc20 silencing supressed cell proliferation and enhanced chemosensitivity to docetaxel (Li et al. 2016), whereas Cdc20 overexpression facilitated docetaxel resistance of castration-resistant prostate cancer cell lines, further emphasising Cdc20 potential as a promising therapeutic target (Wu et al. 2018).

A system review and meta-analysis illustrated Cdc20 elevated expression correlates with poor survival in majority of solid tumours and indicates Cdc20 is both a novel prognostic marker and therapeutic target. However, further investigations to validate the potential mechanism and influence of Cdc20 as well as its prognostic value in human solid tumour pathogenesis are imperative (Wang et al. 2018).

Overall, consistency between findings of associated Cdc20 expression with tumorigenesis and its potential as a valuable prognostic biomarker amongst diverse cancers has been depicted and prospects of Cdc20 as a therapeutic target for cancer considered (Kapanidou et al. 2017). Future investigations ascertaining the precise molecular mechanism in which Cdc20-meditated tumorigenesis evolves are imperative.

Cdh1 is essential for downregulating DNA replication and the mitotic spindle checkpoint, with APC/C-Cdh1 activity in G1 phase averting the accumulation of essential genes and proteins necessary for another round of DNA replication (de Boer et al. 2015).

Cdh1 expression is significantly decreased in diverse cancer types including breast (Fujita et al. 2008a) prostate, ovary, liver, brain, (Bassermann et al. 2008) and colon (Fujita et al. 2008b) tumours with poor patient survival, arguing in favour of Cdh1 function as a tumour suppressor.

Furthermore, APC/Cdh1 has been revealed to target Skp2 for degradation, averting premature Skp2/SCF-meditated destruction of $p27^{Kip1}$, and this subsequently prevents premature entry into S phase, indicating dysregulation of APC/C-Cdh1-dependent proteolysis of substrates is fundamental and likely implicated in tumour initiation (Fujita et al. 2008a, b). Significantly reduced Cdh1 and $p27^{Kip1}$ expression and elevated Skp2 expression in breast and colon cancer tissues have been reported (Fujita et al. 2008a, b). In addition, in non-malignant breast and colon cancer tissues, Cdh1 knockdown by RNAi caused increased cellular proliferation and number of cells in S phase with significantly elevated Skp2 levels and consequential attenuation of $p27^{Kip1}$ levels (Fujita et al. 2008a, b).

Further studies have identified the potential and significance of the physiological tumour suppressor role of Cdh1. In a xenograft breast cancer mouse model, Cdh1 overexpression significantly reduced implanted tumour size, in contrast to Cdh1 depletion by RNAi which accelerated tumour growth (Fujita et al. 2008a). Furthermore, lentiviral-delivered knockdown of Cdh1 by RNAi in primary human fibroblasts and osteosarcoma cells caused aberrant accumulation of numerous APC/C-Cdh1 substrates including Cyclin A, Cyclin B and Aurora A. Subsequent premature and prolonged S phase ensued with significantly delayed mitosis entry, inducing defective chromosome segregation and cytokinesis, collectively stimulating genomic instability (Engelbert et al. 2007). In addition, depletion of Cdh1 in Mouse Embryonic Fibroblasts (MEFs) caused defects in proliferation and accumulation of aberrations including binucleated and multinucleated cells, chromosomal aberrations such as abnormal numbers and complex translocations, and chromosomal missegregation contributing in aneuploidy (García-Higuera et al. 2008).

Furthermore, low expression level of Cdh1 has been associated with poor prognosis in some multiple melanoma patients, with patients demonstrating high Cdh1 expression, exhibiting a significant enrichment of genes related to mature bone marrow plasma cells and Janus Kinase (JAK)/Signal Transducer and Activator of Transcription (STAT) signalling, in comparison to patients with low Cdh1 expression exemplifying significant enrichment in MYC target genes (Lub et al. 2016). Moreover, expression of Cdh1/Fizzy-related protein homolog (Fzr) was reduced in a fully malignant B-cell line as opposed to premalignant cells. Retroviral overexpression of Fzr in B-lymphoma cells caused decreased tumour formation, with tumours that did develop exhibiting diminished or extinguished retroviral Fzr (Wang et al. 2000). Furthermore, a high percentage of malignant tumours showed accumulation of Cdh1 in contrast to most benign and some low-grade tumours which exhibited low Cdh1 expression (Lehman et al. 2007).

Concomitant with Cdh1 downregulation, several APC/C-Cdh1 substrates including Aurora A, Aurora B, Cdc6, Cdc20, Cyclin B, Rad17 and Tpx2 are most frequently overexpressed in human cancers (Carter et al. 2006). Overall, an association between APC/C-Cdh1 downregulation and oncogenesis is evident with reduced APC/C-Cdh1 activity potentially causing the stabilisation and subsequent accumulation of substrates and unscheduled activity that promotes the formation of highly proliferative, genetically unstable and poorly differentiated tumours.

APC/C Inhibitors Contribution to Oncogenic Transformation

APC/C activity requires meticulous regulation executed by an intricate network of signals including endogenous inhibitory proteins. These endogenous inhibitors have evolutionarily developed to ensure appropriate APC/C ligase activity is accomplished with subsequent accurate progression of the cell cycle and chromosomal stability. Defective inhibitory mechanisms of the APC/C participate in human disease development, particularly cancer (Zhang et al. 2014). Fundamental endogenous APC/C inhibitors identified to regulate APC/C activity include Emi1 and SAC core proteins Mad2, BubR1 and Bub3. Deregulation of these inhibitors causes dysregulated APC/C activity and consequential genomic instability indirectly associating the APC/C with cancer. Mad2, BubR1 and Bub3 are three core SAC proteins vital for regulating APC/C activity. These three proteins associate with Cdc20 for the formation of the MCC. This complex monitors chromosome segregation, inhibiting APC/C-Cdc20 activity and averting subsequent degradation of APC/C substrates Cyclin B1 and Securin, consequently delaying metaphase transition to anaphase until all chromosomes have accurately attached to the mitotic spindles. Alternatively, Emi1 crucially functions in regulating the transition from G1 phase to S phase and progression of mitosis through inhibiting the activity of APC/C-Cdh1 (Zhang et al. 2014).

Deregulation of these endogenous APC/C inhibitors has been reported in diverse cancer types with deregulation of Mad2 and Emi1 reported to cause mitotic catastrophe (Hsu et al. 2002; Hernando et al. 2004). The overexpression of Mad2 has been reported in malignant lymphoma (Alizadeh et al. 2000), lung cancer (Garber et al. 2001; Heighway et al. 2002; Kato et al. 2011), hepatocellular carcinoma (Chen et al. 2002; Zhang et al. 2008b), colorectal carcinoma (Rimkus et al. 2007), soft tissue sarcoma (Hisaoka et al. 2008) and gastric cancer (Wang et al. 2009) and correlates to poor prognosis (Hernando et al. 2004; Li et al. 2003; Tanaka et al. 2001; van't Veer et al. 2002). In human fibroblasts and cell lines overexpression of Mad2 can induce the stabilisation of APC/C substrates Cyclin B1 and Securin with a subsequent mitotic delay and an aneuploid cell phenotype observed, indicating upregulation in Mad2 potentially exerts an oncogenic function through producing an overactive MCC with subsequent APC/C activity dysregulation and deficit in cell control that initiates genomic instability (Sotillo et al. 2007). Furthermore, in transgenic mice Mad2 overexpression causes a diverse range of neoplasias including chromosome breaks, anaphase bridges as well as chromosome gain and losses. These results suggest that Mad2 overexpression may exert oncogenic potential by generating a hyperactive mitotic checkpoint to override the critical role of the APC/C in the control of cell cycle and genomic stability (Sotillo et al. 2007). Furthermore, BubR1 overexpression is also observed in diverse cancers (Liu et al. 2009; Lee et al. 2009a; Pinto et al. 2008; Yamamoto et al. 2007; Grabsch et al. 2003; Seike et al. 2002), however the effect of BubR1 overexpression on APC/C substrates stabilisation remains elusive.

Emi1 overexpression has been reported in diverse cancers (Hsu et al. 2002; Lehman et al. 2007; Gütgemann et al. 2008). It has been exemplified that Emi1 through inhibiting APC/C-Cdh1 stimulates an accumulation of APC/C substrate Cyclin A and subsequent S-phase entry in somatic cells. Furthermore, while Emi1 overexpression accelerates S-phase and overrides a G block caused by Cdh1 or pRb overexpression, conversely siRNA knockdown of Emi1 impedes accumulation of Cyclin A and subsequent S-phase entry (Hsu et al. 2002). Furthermore, it has been exemplified that Emi1 overexpression causes severe mitotic catastrophe including chromosome missegregation and concurrent with these phenotypes a stabilisation of APC/C substrate Cyclin A and Aurora A is observed (Margottin Gouget et al. 2003). In addition, in p53-deficient cells, overexpression of Emi1 induces unscheduled cell proliferation, tetraploidy and chromosomal instability. While in p53 wild-type cells this tetraploidy induction by Emi1 overexpression often results in subsequent G1 arrest or apoptotic cell death, in p53-deficient cells the continual unchecked cellular proliferation in conjunction with severe chromosome missegregation likely causes the aneuploidy phenotype exemplified by diverse cancers (Lehman et al. 2006). Overexpression of Emi1 has been reported in HCC and associated with poor survival. In vitro studies revealed Emi1 was highly expressed in HCC cell lines with upregulation of Skp2 and APC/C substrates Cyclin A and Cyclin B, while $p27^{Kip1}$ was downregulated. In addition, siRNA knockout of Emi1 inhibited cell proliferation through impeding S phase and mitosis entry and resulted in the prevention of an accumulation of Skp2, Cyclin A and Cyclin B, while $p27^{Kip1}$ was upregulated. These findings indicate overexpression of Emi1 has oncogenic potential and participates in HCC cell proliferation through APC/C inhibition with subsequent stabilisation of Skp2, Cyclin A and Cyclin B and consequential degradation of $p27^{Kip1}$ that contributes to HCC pathogenesis (Zhao et al. 2013). Furthermore, Emi1 overexpression has been observed in head and neck cancers and correlates with upregulation of APC/C-Cdh1 substrates Cyclin B, Aurora A and Skp2. Head and neck cancers exhibiting high Emi1 expression levels illustrated malignant behaviours including poor differentiation and lymph node metastasis. Additionally, Emi1 knockdown induced downregulation of APC/C substrates Cyclin A, Cyclin B, Aurora A and Skp2, illustrating deregulation of Emi1 contributes to inactivation of APC/C in cancer (Shimizu et al. 2013). Moreover, frequent overexpression of Emi1 and APC/C substrates Securin, Plk1, Aurora A and Skp2 in an abundance of diverse cancer types has been reported in malignant tumours in contrast to benign tumours. Importantly, the APC/C substrates clustered together in numerous tumours with Emi1 elevated expression, indicating a 'mitotic profile' in tumours potentially arises from APC/C dysregulation (Lehman et al. 2007).

Collectively, these findings indicate that deregulation of endogenous APC/C inhibitors can mediate the dysregulation of the APC/C. Overall, whether is it a direct mutation or dysregulation of APC/C subunits, coactivators or inhibitors, APC/C misfunction appears a mutual attribute among diverse cancers.

APC/C as an Attractive Therapeutic Target

The indispensable role of APC/C in regulating mitotic progression, accurate chromosomal segregation and mitotic exit as well as its implication in oncogenesis, renders the APC/C an attractive therapeutic target for cancer treatment. Investigations into discovering and developing small molecule inhibitors to suppress cancer growth or induce cancer death have initiated, which could provide a therapeutic window amongst diverse cancers and is crucial for innovating novel prospective cancer therapeutic strategies (Liu et al. 2019; Wang et al. 2015; Zhou et al. 2016). Antimitotic drugs targeting microtubules including vinca alkaloids and taxanes are used to treat diverse cancers (Montero et al. 2005). However, the clinical efficacy of these small molecules is limited and attributed to the variable responses of the treated cells (Brito and Rieder 2009; Gascoigne and Taylor 2009). In particular, these drugs cause neurotoxicity likely due to perturbation of neuron microtubules (Jordan and Wilson 2004).

In an attempt to develop antimitotic drugs lacking such side effects, investigations were initiated into targeting specific proteins of the mitotic spindle to activate the SAC and subsequently inactivate the APC/C (Huang et al. 2009). In addition, investigations into weakening core proteins of the SAC to impede tumour growth indicate that perturbing the SAC signalling pathway is a promising pool of therapeutic targets for cancer treatment (*reviewed in* Kapanidou and Bolanos-Garcia 2014). These findings instigated the exploration and development of small molecule inhibitors that target the mitotic spindle and SAC kinase catalytic sites, some of which have entered clinical trials. Although these molecules are not neurotoxic, the utilisation of inhibitors that target SAC kinases through functioning as Adenosine Triphosphate (ATP)-binding competitors has proved challenging and of limited success. The inhibitors lack of specificity for the intended drug target potentially causes undesirable side effects and drug resistance of cancer cells to rapidly develop, which remain major concerns (Tsai and Nussinov 2013). Cancer cells can resist such killing by premature mitotic exit before apoptosis is initiated due to a weak checkpoint or rapid slippage (Huang et al. 2009). Slippage is proposed to occur when APC/C-Cdc20-induced background degradation of Cyclin B1 under an active SAC exceeds a certain threshold before cell apoptosis initiation and consequently prompts the cell to exit from mitosis (Brito and Rieder 2006). To improve clinical efficacy and overcome undesirable side effects, developing inhibitors for more specific targets represents a more appropriate approach for treating diverse cancers.

In recent investigations, it has been exemplified that mitotic exit may serve as an appropriate target owing to its pro-apoptotic outcomes of RNAi knockdown of Cdc20 (Huang et al. 2009). Depletion of Cdc20 slowed proteolysis of APC/C substrate Cyclin B1, enabling more time for cell death to be initiated. Such killing of cancer cells was independent of the SAC and could occur by intrinsic apoptosis. Inhibiting mitotic exit causes a permanent mitotic arrest and subsequent mitotic cell death as opposed to other approaches which often cause cell cycle arrest with possible re-entry into a new cell cycle ensuing. These findings depict blocking mitotic exit downstream of the SAC may serve as an improved therapeutic approach for killing apoptosis-resistant, slippage-prone, or SAC defective cancer cells in comparison to antimitotic drugs dependent on targeting the mitotic spindle or the SAC (Huang et al. 2009).

Due to the pivotal role of the APC/C in regulating mitotic progression and exit, renders the APC/C an attractive target for development of mitotic exit inhibitors. As the complex determines substrate specificity, potentially more targeted therapies can be designed and developed. Conversely, the multiplicity and diversity of APC/C substrates can prove challenging for targeted therapy development (Ramanujan and Tiwari 2016). An attractive therapeutic target attribute of APC/C is the modulation of its activity by Cdc20 and Cdh1. The development of inhibitors specifically targeting either of these two APC/C coactivators could reduce the off-target and side effects observed with anti-mitotic agents. Cdc20 is the preferred target as its function and substrates are mainly restricted to mitosis in contrast to Cdh1 which has a more diverse range of functions and substrates (Ramanujan and Tiwari 2016). However, due to Cdh1 implication in maintaining the G0/G1 state and in conjunction with pRb inhibiting transition of G1 to S phase, agonists could be potentially developed to promote these functions. Small compounds stabilising Cdh1 interaction with pRb or preventing its release from pRb could prove valuable in averting proliferation (Ramanujan and Tiwari 2016). The interdependence between Cdh1, SCF, Cdc20 and Emi1 could also possibly be exploited through a combination of antagonists and agonists, signifying future prospects in the design and development of chemotherapeutic agents targeting Cdh1 (Cardozo and Pagano 2007).

Blocking the APC/C through averting Cdc20-dependent mitotic progression with subsequent cell cycle arrest and probable eventual cell death has been the substantial focus in developing inhibitors targeting APC/C. Tosyl-L-arginine methyl ester (TAME) and Apcin are two small size APC/C inhibitors. TAME targets and binds to the APC/C causing a dramatic loss in APC/C activity by perturbing APC/C interaction with its coactivators Cdc20 and Cdh1 (Zeng et al. 2010). TAME imitates the IR tail of Cdc20 and Cdh1, thus specifically interferes with the IR tail-dependent interaction between Cdc20 or Cdh1 and APC/C. Functioning as a Cdc20 and Cdh1 inhibitor, TAME disrupts the coactivators recruitment to the complex impeding APC/C activation and subsequent APC/C substrates degradation. Although Cdh1 and Cdc20 bind to common APC/C sites, only APC/C-Cdc20 is activated by phosphorylation (Zeng et al. 2010). The phosphorylation-independent activity of Cdh1 is attributable to the increased affinity of Cdh1 for unphosphorylated apo APC/C. This increased affinity is caused by more extensive contacts formed between APC/C and Cdh1 in comparison to APC/C and Cdc20. This phenomenon also provides an explanation for TAME exemplifying a greater potency towards APC/C-Cdc20 in contrast to APC/C-Cdh1 (Alfieri et al. 2017).

The precise mechanism in which TAME impedes the activation of APC/C has been further exemplified (Zeng and King 2012). TAME reduces the binding of Cdc20 to the APC/C through inhibiting free Cdc20 binding the complex and actively promoting the dissociation of pre-bound Cdc20 from the APC through inducing auto-ubiquitination of Cdc20 in *Xenopus* extracts. Both aspects of this mechanism are antagonised by the binding of APC/C substrates including Cyclin B1. This phenomenon potentially verifies why in mitotically arrested *Xenopus* extracts in which APC/C substrates are absent, TAME promotes efficient dissociation of Cdc20 from the APC/C, in contrast to the little effect observed in somatic cells in which APC/C substrates seem to promote Cdc20 binding and suppress

Cdc20 auto-ubiquitination (Zeng and King 2012). Nevertheless, TAME inactivates the APC/C and stabilises Cyclin B1 in *Xenopus* extracts by inducing a catalytic defect in the APC/C-Cdc20-Cyclin B1 complex which consequentially slows the initial ubiquitination of unmodified Cyclin B1. As Cyclin B1 becomes ubiquitinated it loses its capability of promoting Cdc20 binding to APC/C in the presence of TAME, the APC/C-Cdc20-substrate-TAME complex remains susceptible to dissociation of Cdc20 as the substrate becomes ubiquitinated, averting the substrate from reaching a threshold of ubiquitination necessary for proteolysis, thus resulting in a persistent mitotic arrest (Zeng and King 2012).

TAME shows low permeability to the cellular membrane, a feature that prompted the search of derivatives of this molecule with better pharmacological properties. These efforts lead to the synthesis of the prodrug pro-TAME, which is processed in vitro by esterases to yield the active molecule TAME (Zeng et al. 2010). ProTAME exhibits the same inhibitory action as TAME in vitro, hindering tumour growth and inducing a mitotic arrest without perturbing the mitotic spindle morphology and function. Intriguingly, the proTAME-induced mitotic arrest is dependent on sustained SAC activity. This is unexpected as treatment with proTAME causes mitotic arrest in cells with normal kinetochore tension development, a condition that ordinarily causes SAC inactivation. The proTAME-induced mitotic arrest in a SAC-dependent manner could be attributable to proTAME treatment possibly instigating defective interactions between kinetochores and microtubules, producing an abnormally high degree of checkpoint signals in contrast to normal metaphase kinetochores. Alternatively, proTAME potentially could impede inactivation of the SAC, regardless of normal interactions between kinetochore and microtubules. The latter condition is more favoured due to the degree of checkpoint dependence exhibited, far exceeding the degree of kinetochore-microtubule perturbation (Zeng et al. 2010). APC/C-dependent ubiquitination or proteolysis is necessary for SAC inactivation. It is postulated and observed that such mutual antagonism between the APC/C and SAC could instigate a positive feedback loop that would augment proTAME inhibitory effects in a SAC dependent manner. This mutual antagonism possibly signifies a system-level behaviour that is controlled by minor abundance alterations in numerous core SAC proteins preceding anaphase (Zeng et al. 2010). Furthermore, it has been exemplified that induction of mitotic arrest could be achieved without complete pharmacological inhibition of the APC/C. This was unexpected as a previous investigation utilising RNAi showed that reduction of Cdc20 to very low levels was required for induction of mitotic arrest and in contrast to proTAME-induced mitotic arrest, was not SAC-dependent (Wolthuis et al. 2008; Huang et al. 2009). Variation in the protein synthesis rates among cells is one potential factor that accounts for the high variability in cellular response to microtubule inhibitors, which could subsequently limit the inhibitors therapeutic effectiveness. In comparison, proTAME-induced mitotic arrest occurs through residual APC/C activity inhibition instead of provoking SAC activation. This mechanism mediating mitotic arrest illustrates reduced dependence on protein synthesis due to the lowered residual APC/C activity rate and consequential lowered protein synthesis required for replenishing APC/C substrates, thus is less prone to mitotic slippage. ProTAME potentially therefore provides enhanced effectiveness in promoting and sustaining mitotic arrest and inducing a greater apoptotic death (Zeng et al. 2010).

More recently, the possibility that APC/C activity is necessary for silencing of the SAC has been further investigated (Lara-Gonzalez and Taylor 2012). It was demonstrated that proTAME does indeed induce a mitotic arrest in a SAC-dependent manner. However, a much simpler explanation has been proposed as to why proTAME-induced mitotic arrest is dependent on the SAC. It has been exemplified that the mitotic arrest induced by treatment with proTAME is attributable to cohesion fatigue induction, a phenotype produced by asynchronous chromatid separation ensuing a prolonged metaphase (Lara-Gonzalez and Taylor 2012). This is a result of the microtubule pulling forces during metaphase eventually overcoming cohesin-based forces holding sister chromatids together (Lara-Gonzalez and Taylor 2012). Subsequent generation of unpaired sisters that are unable to stably attach to microtubules causes reactivation of the SAC (Stevens et al. 2011). Thus, proTAME-induced mitotic arrest dependent on the SAC is an indirect effect of cohesion fatigue. Treatments that delay cells in metaphase including Cdc20 depletion have illustrated the induction of cohesion fatigue (Huang et al. 2009; Stevens et al. 2011; Lara-Gonzalez and Taylor 2012). Thus, it is not surprising that cohesion fatigue can be induced by proTAME. It has been depicted that proTAME inhibits APC/C activity, producing a metaphase delay in which if cohesion fatigue occurs, SAC reactivation arises instigating a prolonged mitotic arrest (Lara-Gonzalez and Taylor 2012). Further investigations identifying and characterising APC/C targets that promote SAC silencing are imperative to ascertain the balance of SAC activating and silencing signals during mitosis, which subsequently can have implications for chemotherapeutic drug development since mitotic exit potentially serves as a valuable therapeutic target for treating cancer.

Furthermore, low concentrations of an APC/C inhibitor may be valuable in combination with microtubule inhibitors to sustain mitotic arrest and enhance cell death. The addition of proTAME to HepG2 cells pre-treated with paclitaxel or an Aurora A inhibitor produced a potent antitumor efficacy by causing a further accumulation of cells in mitosis, inducing significant inhibition of mitotic exit and subsequently substantial apoptotic cancer cell death (Giovinazzi et al. 2013). In addition, treatment of multiple myeloma cell lines with proTAME resulted in growth arrest with subsequent apoptosis. When proTAME was then combined with the microtubule inhibitor vincristine, a significant increase in death of multiple myeloma cell lines and primary cells ensued demonstrating an enhanced anti-myeloma effect. Such findings indicate that APC/C could serve as a promising novel therapeutic target for the treatment of multiple myeloma (Crawford et al. 2016). Furthermore, in vitro studies revealed that proTAME can enhance the effect of commonly used chemotherapeutic drugs Adriamycin (ADM) and Cisplatin (CDDP) in killing osteosarcoma cells, suggesting that the combination of proTAME with chemotherapeutic drugs may constitute an attractive therapeutic approach to treat osteosarcoma (Hu et al. 2014).

While Cdh1 exhibits a potential tumour suppressor role, Cdc20 expression positively correlates with cancer and its depletion impedes tumours growth. Hence, the design and development of an inhibitor specifically targeting APC/C-Cdc20 instead of a pan-APC/C inhibitor like pro-TAME should be more selective to kill cancer cells in those overexpressing Cdc20 (Zhang et al. 2014). In mitosis, APC/C activation requires the binding of Cdc20 which results in the formation of a complex that recognises substrates containing a D-box. The other small compound inhibitor of the APC/C developed to date is Apcin, which directly binds to Cdc20 and prevents the recognition of D-box containing substrates, consequentially competitively impeding APC/C-dependent ubiquitination of APC/C-Cdc20 substrates (Sackton et al. 2014). Apcin binds the pocket defining D-box within Cdc20 WD40 domain, blocking substrate-induced Cdc20 loading onto the APC/C enabling the stabilisation of APC/C substrates containing D-box including Cyclin B1, Cyclin A2 and Securin. However, at low concentrations Apcin is insufficient to inhibit mitotic exit due to substrates outcompeting Apcin binding the pocket, or other mechanisms recruiting substrates to APC/C. Enhanced Apcin efficacy is achieved by combining the compound with TAME (Sackton et al. 2014). Furthermore, the combination of these two inhibitors illustrates greater effectiveness in disrupting APC/C activity than when either compound is used alone, demonstrating synergistic effects in inhibiting APC/C dependent proteolysis and mitotic exit when simultaneously perturbing two protein-protein interactions within the APC/C-Cdc20-substrate complex. This synergistic inhibition of mitotic exit does not depend on the SAC, but likely occurs from direct pharmacological inhibition of the APC/C. As dynamic protein complexes are crucial to ensure homeostasis, these findings depict the possibility of disrupting a protein machine function through simultaneously inhibiting multiple weak protein-protein interactions, and potentially provides a novel therapeutic approach for targeting protein complexes that may otherwise prove particularly challenging to target with a single inhibitor (Sackton et al. 2014).

More recently, it has been shown that Apcin inhibits cell growth and invasion as well as induces significant apoptosis in osteosarcoma cell lines, indicating Cdc20 possibly serves as a valuable therapeutic target for osteosarcoma treatment (Gao et al. 2018). Furthermore, it has been exemplified that myeloma cell lines treated with proTAME caused an accumulation of APC/C substrate Cyclin B1 and increased the number of cells in metaphase. When proTAME was combined with Apcin, an enhanced anti-multiple myeloma effect was observed with greater induction of cell cycle metaphase arrest and subsequent apoptosis, depicting APC/C and Cdc20 could serve as valuable novel therapeutic targets for high-risk multiple myeloma patients (Lub et al. 2016).

Investigations into identifying and developing APC/C inhibitors as promising therapeutic agents which target cells that undergo mitotic slippage following treatment with microtubule inhibitors remains ongoing. More recently, it has been illustrated that when in excess, Mad2 binding motif peptides derived from Cdc20; the endogenous 39 amino acid Tiny yeast comet 1 (Tyc1) protein which contains homology to human p31^{comet}; and a homologous peptide derived from human p31^{comet} can all function as inhibitors of the APC/C (Schuyler et al. 2018). Furthermore, overexpression of these peptides in vivo instigates sensitivity to microtubule inhibitors. These peptides bound and inhibited the APC/C perturbing the ability of the coactivators Cdc20 and Cdh1 to bind to the APC/C. When Cdc20 Mad2-binding motif peptides were co-expressed with Tyc1 an increased sensitivity to microtubule inhibitors was observed in vivo. Although these APC/C inhibitor peptides are too large and functionally too weak to serve as prospective therapeutic agents, these findings indicate that these peptides can all serve as novel valuable molecular tools to investigate potential mechanisms inhibiting APC/C that induce sensitivity to microtubule inhibitors in vivo as a therapeutic strategy towards the unending aim of targeting cancers treated with microtubule inhibitors that undergo mitotic slippage (Schuyler et al. 2018).

Overall, the APC/C serves as a valuable therapeutic target for designing and developing drugs for the treatment of diverse cancers. Through targeting mitotic exit provides an attractive alternative therapeutic strategy in inducing permanent mitotic arrest and subsequent cell death in comparison to other anti-mitotic agents. However, it must be stressed that proTAME is a prodrug that requires activation by esterases to the active impermeable compound TAME, thus upon administration in vivo, active esterases in the blood stream will process majority of the proTAME before it reaches target cells and subsequently no uptake of this drug shall occur (Zeng et al. 2010). As in vivo validation of proTAME is yet to be optimised, the need to design and develop novel more potent APC/C inhibitors that can be validated in vivo is imperative for the treatment of diverse cancers that constitute important yet unmet medical needs.

Concluding Remarks and Perspective Analysis

Renowned for targeting a diverse array of substrates for ubiquitination and subsequent degradation, the APC/C is an indispensable multifunctional E3 ubiquitin-protein ligase accountable for the regulation of numerous crucial and diverse cellular processes. Attributable to its vital functions, the APC/C is therefore tightly regulated by a network of multiple intricate signals involving extensive cooperative interactions that act in a concerted manner to ensure appropriate APC/C activity. The regulation depends on the required APC/C activity at a given point in time, in which interactions can inhibit or activate the complex.

Although since the discovery of the APC/C its crucial implications in cell cycle regulation has been elucidated and its numerous substrates identified, the remarkable versatility and complexity of functions of this E3 ubiquitin-protein ligase elicits the prospect that additional roles and substrates of this complex are yet to be discovered. The emerging utilisation of technologies including single molecule analysis and super-resolution microscopy combined with traditional biochemical, cellular and structural biology approaches will enable to uncover novel aspects of this indispensable complex with an unprecedented level of detail. Further investigations as to whether the subunits exhibit different roles independent of the APC/C holoenzyme are imperative. Future work should also aim to further our understanding of the structural architecture of the APC/C. In particular, the precise location of the lack in well-defined electron density around the anticipated position of Bub3 in the APC/C-MCC assembly and how its mode of action contributes to fine tune APC/C functions require further investigations. Additionally, recent studies by independent groups have elucidated APC/C functions in the cell that are cell cycle-independent including roles in cell differentiation of the nervous system and DNA damage repair. Diverse protein regulators important for the maintenance of a stable genome and implicated in DNA damage repair are Cdh1 substrates. Future work should aim to deepen our understanding of APC/C-Cdh1 physiological role in these cellular processes and if these interactions are confirmed, they will expand further the APC/C repertoire of functions to incorporate crucial roles in apoptosis and senescence.

The discovery of novel APC/C substrates remains an active and substantial focus for research and is fundamental for advancing our understanding and knowledge of its functions, mode of regulation and their implications in health and disease conditions. Within the last two decades, significant advancements in knowledge have provided novel mechanistic insights into the multiple roles of APC/C assembly in chromosome segregation including functions beyond cell cycle regulation. For example, APC/C-Cdh1 is instrumental in the differentiation and function of the nervous system where it plays vital roles in axon growth regulation, synaptic transmission and plasticity, while APC/C-Cdc20 is implicated in neuronal development, regulating dendrite morphogenesis and presynaptic differentiation.

Perhaps, not surprisingly, dysregulated activity of the APC/C appears to be associated with AD pathophysiology. As the APC/C is an indispensable E3 ligase renowned for its capability of stipulating ubiquitination and subsequent degradation of numerous substrates, in conjunction with accumulating evidence indicating a role in vital CNS functions, alterations of the APC/C could affect numerous diverse proteins and their functions and ultimately contribute to the development of neurodegenerative diseases. Accumulating evidence depicts an association between downregulation/inactivation of APC/C-Cdh1 and AD pathological hallmarks, with substantial indication that deregulation of APC/C substrates has significant implications in AD with regards to ectopic cell cycle re-entry, oxidative stress, ecotoxicity and impaired LTP. Aβ oligomers and glutamate excitotoxicity both reduce Cdh1 levels, inactivating APC/C-Cdh1 with subsequent APC/C substrate accumulation. Furthermore, it has been exemplified in AD that Cyclin B1, PFKFB3, and glutaminase are elevated due to APC/C-Cdh1 downregulation. It is probable other APC/C-Cdh1 substrates that are yet to be identified can potentially accumulate in AD due to dysregulation of APC/C activity. Future investigations ascertaining entirely the roles of APC/C-Cdh1 and APC/C-Cdc20 in health and disease are imperative to ascertain whether these complexes can be valuable therapeutic targets for treating AD.

Numerous diverse viruses can affect APC/C functions, suggesting the APC/C is crucially implicated in viral replication and pathogenesis. A common approach amongst some viruses is the disruption of the integrity of the APC/C as exemplified for HCMV, CAV and HPV. Alternatively, the APC/C may be targeted through phosphorylation by HTLV-1 and Ad. ORFV PACR utilises a unique approach incorporating into an otherwise intact APC/C complex through mimicking and competing with APC/C subunit APC11. Although this approach may seem less efficient than targeting substrate recruitment, it elicits the prospect that PACR may redirect APC/C function to fulfil other roles. In the case of oncoviruses, manipulation of the APC/C can lead to premature APC/C-Cdc20 activation which subsequently promotes chromosome instability. For other viruses, APC/C-Cdh1 inhibition stimulates an S-phase like environment while APC/C-Cdc20 inhibition averts the nuclear envelope from reforming, thus enabling nuclear factors to become enriched or accessible for replication of the virus. Intriguingly, viral proteins may be targeted for ubiquitination and degradation by the APC/C, consequently restricting replication of the virus. Numerous HCMV viral proteins contain the consensus D-box, an APC/C recognition signal regularly observed in APC/C substrates. The precise mechanisms in manipulating the APC/C utilised by diverse viruses illustrates variability, as are the effects exhibited by the numerous viral factors. However, as the functional implications of APC/C manipulation by viruses are gradually unveiled, the emerging picture is of a mutual trait between the expression of viral regulators that are crucial for the virus invasive strategy and the control of APC/C functions, which has implications on viral infection.

The identification of diverse viruses and numerous viral protein regulators in manipulating the APC/C, signifies the potential importance of this cell cycle master regulator in virus infection, with more viral factors likely to be discovered. Currently, mechanistic details of how exactly the reported viruses manipulate the APC/C are yet to be defined. The potential implication of virus-mediated APC/C regulation during virus infection remains largely elusive, complicated further by the fact the majority of the viral APC/C regulators identified to date are known to affect other proteins and consequently, other functions. Undoubtedly, further understanding of the role of APC/C manipulation by viruses is imperative. Future investigations should aim to unveil the precise molecular mechanisms these viruses utilise to modulate the APC/C; how exactly the APC/C is implicated in viral replication and pathogenesis; and what are the specific APC/C subunits and APC/C regulators that are targeted by the viruses.

In instances where APC/C potentially restricts replication of the virus, it will be essential to elucidate if the APC/C targets viral or cellular substrates. It is equally important to establish whether the viruses encode adaptor proteins to drive APC/C ubiquitination towards unknown cellular targets in a manner that specific substrates are targeted for destruction by the proteasome, thus selectively redirecting APC/C activity rather than activating or inhibiting the complex. Furthermore, it would be particularly valuable to ascertain whether viral APC/C modulation significantly effects host cell metabolism, favouring an environment that provides not only nucleotides for viral replication, but an extensive repertoire of resources and energy required for viral particle synthesis and assembly.

Equally important is the verification of whether a pseudo-S phase state promoted in otherwise quiescent cells is a mutual trait amongst viral APC/C modulation. It would also be of great significance to determine whether other viruses than those already identified, encode factors modulating APC/C ubiquitin ligase activity. The APC/C has been shown to participate in signalling pathways that implicate pRb and p53. However, further studies should aim to unveil whether some of the viruses known to target pRb and p53 pathways do so through APC/C manipulation as well as elucidate if any viral regulators modulate APC/C activity through targeting APC/C inhibitors including Emi1.

In a broader perspective, future investigations into these viral regulators shall be valuable in further comprehending the biology of the APC/C and ascertaining how exactly these viral regulators interact with the APC/C, which will undeniably further our understanding of the structural architecture, assembly and regulation of the complex. It is likely these studies shall have crucial implications for cancer research, with drugs currently being designed and developed to target the APC/C as a potential therapeutic strategy for treating cancer. These viral APC/C regulators may serve as indispensable tools for identifying APC/C attributes that can be utilised as additional therapeutic targets for innovative development of novel drugs. Furthermore, future studies should analyse whether viral APC/C manipulation can sensitise cancer cells to drugs and/or gamma irradiation and proton beam therapy. Apoptin, a small protein from CAV, can selectively kill tumour cells and for this reason it is currently being investigated as a prospective cancer therapeutic strategy. In addition to this viral protein, it would be of great significance to evaluate the potential utilisation and efficacy of these viral factors as novel drug-delivery systems targeting the SAC-kinetochore-microtubule axis. Overall future investigations should collectively aim to advance our understanding of the intricate interplays amongst the APC/C, virus replication, cell cycle and virus-mediated cancer.

The APC/C is the most complex member of the RING finger ubiquitin E3 ligase family. As the APC/C is instrumental in the regulation of multiple vital cellular processes particularly in cell cycle progression control, timely activation and inactivation of the complex is therefore tightly regulated by a network of sophisticated signals. It is not surprising that when the regulation of the APC/C is dysregulated this instigates genomic instability with the APC/C both directly and indirectly associated with cancer.

Accumulating evidence exemplifies dysregulation of the complex either through mutations in core APC/C subunits, or the deregulation of both activating and inhibiting APC/C regulators are associated with oncogenic transformation. These deregulations have been extensively investigated particularly for deregulations in both APC/C coactivators which show contrasting effects, with Cdc20 and Cdh1 functioning as an oncoprotein and a tumour suppressor, respectively. In addition, APC/C-Cdc20 and APC/C-Cdh1 are active at different phases of the cell cycle which in turn potentially differentially affects carcinogenesis. Moreover, it is very probable the different behaviour exemplified by Cdc20 and Cdh1 potentially originates from their tissue-specific expression, with different levels of expression in different tissues possibly affecting the cellular function of these two APC/C coactivators. Furthermore, Cdh1 can induce Cdc20 degradation, thus the contrasting carcinogenic effects potentially are a result of excess Cdc20 degradation by Cdh1. While the precise molecular details of how tumorigenesis evolves from APC/C deregulation remain to be fully elucidated, ultimately the direct or indirect deregulation of APC/C activity instigates the formation of highly proliferative, genetically unstable and poorly differentiated tumours, in which APC/C dysregulation potentially can serve as a key prognostic marker and therapeutic target for the treatment of cancer. Accumulating evidence illustrates that mutations in several core APC/C subunits occur in numerous cancer tissues. Future studies should aim to elucidate the causal roles of these subunits in cancer both individually and collectively. Future investigations should also aim to further ascertain the precise molecular mechanisms in which Cdc20-mediated and Cdh1-mediated tumorigenesis evolves.

Currently, the development of new anti-mitotic therapies is one of the most active areas of cancer research. Attributable to its instrumental role in regulating cell cycle progression and the mounting evidence implicating the APC/C in tumorigenesis, renders the complex an attractive therapeutic target for treating cancer. In contrast to other anti-mitotic agents that target microtubules and the SAC, targeting mitotic exit provides an attractive innovative novel therapeutic approach in inducing a permanent mitotic arrest and subsequent cell apoptotic death. With other anti-mitotic agents, the clinical efficacy is limited with undesirable effects observed as well as premature mitotic exit and mitotic slippage in cancer cells.

An attractive APC/C therapeutic attribute is its regulation by Cdc20 and Cdh1. Through specifically targeting either of these two coactivators elicits the potential of reducing the off-target and undesirable side effects illustrated with anti-mitotic agents. The inhibitors proTAME and Apcin that target different aspects of APC/C activity, alone illustrate efficient anti-mitotic effects. When these two inhibitors are combined a synergistic inhibition of mitotic exit is observed, depicting the possibility of perturbing a protein machine function through simultaneously inhibiting multiple protein-protein interactions. These two APC/C inhibitors potentially provide an innovative novel therapeutic strategy for targeting protein complexes that may otherwise prove particularly challenging to target with a single inhibitor. However, proTAME requires activation by esterases to render the active impermeable compound TAME. Thus, on administration of proTAME in vivo the majority of the drug will be processed before reaching its target cells and consequentially drug uptake will not occur. Future studies should aim to optimise in vivo validation of proTAME and regardless as to whether this is achievable, the requirement of novel more potent APC/C inhibitors that can be validated in vivo is imperative for the innovate design and development of prospective therapeutic strategies targeting diverse types of cancer.

References

Abrieu A, Magnaghi-Jaulin L, Kahana JA, Peter M, Castro A, Vigneron S, Lorca T, Cleveland DW, Labbe JC (2001) Mps1 is a kinetochore-associated kinase essential for the vertebrate mitotic checkpoint. Cell 106:83–93

Acquaviva C, Herzog F, Kraft C, Pines J (2004) The anaphase promoting complex/cyclosome is recruited to centromeres by the spindle assembly checkpoint. Nat Cell Biol 6:892–898

Ahn J, Kim H, Yoon J, Jang H, Han S, Eun S, Shim H, Kim H, Kim D, Lee J, Lee C, Bae M, Chung K, Jung J, Kim E, Kim S, Chang J, Kim H, Kim J, Lee M, Cho B, Lee J, Bang D (2014) Identification of somatic mutations in EGFR/KRAS/ALK-negative lung adenocarcinoma in never-smokers. Genome Med 6:18

Alfieri C, Chang L, Zhang Z, Yang J, Maslen S, Skehel M, Barford D (2016) Molecular basis of APC/C regulation by the spindle assembly checkpoint. Nature 536:431–436

Alfieri C, Zhang S, Barford D (2017) Visualizing the complex functions and mechanisms of the anaphase promoting complex/cyclosome (APC/C). Open Biol 7:170204

Alizadeh AA, Eisen MB, Davis RE, Ma C, Lossos IS, Rosenwald A, Boldrick JC, Sabet H, Tran T, Yu X, Powell JI, Yang L, Marti GE, Moore T, Hudson J Jr, Lu L, Lewis DB, Tibshirani R, Sherlock G, Chan WC, Greiner TC, Weisenburger DD, Armitage JO, Warnke R, Levy R, Wilson W, Grever MR, Byrd JC, Botstein D, Brown PO, Staudt LM (2000) Distinct types of diffuse large B-cell lymphoma identified by gene expression profiling. Nature 403:503–511

Almeida A (2012) Regulation of APC/C-Cdh1 and its function in neuronal survival. Mol Neurobiol 46:547–554

Almeida A, Bolaños J, Moreno S (2005) Cdh1/Hct1-APC Is Essential for the survival of postmitotic neurons. J Neurosci 25:8115–8121

Almeida A, Moncada S, Bolaños JP (2004) Nitric oxide switches on glycolysis through the AMP protein kinase and 6-phosphofructo-2-kinase pathway. Nat Cell Biol 6:45–51

Amador V, Ge S, Santamaría PG, Guardavaccaro D, Pagano M (2007) APC/C(Cdc20) controls the ubiquitin-mediated degradation of p21 in prometaphase. Mol Cell 27:462

Ang XL, Wade Harper J (2005) SCF-mediated protein degradation and cell cycle control. Oncogene 24:2860–2870

Aristarkhov A, Eytan E, Moghe A, Admon A, Hershko A, Ruderman JV (1996) E2-C, a cyclin-selective ubiquitin carrier protein required for the destruction of mitotic cyclins. Proc Natl Acad Sci USA 93:4294–4299

Arundine M, Tymianski M (2003) Molecular mechanisms of calcium-dependent neurodegeneration in excitotoxicity. Cell Calcium 34:325–337

Aulia S, Tang B (2006) Cdh1-APC/C, cyclin B-Cdc2, and Alzheimer's disease pathology. Biochem and Biophys Res Comm 339:1–6

Baek K, Park H, Kang C, Kim S, Jeong S, Hong E, Park J, Sung Y, Suzuki T, Kim C, Lee C (2006) Overexpression of Hepatitis C Virus NS5A protein induces chromosome instability via mitotic cell cycle dysregulation. J Mol Biol 359:22–34

Bain M, Sinclair J (2007) The S phase of the cell cycle and its perturbation by human cytomegalovirus. Rev Med Virol 17:423–434

Barford D (2011) Structural insights into anaphase-promoting complex function and mechanism. Philos Trans R Soc London B Biol Sci 366:3605–3624

Barford D (2015) Understanding the structural basis for controlling chromosome division. Philos Trans A Math Phys Eng Sci 373:pii:20130392

Baron AP, von Schubert C, Cubizolles F, Siemeister G, Hitchcock M, Mengel A, Schröder J, Fernández-Montalván A, von Nussbaum F, Mumberg D, Nigg EA (2016) Probing the catalytic functions of Bub1 kinase using the small molecule inhibitors BAY-320 and BAY-524. eLife 5. pii:e12187

Bashir T, Dorrello NV, Amador V, Guardavaccaro D, Pagano M (2004) Control of the SCF (Skp2-Cks1) ubiquitin ligase by the APC/C(Cdh1) ubiquitin ligase. Nature 428:190–193
Bassermann F, Frescas D, Guardavaccaro D, Busino L, Peschiaroli A, Pagano M (2008) The Cdc14B-Cdh1-Plk1 axis controls the G2 DNA-damage-response checkpoint. Cell 134:256–267
Bedford L, Lowe J, Dick LR, Mayer RJ, Brownell JE (2011) Ubiquitin-like protein conjugation and the ubiquitin-proteasome system as drug targets. Nat Rev Drug Discov 10:29–46
Bellanger S, Blachon S, Mechali F, Bonne-Andrea C, Thierry F (2005) High-risk but not low-risk HPV E2 proteins bind to the APC activators Cdh1 and Cdc20 and cause genomic instability. Cell Cycle 4:1608–1615
Benanti J, Matyskiela ME, Morgan DO, Toczyski DP (2009) Functionally distinct isoforms of Cik1 are differentially regulated by APC/C-mediated proteolysis. Mol Cell 33:581–590
Bharadwaj R, Yu H (2004) The spindle checkpoint, aneuploidy, and cancer. Oncogene 23:2016–2027
Bhattacharjee B, Renzette N, Kowalik T (2012) Genetic Analysis of cytomegalovirus in malignant gliomas. J Virol 86:6815–6824
Bliss T, Cooke S (2011) Long-term potentiation and long-term depression: a clinical perspective. Clinics 66:3–17
Bobo-Jiménez V, Delgado-Esteban M, Angibaud J, Sánchez-Morán I, de la Fuente A, Yajeya J, Nägerl U, Castillo J, Bolaños J, Almeida A (2017) APC/CCdh1-Rock2 pathway controls dendritic integrity and memory. Proc Natl Acad Sci USA 114:4513–4518
Bolanos-Garcia VM, Blundell TL (2011) BUB1 and BUBR1: multifaceted kinases of the cell cycle. Trends Biochem Sci 36:141–150
Boxus M, Willems L (2009) Mechanisms of HTLV-1 persistence and transformation. Br J Cancer 101:1497–1501
Braunstein I, Miniowitz S, Moshe Y, Hershko A (2007) Inhibitory factors associated with anaphase-promoting complex/cylosome in mitotic checkpoint. Proc Natl Acad Sci USA 104:4870–4875
Brito D, Rieder C (2006) Mitotic checkpoint slippage in humans occurs via cyclin B destruction in the presence of an active checkpoint. Curr Biol 16:1194–1200
Brito D, Rieder C (2009) The ability to survive mitosis in the presence of microtubule poisons differs significantly between human nontransformed (RPE-1) and cancer (U2OS, HeLa) cells. Cell Motil Cytoskeleton 66:437–447
Britt W (2008) Manifestations of human cytomegalovirus infection: proposed mechanisms of acute and chronic disease. Curr Top Microbiol Immunol 325:417–470
Brune W (2011) Inhibition of programmed cell death by cytomegaloviruses. Virus Res 157:144–150
Budhavarapu VN, White ED, Mahanic CS, Chen L, Lin FT, Lin WC (2012) Regulation of E2F1 by APC/C Cdh1 via K11 linkage-specific ubiquitin chain formation. Cell Cycle 11:2030–2038
Burton JL, Solomon MJ (2007) Mad3p, a pseudosubstrate inhibitor of APCCdc20 in the spindle assembly checkpoint. Genes Dev 21:655–667
Buschhorn BA, Petzold G, Galova M, Dube P, Kraft C, Herzog F, Stark H, Peters JM (2011) Substrate binding on the APC/C occurs between the coactivator Cdh1 and the processivity factor Doc1. Nat Struct Mol Biol 18:6–13
Caffarelli N, Fehr A, Yu D (2013) Cyclin A degradation by primate cytomegalovirus protein pUL21a counters its innate restriction of virus replication. PLoS Pathog 9:e1003825
Calabrese V, Scapagnini G, Ravagna A, Colombrita C, Spadaro F, Butterfield DA, Giuffrida Stella AM (2004) Increased expression of heat shock proteins in rat brain during aging: relationship with mitochondrial function and glutathione redox state. Mech Ageing Dev 125:325–335
Caposio P, Orloff SL, Streblow DN (2011) The role of cytomegalovirus in angiogenesis. Virus Res 157:204–211
Cardozo T, Pagano M (2004) The SCF ubiquitin ligase: insights into a molecular machine. Nat Rev Mol Cell Biol 5:739–751

Cardozo T, Pagano M (2007) Wrenches in the works: drug discovery targeting the SCF ubiquitin ligase and APC/C complexes. BMC Biochem 8:S9

Carroll CW, Morgan DO (2005) Enzymology of the anaphase-promoting complex. Methods Enzymol 398:219–230

Carter S, Eklund A, Kohane I, Harris L, Szallasi Z (2006) A signature of chromosomal instability inferred from gene expression profiles predicts clinical outcome in multiple human cancers. Nat Genet 38:1043–1048

Chaabane W, Ghavami S, Małecki A, Łos M (2017) Human Gyrovirus-Apoptin interferes with the cell cycle and induces G2/M arrest prior to apoptosis. Arch Immunol Ther Exp 65:545–552

Chabes AL, Pfleger CM, Kirschner MW, Thelander L (2003) Mouse ribonucleotide reductase R2 protein: A new target for anaphase-promoting complex-Cdh1-mediated proteolysis. Proc Natl Acad Sci USA 100:3925–3929

Chang DZ, Ma Y, Ji B, Liu Y, Hwu P, Abbruzzese JL, Logsdon C, Wang H (2012) Increased CDC20 expression is associated with pancreatic ductal adenocarcinoma differentiation and progression. J Hematol Oncol 5:15

Chang L, Zhang Z, Yang J, McLaughlin S, Barford D (2015) Atomic structure of the APC/C and its mechanism of protein ubiquitination. Nature 522:450–454

Chang LF, Zhang Z, Yang J, McLaughlin SH, Barford D (2014) Molecular architecture and mechanism of the anaphase-promoting complex. Nature 513:388–393

Chao W, Kulkarni K, Zhang Z, Kong E, Barford D (2012) Structure of the mitotic checkpoint complex. Nature 484:208–213

Chen X, Cheung ST, So S, Fan ST, Barry C, Higgins J, Lai KM, Ji J, Dudoit S, Ng IO, Van De Rijn M, Botstein D, Brown PO (2002) Gene expression patterns in human liver cancers. Mol Biol Cell 13:1929–1939

Cho HJ, Lee EH, Han SH, Chung HJ, Jeong JH, Kwon J, Kim H (2012) Degradation of human RAP80 is cell cycle regulated by Cdc20 and Cdh1 ubiquitin ligases. Mol Cancer Res 10:615–625

Choi JW, Kim Y, Lee JH, Kim YS (2013) High expression of spindle assembly checkpoint proteins CDC20 and MAD2 is associated with poor prognosis in urothelial bladder cancer. Virchows Arch 463:681–687

Chow C, Wong N, Pagano M, Lun SW, Nakayama KI, Nakayama K, Lo KW (2012) Regulation of APC/CCdc20 activity by RASSF1A-APC/CCdc20 circuitry. Oncogene 31:1975–1987

Christensen KL, Brennan JD, Aldridge CS, Ford HL (2007) Cell cycle regulation of the human Six1 homeoprotein is mediated by APC(Cdh1). Oncogene 26:3406–3414

Chun AC, Kok KH, Jin DY (2013) REV7 is required for anaphase-promoting complex-dependent ubiquitination and degradation of translation DNA polymerase REV1. Cell Cycle 12:365–378

Chung E, Chen RH (2003) Phosphorylation of Cdc20 is required for its inhibition by the spindle checkpoint. Nat Cell Biol 5:748–753

Chung E, Chen RH (2002) Spindle checkpoint requires Mad1-bound and Mad1-free Mad2. Mol Biol Cell 13:1501–1511

Ciosk R, Zachariae W, Michaelis C, Shevchenko A, Mann M, Nasmyth K (1998) An ESP1/PDS1 complex regulates loss of sister chromatid cohesion at the metaphase to anaphase transition in yeast. Cell 93:1067–1076

Ciossani G, Overlack K, Petrovic A, Huis In't Veld PJ, Koerner C, Wohlgemuth S, Maffini S, Musacchio A (2018) The kinetochore proteins CENP-E and CENP-F directly and specifically interact with distinct BUB mitotic checkpoint Ser/Thr kinases. J Biol Chem 293:10084–10101

Clark E, Spector D (2015) Studies on the contribution of human cytomegalovirus UL21a and UL97 to viral growth and inactivation of the anaphase-promoting complex/cyclosome (APC/C) E3 ubiquitin ligase reveal a unique cellular mechanism for downmodulation of the APC/C subunits APC1, APC4, and APC5. J Virol 89:6928–6939

Clute P, Pines J (1999) Temporal and spatial control of cyclin B1 destruction in metaphase. Nat Cell Biol 1:82–87

Cohen-Fix O, Peters JM, Kirschner MW, Koshland D (1996) Anaphase initiation in *Saccharomyces cerevisiae* is controlled by the APC-dependent degradation of the anaphase inhibitor Pds1p. Genes Dev 10:3081–3093

Colombo SL, Palacios-Callender M, Frakich N, Carcamo S, Kovacs I, Tudzarova S, Moncada S (2011) Molecular basis for the differential use of glucose and glutamine in cell proliferation as revealed by synchronized HeLa cells. Proc Natl Acad Sci USA 108:21069–21074

Coster G, Hayouka Z, Argaman L, Strauss C, Friedler A, Brandeis M, Goldberg M (2007) The DNA damage response mediator MDC1 directly interacts with the anaphase-promoting complex/cyclosome. J Biological Chem 282:32053–32064

Cotsiki M, Lock R, Cheng Y, Williams G, Zhao J, Perera D, Freire R, Entwistle A, Golemis E, Roberts T, Jat P, Gjoerup O (2004) Simian virus 40 large T antigen targets the spindle assembly checkpoint protein Bub1. Proc Nat Acad Sci USA 101:947–952

Cotto-Rios XM, Jones MJ, Busino L, Pagano M, Huang TT (2011) APC/CCdh1-dependent proteolysis of USP1 regulates the response to UV-mediated DNA damage. J Cell Biol 194:177–186

Craney A, Kelly A, Jia L, Fedrigo I, Yu H, Rape M (2016) Control of APC/C-dependent ubiquitin chain elongation by reversible phosphorylation. Proc Nat Acad Sci USA 113:1540–1545

Crawford L, Anderson G, Johnston C, Irvine A (2016) Identification of the APC/C co-factor FZR1 as a novel therapeutic target for multiple myeloma. Oncotarget 7:70481–70493

Cui Y, Cheng X, Zhang C, Zhang Y, Li S, Wang C, Guadagno TM (2010) Degradation of the human mitotic checkpoint kinase Mps1 is cell cycle-regulated by APC-cCdc20 and APC-cCdh1 ubiquitin ligases. J Biol Chem 285:32988–32998

da Fonseca PC, Kong EH, Zhang Z, Schreiber A, Williams MA, Morris EP, Barford D (2011) Structures of APC/C(Cdh1) with substrates identify Cdh1 and Apc10 as the D-box co-receptor. Nature 470:274–278

Danen-van Oorschot A, Zhang Y, Leliveld S, Rohn J, Seelen M, Bolk M, van Zon A, Erkeland S, Abrahams J, Mumberg D, Noteborn M (2003) Importance of nuclear localization of apoptin for tumor-specific induction of apoptosis. J Biol Chem 278:27729–27736

De Antoni A, Pearson CG, Cimini D, Canman JC, Sala V, Nezi L, Mapelli M, Sironi L, Faretta M, Salmon ED, Musacchio A (2005) The Mad1/Mad2 complex as a template for Mad2 activation in the spindle assembly checkpoint. Curr Biol 15:214–225

de Boer H, Guerrero Llobet S, van Vugt M (2015) Controlling the response to DNA damage by the APC/C-Cdh1. Cell Mol Life Sci 73:949–960

DeCaprio J (2009) How the Rb tumor suppressor structure and function was revealed by the study of Adenovirus and SV40. Virol 384:274–284

De Felice F, Velasco P, Lambert M, Viola K, Fernandez S, Ferreira S, Klein W (2007) Aβ oligomers induce neuronal oxidative stress through an N-Methyl-D-aspartate receptor-dependent mechanism that is blocked by the alzheimer drug memantine. J Biol Chem 282:11590–11601

Delgado-Esteban M, García-Higuera I, Maestre C, Moreno S, Almeida A (2013) APC/C-Cdh1 coordinates neurogenesis and cortical size during development. Nat Commun 4:2879

den Elzen N, Pines J (2001) Cyclin A is destroyed in prometaphase and can delay chromosome alignment and anaphase. J Cell Biol 153:121–136

Derive N, Landmann C, Montembault E, Claverie M, Pierre-Elies P, Goutte-Gattat D, Founounou N, McCusker D, Royou A (2015) Bub3–BubR1-dependent sequestration of Cdc20 Fizzy at DNA breaks facilitates the correct segregation of broken chromosomes. J Cell Biol 211:517–532

Deshaies RJ, Joazeiro CA (2009) RING domain E3 ubiquitin ligases. Annu Rev Biochem 78:399–434

Diaz-Martinez LA, Tian W, Li B, Warrington R, Jia L, Brautigam CA, Luo X, Yu H (2015) The Cdc20-binding Phe box of the spindle checkpoint protein BubR1 maintains the mitotic checkpoint complex during mitosis. J Biol Chem 290:2431–2443

Di Fiore B, Davey NE, Hagting A, Izawa D, Mansfeld J, Gibson TJ, Pines J (2015) The ABBA motif binds APC/C activators and is shared by APC/C substrates and regulators. Dev Cell 32:358–372

Di Fiore B, Wurzenberger C, Davey NE, Pines J (2016) The mitotic checkpoint complex requires an evolutionary conserved cassette to bind and inhibit active APC/C. Mol Cell 64:1144–1153

Ding ZY, Wu HR, Zhang JM, Huang GR, Ji DD (2014) Expression characteristics of CDC20 in gastric cancer and its correlation with poor prognosis. Int J of Cli Exp Pathol 7:722–727

Dong S, Huang F, Zhang H, Chen Q (2019) Overexpression of BUB1B, CCNA2, CDC20, and CDK1 in tumor tissues predicts poor survival in pancreatic ductal adenocarcinoma. Biosci Rep 39:BSR20182306

Dong X, Wang Y, Qin Z (2009) Molecular mechanisms of excitotoxicity and their relevance to pathogenesis of neurodegenerative diseases. Acta Pharmacol Sin 30:379–387

Donzelli M, Squatrito M, Ganoth D, Hershko A, Pagano M, Draetta GF (2002) Dual mode of degradation of Cdc25 A phosphatase. EMBO J 21:4875–4884

Drouet Y, Treilleux I, Viari A, Léon S, Devouassoux-Shisheboran M, Voirin N, de la Fouchardière C, Manship B, Puisieux A, Lasset C, Moyret-Lalle C (2018) Integrated analysis highlights APC11 protein expression as a likely new independent predictive marker for colorectal cancer. Sci Rep 8:7386

Engelbert D, Schnerch D, Baumgarten A, Wäsch R (2007) The ubiquitin ligase APCCdh1 is required to maintain genome integrity in primary human cells. Oncogene 27:907–917

Enquist-Newman M, Sullivan M, Morgan DO (2008) Modulation of the mitotic regulatory network by APC-dependent destruction of the Cdh1 inhibitor Acm1. Mol Cell 30:437–446

Fehr A, Gualberto N, Savaryn J, Terhune S, Yu D (2012) Proteasome-dependent disruption of the E3 ubiquitin ligase anaphase-promoting complex by HCMV protein pUL21a. PLoS Pathog 8: e1002789

Fehr A, Yu D (2013) Control the host cell cycle: viral regulation of the anaphase-promoting complex. J Virol 87:8818–8825

Fehr A, Yu D (2010) Human cytomegalovirus early protein pUL21a promotes efficient viral DNA synthesis and the late accumulation of immediate-early transcripts. J Virol 85:663–674

Fehr A, Yu D (2009) Human cytomegalovirus gene UL21a Encodes a short-lived cytoplasmic protein and facilitates virus replication in fibroblasts. J Virol 84:291–302

Feine O, Zur A, Mahbubani H, Brandeis M (2007) Human Kid is degraded by the APC/C(Cdh1) but not by the APC/C(Cdc20). Cell Cycle 6:2516–2523

Fleming S, Wise L, Mercer A (2015) Molecular genetic analysis of orf virus: a poxvirus that has adapted to skin. Viruses 7:1505–1539

Foe I, Toczyski D (2011) Structural biology: a new look for the APC. Nature 470:182–183

Fonseca R, Vabulas RM, Hartl FU, Bonhoeffer T, Nagerl UV (2006) A balance of protein synthesis and proteasome-dependent degradation determines the maintenance of LTP. Neuron 52:239–245

Fortunato EA, Dell'Aquila ML, Spector DH (2000) Specific chromosome 1 breaks induced by human cytomegalovirus. Proc Natl Acad Sci USA 97:853–858

Frye JJ, Brown NG, Petzold G, Watson ER, Grace CR, Nourse A, Jarvis MA, Kriwacki RW, Peters JM, Stark H, Schulman BA (2013) Electron microscopy structure of human APC/C (CDH1)-EMI1 reveals multimodal mechanism of E3 ligase shutdown. Nat Struct Mol Biol 20:827–835

Fu AK, Hung KW, Fu WY, Shen C, Chen Y, Xia J, Lai KO, Ip NY (2011) APC(Cdh1) mediates EphA4-dependent downregulation of AMPA receptors in homeostatic plasticity. Nat Neurosci 14:181–189

Fuchsberger T, Lloret A, Viña J (2017) New functions of APC/C ubiquitin ligase in the nervous system and its role in alzheimer's disease. Int J Mol Sci 18:pii:E1057

Fuchsberger T, Martínez-Bellver S, Giraldo E, Teruel-Martí V, Lloret A, Viña J (2016) Aβ induces excitotoxicity mediated by APC/C-Cdh1 depletion that can be prevented by glutaminase inhibition promoting neuronal survival. Sci Rep 6:31158

Fujita T, Liu W, Doihara H, Date H, Wan Y (2008a) Dissection of the APCCdh1-Skp2 cascade in breast cancer. Clin Cancer Res 14:1966–1975

Fujita T, Liu W, Doihara H, Wan Y (2008b) Regulation of Skp2-p27 Axis by the Cdh1/anaphase-promoting complex pathway in colorectal tumorigenesis. Am J Pathol 173:217–228

Gao D, Inuzuka H, Korenjak M, Tseng A, Wu T, Wan L, Kirschner M, Dyson N, Wei W (2009) Cdh1 regulates cell cycle through modulating the claspin/Chk1 and the Rb/E2F1 pathways. Mol Biol Cell 20:3305–3316

Gao Y, Zhang B, Wang Y, Shang G (2018) Cdc20 inhibitor apcin inhibits the growth and invasion of osteosarcoma cells. Oncol Rep 40:841–848

Garber ME, Troyanskaya OG, Schluens K, Petersen S, Thaesler Z, Pacyna-Gengelbach M, van de Rijn M, Rosen GD, Perou CM, Whyte RI, Altman RB, Brown PO, Botstein D, Petersen I (2001) Diversity of gene expression in adenocarcinoma of the lung. Proc Nat Acad Sci USA 98:13784–13789

García-Higuera I, Manchado E, Dubus P, Cañamero M, Méndez J, Moreno S, Malumbres M (2008) Genomic stability and tumour suppression by the APC/C cofactor Cdh1. Nat Cell Biol 10:802–811

Gascoigne K, Taylor S (2009) How do anti-mitotic drugs kill cancer cells? J Cell Sci 122:2579–2585

Geley S, Kramer E, Gieffers C, Gannon J, Peters JM, Hunt T (2001) Anaphase-promoting complex/cyclosome-dependent proteolysis of human cyclin A starts at the beginning of mitosis and is not subject to the spindle assembly checkpoint. J Cell Biol 153:137–148

Gella A, Durany N (2009) Oxidative stress in Alzheimer disease. Cell Adh Migr 3:88–93

Giovinazzi S, Bellapu D, Morozov V, Ishov A (2013) Targeting mitotic exit with hyperthermia or APC/C inhibition to increase paclitaxel efficacy. Cell Cycle 12:2598–2607

Glotzer M, Murray AW, Kirschner MW (1991) Cyclin is degraded by the ubiquitin pathway. Nature 349:132–138

Goodman RH, Smolik S (2000) CBP/p300 in cell growth, transformation and development. Genes Dev 14:1553–1577

Grabsch H, Takeno S, Parsons WJ, Pomjanski N, Boecking A, Gabbert HE, Mueller W (2003) Overexpression of the mitotic checkpoint genes BUB1, BUBR1, and BUB3 in gastric cancer-association with tumour cell proliferation. J Pathol 200:16–22

Gujar A, Yano H, Kim A (2015) The CDC20-APC/SOX2 signaling axis: an achilles' heel for glioblastoma. Mol Cellular Oncol 3:pe.1075644

Gurden MD, Holland AJ, van Zon W, Tighe A, Vergnolle MA, Andres DA, Spielmann HP, Malumbres M, Wolthuis RM, Cleveland DW, Taylor SS (2010) Cdc20 is required for the post-anaphase, KEN-dependent degradation of centromere protein F. J Cell Sci 123:321–330

Gütgemann I, Lehman NL, Jackson PK, Longacre TA (2008) Emi1 protein accumulation implicates misregulation of the anaphase promoting complex/cyclosome pathway in ovarian clear cell carcinoma. Mod Pathol 21:445–454

Gutierrez GJ, Tsuji T, Chen M, Jiang W, Ronai ZA (2010) Interplay between Cdh1 and JNK activity during the cell cycle. Nat Cell Biol 12:686–695

Hagting A, Den Elzen N, Vodermaier HC, Waizenegger IC, Peters JM, Pines J (2002) Human securin proteolysis is controlled by the spindle checkpoint and reveals when the APC/C switches from activation by Cdc20 to Cdh1. J Cell Biol 157:1125–1137

Hames RS, Wattam SL, Yamano H, Bacchieri R, Fry AM (2001) APC/C-mediated destruction of the centrosomal kinase Nek2A occurs in early mitosis and depends upon a cyclin A-type D-box. EMBO J 20:7117–7127

Hardwick KG, Weiss E, Luca FC, Winey M, Murray AW (1996) Activation of the budding yeast spindle assembly checkpoint without mitotic spindle disruption. Science 273:953–956

Harley ME, Allan LA, Sanderson HS, Clarke PR (2010) Phosphorylation of Mcl-1 by CDK1-cyclin B1 initiates its Cdc20-dependent destruction during mitotic arrest. EMBO J 29:2407–2420

Harper JW, Burton JL, Solomon MJ (2002) The anaphase-promoting complex: it's not just for mitosis any more. Genes Dev 16:2179–2206

Hayes MJ, Kimata Y, Wattam SL, Lindon C, Mao G, Yamano H, Fry AM (2006) Early mitotic degradation of Nek2A depends on Cdc20-independent interaction with the APC/C. Nat Cell Biol 8:607–614

He J, Chao WC, Zhang Z, Yang J, Cronin N, Barford D (2013) Insights into degron recognition by APC/C coactivators from the structure of an Acm1-Cdh1 complex. Mol Cell 50:649–656

Heighway J, Knapp T, Boyce L, Brennand S, Field JK, Betticher DC, Ratschiller D, Gugger M, Donovan M, Lasek A, Rickert P (2002) Expression profiling of primary non-small cell lung cancer for target identification. Oncogene 21:7749–7763

Heilman D, Teodoro J, Green M (2006) Apoptin nucleocytoplasmic shuttling is required for cell type-specific localization, apoptosis, and recruitment of the anaphase-promoting complex/cyclosome to PML bodies. J Virol 80:7535–7545

Hein J, Boichuk S, Wu J, Cheng Y, Freire R, Jat P, Roberts T, Gjoerup O (2008) Simian virus 40 large T antigen disrupts genome integrity and activates a DNA damage response via Bub1 binding. J Virol 83:117–127

Hein JB, Hertz EPT, Garvanska DH, Kruse T, Nilsson J (2017) Distinct kinetics of serine and threonine dephosphorylation are essential for mitosis. Nat Cell Biol 19:1433–1440

Hein JB, Nilsson J (2014) Stable MCC binding to the APC/C is required for a functional spindle assembly checkpoint. EMBO Rep 15:264–272

Hernando E, Nahle Z, Juan G, Diaz-Rodriguez E, Alaminos M, Hemann M, Michel L, Mittal V, Gerald W, Benezra R (2004) Rb inactivation promotes genomic instability by uncoupling cell cycle progression from mitotic control. Nature 430:797–802

Herrero-Mendez A, Almeida A, Fernández E, Maestre C, Moncada S, Bolaños J (2009) The bioenergetic and antioxidant status of neurons is controlled by continuous degradation of a key glycolytic enzyme by APC/C–Cdh1. Nat Cell Biol 11:747–752

Hewitt L, Tighe A, Santaguida S, White A, Jones C, Musacchio A, Green S, Taylor S (2010) Sustained Mps1 activity is required in mitosis to recruit O-Mad2 to the Mad1–C-Mad2 core complex. J Cell Biol 190:25–34

Hisaoka M, Matsuyama A, Hashimoto H (2008) Aberrant MAD2 expression in soft-tissue sarcoma. Pathol Int 58:329–333

Höckner S, Neumann-Arnold L, Seufert W (2016) Dual control by Cdk1 phosphorylation of the budding yeast APC/C ubiquitin ligase activator Cdh1. Mol Biol Cell 27:2198–2212

Hsu JY, Reimann JD, Sørensen CS, Lukas J, Jackson PK (2002) E2F-dependent accumulation of hEmi1 regulates S phase entry by inhibiting APC(Cdh1). Nat Cell Biol 4:358–366

Hu K, Liao D, Wu W, Han A, Shi H, Wang F, Wang X, Zhong L, Duan T, Wu Y, Cao J, Tang J, Sang Y, Wang L, Lv X, Xu S, Zhang R, Deng W, Li S, Zeng Y, Kang T (2014) Targeting the anaphase-promoting complex/cyclosome (APC/C)- bromodomain containing 7 (BRD7) pathway for human osteosarcoma. Oncotarget 5:3088–3100

Hu D, Qiao X, Wu G, Wan Y (2011) The emerging role of APC/CCdh1 in development. Semin Cell Dev Biol 22:579–585

Huang JN, Park I, Ellingson E, Littlepage LE, Pellman D (2001) Activity of the APC(Cdh1) form of the anaphase-promoting complex persists until S phase and prevents the premature expression of Cdc20p. J Cell Biol 154:85–94

Huang H, Shi J, Orth J, Mitchison T (2009) Evidence that mitotic exit is a better cancer therapeutic target than spindle assembly. Cancer Cell 16:347–358

Huang NJ, Zhang L, Tang W, Chen C, Yang CS, Kornbluth S (2012) The Trim39 ubiquitin ligase inhibits APC/CCdh1-mediated degradation of the Bax activator MOAP-1. J Cell Biol 197:361–367

Huang W, Zhang X, Chen W (2016) Role of oxidative stress in Alzheimer's disease. Biomed Rep 4:519–522

Hume A, Finkel J, Kamil J, Coen D, Culbertson M, Kalejta R (2008) Phosphorylation of retinoblastoma protein by viral protein with cyclin-dependent kinase function. Science 320:797–799

Hume AJ, Kalejta RF (2009) Regulation of the retinoblastoma proteins by the human herpesviruses. Cell Div 4:1

Hyun SY, Sarantuya B, Lee HJ, Jang YJ (2013) APC/C(Cdh1)-dependent degradation of Cdc20 requires a phosphorylation on CRY-box by Polo-like kinase-1 during somatic cell cycle. Biochem Biophys Res Comm 436:12–18

Ichim G, Mola M, Finkbeiner MG, Cros MP, Herceg Z, Hernandez-Vargas H (2014) The histone acetyltransferase component TRRAP is targeted for destruction during the cell cycle. Oncogene 33:181–192

Irniger S, Nasmyth K (1997) The anaphase-promoting complex is required in G1 arrested yeast cells to inhibit B-type cyclin accumulation and to prevent uncontrolled entry into S-phase. J Cell Sci 110:1523–1531

Izawa D, Pines J (2012) Mad2 and the APC/C compete for the same site on Cdc20 to ensure proper chromosome segregation. J Cell Biol 199:27–37

Izawa D, Pines J (2015) The mitotic checkpoint complex binds a second CDC20 to inhibit active APC/C. Nature 517:631–634

Jeng JC, Lin YM, Lin CH, Shih HM (2009) Cdh1 controls the stability of TACC3. Cell Cycle 8:3537–3544

Jin J, Cardozo T, Lovering RC, Elledge SJ, Pagano M, Harper JW (2004) Systematic analysis and nomenclature of mammalian F-box proteins. Genes Dev 18:2573–2580

Jordan M, Wilson L (2004) Microtubules as a target for anticancer drugs. Nat Rev Cancer 4:253–265

Juhlin C, Goh G, Healy J, Fonseca A, Scholl U, Stenman A, Kunstman J, Brown T, Overton J, Mane S, Nelson-Williams C, Bäckdahl M, Suttorp A, Haase M, Choi M, Schlessinger J, Rimm D, Höög A, Prasad M, Korah R, Larsson C, Lifton R, Carling T (2015) Whole-exome sequencing characterizes the landscape of somatic mutations and copy number alterations in adrenocortical carcinoma. J Clin Endocrinol Metab 100:E493–E502

Kapanidou M, Bolanos-Garcia V (2014) Spindle assembly checkpoint (SAC): more new targets for anti-cancer drug therapies. Adv Cancer Drug Targets 2:54–79

Kapanidou M, Curtis N, Bolanos-Garcia V (2017) Cdc20: at the crossroads between chromosome segregation and mitotic exit. Trends Biochem Sci 42:193–205

Kapanidou M, Lee S, Bolanos-Garcia V (2015) BubR1 kinase: protection against aneuploidy and premature aging. Trends Mol Med 21:364–372

Karamysheva Z, Diaz-Martinez LA, Crow SE, Li B, Yu H (2009) Multiple anaphase-promoting complex/cyclosome degrons mediate the degradation of human Sgo1. J Biol Chem 284:1772–1780

Karra H, Repo H, Ahonen I, Loyttyniemi E, Pitkanen R, Lintunen M, Kuopio T, Soderstrom M, Kronqvist P (2014) Cdc20 and securin overexpression predict short-term breast cancer survival. Brit J Cancer 110:2905–2913

Kato T, Daigo Y, Aragaki M, Ishikawa K, Sato M, Kaji M (2012) Overexpression of CDC20 predicts poor prognosis in primary non-small cell lung cancer patients. J Surg Oncol 106:423–430

Kato T, Daigo Y, Aragaki M, Ishikawa K, Sato M, Kondo S, Kaji M (2011) Overexpression of MAD2 predicts clinical outcome in primary lung cancer patients. Lung Cancer 74:124–131

Ke PY, Chang ZF (2004) Mitotic degradation of human thymidine kinase 1 is dependent on the anaphase-promoting complex/cyclosome-CDH1-mediated pathway. Mol Cell Biol 24:514–526

Ke PY, Hu CM, Chang YC, Chang ZF (2007) Hiding human thymidine kinase 1 from APC/C-mediated destruction by thymidine binding. FASEB J 21:1276–1284

Ke PY, Kuo YY, Hu CM, Chang ZF (2005) Control of dTTP pool size by anaphase promoting complex/cyclosome is essential for the maintenance of genetic stability. Genes Dev 19:1920–1933

Kim Y, Choi JW, Lee JH, Kim YS (2014) MAD2 and CDC20 are upregulated in high-grade squamous intraepithelial lesions and squamous cell carcinomas of the uterine cervix. Int J Gynecol Pathol 33:517–523

Kim S, Park S, Yong H, Famulski J, Chae S, Lee J, Kang C, Saya H, Chan G, Cho H (2008) HBV X protein targets hBubR1, which induces dysregulation of the mitotic checkpoint. Oncogene 27:3457–3464

Kim AH, Puram SV, Bilimoria PM, Ikeuchi Y, Keough S, Wong M, Rowitch D, Bonni A (2009) A centrosomal Cdc20-APC pathway controls dendrite morphogenesis in postmitotic neurons. Cell 136:322–336

Kimata Y, Trickey M, Izawa D, Gannon J, Yamamoto M, Yamano H (2008) A mutual inhibition between APC/C and its substrate Mes1 required for meiotic progression in fission yeast. Dev Cell 14:446–454

Kleinberger T (2015) Mechanisms of cancer cell killing by the adenovirus E4orf4 protein. Viruses 7:2334–2357

Klitzing CV, Huss R, Illert AL, Fröschl A, Wötzel S, Peschel C, Bassermann F, Duyster J (2011) APC/C(Cdh1)-mediated degradation of the F-box protein NIPA is regulated by its association with Skp1. PLoS ONE 6:e28998

Ko N, Nishihama R, Tully GH, Ostapenko D, Solomon MJ, Morgan DO, Pringle JR (2007) Identification of yeast IQGAP (Iqg1p) as an anaphase-promoting-complex substrate and its role in actomyosin-ring-independent cytokinesis. Mol Biol Cell 18:5139–5153

Komander D, Rape M (2012) The ubiquitin code. Ann Rev Biochem 81:203–229

Konishi Y, Stegmuller J, Matsuda T, Bonni S, Bonni A (2004) Cdh1-APC controls axonal growth and patterning in the mammalian brain. Science 303:1026–1030

Kornitzer D, Sharf R, Kleinberger T (2001) Adenovirus E4orf4 protein induces PP2A-dependent growth arrest in *S. cerevisiae* and interacts with the anaphase-promoting complex/cyclosome. J Cell Biol 154:331–344

Kowall N, Beal M (1991) Glutamate-, glutaminase-, and taurine-immunoreactive neurons develop neurofibrillary tangles in Alzheimer's disease. Ann Neurol 29:162–167

Kraft C, Herzog F, Gieffers C, Mechtler K, Hagting A, Pines J, Peters JM (2003) Mitotic regulation of the human anaphase-promoting complex by phosphorylation. EMBO J 22:6598–6609

Kramer ER, Scheuringer N, Podtelejnikov AV, Mann M, Peters JM (2000) Mitotic regulation of the APC activator proteins CDC20 and CDH1. Mol Biol Cell 11:1555–1569

Kucharski T, Gamache I, Gjoerup O, Teodoro J (2011) DNA damage response signaling triggers nuclear localization of the chicken anemia virus protein apoptin. J Virol 85:12638–12649

Kucharski T, Ng T, Sharon D, Navid-Azarbaijani P, Tavassoli M, Teodoro J (2016) Activation of the chicken anemia virus apoptin protein by Chk1/2 phosphorylation is required for apoptotic activity and efficient viral replication. J Virol 90:9433–9445

Kuczera T, Stilling R, Hsia H, Bahari-Javan S, Irniger S, Nasmyth K, Sananbenesi F, Fischer A (2010) The anaphase promoting complex is required for memory function in mice. Learn Mem 18:49–57

Lai F, Hu K, Wu Y, Tang J, Sang Y, Cao J, Kang T (2012) Human KIAA1018/FAN1 nuclease is a new mitotic substrate of APC/C(Cdh1). Chin J Cancer 31:440–448

Lara-Gonzalez P, Taylor S (2012) Cohesion fatigue explains why pharmacological inhibition of the APC/C induces a spindle checkpoint-dependent mitotic arrest. PLoS ONE 7:e49041

Lara-Gonzalez P, Westhorpe FG, Taylor SS (2012) The spindle assembly checkpoint. Curr Biol 22:R966–R980

Lasorella A, Stegmuller J, Guardavaccaro D, Liu G, Carro MS, Rothschild G, de la Torre-Ubieta L, Pagano M, Bonni A, Iavarone A (2006) Degradation of Id2 by the anaphase-promoting complex couples cell cycle exit and axonal growth. Nature 442:471–474

Lee YK, Choi E, Kim MA, Park PG, Park NH, Lee H (2009a) BubR1 as a prognostic marker for recurrence-free survival rates in epithelial ovarian cancers. Brit J Cancer 101:504–510

Lee J, Kim J, Barbier V, Fotedar A, Fotedar R (2009b) DNA Damage Triggers p21WAF1-dependent Emi1 down-regulation that maintains G2 arrest. Mol Biol Cell 20:1891–1902

Lee SJ, Rodriguez-Bravo V, Kim H, Datta S, Foley EA (2017) The PP2AB56 phosphatase promotes the association of Cdc20 with APC/C in mitosis. J Cell Sci 130:1760–1771

Lehman N, Tibshirani R, Hsu J, Natkunam Y, Harris B, West R, Masek M, Montgomery K, van de Rijn M, Jackson P (2007) Oncogenic regulators and substrates of the anaphase promoting complex/cyclosome are frequently overexpressed in malignant tumors. Am J Pathol 170:1793–1805

Lehman NL, Verschuren EW, Hsu JY, Cherry AM, Jackson PK (2006) Overexpression of the anaphase promoting complex/cyclosome inhibitor Emi1 leads to tetraploidy and genomic instability of p53-deficient cells. Cell Cycle 5:1569–1573

Li GQ, Li H, Zhang HF (2003) Mad2 and p53 expression profiles in colorectal cancer and its clinical significance. World J Gastroenterol 9:1972–1975

Li J, Gao J, Du J, Huang Z, Wei L (2014) Increased CDC20 expression is associated with development and progression of hepatocellular carcinoma. Int J Oncol 45:1547–1555

Li K, Mao Y, Lu L, Hu C, Wang D, Si-Tu J, Lu M, Peng S, Qiu J, Gao X (2016) Silencing of CDC20 suppresses metastatic castration-resistant prostate cancer growth and enhances chemosensitivity to docetaxel. Int J Oncol 49:1679–1685

Li M, Shin Y, Hou L, Huang X, Wei Z, Klann E, Zhang P (2008) The adaptor protein of the anaphase promoting complex Cdh1 is essential in maintaining replicative lifespan and in learning and memory. Nat Cell Biol 10:1083–1089

Li B, Sun M, Zhang L, Takahashi S, Ma W, Vinade L, Kulkarni A, Brady R, Pant H (2001) Regulation of NMDA receptors by cyclin-dependent kinase-5. Proc Natl Acad Sci USA 98:12742–12747

Li R, Wan B, Zhou J, Wang Y, Luo T, Gu X, Chen F, Yu L (2012) APC/C(Cdh1) targets brain-specific kinase 2 (BRSK2) for degradation via the ubiquitin-proteasome pathway. PLoS ONE 7:e45932

Li L, Zhou Y, Wang GF, Liao SC, Ke YB, Wu W, Li XH, Zhang RL, Fu YC (2011) Anaphase-promoting complex/cyclosome controls HEC1 stability. Cell Prolif 44:1–9

Liang M, Geisbert T, Yao Y, Hinrichs S, Giam C (2002) Human T-Lymphotropic virus type 1 oncoprotein tax promotes S-phase entry but blocks mitosis. J Virol 76:4022–4033

Lim HJ, Dimova NV, Tan MK, Sigoillot FD, King RW, Shi Y (2013) The G2/M regulator histone demethylase PHF8 is targeted for degradation by the anaphase-promoting complex containing CDC20. Mol Cell Biol 33:4166–4180

Lindberg J, Mills I, Klevebring D, Liu W, Neiman M, Xu J, Wikström P, Wiklund P, Wiklund F, Egevad L, Grönberg H (2013) The mitochondrial and autosomal mutation landscapes of prostate cancer. Eur Urol 63:702–708

Lindon C, Pines J (2004) Ordered proteolysis in anaphase inactivates Plk1 to contribute to proper mitotic exit in human cells. J Cell Biol 164:233–241

Liot C, Seguin L, Siret A, Crouin C, Schmidt S, Bertoglio J (2011) APC(cdh1) mediates degradation of the oncogenic Rho-GEF Ect2 after mitosis. PLoS ONE 6:e23676

Lipkowitz S, Weissman AM (2011) RINGs of good and evil: RING finger ubiquitin ligases at the crossroads of tumour suppression and oncogenesis. Nat Rev Cancer 11:629–643

Litchfield K, Summersgill B, Yost S, Sultana R, Labreche K, Dudakia D, Renwick A, Seal S, Al-Saadi R, Broderick P, Turner N, Houlston R, Huddart R, Shipley J, Turnbull C (2015) Whole-exome sequencing reveals the mutational spectrum of testicular germ cell tumours. Nat Comm 6:5973

Littlepage LE, Ruderman JV (2002) Identification of a new APC/C recognition domain, the A box, which is required for the Cdh1-dependent destruction of the kinase Aurora-A during mitotic exit. Genes Dev 16:2274–2285

Liu AW, Cai J, Zhao XL, Xu AM, Fu HQ, Nian H, Zhang SH (2009) The clinicopathological significance of BUBR1 overexpression in hepatocellular carcinoma. J Clin Pathol 62:1003–1008

Liu X, Chen Y, Li Y, Petersen R, Huang K (2019) Targeting mitosis exit: a brake for cancer cell proliferation. Biochim Biophys Acta Rev Cancer 1871:179–191

Liu J, Grimison B, Maller JL (2007) New insight into metaphase arrest by cytostatic factor: from establishment to release. Oncogene 26:1286–1289

Liu B, Hong S, Tang Z, Yu H, Giam C (2004) HTLV-I Tax directly binds the Cdc20-associated anaphase-promoting complex and activates it ahead of schedule. Proc Natl Acad Sci USA 102:63–68

Liu B, Liang M, Kuo Y, Liao W, Boros I, Kleinberger T, Blancato J, Giam C (2003) Human T-Lymphotropic Virus Type 1 oncoprotein tax promotes unscheduled degradation of Pds1p/securin and Clb2p/Cyclin B1 and causes chromosomal instability. Mol Cell Biol 23:5269–5281

London N, Biggins S (2014) Signalling dynamics in the spindle checkpoint response. Nat Rev Mol Cell Biol 15:736–747

Los M, Panigrahi S, Rashedi I, Mandal S, Stetefeld J, Essmann F, Schulze-Osthoff K (2009) Apoptin, a tumor-selective killer. Biochim Biophys Acta Mol Cell Res 1793:1335–1342

Lu L, Hu S, Wei R, Qiu X, Lu K, Fu Y, Li H, Xing G, Li D, Peng R, He F, Zhang L (2013) The HECT type ubiquitin ligase NEDL2 is degraded by anaphase-promoting complex/cyclosome (APC/C)-Cdh1, and its tight regulation maintains the metaphase to anaphase transition. J Biol Chem 288:35637–35650

Lub S, Maes A, Maes K, De Veirman K, De Bruyne E, Menu E, Fostier K, Kassambara A, Moreaux J, Hose D, Leleu X, King R, Vanderkerken K, Van Valckenborgh E (2016) Inhibiting the anaphase promoting complex/cyclosome induces a metaphase arrest and cell death in multiple myeloma cells. Oncotarget 7:4062–4076

Luetjens C, Bui N, Sengpiel B, Münstermann G, Poppe M, Krohn A, Bauerbach E, Krieglstein J, Prehn J (2000) Delayed mitochondrial dysfunction in excitotoxic neuron death: cytochrome-crelease and a secondary increase in superoxide production. J Neurosci 20:5715–5723

Maestre C, Delgado-Esteban M, Gomez-Sanchez J, Bolaños J, Almeida A (2008) Cdk5 phosphorylates Cdh1 and modulates cyclin B1 stability in excitotoxicity. EMBO J 27:2736–2745

Manchado E, Guillamot M, de Carcer G, Eguren M, Trickey M, Garcia-Higuera I, Moreno S, Yamano H, Canamero M, Malumbres M (2010) Targeting mitotic exit leads to tumor regression in vivo: modulation by Cdk1, Mastl, and the PP2A/B55alpha, delta phosphatase. Cancer Cell 18:641–654

Mao D, Gujar A, Mahlokozera T, Chen I, Pan Y, Luo J, Brost T, Thompson E, Turski A, Leuthardt E, Dunn G, Chicoine M, Rich K, Dowling J, Zipfel G, Dacey R, Achilefu S, Tran D, Yano H, Kim AA (2015) CDC20-APC/SOX2 signaling axis regulates human glioblastoma stem-like cells. Cell Rep 11:1809–1821

Mao Y, Li K, Lu L, Si-tu J, Lu M, Gao X (2016) Overexpression of Cdc20 in clinically localized prostate cancer: relation to high Gleason score and biochemical recurrence after laparoscopic radical prostatectomy. Cancer Biomark 16:351–358

Maoz T, Koren R, Ben-Ari I, Kleinberger T (2005) YND1 interacts with CDC55 and is a novel mediator of E4orf4-induced toxicity. J Biol Chem 280:41270–41277

Margottin-Goguet F, Hsu JY, Loktev A, Hsieh HM, Reimann JD, Jackson PK (2003) Prophase destruction of Emi1 by the $SCF^{\beta TrCP/Slimb}$ ubiquitin ligase activates the anaphase promoting complex to allow progression beyond prometaphase. Dev Cell 4:813–826

Martin-Lluesma S, Schaeffer C, Robert E, van Breugel P, Leupin O, Hantz O, Strubin M (2008) Hepatitis B virus X protein affects S phase progression leading to chromosome segregation defects by binding to damaged DNA binding protein 1. Hepatology 48:1467–1476

Marucci G, Morandi L, Magrini E, Farnedi A, Franceschi E, Miglio R, Calo D, Pession A, Foschini MP, Eusebi V (2008) Gene expression profiling in glioblastoma and immunohistochemical evaluation of IGFBP-2 and CDC20. Virchows Arch 453:599–609

Matyskiela ME, Morgan DO (2009) Analysis of activator-binding sites on the APC/C supports a cooperative substrate-binding mechanism. Mol Cell 34:68–80

McGarry TJ, Kirschner MW (1998) Geminin, an inhibitor of DNA replication, is degraded during mitosis. Cell 93:1043–1053

McLean JR, Chaix D, Ohi MD, Gould KL (2011) State of the APC/C: organization, function, and structure. Crit Rev Biochem Mol Biol 46:118–136

Meghini F, Martins T, Tait X, Fujimitsu K, Yamano H, Glover D, Kimata Y (2016) Targeting of Fzr/Cdh1 for timely activation of the APC/C at the centrosome during mitotic exit. Nat Comm 7:12607

Menyhárt O, Nagy Á, Győrffy B (2018) Determining consistent prognostic biomarkers of overall survival and vascular invasion in hepatocellular carcinoma. R Soc Open Sci 5:181006

Metzger MB, Hristova VA, Weissman AM (2012) HECT and RING finger families of E3 ubiquitin ligases at a glance. J Cell Sci 125:531–537

Meyer SK, Dunn M, Vidler DS, Porter A, Blain PG, Jowsey PA (2019) Phosphorylation of MCPH1 isoforms during mitosis followed by isoform-specific degradation by APC/C-CDH1. FASEB J 33:2796–2808

Michaelis C, Ciosk R, Nasmyth K (1997) Cohesins: chromosomal proteins that prevent premature separation of sister chromatids. Cell 91:35–45

Miller JJ, Summers MK, Hansen DV, Nachury MV, Lehman NL, Loktev A, Jackson PK (2006) Emi1 stably binds and inhibits the anaphase-promoting complex/cyclosome as a pseudosubstrate inhibitor. Genes Dev 20:2410–2420

Miron M, Blanchette P, Groitl P, Dallaire F, Teodoro J, Li S, Dobner T, Branton P (2008) Localization and importance of the adenovirus E4orf4 protein during lytic infection. J Virol 83:1689–1699

Mo M, Fleming S, Mercer A (2009) Cell cycle deregulation by a poxvirus partial mimic of anaphase-promoting complex subunit 11. Proc Natl Acad Sci USA 106:19527–19532

Mo M, Fleming S, Mercer A (2010a) Changing pace: viral mimicry of an anaphase promoting complex subunit. Cell Cycle 9:632–633

Mo M, Fleming S, Mercer A (2010b) Orf virus cell cycle regulator, PACR, competes with subunit 11 of the anaphase promoting complex for incorporation into the complex. J Gen Virol 91:3010–3015

Mo M, Shahar S, Fleming S, Mercer A (2012) How viruses affect the cell cycle through manipulation of the APC/C. Trends Microbiol 20:440–448

Mondal G, Sengupta S, Panda CK, Gollin SM, Saunders WS, Roychoudhury S (2007) Overexpression of Cdc20 leads to impairment of the spindle assembly checkpoint and aneuploidization in oral cancer. Carcinogenesis 28:81–92

Montero A, Fossella F, Hortobagyi G, Valero V (2005) Docetaxel for treatment of solid tumours: a systematic review of clinical data. Lancet Oncol 6:229–239

Morrow CJ, Tighe A, Johnson VL, Scott MI, Ditchfield C, Taylor SS (2005) Bub1 and aurora B cooperate to maintain BubR1-mediated inhibition of APC/CCdc20. J Cell Sci 118:3639–3652

Mosch B, Morawski M, Mittag A, Lenz D, Tarnok A, Arendt T (2007) Aneuploidy and DNA replication in the normal human brain and alzheimer's disease. J Neurosci 27:6859–6867

Moura IM, Delgado ML, Silva PM, Lopes CA, do Amaral JB, Monteiro LS, Bousbaa H (2014) High CDC20 expression is associated with poor prognosis in oral squamous cell carcinoma. J Oral Pathol Med 43:225–231

Mu Y, Gage F (2011) Adult hippocampal neurogenesis and its role in Alzheimer's disease. Mol Neurodegener 6:85

Mui M, Roopchand D, Gentry M, Hallberg R, Vogel J, Branton P (2010) Adenovirus protein E4orf4 induces premature APCCdc20 activation in *Saccharomyces cerevisiae* by a protein phosphatase 2A-dependent mechanism. J Virol 84:4798–4809

Musacchio A (2011) Spindle assembly checkpoint: the third decade, philosophical transactions of the royal society of London. Series B Biol Sci 366:3595–3604

Musacchio A, Salmon ED (2007) The spindle-assembly checkpoint in space and time. Nat Rev Mol Cell Biol 8:379–393

Nagy V, Dikic I (2010) Ubiquitin ligase complexes: from substrate selectivity to conjugational specificity. Biol Chem 391:163–169

Nakayama KI, Nakayama K (2006) Ubiquitin ligases: cell-cycle control and cancer. Nat Rev Cancer 6:369–381

Nalepa G, Rolfe M, Harper JW (2006) Drug discovery in the ubiquitin-proteasome system. Nat Rev Drug Discov 5:596–613

Naoe H, Araki K, Nagano O, Kobayashi Y, Ishizawa J, Chiyoda T, Shimizu T, Yamamura K, Sasaki Y, Saya H, Kuninaka S (2010) The anaphase-promoting complex/cyclosome activator Cdh1 modulates Rho GTPase by targeting p190 RhoGAP for degradation. Mol Cell Biol 30:3994–4005

Nguyen D, Alavi M, Kim K, Kang T, Scott R, Noh Y, Lindsey J, Wissinger B, Ellisman M, Weinreb R, Perkins G, Ju W (2011) A new vicious cycle involving glutamate excitotoxicity, oxidative stress and mitochondrial dynamics. Cell Death Dis 2:e240

Nguyen HG, Chinnappan D, Urano T, Ravid K (2005) Mechanism of Aurora-B degradation and its dependency on intact KEN and A-boxes: identification of an aneuploidy-promoting property. Mol Cell Biol 25:4977–4992

Nishimura K, Oki T, Kitaura J, Kuninaka S, Saya H, Sakaue-Sawano A, Miyawaki A, Kitamura T (2013) APC(CDH1) targets MgcRacGAP for destruction in the late M phase. PLoS ONE 8: e63001

Noteborn M (2004) Chicken anemia virus induced apoptosis: underlying molecular mechanisms. Vet Microbiol 98:89–94

Noteborn M (2009) Proteins selectively killing tumor cells. Eur J Pharmacol 625:165–173

Ohoka N, Sakai S, Onozaki K, Nakanishi M, Hayashi H (2010) Anaphase-promoting complex/cyclosome-cdh1 mediates the ubiquitination and degradation of TRB3. Biochem Biophys Res Commun 392:289–294

Ostapenko D, Solomon MJ (2011) Anaphase promoting complex-dependent degradation of transcriptional repressors Nrm1 and Yhp1 in *Saccharomyces cerevisiae*. Mol Biol Cell 22:2175–2184

Pandey V, Kumar V (2012) HBx protein of hepatitis B virus promotes reinitiation of DNA replication by regulating expression and intracellular stability of replication licensing factor CDC6. J Biol Chem 287:20545–20554

Park K, Choi S, Eom M, Kang Y (2005) Downregulation of the anaphase-promoting complex (APC)7 in invasive ductal carcinomas of the breast and its clinicopathologic relationships. Breast Cancer Res 7:R238–R247

Park HJ, Costa RH, Lau LF, Tyner AL, Raychaudhuri P (2008) Anaphase-promoting complex/cyclosome-CDH1-mediated proteolysis of the forkhead box M1 transcription factor is critical for regulated entry into S phase. Mol Cell Biol 28:5162–5171

Parmar MB, KC RB, Löbenberg R, Uludağ H (2018) Additive polyplexes to undertake siRNA therapy against CDC20 and survivin in breast cancer cells. Biomacromolecules 19:4193–4206

Parmar M, Meenakshi Sundaram D, KC R, Maranchuk R, Montazeri Aliabadi H, Hugh J, Löbenberg R, Uludağ H (2018) Combinational siRNA delivery using hyaluronic acid modified amphiphilic polyplexes against cell cycle and phosphatase proteins to inhibit growth and migration of triple-negative breast cancer cells. Acta Biomater 66:294–309

Passmore LA, Booth CR, Vénien-Bryan C, Ludtke SJ, Fioretto C, Johnson LN, Chiu W, Barford D (2005) Structural analysis of the anaphase-promoting complex reveals multiple active sites and insights into polyubiquitylation. Mol Cell 20:855–866

Patel D, McCance D (2010) Compromised spindle assembly checkpoint due to altered expression of Ubch10 and Cdc20 in human papillomavirus Type 16 E6- and E7-expressing keratinocytes. J Virol 84:10956–10964

Penas C, Govek E, Fang Y, Ramachandran V, Daniel M, Wang W, Maloof M, Rahaim R, Bibian M, Kawauchi D, Finkelstein D, Han J, Long J, Li B, Robbins D, Malumbres M, Roussel M, Roush W, Hatten M, Ayad N (2015) Casein kinase 1δ Is an APC/CCdh1 substrate that regulates cerebellar granule cell neurogenesis. Cell Rep 11:249–260

Penrose K, McBride A (2000) Proteasome-mediated degradation of the papillomavirus E2-TA protein is regulated by phosphorylation and can modulate viral genome copy number. J Virol 74:6031–6038

Peters JM (2006) The anaphase promoting complex/cyclosome: a machine designed to destroy. Nature Rev Mol Cell Biol 7:644–656

Petersen B, Wagener C, Marinon F, Kramer E, Melixetian M, Denchi E, Gieffers C, Matteucci C, Peters J, Helin K (2000) Cell cycle- and cell growth-regulated proteolysis of mammalian CDC6 is dependent on APC-CDH1. Genes Dev 14:2330–2343

Petroski MD, Deshaies RJ (2005) Function and regulation of cullin-RING ubiquitin ligases. Nature Rev Mol Cell Biol 6:9–20

Pfleger CM, Kirschner MW (2000) The KEN box: an APC recognition signal distinct from the D box targeted by Cdh1. Genes Dev 14:655–665

Pick J, Malumbres M, Klann E (2012) The E3 ligase APC/C-Cdh1 is required for associative fear memory and long-term potentiation in the amygdala of adult mice. Learn Mem 20:11–20

Pickart CM (2001) Mechanisms underlying ubiquitination. Ann Rev Biochem 70:503–533

Pines J (2011) Cubism and the cell cycle: the many faces of the APC/C. Nat Rev Mol Cell Biol 12:427–438

Ping Z, Lim R, Bashir T, Pagano M, Guardavaccaro D (2012) APC/C (Cdh1) controls the proteasome-mediated degradation of E2F3 during cell cycle exit. Cell Cycle 11:1999–2005

Pinto M, Vieira J, Ribeiro FR, Soares MJ, Henrique R, Oliveira J, Jerónimo Teixeira MR (2008) Overexpression of the mitotic checkpoint genes BUB1 and BUBR1 is associated with genomic complexity in clear cell kidney carcinomas. Cell Oncol 30:389–395

Powers C, DeFilippis V, Malouli D, Fruh K (2008) Cytomegalovirus immune evasion. Curr Top Microbiol Immunol 325:333–359

Prasetyo A, Kamahora T, Kuroishi A, Murakami K, Hino S (2009) Replication of chicken anemia virus (CAV) requires apoptin and is complemented by VP3 of human torque teno virus (TTV). Virol 385:85–92

Puram SV, Bonni A (2011) Novel functions for the anaphase-promoting complex in neurobiology. Semin Cell Dev Biol 22:586–594

Qi W, Yu H (2007) KEN-box-dependent degradation of the Bub1 spindle checkpoint kinase by the anaphase-promoting complex/cyclosome. J Biol Chem 282:3672–3679

Qian Z, Leung-Pineda V, Xuan B, Piwnica-Worms H, Yu D (2010) Human cytomegalovirus protein pUL117 targets the mini-chromosome maintenance complex and suppresses cellular DNA synthesis. PLoS Pathog 6:e1000814

Qiu L, Tan X, Lin J, Liu R, Chen S, Geng R, Wu J, Huang W (2017) CDC27 induces metastasis and invasion in colorectal cancer via the promotion of epithelial-to-mesenchymal transition. J Cancer 8:2626–2635

Qiu L, Wu J, Pan C, Tan X, Lin J, Liu R, Chen S, Geng R, Huang W (2016) Downregulation of CDC27 inhibits the proliferation of colorectal cancer cells via the accumulation of p21Cip1/Waf1. Cell Death Dis 7:e2074

Rahimi H, Ahmadzadeh A, Yousef-amoli S, Kokabee L, Shokrgozar M, Mahdian R, Karimipoor M (2015) The expression pattern of APC2 and APC7 in various cancer cell lines and AML patients. Adv Med Sci 60:259–263

Ramanujan A, Tiwari S (2016) APC/C and retinoblastoma interaction: cross-talk of retinoblastoma protein with the ubiquitin proteasome pathway. Biosci Rep 36:e00377

Ranganathan P, Clark P, Kuo J, Salamat M, Kalejta R (2011) Significant association of multiple human cytomegalovirus genomic loci with glioblastoma multiforme samples. J Virol 86:854–864

Rape M (2010) Assembly of k11-linked ubiquitin chains by the anaphase-promoting complex. Subcell Biochem 54:107–115

Rape M, Kirschner MW (2004) Autonomous regulation of the anaphase-promoting complex couples mitosis to S-phase entry. Nature 432:588–595

Reimann JD, Freed E, Hsu JY, Kramer ER, Peters JM, Jackson PK (2001) Emi1 is a mitotic regulator that interacts with Cdc20 and inhibits the anaphase promoting complex. Cell 105:645–655

Reimann E, Kõks S, Ho X, Maasalu K, Märtson A (2014) Whole exome sequencing of a single osteosarcoma case¿integrative analysis with whole transcriptome RNA-seq data. Hum Genomics 8:20

Rimkus C, Friederichs J, Rosenberg R, Holzmann B, Siewert JR, Janssen KP (2007) Expression of the mitotic checkpoint gene MAD2L2 has prognostic significance in colon cancer. Int J Cancer 120:207–211

Rodriguez-Rodriguez P, Almeida A, Bolaños J (2013) Brain energy metabolism in glutamate-receptor activation and excitotoxicity: Role for APC/C-Cdh1 in the balance glycolysis/pentose phosphate pathway. Neurochem Int 62:750–756

Rotin D, Kumar S (2009) Physiological functions of the HECT family of ubiquitin ligases. Nature Rev Mol Cell Biol 10:398–409

Sackton K, Dimova N, Zeng X, Tian W, Zhang M, Sackton T, Meaders J, Pfaff K, Sigoillot F, Yu H, Luo X, King R (2014) Synergistic blockade of mitotic exit by two chemical inhibitors of the APC/C. Nature 514:646–649

Sacristan C, Kops GJ (2015) Joined at the hip: kinetochores, microtubules, and spindle assembly checkpoint signaling. Trends Cell Biol 25:21–28

Sandhu SK, Kaur G (2002) Alterations in oxidative stress scavenger system in aging rat brain and lymphocytes. Biogerontology 3:161–173

Sansregret L, Patterson J, Dewhurst S, López-García C, Koch A, McGranahan N, Chao W, Barry D, Rowan A, Instrell R, Horswell S, Way M, Howell M, Singleton M, Medema R, Nurse P, Petronczki M, Swanton C (2017) APC/C dysfunction limits excessive cancer chromosomal instability. Cancer Discov 7:218–233

Sasaki T, Senda M, Kim SN, Kojima S, Kubodera A (2001) Age-related changes of glutathione content, glucose transport and metabolism, and mitochondrial electron transfer function in mouse brain. Nucl Med Biol 28:25–31

Saxena N, Kumar V (2014) The HBx oncoprotein of hepatitis B virus deregulates the cell cycle by promoting the intracellular accumulation and re-compartmentalization of the cellular deubiquitinase USP37. PLoS ONE 9:e111256

Schreiber A, Stengel F, Zhang Z, Enchev RI, Kong EH, Morris EP, Robinson CV, da Fonseca PC, Barford D (2011) Structural basis for the subunit assembly of the anaphase-promoting complex. Nature 470:227–232

Schuyler S, Wu Y, Chen H, Ding Y, Lin C, Chu Y, Chen T, Liao L, Tsai W, Huang A, Wang L, Liao T, Jhuo J, Cheng V (2018) Peptide inhibitors of the anaphase promoting-complex that cause sensitivity to microtubule poison. PLoS ONE 13:e0198930

Schwab M, Neutzner M, Möcker D, Seufert W (2001) Yeast Hct1 recognizes the mitotic cyclin Clb2 and other substrates of the ubiquitin ligase APC. EMBO J 20:5165–5175

Sedgwick GG, Hayward DG, Di Fiore B, Pardo M, Yu L, Pines J, Nilsson J (2013) Mechanisms controlling the temporal degradation of Nek2A and Kif18A by the APC/C-Cdc20 complex. EMBO J 32:303–314

Seike M, Gemma A, Hosoya Y, Hosomi Y, Okano T, Kurimoto F, Uematsu K, Takenaka K, Yoshimura A, Shibuya M, Ui-Tei K, Kudoh S (2002) The promoter region of the human BUBR1 gene and its expression analysis in lung cancer. Lung Cancer 38:229–234

Seki A, Fang G (2007) CKAP2 is a spindle-associated protein degraded by APC/C-Cdh1 during mitotic exit. J Biol Chem 282:15103–15113

Shi R, Sun Q, Sun J, Wang X, Xia W, Dong G, Wang A, Jiang F, Xu L (2017) Cell division cycle 20 overexpression predicts poor prognosis for patients with lung adenocarcinoma. Tumor Biol 39:101042831769223

Shimizu N, Nakajima N, Tsunematsu T, Ogawa I, Kawai H, Hirayama R, Fujimori A, Yamada A, Okayasu R, Ishimaru N, Takata T, Kudo Y (2013) Selective enhancing effect of early mitotic inhibitor 1 (Emi1) depletion on the sensitivity of doxorubicin or X-ray treatment in human cancer cells. J Biol Chem 288:17238–17252

Simpson-Lavy KJ, Sajman J, Zenvirth D, Brandeis M (2009) APC/CCdh1 specific degradation of Hsl1 and Clb2 is required for proper stress responses of *S. cerevisiae*. Cell Cycle 8:3003–3009

Sivakumar S, Gorbsky GJ (2015) Spatiotemporal regulation of the anaphase-promoting complex in mitosis. Nat Rev Mol Cell Biol 16:82–94

Skaar JR, Pagano M (2009) Control of cell growth by the SCF and APC/C ubiquitin ligases. Curr Opin Cell Biol 21:816–824

Song L, Rape M (2010) Regulated degradation of spindle assembly factors by the anaphase-promoting complex. Mol Cell 38:369–382

Soroceanu L, Cobbs C (2011) Is HCMV a tumor promoter? Virus Res 157:193–203

Sotillo R, Hernando E, Diaz-Rodriguez E, Teruya-Feldstein J, Cordon-Cardo C, Lowe SW, Benezra R (2007) Mad2 overexpression promotes aneuploidy and tumorigenesis in mice. Cancer Cell 11:9–23

Stegmuller J, Konishi Y, Huynh MA, Yuan Z, Dibacco S, Bonni A (2006) Cell-intrinsic regulation of axonal morphogenesis by the Cdh1-APC target SnoN. Neuron 50:389–400

Stevens D, Gassmann R, Oegema K, Desai A (2011) Uncoordinated loss of chromatid cohesion is a common outcome of extended metaphase arrest. PLoS ONE 6:e22969

Stewart S, Fang G (2005) Destruction box-dependent degradation of aurora B is mediated by the anaphase-promoting complex/cyclosome and Cdh1. Cancer Res 65:8730–8735

Sugimoto N, Kitabayashi I, Osano S, Tatsumi Y, Yugawa T, Narisawa-Saito M, Matsukage A, Kiyono T, Fujita M (2008) Identification of novel human Cdt1-binding proteins by a proteomics approach: proteolytic regulation by APC/CCdh1. Mol Biol Cell 19:1007–1021

Sun J, Karoulia Z, Wong EY, Ahmed M, Itoh K, Xu PX (2013) The phosphatase-transcription activator EYA1 is targeted by anaphase-promoting complex/Cdh1 for degradation at M-to-G1 transition. Mol Cell Biol 33:927–936

Takahashi A, Imai Y, Yamakoshi K, Kuninaka S, Ohtani N, Yoshimoto S, Hori S, Tachibana M, Anderton E, Takeuchi T, Shinkai Y, Peters G, Saya H, Hara E (2012) DNA damage signaling triggers degradation of histone methyltransferases through APC/C(Cdh1) in senescent cells. Mol Cell 45:123–131

Tan C, Teissier S, Gunaratne J, Quek L, Bellanger S (2015) Stranglehold on the spindle assembly checkpoint: the human papillomavirus E2 protein provokes BUBR1-dependent aneuploidy. Cell Cycle 14:1459–1470

Tanaka K, Nishioka J, Kato K, Nakamura A, Mouri T, Miki C, Kusunoki M, Nobori T (2001) Mitotic checkpoint protein hsMAD2 as a marker predicting liver metastasis of human gastric cancers. Jpn J Cancer Res 92:952–958

Tang Z, Shu H, Oncel D, Chen S, Yu H (2004) Phosphorylation of Cdc20 by Bub1 provides a catalytic mechanism for APC/C inhibition by the spindle checkpoint. Mol Cell 16:387–397

Taniguchi K, Momiyama N, Ueda M, Matsuyama R, Mori R, Fujii Y, Ichikawa Y, Endo I, Togo S, Shimada H (2008) Targeting of CDC20 via small interfering RNA causes enhancement of the cytotoxicity of chemoradiation. Anticancer Res 28:1559–1563

Teng FY, Tang BL (2005) APC/C regulation of axonal growth and synaptic functions in postmitotic neurons: the Liprin-alpha connection. Cell Molecular Life Sci 62:1571–1578

Teodoro J, Heilman D, Parker A, Green M (2004) The viral protein Apoptin associates with the anaphase-promoting complex to induce G2/M arrest and apoptosis in the absence of p53. Genes Dev 18:1952–1957

Thornton BR, Ng TM, Matyskiela ME, Carroll CW, Morgan DO, Toczyski DP (2006) An architectural map of the anaphase-promoting complex. Genes Dev 20:449–460

Tipton AR, Ji W, Sturt-Gillespie B, Bekier ME 2nd, Wang K, Taylor WR, Liu ST (2013) Monopolar spindle 1 (MPS1) kinase promotes production of closed MAD2 (C-MAD2) conformer and assembly of the mitotic checkpoint complex. J Biol Chem 288:35149–35158

Touati SA, Kataria M, Jones AW, Snijders AP, Uhlmann F (2018) Phosphoproteome dynamics during mitotic exit in budding yeast. EMBO J 37:pii:e98745

Tran K, Kamil J, Coen D, Spector D (2010) Inactivation and disassembly of the anaphase-promoting complex during human cytomegalovirus infection is associated with degradation of the APC5 and APC4 subunits and does not require UL97-mediated phosphorylation of Cdh1. J Virol 84:10832–10843

Tran K, Mahr J, Choi J, Teodoro J, Green M, Spector D (2007) Accumulation of substrates of the anaphase-promoting complex (APC) during human cytomegalovirus infection is associated with the phosphorylation of Cdh1 and the dissociation and relocalization of APC subunits. J Virol 82:529–537

Tsai C, Nussinov R (2013) The molecular basis of targeting protein kinases in cancer therapeutics. Semin Cancer Biol 23:235–242

Tu S, Okamoto S, Lipton S, Xu H (2014) Oligomeric Aβ-induced synaptic dysfunction in Alzheimer's disease. Mol Neurodegener 9:48

Tung JJ, Padmanabhan K, Hansen DV, Richter JD, Jackson PK (2007) Translational unmasking of Emi2 directs cytostatic factor arrest in meiosis II. Cell Cycle 6:725–731

Turnell A, Mymryk J (2006) Roles for the coactivators CBP and p300 and the APC/C E3 ubiquitin ligase in E1A-dependent cell transformation. Brit J Cancer 95:555–560

Turnell A, Stewart G, Grand R, Rookes S, Martin A, Yamano H, Elledge S, Gallimore P (2005) The APC/C and CBP/p300 cooperate to regulate transcription and cell-cycle progression. Nature 438:690–695

Uhlmann F, Lottspeich F, Nasmyth K (1999) Sister-chromatid separation at anaphase onset is promoted by cleavage of the cohesin subunit Scc1. Nature 400:37–42

Vanoosthuyse V, Meadows J, van der Sar S, Millar J, Hardwick K (2009) Bub3p facilitates spindle checkpoint silencing in fission yeast. Mol Biol Cell 20:5096–5105

van Roessel P, Elliott DA, Robinson IM, Prokop A, Brand AH (2004) Independent regulation of synaptic size and activity by the anaphase-promoting complex. Cell 119:707–718

van't Veer LJ, Dai H, van de Vijver MJ, He YD, Hart AA, Mao M, Peterse HL, van der Kooy K, Marton MJ, Witteveen AT (2002) Gene expression profiling predicts clinical outcome of breast cancer. Nature 415:530–536

Veas-Pérez de Tudela M, Maestre C, Delgado-Esteban M, Bolaños J, Almeida A (2015) Cdk5-mediated inhibition of APC/C-Cdh1 switches on the cyclin D1-Cdk4-pRb pathway causing aberrant S-phase entry of postmitotic neurons. Sci Rep 5:18180

Visintin R, Prinz S, Amon A (1997) CDC20 and CDH1: a family of substrate-specific activators of APC-dependent proteolysis. Science 278:460–463

Visintin C, Tomson BN, Rahal R, Paulson J, Cohen M, Taunton J, Amon A, Visintin R (2008) APC/C-Cdh1-mediated degradation of the Polo kinase Cdc5 promotes the return of Cdc14 into the nucleolus. Genes Dev 22:79–90

Vodermaier HC, Gieffers C, Maurer-Stroh S, Eisenhaber F, Peters JM (2003) TPR subunits of the anaphase-promoting complex mediate binding to the activator protein CDH1. Curr Biol 13:1459–1468

Vomaske J, Varnum S, Melnychuk R, Smith P, Pasa-Tolic L, Shutthanandan J, Streblow D (2010) HCMV pUS28 initiates pro-migratory signaling via activation of Pyk2 kinase. Herpesviridae 1:2

Wamelink M, Struys E, Jakobs C (2008) The biochemistry, metabolism and inherited defects of the pentose phosphate pathway: a review. J Inherit Metab Dis 31:703–717

Wan L, Tan M, Yang J, Inuzuka H, Dai X, Wu T, Liu J, Shaik S, Chen G, Deng J, Malumbres M, Letai A, Kirschner M, Sun Y, Wei W (2014) APCCdc20 suppresses apoptosis through targeting bim for ubiquitination and destruction. Dev Cell 29:377–391

Wang S, Chen B, Zhu Z, Zhang L, Zeng J, Xu G, Liu G, Xiong D, Luo Q, Huang Z (2018) CDC20 overexpression leads to poor prognosis in solid tumors. Medicine 97:e13832

Wang CX, Fisk BC, Wadehra M, Su H, Braun J (2000) Overexpression of murine fizzy-related (fzr) increases natural killer cell-mediated cell death and suppresses tumor growth. Blood 96:259–263

Wang R, Li KM, Zhou CH, Xue JL, Ji CN, Chen JZ (2011) Cdc20 mediates D-box-dependent degradation of Sp100. Biochem Biophys Res Commun 415:702–706

Wang Q, Moyret-Lalle C, Couzon F, Surbiguet-Clippe C, Saurin J, Lorca T, Navarro C, Puisieux A (2003) Alterations of anaphase-promoting complex genes in human colon cancer cells. Oncogene 22:1486–1490

Wang L, Yin F, Du Y, Du W, Chen B, Zhang Y, Wu K, Ding J, Liu J, Fan D (2009) MAD2 as a key component of mitotic checkpoint: a probable prognostic factor for gastric cancer. Am J Clin Pathol 131:793–801
Wang Y, Zhan Q (2007) Cell cycle-dependent expression of centrosomal ninein-like protein in human cells is regulated by the anaphase-promoting complex. J Biol Chem 282:17712–17719
Wang L, Zhang J, Wan L, Zhou X, Wang Z, Wei W (2015) Targeting Cdc20 as a novel cancer therapeutic strategy. Pharmacol Thera 151:141–151
Wasch R, Robbins JA, Cross FR (2010) The emerging role of APC/CCdh1 in controlling differentiation, genomic stability and tumor suppression. Oncogene 29:1–10
Wei W, Ayad NG, Wan Y, Zhang GJ, Kirschner MW, Kaelin WG Jr (2004) Degradation of the SCF component Skp2 in cell-cycle phase G1 by the anaphase-promoting complex. Nature 428:194–198
Whyte P, Buchkovich K, Horowitz J, Friend S, Raybuck M, Weinberg R, Harlow E (1988) Association between an oncogene and an anti-oncogene: the adenovirus E1A proteins bind to the retinoblastoma gene product. Nature 334:124–129
Wickliffe KE, Williamson A, Meyer HJ, Kelly A, Rape M (2011) K11-linked ubiquitin chains as novel regulators of cell division. Trends Cell Biol 21:656–663
Wiebusch L, Bach M, Uecker R, Hagemeier C (2005) Human cytomegalovirus inactivates the G0/G1-APC/C ubiquitin ligase by Cdh1 dissociation. Cell Cycle 4:1435–1439
Wiebusch L, Hagemeier C (2010) p53- and p21-dependent premature APC/C–Cdh1 activation in G2 is part of the long-term response to genotoxic stress. Oncogene 29:3477–3489
Williams SJ, Abrieu A, Losada A (2017) Bub1 targeting to centromeres is sufficient for Sgo1 recruitment in the absence of kinetochores. Chromosoma 126:279–286
Wolthuis R, Clay-Farrace L, van Zon W, Yekezare M, Koop L, Ogink J, Medema R, Pines J (2008) Cdc20 and Cks direct the spindle checkpoint-independent destruction of cyclin A. Mol Cell 30:290–302
Woodbury EL, Morgan DO (2007) Cdk and APC activities limit the spindle-stabilizing function of Fin1 to anaphase. Nat Cell Biol 9:106–112
Wu F, Dai X, Gan W, Wan L, Li M, Mitsiades N, Wei W, Ding Q, Zhang J (2017) Prostate cancer-associated mutation in SPOP impairs its ability to target Cdc20 for poly-ubiquitination and degradation. Cancer Lett 385:207–214
Wu WJ, Hu KS, Wang DS, Zeng ZL, Zhang DS, Chen DL, Bai L, Xu RH (2013) CDC20 overexpression predicts a poor prognosis for patients with colorectal cancer. J Transl Med 11:142
Wu F, Lin Y, Cui P, Li H, Zhang L, Sun Z, Huang S, Li S, Huang S, Zhao Q, Liu Q (2018) Cdc20/p55 mediates the resistance to docetaxel in castration-resistant prostate cancer in a Bim-dependent manner. Cancer Chemother Pharmacol 81:999–1006
Wu S, Wang W, Kong X, Congdon LM, Yokomori K, Kirschner MW, Rice JC (2010) Dynamic regulation of the PR-Set7 histone methyltransferase is required for normal cell cycle progression. Genes Dev 24:2531–2542
Xie C, Powell C, Yao M, Wu J, Dong Q (2014) Ubiquitin-conjugating enzyme E2C: a potential cancer biomarker. Int J Biochem Cell Biol 47:113–117
Xin Y, Ning S, Zhang L, Cui M (2018) CDC27 facilitates gastric cancer cell proliferation, invasion and metastasis via twist-induced epithelial-mesenchymal transition. Cellular Physiol Biochem 50:501–511
Yamada M, Watanabe K, Mistrik M, Vesela E, Protivankova I, Mailand N, Lee M, Masai H, Lukas J, Bartek J (2013) ATR-Chk1-APC/CCdh1-dependent stabilization of Cdc7-ASK (Dbf4) kinase is required for DNA lesion bypass under replication stress. Genes Dev 27:2459–2472
Yamaguchi M, VanderLinden R, Weissmann F, Qiao R, Dube P, Brown NG, Haselbach D, Zhang W, Sidhu SS, Peters JM, Stark H, Schulman BA (2016) Cryo-EM of mitotic checkpoint complex-bound APC/C reveals reciprocal and conformational regulation of ubiquitin ligation. Mol Cell 63:593–607

Yamamoto Y, Matsuyama H, Chochi Y, Okuda M, Kawauchi S, Inoue R, Furuya T, Oga A, Naito K, Sasaki K (2007) Overexpression of BUBR1 is associated with chromosomal instability in bladder cancer. Cancer Genet Cytogenet 174:42–47

Yang Y, Kim AH, Yamada T, Wu B, Bilimoria PM, Ikeuchi Y, de la Iglesia N, Shen J, Bonni AA (2009) Cdc20-APC ubiquitin signaling pathway regulates presynaptic differentiation. Science 326:575–578

Ye Y, Rape M (2009) Building ubiquitin chains: E2 enzymes at work. Nature Rev Mol Cell Biol 10:755–764

Yi J, Wei X, Li X, Wan L, Dong J, Wang R (2018) A genome-wide comprehensive analysis of alterations in driver genes in non-small-cell lung cancer. Anticancer Drugs 29:10–18

Yu H (2007) Cdc20: a WD40 activator for a cell cycle degradation machine. Mol Cell 27:3–16

Yu H (2002) Regulation of APC-Cdc20 by the spindle checkpoint. Curr Opin Cell Biol 14:706–714

Yu Y, Munger K (2013) Human papillomavirus type 16 E7 oncoprotein inhibits the anaphase promoting complex/cyclosome activity by dysregulating EMI1 expression in mitosis. Virol 446:251–259

Yudkovsky Y, Shteinberg M, Listovsky T, Brandeis M, Hershko A (2000) Phosphorylation of Cdc20/fizzy negatively regulates the mammalian cyclosome/APC in the mitotic checkpoint. Biochem Biophys Res Commun 271:299–304

Zeng X, King R (2012) An APC/C inhibitor stabilizes cyclin B1 by prematurely terminating ubiquitination. Nat Chem Biol 8:383–392

Zeng X, Sigoillot F, Gaur S, Choi S, Pfaff KL, Oh DC, Hathaway N, Dimova N, Cuny GD, King RW (2010) Pharmacologic inhibition of the anaphase-promoting complex induces a spindle checkpoint-dependent mitotic arrest in the absence of spindle damage. Cancer Cell 18:382–395

Zhang S, Chang L, Alfieri C, Zhang Z, Yang J, Maslen S, Skehel M, Barford D (2016) Molecular mechanism of APC/C activation by mitotic phosphorylation. Nature 533:260–264

Zhang J, Cicero S, Wang L, Romito-DiGiacomo R, Yang Y, Herrup K (2008a) Nuclear localization of Cdk5 is a key determinant in the postmitotic state of neurons. Proc Natl Acad Sci USA 105:8772–8777

Zhang SH, Xu AM, Chen XF, Li DH, Sun MP, Wang YJ (2008b) Clinicopathologic significance of mitotic arrest defective protein 2 overexpression in hepatocellular carcinoma. Hum Pathol 39:1827–1834

Zhang J, Li H, Zhou T, Zhou J, Herrup K (2012) Cdk5 levels oscillate during the neuronal cell cycle. J Biol Chem 287:25985–25994

Zhang L, Park CH, Wu J, Kim H, Liu W, Fujita T, Balasubramani M, Schreiber EM, Wang XF, Wan Y (2010) Proteolysis of Rad17 by Cdh1/APC regulates checkpoint termination and recovery from genotoxic stress. EMBO J 29:1726–1737

Zhang J, Wan L, Dai X, Sun Y, Wei W (2014) Functional characterization of anaphase promoting complex/cyclosome (APC/C) E3 ubiquitin ligases in tumorigenesis. Biochim et Biophys Acta 1845:277–293

Zhao WM, Coppinger JA, Seki A, Cheng XL, Yates JR, Fang G (2008) RCS1, a substrate of APC/C, controls the metaphase to anaphase transition. Proc Natl Acad Sci USA 105:13415–13420

Zhao WM, Fang G (2005) Anillin is a substrate of anaphase-promoting complex/cyclosome (APC/C) that controls spatial contractility of myosin during late cytokinesis. J Biol Chem 280:33516–33524

Zhao Y, Tang Q, Ni R, Huang X, Wang Y, Lu C, Shen A, Wang Y, Li C, Yuan Q, Chen H, Cheng C, He S (2013) Early mitotic inhibitor-1, an anaphase-promoting complex/cyclosome inhibitor, can control tumor cell proliferation in hepatocellular carcinoma: correlation with Skp2 stability and degradation of p27Kip1. Hum Path 44:365–373

Zhou Z, He M, Shah A, Wan Y (2016) Insights into APC/C: from cellular function to diseases and therapeutics. Cell Div 11:9

Zhou J, Zhang S, Fu G, He Z, Xu Y, Ye W, Chen Z (2018) Overexpression of APC11 predicts worse survival in lung adenocarcinoma. OncoTargets Ther 11:7125–7132

Zhu Y, Carvey PM, Ling Z (2006) Age-related changes in glutathione and glutathione-related enzymes in rat brain. Brain Res 1090:35–44

Zhuang L, Yang Z, Meng Z (2018) Upregulation of BUB1B, CCNB1, CDC7, CDC20, and MCM3 in tumor tissues predicted worse overall survival and disease-free survival in hepatocellular carcinoma patients. Biomed Res Int 2018:1–8

Zich J, Hardwick KG (2010) Getting down to the phosphorylated 'nuts and bolts' of spindle checkpoint signalling. Trends Biochem Sci 35:18–27

Chapter 19
TRiC/CCT Chaperonin: Structure and Function

Mingliang Jin, Caixuan Liu, Wenyu Han and Yao Cong

Abstract The eukaryotic group II chaperonin TRiC/CCT assists the folding of 10% of cytosolic proteins including many key structural and regulatory proteins. TRiC plays an essential role in maintaining protein homeostasis, and dysfunction of TRiC is closely related to human diseases including cancer and neurodegenerative diseases. TRiC consists of eight paralogous subunits, each of which plays a specific role in the assembly, allosteric cooperativity, and substrate recognition and folding of this complex macromolecular machine. TRiC-mediated substrate folding is regulated through its ATP-driven conformational changes. In recent years, progresses have been made on the structure, subunit arrangement, conformational cycle, and substrate folding of TRiC. Additionally, accumulating evidences also demonstrate the linkage between TRiC oligomer or monomer and diseases. In this review, we focus on the TRiC structure itself, TRiC assisted substrate folding, TRiC and disease, and the potential therapeutic application of TRiC in various diseases.

Keywords Chaperonin · TRiC/CCT · Structure · Function · Substrate folding · ATP-driven conformational changes · Cryo-EM

M. Jin · C. Liu · W. Han · Y. Cong (✉)
National Center for Protein Science Shanghai, State Key Laboratory of Molecular Biology, CAS Center for Excellence in Molecular Cell Science, Shanghai Institute of Biochemistry and Cell Biology, Chinese Academy of Sciences, University of Chinese Academy of Sciences, Shanghai, China
e-mail: cong@sibcb.ac.cn

M. Jin
e-mail: jinmingliang2013@sibcb.ac.cn

C. Liu
e-mail: liucaixuan2015@sibcb.ac.cn

W. Han
e-mail: hanwenyu6@sibcb.ac.cn

Y. Cong
Shanghai Science Research Center, Chinese Academy of Sciences, Shanghai, China

J. R. Harris and J. Marles-Wright (eds.), *Macromolecular Protein Complexes II: Structure and Function*, Subcellular Biochemistry 93,
https://doi.org/10.1007/978-3-030-28151-9_19

Introduction

Many proteins need assistance to fold into their native state and achieve functional activity after release from ribosome into the crowded environment of the cell (Fig. 19.1) (Kim et al. 2013). Correspondingly, cells have evolved a system of molecular chaperones and quality control machineries, often called the "protein homeostasis" network. Proteins from this network function to bind unfolded or misfolded polypeptides and induce their folding, sequester them, or facilitate their degradation (Kim et al. 2013; Willison 2018). The chaperonins, a critical group of molecular chaperones, are large double-ring complexes of 800–1000 kDa built of 7–9 subunits per ring, and each ring provides a central cavity for ATP-driven chaperonin-mediated protein folding cycle (Hartl and Hayer-Hartl 2002; Horwich et al. 2007; Willison 2018).

Chaperonins have been classified into two groups (Horwich and Willison 1993), including group I chaperonins presented in prokaryotes (GroEL/GroES), mitochondria (Hsp60/Hsp10), and chloroplasts (Rubisco subunit binding protein), and group II chaperonins presented in archaea (thermosome, Mm-cpn) and eukaryotes (TCP-1 ring complex, TRiC, or chaperonin containing TCP-1, CCT) (Fig. 19.2a–c). Group I chaperonin consists of a detachable co-chaperonin, while group II chaperonin has a build-in lid in the apical domain that can shut close the chamber (Fig. 19.2a–c). The archaea group II chaperonin has eight or nine subunits per ring, while the eukaryotic group II chaperonin has eight paralogous subunits. The eukaryotic group II chaperonin TRiC/CCT (~960 kDa) comprises of two hetero-oligomeric stacked rings with eight distinct subunits each in a certain arrangement. The paralogous subunits share about 23–35% and 27–39% sequence identity for yeast and bovine TRiC, respectively (Frydman et al. 1992; Ursic and Culbertson 1991). An individual TRiC subunit has three domains: an equatorial domain (E domain) that contains the ATP-binding site and forms the intra- and inter-ring contacts, an apical domain (A domain) that comprises the interaction sites with target proteins, and an intermediate domain (I domain) that connects the other two domains (Fig. 19.2d) (Ditzel et al. 1998; Waldmann et al. 1995; Klumpp et al. 1997).

Coordinated with its ATP-driven conformational cycle, TRiC can assist the folding of approximately 10% of cytosolic proteins, including the cytoskeleton proteins actin and tubulin, which are also redundant substrates of TRiC (Khabirova et al. 2014; Waldmann et al. 1995; Llorca et al. 1999, 2000), cell cycle regulator CDC20 (Camasses et al. 2003), and G-protein signaling related element Gβ (Plimpton et al. 2015). TRiC is also essential for the folding of many proteins involved in oncogenesis, such as p53, VHL tumor suppressor, and STAT3 (Trinidad et al. 2013; Kasembeli et al. 2014; McClellan et al. 2005), and for the folding of retinal developmental factors, including transducin α and PEX7 (Tables 19.1 and 19.2) (Hunziker et al. 2016; van den Brink et al. 2003). Hence, dysfunction of TRiC is closely related to cancer and neurodegenerative diseases

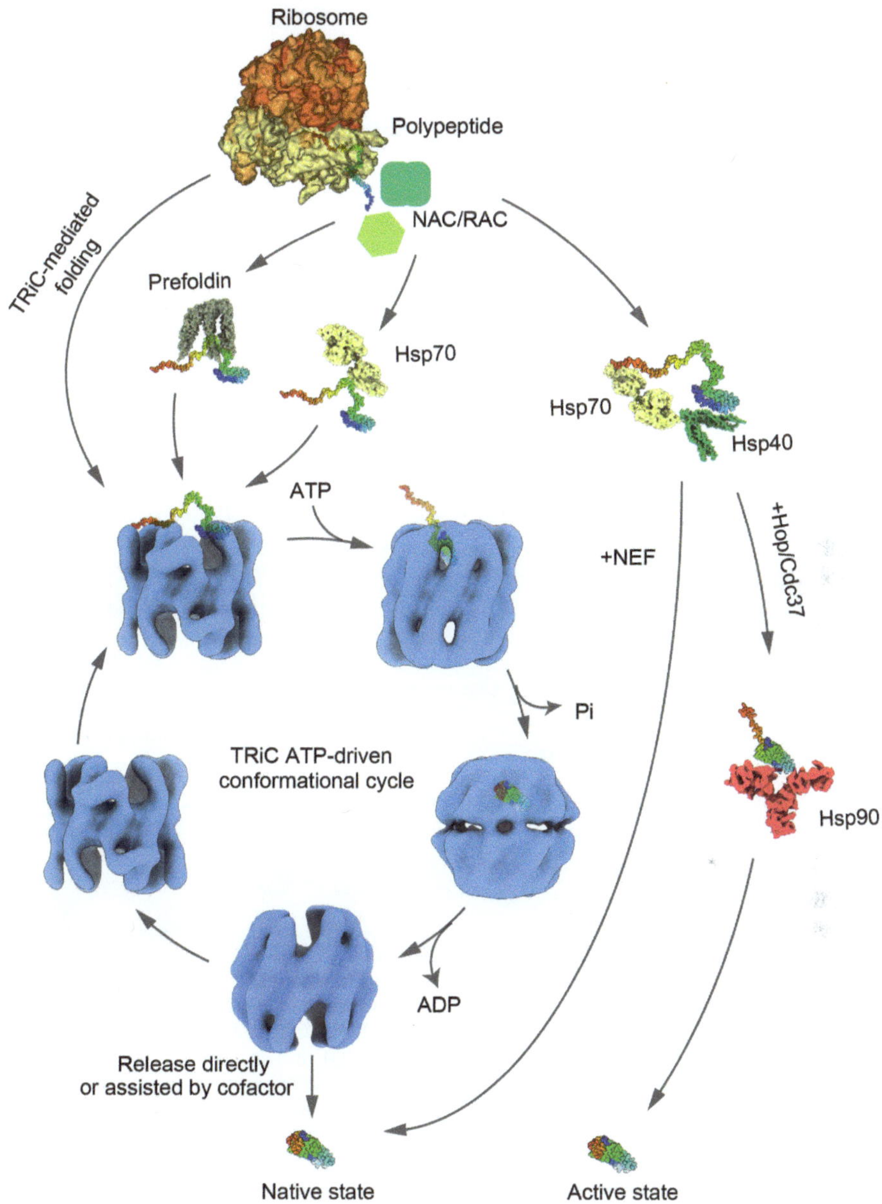

Fig. 19.1 Proposed TRiC-mediated substrate folding pathway. The scheme shows the downstream protein folding pathway from the ribosome. Hsp70 serves as the hub of the chaperone network, which cooperates with Hsp40s and nucleotide exchange factors (NEFs) to fold ~20% of the proteome. TRiC binds and helps fold ~10% of cytosolic proteins directly or with the assistance of cofactors (prefoldin, Hsp70) for efficient substrate delivery. TRiC-mediated substrate folding is regulated through its ATP-driven conformational changes. Then substrate is released from the central cavity of TRiC to achieve its native state, which sometimes needs the help of cofactor, such as Phlp1. In addition, eukaryotes also engage the Hsp90 chaperone system to catalyze the activation of metastable proteins *via* the Hsp70 system and the co-chaperone Hop

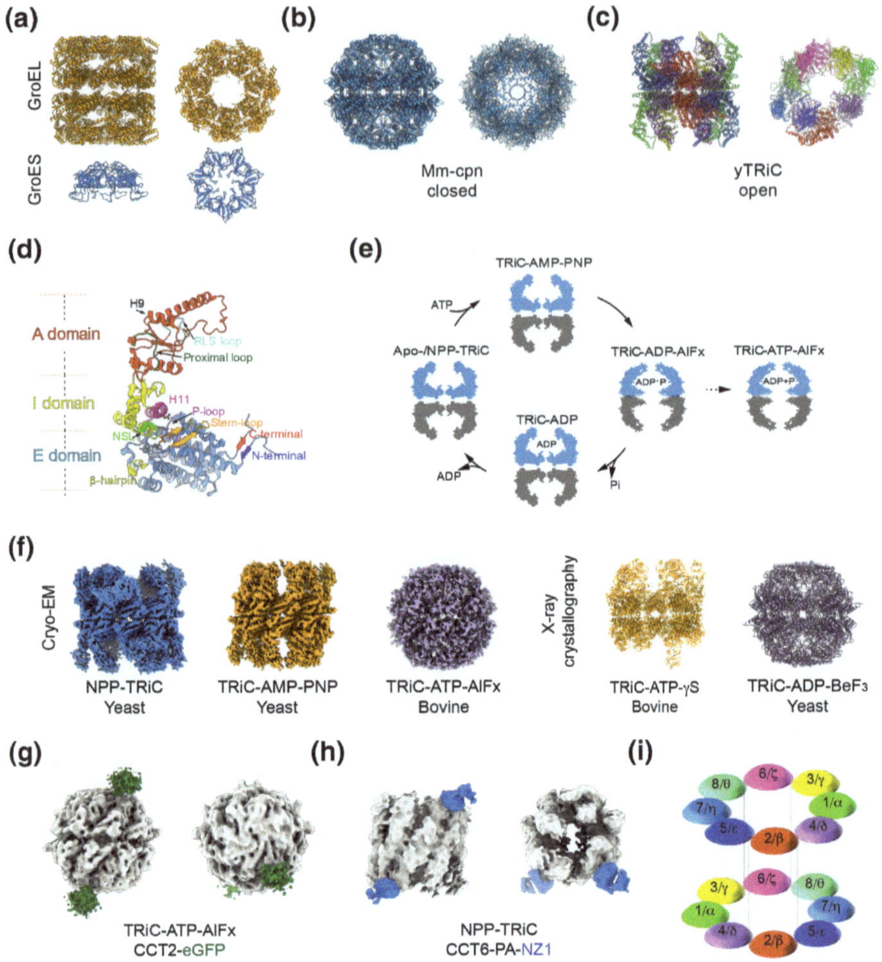

Fig. 19.2 TRiC structural studies and subunit arrangement. **a** The crystal structures of GroEL (PDB ID: 1SS8, orange) and co-chaperone GroES (PDB ID: 1PCQ, cyan). **b** The crystal structure of archaea group II chaperonin Mm-cpn (PDB ID: 3RUQ). **c** The cryo-EM-structure-derived pseudo atomic model of yeast TRiC at open NPP state with different subunits in distinct colors (PDB ID: 5GW4). **d** The three domains of a TRiC subunit, with the key structural elements labeled. **e** A cartoon diagram illustrating TRiC ATPase cycle. **f** Available relative high-resolution structures of TRiC at different nucleotide states, including the cryo-EM structures of NPP-TRiC (EMDB: 9540), TRiC-AMP-PNP (EMDB: 9541), and TRiC-ATP-AlFx (EMDB: 5415), X-ray structures of TRiC-ATP-γS (PDB: 4B2T) and TRiC-ADP-BeF_3 (PDB: 4V8R). **g** Illustration of using YISEL method to identify the location of CCT2 subunit in TRiC complex. The map of the closed yeast TRiC is in grey, with eGFP (dark green) tagged to CCT2 subunits (one in each ring). **h** Illustration of using YISPANL method to identify the location of CCT6 subunit in TRiC complex. The map of open yeast TRiC is in grey, with NZ-1 fab (sky blue) tagged to CCT6 subunits. **i** Determined TRiC subunit arrangement

Table 19.1 Involvement of TRiC in diseases

Disease		TRiC or subunits	Biological behavior	Substrate	References
Neuro-related diseases	Huntington's disease	CCT1, CCT5	Inhibit aggregation	Huntingtin	Shahmoradian et al. (2013), Darrow et al. (2015), Sontag et al. (2013)
	Parkinson's disease	CCT2, CCT3, CCT6	Inhibit amyloid fiber assembly of α-synuclein A53T	α-synuclein A53T	Sot et al. (2017)
	Alzheimer's disease	TRiC	Inhibit Aβ toxicity	Aβ peptide	Khabirova et al. (2014)
	Cockayne syndrome	CCT4, CCT5	Interaction	CSA	Pines et al. (2018)
	Hereditary sensory neuropathy	CCT4	C450Y (Mutation)		Lee et al. (2003)
	Mutilating sensory neuropathy	CCT5	H147R (Mutation)		Bouhouche et al. (2006)
	Neuronal Apoptosis	CCT8	up-regulated		Wu et al. (2015b)
Eye-related diseases	Leber Congenital amaurosis	CCT2	T400P, R516H (Mutation)		Minegishi et al. (2016)
	Achromatopsia	TRiC	Folding	Transducin α	Kohl et al. (2002), Koulikovska et al. (2005), Farr et al. (1997)
	Refsum disease	TRiC	Folding	Peroxisomal targeting signal 2 receptor protein	Siegers et al. (2003), van den Brink et al. (2003), Erdo et al. (2004)
	Usher syndrome	TRiC	Folding	Myosin	Srikakulam and Winkelmann (1999), Adato et al. (1997), Erdo et al. (2004)

(continued)

Table 19.1 (continued)

Disease		TRiC or subunits	Biological behavior	Substrate	References
Cancer	Breast cancer	CCT1, CCT2, CCT5	Up-regulated	p53	Guest et al. (2015), Ooe et al. (2007)
	Colorectal cancer	CCT2, CCT6	Up-regulated		Qian-Lin et al. (2010), Coghlin et al. (2006)
	Glioma	CCT8	Up-regulated		Qiu et al. (2015)
	Hepatocellular cancer	CCT3, CCT8	Up-regulated		Zhang et al. (2016), Wei et al. (2016), Cui et al. (2015), Huang et al. (2014), Yokota et al. (2001)
	Lymphoma	CCT1, CCT8	Up-regulated		Yin et al. (2016), Xudong Jiang et al. (2015)
	Small cell lung cancer	CCT2	Up-regulated		Carr et al. (2017)
	SC/ASC, AC	CCT2	Up-regulated		Zou et al. (2013)
	Uterine Cancer	CCT2	Interaction	β-tubulin	Lin et al. (2009)

such as Huntington's, Alzheimer's, and Parkinson's diseases (Chen et al. 2014; Bassiouni et al. 2016).

Our knowledge of the TRiC complex has been summarized in a large number of superb reviews (Willison 2018a, b; Balchin et al. 2016; Lopez et al. 2015; Broadley and Hartl 2009; Dunn et al. 2001; Skjaerven et al. 2015; Fernandez-Fernandez et al. 2016). Our current review will mainly focus on the recent progresses on the structure and function of TRiC complex and TRiC related diseases.

TRiC Molecular Structure

Structural Studies of TRiC and Its ATP-Driven Conformational Cycle

Purification of such a large eukaryotic complex as TRiC is usually challenging. There are several established purification protocols for direct purification of TRiC from mammalian tissues or cells (Lewis et al. 1992; Frydman et al. 1992; Knee et al. 2013; Gao et al. 1992), as well as from yeast (Pappenberger et al. 2006; Zang et al. 2016). In addition, a method was also reported to reconstitute human TRiC by co-expressing its eight distinct subunits in BHK-21 cells (Machida et al. 2012).

Table 19.2 Involvement of TRiC in biological system/process

Biological system/ Process	Organism	Related subunits	Substrate	References
MCC assembly	Human	TRiC	CDC20	Kaisari et al. (2017)
Autophagy	Human	CCT2, CCT5, CCT7		Pavel et al. (2016)
Sarcomere Z-disk	*Dario rerio*	CCT5 (Mutation G422V)		Berger et al. (2018)
Invasion	*C. elegans*	CCT5		Matus et al. (2010)
Cell polarity	Yeast	TRiC		Liu et al. (2010)
Morphogenesis	Yeast	CCT8		Rademacher et al. (1998)
Spermatogenesis	*Flatworm*	CCT1, CCT3, CCT4, CCT8		Counts et al. (2017)
Fibroblast motility	Human	CCT7		Satish et al. (2010)
Proteostatic control of telomerase function	Human	TRiC	TCAB1	Freund et al. (2014)
Cancer signaling	Human	CCT3	STAT3	Kasembeli et al. (2014)
Apoptosis/ autophagy/cell motility	Human	TRiC		Fontanella et al. (2010)
Myeloid leukemia	Human	CCT6, CCT8	AML1-ETO	Roh et al. (2016a, b)
Atherosclerosis development	Human	TRiC	LOX-1	Bakthavatsalam et al. (2014)
Mediating cellular responses to oxygen	Human	TRiC	PHD3	Masson et al. (2004)
Chromatin remodeling	Human	TRiC	HDACs	Guenther et al. (2002)
Cilia biogenesis	*Tetrahymena*	CCT1, CCT3, CCT4, CCT7		Soues et al. (2003), Seixas et al. (2010), Cyrne et al. (1996)
Proper formation of microvilli in intestinal cells	*C. elegans*	TRiC		Saegusa et al. (2014)
Lifespan extension	*C. elegans*	CCT8		Noormohammadi et al. (2016)
Fragile X mental retardation	*Drosophila*	CCT3, CCT4, CCT7		Monzo et al. (2010)
Stem cell function	*Arabidopsis*	CCT8		Xu et al. (2011)
TFIID assembly	Human	TRiC	TAF5	Antonova et al. (2018)

Chaperonins are molecular machines that assist protein folding by undergoing ATP-dependent conformational changes that are coordinated in time and space owing to complex allosteric regulation (Fig. 19.2e). Since the early days, both negative staining-electron microscopy (NS-EM) and cryo-electron microscopy (cryo-EM) have been used to investigate the complex structure itself and the ATP-induced conformational changes of TRiC (Lewis et al. 1992; Frydman et al. 1992; Gao et al. 1992; Llorca et al. 1998). Moreover, small angle X-ray scattering (SAXS) analysis also suggested that ATP hydrolysis is needed for the lid closure of TRiC (Meyer et al. 2003).

Furthermore, reference-free 2D analysis of cryo-EM data showed the asymmetric nature among the eight distinct subunits of bovine TRiC (Rivenzon-Segal et al. 2005). A cryo-EM structural study revealed that the inter-domain motions lead to lid closure in bovine TRiC, which is radically different from those of group I chaperonins despite their overall structural similarity (Booth et al. 2008). Our previous asymmetric cryo-EM 3D reconstructions on TRiC in different nucleotide states demonstrated that TRiC changes its intra- and inter-ring subunit interaction pattern during the ATPase cycle, with all the subunits in each ring highly asymmetric in the apo state, whereas all the nucleotide-containing states tend to be more symmetrical. Moreover, the eight subunits in each ring form four subunits pairs, including CCT2-CCT4, CCT1-CCT3, CCT6-CCT8, and CCT7-CCT5; and there exists a two fold axis between its two rings, resulting in two homotypic subunit interactions across the rings (Cong et al. 2010, 2012).

We recently reconstructed two cryo-EM structures of yeast TRiC in a newly captured nucleotide partially preloaded (NPP) state (termed NPP-TRiC) and in the ATP-bound state at ~4.6 Å resolution (Zang et al. 2016). Our NPP-TRiC map showed that CCT2 subunit-pair forms an unexpected Z-shaped feature (red subunit pair, Fig. 19.2c), which conformation is different from that of bovine TRiC in the apo state, whose CCT2 subunit is intrinsically dynamic with its A domain mostly missing (Cong et al. 2010, 2012). ATP binding could induce a dramatic conformational change of CCT2 subunit, which could be either unbent or stabilized for yeast or bovine TRiC, respectively, constructing a typical inward tilting conformation as the other subunits. Strikingly, for yeast TRiC, CCT3, CCT6, and CCT8 subunits have preloaded nucleotide from the environment of the yeast cell (Zang et al. 2016). Overall, TRiC has evolved into a complex that is structurally divided into two sides and reveals a staggered ATP binding mechanism. Interestingly, there are extra densities blocking the two chambers of yeast TRiC in both NPP-TRiC and TRiC-AMP-PNP maps as well as in the bovine apo-TRiC and TRiC-AMP-PNP maps, which may be correlated to the N- and C- termini and might be involved in the complex assembly and allosteric regulation (Cong et al. 2012; Zang et al. 2016).

There are biochemical investigations and computational analyses on the subunit specificity of TRiC in ATP usage. It has also been suggested the ATP binding affinity varies among the eight distinct subunits of TRiC. A recent report has shown that within one ring of bovine TRiC, only four subunits (CCT1, CCT2, CCT4, and CCT5) bind ATP at its physiological concentration (Reissmann et al. 2012). Additionally, it has also been reported that the most dramatic phenotypic effects are associated with mutations in the nucleotide-binding pockets of CCT4, CCT2, CCT5, and CCT7 of yeast TRiC; however, similar mutations do not markedly affect CCT3, CCT6, and CCT8 (Amit et al. 2010), which is consistent with our observation that there are preloaded nucleotides in these three subunits (CCT3/6/8) (Zang et al. 2016). Moreover, equivalent mutations in the eight subunits of TRiC at the positions involved in ATP hydrolysis led to dramatically different phenotypes, indicating each subunit may have different function and play distinct role in TRiC ATPase hydrolysis (Amit et al. 2010). By measuring emissions from chaperonin-bound fluorescent nucleotides, it was suggested that TRiC either hydrolyzes no ATP or hydrolyzes about four ATPs in each of the two rings at the same time (Jiang et al. 2011). By measuring the amount of phosphate generated by ATP hydrolysis as a function of time, it has been indicated that TRiC is in an equilibrium with a large number of conformational states, and that "conformational selection" by ATP takes place before hydrolysis (Korobko et al. 2016). Diffracted X-ray tracking experiment by characterizing *Chaetomium thermophilum* TRiC variants containing ATPase-deficient subunits showed that ATP-induced motion in a ring is in an asymmetric manner (Yamamoto et al. 2017). All the above studies show that TRiC subunits consume ATP in an asymmetric manner, coordinated in time and space owing to complex allosteric regulations, and each subunit has its own functional specificity. However, the underling structural basis of this specificity and the evolutional benefit need further investigation.

TRiC Subunit Arrangement

To better understand the subunit specificity in ATP usage and allosteric regulation, as well as in substrate recognition and folding, it is essential to address the long-standing question regarding the subunit arrangement of the TRiC complex. In the early studies, TRiC micro-complexes (comprised of subsets of the constitutively expressed TRiC subunits) were examined and implied the unique topology of TRiC subunits (Liou and Willison 1997). In the past decade, extensive efforts have been made to determine the subunit arrangement of this complex molecular machine (Fig. 19.2f) (Dekker et al. 2011; Munoz et al. 2011; Cong et al. 2010; Liou and Willison 1997; Kalisman et al. 2012). Recently, the arrangement has been better determined by chemical cross-linking and mass spectrometry (XL-MS) (Kalisman et al. 2012; Leitner et al. 2012). Moreover, computational approaches have been used to allocate the subunits in the X-ray structures of yeast TRiC in the presence of ATP-BeFx and bovine TRiC in the presence of ATP-γS, which confirmed the XL-MS result (Kalisman et al. 2013). However, due to the structural similarity among TRiC subunits and resolution limitation, the conclusions have been inconsistent across studies (Dekker et al. 2011; Munoz et al. 2011; Cong et al. 2010; Liou and Willison 1997; Kalisman et al. 2012). Recently, combining the cryo-EM reconstruction of yeast TRiC at the open NPP state with our recently developed yeast internal-subunit eGFP labeling (YISEL) strategy and yeast inner-subunit

PA–NZ-1 labeling (YISPANL) strategy, we unambiguously identified the subunit locations in the open-state TRiC (Fig. 19.2g–i) (Zang et al. 2018; Wang et al. 2018), and the subunit ordering determined in this way is in agreement with the previous XL-MS result (Kalisman et al. 2012; Leitner et al. 2012).

Structural Study on TRiC with Substrate

TRiC plays an important role in the maintenance of protein homeostasis. Therefore, capturing high-resolution structural information of TRiC in complex with substrate is of great significance for our understanding of the mechanism of how TRiC recognizes and folds substrate proteins through its ATP-driven conformational cycle. However, due to the relative low binding efficiency between TRiC and the pretreated unfolded substrates (by applying urea or GuHCl in vitro), and the potential conformational heterogeneity of bound substrate, so far the available structural information of TRiC-substrate remains limited.

Along this line, a number of efforts have been made to investigate the TRiC-substrate complex mainly using cryo-EM. Early cryo-EM studies of TRiC in complex with denatured actin (Llorca et al. 1999) or denatured tubulin (Llorca et al. 2000) in the open state indicated that actin binds below the helical protrusions with multiple TRiC subunits, while tubulin interacts directly with the helical protrusion of two TRiC subunits. Subsequently, structural and functional analyses of Huntingtin-TRiC complex uncovered the A domains of CCT1 and CCT4 involve in the inhibition of the aggregation process of Huntingtin, the causative agent of Huntington's disease (Tam et al. 2006; Darrow et al. 2015; Sontag et al. 2013; Shahmoradian et al. 2013). Recently, the cryo-EM structure of cancer-related protein AML1-ETO in complex with human TRiC revealed the subunit-specific interaction, primarily through its DNA binding domain with two TRiC subunits (Fig. 19.3a) (Roh et al. 2016a, b).

There are few clear definitions from the common sequence or structural features of TRiC substrates except for WD40-domain containing proteins, which have been described to be an important family of TRiC substrates (Lopez et al. 2015; Willison 2018b). The WD40-domain containing proteins usually consist of a 7-bladed-β-propeller and could coordinate multi-protein complex assemblies, where the repeating units serve as a rigid scaffold for protein interactions. A number of TRiC substrates have been found to have the WD40 β-propeller features such as TCAB1, CDC20, and Gβ (Freund et al. 2014; Kaisari et al. 2017; Plimpton et al. 2015; Miyata et al. 2014). To overcome the low binding efficiency problem, researchers purified the TRiC substrate, human Gβ, from insect cells. In this way, the potentially partially folded human Gβ may retrieve the intact insect TRiC from the cells. Then the cryo-EM structures of Gβ-TRiC and PhLP1-Gβ-TRiC complexes were reconstructed, revealing a mechanism for Gβ folding and Gβγ dimer assembly assisted by TRiC (Plimpton et al. 2015). Recently, it has been reported that TRiC

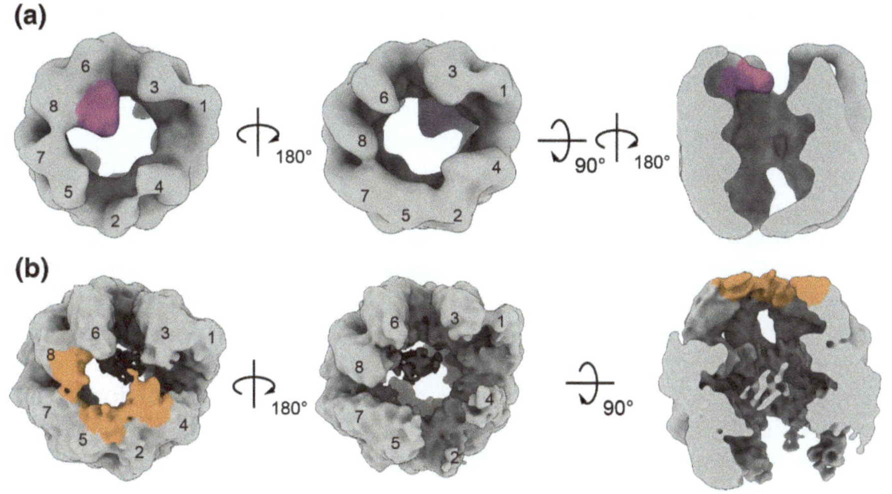

Fig. 19.3 Cryo-EM structures of TRiC in complex with substrate. **a** Cryo-EM structure of TRiC-AML1-175 (EMDB: 6227, the map is rendered according to the recommended threshold). TRiC density is colored in grey and AML1-175 in purple. Subunits are also labeled. **b** Cryo-EM structure of TRiC-actin (EMDB: 0017, the map is rendered according to the recommended threshold). TRiC density is colored in grey and actin density in orange

could capture TAF5, assist its WD40 domain folding and subsequent handover to TAF6-TAF9 and ultimately holo-TFIID formation (Antonova et al. 2018).

There are also X-ray structures of actin-TRiC in the presence of ADP-BeFx in the closed state at 3.8 Å resolution, revealing residual density in one of the TRiC cavities as unfolded actin interacting with multiple TRiC subunits (Dekker et al. 2011), and of tubulin-TRiC with ATP-γS in the open state at 5.5 Å resolution, which also allocated part of the substrate density contacting with three TRiC subunits (Munoz et al. 2011). A recent study combining cryo-EM, fluorescence correlation spectroscopy (FCS), and hydrogen/deuterium exchange coupled to mass spectrometry (H/DX-MS) determined the conformational progression of actin as it is folded by bovine TRiC. This analysis revealved that actin binds to five TRiC subunits (CCT2/4/8/7/5), and captured the dynamic interaction locations through the ATP-driven TRiC-mediated actin folding cycle (Fig. 19.3b) (Balchin et al. 2018).

Key Structural Elements of Group II Chaperonin

To-date, many key structural elements of group II chaperonins involved in nucleotide consumption, subunit coordination, and substrate recognition have been delineated (Fig. 19.2d). The GDGTT motif, called the p-loop and located in the

E domain of individual subunits of group II chaperonins, involves ATP binding and is strictly conserved in all chaperonins (Amit et al. 2010). The nucleotide-sensing loop (NSL) located in the I domain monitors the presence of the γ-phosphate of ATP (Pereira et al. 2012; Zang et al. 2016). The β-hairpin in the E domain of almost every archaea and eukaryotic group II chaperonin subunit has been suggested to play a role covering the entrance of the nucleotide pocket (Pereira et al. 2012; Zang et al. 2016). Between two adjacent subunits of group II chaperonins within a ring, there are contacts linking their E domains *via* a mixed 4-stranded β-sheet involving the N- and C- termini of one subunit and the stem-loop of the neighboring subunit that communicates with its nucleotide pocket (Zhang et al. 2010; Pereira et al. 2010; Ditzel et al. 1998). In addition, the proximal loop (PL) in the A domain recognizes substrate together with its neighboring α-helix 9 (H9), while the release loop for substrate (RLS) is responsible for evicting bound substrate into the TRiC chamber during TRiC ring closure (Lopez et al. 2015; Joachimiak et al. 2014). Still, it needs further examination on how these structural elements coordinate together to perform allosteric conformational transitions in ring closure and substrate folding.

Structure of TRiC with Co-chaperone

Although TRiC does not need a co-chaperone to cover the central cavity forming a closed ring, it has co-chaperone or cofactors to deliver unfolded/partially folded substrates, or to accelerate the substrate folding process (Fig. 19.1). There are several cofactors for TRiC: (1) prefoldin, a hexameric chaperone, binds specifically to chaperonin and transfers target protein (mainly actin) to it (Vainberg et al. 1998); (2) phosducin-like protein, including three known families of phosducin-like proteins participating in Gβ-protein signaling, which have also been implicated in other process, such as the folding of actin and tubulin (Plimpton et al. 2015; Stirling et al. 2006; Humrich et al. 2005); (3) Hsp70, another heat-shock protein that was found to be involved in substrate delivery to TRiC (Cuellar et al. 2008; Melville et al. 2003). Therefore, understanding the interaction mechanism of TRiC with cofactors may provide a thorough picture of TRiC-assisted substrate folding. The cryo-EM map of the TRiC-PFD complex indicates that PFD (human) binds to both rings of TRiC (bovine) in a unique configuration, suggesting a substrate handoff mechanism from PFD to TRiC (Martin-Benito et al. 2002). The cryo-EM structure of bovine TRiC-Hsc70 also supports the notion that Hsc70 may deliver the unfolded proteins to the substrate receiving region of TRiC, and it has been postulated that in eukarya, TRiC and Hsc70 evolved as a concerted action that makes the folding task more efficient (Cuellar et al. 2008). Recently, a cryo-EM structure of insect TRiC in complex with the co-chaperone phosducin-like protein 1 (PhLP1, human) and substrate Gβ (human) was reported. This study indicated that PhLP1 binding could stabilize the Gβ fold, disrupting interactions with TRiC and releasing a PhLP1-Gβ dimer for assembly with Gγ (Plimpton et al. 2015). Still, there are remaining obscure questions regarding the role that co-chaperone plays in cooperating with

TRiC that need to be addressed, such as in what situation co-chaperone binds substrate, and when and how co-chaperone delivers substrate to TRiC or helps release substrate from TRiC.

Cryo-ET of TRiC with Substrate

It is exciting to directly visualize locations of TRiC in the cell and to understand how TRiC assists substrate folding in situ. In recent years, cryo-electron tomography (cryo-ET) has emerged as a powerful tool for direct visualization of the molecular organization of cellular landscapes, with the potential to reach near-atomic resolution (Beck and Baumeister 2016; Mosalaganti et al. 2018; Guo et al. 2018). Cryo-ET has been used to visualize the interacting location of TRiC on the mutant huntingtin exon fibril, containing an expanded polyglutamine tract with 51 residues (mhttQ51) and resolved a 3D structure of TRiC interacting with mhttQ51, which provided a structural description for TRiC's inhibition of mhttQ51 aggregation in vitro (Shahmoradian et al. 2013). Recently, using cryo-ET and sub-tomogram averaging, the structure and cellular interactions of poly-GA aggregates (poly-Gly-Ala aggregates resulting from aberrant translation of an expanded GGGGCC repeat in C9orf72) within intact neurons was visualized, in which TRiC molecules were also found to locate between poly-GA ribbons (Guo et al. 2018). These studies pave the way for further understanding of the function of TRiC in substrate folding or its role in hindering substrate protein aggregation.

Structure of Mutated TRiC and Homo-Oligomer of TRiC Single Subunit

TRiC consists of eight distinct subunits which bring extra complexity to its architecture and functional specificity. Each subunit of TRiC may play a distinct role in the assembly, allosteric coordination, and substrate processing of TRiC complex (Spiess et al. 2004). Some TRiC subunits are able to form a biologically active homo-oligomer when respectively expressed in *Escherichia coli*. It has been shown that human CCT4 and CCT5 could form biologically active homo-oligomer independently, and a cryo-EM reconstruction of CCT5 homo-oligomeric ring suggests an overall TRiC-like structure (Sergeeva et al. 2013). Intriguingly, similar to WT TRiC, the synthetic CCT5 homo-oligomer complex was found to be able to cap mHTT fibrils at their tips and inhibit mHTT aggregation, revealed by cryo-ET analysis (Darrow et al. 2015), and it is also able to promote the disassembly of the mitotic checkpoint complex (MCC) in the same fashion as TRiC (Kaisari et al. 2017). Moreover, two mutations, C450Y of CCT4 and H147R of CCT5, were identified in hereditary sensory neuropathies (HSNs) (Bouhouche et al. 2006;

Lee et al. 2003). NS-EM study revealed that CCT4-C450Y forms few ring-shaped species, whereas WT CCT4, CCT5-H147R, and WT CCT5 can form similar ring structures (Sergeeva et al. 2014). Recently, the X-ray structures of human CCT5 and CCT5-H147R recombinant complexes were resolved, respectively, which demonstrate the mutation may impact the allosteric cooperativity of TRiC (Pereira et al. 2017).

Functions of TRiC

Substrate Folding Assisted by TRiC

Although the mechanism of TRiC-substrate recognition and binding need further investigation, it has been demonstrated that substrate motifs are recognized by a cleft formed between the H9 and PL in the A domains of TRiC individual subunits (Fig. 19.2d), and this cleft contains subunit-specific patterns of polar and hydrophobic residues (Joachimiak et al. 2014). Thus, the diversification of TRiC subunits, especially in the A domain regions, enables TRiC to recognize and bind diverse substrates (Yamamoto et al. 2017). Considering that there are usually multiple TRiC subunits involving in substrate binding (Lopez et al. 2015), this explains why TRiC can assist up to 10% of cytosolic proteins to fold in the cell. Moreover, TRiC plays an important role in the folding of newly synthesized proteins (Frydman et al. 1994; Yam et al. 2008), but it can also prevent the aggregation of proteins with polyglutamine regions (Kitamura et al. 2006; Tam et al. 2006), thus potentially contributing to the suppression of mis-folding in diseases such as Huntington's, Parkinson's, and Alzheimer's diseases. Still, for the folding state changes of substrates along with the ATP-driven ring closure process, TRiC remains to be further explored.

TRiC and Diseases

TRiC and Neurodegenerative Diseases

Many neurodegenerative disorders, such as Alzheimer's, Parkinson's, and polyglutamine diseases, as well as amyotrophic lateral sclerosis (ALS), are characterized by conformational changes in proteins that result in misfolding, aggregation and intra- or extra-neuronal accumulation of amyloid fibrils, molecular chaperones including TRiC provide the first line of defense (Table 19.1) (Muchowski and Wacker 2005; Roh et al. 2016b). Recent studies have investigated the role of TRiC in neurodegenerative diseases (Voisine et al. 2010; Liebman and Meredith 2010). It has been biochemically shown that TRiC can inhibit several aggregation variants both in vitro and in vivo (Kitamura et al. 2006; Tam et al. 2006; Behrends et al. 2006) and in cooperation with the Hsp70 system (Behrends et al. 2006). Moreover, a cryo-ET analysis provided a structural description for TRiC's inhibition of mhttQ51 aggregation in vitro (Shahmoradian et al. 2013).

Moreover, genetic defects in CSA, a 7-bladed WD40 protein, mostly give rise to Cockayne syndrome, which is characterized by premature aging, progressive mental and sensorial retardation, microcephaly, severe growth failure, and cutaneous photosensitivity. TRiC has been shown to interact with CSA through its WD40 domain, thereby regulating CSA stability (Pines et al. 2018). In addition, several individual subunits of TRiC have been reported to be related to hereditary sensory neuropathy, neuronal apoptosis or neurological defects (Lee et al. 2003; Bouhouche et al. 2006; Wu et al. 2015b; Matsuda and Mishina 2004) (described elsewhere in this chapter).

TRiC and Eye Related Diseases

It had been reported that several retinal developmental factors are the client of TRiC, such as myosin, the cause of Usher syndrome (Bonné-Tamir et al. 1994; Adato et al. 1997), transducin α, the cause of Achromatopsia (Kohl et al. 2002), and peroxisomal targeting signal 2 receptor protein, the cause of Refsum disease (Srikakulam and Winkelmann 1999; Farr et al. 1997).

Several studies suggested an essential role of TRiC in ciliogenesis. The most abundant clients of TRiC are represented by cytoskeletal proteins from the actin and tubulin families (Gao et al. 1992; Grantham et al. 2006; Hanafy et al. 2004; Vinh and Drubin 1994; Stemp et al. 2005; Chen et al. 1994). It has been proven that TRiC plays an essential role in the biogenesis of vertebrate photoreceptor sensory cilia (Sinha et al. 2014). Additionally, suppressing TRiC activity in mouse photoreceptors results in the malformation of the retinal outer segment, a cellular compartment responsible for light detection, and triggers rapid retinal degeneration (Posokhova et al. 2011). The activity of TRiC is also essential for the assembly of the BBSome complex (Zhang et al. 2012), a complex thought to control trafficking of molecules into the cilium (Nachury et al. 2007; Loktev et al. 2008). As a result, knockdown of TRiC in zebrafish results in a Bardet-Biedl-like syndrome, a human genetic disorder resulting in obesity, retinal degeneration, polydactyly, and nephropathy (Seo et al. 2010).

Gβ, as part of G$\beta\gamma$ (G protein $\beta\gamma$) and RGS9-Gβ_5 heterodimers, plays an essential role in photo-transduction. Both are obligate dimers that rely on TRiC and its co-chaperone PhLP1 to form a complex from their nascent polypeptides (Tracy et al. 2015; Willardson and Howlett 2007). Moreover, two mutations T400P and R516H in CCT2 subunit has been reported to be related to Leber congenital amaurosis (LCA), a hereditary early-onset retinal dystrophy that is accompanied by severe macular degeneration (Minegishi et al. 2016).

TRiC and Cancer

TRiC has been found to be related to oncogenesis, probably through the interaction of TRiC with oncogenic clients which modulate the cancer cells (Roh et al. 2015). Studies have indicated that both TRiC and Hsp70 are required for correct folding of newly translated von Hippel-Lindau (VHL), which is coupled to assembly of a ternary complex with its partner proteins elongin BC (McClellan et al. 2005; Feldman et al. 2003; Melville et al. 2003; Feldman et al. 1999). Loss of VHL function is associated with a number of inherited and spontaneous tumors (Kaelin 2002). Moreover, TRiC also contributes to the folding and function of STAT3, a transcription factor for the transmission of peptide hormone signals from receptors on the extracellular membrane to nucleus, which is up-regulated in many pathological conditions, including cancer and inflammatory diseases (Kasembeli et al. 2014).

p53 is a transcription factor that mediates tumor suppressor responses. Correct folding of the p53 protein is essential for these activities, and point mutations that induce conformational instability of p53 are frequently found in cancers. It has been shown that folding of WT p53 is promoted by the interaction with TRiC. Depletion of TRiC in cells results in the accumulation of misfolded p53, leading to a reduction in p53-dependent gene expression. Although the most straightforward model for the role of TRiC is to assist in the correct folding of newly synthesized p53 (Malcikova et al. 2010), it is also possible that TRiC binding to p53 prevents its aggregation (Trinidad et al. 2013).

TRiC expression is closely correlated with cell growth and is markedly enhanced at early S phase of the cell cycle in mouse and human cultured cells (Yokota et al. 1999). Furthermore, specific TRiC subunits have been found to be up-regulated in the development of diverse cancers (Table 19.1). For instance, CCT1 and CCT2 have been reported to be essential for survival and proliferation of breast cancer (Guest et al. 2015). The expression levels of CCT3 mRNA and protein are up-regulated in hepatocellular carcinoma cells (HCC) cell lines, and over expression of CCT3 in the nuclei of cancerous cells is associated with HCC progression. CCT3 may be a target that affects the activation of STAT3 in HCC (Cui et al. 2015).

CCT8 has been reported to be dysregulated in several tumor tissues. In glioma, high expression of CCT8 is significantly associated with shorter overall survival. CCT8 can regulate the proliferation and invasion of glioblastomas (Qiu et al. 2015). Similarly, CCT8 expression is increased in hepatocellular carcinoma specimens, and knockdown of CCT8 can inhibit the proliferation of HCC (Huang et al. 2014). Also, CCT8 is highly expressed in proliferating germinal center cells and in progressive lymphomas. High expression of CCT8 is significantly associated with shorter overall survival in patients with diffuse large B-cell lymphoma (Yin et al. 2016). Further study is needed to address the following questions: whether the TRiC subunits in cancer act as a monomer, a homo-oligomer, or a TRiC complex; if it functions as a monomer, does it work as a chaperone with ATPase activity or just transiently interacts with a certain substrate? Hence, elucidating the roles of TRiC in oncogenesis may provide new strategies for related cancer therapy.

Potential Therapeutic Applications of TRiC

TRiC Subunit Can Serve as a Biomarker for Related Disease

Recently, accumulating evidences have suggested the relationship between specific TRiC subunits and disease, and several TRiC subunits have been found to be over-expressed in certain diseases (Table 19.1). Therefore, over-expressed specific TRiC subunit could be used as a clinical marker for related disease.

The expression of CCT5 along with RGS3 and YKT6 genes are up-regulated in p53-mutated breast tumors and are associated with a resistance to docetaxel, and this might be clinically useful in identifying the subset of breast cancer patients who may or may not benefit from docetaxel treatment. Treatment of human breast cancer cell line MCF-7 with siRNA specific for CCT5 resulted in a significant enhancement of docetaxel-induced apoptosis and this may provide new insights into the molecular mechanism of resistance to docetaxel (Ooe et al. 2007). It has been reported that CCT5 also shows higher expression in non-small cell lung cancer (NSCLC) tissues and could be used as a biomarker in the diagnosis of NSCLC in an early stage (Gao et al. 2017). Moreover, the expression of CCT2 has been shown to be elevated in urine of patients with type 2 diabetic mellitus in the hyperfiltration stage. Hyperfiltration in early diabetic nephropathy (DN) may be detectable by measuring urine CCT2, which can be a novel and valuable biomarker for clinical evaluation (Wu et al. 2015a). Increased CCT7 has been observed in fibrous diseases such as Dupuytren's contracture (Satish et al. 2013) and skin contractive scars (Satish et al. 2010), and CCT7 has been indicated to be a unique potential marker for active fibroblasts (Bai et al. 2015).

Potential Application of TRiC in Preventing MHtt (Mutant Huntingtin) Aggregation

TRiC has been suggested to be able to prevent mHtt aggregation in cells (Behrends et al. 2006; Kitamura et al. 2006; Tam et al. 2006, 2009), and the underlying inhibition mechanism of TRiC against mhttQ51 aggregation *in vitro* has been investigated (Shahmoradian et al. 2013). Besides, the CCT5 homo-oligomer can cap mHtt fibrils in a similar fashion to the TRiC complex. These results suggest the option of exploring the use of the TRiC complex or CCT5 as a reagent to prevent mHtt aggregation, which may contribute to the development of HD therapy (Darrow et al. 2015). Moreover, it has been reported that the CCT1 subunit is sufficient to inhibit aggregation and reduce mHtt-mediated toxicity in mouse N2a neuronal cells (Tam et al. 2006), and the exogenous delivery of the 20 kDa recombinant apical domain of CCT1 (ApiCCT1) can inhibit aggregation of recombinant mHtt *in vitro* (Sontag et al. 2013). More remarkably, applying ApiCCT1 to the cortical compartment of Htt disease mouse model can rescue impaired anterograde transport of brain-derived neurotrophic factor (BDNF) (Zhao et al. 2016). Therefore, there is a potential application for ApiCCT1 to be explored in the treatment of HD.

TRiC Is a Potential Target for Therapeutic Intervention

Cryo-EM analysis revealed that the CCT2 subunit of TRiC interacts with β-tubulin (Llorca et al. 2001). Disrupting the constitutively associated β-tubulin-CCT2 complex can cause severe cell apoptosis in multidrug-resistant MES-SA/Dx5 cancer cells (Harker and Sikic 1985). This finding suggests that the β-tubulin-CCT2 complex may serve as an effective chemotherapeutic target for treating clinical tubulin-binding agent-resistant tumor (Lin et al. 2009).

Breast cancer is the leading cause of death in women between the ages of 35 and 54 (Fiorica 1992). The CCT2 subunit was identified as the intracellular target of CT20p, a peptide that displays cancer (include breast cancer) specific cytotoxicity. The susceptibility of breast cancer cell to CT20p would be increased when over-expressing CCT2. TRiC is thus a potential target for therapeutic intervention for breast cancer treatment (Bassiouni et al. 2016). SMADs are the main signal transducers for receptors of the TGF-β superfamily, which are critically important for regulating cell development and growth. CCT6 has been shown to be the SMAD2 blocker, and silencing CCT6 could result in efficient suppression of metastasis *in vivo* and significantly prolonged the survival of tumor-bearing mice (Ying et al. 2017).

It has been reported that the homozygous cct2-L394H-7del mutation in the zebrafish leads to a small eye phenotype, and injection of RNA encoding wild-type human CCT2 could rescue the small eye phenotype (Minegishi et al. 2018). Also, it has been suggested that two mutations, T400P and R516H, in the CCT2 subunit can evoke leber congenital amaurosis (LCA) (Minegishi et al. 2016). This indicates a potential novel direction for the treatment of the LCA eye disease.

Perspective

In recent years, accumulating evidence has suggested the linkage between TRiC and certain diseases; still, there is a long way to go to fully understand the underlying molecular mechanism. Future atomic-resolution structural details and biochemical analysis of the eukaryotic group II chaperonin TRiC/CCT, may offer the structural basis of its subunit specificity, and provide more mechanistic insights into the allosteric network of TRiC and how this network regulates TRiC conformational transitions. Additionally, how the coordinated conformational transitions of TRiC correlate with their productive substrate folding is yet to be further elucidated (Fig. 19.4). Moreover, with the development of computational method in cryo-EM, such as manifold embedding algorithm (Frank and Ourmazd 2016), the continuous conformational landscape of TRiC may be delineated, and its free-energy landscape could be mapped out. Furthermore, the in vivo study of TRiC in the native environment by cryo-ET and sub-tomogram averaging could enable us to directly visualize the process of TRiC-assisted substrate folding in the cell (Guo et al. 2018; Shahmoradian et al. 2013). Eventually, a combination of all the information on this complex macromolecular machine will facilitate therapeutic strategy development against TRiC-related diseases (Fig. 19.4).

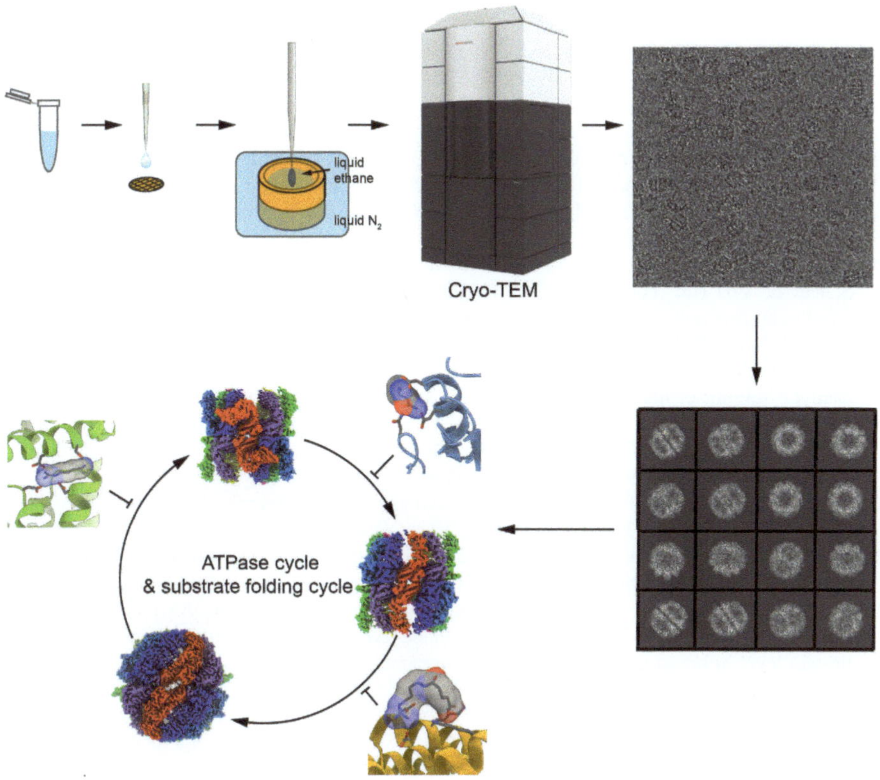

Fig. 19.4 Potential therapeutic development based on high-resolution cryo-EM structural studies of TRiC. Illustration of cryo-EM sample preparation, data collection, 3D reconstruction, model building, and structure-based drug development. The high-resolution reconstruction of an ensemble of TRiC structures in its substrate folding cycle regulated by its ATP-driven conformational cycle, will provide a complete picture of the mechanism of TRiC-assisted substrate folding. This structural information will enable us to design drugs to inhibit the allosteric transition at any desired step, and to block the reorganization of TRiC with specific substrate so as to prevent the development of related diseases such as cancer

Author Declaration

During the page proof process, a cryo-EM study on TRiC in complex with substrate mLST8 was published, revealing the function of TRiC in the folding of mLST8 and in the preparation for the assembly of mLST8 into mTOR complexes (Cuéllar et al. 2019). Moreover, another study resolved the cryo-EM structure of TRiC-PFD complex, established the structural and functional basis for TRiC and PFD cooperation in protein folding, essential for cellular proteostasis (Gestaut et al. 2019).

References

Adato A, Weil D, Kalinski H, Pel-Or Y, Ayadi H, Petit C, Korostishevsky M, Bonne-Tamir B (1997) Mutation profile of all 49 exons of the human myosin VIIA gene, and haplotype analysis, in Usher 1B families from diverse origins. Am J Hum Genet 61(4):813–821. https://doi.org/10.1086/514899

Amit M, Weisberg SJ, Nadler-Holly M, McCormack EA, Feldmesser E, Kaganovich D, Willison KR, Horovitz A (2010) Equivalent mutations in the eight subunits of the chaperonin CCT produce dramatically different cellular and gene expression phenotypes. J Mol Biol 401 (3):532–543. https://doi.org/10.1016/j.jmb.2010.06.037

Antonova SV, Haffke M, Corradini E, Mikuciunas M, Low TY, Signor L, van Es RM, Gupta K, Scheer E, Vos HR, Tora L, Heck AJR, Timmers HTM, Berger I (2018) Chaperonin CCT checkpoint function in basal transcription factor TFIID assembly. Nat Struct Mol Biol. https://doi.org/10.1038/s41594-018-0156-z

Bai XC, Rajendra E, Yang G, Shi Y, Scheres SH (2015) Sampling the conformational space of the catalytic subunit of human gamma-secretase. Elife 4. https://doi.org/10.7554/elife.11182

Bakthavatsalam D, Soung RH, Tweardy DJ, Chiu W, Dixon RA, Woodside DG (2014) Chaperonin-containing TCP-1 complex directly binds to the cytoplasmic domain of the LOX-1 receptor. FEBS Lett 588(13):2133–2140. https://doi.org/10.1016/j.febslet.2014.04.049

Balchin D, Hayer-Hartl M, Hartl FU (2016) In vivo aspects of protein folding and quality control. Science 353(6294):aac4354. https://doi.org/10.1126/science.aac4354

Balchin D, Milicic G, Strauss M, Hayer-Hartl M, Hartl FU (2018) Pathway of actin folding directed by the eukaryotic chaperonin TRiC. Cell. https://doi.org/10.1016/j.cell.2018.07.006

Bassiouni R, Nemec KN, Iketani A, Flores O, Showalter A, Khaled AS, Vishnubhotla P, Sprung RW Jr, Kaittanis C, Perez JM, Khaled AR (2016) Chaperonin containing TCP-1 protein level in breast cancer cells predicts therapeutic application of a cytotoxic peptide. Clin Cancer Res 22(17):4366–4379. https://doi.org/10.1158/1078-0432.CCR-15-2502

Beck M, Baumeister W (2016) Cryo-Electron tomography: can it reveal the molecular sociology of cells in atomic detail? Trends Cell Biol 26(11):825–837. https://doi.org/10.1016/j.tcb.2016.08.006

Behrends C, Langer CA, Boteva R, Bottcher UM, Stemp MJ, Schaffar G, Rao BV, Giese A, Kretzschmar H, Siegers K, Hartl FU (2006) Chaperonin TRiC promotes the assembly of polyQ expansion proteins into nontoxic oligomers. Mol Cell 23(6):887–897. https://doi.org/10.1016/j.molcel.2006.08.017

Berger J, Berger S, Li M, Jacoby AS, Arner A, Bavi N, Stewart AG, Currie PD (2018) In vivo function of the chaperonin TRiC in alpha-actin folding during sarcomere assembly. Cell Rep 22(2):313–322. https://doi.org/10.1016/j.celrep.2017.12.069

Bonné-Tamir B, Korostishevsky M, Kalinsky H, Seroussi E, Beker R, Weiss S, Godel V (1994) Genetic mapping of the gene for usher syndrome: linkage analysis in a large samaritan kindred. Genomics 20(1):36–42. https://doi.org/10.1006/geno.1994.1124

Booth CR, Meyer AS, Cong Y, Topf M, Sali A, Ludtke SJ, Chiu W, Frydman J (2008) Mechanism of lid closure in the eukaryotic chaperonin TRiC/CCT. Nat Struct Mol Biol 15(7):746–753. https://doi.org/10.1038/nsmb.1436

Bouhouche A, Benomar A, Bouslam N, Chkili T, Yahyaoui M (2006) Mutation in the epsilon subunit of the cytosolic chaperonin-containing t-complex peptide-1 (Cct5) gene causes autosomal recessive mutilating sensory neuropathy with spastic paraplegia. J Med Genet 43 (5):441–443. https://doi.org/10.1136/jmg.2005.039230

Broadley SA, Hartl FU (2009) The role of molecular chaperones in human misfolding diseases. FEBS Lett 583(16):2647–2653. https://doi.org/10.1016/j.febslet.2009.04.029

Camasses A, Bogdanova A, Shevchenko A, Zachariae W (2003) The CCT Chaperonin promotes activation of the anaphase-promoting complex through the generation of functional Cdc20. Mol Cell 12(1):87–100. https://doi.org/10.1016/s1097-2765(03)00244-2

Carr AC, Khaled AS, Bassiouni R, Flores O, Nierenberg D, Bhatti H, Vishnubhotla P, Perez JM, Santra S, Khaled AR (2017) Targeting chaperonin containing TCP1 (CCT) as a molecular therapeutic for small cell lung cancer. Oncotarget 8(66):110273–110288. https://doi.org/10.18632/oncotarget.22681

Chen X, Sullivan DS, Huffaker TC (1994) Two yeast genes with similarity to TCP-1 are required for microtubule and actin function in vivo. Proc Natl Acad Sci USA 91(19):9111–9115. https://doi.org/10.1073/pnas.91.19.9111

Chen L, Zhang Z, Qiu J, Zhang L, Luo X, Jang J (2014) Chaperonin CCT-mediated AIB1 folding promotes the growth of ERalpha-positive breast cancer cells on hard substrates. PLoS ONE 9 (5):e96085. https://doi.org/10.1371/journal.pone.0096085

Coghlin C, Carpenter B, Dundas SR, Lawrie LC, Telfer C, Murray GI (2006) Characterization and over-expression of chaperonin t-complex proteins in colorectal cancer. J Pathol 210(3):351–357. https://doi.org/10.1002/path.2056

Cong Y, Baker ML, Jakana J, Woolford D, Miller EJ, Reissmann S, Kumar RN, Redding-Johanson AM, Batth TS, Mukhopadhyay A, Ludtke SJ, Frydman J, Chiu W (2010) 4.0-A resolution cryo-EM structure of the mammalian chaperonin TRiC/CCT reveals its unique subunit arrangement. Proc Nat Acad Sci USA 107(11):4967–4972. https://doi.org/10.1073/pnas.0913774107

Cong Y, Schroder GF, Meyer AS, Jakana J, Ma B, Dougherty MT, Schmid MF, Reissmann S, Levitt M, Ludtke SL, Frydman J, Chiu W (2012) Symmetry-free cryo-EM structures of the chaperonin TRiC along its ATPase-driven conformational cycle. EMBO J 31(3):720–730. https://doi.org/10.1038/emboj.2011.366

Counts JT, Hester TM, Rouhana L (2017) Genetic expansion of chaperonin-containing TCP-1 (CCT/TRiC) complex subunits yields testis-specific isoforms required for spermatogenesis in planarian flatworms. Mol Reprod Dev 84(12):1271–1284. https://doi.org/10.1002/mrd.22925

Cuellar J, Martin-Benito J, Scheres SH, Sousa R, Moro F, Lopez-Vinas E, Gomez-Puertas P, Muga A, Carrascosa JL, Valpuesta JM (2008) The structure of CCT-Hsc70 NBD suggests a mechanism for Hsp70 delivery of substrates to the chaperonin. Nat Struct Mol Biol 15(8):858–864. https://doi.org/10.1038/nsmb.1464

Cuéllar J, Ludlam WG, Tensmeyer NC, Aoba T, Dhavale M, Santiago C, Bueno-Carrasco MT, Mann MJ, Plimpton RL, Makaju A, Franklin S, Willardson BM, Valpuesta JM (2019) Structural and functional analysis of the role of the chaperonin CCT in mTOR complex assembly. Nat Commun 10(1)

Cui X, Hu ZP, Li Z, Gao PJ, Zhu JY (2015) Overexpression of chaperonin containing TCP1, subunit 3 predicts poor prognosis in hepatocellular carcinoma. World J Gastroenterol 21 (28):8588–8604. https://doi.org/10.3748/wjg.v21.i28.8588

Cyrne L, Guerreiro P, Cardoso AC, RodriguesPousada C, Soares H (1996) The Tetrahymena chaperonin subunit CCT eta gene is coexpressed with CCT gamma gene during cilia biogenesis and cell sexual reproduction. FEBS Lett 383(3):277–283. https://doi.org/10.1016/0014-5793(96)00240-2

Darrow MC, Sergeeva OA, Isas JM, Galaz-Montoya JG, King JA, Langen R, Schmid MF, Chiu W (2015) Structural mechanisms of mutant huntingtin aggregation suppression by the synthetic chaperonin-like CCT5 complex explained by cryoelectron tomography. J Biol Chem 290 (28):17451–17461. https://doi.org/10.1074/jbc.M115.655373

Dekker C, Roe SM, McCormack EA, Beuron F, Pearl LH, Willison KR (2011) The crystal structure of yeast CCT reveals intrinsic asymmetry of eukaryotic cytosolic chaperonins. EMBO J 30(15):3078–3090. https://doi.org/10.1038/emboj.2011.208

Ditzel L, Lowe J, Stock D, Stetter KO, Huber H, Huber R, Steinbacher S (1998) Crystal structure of the thermosome, the archaeal chaperonin and homolog of CCT. Cell 93(1):125–138. https://doi.org/10.1016/s0092-8674(00)81152-6

Dunn AY, Melville MW, Frydman J (2001) Review: cellular substrates of the eukaryotic chaperonin TRiC/CCT. J Struct Biol 135(2):176–184. https://doi.org/10.1006/jsbi.2001.4380

Erdo F, Trapp T, Mies G, Hossmann KA (2004) Immunohistochemical analysis of protein expression after middle cerebral artery occlusion in mice. Acta Neuropathol 107(2):127–136. https://doi.org/10.1007/s00401-003-0789-8

Farr GW, Scharl EC, Schumacher RJ, Sondek S, Horwich AL (1997) Chaperonin-mediated folding in the eukaryotic cytosol proceeds through rounds of release of native and nonnative forms. Cell 89(6):927–937. https://doi.org/10.1016/S0092-8674(00)80278-0

Feldman DE, Thulasiraman V, Ferreyra RG, Frydman J (1999) Formation of the VHL-elongin BC tumor suppressor complex is mediated by the chaperonin TRiC. Mol Cell 4(6):1051–1061

Feldman DE, Spiess C, Howard DE, Frydman J (2003) Tumorigenic mutations in VHL disrupt folding in vivo by interfering with chaperonin binding. Mol Cell 12(5):1213–1224

Fernandez-Fernandez MR, Sot B, Valpuesta JM (2016) Molecular chaperones: functional mechanisms and nanotechnological applications. Nanotechnology 27(32):324004. https://doi.org/10.1088/0957-4484/27/32/324004

Fiorica JV (1992) Breast disease. Curr Opin Obstet Gynecol 4(6):897–903

Fontanella B, Birolo L, Infusini G, Cirulli C, Marzullo L, Pucci P, Turco MC, Tosco A (2010) The co-chaperone BAG3 interacts with the cytosolic chaperonin CCT: new hints for actin folding. Int J Biochem Cell Biol 42(5):641–650. https://doi.org/10.1016/j.biocel.2009.12.008

Frank J, Ourmazd A (2016) Continuous changes in structure mapped by manifold embedding of single-particle data in cryo-EM. Methods 100:61–67. https://doi.org/10.1016/j.ymeth.2016.02.007

Freund A, Zhong FL, Venteicher AS, Meng Z, Veenstra TD, Frydman J, Artandi SE (2014) Proteostatic control of telomerase function through TRiC-mediated folding of TCAB1. Cell 159(6):1389–1403. https://doi.org/10.1016/j.cell.2014.10.059

Frydman J, Nimmesgern E, Erdjument-Bromage H, Wall JS, Tempst P, Hartl FU (1992) Function in protein folding of TRiC, a cytosolic ring complex containing TCP-1 and structurally related subunits. EMBO J 11(13):4767–4778. https://doi.org/10.1002/j.1460-2075.1992.tb05582.x

Frydman J, Nimmesgern E, Ohtsuka K, Hartl FU (1994) Folding of nascent polypeptide chains in a high molecular mass assembly with molecular chaperones. Nature 370(6485):111–117. https://doi.org/10.1038/370111a0

Gao YJ, Thomas JO, Chow RL, Lee GH, Cowan NJ (1992) A cytoplasmic chaperonin that catalyzes beta-actin folding. Cell 69(6):1043–1050. https://doi.org/10.1016/0092-8674(92)90622-J

Gao HJ, Zheng M, Sun SJ, Wang HW, Yue ZG, Zhu Y, Han XC, Yang JQ, Zhou YQ, Cai YR, Hu WN (2017) Chaperonin containing TCP1 subunit 5 is a tumor associated antigen of non-small cell lung cancer. Oncotarget 8(38):64170–64179. https://doi.org/10.18632/oncotarget.19369

Gestaut D, Roh SH, Ma B, Pintilie G, Joachimiak LA, Leitner A, Walzthoeni T, Aebersold R, Chiu W, Frydman J (2019) The chaperonin TRiC/CCT associates with prefoldin through a conserved electrostatic interface essential for cellular proteostasis. Cell 177(3):751–765.e15

Grantham J, Brackley KI, Willison KR (2006) Substantial CCT activity is required for cell cycle progression and cytoskeletal organization in mammalian cells. Exp Cell Res 312(12):2309–2324. https://doi.org/10.1016/j.yexcr.2006.03.028

Guenther MG, Yu JJ, Kao GD, Yen TJ, Lazar MA (2002) Assembly of the SMRT-histone deacetylase 3 repression complex requires the TCP-1 ring complex. Genes Dev 16(24):3130–3135. https://doi.org/10.1101/gad.1037502

Guest ST, Kratche ZR, Bollig-Fischer A, Haddad R, Ethier SP (2015) Two members of the TRiC chaperonin complex, CCT2 and TCP1 are essential for survival of breast cancer cells and are linked to driving oncogenes. Exp Cell Res 332(2):223–235. https://doi.org/10.1016/j.yexcr.2015.02.005

Guo Q, Lehmer C, Martinez-Sanchez A, Rudack T, Beck F, Hartmann H, Perez-Berlanga M, Frottin F, Hipp MS, Hartl FU, Edbauer D, Baumeister W, Fernandez-Busnadiego R (2018) In situ structure of neuronal C9orf72 Poly-GA aggregates reveals proteasome recruitment. Cell. https://doi.org/10.1016/j.cell.2017.12.030

Hanafy KA, Martin E, Murad F (2004) CCTeta, a novel soluble guanylyl cyclase-interacting protein. J Biol Chem 279(45):46946–46953. https://doi.org/10.1074/jbc.M404134200

Harker WG, Sikic BI (1985) Multidrug (pleiotropic) resistance in doxorubicin-selected variants of the human sarcoma cell line MES-SA. Cancer Res 45(9):4091–4096

Horwich AL, Willison KR (1993) Protein folding in the cell: functions of two families of molecular chaperone, hsp 60 and TF55-TCP1. Philos Trans R Soc Lond B Biol Sci 339 (1289):313–325; discussion 325–316. https://doi.org/10.1098/rstb.1993.0030

Huang X, Wang X, Cheng C, Cai J, He S, Wang H, Liu F, Zhu C, Ding Z, Huang X, Zhang T, Zhang Y (2014) Chaperonin containing TCP1, subunit 8 (CCT8) is upregulated in hepatocellular carcinoma and promotes HCC proliferation. APMIS 122(11):1070–1079. https://doi.org/10.1111/apm.12258

Humrich J, Bermel C, Bunemann M, Harmark L, Frost R, Quitterer U, Lohse MJ (2005) Phosducin-like protein regulates G-protein betagamma folding by interaction with tailless complex polypeptide-1alpha: dephosphorylation or splicing of PhLP turns the switch toward regulation of Gbetagamma folding. J Biol Chem 280(20):20042–20050. https://doi.org/10.1074/jbc.M409233200

Hunziker M, Barandun J, Petfalski E, Tan D, Delan-Forino C, Molloy KR, Kim KH, Dunn-Davies H, Shi Y, Chaker-Margot M, Chait BT, Walz T, Tollervey D, Klinge S (2016) UtpA and UtpB chaperone nascent pre-ribosomal RNA and U3 snoRNA to initiate eukaryotic ribosome assembly. Nat Commun 7:12090. https://doi.org/10.1038/ncomms12090

Jiang Y, Douglas NR, Conley NR, Miller EJ, Frydman J, Moerner WE (2011) Sensing cooperativity in ATP hydrolysis for single multisubunit enzymes in solution. Proc Natl Acad Sci USA 108(41):16962–16967. https://doi.org/10.1073/pnas.1112244108

Jiang XD, Mao WJ, Yang ZY, Zeng J, Zhang Y, Song Y, Kong Y, Ren SY, Zuo YF (2015) Silencing P2X7 receptor downregulates the expression of TCP-1 involved in lymphoma lymphatic metastasis. Oncotarget 6(39):42105–42117. https://doi.org/10.18632/oncotarget.5870

Joachimiak LA, Walzthoeni T, Liu CW, Aebersold R, Frydman J (2014) The structural basis of substrate recognition by the eukaryotic chaperonin TRiC/CCT. Cell 159(5):1042–1055. https://doi.org/10.1016/j.cell.2014.10.042

Kaelin WG Jr (2002) Molecular basis of the VHL hereditary cancer syndrome. Nat Rev Cancer 2 (9):673–682. https://doi.org/10.1038/nrc885

Kaisari S, Sitry-Shevah D, Miniowitz-Shemtov S, Teichner A, Hershko A (2017) Role of CCT chaperonin in the disassembly of mitotic checkpoint complexes. Proc Natl Acad Sci USA 114 (5):956–961. https://doi.org/10.1073/pnas.1620451114

Kalisman N, Adams CM, Levitt M (2012) Subunit order of eukaryotic TRiC/CCT chaperonin by cross-linking, mass spectrometry, and combinatorial homology modeling. Proc Natl Acad Sci USA 109(8):2884–2889. https://doi.org/10.1073/pnas.1119472109

Kalisman N, Schroder GF, Levitt M (2013) The crystal structures of the eukaryotic chaperonin CCT reveal its functional partitioning. Structure 21(4):540–549. https://doi.org/10.1016/j.str.2013.01.017

Kasembeli M, Lau WC, Roh SH, Eckols TK, Frydman J, Chiu W, Tweardy DJ (2014) Modulation of STAT3 folding and function by TRiC/CCT chaperonin. PLoS Biol 12(4):e1001844. https://doi.org/10.1371/journal.pbio.1001844

Khabirova E, Moloney A, Marciniak SJ, Williams J, Lomas DA, Oliver SG, Favrin G, Sattelle DB, Crowther DC (2014) The TRiC/CCT chaperone is implicated in Alzheimer's disease based on patient GWAS and an RNAi screen in Abeta-expressing Caenorhabditis elegans. PLoS One 9(7):e102985. https://doi.org/10.1371/journal.pone.0102985

Kim YE, Hipp MS, Bracher A, Hayer-Hartl M, Hartl FU (2013) Molecular chaperone functions in protein folding and proteostasis. Annu Rev Biochem 82(82):323–355. https://doi.org/10.1146/annurev-biochem-060208-092442

Kitamura A, Kubota H, Pack CG, Matsumoto G, Hirayama S, Takahashi Y, Kimura H, Kinjo M, Morimoto RI, Nagata K (2006) Cytosolic chaperonin prevents polyglutamine toxicity with altering the aggregation state. Nat Cell Biol 8(10):1163–1170. https://doi.org/10.1038/ncb1478

Klumpp M, Baumeister W, Essen LO (1997) Structure of the substrate binding domain of the thermosome, an archaeal group II chaperonin. Cell 91(2):263–270. https://doi.org/10.1016/S0092-8674(00)80408-0

Knee KM, Sergeeva OA, King JA (2013) Human TRiC complex purified from HeLa cells contains all eight CCT subunits and is active in vitro. Cell Stress Chaperones 18(2):137–144. https://doi.org/10.1007/s12192-012-0357-z

Kohl S, Baumann B, Rosenberg T, Kellner U, Lorenz B, Vadalà M, Jacobson SG, Wissinger B (2002) Mutations in the cone photoreceptor G-protein alpha-subunit gene GNAT2 in patients with achromatopsia. Am J Hum Genet 71(2):422–425. https://doi.org/10.1086/341835

Korobko I, Nadler-Holly M, Horovitz A (2016) Transient kinetic analysis of ATP hydrolysis by the CCT/TRiC chaperonin. J Mol Biol 428(22):4520–4527. https://doi.org/10.1016/j.jmb.2016.09.017

Koulikovska M, Podskochy A, Fagerholm P (2005) The expression pattern of the subunit of chaperonin containing T-complex polypeptide 1 and its substrate, alpha-smooth muscle actin, during corneal wound healing. Acta Ophthalmol Scand 83(5):543–548. https://doi.org/10.1111/j.1600-0420.2005.00482.x

Lee MJ, Stephenson DA, Groves MJ, Sweeney MG, Davis MB, An SF, Houlden H, Salih MAM, Timmerman V, de Jonghe P, Auer-Grumbach M, Di Maria E, Scaravilli F, Wood NW, Reilly MM (2003) Hereditary sensory neuropathy is caused by a mutation in the delta subunit of the cytosolic chaperonin-containing t-complex peptide-1 (Cct4) gene. Hum Mol Genet 12(15):1917–1925. https://doi.org/10.1093/hmg/ddg198

Leitner A, Joachimiak LA, Bracher A, Monkemeyer L, Walzthoeni T, Chen B, Pechmann S, Holmes S, Cong Y, Ma B, Ludtke S, Chiu W, Hartl FU, Aebersold R, Frydman J (2012) The molecular architecture of the eukaryotic chaperonin TRiC/CCT. Structure 20(5):814–825. https://doi.org/10.1016/j.str.2012.03.007

Lewis VA, Hynes GM, Zheng D, Saibil H, Willison K (1992) T-complex polypeptide-1 is a subunit of a heteromeric particle in the eukaryotic cytosol. Nature 358(6383):249–252. https://doi.org/10.1038/358249a0

Liebman SW, Meredith SC (2010) Protein folding: sticky N17 speeds huntingtin pile-up. Nat Chem Biol 6(1):7–8. https://doi.org/10.1038/nchembio.279

Lin YF, Tsai WP, Liu HG, Liang PH (2009) Intracellular beta-tubulin/chaperonin containing TCP1-beta complex serves as a novel chemotherapeutic target against drug-resistant tumors. Cancer Res 69(17):6879–6888. https://doi.org/10.1158/0008-5472.CAN-08-4700

Liou AK, Willison KR (1997) Elucidation of the subunit orientation in CCT (chaperonin containing TCP1) from the subunit composition of CCT micro-complexes. EMBO J 16 (14):4311–4316. https://doi.org/10.1093/emboj/16.14.4311

Liu BD, Larsson L, Caballero A, Hao XX, Oling D, Grantham J, Nystrom T (2010) The polarisome is required for segregation and retrograde transport of protein aggregates. Cell 140 (2):257–267. https://doi.org/10.1016/j.cell.2009.12.031

Llorca O, Smyth MG, Marco S, Carrascosa JL, Willison KR, Valpuesta JM (1998) ATP binding induces large conformational changes in the apical and equatorial domains of the eukaryotic chaperonin containing TCP-1 complex. J Biol Chem 273(17):10091–10094. https://doi.org/10.1074/jbc.273.17.10091

Llorca O, McCormack EA, Hynes G, Grantham J, Cordell J, Carrascosa JL, Willison KR, Fernandez JJ, Valpuesta JM (1999) Eukaryotic type II chaperonin CCT interacts with actin through specific subunits. Nature 402(6762):693–696. https://doi.org/10.1038/45294valpuesta 1999.pdf

Llorca O, Martin-Benito J, Ritco-Vonsovici M, Grantham J, Hynes GM, Willison KR, Carrascosa JL, Valpuesta JM (2000) Eukaryotic chaperonin CCT stabilizes actin and tubulin folding intermediates in open quasi-native conformations. EMBO J 19(22):5971–5979. https://doi.org/10.1093/emboj/19.22.5971

Llorca O, Martin-Benito J, Gomez-Puertas P, Ritco-Vonsovici M, Willison KR, Carrascosa JL, Valpuesta JM (2001) Analysis of the interaction between the eukaryotic chaperonin CCT and its substrates actin and tubulin. J Struct Biol 135(2):205–218. https://doi.org/10.1106/jsbi.2001.4359

Loktev AV, Zhang Q, Beck JS, Searby CC, Scheetz TE, Bazan JF, Slusarski DC, Sheffield VC, Jackson PK, Nachury MV (2008) A BBSome subunit links ciliogenesis, microtubule stability, and acetylation. Dev Cell 15(6):854–865. https://doi.org/10.1016/j.devcel.2008.11.001

Lopez T, Dalton K, Frydman J (2015) The mechanism and function of group ii chaperonins. J Mol Biol 427(18):2919–2930. https://doi.org/10.1016/j.jmb.2015.04.013

Machida K, Masutani M, Kobayashi T, Mikami S, Nishino Y, Miyazawa A, Imataka H (2012) Reconstitution of the human chaperonin CCT by co-expression of the eight distinct subunits in mammalian cells. Protein Expr Purif 82(1):61–69. https://doi.org/10.1016/j.pep.2011.11.010

Malcikova J, Tichy B, Damborsky J, Kabathova J, Trbusek M, Mayer J, Pospisilova S (2010) Analysis of the DNA-binding activity of p53 mutants using functional protein microarrays and its relationship to transcriptional activation. Biol Chem 391(2–3):197–205. https://doi.org/10.1515/BC.2010.027

Martin-Benito J, Boskovic J, Gomez-Puertas P, Carrascosa JL, Simons CT, Lewis SA, Bartolini F, Cowan NJ, Valpuesta JM (2002) Structure of eukaryotic prefoldin and of its complexes with unfolded actin and the cytosolic chaperonin CCT. EMBO J 21(23):6377–6386. https://doi.org/10.1093/emboj/cdf640

Masson N, Appelhoff RJ, Tuckerman JR, Tian YM, Demol H, Puype M, Vandekerckhove J, Ratcliffe PJ, Pugh CW (2004) The HIF prolyl hydroxylase PHD3 is a potential substrate of the TRiC chaperonin. FEBS Lett 570(1–3):166–170. https://doi.org/10.1016/j.febslet.2004.06.040

Matsuda N, Mishina M (2004) Identification of chaperonin CCT gamma subunit as a determinant of retinotectal development by whole-genome subtraction cloning from zebrafish no tectal neuron mutant. Development 131(9):1913–1925. https://doi.org/10.1242/dev.01085

Matus DQ, Li XY, Durbin S, Agarwal D, Chi Q, Weiss SJ, Sherwood DR (2010) In vivo identification of regulators of cell invasion across basement membranes. Sci Signal 3(120): ra35. https://doi.org/10.1126/scisignal.2000654

McClellan AJ, Scott MD, Frydman J (2005) Folding and quality control of the VHL tumor suppressor proceed through distinct chaperone pathways. Cell 121(5):739–748. https://doi.org/10.1016/j.cell.2005.03.024

Melville MW, McClellan AJ, Meyer AS, Darveau A, Frydman J (2003) The Hsp70 and TRiC/CCT chaperone systems cooperate in vivo to assemble the Von Hippel-Lindau tumor suppressor complex. Mol Cell Biol 23(9):3141–3151. https://doi.org/10.1128/mcb.23.9.3141-3151.2003

Meyer AS, Gillespie JR, Walther D, Millet IS, Doniach S, Frydman J (2003) Closing the folding chamber of the eukaryotic chaperonin requires the transition state of ATP hydrolysis. Cell 113(3):369–381. https://doi.org/10.1016/S0092-8674(03)00307-6

Minegishi Y, Sheng X, Yoshitake K, Sergeev Y, Iejima D, Shibagaki Y, Monma N, Ikeo K, Furuno M, Zhuang W, Liu Y, Rong W, Hattori S, Iwata T (2016) CCT2 mutations evoke leber congenital amaurosis due to chaperone complex instability. Sci Rep 6:33742. https://doi.org/10.1038/srep33742

Minegishi Y, Nakaya N, Tomarev SI (2018) Mutation in the Zebrafish cct2 gene leads to abnormalities of cell cycle and cell death in the retina: a model of CCT2-related leber congenital amaurosis. Invest Ophthalmol Vis Sci 59(2):995–1004. https://doi.org/10.1167/iovs.17-22919

Miyata Y, Shibata T, Aoshima M, Tsubata T, Nishida E (2014) The molecular chaperone TRiC/CCT binds to the Trp-Asp 40 (WD40) repeat protein WDR68 and promotes its folding, protein kinase DYRK1A binding, and nuclear accumulation. J Biol Chem 289(48):33320–33332. https://doi.org/10.1074/jbc.M114.586115

Monzo K, Dowd SR, Minden JS, Sisson JC (2010) Proteomic analysis reveals CCT is a target of Fragile X mental retardation protein regulation in Drosophila. Dev Biol 340(2):408–418. https://doi.org/10.1016/j.ydbio.2010.01.028

Mosalaganti S, Kosinski J, Albert S, Schaffer M, Strenkert D, Salome PA, Merchant SS, Plitzko JM, Baumeister W, Engel BD, Beck M (2018) In situ architecture of the algal nuclear pore complex. Nat Commun 9(1):2361. https://doi.org/10.1038/s41467-018-04739-y

Muchowski PJ, Wacker JL (2005) Modulation of neurodegeneration by molecular chaperones. Nat Rev Neurosci 6:11. https://doi.org/10.1038/nrn1587

Munoz IG, Yebenes H, Zhou M, Mesa P, Serna M, Park AY, Bragado-Nilsson E, Beloso A, de Carcer G, Malumbres M, Robinson CV, Valpuesta JM, Montoya G (2011) Crystal structure of the open conformation of the mammalian chaperonin CCT in complex with tubulin. Nat Struct Mol Biol 18(1):14–19. https://doi.org/10.1038/nsmb.1971

Nachury MV, Loktev AV, Zhang Q, Westlake CJ, Peranen J, Merdes A, Slusarski DC, Scheller RH, Bazan JF, Sheffield VC, Jackson PK (2007) A core complex of BBS proteins cooperates with the GTPase Rab8 to promote ciliary membrane biogenesis. Cell 129(6):1201–1213. https://doi.org/10.1016/j.cell.2007.03.053

Noormohammadi A, Khodakarami A, Gutierrez-Garcia R, Lee HJ, Koyuncu S, Konig T, Schindler C, Saez I, Fatima A, Dieterich C, Vilchez D (2016) Somatic increase of CCT8 mimics proteostasis of human pluripotent stem cells and extends *C. elegans* lifespan. Nat Commun 7:13649. https://doi.org/10.1038/ncomms13649

Ooe A, Kato K, Noguchi S (2007) Possible involvement of CCT5, RGS3, and YKT6 genes up-regulated in p53-mutated tumors in resistance to docetaxel in human breast cancers. Breast Cancer Res Treat 101(3):305–315. https://doi.org/10.1007/s10549-006-9293-x

Pappenberger G, McCormack EA, Willison KR (2006) Quantitative actin folding reactions using yeast CCT purified via an internal tag in the CCT3/gamma subunit. J Mol Biol 360(2):484–496. https://doi.org/10.1016/j.jmb.2006.05.003

Pavel M, Imarisio S, Menzies FM, Jimenez-Sanchez M, Siddiqi FH, Wu X, Renna M, O'Kane CJ, Crowther DC, Rubinsztein DC (2016) CCT complex restricts neuropathogenic protein aggregation via autophagy. Nat Commun 7:13821. https://doi.org/10.1038/ncomms13821

Pereira JH, Ralston CY, Douglas NR, Meyer D, Knee KM, Goulet DR, King JA, Frydman J, Adams PD (2010) Crystal structures of a group II chaperonin reveal the open and closed states associated with the protein folding cycle. J Biol Chem 285(36):27958–27966. https://doi.org/10.1074/jbc.M110.125344

Pereira JH, Ralston CY, Douglas NR, Kumar R, Lopez T, McAndrew RP, Knee KM, King JA, Frydman J, Adams PD (2012) Mechanism of nucleotide sensing in group II chaperonins. EMBO J 31(3):731–740. https://doi.org/10.1038/emboj.2011.468

Pereira JH, McAndrew RP, Sergeeva OA, Ralston CY, King JA, Adams PD (2017) Structure of the human TRiC/CCT Subunit 5 associated with hereditary sensory neuropathy. Sci Rep 7(1):3673. https://doi.org/10.1038/s41598-017-03825-3

Pines A, Dijk M, Makowski M, Meulenbroek EM, Vrouwe MG, van der Weegen Y, Baltissen M, French PJ, van Royen ME, Luijsterburg MS, Mullenders LH, Vermeulen M, Vermeulen W, Pannu NS, van Attikum H (2018) TRiC controls transcription resumption after UV damage by regulating Cockayne syndrome protein A. Nat Commun 9(1):1040. https://doi.org/10.1038/s41467-018-03484-6

Plimpton RL, Cuellar J, Lai CW, Aoba T, Makaju A, Franklin S, Mathis AD, Prince JT, Carrascosa JL, Valpuesta JM, Willardson BM (2015) Structures of the Gbeta-CCT and PhLP1-Gbeta-CCT complexes reveal a mechanism for G-protein beta-subunit folding and Gbetagamma dimer assembly. Proc Natl Acad Sci USA 112(8):2413–2418. https://doi.org/10.1073/pnas.1419595112

Posokhova E, Song H, Belcastro M, Higgins L, Bigley LR, Michaud NA, Martemyanov KA, Sokolov M (2011) Disruption of the chaperonin containing TCP-1 function affects protein networks essential for rod outer segment morphogenesis and survival. Mol Cell Proteomics 10 (1):M110 000570. https://doi.org/10.1074/mcp.m110.000570

Qian-Lin Z, Ting-Feng W, Qi-Feng C, Min-Hua Z, Ai-Guo L (2010) Inhibition of cytosolic chaperonin CCTzeta-1 expression depletes proliferation of colorectal carcinoma in vitro. J Surg Oncol 102(5):419–423. https://doi.org/10.1002/jso.21625

Qiu X, He X, Huang Q, Liu X, Sun G, Guo J, Yuan D, Yang L, Ban N, Fan S, Tao T, Wang D (2015) Overexpression of CCT8 and its significance for tumor cell proliferation, migration and invasion in glioma. Pathol Res Pract 211(10):717–725. https://doi.org/10.1016/j.prp.2015.04.012

Rademacher F, Kehren V, Stoldt VR, Ernst JF (1998) A Candida albicans chaperonin subunit (CaCct8p) as a suppressor of morphogenesis and Ras phenotypes in C-albicans and Saccharomyces cerevisiae. Microbiol-Uk 144:2951–2960. https://doi.org/10.1099/00221287-144-11-2951

Reissmann S, Joachimiak LA, Chen B, Meyer AS, Nguyen A, Frydman J (2012) A gradient of ATP affinities generates an asymmetric power stroke driving the chaperonin TRIC/CCT folding cycle. Cell reports 2(4):866–877. https://doi.org/10.1016/j.celrep.2012.08.036

Rivenzon-Segal D, Wolf SG, Shimon L, Willison KR, Horovitz A (2005) Sequential ATP-induced allosteric transitions of the cytoplasmic chaperonin containing TCP-1 revealed by EM analysis. Nat Struct Mol Biol 12(3):233–237. https://doi.org/10.1038/nsmb901

Roh SH, Kasembeli M, Bakthavatsalam D, Chiu W, Tweardy DJ (2015) Contribution of the type II chaperonin, TRiC/CCT, to oncogenesis. Int J Mol Sci 16(11):26706–26720. https://doi.org/10.3390/ijms161125975

Roh SH, Kasembeli M, Galaz-Montoya JG, Trnka M, Lau WC, Burlingame A, Chiu W, Tweardy DJ (2016a) Chaperonin TRiC/CCT modulates the folding and activity of leukemogenic fusion oncoprotein AML1-ETO. J Biol Chem 291(9):4732–4741. https://doi.org/10.1074/jbc.M115.684878

Roh SH, Kasembeli MM, Galaz-Montoya JG, Chiu W, Tweardy DJ (2016b) Chaperonin TRiC/CCT recognizes fusion oncoprotein AML1-ETO through subunit-specific interactions. Biophys J 110(11):2377–2385. https://doi.org/10.1016/j.bpj.2016.04.045

Saegusa K, Sato M, Sato K, Nakajima-Shimada J, Harada A, Sato K (2014) Caenorhabditis elegans chaperonin CCT/TRiC is required for actin and tubulin biogenesis and microvillus formation in intestinal epithelial cells. Mol Biol Cell 25(20):3095–3104. https://doi.org/10.1091/mbc.E13-09-0530

Satish L, Johnson S, Wang JH, Post JC, Ehrlich GD, Kathju S (2010) Chaperonin containing T-complex polypeptide subunit eta (CCT-eta) is a specific regulator of fibroblast motility and contractility. PLoS ONE 5(4):e10063. https://doi.org/10.1371/journal.pone.0010063

Satish L, O'Gorman DB, Johnson S, Raykha C, Gan BS, Wang JH, Kathju S (2013) Increased CCT-eta expression is a marker of latent and active disease and a modulator of fibroblast contractility in Dupuytren's contracture. Cell Stress Chaperones 18(4):397–404. https://doi.org/10.1007/s12192-012-0392-9

Seixas C, Cruto T, Tavares A, Gaertig J, Soares H (2010) CCTalpha and CCTdelta chaperonin subunits are essential and required for cilia assembly and maintenance in Tetrahymena. PLoS ONE 5(5):e10704. https://doi.org/10.1371/journal.pone.0010704

Seo S, Baye LM, Schulz NP, Beck JS, Zhang Q, Slusarski DC, Sheffield VC (2010) BBS6, BBS10, and BBS12 form a complex with CCT/TRiC family chaperonins and mediate BBSome assembly. Proc Natl Acad Sci USA 107(4):1488–1493. https://doi.org/10.1073/pnas.0910268107

Sergeeva OA, Chen B, Haase-Pettingell C, Ludtke SJ, Chiu W, King JA (2013) Human CCT4 and CCT5 chaperonin subunits expressed in *Escherichia coli* form biologically active homo-oligomers. J Biol Chem 288(24):17734–17744. https://doi.org/10.1074/jbc.M112.443929

Sergeeva OA, Tran MT, Haase-Pettingell C, King JA (2014) Biochemical characterization of mutants in chaperonin proteins CCT4 and CCT5 associated with hereditary sensory neuropathy. J Biol Chem 289(40):27470–27480. https://doi.org/10.1074/jbc.M114.576033

Shahmoradian SH, Galaz-Montoya JG, Schmid MF, Cong Y, Ma B, Spiess C, Frydman J, Ludtke SJ, Chiu W (2013) TRiC's tricks inhibit huntingtin aggregation. Elife 2:e00710. https://doi.org/10.7554/eLife.00710

Siegers K, Bolter B, Schwarz JP, Bottcher UMK, Guha S, Hartl FU (2003) TRiC/CCT cooperates with different upstream chaperones in the folding of distinct protein classes (Retracted Article. See vol 27, 301 p, 2008). EMBO J 22(19):5230–5240. https://doi.org/10.1093/emboj/cdg483

Sinha S, Belcastro M, Datta P, Seo S, Sokolov M (2014) Essential role of the chaperonin CCT in rod outer segment biogenesis. Invest Ophth Vis Sci 55(6):3775–3784. https://doi.org/10.1167/iovs.14-13889

Skjaerven L, Cuellar J, Martinez A, Valpuesta JM (2015) Dynamics, flexibility, and allostery in molecular chaperonins. FEBS Lett 589(19 Pt A):2522–2532. https://doi.org/10.1016/j.febslet.2015.06.019

Sontag EM, Joachimiak LA, Tan Z, Tomlinson A, Housman DE, Glabe CG, Potkin SG, Frydman J, Thompson LM (2013) Exogenous delivery of chaperonin subunit fragment ApiCCT1 modulates mutant Huntingtin cellular phenotypes. Proc Natl Acad Sci USA 110(8):3077–3082. https://doi.org/10.1073/pnas.1222663110

Sot B, Rubio-Munoz A, Leal-Quintero A, Martinez-Sabando J, Marcilla M, Roodveldt C, Valpuesta JM (2017) The chaperonin CCT inhibits assembly of alpha-synuclein amyloid fibrils by a specific, conformation-dependent interaction. Sci Rep 7:40859. https://doi.org/10.1038/srep40859

Soues S, Kann M-L, Fouquet J-P, Melki R (2003) The cytosolic chaperonin CCT associates to cytoplasmic microtubular structures during mammalian spermiogenesis and to heterochromatin in germline and somatic cells. Exp Cell Res 288(2):363–373. https://doi.org/10.1016/s0014-4827(03)00248-9

Spiess C, Meyer AS, Reissmann S, Frydman J (2004) Mechanism of the eukaryotic chaperonin: protein folding in the chamber of secrets. Trends Cell Biol 14(11):598–604. https://doi.org/10.1016/j.tcb.2004.09.015

Srikakulam R, Winkelmann DA (1999) Myosin II folding is mediated by a molecular chaperonin. J Biol Chem 274(38):27265–27273. https://doi.org/10.1074/jbc.274.38.27265

Stemp MJ, Guha S, Hartl FU, Barral JM (2005) Efficient production of native actin upon translation in a bacterial lysate supplemented with the eukaryotic chaperonin TRiC. Biol Chem 386(8):753–757. https://doi.org/10.1515/BC.2005.088

Stirling PC, Cuellar J, Alfaro GA, El Khadali F, Beh CT, Valpuesta JM, Melki R, Leroux MR (2006) PhLP3 modulates CCT-mediated actin and tubulin folding via ternary complexes with substrates. J Biol Chem 281(11):7012–7021. https://doi.org/10.1074/jbc.M513235200

Tam S, Geller R, Spiess C, Frydman J (2006) The chaperonin TRiC controls polyglutamine aggregation and toxicity through subunit-specific interactions. Nat Cell Biol 8(10):1155–1162. https://doi.org/10.1038/ncb1477

Tam S, Spiess C, Auyeung W, Joachimiak L, Chen B, Poirier MA, Frydman J (2009) The chaperonin TRiC blocks a huntingtin sequence element that promotes the conformational switch to aggregation. Nat Struct Mol Biol 16(12):1279–1285. https://doi.org/10.1038/nsmb.1700

Tracy CM, Kolesnikov AV, Blake DR, Chen CK, Baehr W, Kefalov VJ, Willardson BM (2015) Retinal cone photoreceptors require phosducin-like protein 1 for G protein complex assembly and signaling. PLoS ONE 10(2):e0117129. https://doi.org/10.1371/journal.pone.0117129

Trinidad AG, Muller PA, Cuellar J, Klejnot M, Nobis M, Valpuesta JM, Vousden KH (2013) Interaction of p53 with the CCT complex promotes protein folding and wild-type p53 activity. Mol Cell 50(6):805–817. https://doi.org/10.1016/j.molcel.2013.05.002

Ursic D, Culbertson MR (1991) The yeast homolog to mouse Tcp-1 affects microtubule-mediated processes. Mol Cell Biol 11(5):2629–2640. https://doi.org/10.1128/mcb.11.5.2629

Vainberg IE, Lewis SA, Rommelaere H, Ampe C, Vandekerckhove J, Klein HL, Cowan NJ (1998) Prefoldin, a chaperone that delivers unfolded proteins to cytosolic chaperonin. Cell 93(5):863–873. https://doi.org/10.1016/S0092-8674(00)81446-4

van den Brink DM, Brites P, Haasjes J, Wierzbicki AS, Mitchell J, Lambert-Hamill M, de Belleroche J, Jansen GA, Waterham HR, Ronald Wanders JA (2003) Identification of PEX7 as the Second Gene Involved in Refsum Disease. Am J Human Genet 72(2):471–477. https://doi.org/10.1086/346093

Vinh DBN, Drubin DG (1994) A yeast Tcp-1-like protein is required for actin function in-vivo. P Natl Acad Sci USA 91(19):9116–9120. https://doi.org/10.1073/pnas.91.19.9116

Voisine C, Pedersen JS, Morimoto RI (2010) Chaperone networks: tipping the balance in protein folding diseases. Neurobiol Dis 40(1):12–20. https://doi.org/10.1016/j.nbd.2010.05.007

Waldmann T, Lupas A, Kellermann J, Peters J, Baumeister W (1995) Primary structure of the thermosome from thermoplasma-acidophilum. Biol Chem H-S 376(2):119–126. https://doi.org/10.1515/bchm3.1995.376.2.119

Wang H, Han W, Takagi J, Cong Y (2018) Yeast inner-subunit PA–NZ-1 labeling strategy for accurate subunit identification in a macromolecular complex through Cryo-EM analysis. J Mol Biol 430(10):1417–1425. https://doi.org/10.1016/j.jmb.2018.03.026

Wei PL, Huang CY, Tai CJ, Batzorig U, Cheng WL, Hunag MT, Chang YJ (2016) Glucose-regulated protein 94 mediates metastasis by CCT8 and the JNK pathway in hepatocellular carcinoma. Tumour Biol 37(6):8219–8227. https://doi.org/10.1007/s13277-015-4669-3

Willardson BM, Howlett AC (2007) Function of phosducin-like proteins in G protein signaling and chaperone-assisted protein folding. Cell Signal 19(12):2417–2427. https://doi.org/10.1016/j.cellsig.2007.06.013

Willison KR (2018a) The structure and evolution of eukaryotic chaperonin-containing TCP-1 and its mechanism that folds actin into a protein spring. Biochem J 475(19):3009–3034. https://doi.org/10.1042/bcj20170378

Willison KR (2018b) The substrate specificity of eukaryotic cytosolic chaperonin CCT. Philos Trans R Soc Lond B Biol Sci 373(1749). https://doi.org/10.1098/rstb.2017.0192

Wu CZ, Chang LC, Lin YF, Hung YJ, Pei D, Chen JS (2015a) Chaperonin-containing t-complex protein-1 subunit beta as a possible biomarker for the phase of glomerular hyperfiltration of diabetic nephropathy. Dis Markers 2015:548101. https://doi.org/10.1155/2015/548101

Wu X, Zhang H, Chen D, Song Y, Qian R, Chen C, Mao X, Chen X, Zhang W, Shao B, Shen J, Yan Y, Wu X, Liu Y (2015b) Up-Regulation of CCT8 related to neuronal apoptosis after traumatic brain injury in adult rats. Neurochem Res 40(9):1882–1891. https://doi.org/10.1007/s11064-015-1683-1

Xu XM, Wang J, Xuan ZY, Goldshmidt A, Borrill PGM, Hariharan N, Kim JY, Jackson D (2011) Chaperonins facilitate KNOTTED1 cell-to-cell trafficking and stem cell function. Science 333(6046):1141–1144. https://doi.org/10.1126/science.1205727

Yam AY, Xia Y, Lin HT, Burlingame A, Gerstein M, Frydman J (2008) Defining the TRiC/CCT interactome links chaperonin function to stabilization of newly made proteins with complex topologies. Nat Struct Mol Biol 15(12):1255–1262. https://doi.org/10.1038/nsmb.1515

Yamamoto YY, Uno Y, Sha E, Ikegami K, Ishii N, Dohmae N, Sekiguchi H, Sasaki YC, Yohda M (2017) Asymmetry in the function and dynamics of the cytosolic group II chaperonin CCT/TRiC. PLoS ONE 12(5):e0176054. https://doi.org/10.1371/journal.pone.0176054

Yin H, Miao X, Wu Y, Wei Y, Zong G, Yang S, Chen X, Zheng G, Zhu X, Guo Y, Li C, Chen Y, Wang Y, He S (2016) The role of the Chaperonin containing t-complex polypeptide 1, subunit 8 (CCT8) in B-cell non-Hodgkin's lymphoma. Leuk Res 45:59–67. https://doi.org/10.1016/j.leukres.2016.04.010

Ying Z, Tian H, Li Y, Lian R, Li W, Wu S, Zhang HZ, Wu J, Liu L, Song J, Guan H, Cai J, Zhu X, Li J, Li M (2017) CCT6A suppresses SMAD2 and promotes prometastatic TGF-beta signaling. J Clin Invest 127(5):1725–1740. https://doi.org/10.1172/JCI90439

Yokota S, Yamamoto Y, Shimizu K, Momoi H, Kamikawa T, Yamaoka Y, Yanagi H, Yura T, Kubota H (2001) Increased expression of cytosolic chaperonin CCT in human hepatocellular and colonic carcinoma. Cell Stress Chaperon 6(4):345–350. https://doi.org/10.1379/1466-1268(2001)006<0345:Ieoccc>2.0.Co;2

Yokota S, Yanagi H, Yura T, Kubota H (1999) Cytosolic chaperonin is up-regulated during cell growth - preferential, expression and binding to tubulin at G(1)/S transition through early S phase. J Biol Chem 274(52):37070–37078. https://doi.org/10.1074/jbc.274.52.37070

Zang Y, Jin M, Wang H, Cui Z, Kong L, Liu C, Cong Y (2016) Staggered ATP binding mechanism of eukaryotic chaperonin TRiC (CCT) revealed through high-resolution cryo-EM. Nat Struct Mol Biol 23(12):1083–1091. https://doi.org/10.1038/nsmb.3309

Zang Y, Wang H, Cui Z, Jin M, Liu C, Han W, Wang Y, Cong Y (2018) Development of a yeast internal-subunit eGFP labeling strategy and its application in subunit identification in eukaryotic group II chaperonin TRiC/CCT. Sci Rep 8(1):2374. https://doi.org/10.1038/s41598-017-18962-y

Zhang J, Baker ML, Schroder GF, Douglas NR, Reissmann S, Jakana J, Dougherty M, Fu CJ, Levitt M, Ludtke SJ, Frydman J, Chiu W (2010) Mechanism of folding chamber closure in a group II chaperonin. Nature 463(7279):379–383. https://doi.org/10.1038/nature08701

Zhang Q, Yu D, Seo S, Stone EM, Sheffield VC (2012) Intrinsic protein-protein interaction-mediated and chaperonin-assisted sequential assembly of stable bardet-biedl syndrome protein complex, the BBSome. J Biol Chem 287(24):20625–20635. https://doi.org/10.1074/jbc.M112.341487

Zhang Y, Wang Y, Wei Y, Wu J, Zhang P, Shen S, Saiyin H, Wumaier R, Yang X, Wang C, Yu L (2016) Molecular chaperone CCT3 supports proper mitotic progression and cell proliferation in hepatocellular carcinoma cells. Cancer Lett 372(1):101–109. https://doi.org/10.1016/j.canlet.2015.12.029

Zhao X, Chen XQ, Han E, Hu Y, Paik P, Ding Z, Overman J, Lau AL, Shahmoradian SH, Chiu W, Thompson LM, Wu C, Mobley WC (2016) TRiC subunits enhance BDNF axonal transport and rescue striatal atrophy in Huntington's disease. Proc Natl Acad Sci USA 113(38):E5655–E5664. https://doi.org/10.1073/pnas.1603020113

Zou Q, Yang ZL, Yuan Y, Li JH, Liang LF, Zeng GX, Chen SL (2013) Clinicopathological features and CCT2 and PDIA2 expression in gallbladder squamous/adenosquamous carcinoma and gallbladder adenocarcinoma. World J Surg Oncol 11:143. https://doi.org/10.1186/1477-7819-11-143

Index

J. R. Harris and J. Marles-Wright (eds.), *Macromolecular Protein Complexes II: Structure and Function*, Subcellular Biochemistry 93,
https://doi.org/10.1007/978-3-030-28151-9

FSC
www.fsc.org
MIX
Papier | Fördert gute Waldnutzung
FSC® C083411

Kolibri 360 GmbH
Ferdinand-Jühlke-Straße 7
99095 Erfurt, Deutschland
produktsicherheit@kolibri360.de